90

D1765823

HYBRIDIZATION

and the

Flora of the British Isles

HYBRIDIZATION
and the
Flora of the British Isles

Edited by

C. A. STACE

Reader in Plant Taxonomy
University of Leicester

1975

Published in collaboration with
THE BOTANICAL SOCIETY OF THE BRITISH ISLES
by
ACADEMIC PRESS
LONDON · NEW YORK · SAN FRANCISCO

ACADEMIC PRESS INC. (LONDON) LTD.
24/28 Oval Road,
London NW1

United States Edition published by
ACADEMIC PRESS INC.
111 Fifth Avenue
New York, New York 10003

Library of Congress Catalog Card Number: 74-5671
ISBN: 0-12-661650-7

PRINTED IN GREAT BRITAIN BY
THE WHITEFRIARS PRESS LTD., LONDON AND TONBRIDGE

List of Contributors

BENOIT, P. M. "Pencarreg", Barmouth, Merionethshire.

BORRILL, M. Welsh Plant Breeding Station, Plas Gogerddan, nr. Aberystwyth, Cardiganshire.

BRADSHAW, A. D. Department of Botany, University of Liverpool.

CALLOW, R. S. Department of Botany, University of Manchester.

CANDLISH, P. A. "Rockside", The Drive, Woodhouse Eaves, Leicestershire.

COOK, C. D. K. Botanischer Garten und Institut für Systematische Botanik der Universität, Zurich, Switzerland.

COUSENS, J. E. Department of Forestry and Natural Resources, University of Edinburgh.

CRABBE, J. A. Department of Botany, British Museum (Natural History), London S.W.7.

CRACKLES, F. E. 143 Holmgarth Drive, Bellfield Avenue, Kingston-upon-Hull, Yorkshire.

CRISP, P. C. National Vegetable Research Station, Wellesbourne, Warwickshire.

CURRAN, P. L. Faculty of Agriculture, University College, Dublin 9.

DAKER, M. G. Paediatric Research Unit, Guy's Hospital Medical School, London S.E.1.

DALBY, D. H. Department of Botany, Imperial College of Science and Technology, London S.W.7.

DANDY, J. E. British Museum (Natural History), Tring, Hertfordshire.

DUCKETT, J. G. School of Plant Biology, University College of North Wales, Bangor, Caernarvonshire.

ELLIS, J. R. Department of Botany and Microbiology, University College, London W.C.1.

FERGUSON, I. K. Royal Botanic Gardens, Kew, Richmond, Surrey.

FERGUSON, L. F. Royal Botanic Gardens, Kew, Richmond, Surrey.

FINCH, R. A. Plant Breeding Institute, Trumpington, Cambridge.

GILL, J. J. B. Department of Genetics, University of Liverpool.

GOODWAY, K. M. Department of Biology, University of Keele, Staffordshire.

GREEN, P. S. Royal Botanic Gardens, Kew, Richmond, Surrey.

HALLIDAY, G. Department of Biological Sciences, University of Lancaster.

HARBERD, D. J. School of Agricultural Sciences, University of Leeds.

HARLEY, R. M. Royal Botanic Gardens, Kew, Richmond, Surrey.

HEPPER, F. N. Royal Botanic Gardens, Kew, Richmond, Surrey.

HESLOP-HARRISON, Y. Royal Botanic Gardens, Kew, Richmond, Surrey.

HUBBARD, C. E. Royal Botanic Gardens, Kew, Richmond, Surrey.

HUNT, P. F. School of Architecture, Thames Polytechnic, London W.6.

JERMY, A. C. Department of Botany, British Museum (Natural History), London S.W.7.

JONES, B. M. G. Department of Botany, Royal Holloway College, Englefield Green, Surrey.

JONES, E. M. 17 Bankfield Drive, Bramcote Hills, Beeston, Nottingham.

JONES, K. Royal Botanic Gardens, Kew, Richmond, Surrey.

KAY, Q. O. N. Department of Botany, University College of Swansea, Swansea.

LEAN, A. S. "Kyloe", Plain Road, Smeeth, Ashford, Kent.

LEWIN, R. A. Scripps Institution of Oceanography, La Jolla, California, U.S.A.

LEWIS, E. J. Welsh Plant Breeding Station, Plas Gogerddan, nr. Aberystwyth, Cardiganshire.

LOUSLEY, J. E. 7 Penistone Road, London S.W.16.

LOVIS, J. D. Department of Plant Sciences, University of Leeds.

McCLINTOCK, D. "Bracken Hill", Platt, Sevenoaks, Kent.

MARCHANT, C. J. Botanical Garden, University of British Columbia, Vancouver, Canada.

MATFIELD, B. 27 Grove Wood Close, Chorleywood, Hertfordshire.

MEIKLE, R. D. Royal Botanic Gardens, Kew, Richmond, Surrey.

MELDERIS, A. Department of Botany, British Museum (Natural History), London S.W.7.

MELVILLE, R. Royal Botanic Gardens, Kew, Richmond, Surrey.

MILLER, D. J. 14 Spring Road, Nahant, Massachusetts, U.S.A.

MOORE, D. M. Department of Botany, University of Reading.

MORTON, J. K. Department of Biology, University of Waterloo, Waterloo, Ontario, Canada.

NEWTON, A. 11 Kensington Gardens, Hale, Cheshire.

OCKENDON, D. J. National Vegetable Research Station, Wellesbourne, Warwickshire.

PAGE, C. N. Royal Botanic Garden, Edinburgh 3.

PARKER, J. S. Department of Plant Biology and Microbiology, Queen Mary College, London E.1.

PERRING, F. H. Monks Wood Experimental Station, Abbots Ripton, Huntingdon.

PIGOTT, C. D. Department of Biological Sciences, University of Lancaster.

PIKE, A. E. S. 17 Wentworth Drive, Seal Road, Bramhall, Cheshire.

PRIME, C. T. 7 Westview Road, Warlingham, Surrey.

PRITCHARD, N. M. Department of Botany, University of Aberdeen.

PROCTOR, M. C. F. Department of Biological Sciences, University of Exeter.

RATTER, J. R. Royal Botanic Garden, Edinburgh 3.

RICHARDS, A. J. Department of Plant Biology, University of Newcastle-upon-Tyne.

ROBERTS, R. H. 51 Belmont Road, Bangor, Caernarvonshire.

ROBSON, N. K. B. Department of Botany, British Museum (Natural History), London S.W.7.

ROGERS, S. Department of Biology, Queen Elizabeth College, London W.8.

SELL, P. D. Botany School, University of Cambridge.

SHIMWELL, D. W. Department of Geography, University of Manchester.

SHIVAS, M. G. Department of Plant Biology, University of Newcastle-upon-Tyne.

SLEDGE, W. A. Department of Plant Sciences, University of Leeds.

SLEEP, A. Department of Plant Sciences, University of Leeds.

SMITH, A. J. E. School of Plant Biology, University College of North Wales, Bangor, Caernarvonshire.

SMITH, P. M. Department of Botany, University of Edinburgh.

STACE, C. A. Department of Botany, University of Leicester.

STYLES, B. T. Department of Forestry, University of Oxford.

TIMSON, J. Department of Medical Genetics, University of Manchester.

TRUEMAN, I. C. Department of Biological Sciences, Wolverhampton Polytechnic.

TUTIN, T. G. Department of Botany, University of Leicester.

UBSDELL, R. Department of Botany, University of Oxford.

VALENTINE, D. H. Department of Botany, University of Manchester.

WALKER, S. School of Medicine, University of Liverpool.

WALLACE, E. C. 2 Strathearn Road, Sutton, Surrey.

WALTERS, S. M. University Botanic Garden, Cambridge.

WEBB, D. A. School of Botany, Trinity College, Dublin 2.

WEST, C. 96 New Road, Ditton, Maidstone, Kent.

WILLIAMS, J. T. Department of Botany, University of Birmingham.

WILLIS, A. J. Department of Botany, University of Sheffield.

YEO, P. F. University Botanic Garden, Cambridge.

YOUNG, D. P. (Deceased 18th March, 1972).

Preface

Hybrids have in the past provided a source of as much interest and controversy, and of as much information and perplexity, as any other aspect of plant taxonomy, and today arguments still continue concerning their interpretation, significance and treatment.

Although a great deal of information on hybrids, both descriptive and experimental, exists, our knowledge in terms of the proportion of known hybrids properly investigated is still rudimentary. Nevertheless, the volume of relevant literature is vast, and moreover is very widely scattered in books and journals, besides the unknown quantities of unpublished data in theses and field and laboratory note-books, etc. Thus the broad purpose of this book is to bring together as much as possible of this information for the first time.

In the Systematic section the accounts have been made as authoritative as possible by the collaboration of a total of 86 specialists, who in many cases have enriched the accounts for which they are responsible by the addition of unpublished information obtained by themselves or by various colleagues and correspondents. In editing the accounts I have tried to attain a high degree of consistency without altering the style or precise meanings of the authors. Thus in some cases minor inconsistencies persist, but I hope they will be excused.

The Introductory section is an attempt to provide a broad review of the whole subject of hybridization as a background to the systematic accounts, and to prevent too narrow an approach by placing the work on British plants in its world-wide perspective. I have tried where appropriate to make it a practical aid to those wishing to discover, recognize and study hybrids, and have utilized familiar British and Continental examples wherever possible. Although other extensive general discussions of the subject exist most of them are now somewhat out of date, and almost all of them rely heavily on American examples. The treatment of the various topics is admittedly somewhat uneven, but I trust not illogically so; for instance I have avoided too long a discussion of those subjects for which good modern reviews exist.

It is hoped that this book will serve three main purposes: a reference work for professional botanists for both research and teaching; a source of information for the field-botanist, who might find a greater measure of interest accorded his excursions, especially in the present era when the traditional search for rarities is relatively out of fashion and very undesirable; and a stimulus to all botanists, both amateur and professional, to investigate those hybrid situations about which it can be seen that too little is known. There is a tremendous scope for amateurs and professionals alike, and it is to be hoped that hybrids may form one basis for their continuing collaboration, a collaboration which has been a major feature of the Botanical Society of the British Isles for over a century.

February, 1975   C. A. S.

ix

Acknowledgments

The publication of this volume has been made possible by the generous collaboration of the 86 contributors listed on pp. v–viii. Many of them have answered numerous editorial queries and several continued to send the results of their current research as it became available, so that the accounts could be made as up to date as possible. In addition to the genera for which they were primarily responsible a number of the contributors supplied valuable information on other taxa. The following correspondents also provided useful data on a wide range of topics: J. Anthony, G. H. Ballantyne, J. P. M. Brenan, R. K. Brummitt, E. R. Bullard, W. A. Charlton, A. O. Chater, T. Edmondson, T. T. Elkington, G. Ellis, B. E. M. Garratt, M. B. Gerrans, G. G. Graham, N. Hamilton, V. H. Heywood, A. G. Kenneth, A. C. Leslie, I. H. McNaughton, J. R. Matthews, K. G. Messenger, P. H. Raven, P. W. Richards, A. W. Robson, M. J. P. Scannell, A. R. Smith, U. K. Smith, W. T. Stearn, E. Steiner, A. McG. Stirling, W. Stubbe, P. J. O. Trist, M. McC. Webster and D. L. Wigston.

In particular, I should like to express my gratitude to Professor D. H. Valentine, for his help and encouragement during numerous discussions, for reading the draft of the Introductory section and suggesting many improvements, and for writing the brief Foreword. Finally I wish to thank Mr J. C. Gardiner for valuable advice on general matters (particularly during the early stages), Academic Press for ever-ready advice and assistance, and my wife for preparing almost all of the typescript. Mrs I. Dingwall helped by kindly preparing the indices.

C. A. S.

Foreword

In a paper which I gave in 1950 at the second B.S.B.I. Conference in London, I said "I think it would be of use if the Society were to sponsor the compilation and publication of a list of the interspecific hybrids which occur in Britain and of their geographical distribution". These were prophetic words. At a much later, informal meeting of University Lecturers, held in Cambridge in 1969, Dr S. M. Walters revived the proposal; and very soon afterwards it was taken up by Dr Stace. The result now lies before you; and I think it will be agreed that it is very satisfactory. The introductory essay on hybridization presents an excellent and comprehensive review of the subject; and the text which follows is a tribute to Dr Stace's zeal in obtaining contributors and to his editorial skill. I am sure that this book is a worthy addition to the roll of standard works on British Botany. I am sure too that it will have a wide circulation, and stimulate interest in problems of hybridization the world over.

D. H. Valentine

Professor of Botany,
University of Manchester

Contents

SECTION A. INTRODUCTORY (by C. A. Stace)

SECTION B. SYSTEMATIC

Section A

Introductory

1. Preamble

Much of biology today is built around the species concept. Although man has recognized species as such from the very beginning (for species recognition is essential in detecting food, predators, etc.), it was the nineteenth century theories of evolution and the twentieth century studies of genetics and cytology that gave the species its present dominant position in biological thinking. The species, either as a concept or as a group of organisms, is utilized nowadays by scientists of all sorts as the standard biological unit. History has shown that even those disciplines which start by emphasizing the universal application of their principles sooner or later turn to investigating interspecific differences. Modern examples of this are seen in metabolic biochemistry and in cell ultrastructure.

The choice of the species rather than any other taxonomic rank, all of which equally figure in the hierarchy of all groups of organisms, is surely due to the belief that there is some sort of common denominator which renders the species of different phyla comparable. Certainly there is no way yet of comparing meaningfully the families of, for example, insects with those of algae or flowering plants. Perhaps in the future we shall be able to quote the percentage of DNA in common as a measure of taxonomic rank, but meanwhile biologists have to make do with more subjective means of estimating rank, and only in the case of the species (and some lower taxa) are these means comparable in all living organisms. The special feature of species is, of course, that they are the units of evolution, in the sense that they are the sum of the interbreeding populations which evolve as a single unit. When the populations behave as completely separate units they become recognized as separate species. There are many exceptions to this greatly over-simplified species concept, but it is certainly true that the vast majority of species recognized today conform very closely to it, i.e. the morphological or traditional species can in most cases be loosely equated with the biological or evolutionary species.

Interspecific hybrids, with which this book mainly deals, are results of situations where the morphological and biological species concepts do not

1

exactly coincide. In some instances the two species have not become sufficiently distinct to prevent hybridization; in others hybridization is a sign that the distinctness which once existed is breaking down. These two situations were described by Mayr (1942) as primary and secondary intergradation respectively. In yet other cases, of course, hybridization occurs between normally well-isolated species and does not lead to a breakdown of the specific boundaries. Although an enormous number of hybrids is now known they have been used not as a sign that the species concept is not a valid one, but rather as exceptions that prove the rule, and of course as valuable tools in studying interspecific relationships.

Perhaps because they are still considered to be somewhat enigmatic or exceptional, hybrids have not been given sufficiently detailed treatment in most taxonomic works; where space is at a premium hybrids have often been one of the first topics to be sacrificed. For instance, in the *Flora of the British Isles* (Clapham *et al.*, 1952, 1962) hybrids "have as far as possible been mentioned", but descriptions are provided only "where the hybrid is common . . ., where it is a highly distinct plant . . ., or where it is liable to lead to confusion between species . . .". In *Flora Europaea* (Tutin *et al.*, 1964) "only those few hybrids which reproduce vegetatively and are frequent over a reasonably large area are described", although "other common hybrids may be mentioned . . . in notes". These criteria have been discussed by Valentine (1963). When it is realized that many major crops, e.g. wheat, are of hybrid derivation, and that hybrids are being more and more frequently used in agriculture and elsewhere, it becomes clear that their detailed study and precise documentation is of great importance. Although an enormous amount of information exists it is very widely scattered, and unless it is brought together in a single text it is difficult to judge its overall significance and to detect gaps in our knowledge. This is a priority in an age when many sources of information are being lost by the development of natural resources, and when a correct assessment of the areas urgently requiring conservation is absolutely vital.

That only the flora of the British Isles is covered in detail in this book is obvious from its title. Two other deliberate limitations, its restriction to vascular plants and (largely) to interspecific hybrids, will become clear in the next few pages.

2. What is a hybrid?

A hybrid may be defined simply as the offspring of two dissimilar individuals, but in order to have any real meaning it becomes necessary to define what is meant by "two dissimilar individuals". In fact the phrase is used for a whole range of situations ranging from two distinct species to virtually any two organisms, according to the context. Thus, in the latter extreme, almost any organism formed by cross-fertilization is a hybrid. This is the definition of the geneticist, who is concerned with pure lines and individual genes, and to whom the mating of two individuals showing any degree of genetic difference is a hybridization. Taxonomists are concerned with taxa, i.e. genera, species, varieties, etc., and a taxonomic hybrid therefore involves a cross between two individuals of different taxa. If the taxa are two different species one talks of an interspecific hybrid; if two varieties of a single species, an intervarietal hybrid; and so on.

Darlington (1937), in a note with the same title as this chapter, maintained that "a hybrid is a zygote produced by the union of dissimilar gametes (or which by mutation has the character of such a zygote). Whether the gametes come from similar or dissimilar parents does not signify". It is doubtful whether this broadest of all possible definitions (by which even the offspring of a self-fertilization of a pure-bred individual could be a hybrid) has any practical value at all. His assertion that "Systematists have generally been content to use it [the term hybrid] as a label for misfits" was even in 1937 blatantly untrue, and probably simply reflects a distaste for taxonomy.

Stebbins (1959) defined a hybrid as a cross "between individuals belonging to separate populations which have different adaptive norms". This is an evolutionary definition, for while it attempts to limit the term to cases where the two entities differ adaptively it avoids a mention of any taxonomic ranks or of the level of reproductive isolation. It is not a taxonomic definition because, whereas many populations with different adaptive norms are recognized as distinct species, subspecies, varieties, etc., many are not, and many hybrids in Stebbins' sense are therefore not hybrids to a taxonomist. For this reason Wagner (1968) suggested adding to Stebbins' definition "and which would be separated in ordinary taxonomic practice as readily defined phenetic species".

It may be desirable to retain the term hybrid in a flexible sense, with a precise meaning dependent upon the context. For solely taxonomic purposes the loose definition given above, a cross between individuals of different taxa, is probably as useful and meaningful as any that could be constructed. It is certainly simpler than most, and has the advantage that it is equally applicable whatever the concept of the various taxonomic levels.

Apart from these problems concerning the level of hybridization there are others involving its degree and its age. There are many known species which have in the past crossed with another species, but in which the results of hybridization have become enormously diluted in time due to the repeated backcrossing of the F_1 hybrids with the first species (introgression). Although such introgressed individuals often scarcely differ from the unadulterated species, the influence of the second species may still be detectable in certain characteristics, and it is difficult in many cases to decide whether to recognize such plants as variants of the first species or as hybrids between the two. Webb (1951) unequivocally opted for the first alternative: "if species A, over most of its area, comprises a proportion of individuals with certain marks of species B, then we must accept this as part of its pattern of variation and call them by the name of A, even if we believe that there has been interbreeding at some time".

This is clearly the common-sense judgement in those cases where introgression has played an extensive role in the pattern of variation of a species; indeed, there are undoubtedly a great many species which, quite unknown to us, owe their present range of variation to past interspecific hybridization. But where introgression is apparent in only a part of the range of the species it might be more useful to recognize this by treating the introgressants as hybrids, which they are, even though they fall much closer to one parent than to the other.

The difficulty of classifying plants of ancient hybrid origin ("historical hybrids") arises in other situations, the best-known and most widespread being the case of amphidiploids. Amphidiploids are allopolyploids which have arisen from diploid hybrids by a doubling of the chromosome number, so that although polyploid they behave as if diploid. Because of this they share all the characteristics of normal diploid species, and are universally treated as such, even though they might have arisen from hybrids in a single step or generation. Many species are known to be amphidiploids, and their parental species are known or surmised, but in other cases the parents are not known (and indeed may be extinct or have evolved into something very different from the original). In general it is true that the parents are more often known where the amphidiploid is of more recent origin (e.g. *Spartina "townsendii", Senecio cambrensis, Primula "kewensis", Tragopogon mirus,* etc.). A further reason for considering such plants as species rather than hybrids is that hybrids between them and their diploid parents are usually highly sterile.

Other historical hybrids have arisen by the stabilization of various F_2 or later segregants from an F_1 hybrid, of well-defined segments of a hybrid complex, or even of F_1 hybrids themselves. In the last case the hybrids are frequently highly or wholly sterile, but have spread vegetatively and come to occupy habitats and geographical areas different from those of the parents. For instance *Circaea* x *intermedia (C. alpina* x *C. lutetiana)* is far more widespread in Britain than is *C. alpina,* and it occupies habitats not exploited by *C. lutetiana. Symphytum* x *uplandicum (S. asperum* x *S. officinale)* is more widespread than either parent, and, unlike them, is known to hybridize with *S. tuberosum. Equisetum* x *moorei* and *E.* x *trachyodon* both occur in places well outside the ranges of one or both their putative parents. In several genera undoubted hybrids occur whose parents are unknown (e.g. x *Asplenophyllitis, Salix*). The genus *Mentha* furnishes many examples of hybrid taxa which, like those above, have often been treated as species. In *Flora Europaea* Harley (1973) numbered three *Mentha* hybrids as though they were species. In addition *M. scotica* (Graham, 1958) is a fertile plant and opinions differ as to whether it should be recognized as a species, a variant of *M. spicata* (which is itself a segmental allopolyploid), or a variant of *M. spicata* x *M. suaveolens (M.* x *villosa)*. Despite the sterility of many *Mentha* hybrids some have become very widespread as a result of their vigorous vegetative propagation and often deliberate cultivation. It would seem that hybrids in several other genera, e.g. *Potamogeton,* should merit similar attention in Floras.

Hybrid complexes were defined by Grant (1953) as "groups of species in which hybridization has obscured the morphological discontinuities between the basic diploid types". Many examples are known. Grant classified them into five sorts: homogamic complexes, in which the hybrid derivatives are normal sexual diploids; polyploid complexes, in which the derivatives are sexual polyploids; heterogamic complexes, in which they are permanent heterozygotes for structural hybridity; agamic complexes, where they are partly or wholly agamospermous; and clonal complexes, where reproduction is mainly or entirely vegetative. As might be expected, cases are known where the actual situation combines two or more of the above categories (e.g. *Mentha*, with both polyploid and clonal hybrid complexes) or is otherwise intermediate in nature.

In the genus *Euphrasia* (Yeo, 1956) there are cases where diploid hybrids have arisen from crosses between diploid species and tetraploid ones, via the production of occasional haploid gametes by triploid F_1 plants. Where these diploid hybrids are fairly uniform and form extensive populations which occupy habitats different from the parents Yeo suggests they should be given specific status. Such is the postulated origin of *E. vigursii* (from *E. anglica* x *E. micrantha*), but similar plants derived from *E. anglica* x *E. nemorosa* and *E. anglica* x *E. brevipila* have not received a name.

Plants of presumed hybrid origin are also sometimes treated, for convenience, as infraspecific taxa. *Montia fontana* subsp. *variabilis* is thought to consist of hybrids between subsp. *fontana* and subsp. *intermedia* (Walters, 1953), and similarly *Rhinanthus minor* subsp. *lintonii* of hybrids between subsp. *borealis* and both subsp. *monticola* and subsp. *stenophyllus* (Sell, 1967). In some Floras which recognize *Medicago sativa* and *M. falcata* as subspecies of *M. sativa*, their hybrid (*M.* x *varia*) is treated as a third subspecies.

Many apomictic taxa ("microspecies", "agamospecies") are undoubtedly of hybrid origin, and a high proportion have odd numbers of chromosome sets (triploid, pentaploid, etc.). In genera where the apomictic taxa are given binomials, as if species, those which are known to be derived by hybridization are treated identically with the others. It is quite likely, of course, that the majority of such taxa are hybrid derivatives which have escaped sterility by means of apomixis. The genus *Sorbus* provides good examples of agamospecies which are certainly of hybrid origin. For instance the numerous species in the *S. latifolia* agg. (42 are listed by Warburg and Kárpáti (1968)) are probably all apomictic and derived from hybrids between the sexual *S. torminalis* and various sexual taxa in the *S. aria* aggregate. In *Alchemilla* there are several species which are intermediate between *A. alpina* agg. and *A. vulgaris* agg. and which might well have arisen from them, although so far as is known both these aggregates are now entirely apomictic. Many other examples from well-known apomictic genera such as *Rubus, Hieracium* and *Taraxacum* could be cited.

In a pioneer series of investigations started in 1913, Blackburn and Harrison (1924) found that many of the British species of *Rosa* recognized at that time were hybrids. Using the terminology of Jeffrey (1914), they distinguished two sorts: phenhybrids, whcih are fairly obvious hybrids (F_1 or later generations or backcrosses) of recent origin; and crypthybrids, which are stabilized ancient hybrids often recognized as distinct species. Even the crypthybrids have a somewhat disturbed meiosis and reduced fertility and, as many of the phenhybrids have become fertile owing to a doubling of the chromosome number, fertility is not a reliable measure of the age of the hybridization. In this genus the tendency in Britain has been for the hybrids of both sorts to be treated taxonomically as such, rather than as species, so that the number of British species in the genus dropped from over 100 recognized in 1910 (Wolley-Dod, 1910) to 17 in 1930 (Wolley-Dod, 1930–31). R. Melville (1972, pers. comm.) considers that 8 of the 19 species in section Caninae which are recorded from Britain by Klášterský (1968) are hybrids, and he treats them as such in his account of *Rosa* for this book. Many of the infraspecific taxa of Wolley-Dod are dealt with similarly.

In short, plants of hybrid derivation (either recent or ancient) have often been considered as species when they have developed a distributional, morphological or genetical set of characteristics which is no longer strictly related to that of their parents. The taxonomic treatment of such plants (apart from the amphidiploids) has, however, often been very uneven, and decisions on their recognition as species *vis-à-vis* hybrids are still largely subjective. Because of these uncertainties all such plants with a known origin have been dealt with in this book as hybrids.

3. Hybridization as a species criterion

It was mentioned in Chapter 1 that the ability to interbreed is an essential part of the biological species concept, with which the morphological species is often synonymous. Where the biological and morphological species are found to have different circumscriptions there arises the possibility that the limits of the latter might be altered to coincide with those of the former, for it would certainly be convenient if all species were composed of potentially interbreeding individuals, none of which was capable of crossing with another species. Such inclinations are encouraged by the discovery that, in any given group, the degree of morphological similarly between taxa is usually more or less proportional to the degree of interfertility, viz. the more alike in appearance are two organisms the more likely it is that they can interbreed and, if hybrids are formed, the more likely it is that such hybrids are fertile.

The formation of sterile hybrids between such idealized species might be quite cheerfully tolerated, because such hybrids are usually easily recognized and they scarcely alter the evolutionary status of a species, but it could be argued that the formation of fertile hybrids between two taxa should be used as an indication that those taxa represent a single species. That this is a wholly unrealistic argument is shown by three major considerations.

In the first place it is not possible to differentiate between fertile and sterile hybrids because every intermediate condition exists, and intermediates probably outnumber hybrids exhibiting the extreme conditions. Thus an arbitrary distinction would have to be made between sterile and fertile hybrids.

Secondly, although in any one group the degree of morphological similarity is likely to be proportional to the interbreedability, overall this is not so. In many groups fully fertile hybrids occur in nature between taxa of vastly different appearance, and in others there are strong sterility barriers between taxa which are hardly, if at all, morphologically separable. Species delimited on the basis of breeding behaviour as the major criterion would thus be very varied assemblages, which would greatly debase the value of the species in practical terms, i.e. it would not be possible to recognize species visually, and the limits of species which had not been studied experimentally or in the field would be quite unknown.

Thirdly, there is a great difference between the ability of two species to hybridize and the occurrence of actual hybrids in the field, because there are many barriers to hybridization apart from the genetic incompatibility of the gametes. Many of these barriers which are fully effective in nature can be overcome experimentally, and so it would be necessary to prescribe a set of conditions in which to test for hybridization.

Nevertheless there are many taxonomists who place a great deal of emphasis on the intersterility of species, and it is in fact still possible to encounter biologists who are genuinely surprised that taxonomists could hold the opinion that separate species can produce highly fertile hybrids. The varying taxonomic importance placed on the formation of fertile hybrids has given rise to many differences of opinion concerning species limits. For instance *Calystegia sepium* and *C. silvatica* form fertile hybrids in the wild, and for this reason are considered by some workers to be subspecies of a single species. *Medicago sativa* and *M. falcata*, and *Silene vulgaris* and *S. maritima,* are two other well-known examples.

Differences of opinion on such matters will doubtless always exist, but it should be possible to minimize them by sensible compromise. The species is by present-day consensus a unit of practical value, visually recognizable but wherever possible also of evolutionary significance. Thus morphological and genetical data should both be used in its recognition, and neither to the subjugation of the other. The evidence of interfertility between two taxa should be used where possible in relation to the interfertility between other closely related taxa. Hence, in the *Calystegia* case above, if many of the taxa in the genus are interfertile it is more useful to maintain *C. sepium* and *C. silvatica* as distinct species, to avoid a situation where there are very few polymorphic species with numerous infraspecific taxa. But if *C. sepium* and *C. silvatica* are the only two interfertile species it would be better to recognize their greater degree of relationship by relegating them to subspecific rank. In this example the necessary information concerning the other species is not yet available (Stace, 1961).

The non-coincidence in many cases of the limit of the interbreeding unit and the morphologically defined species has led to attempts by ecological geneticists to construct a hierarchy of terms (alternative to genus, species, etc.) to describe degrees of interfertility among plants without reference to visual appearance. The earliest and most used system is that of Turesson (1922; and later amendments by Turesson and others), who employed the terms ecotype, ecospecies and coenospecies. Danser (1929) similarly used three terms (convivium, commiscuum and comparium) but they were defined in such a way that, although Turesson and Danser between them described only four different levels of the genecological hierarchy, none of the six terms is synonymous with any other. Danser's terminology never became popular and is rarely encountered nowadays. The four levels of the hierarchy can all be defined by means of the deme terminology of Gilmour and Gregor (1939), as was demonstrated by Gilmour and Heslop-Harrison (1954). The deme terminology, despite its several advantages, has not been widely adopted, although Briggs and Walters (1969) recently utilized it extensively in their textbook for university students. For ease of reference the definitions of the ten terms are compared in Fig. 1.

	Turesson	Danser	Gilmour and Gregor
A group of individuals occupying a specific ecological habitat and forming an interbreeding population which differs genetically from other such populations.	ECOTYPE	CONVIVIUM	GENECODEME
All individuals capable of hybridization among one another to give hybrids showing complete fertility.	ECOSPECIES		HOLOGAMODEME
All individuals capable of hybridization among one another to give hybrids showing some degree of fertility.	COENOSPECIES	COMMISCUUM	COENOGAMODEME
All individuals capable of hybridization among one another.		COMPARIUM	SYNGAMODEME

Fig. 1. The genecological hierarchies of Turesson, Danser, and Gilmour and Gregor compared.

The ideal species is the syngamodeme or, if hybridization to produce wholly sterile offspring is to be permitted, the coenogamodeme. But a great many plants recognized as species are hologamodemes (i.e. able to form fertile hybrids with others), and a not inconsiderable number genecodemes (able to form hybrids showing complete fertility).

4. Frequency and level of hybridization

It is evident from what has been said in the previous chapter that hybridization is on the whole more frequent the more similar in morphology are the plants concerned, but that there is no absolute correlation between these features and therefore it is not possible to define any of the orthodox taxonomic levels on this basis. Thus the taxonomic level to which hybridization extends varies from one group of organisms to another. The topic can be conveniently dealt with in three parts: intraspecific, interspecific, and supraspecific.

INTRASPECIFIC HYBRIDS

Of the traditionally recognized infraspecific categories the lowest ones in the hierarchy (variety, subvariety, form, subform) represent taxa differing from each other in relatively minor characteristics. Usually no sterility barriers exist between the varieties, for example, of one species and, as there are often no marked geographical or ecological differences between them, hybrids occur frequently. Such hybrids are fully fertile and the characters separating the parental taxa mostly segregate in a simple way. Similarly the subspecies in its usual connotation, i.e. a geographical race, differs from other conspecific subspecies in rather few, often quantitative characters, and by their very definition such subspecies can hybridize wherever their geographical ranges meet or overlap. Thus a catalogue of these inter-subspecific hybridizations would be of as little consequence as one of those between varieties or forms.

In more recent years there has, however, been a debasement of the subspecies concept to the extent that various other sorts of taxa (not geographical races) are now similarly classified. This most commonly applies to various levels of a polyploid series where the degree of morphological difference is very small ("semicryptic polyploidy"—Davis and Heywood, 1963). In such cases it is by no means certain that hybrids between them occur in nature, and even when they do they would be expected to be sterile, so that documentation of them would be as worthwhile as that of interspecific hybrids. This view is strengthened by the fact that many semicryptic polyploids are treated as species by some authors, e.g. the three British cytodemes of *Polypodium vulgare*. For these reasons hybrids between such taxa are included in this book, but hybrids between geographical races recognized as subspecies and between varieties or lower taxa are not. Examples of hybrids between semicryptic polyploids treated as subspecies are to be found under *Asplenium trichomanes* and *Eleocharis palustris*, etc.

Sterility barriers at the intraspecific level are not confined to cases of polyploidy, for there are many examples of species exhibiting inter-race sterility at one ploidy level. The topic appears not to have received much attention in Britain, but there are many examples from elsewhere, e.g. *Galeopsis tetrahit* (Muntzing, 1929), *Elymus glaucus* (Snyder, 1951), *Epilobium hirsutum* (Lehmann, 1941). In these cases sterility may be due to the mutation of particular genes, to minor chromosomal rearrangements, or to cytoplasmic incompatibility. In intermediate cases intraspecific sterility occurs between races with different chromosome numbers close to one ploidy level. In *Vicia sativa*, for example, several of the subspecies are represented by plants with different chromosome numbers (e.g. subsp. *amphicarpa* $2n$ = 10, 12 or 14; subsp. *segetalis* $2n$ = 10 or 12); hybrids between plants of the same subspecies with different chromosome numbers are far more sterile than are hybrids between plants of different subspecies but with the same chromosome number (Hollings and Stace, 1974 and unpublished). In this case, of course, the subspecies are based on morphological rather than cytological criteria.

INTERSPECIFIC HYBRIDS

Interspecific hybrids naturally attract far more attention than intraspecific hybrids. Among the vascular plants they are very numerous and exhibit every grade of fertility. Knobloch (1972) stated that he had catalogued 23,675 in the flowering plants (i.e. angiosperms and gymnosperms), but this figure must be used with great caution. Many of the total have been shown to be fanciful identifications, and no doubt even more will prove to be so; on the other hand there are many omissions of known hybrids (e.g. all those in the Combretaceae) and there are undoubtedly many hundreds of unrecognized hybrids in less well worked regions of the world, especially the tropics. Whether the errors will compensate the omissions or one will greatly exceed the other is of course quite unknown. Nevertheless the total figure Knobloch gave does help to give an impression of the magnitude of the problem set against a background of perhaps 250,000–300,000 described species. Of his 23,675, over 10,000 belong to the Orchidaceae, 2400 to the Gramineae (see also Knobloch, 1968), 2242 to the Compositae, and 1478 to the Rosaceae; no other families reach 600. The most important unknown in Knobloch's data is the number of purely artificial crosses, for he has combined natural and artificial hybrids without discrimination. It would be reasonable to assume that at least 50% of them, and probably very considerably more, are not known in the wild, and this throws a rather different light on the situation. In addition the figures for the Orchidaceae should be radically revised; according to P. F. Hunt (see below under Orchidaceae) the total number of hybrids is now about 45,000, almost all the increase being due to additional artificial hybrids.

In many cases the number of hybrids is roughly proportional to the number of species in a given family, but there are some notable exceptions. The Betulaceae, Onagraceae, Orchidaceae, Pinaceae, Rosaceae and Salicaceae are among those conspicuously over-represented. A large number of families is under-represented, but in most cases probably due to a lack of information; the Labiatae and Leguminosae are perhaps reliably so.

Among the numerous other trends which have been noted by various authors, hybrids do appear to be much commoner among perennials than annuals. This, however, seems to be simply due to the floral biology of the two groups, for autogamy is far more frequent in annuals. Hybrids between annuals and perennials are not rare, e.g. *Festuca* x *Vulpia; Senecio squalidus* x *S. vulgaris*, and Knobloch (1972) quoted a cross in the Compositae between an annual and a shrub. Clifford (1961) discussed a number of other possible correlations, and showed how difficult it is to make valid generalizations.

Natural hybrids between three or more species are fairly common in those genera which produce fertile binary hybrids. In the British flora there are fairly well substantiated examples in x *Dactyloglossum, Epilobium, Mentha, Rosa, Salix, Symphytum* and *Ulmus.* R. D. Meikle and R. Melville, in their accounts for this book of *Salix* and *Ulmus* respectively, each mention one hybrid involving four species, but it is probably fair to say that some authorities would query the identity of these. In the genus *Salix* extensive artificial hybridizations have been carried out in Sweden (see Nilsson, 1954), and a hybrid involving 13 different species, eight of them British, has been obtained (Fig. 2).

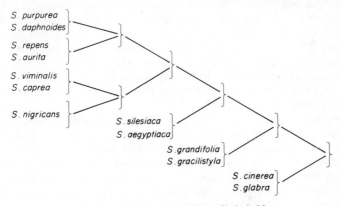

Fig. 2. Nilsson's 13-species *Salix* hybrid.

Interspecific hybrids occur in all four divisions of the pteridophytes (Psilophyta, Lepidophyta, Calamophyta and Filicophyta), and in the last two (horsetails and ferns respectively) they are very widespread.* Despite their free-living gametophytes (whose direct method of fertilization probably contributes to the frequency of hybrids), pteridophytes have a large dominant sporophyte comparable to that of the flowering plants, and they are therefore included with them in this book.

The situation in the bryophytes is somewhat different, since the dominant phase of the life-cycle is gametophytic. A successful hybridization is followed by the production of a hybrid sporophyte (capsule) retained upon the maternal gametophyte, and several reports of moss capsules (often with abortive spores) which are intermediate between the species bearing them and another related one have been made. Examples are found in the British genera *Funaria, Grimmia, Orthotrichum* and *Weissia.* Since these four are currently placed in different families it seems likely that moss hybrids are more widespread than often realized, and the lack of records in the liverworts might simply be the

* See also Knobloch (1973).

result of their not having been recognized, as the sporophyte of liverworts possesses relatively very few diagnostic characters. Nicholson (1905, 1906) reported in some detail the discovery of *Weissia crispa* (*W. longifolia* Mitt.) x *W. crispata* (= *Astomum* x *nicholsonii* Roth) in Dorset, Kent and Sussex, and of *W. crispa* x *W. microstoma* in Sussex. In the first case reciprocal hybrids were found, and one gametophyte of *W. crispa* possessed two capsules of pure *W. crispa* and one of the hybrid.

If hybrid moss plants (gametophytes) are to occur the hybrid capsule must have a fairly regular meiosis and produce viable spores. Few cases of this are known. Nicholson (1905) found that the hybrid *W. crispa* female x *W. crispata* male produced some apparently normal spores, but he was not able to germinate them and he could not find hybrid gametophytes in the wild. The reciprocal hybrid produced no viable spores.

In the algae and fungi far less is known about hybridization, and indeed the species concept is far less well established, but there are reliable records of natural hybrids from various algal groups such as the Phaeophyta (about four British species of *Fucus* in various combinations), Charophyta (where Druce, 1928, recorded *Chara contraria* x *C. hispida* on good authority), and Chlorophyta (*Chlamydomonas, Spirogyra, Mougeotia*, etc.). In some of these genera, as well as in *Eudorina* (Chlorophyta) and others, artificial hybrids have been synthesized.[*]

Hybrids have similarly been synthesized in many lower fungi, principally Ascomycetes and some Phycomycetes, for genetic purposes. Emerson and Wilson (1954) produced hybrids between *Allomyces arbuscula* and *A. macrogynus* and found that the fungus previously known as *A. javanicus* var. *javanicus* was in fact the natural hybrid between them. By means of tetrasporic analysis Parker-Rhodes (1950) showed that hybrids between *Psilocybe bullacea* and *P. coprophila* occur on Skokholm Island, and he named them as a new species *P. scocholmica* (Basidiomycetes). Culberson and Hale (1973) concluded from chemical evidence that some present-day species of the lichen genus *Parmelia* have arisen from ancient hybridizations. Because of the relative lack of information concerning hybrids in non-vascular plants, and because of the rather different principles which can be applied to them, the systematic part of this book is confined to the vascular plants. A review of hybrids in the lower plants is clearly much needed.

SUPRASPECIFIC HYBRIDS

At the supraspecific level hybrids are much less frequent, and because of this it has very often been suggested that the occurrence of hybrids between species of different genera indicates that the genera should be combined. This is no more valid an argument than that concerning the limits of species, and the criticisms levelled against it there are equally applicable here. Kruckeberg (1962) surveyed the characters of *Silene, Lychnis* and *Melandrium* and was able to synthesize hybrids between species in all three generic combinations. He pointed out that several of his hybrids were intergeneric or intrageneric according to the variously

[*] See also Rueness (1973).

recognized limits of the three genera, and quite rightly concluded that "the ability to make successful hybridizations between species of related genera is not sufficient in itself to cause the joining of those species into a single genus". In his interesting discussion of generic limits in the grass tribes Festuceae and Hordeeae, Stebbins (1956) suggested that the evidence from intergeneric hybrids was in fact supported by morphological criteria, and that many genera in these tribes should be united, but his arguments were to some extent clouded by the then contemporary practice of placing *Lolium* in the Hordeeae. A more recent discussion has been provided by Terrell (1966).

Knobloch (1972) listed all the recorded intergeneric hybrids that he could trace in the flowering plants, but, as noted previously, his list fails to discriminate between natural and artificial crosses. His total of 2993 combinations between species of different genera (not different generic combinations) is dispersed in 45 different families. Of the 2993 about 1869 occur in the Orchidaceae and over 800 in the Gramineae, leaving only just over 300 others. In fact Knobloch misinterpreted Adams and Anderson's (1958) data, which he used for the Orchidaceae: the figure of 62 Knobloch gave for bigeneric hybrids should have read *c* 642. Thus the total number 2993 becomes 3573, of which about 2449 (not about 1869) occur in the Orchidaceae. This figure is now greatly out of date. As with the case of interspecific hybrids well over 50% of Knobloch's total are solely artificial hybrids, and many of the others (e.g. *Lens* x *Vicia*) are based on erroneous determinations.

Hybrids involving three or more genera were cited by Knobloch (1972) only in the Cactaceae (one artificial quadrigeneric combination), Rutaceae (two different artificial trigeneric combinations), and Orchidaceae, although earlier (Knobloch, 1968) he had listed several artificial trigeneric hybrids in the Gramineae. Hybridization in the Orchidaceae has been surveyed by Adams and Anderson (1958), who distinguished carefully between natural and artificial hybrids. They recorded 11 different trigeneric and two different quadrigeneric combinations, all of which were artificial only. Since the 1958 review many other tri- and quadrigeneric orchid hybrids have been synthesized, and in 1970 the first quinquigeneric hybrid, x *Rothara*, was announced (P. F. Hunt, pers. comm., 1973). Thus it appears that no natural hybrids are known to involve more than two genera of flowering plants.

The great majority of intergeneric hybrids involve obviously closely related genera. In the Gramineae all natural hybrids, as far as I am aware, involve genera within a single tribe, although successful inter-tribal crosses have often been carried out artificially (e.g. *Bromus* x *Festuca*). In the Orchidaceae all the British hybrids are similarly confined to a single tribe, but according to Solbrig (1970, p. 90) artificial hybrids can be made between orchids of different subfamilies.

In other families as well artificial crossing techniques have produced some spectacular hybrid combinations which field botanists would scarcely have thought likely. Amongst the earliest of these was the cabbage–radish hybrid (*Brassica* x *Raphanus*) (Karpechenko, 1927), with its exciting culinary prospects (as yet hardly realized). More recent examples are the apple–pear (*Malus* x *Pyrus*) hybrids (Crane and Marks, 1952; Williams, 1959; Gorshkov, 1962) and various combinations in *Ribes* (gooseberry x blackcurrant) and *Prunus* (e.g. plum, cherry and peach in all three combinations) (Gorshkov, 1962; Yenikeyev, 1966), etc. Whether any of these will become rivals to their parents in horticultural value remains to be seen; more likely they will prove interesting diversions like the loganberry (*Rubus idaeus* x *R. vitifolius*), which arose by chance in a garden in 1881. Many details concerning the production of "wide hybrids" in economically important plants (particularly cereals, but also in tree-fruits, soft-fruits, vegetables, forest-trees, legumes, cotton, *Helianthus*, *Nicotiana* and melons, etc.) are to be found in Tsitsin (1962).

One inter-familial hybrid, x *Veronicena* Moldenke (*Verbena*, Verbenaceae, x *Veronica*, Scrophulariaceae) was described in 1955, but this was based on a plant reported by Haartman (1751), and is almost certainly an error, as, without doubt, are several fanciful records of inter-familial fern hybrids. The ability to generate, in microculture conditions, cell-hybrids between organisms of vastly different affinity is mentioned in Chapter 5.

In the British flora authenticated natural intergeneric hybrids are known only in the Aspleniaceae (*Asplenium* x *Phyllitis*), Compositae (*Anthemis* x *Tripleurospermum*; *Conyza* x *Erigeron*), Gramineae (*Agropyron* x *Hordeum*; *Agrostis* x *Polypogon*; *Ammophila* x *Calamagrostis*; *Festuca* x *Lolium*; *Festuca* x *Vulpia*), Orchidaceae (*Anacamptis* x *Gymnadenia*; *Coeloglossum* x *Dactylorhiza*; *Coeloglossum* x *Gymnadenia*; *Dactylorhiza* x *Gymnadenia*; *Gymnadenia* x *Pseudorchis*), and Rosaceae (*Crataegus* x *Mespilus*).

Intergeneric hybrids in the ferns (such as *Asplenium* x *Phyllitis* mentioned above) are probably as frequent as in the flowering plants, and it is likely that the same is true of the bryophytes. Reciprocal hybrids between *Funaria* and *Physcomitrium* (both Funariaceae) have been frequently found on the Continent, and have been made artificially and used in pioneer studies on cytoplasmic inheritance, polyploidy, and other phenomena. Nicholson (1910) found hybrid sporophytes of *Trichostomum flavovirens* (*Tortella flavovirens* (Bruch) Broth.) female x *Weissia crispa* (*W. longifolia* Mitt.) male (both Pottiaceae) in Sussex. Putative intergeneric hybrids have also been reported in the algae, e.g. *Ascophyllum* x *Fucus* (both Fucaceae), which has also been synthesized artificially.

ANIMAL HYBRIDS

A brief comment on hybridization in animals is relevant here. There is no doubt that hybridization is far less frequent in animals than in plants, and that the hybrids, when formed, are far less often fertile. Their rarity is usually attributed to the highly developed behavioural patterns in animals which restrict mating mostly within single species. Moreover in the largest group, the insects, the genitalia are often highly evolved to form an intricate sort of lock and key mechanism. When hybrids are formed (often artificially, e.g. mule and hinny, the reciprocal hybrids of the horse x donkey) they are usually sterile, so that there must be particularly well-developed internal sterility barriers as well. These are often said to be related to the XY sex-determining mechanism. Whatever their causes, the result is that in animals the biological and morphological species much more frequency coincide than in plants, and the temptation to define species on the fertility–sterility criterion is much stronger. There are several groups of insects, for example, in which the species can only be identified when dead by examination of the genitalia, and sometimes this in one sex only.

Nevertheless, in more recent years, more and more examples of animal hybridization have come to light. In the fishes hybrids are known at the inter-ordinal level (higher than that in any plants), and fertile interspecific hybrids showing introgression are now known in many groups ranging from birds to various invertebrates.

The situation in animals supports the belief that there is no absolute correlation between the rates of evolution of different sorts of isolating mechanisms. Usually morphological and physiological characters have evolved at about the same pace, but sometimes one or the other is left behind, so to speak. Thus one should not be surprised to discover that intergeneric orchid hybrids are fertile, or that one cannot distinguish morphologically between two intersterile chromosome-races, or that in many insects the primary sexual characters have been the only morphological features to parallel the physiological ones.

5. Other sorts of hybrids

The terms hybrid and hybridization are used in senses other than those concerned with the sexual union of two different organisms, and as these other sorts of hybrids are all relevant to taxonomic problems they will be mentioned here.

INTERNAL OR STRUCTURAL HYBRIDITY

Many plants are known in which structural alterations have taken place in the chromosomes, involving a change in the relative positions of various parts of the chromosomes quite apart from any gene mutations which might have taken place. The best understood cases involve inversions, where part of a chromosome has become inverted, and translocations, where part of a chromosome has become detached and has jointed to another chromosome of the same set (usually reciprocally). Such structural changes only become discernible in the heterozygous condition, when distinctive configurations appear at meiosis as the homologous parts of chromosomes pair. In the case of inversion heterozygotes the paired chromosomes form loops, and in the case of translocation heterozygotes multivalents are formed. Such heterozygotes are known as structural hybrids, and they have important evolutionary consequences.

In the first place they have the effect of keeping favourable gene combinations together in a definable segment of a chromosome, which enables

the plants immediately to take advantage of any environmental situation which might particularly favour that gene combination. Unfavourable genes of course equally occur in these segments, but if they are recessive and in the heterozygous condition they are preserved from selection pressures. For this reason hetero-zygotic inversions and translocations are particularly common. The cytological and genetical phenomena behind these situations have been much studied and are understood in considerable detail; there is no need or space for them to be discussed here. In various species of the genus *Oenothera* ($2n = 14$) all the individuals exist as complex translocation heterozygotes which at meiosis form multivalents of varying size (often 14-valents). This, together with its systems of gametic and zygotic lethality, has made *Oenothera* a unique genus whose peculiarities must be fully grasped before its mode of speciation and the consequences of its hybridization can be properly understood. Cleland (1972) has recently written an excellent monograph on this topic.

In species which frequently produce chromosomal inversions or transloca-tions, populations which become isolated reproductively will tend to build up a stock of different structural alterations, and this can lead to intersterility between the races, finally even resulting in speciation. Probably the best-known example of this in the flowering plants is in *Crepis,* where it is thought that one chromosome pair of *C. neglecta* ($2n = 8$) became so shortened by unequal reciprocal translocations that its loss was not deleterious and resulted in the formation of *C. fuliginosa* ($2n = 6$) (Tobgy, 1943). Other taxa where there has been a reduction in chromosome number might have arisen in the same way, e.g. *Vicia sativa* (Hollings and Stace, 1974). These, however, are extreme cases, for most chromosome races differing in their pattern of translocations have the same chromosome number.

It seems that by themselves inversions are much less effective than translocations in promoting racial intersterility, and that they have rarely, if ever, caused speciation. On the other hand many interspecific hybrids which are sterile show apparently normal chromosome pairing, and it is possible that the chromosomes of such species pairs differ by very small structural alterations such as inversions or translocations. This condition has been described by Stebbins (1945) as cryptic structural hybridity, and the whole topic of structural hybridity was reviewed by him five years later (Stebbins, 1950).

CELL AND SOMATIC HYBRIDIZATION

In 1960 it was reported by Barski *et al.,* that genetically different mouse cells grown in culture conditions could on occasion fuse together to form cells with a single nucleus containing the chromosomes of both cell-types. Later on it was shown by others that the yield of fused cells could be increased about a hundredfold by the addition of various viruses, especially Sendai virus. Using this technique Harris and Watkins (1965) were able to obtain cell-hybrids between different species of mammals such as man and mouse, and since then even wider hybridizations have been obtained, e.g. mammals with birds and amphibians. The subject has developed greatly in the last ten years, and two extensive reviews have already appeared (Harris, 1970; Ephrussi, 1972). Ephrussi concluded that one can now effect "the production of practically any hybrid one wishes to have for any purpose", but he was presumably confining himself to the vertebrates as relatively little work has been done outside that group.

It has been known for several years that plant protoplasts (stripped of their cell walls) could be fused together in certain circumstances. Power *et al.* (1970) were able to produce interspecific cell-hybrids without the use of viruses by immersion in sodium nitrate solution followed by low-speed centrifugation.* Because of the totipotency of plant cells (a feature not shared by animal cells) it was suggested that the fusion could be taken further than in animals, and whole hybrid plants formed, and recently Carlson *et al.* (1972) have announced that such plants have been raised. *Nicotiana glauca* (2*n* = 24) and *N. langsdorffii* (2*n* = 18) were used as parents, and cell-hybrids were made from leaf-mesophyll cells using a similar technique to that above. The cell-hybrids developed into callus tissue and then leafy shoots, after which they were grafted on to stocks of *N. glauca* for further development to the fruiting stage. Sterile F_1 hybrids and fertile amphidiploids (2*n* = 42) derived from these two species are known. The somatic hybrids resemble the sexual amphidiploids in morphology, anatomy, cytology, peroxidase isozyme patterns and chromosome number, and they are similarly fertile.

The future prospects of these techniques are exciting. The formation of somatic hybrids might have two important advantages over that of sexual hybrids: it is possible that "hybrids" can be made between much more distantly related taxa (by analogy with animal cell-hybrids); and the hybrid obtained is a fertile amphidiploid from the start, whereas most sexual hybrids are sterile to varying degrees.

Whether somatic hybrids have any taxonomic implications, and whether they will develop into important tools for the study of crop improvement, evolution, differentiation or other fields, remains to be seen. It is not yet known whether plant somatic hybrids can be obtained between very widely differing taxa. In such animal cell-hybrids a certain instability remains; for example chromosomes are usually selectively lost from the hybrid nucleus, usually those of one parent at the expense of those of the other. Nevertheless such cells are true hybrids, for the genes of the two parents show dominance and complementarity as in heterozygous whole organisms, and the DNA–RNA–protein coding system can incorporate elements from different parents.

It is perhaps not out of place here to draw attention to sexual recombination in bacteria where, by various different means, foreign DNA is introduced into a bacterial cell. Its subsequent incorporation in some ways bears a closer resemblance to cell-hybridity than to sexual reproduction, for example in the apparently broad taxonomic spectrum over which it can occur. Certain aspects of viral infection, diploidization of heterothallic higher fungi and other phenomena may be similarly related.†

* See also Kao and Michayluk (1974).
† See also Carlile (1973) and Ling and Ling (1974).

DNA HYBRIDIZATION

The DNA (genetic code) of all organisms except some viruses exists in a double-stranded form, the two sister strands being precisely complementary to each other. It is possible to dissociate these two sister strands, and to promote their reunion, and these techniques are widely used in order to test for DNA similarities between different sorts of organisms. Broadly speaking the degree of association which occurs between the DNA from different sources is proportional to the similarity of the molecules, and this is used as an assessment of taxonomic similarity. An outline of the techniques involved, examples of results obtained and further references were given by Kohne (1968).

Just as cell-hybrids can be obtained between more widely dissimilar taxa than sexual hybrids, so DNA hybridization can encompass an even wider spectrum, since at this molecular level the genetic material of all organisms is essentially the same. In fact there seems to be no limit to the scope for DNA hybridization, and taxa as wide apart as vascular plants, fungi and mammals can be directly compared.

GRAFT-HYBRIDS

The art of grafting has a very long history, and it has long been known that the ability to graft the scion of one plant on to the stock of another is some measure of taxonomic similarity, for successful grafts always involve relatively closely related plants (e.g. apple–pear, *Castanea–Quercus*). However, the stock and scion are joined only at the point of union of the graft, and the branches produced by each part are usually wholly typical of that part alone, although in some cases modifications of the scion attributable to the characteristics of the stock are apparent (e.g. see Yablokov, 1962). Sometimes, however, buds may arise from the point of the graft-union itself, so that the apical meristem combines tissues from each, either anticlinally or periclinally separated. The branches which develop from this bud are similarly mixed, and show intriguing combinations of the two parents. The most interesting cases are periclinal, where the tissue of one parent is covered by a few cell-layers of the other. The most famous example of this is the *Cytisus purpureus–Laburnum anagyroides* graft-hybrid, which arose accidentally in France in 1825 but which, despite numerous attempts, has never been resynthesized. The hybrid is not at all rare in gardens, but all plants arose from the single original plant. Sneath (1968) has carried out a numerical taxonomic study of a tree grown at Leicester.

Some early theories attempted to account for graft-hybrids by a form of cell hybridization, whereby vegetative cells, including their nuclei, fused, but this has never been demonstrated and such an occurrence is now considered to be unlikely. Since the term graft-hybrid came to be particularly associated with theories of cell-fusion many authors prefer the use of the term graft-chimera, but unfortunately the word chimera (like "hybrid") has also been used in a variety of senses.

Many graft-hybrids are unstable in that some buds, having somehow lost the tissues of one parent, develop into branches purely of the other parent. Thus the *Cytisus–Laburnum* graft-hybrid frequently produces pure *Laburnum* branches. Graft-hybrids are either sterile or produce seeds which are purely of the parent occupying the core of the graft-hybrid (*Laburnum* in the example above).

Sexual hybrids between *Cytisus purpureus* and *Laburnum anagyroides* have not been obtained, but they exist between the parents of other graft-hybrids, e.g. *Crataegus monogyna* x *Mespilus germanica*, and usually differ noticeably from the graft-hybrids in morphological characters. Their vegetative stability and the production of hybrid progeny sets them well apart. There are at least four known hawthorn–medlar graft-hybrids which differ in rather minor morphological characters as well as in stability. The two best-known ones (see under *Crataegus* x *Mespilus*) arose in 1899 as two branches from a single graft-union. Useful reviews are given by Bean (1911), Daniel (1914–15) and Neilson-Jones (1969).

6. Historical aspects

Hybridization has a long and fascinating history which has been very well covered by two books: Roberts (1929) dealt with the subject up to the rediscovery of Mendelism in 1900, although he provided little information on work before that of Kölreuter in the 1760s; and Zirkle (1935) reviewed the period up to Kölreuter's contributions.

A study of these books shows clearly that it is not easy to quote the year in which a botanist or gardener first scientifically carried out an interspecific hybridization, for the operation was carried out many times with varying degrees of accident or deliberation, and with varying but increasing degrees of comprehension of the process and its results.

Linnaeus reported on hybridization experiments which he carried out in 1758 in crossing *Tragopogon porrifolius* and *T. pratensis*, and this is often claimed as the earliest deliberate interspecific hybrid. His report was translated into English, with annotations, by J. E. Smith (Linnaeus, 1786). However, other artificial hybrids had been obtained before then, the first perhaps in 1717 by T. Fairchild, who crossed Carnation (*Dianthus caryophyllus*) with Sweet William (*D. barbatus*). Also before Linnaeus' book Haartman (1751) had published what is considered the first treatise on plant hybridization, entitled simply *Plantae Hybridae*. It listed 100 plants which were considered to be of hybrid origin, though which of them were so is a matter for speculation. Among the 100 were *Trifolium hybridum*, said to be *T. pratense* x *T. repens*, a reputedly interfamilial cross, *Verbena officinalis* x *Veronica maritima*, and the well-known hybrid lucerne, *Medicago falcata* x *M. sativa*.

The development of our knowledge of hybrids can be conveniently divided into three phases: firstly, the discovery of sex in plants and the demonstration that artificial intraspecific and interspecific hybrids could be obtained (this period ended in the 1760s, with Kölreuter's contributions), i.e. that period covered by Zirkle's book; secondly, a period of extensive artificial hybridization and the demonstration of the existence of widespread natural hybridization (ending about 1900 with the development of genetics), i.e. the period covered by the bulk of Roberts' book; and thirdly, investigations of the cytological, genetical and evolutionary aspects of hybrids, and their widespread use as a technique in plant breeding. It is arguable that the third phase has passed and another arrived; how one should describe this is probably best left until it also has finished, but a more precise understanding of the overall role of hybridization in evolution at the species level and above, and the development of somatic hybridization, might prove to be major features.

Undoubtedly the most comprehensive treatise on hybrids that has appeared is *Die Pflanzen-Mischlinge* (Focke, 1881). The main part of the book is concerned with a systematic treatment of hybrids, under families and genera, from Ranunculaceae to Fucaceae, but there are in addition six shorter review sections: a historical account; the occurrence and formation of hybrids; the characteristics of hybrids; the nomenclature of hybrids; hybrids in relation to nature and mankind; and topics related to hybrids such as xenia and graft-hybrids. This work of 569 pages amounted to a thorough survey of all aspects of hybrids then under consideration, and it provided a reference-point for future work over several decades. When one remembers that it was written before the days of genetics and chromosomes, let alone cell biology and biochemistry, its true status becomes clear.

According to the chronological classification set out above, *Die Pflanzen-Mischlinge* appeared towards the end of the phase of realization that hybridization was widespread in the wild. Nevertheless, among many taxonomists, a considerable amount of reluctance to accept this fact was apparent. In Britain particularly there was still much argument concerning not only the hybrid nature of many taxa but also the general notion that hybrids were at all frequent and ever fertile, and this extended into the present century. A few examples will demonstrate this point.

Among the well-known botanists who claimed or implied at various times that natural hybrids were always intermediate in appearance, sterile, and sporadic in occurrence were J. E. Smith, H. C. Watson, J. D. Hooker, N. E. Brown and C. B. Clarke. Hooker (1853, p. xv) stated that "no hybrid has ever afforded a character foreign to that of its parents", and that "hybrids generally are constitutionally weak, and almost invariably barren". He used this as "the most satisfactory proof we can adduce of hybridization being powerless as an agent in producing species." The extremely sceptical views of J. E. Smith concerning *Salix* hybrids and of N. E. Brown concerning *Epilobium* hybrids are quoted below in the accounts of those genera, and it seems that C. B. Clarke was of a similar opinion. He (Clarke, 1891) said of Haussknecht's (1884) *Monographie der Gattung Epilobium* "his varieties and hybrids are altogether beyond me", which with other disparaging remarks sparked off a lively controversy between him and E. S. Marshall. Marshall (1891) said, "I cannot see why it should be inconceivable for insects to do unconsciously what all are agreed that florists do consciously and more clumsily". However, Clarke (1892a) considered that most putative hybrids could be more satisfactorily explained as variants of one of the claimed parents or as intermediates between extremes of one species: "Some hybrid-monger gets an example that has glabrous leaves, but acute sepals, and at once describes it as the new 'hybrid' " between what were only "two of the hybrid-monger's own diagnoses". Marshall (1892), talking of hybrids in *Salix* and *Epilobium*, was in "no more doubt about their frequency than about the fact that two and two make four". Clarke (1892b) clearly considered that fertility in intermediates showed that they were not interspecific hybrids but "should be called only crosses", and he even doubted "if observations or even experiments in botanic intercrossing of very closely allied (dubious) wild species are very promising of theoretic results". Marshall's (1893) later note and Beeby's (1892) intervention seem to have ended this personal argument, but much later reviewers such as Heiser (1949) still considered it necessary to cite evidence for hybridization. Perhaps we may leave the last word with Anderson (1949): "Those who have pioneered in the analysis of introgression are sometimes accused of 'seeing hybrids under every bush'. The truth of the matter is that, in certain groups of plants and animals, the results of hybridization are more widespread than had previously been suspected by most biologists . . .".

The surveys of hybrids and hybridization, which began to appear in some numbers towards the end of the nineteenth century, have been discussed in some detail by Focke (1881) and Roberts (1929), and will not be listed here. The following geographically based accounts, among others, must have had considerable influence on field-workers in the areas concerned. Neilreich (1852) presented a rather short annotated list of the hybrids found in the vicinity of Vienna, Austria, and Lasch (1857) catalogued the hybrids of the province of Brandenburg, E. Germany, his list being based on an earlier prodromus (Lasch, 1831). Brügger (1880) provided a long systematic discussion of the hybrids found in the wild in Switzerland, including comments on variability, distribution and experimental work. Camus (1907) presented to a genetical conference in 1906 a most interesting discussion on European natural hybrids, including notes on their relative frequency in various families. Among his general points is the comment that "clearings in woods are favourable places" for hybrids, easily pre-dating Wiegland's (1935) widely quoted paper on the subject, but Camus misinterpreted the significance of this (see Chapter 9). It is most unfortunate that the *Analytical catalogue of spontaneous hybrids of European plants* which he had prepared for publication (and which had been the result of 30 years' work) was "too long to be embodied in a Report of the Conference"; I have not detected it published elsewhere. At the same conference Lynch (1907) gave a selected list of Continental wild hybrids which would be suitable for experimental work; it is a useful compilation from various Floras, etc., published since Focke's (1881) monograph. MacDougal (1907) gave a list of known natural hybrids in North American flowering plants, amounting at that time to only 117. Leveillé's (1917) list of French hybrids was accompanied by very few comments and, as he said in his preface, "Beaucoup ne sont pas hybrides".

Guétrot (1927–31) embarked on an over-ambitious series of publications which was aimed at producing a résumé of all the known data concerning the *Plantes hybrides de France*. Parts 1 and 2 (1925–26) appeared in 1927 as one volume and included descriptions of 69 hybrids; Parts 3 and 4 (1927–28), similarly in one volume and published in 1929, contained 32 hybrids; and the third volume (Parts 5–7, 1929–31) appeared in 1931 and was devoted to a treatment of the genus *Saxifraga* by D. Luizet. The coverage of the hybrids is very varied, in some cases detailed and authoritative, in others superficial and uncritical, but there is much of interest in the work, both in the systematic treatment and in the various supplementary chapters on biography, nomenclature and some other aspects. Unfortunately Guétrot's views on hybrid nomenclature were eccentric, and involved changes to a good proportion of the binomials already in existence. I have not traced any volumes subsequent to the three above, and have seen nothing of the promised journal *Revue internationale d'hybridologie végétale*. Cockayne (1923), and Cockayne and Allan (1934) discussed the information which had accumulated on wild hybrids in New Zealand, much of it due to their own work from 1912 onwards. Work in Australia has not kept pace with that in New Zealand (for obvious reasons), but Clifford (1963) has summarized some salient aspects.

Concerning the British flora, Beeby (1892) made a list of hybrids recorded by Focke (1881) and which he thought should be sought in Britain. Many of them are now known in Britain, but many are not, although several of these are false identifications. In 1906 Lynch (1907) listed hybrids which have been recorded from Britain, with contributions by E. F. Linton and W. M. Rogers for *Salix* and *Rubus* respectively. The list was not complete, but that produced by Linton (1907) just after was more so, and also provided notes on frequency, fertility and related matters. Some of Linton's prefatory comments are as applicable today as then. He spoke of "the danger . . . of too much being made of hybrids, and hybridity being set up as the chief factory of variation in all genera where great variety occurs". Certainly in the first part of the present century a hybrid origin was attributed to far too many taxa, as Linton's list shows.

Linton also made reference to the *London Catalogue of British Plants*, which had for the first time in the then current (9th) edition treated putative hybrids as such by the use of a multiplication sign. He was of the opinion that many other hybrids were still treated as species and varieties, and history has shown him to be correct. The 7th edition (Watson, 1874), the last edited by H. C. Watson, contained eight hybrids, the 9th edition (Hanbury, 1895) about 145, and the 11th edition (Hanbury, 1925), which was the last of all, about 402. Druce's (1928) *British Plant List* contained "over 500" hybrids, but Druce was noted for his over-willingness to identify plants as hybrids, and many of his determinations are now known to be errors. Fortunately he preserved vouchers of most of his important records; if his predecessors all had done likewise many of them would not have been spared the criticisms now often levelled at Druce. Despite the errors included in Druce's total, Dandy's (1958) *List of British Vascular Plants* contains 538 hybrids, and the present book adds to that number.*

* 626 are accepted in this book, with a further 122 possible ones.

Praeger (1951) published an annotated list of Irish hybrids, which contains much of interest, but it is incomplete and very uncritical, and there are numerous errors. Webb (1951), largely prompted by Praeger's paper, provided a commentary on certain aspects of Irish hybrids.

Far more work has been carried out on the hybrids of a few commercially important plant groups than on all the others combined. The Gramineae, in particular, have received an enormous amount of attention, and Ullmann (1936), Myers (1947), Carnahan and Hill (1961) and Knobloch (1968) have provided valuable reviews. Useful national surveys also exist for Denmark (Anderson, 1931) and France (Camus, 1957, 1958; Hansen, 1959). Among other groups the reviews of Johnson (1939) and Richens (1945) on forest trees, of Garay and Sweet (1974) on orchids, and of Yarnell (1956) on crucifers, should be mentioned.

Arguments which have taken place in the last 50 years have mostly concerned the hybrid origin or otherwise of particular taxa, or the overall significance of hybridization in plant evolution. The notions on the latter topic held by some early workers were crude, often involving the crossing of two species to produce a third in a single step, e.g. *Trifolium hybridum (T. pratense* x *T. repens), Epilobium roseum (E. adnatum* x *E. montanum),* and *Primula elatior (P. veris* x *P. vulgaris).* Reviews of the status of plant hybridization, with particular reference to evolution, have been made by many leading evolutionists, cytogeneticists and taxonomists. A chronological study of their publications clearly shows how ideas have advanced as detailed experimental results have replaced subjective intuition, and as the number of situations investigated has increased. The following is a small selection of some of the prominent milestones.

In July 1899 the Royal Horticultural Society sponsored an "International Conference on hybridisation (the cross-breeding of species) and on the cross-breeding of varieties", which was fully reported in a volume of their journal devoted to the conference (Wilks, 1900). Much of the conference, of course, comprised discussions of artificial hybridization in horticultural groups such as orchids, ferns, etc., but of general interest are articles by Wilson (1900), which analysed in considerable detail the morphology of a number of cultivated hybrids, and Rolfe (1900), which surveyed the state of play regarding hybrids and hybridization from a taxonomic viewpoint. The timing of the symposium was, in retrospect, particularly appropriate, coming only one year before the rediscovery of Mendelism, shortly after which two further conferences were held. In 1902 the Horticultural Society of New York staged a conference on "Plant Breeding and Hybridization", whose proceedings were published as Volume 1 of the Memoirs of the Society (not seen), and in 1906 the Royal Horticultural Society sponsored in London "The Third International Conference 1906 on Genetics" (Wilks, 1907). The differences in emphasis between the reports of the 1899 and 1906 conferences are very revealing, and reflect as well as anything the actual impact of Mendelism on botanists and gardeners. Apart from various papers mentioned elsewhere in this Introduction the article by Rosenberg (1907) on *Drosera* cytology is particularly important.

After about 1910 the tone of most scientific works concerning hybrids changed, for botanists then made serious attempts to utilize contemporary developments in genetics in explaining the relevance of hybridization to the evolution of species. Winge (1917, 1932) wrote masterly and now classic accounts of the origin of new fertile species from sterile hybrids by polyploidization, a concept which has become basic in modern theories of plant evolution.

Lotsy (1916) provided a detailed argument claiming that hybridization was the dominant factor in plant evolution. His book was described by Allan (1937) as "that delightful provocative small volume" which "cast aside the results of the heavy labours of years on phylogeny as 'a product of phantastic speculations'." His other publications are equally provocative, e.g. the one (Lotsy and Goddijn, 1928) describing trips to South Africa to study "the bearing of hybridisation upon evolution", mainly in humans and succulent dicotyledons. Almqvist (1926) concluded that in Sweden, on the other hand, "nature at present is standing rather still", since new, fertile species clearly derived from hybridization were rare, despite the frequency of hybrids as a whole. Ostenfeld (1928) similarly arrived "at the general conclusion that species-hybrids in nature *apparently* do not play any role worth mentioning, as the species keep constant and the hybrids soon disappear." But he demonstrated that this was not always the case, and outlined examples of amphidiploidy and fertile hybrid segregates which indicate that new species did arise by hybridization. His was a very successful, reasoned discussion of the state of knowledge at that time.

Renner's (1929) review is in some ways complementary to Ostenfeld's article, as it sets out in detail a summary of cytological and experimental work thitherto carried out on species hybrids, but the section "Die Artbastarde in der Natur" occupies only one of the 161 pages, in marked contrast to the emphasis of the contemporary research being carried out by the genecologists, such as Turesson, mentioned in Chapter 3. Useful abstracts of classic work on *Erophila, Funaria, Epilobium* and *Oenothera*, etc., and an extensive bibliography, were included by Renner. Watson (1932) touched on many similar points. Exhaustive reviews of the cytological analysis of hybrids were produced by Sax (1935) and Stebbins (1945), providing a wealth of data on which to base evolutionary theory. Allan's (1937, 1949) papers in the same journal provide a similar background on many other aspects of hybridization. A useful historical account of plant breeding work in the U.S.S.R. is given by Goryunov (1962).

One of the major new concepts concerning hybridization which has emerged in the past 30 years is that of introgressive hybridization, or introgression, a term first introduced by Anderson and Hubricht (1938) to describe the gradual infiltration of the germplasm of one species into that of another by means of hybridization followed by successive backcrosses. Important reviews have been made by Heiser (1949) and Anderson (1949, 1953), and the phenomenon has been shown to have a very considerable evolutionary significance which has been taken into full account by subsequent authors, e.g. Baker (1951), Dillemann (1954) and Stebbins (1950, 1959).* The papers of Baker and Dillemann are of

---

* See also Brobor (1973) and Heiser (1973).

particular interest to British botanists, as they are well composed and are among the very few not based largely on North American work.

Whether any more major new concepts will arise from the vast amount of work being pursued at present remains to be seen.

7. Hybrid nomenclature

This topic has promoted an enormous amount of discussion and argument which is in a sense unfortunate for many would consider that it is not the most urgent or vital of the problems which confront students of hybrids.

Differences of opinion concern two major aspects: firstly, broadly the same points that were discussed in Chapter 2 under "What is a hybrid?"; and secondly, whether or not accepted interspecific hybrids should receive a name in addition to their "formula". It is not profitable to reiterate all the pros and cons, but broadly speaking the main arguments against the use of a name for a hybrid are that hybrids are not taxa but crosses between taxa; that it causes an over-encumbrance of the literature with extra names; and that many hybrids exist in many different guises (backcrosses, segregates, variants derived from different variants of the parents, reciprocal crosses, etc.), i.e. they may be multiform rather than pauciform, and one name does not adequately cover them all. On the other hand it is undoubtedly more convenient to refer to a plant by a single name. A name is subject to fewer changes than a formula—on average half the number of nomenclatural changes and far fewer taxonomic changes; unless hybrids are given names those of unknown parentage have no name or formula; and if there is no system of hybrid nomenclature it is difficult to refer to the different variants of a hybrid. More details of the sorts of proposals forwarded can be found in Rickett and Camp (1948), Little (1960), Wagner (1968) and Baum (1969).

Whether or not one approves of hybrid names there are many thousands of them, and more appear almost daily. The application of the names of hybrids, like those of any other plants, is governed by the *International Code of Botanical Nomenclature* (Stafleu et al., 1972) and the *International Code of Nomenclature of Cultivated Plants* (Gilmour et al., 1969). Recognizing that differences of opinion exist, the *Codes* sanction the use of a special name for an interspecific hybrid "whenever it seems useful or necessary" (which to some would be always), but do not make it obligatory. On the subject of "what is a hybrid?" the *Codes* are much less helpful. They simply state that hybrids and "putative hybrids" should be treated similarly, and they allow "amphidiploids, and similar polyploids treated as species" to be named exactly as species, implying that diploids of hybrid origin must always be named as hybrids. The main provisions of the *Codes* are as follows:

Interspecific Hybrids (i.e. within one genus)

1. Interspecific hybrids are designated by a formula consisting of the names of the two parents connected by a multiplication sign, e.g. *Calystegia sepium* × *C.*

silvatica. The names should be in alphabetical order, or the female parent (if known) may be placed first.

2. Interspecific hybrids may in addition be designated by a binary name, which consists of the genus name and a hybrid epithet separated by a multiplication sign, e.g. (in the above example) *Calystegia* x *lucana.*

3. The binary name is equivalent in rank to a species, and the same regulations govern its formation, publication, priority and use. It is followed by its authority, e.g. *C.* x *lucana* (Tenore) G. Don. If it was first published as a species no new combination is involved when the multiplication is added to denote hybrid origin. The example above was considered by Don to be a species, *C. lucana* (Tenore) G. Don. To be validly published the name must be accompanied by a latin description and a type specimen cited, etc. It covers all hybrids derived from the same interspecific cross, e.g. backcrosses, segregates, polytopic origins, etc.

4. Hybrids between infraspecific taxa of the same species are designated by a formula and (if required) by a name (without a multiplication sign) at the same rank as that of its parents. Thus (if the two above species are considered subspecies of one species) *Calystegia sepium* subsp. *sepium* x *C. sepium* subsp. *silvatica* = *C. sepium* subsp. *lucana.*

5. Variants of an interspecific hybrid may be described as nothomorphs (abbreviated nm.), which is equivalent in rank to a variety, e.g. *Hypericum* x *desetangsii* nm. *desetangsii* and nm. *carinthiacum.* Variants of an interspecific hybrid which have been described as subspecies, varieties, forms, etc. are treated as having been described as nothomorphs, and substitution of the abbreviation nm. does not constitute a new combination.

6. Variants of cultivated interspecific hybrids may similarly be classed as cultivars, e.g. *Viburnum* x *bodnantense* 'Dawn'.

Intergeneric Hybrids

1. Intergeneric hybrids are designated at the generic level by a formula consisting of the generic names of the two parents connected by a multiplication sign, e.g. *Cattleya* x *Epidendrum.* The names should be in alphabetical order.

2. Intergeneric hybrids may also be designated at the generic level by a hybrid-genus name, which in the case of a bigeneric hybrid is constructed from parts of the two parental names joined together and preceded by a multiplication sign, e.g. (in the above example) x *Epicattleya.* In the case of a quadrigeneric (or higher) hybrid the name is derived from that of an eminent botanist, with the ending-*ara,* e.g. x *Rothara* (= *Brassavola* x *Cattleya* x *Epidendrum* x *Laelia* x *Sophronitis*). In the case of a trigeneric hybrid either method of naming may be used.

3. The hybrid-genus name is equivalent in rank to a genus, but to be validly published it merely requires mention of its parental genera, i.e. no description or designation of a type is needed. It is followed by the authority, e.g. x *Epicattleya* Rosita. It covers all hybrids between any species of its parental genera.

4. Hybrids between infrageneric taxa (e.g. subgenera) of the same genus may be named in the same way as a hybrid-genus, e.g. *Iris* subgen. x *Regeliocyclus* (= *Iris* subgen. *Regelia* x subgen. *Oncocyclus*).

5. Intergeneric hybrids are designated at the specific level by a formula consisting of the binary names of the two parents connected by a multiplication sign, e.g. *Festuca pratensis* x *Lolium perenne.* They may also be designated by a binary name consisting of the hybrid-genus name followed by a new epithet, e.g. (in the above example) x *Festulolium loliaceum.*

The idea of naming intergeneric hybrids at the genus level by merely citing the parents is very attractive to many botanists, particularly those not expert in constructing latin descriptions, and some would suggest that the idea ought to be extended to the naming of individual species-hybrids. There are, however, very real disadvantages in such a short-cut method. For example it becomes much easier and thus far more tempting to name hybrids, and in some cases hybrid-generic names have been coined by non-taxonomists without any intention of formal naming. This has led to the existence of several synonyms for many well-known hybrids, e.g. x *Elyhordeum,* x *Elymordeum* and x *Hordelymus* for *Elymus* x *Hordeum.* Secondly, since a hybrid-genus name is built up from parts of the two parental genus names, the former has to be changed whenever either of the latter is changed (whether for taxonomic or nomenclatural reasons). Hence the hybrids between *Coeloglossum* and the various marsh orchids have gone under the names x *Orchicoeloglossum* or x *Dactyloglossum* according to the genus (*Orchis* or *Dactylorhiza*) assigned to the latter parent. Clearly the extension of these rules to species-hybrids would immediately necessitate coining a very large number of new names.

In the past many other methods of designating hybrids have been used. For instance Focke (1881), who preferred to use formulae, often substituted various symbols for the multiplication sign to indicate degrees of fertility, intermediacy, certainty of the parentage, etc., and more modern workers have sometimes advocated similar procedures (see Terrell, 1963). A cruder method was to describe a hybrid plant as super-X or super-Y, according to the parental characters considered to predominate. Frequently a hybrid has been designated by a hyphenated specific name, such as *Verbascum nigro-pulverulentum*; according to the *Codes* these are to be considered formulae, not hybrid names.

Most of these methods not sanctioned by the *Codes* are either too imprecise or too complicated for general use. It is generally agreed that at least some plants of hybrid origin (e.g. amphidiploids; hybrids of unknown, uncertain or disputed parentage as in *Salix, Ulmus, Potamogeton, Rosa,* etc.) must be given names, and unless the rules governing them are kept as simple as is compatible with utility they will be abused, and a chaotic nomenclatural situation will arise.

Finally, a note concerning the nomenclature of graft-hybrids (governed by the *Code* for cultivated plants) is necessary. Graft-hybrid nomenclature is largely analogous to that of sexual hybrids, except that a plus sign (+) is used instead of a multiplication sign (x), e.g. *Syringa + correlata,* + *Laburnocytisus adamii.* The name of a graft-hybrid cannot be the same as that of a sexual hybrid between the same taxa. Thus the sexual hybrid *Crataegus* x *Mespilus* is x *Crataemespilus,* and the graft-hybrid is + *Crataegomespilus*. When naming graft-hybrids (which are relatively rare plants) it is necessary only to cite the names of the parents, whether at the generic or specific level.

8. Natural hybridization

Species are normally prevented from crossing by isolating mechanisms or breeding barriers, the breakdown of which allows hybridization to take place. Isolating mechanisms are of many different sorts, but they have been classified admirably by numerous authors who have cited examples of species apparently isolated by each of the various categories (e.g. Dobzhansky, 1937; Stebbins, 1950; Dobzhansky, 1951; Baker, 1951; Riley, 1952; Davis and Heywood, 1963; Solbrig, 1968; Levin, 1971). It should be borne in mind, however, that in nature most species are isolated by a combination of several sorts of barrier, and that those separated by only one sort (unless it is a strong, genetically controlled cross-incompatibility) are those most likely to produce hybrids in the wild. The well-known species-pair *Geum rivale* and *G. urbanum* are separated in Britain by geographical, ecological, seasonal and ethological isolating mechanisms, but genetic incompatibility is relatively weak and hybrids are common.

Isolating mechanisms preventing the formation of hybrids are often said to be of two main sorts: external and internal. In the following list of prezygotic mechanisms the first six are largely of the former type and the last three of the latter, but the distinction is of limited value because most external barriers rely to varying degrees on internal features as well.

GEOGRAPHICAL ISOLATION

Several allopatric pairs or groups of species have been shown experimentally to be interfertile, i.e. they largely or wholly owe their present genetic isolation to their geographical separation. Many examples are known, such as *Platanus occidentalis* and *P. orientalis* from North America and the Mediterranean respectively, and *Catalpa ovata* and *C. bignonioides** from China and North America respectively. The disjunctions are usually much narrower or only partial, however, e.g. *Primula elatior* and *P. vulgaris* in England (as well as in Europe generally). Of the 17 European species of *Antirrhinum* 14 are confined mostly to different small areas of Spain and Portugal (Webb, 1972), and similar statements can be made about other genera, such as *Ulex* and *Sempervivum*, etc. At least some of the species in these three genera are known to be interfertile. In many instances geographically separated taxa which can interbreed are recognized only at the subspecific level. Examples which have been investigated in Britain in recent years are to be found in *Gentianella*, *Anthyllis* and *Dactylorhiza*, among others.

* (Bignoniaceae)

In order for geographical isolation to break down naturally either one or both species must extend their ranges to become sympatric, or chance, long-range pollen dispersal must occasionally take place. Indeed, the possession of an outbreeding system coupled with powers of long-range pollen dispersal has probably contributed to the wide distribution of many taxa, e.g. several *Pinus* species, which might otherwise have been subject to speciation or subspeciation.

Man has often broken geographical barriers by transporting plants, intentionally or otherwise, into areas where other compatible species exist. Thus one might expect to find a high incidence of hybrids involving alien, naturalized species with native ones. Although no detailed analyses appear to have been undertaken there are many examples of this phenomenon in the British flora, e.g. *Heracleum mantegazzianum* x *H. sphondylium*, *Calystegia sepium* x *C. silvatica*, *Senecio squalidus* x *S. vulgaris* and *S. viscosus*, *Linaria purpurea* x *L. repens*, *Juncus effusus* and *J. inflexus* x *J. pallidus*, *Senecio cinerea* x *S. jacobaea*, *Conyza canadensis* x *Erigeron acer*, *Tragopogon porrifolius* x *T. pratensis*, *Endymion hispanicus* x *E. non-scripta*, *Spartina alterniflora* x *S. maritima* and *Raphanus raphanistrum* x *R. sativus*, besides other cases in genera such as *Epilobium*, *Rumex*, *Rosa* and *Verbascum* where hybridization is also widespread among native species. Allan (1937) and Almqvist (1926) have previously commented on this aspect of hybridization in other areas, and Clifford (1963) drew attention to the fact that the *Juncus* example above is reversed in Australia, where *J. pallidus* is native but the other two naturalized. Allan (1940) later, however, emphasized the relative scarcity in New Zealand of hybrids among naturalized species.

ECOLOGICAL ISOLATION

Interfertile species occupying different ecological niches in the same broad geographical zone have been used as experimental material in many classical genecological studies. The best-known example is *Geum rivale* and *G. urbanum*, which are characteristic of ditches and marshes and of woods and hedgerows respectively (Marsden-Jones, 1930), but there are many other equally clear cases, e.g. *Primula veris* and *P. vulgaris*, *Silene maritima* and *S. vulgaris*, *Silene alba* and *S. dioica*, *Hypericum maculatum* and *H. perforatum*, *Spergularia marina* and *S. rupicola*, *Juncus effusus* and *J. inflexus*, *Galium saxatile* and *G. sterneri*, *Alopecurus geniculatus* and *A. pratensis*, etc. Moreover subspecies within various species, in Britain at least, are often ecologically rather than geographically separated, e.g. *Rhinanthus minor*, *Melampyrum pratense*. It is not to be denied that ecological separation is a relatively weak barrier to hybridization, and in all the above examples the habitat requirements of the two species do overlap, or the different habitats may come into very close contact, and hybrids have been found, sometimes commonly but sometimes over only part of the area of contact of the two species, e.g. *Scutellaria galericulata* x *S. minor*. In addition, disturbance by man has often caused intermingling of the habitats and has increased hybridization, particularly (among the above) of *Silene alba* x *S. dioica* and *Geum rivale* x *G. urbanum*.

SEASONAL ISOLATION

Of the many examples known of species which are separated by different flowering times the most marked are those in temperate climates where one species is vernal and the other autumnal in flowering, e.g. *Ulmus* (Santamour, 1972), various corm- and bulb-bearing genera such as *Galanthus*, *Crocus* and *Cyclamen*, and *Madia* (Compositae), but less extreme cases are much more frequent. The three common species of *Juncus* subgenus *Genuini*, *J. conglomeratus*, *J. effusus* and *J. inflexus*, flower successively during the early summer at approximately two-week intervals, but the flowering seasons often extend for more than two weeks so that there is considerable overlap between the first two and second two species, but much less or none between *J. conglomeratus* and *J. inflexus*. Of the three possible hybrids *J. conglomeratus* x *J. inflexus* is the only one not known to occur in the wild. Three of the four British species of *Vulpia* are known to hybridize with *Festuca rubra*. The absence of hybrids involving the fourth species, *V. ambigua*, might well be due to its distinctly earlier flowering. In Britain *Ophrys sphegodes* flowers well before *O. apifera* and *O. fuciflora*, and in the Mediterranean area *O. fusca* usually well before *O. lutea*, but overlaps and therefore hybrids occur. In Denmark the hybrid between *Sambucus nigra* and *S. racemosa* is very rare largely because of the earlier flowering of the latter by 7–8 weeks (Winge, 1944).

Unless the seasonal difference is extremely well marked it is likely to be only a weak isolating mechanism, as vicissitudes of the weather frequently cause plants to flower well outside their normal periods. Moreover different weather patterns in different areas can give rise to lesser or greater degrees of isolation. Agnew (1968) suggested that the presence of *J. conglomeratus* x *J. effusus* only in upland areas of Wales was because of the shorter and more overlapping flowering seasons of the two species there. The discovery of hybrid populations in coastal habitats in eastern Scotland (C. A. Stace, unpublished) supports this suggestion.

TEMPORAL ISOLATION

Differences of flowering time within the day have frequently been reported between species which are interfertile, but it is to be expected that they very rarely, if ever, cause effective isolation. *Silene alba* is nocturnal and *S. dioica* diurnal in anthesis, and a similar relationship exists between various European species of *Matthiola*. Philipson (1937) found that *Agrostis stolonifera* shed its pollen in the morning and *A. tenuis* in the afternoon, and the same has been claimed for *Alisma lanceolatum* and *A. plantago-aquatica*, but in both cases overlaps occur and there must be a considerable amount of pollen carried over either in the air or on insect vectors from one period to the next. This might have the effect of confining hybridization to one direction. Species pairs in which one is completely nocturnal and the other completely diurnal, and which are therefore pollinated by different insects, seem to stand the best chance of continued isolation. In the *Silene* example above the flowering times overlap and many sorts of insects visit both species; hybrids are common.

ETHOLOGICAL ISOLATION

This refers to the considerable degree of fidelity to one particular species which is shown by some pollinators, and is of two distinct sorts. In the first place different but related species of plant are often adapted for pollination by different species or groups of animal. In America several remarkable cases have been investigated, e.g. *Aquilegia formosa* (humming-birds) and *A. pubescens* (hawkmoths), *Mimulus cardinalis* (humming-birds) and *M. lewisii* (bees), and *Penstemon centranthifolius* (Scrophulariaceae) (humming-birds) and *P. grinnellii* (bees). In Europe the best-known examples are found in the Orchidaceae, particularly the "insect-orchids" of the genus *Ophrys* where the flowers resemble the females of various Hymenoptera and so attract the males of the same species ("pseudo-copulation"). Of the two species of *Geum* mentioned previously *G. rivale* is largely pollinated by bees and *G. urbanum* by smaller insects, mostly hover-flies and other Diptera.

The second sort of ethological separation is shown by the constancy of one individual pollinator for one sort of flower. This constancy may extend for a single forage, a single day or throughout the life of the individual. Thus related species with the same species of pollinator might to some extent remain isolated. Mather (1947) found that in mixed cultivated beds of *Antirrhinum majus* and *A. glutinosum* few hybrids were produced, owing to the constancy of the pollinating bees, and McNaughton and Harper (1960a) reported a similar result among several species of *Papaver*.

Obviously the degree of isolation reflects the degree of fidelity of the pollinators, and it is very unlikely that either is ever complete. Moreover, species which are mainly visited by distinct pollinators may also be visited by less discriminating ones. In the *Aquilegia* example above, for instance, both species are occasionally visited by bumble-bees, which probably account for most of the naturally-occurring hybrids (Grant, 1952). Where one species is much less common than the other ethological isolation also tends to break down, because the pollinator of the rarer plant is forced to turn its attentions to the commoner one as well.

MECHANICAL ISOLATION

This is found in plants with elaborate floral structures which physically prevent pollination from one species to another by the different relative positions of the floral parts, or by the different shapes of orifices through which the pollinator or pollen must pass. There are very well-known examples in the Orchidaceae and Asclepiadaceae, and probably many others in various tropical families. It is often associated with ethological isolation, and frequently difficult to distinguish from it.

BREEDING BEHAVIOURAL ISOLATION

The breeding behaviour of a plant determines its degree of outbreeding, and hence to some extent the amount of hybridization. The possibilities of hybridization are greatest in dioecious or self-incompatible species, for these have to rely for fertilization on pollen from other plants. In such plants hybridization would be expected less often where the species is common and plenty of compatible pollen of the same species is available. Dillemann (1954, etc.) found that in France *Linaria repens* and *L. vulgaris* (both of which are self-sterile) seldom form hybrids when they are abundant, but that hybrids are frequent where one species is common and the other rare, often due to recent introduction. Dillemann (1948) also found that seed from various self-incompatible species of *Linaria* grown in Botanic Gardens (where usually few plants are present, but related species are often grown adjacent) was frequently of hybrid origin. Clifford (1961), however, claimed that there is no overall correlation between breeding behaviour and the occurrence or frequency of hybrids.

At the opposite end of the scale autogamous plants obviously less often hybridize, and autogamy may well be a strong isolating mechanism. McNaughton and Harper (1960a) concluded that the rarity of hybrids in *Papaver* was in part due to the usual self-fertilization of several of the species. In the *Vicia sativa* aggregate autogamy is the rule, for the anthers dehisce and the pollen germinates on the stigma of the flower before the latter opens. This is undoubtedly the major or only barrier to hybridization between the various segregates, and perhaps between the aggregate and *V. lutea* (which can be artificially crossed) (C. A. Stace and E. Hollings, unpublished). Autogamy in *Vulpia membranacea* probably reduces the number of hybridizations with *Festuca rubra,* for the artificial hybrid is easily made (Stace and Cotton, 1974).

Plants with cleistogamous flowers are extreme examples of the effect of autogamy as an isolating mechanism, and could also be regarded as cases showing mechanical isolation. Although most seeds produced by many species of violets (*Viola* spp.) arise from cleistogamous flowers (Valentine, 1941), hybrids are not rare and presumably come from occasional pollinations of the chasmogamous flowers. In *Vicia sativa* several races (usually known as subsp. *amphicarpa*) bear cleistogamous subterranean flowers, reinforcing the isolation due to normal aerial autogamy.

Plants reproducing mainly or wholly vegetatively or apomictically are also to some extent protected from hybridization. In many apomictic groups large numbers of "species" are in fact very closely related, and, if apomixis is facultative, hybridization can occur in many combinations. The best example in Britain is the crossing of the sexual *Rubus idaeus* and *R. ulmifolius* with a very large number of apomictic *Rubus* microspecies. R. Melville (pers. comm., 1972) considers that in *Rosa* the facts that the male parent contributes only one-fifth of the chromosomes to the F_1, and that species within the *Caninae* are closely related, similarly allows hybridization in large numbers of combinations.

GAMETOPHYTIC ISOLATION

Under this heading can be grouped a number of processes which block hybridization between the time that the pollen is deposited upon the stigma and the time the pollen-tube reaches the egg (Riley, 1952). The collective phenomenon is also known as pollen incompatibility. Examples have been found which show blockage at almost every conceivable intermediate stage: pollen fails to germinate; pollen-tube grows too slowly, especially in relation to the rate of growth of conspecific pollen-tubes; pollen-tube dies in the stigma or style, often indicated by bursting (perhaps due to osmotic pressure differences); pollen-tube is too short or too wide; pollen-tube reaches egg but does not release gametes. These processes have been subjected to a considerable amount of investigation and there are several well-composed discussions of them in the literature (e.g. Watkins, 1932; Dobzhansky, 1937; Thompson, 1940; Blakeslee, 1945; Dobzhansky, 1951; Riley, 1952; Levin, 1971). Most of the work has been carried out with economically important genera, such as cereals, or on traditional genetical material, e.g. *Datura.*

Some of the processes of gametophytic isolation are related to those of self-incompatibility, as is shown by the fact that self-compatible species can frequently hybridize with related self-incompatible species when the former are the female parents, but not when they are the male parents. This is known as unilateral incompatibility (Harrison and Darby, 1955; Lewis and Crowe, 1958).

Sometimes the success of hybridization is related to a simple mechanical feature such as style-length. For instance, in the crosses between *Polemonium mexicanum* and *P. pauciflorum,* Ostenfeld (1929) found that hybrids were obtained only when *P. mexicanum* was used as the female parent; this has styles eight times shorter than those of *P. pauciflorum,* and its pollen cannot penetrate the long styles of the latter. In other cases the ploidy level of the parents is of importance. Watkins (1932) found that if the normal ploidy ratio of pollen : style (1 : 2) is lowered (e.g. 2 : 2, as in a female diploid × male tetraploid) the pollen-tubes are usually unsuccessful in negotiating the style, but they are reasonably successful in the reciprocal cross (which gives 1 : 4). This phenomenon has been demonstrated in several genera. Similarly, the same reciprocal differences can result from the polyploid pollen-tubes being too wide for the diploid style, as has been found in *Nicotiana* (Solanaceae). Nevertheless there are exceptions; in *Prunus,* hybrids between diploid and tetraploid cherries are much more easily obtained if the diploid is used as female (Crane and Lawrence, 1956). The situation regarding crosses between diploids and their autotetraploids has been discussed by Valentine and Woodell (1961).

The rate of growth of pollen-tubes has often been found to be lower in styles of foreign species than in those of their own species, which accounts for the usual relative failure of hybridization in mixed pollinations, a phenomenon known as certation. On the other hand mixed pollination can sometimes result in a greater success in hybridization than interspecific pollination alone. There is obviously still a great deal to learn concerning the physiological basis of such phenomena.

GAMETIC ISOLATION

In some cases the pollen-tube reaches the megaspore and breaks open, but the gametes do not fuse with the egg nucleus and/or primary endosperm nucleus.

Sometimes the male gametes remain outside the megaspore, in other cases they enter it but fail to fuse, often being digested or ejected. In general, work on gametic isolation has paralleled that on gametophytic isolation, having being carried out with the same sort of material and often by the same workers.

9. Establishment of hybrids

In the previous chapter the various barriers to hybridization up to the formation of a hybrid zygote were considered; in this chapter the postzygotic development up to the formation of a viable hybrid progeny is briefly discussed. Hybrid failure may occur at any stage from the zygote onwards, and the various isolating mechanisms involved are again conveniently classified on a sequential basis; three are recognized here. In addition to the reviews quoted at the beginning of the previous chapter there are useful discussions of postzygotic failures by Watkins (1932), Thompson (1940), Blakeslee (1945), Brink and Cooper (1947), Stebbins (1958) and Poddubnaya-Arnol'di (1962).*

SEED INCOMPATIBILITY

Cases where embryo death occurs between zygote formation and seed maturity have been called seed-incompatibility by Valentine (1956), who has studied the phenomenon in some detail in *Primula*. Valentine (1961) used the strength of interspecific seed incompatibility as an indication of phylogenetic affinity in this genus, although Eaton (1973) has since expressed doubts on the extent of its value for such purposes.

In the earliest cases of failure the zygote divides to form a few-celled embryo (say of four or eight cells) and then aborts. Later embryo death can be at particular, definable stages, such as at the differentiation of the apical meristems, or when fully mature but with no subsequent germination, or may occur at less definite times.

Stebbins (1958) classified the causes of postzygotic death as incompatibilities between (a) the chromosomal material of the two parents, (b) the chromosomal material of one parent and the cytoplasm of the other, and (c) the hybrid embryo and the endosperm or the maternal tissue. This scheme cuts across the sequential system adopted here since the former two modes of breakdown can take effect at any postzygotic stage, even up to failures in the F_2 or later generations. But it is probably more fundamental because, in the same mating combination, failures can be found at different stages, probably all with the same underlying causes.

Incompatibilities between the hybrid embryo and surrounding tissue obviously occur only in the pre-germination stage, and many examples are

* See also discussion of chromosome elimination in hybrids by Davies (1974).

known. When the endosperm is involved the embryo usually passes the few-celled stage and its death during later growth is preceded by abortion of or abnormalities in the endosperm. There is evidence for competition for food, in some such cases, between the maternal tissue and the endosperm. This physiological imbalance is illustrated, for example, by abnormal development of the antipodal cells or of the inner integument, resulting in starvation of the endosperm and therefore of the embryo.

The distinction between inter-chromosomal and chromosomal–cytoplasmic incompatibility may be difficult to recognize in practice, but generally the latter is marked by reciprocal hybrid differences while the former is not. Stebbins (1958) considered interchromosomal incompatibility to be due either to general effects or to effects of specific genes, and examples of both are known. Incompatibility between the maternal cytoplasm and the chromosomes of the male parent has been studied in great detail in *Epilobium* (for references see under that genus), where it may take effect during seed-formation or at various post-germination stages.

HYBRID INVIABILITY

F_1 hybrids may die during or just after germination, or as soon as the seed food-reserves are used up, or at any later stages before reproduction. Often they are perfectly viable in the absence of competition, but soon killed off by neighbouring plants in the wild. Finally, the F_1 may reach maturity but produce few or no flowers.

The reasons behind such failures are presumably the same as the first two of Stebbins' three causes mentioned above. Sears (1944) found three separate alleles concerned with death of seedlings of F_1 hybrids between various species of *Aegilops* and *Triticum* (Gramineae); they acted at two distinct stages of development. In *Epilobium*, chromosomal–cytoplasmic incompatibility often results in odd-looking plants with varying degrees of chlorosis and stunted growth. Frequently the plants have twisted, bullate leaves and resemble virus-infected or otherwise diseased plants. Chlorotic seedlings are very commonly encountered in the offspring of artificial hybridization in many other genera, and McNaughton and Harper (1960b) reported plants with "aberrant morphology and a virus-like syndrome" among F_1 *Papaver dubium* x *P. rhoeas*. In *Brassica napus* x *B. rapa* (see below under *Brassica*) the roots may develop "hybridization nodules", which resemble club-root infections but lack the pathogen. The hybrids are, however, perfectly viable.

Several examples are known in which viable hybrid seed has been collected in the wild from the maternal parent, and grown in cultivation, but the F_1 hybrid itself has not been found in the wild. It is impossible to know in such cases whether the seed would have germinated in nature, and, if so, whether it would have produced a mature F_1 plant. In the British Isles at least two instances have been recorded. *Poa annua* x *P. supina* caryopses have been collected in Guernsey, and *Anthemis arvensis* x *Tripleurospermum inodorum* cypselae on an isolated plant of the former species in Oxfordshire; in neither case is the F_1 hybrid recorded from the British Isles, but the second one has been reported from

Germany. In Poland Skalínska (1971) found heptaploid caryopses on plants of *Festuca rubra*, although all the populations sampled were either hexaploid or octoploid.

NON-FITNESS OF F_1 HYBRIDS

As has been mentioned previously, it has frequently been found that whereas F_1 plants are easy to rear in cultivation, seemingly without special attention, they have not been found in nature. In these cases the hybrids are physiologically completely viable, but fail to survive the extra pressures imposed by the communities in which they occur. This sort of isolating mechanism is therefore equivalent to several of those in Chapter 8, but here the barriers are acting not to keep the species apart but to disfavour the hybrid offspring. Very often the same barriers act at both levels. To exemplify these situations the action of three barriers (ecological, seasonal and ethological) will be mentioned.

In most cases hybrids are to some extent intermediate between their two parents in many characteristics, and if the parents have distinctly different ecological tolerances it is likely that the hybrids will be different again. They might tend to occupy intermediate habitats, overlapping those of one or both parents or not, or occupy a habitat similar to one of the parents, or demand a habitat quite different from that of either parent. Thus, however vigorous the hybrid, its occurrence in the wild will be limited by the availability of suitable niches, just as is the distribution of any other plant.

Cases where the hybrid closely resembles one of its parents in ecological preferences are naturally more favourable to hybrid survival whenever the parental habitats are distinct and without intermediates, and one might expect to find a considerable number of examples of such situations. However, in the great majority of cases, the detailed investigations necessary to determine this have simply not been carried out. Even where much work has been done on the cytology, genetics and morphology of hybrids very rarely have autecological analyses been made of both parents and hybrids. For instance, in the well-known *Geum* example, hybrid plants more often appear to grow in places favoured by *G. rivale* than in those by *G. urbanum*, but no detailed data are available (see Marsden-Jones, 1930). Indeed, as it is difficult to distinguish F_1 hybrids from some F_2 segregates, the actual range of ecological tolerance of the F_1 is not known even in terms of subjective observation. Valentine (1970) has discussed some further aspects of hybridization at zones of vegetational transition (ecotones).

It has been realized for a long time that the widespread disturbance of natural vegetation by man's various activities has frequently provided habitats suitable for hybrid colonization, both because some of the habitats so created are new sorts and because many of them are open, i.e. relatively free from competition. Wiegland (1935) was one of the first to discuss this view in some detail, and the concept has been developed a great deal by Anderson (1948, 1949, 1953a). There are many examples to illustrate this idea; the *Silene* and *Geum* hybrids have been mentioned previously, and the point is made in more detail below under *Epilobium*. Anderson spoke of "hybridization of the habitat" as a requisite before the establishment of hybrid populations could become possible.

Of course man is not the only agency capable of creating new habitats. Ice-ages, fires, land-slips, damage by the sea, volcanic eruptions, sudden large-scale attacks by predators or parasites and other natural catastrophes must also be considered to provide opportunities for hybrid establishment. C. D. K. Cook (see below under *Sparganium*) has found *Sparganium angustifolium* x *S. emersum* on a Scottish shingle-beach, a habitat not exploited by either parent.

Several of the above phenomena are relatively localized in the world, if not permanently then at any given period of time, and the resultant establishment of hybrids then often becomes equally localized. For instance Briggs and Walters (1969) pointed out the abundance of hybrid populations of *Geum rivale* x *G. urbanum* in Britain compared with the situation in Poland, and related this to differences in the vegetational history of the two countries which have resulted in the loss of almost all unadulterated woodland in Britain, but not so in Poland. The prevalence of forest fires in some parts of the world, e.g. the Mediterranean, where they are caused by man, and large areas of Africa, where they are natural (Exell and Stace, 1972), provides another example of local habitat changes. Allan (1937) quoted Cockayne's work on *Nothofagus* (Fagaceae) in New Zealand, where hybrids may occur "in great profusion" after forest fires and "the reinstated forest will certainly contain more hybrid trees than did the former community". Rattenbury (1962) also discussed the high incidence of hybridization in New Zealand, but related it to other factors. All these situations give rise to the well-established principle that the existence of hybrids in one area does not necessarily mean that they occur wherever the two parents grow together.

Briggs and Walters (1969) also suggested that introgression, which has been much more extensively studied in North America than in Europe and of which many more examples are known there, might in fact be much more frequent there because of the very recent widespread destruction of natural vegetation. In Europe the richest vegetational zone (the Mediterranean) has been cultivated extensively for thousands of years, not just for 200. Nevertheless Davis and Heywood (1963) quite rightly point out that introgression has been relatively little studied in Eurasia, a situation which ought to be remedied. The significance of "hybridized habitats" becomes much more important in the F_2 and later generations, where the range of variation is so much greater, and will be mentioned again in the next chapter.

The need for habitats suitable for the growth of F_1 hybrids must be viewed also from the temporal point of view. In some cases the hybrids, once well established, are vigorous enough to withstand strong competition, and thus can persist after the "new" habitats have been swamped by the old ones during the normal process of succession. In other cases they cannot, and perish. The need for the continued existence of "new" habitats is most acute in the case of annual hybrids, which need to be created anew annually, e.g. various *Senecio* hybrids, but is probably not a limiting factor in their occurrence since in most cases their parents are equally dependent upon open conditions. Many vigorous perennial hybrids, e.g. colonies of *Juncus balticus* x *J. effusus* and *J. balticus* x *J. inflexus* on the Lancashire coast, as well as obvious examples of trees and shrubs, have been known to persist for a great many years, while others are relatively short-lived and sporadic in appearance.

Levin (1971) has discussed hybridization between species with partial seasonal isolation where the hybrids have yet a different flowering period. The hybrid *Phlox glaberrima* subsp. *interior* x *P. maculata* subsp. *maculata* (Polemoniaceae) flowers later than either parent, but in its natural habitat it finds plenty of pollinators able to exploit the prolonged flowering season. Clausen (1951), on the other hand, found that the hybrid *Layia gaillardioides* x *L. hieracioides* (Compositae), which again flowers later than either parent, was at a distinct disadvantage because it was not adapted to the conditions of drought which followed the normal flowering season. Thus in the latter case there is a seasonal barrier to the success of hybrids.

A similar principle can be extended to ethological barriers, as shown by the *Aquilegia* and *Penstemon* examples mentioned in Chapter 8. The hybrid between *P. centranthifolius* (pollinated by humming-birds) and *P. grinnellii* (bees) has flowers which are more or less intermediate in appearance between those of its parents, and which resemble in many respects those of a third species, *P. spectabilis*. This species is pollinated chiefly by large wasps, which are similarly attracted to the flowers of the hybrid (Straw, 1955). The flowers of the hybrid between *Aquilegia formosa* (humming-birds) and *A. pubescens* (hawkmoths) also have more or less intermediate flowers, but in this case they are not adapted to pollination by a third group of pollinators. Instead, they are visited occasionally by the pollinators of both parents, so the hybrid tends to backcross with the parents rather than to remain a distinct entity (Grant, 1952). Levin (1971) has called this "hybrid floral isolation".

10. Hybrid sterility and fertility

This is the third and last chapter dealing with the various stages at which isolating mechanisms can take effect; the references given at the beginning of the previous two chapters are mostly appropriate here too. This chapter covers those breeding barriers which occur after the F_1 hybrid has reached the flowering stage. The three barriers here recognized may be termed F_1 hybrid sterility, F_2 hybrid inviability or sterility, and non-fitness of F_2 and backcross hybrids, and they are treated here together with other topics related to hybrid fertility.

As has often been pointed out, sterility in hybrids is not correlated in any way with their vigour. In a great many cases very strong-growing hybrids are completely sterile; indeed, because seeds are not formed, sterile hybrids may possess a far longer flowering period and vegetative growth phase than their parents. This is most noticeable in annuals, where normally seed-production involves a diversion of energy resources from all other parts of the plant and so causes its death. As an example one can cite the artificial *Vulpia geniculata* x *V. membranacea,* which continues to produce new flowers and retain green leaves long after both its parents have died (R. Cotton and C. A. Stace, unpublished). This is akin to the gardening practice of removing all flowers as soon as they are withered in order to extend the flowering period. In other cases strong-growing hybrids can be attributed to hybrid vigour, or positive heterosis, a phenomenon which is discussed in Chapter 12 and is apparently not related to fertility.

Although even today it is often said that interspecific hybrids are usually or mostly sterile, many show a great deal of fertility and it could be argued that the majority are fertile to some degree. In fact every stage from complete sterility to complete fertility exists, and it is difficult to make generalizations. For example most grass hybrids are sterile, but many of those in the genus *Lolium* are not; similarly most British orchid hybrids are fertile, but some (e.g. *Dactylorhiza fuchsii* x *D. maculata*) are highly sterile. Moreover it is sometimes difficult to determine the degree of fertility. In certain *Epilobium* hybrids individuals can be self-compatible or self-incompatible; in the latter case the plants appear sterile if they happen not to be pollinated by a compatible pollen source, although they are potentially fertile. In many other cases the true fertility of a hybrid varies considerably from individual to individual, probably resulting from the different parental strains from which they were derived, e.g. *Juncus effusus* x *J. inflexus.* Sometimes the cause is more obvious, as in *Betula pendula* (diploid) x *B. pubescens* (tetraploid). Here, some hybrids are triploid and sterile but others, equally intermediate in appearance, are tetraploid and fertile, having presumably arisen from the production of unreduced, diploid gametes by *B. pendula.*

F_1 HYBRID STERILITY

Dobzhansky (1937), followed by Thompson (1946) and Stebbins (1950), recognized two basic causes of hybrid sterility: genic and chromosomal. Genic sterility is caused by particular cytoplasmic or chromosomal genes which govern interspecific incompatibility and whose action is delayed until the reproductive phase of the F_1. Hybrid failure may take effect before meiosis occurs, during it or after it. Premeiotic events include abnormal floral development, such as the conversion of stamens to petal-like organs, a failure to organize the anther tissue and the consequent lack of differentiation of pollen mother cells, or degeneration of the pollen mother cells. In various *Luzula* crosses (mostly intercontinental) Nordenskiöld (1956) commonly observed degeneration of the pollen mother cells, and in some cases the flowers were not properly developed. Cook (1970) described *Ranunculus* subgen. *Batrachium* hybrids where the reproductive organs abort before the archesporial cells are formed, and in *Sorbus aucuparia* x *S. intermedia* (see below under *Sorbus*) pollen mother cells degenerate into a plasmodial condition. In *Galium saxatile* x *G. sterneri* (see below under *Galium*) the flower-buds may abort at an early stage.

Many genic effects on meiosis are now known, covering all aspects of the process; Solbrig (1968) usefully listed 15 such events, ranging from asynapsis to non-separation, and Rees (1961) surveyed the topic in considerable detail.

Postmeiotic effects are equally numerous, again involving all stages of development from the degeneration of the pollen or megaspores before they undergo any further development to the abortion of the F_2 embryo. The immediate physiological causes of many of these phenomena are probably similar to those preventing the formation of F_1 hybrids, but they are delayed a whole generation. For instance post-fertilization failure of the F_2 zygote may or may not be connected with prior endosperm degeneration. Pre-fertilization failures do not always run exactly parallel in the male and female organs.

Chromosomal sterility is the failure of synapsis during meiotic prophase I due to varying degrees of non-homology of the chromosomes of the two parental species, and is an extremely common form of isolating mechanism in plants. Non-homology is caused by rearrangements of the chromosomal material of one parent *vis-à-vis* the other (inversions and translocations), or the loss or gain of extra material (deletions and duplications). In its most obvious form it involves hybridization between species with different chromosome numbers. In these cases, if one species has originated fairly recently from the other, a good degree of chromosome pairing may take place (e.g. *Crepis fuliginosa* x *C. neglecta*—see Tobgy, 1943), and a study of meiosis can show the degree of chromosome homology between the parents. If the parental relationship is more remote, and many differences have accumulated, there may be no pairing at all. In the most puzzling cases chromosome pairing appears complete, but the microspores and megaspores are non-functional. This is usually explained by the concept of cryptic structural hybridity (Stebbins, 1945), whereby minute inversions or translocations have occurred followed by nearly, but not quite, identical re-inversions or re-translocations, so that small parts of chromosomes are misplaced; this is sufficient to create inviability of the meiotic products without upsetting the gross features of synapsis.

Unpaired chromosomes (univalents) show erratic behaviour at the first division of meiosis. They may be included at random in either daughter nucleus, or become left out of both. In other cases they divide mitotically at anaphase I and one chromatid becomes incorporated in each daughter cell; in these cases their behaviour at the second meiotic division remains erratic. The result in almost all cases is that the chromosome numbers of the microspores and megaspores are irregular, and these duplications and deficiencies are usually lethal. Occasionally, by chance, viable gametes are produced, thus resulting in a low degree of hybrid fertility. In some cases irregular numbers of chromosomes can be accommodated in the hybrids without complete loss of fertility. A great many hybrids show such a variation in chromosome number. Fothergili (1938) found that in *Viola lutea* x *V. tricolor* the chromosome number tends to increase due to the double division of univalents at meiosis, and moreover the increase in chromosome number leads to greater vigour and fertility of the hybrids.

Clearly the study of meiosis in hybrids, whether they are highly fertile or highly sterile, can throw a great deal of light on the causes of sterility and on the parental inter-relationship, and it is has become one of the most powerful tools of the biosystematist. Nevertheless some care is necessary in the interpretation of the data. In particular it is often assumed that, in a diploid hybrid between diploid parents which show only bivalent formation, pairing only occurs between chromosomes of different parents (heterogenetic pairing), but in fact this is known not to be so. The frequency of homogenetic pairing is not always easy to ascertain, but it is essential that it is so before an accurate assessment of the inter-parental relationships can be obtained from pairing data. This problem is much more acute in the case of polyploids, and will be mentioned again later. Secondly, failure to pair can be due to relatively minor features which might exist in hybrids between closely related species, and in such cases the low degree of pairing does not indicate a remote parental relationship. This point becomes especially important in hybrids which combine chromosomal sterility with genic sterility acting upon meiosis itself. Sax (1935) and Stebbins (1945) have surveyed this topic in some detail; nowadays many more cases could be cited but the principles remain the same.

Although Stebbins (1950) argued that Dobzhansky's classification of sterility into genic and chromosomal types (which was first suggested in 1933) was the most fundamental, Stebbins (1958) later changed his mind and gave reasons for preferring a system derived from the ideas of Federley, Renner and Müntzing in 1928–30, whereby haplontic or gametic, and diplontic or zygotic, sterility is recognized. The latter system has also been adapted by Solbrig (1968) and others. One of the main objections to the genic/chromosomal classification is the practical difficulty of distinguishing between genic sterility and such chromosomal sterility as is caused by cryptic structural hybridity. If artificial tetraploids are created from the sterile hybrid the effects of genic sterility will still be apparent, but in the case of chromosomal sterility the doubled chromosome number will enable homologous pairing to take place and hence restore fertility to the hybrid. However, it is not always easy to synthesize such tetraploids, and in any case the situation is not quite as simple as here suggested (see Stebbins, 1958).

Haplontic sterility affects the haplophase of the F_1 hybrid, so that the microspores and megaspores or the gametes are non-functional, while diplontic sterility affects either the diplophase of the F_1 (somatic or sporogenous tissue) or the diploid F_2 hybrid embryo. Diplontic sterility is mostly genic in nature and commoner in animals than in plants. In some cases it results from an incompatibility between the chromosomes of one parent and the cytoplasm of the other, and in the genic/chromosomal classification above this could be considered a third category. Haplontic sterility is rare in animals but many botanical examples are known, most of which are due to chromosomal sterility.

It is possible to have both genic and chromosomal, and both haplontic and diplontic, sterility in a single hybrid, which suggests that the wisdom of defending or adhering to one classification rather than the other is questionable.

F_2 HYBRID INVIABILITY OR STERILITY

There is a number of examples known where the individuals of F_2 or subsequent generations show general weakness or markedly reduced fertility, even where the

F_1 hybrids are fully viable and fertile. It appears that the sterility mechanisms involved are similar to those concerned with F_1 inviability or sterility, but that they take effect in a later generation. It is to be expected that the range of fertility might be wider in the F_2 than in the F_1, and many examples are known where selection for greater fertility in F_2 and successive generations occurs, but this does not account for the cases being considered here where all or nearly all of the F_2 are markedly less fertile. Examples of such an isolating mechanism are *Populus alba* x *P. grandidentata, Larix gmelinii* x *L. kaempferi, Sambucus nigra* x *S. racemosa* and others discussed by Stebbins (1950, 1958).

POLYPLOIDY AND AMPHIDIPLOIDY

Amphidiploids (or amphiploids) are polyploids which have arisen by hybridization and which behave as diploids. The explanation of the diploid behaviour of many polyploids was one of the major achievements of cytogenetical work in the first quarter of this century, and in the subsequent half-century it has continued to provide answers to many taxonomic problems. In particular the artificial re-synthesis of natural amphidiploids, first carried out by Müntzing (1930), has been repeated many times. There are many reviews, e.g. Winge (1917, 1932), Goodspeed and Bradley (1942), Clausen *et al.* (1945), and Stebbins (1947, 1950).

In most but not all cases amphidiploids arise from sterile, diploid F_1 hybrids by a doubling of the chromosome number. If two parental diploid species are designated AA and BB, to indicate their genomes, their F_1 hybrid is AB. A tetraploid derived from this is AABB, and is usually known as an allotetraploid to distinguish it from the opposite cases (autotetraploid) where the chromosome number of a single diploid species has doubled (AAAA or BBBB). Because of the presence of four homologous sets of chromosomes and the usually subsequent formation of multivalents at meiosis, autotetraploids are usually highly infertile. In the allotetraploid the degree of fertility will be largely dependent upon the degree of homology between the A and B genomes. If in the F_1 hybrid (AB) there is no heterogenetic (A–B) pairing, in the allotetraploid (AABB) only homogenetic pairing (A–A and B–B) occurs, which leads to full fertility, i.e. diploid behaviour. Such allopolyploids are known as amphidiploids. (The terms homogenetic and heterogenetic pairing (Waddington, 1939) are to be preferred to autosyndetic and allosyndetic pairing because the latter have been used in two different, opposite senses.)

Amphidiploids are thus extreme sorts of allopolyploids, i.e. the extreme opposites of autopolyploids. In between these two extremes is a whole range of intermediates, some lying close to one extreme or the other but some occupying the half-way condition, where some heterogenetic (A–B) pairing occurs in the F_1, and therefore some also in the polyploid, resulting in the latter in the formation at meiosis of some multivalents as well as the bivalents formed by homogenesis. Such intermediates are known as segmental allopolyploids, and their genomic constitution can be written $AAA'A'$.

From what has been said concerning intraspecific sterility barriers and the impossibility of defining species on the fertility criterion, it will be clear that segmental allopolyploids and amphidiploids may involve in their ancestry different strains of one species or different species, but the closer to the amphidiploid condition an allopolyploid is, the more likely it is that different species are involved. The exact relative frequencies of autopolyploids, segmental allopolyploids and amphidiploids in nature is uncertain, but if the definitions of the extremes are drawn tightly most wild polyploids are probably segmental allopolyploids. In this account I shall use the term amphidiploid to cover true amphidiploids and also segmental allopolyploids which are derived from parents usually recognized as separate species.

Allotetraploids arise by chromosome doubling in one of several ways. Basically they can be formed by somatic or meiotic processes. In the former case the chromosome number of the F_1 zygote or very young embryo, or of a bud on an established F_1 plant, doubles because a mitotic chromosome division is not followed by nuclear division. This gives rise to a plant or a shoot which produces flowers with diploid microspores and megaspores, and hence tetraploid seeds. The well-known allotetraploid derived from *Primula floribunda* x *P. verticillata* arose in this way (Newton and Pellew, 1929). In the latter case, on the other hand, the F_1 hybrid remains diploid but it produces diploid microspores and megaspores owing to premeiotic doubling or to non-reduction at meiosis. The equally well-known allotetraploid derived from *Brassica oleracea* x *Raphanus sativus* originated thus (Karpechenko, 1927). Many different factors are known which cause mitotic or meiotic misdivision of cells to produce daughter cells with twice as many chromosomes as expected. In F_1 hybrids the upsets caused by non-homologies frequently result in a complete breakdown of meiosis, but sometimes viable microspores or megaspores are produced without a division occurring and containing the whole of the diploid chromosome complement.

Such phenomena are not rare even in pure species; when large progenies are analysed autotriploids (resulting from the fusion of one haploid and one unreduced diploid gamete) and even autotetraploids (from two unreduced gametes) are often found in odd individuals. This sort of behaviour also accounts for the variation in chromosome number found in some hybrids. For example the hybrid *Festuca pratensis* x *Lolium perenne* exists in three cytotypes: the expected diploid (FL) and two sorts of triploids (FFL and FLL), the latter two corresponding to derivations from unreduced *Festuca* and *Lolium* gametes respectively (Gymer and Whittington, 1973b). Similarly the hybrid *Ammophila arenaria* x *Calamagrostis epigeios* exists in three cytotypes of a comparable nature, and *Betula pendula* (diploid) x *B. pubescens* (tetraploid) may also be triploid or tetraploid (and accordingly sterile or fertile) depending upon whether or not the *B. pendula* gamete is reduced.

In *Delphinium* hybridization between a diploid *(D. nudicaule)* and a tetraploid *(D. elatum)* species, in which the former contributed unreduced gametes, has given rise to a new, fertile, allotetraploid species, *D. ruysii* (Lawrence, 1936). Allotetraploids can also arise from hybridization between two autotetraploids. Usually autotetraploids are largely sterile due to the formation of multivalents at meiosis, but in many cases some viable (diploid) gametes are formed, and sometimes, due to the presence of multivalent suppressor mechanisms, bivalents alone form and therefore a full complement of viable gametes results. Bivalent-forming autotetraploids which can hybridize to form allotetraploids are known in *Erucastrum* (D. J. Harberd, pers. comm. 1973).

When levels higher than tetraploidy are involved more possibilities offer themselves. A hexaploid might, for example, be an amphidiploid (AABBCC), an autohexaploid (AAAAAA), a segmental allohexaploid (AAA′A′BB or AAA′A′A″A″, etc.), or an autoallohexaploid (AAAABB). Amphidiploid allo-hexaploids (AABBCC) arise by the doubling of the chromosome number of a sterile allotriploid (ABC), which was in turn derived from the hybridization of an allotetraploid (AABB) with a third diploid (CC). Octoploids and other higher polyploids are derived by mechanisms comparable to those producing tetraploids and hexaploids.

Rather than question the possibility that amphidiploids might arise naturally, one might wonder why they do not arise more frequently. It is likely that the factors discussed under "non-fitness of F_1 hybrids" in Chapter 9 would operate. Nevertheless some amphidiploids have had a polyphyletic origin. Perhaps the best documented case is in *Tragopogon*, where, in North America, two amphidiploids are known: *T. mirus* (from *T. dubius* x *T. porrifolius*) and *T. miscellus (T. dubius* x *T. pratensis).* In the former case detailed cytological study has shown that there have been at least three different independent origins (Ownbey and McCollum, 1954). In addition it is possible that the amphidiploid *Spartina "townsendii"* has arisen independently in England and south-western France, and disagreements as to the origin of species such as *Poa annua* (Tutin, 1957; Koshy, 1968) might be similarly explained. In the case of *Galeopsis tetrahit*, which arose in a rather complicated way from *G. pubescens* x *G. speciosa* (Müntzing, 1930), a separate independent origin from the same original parents has produced a taxon sufficiently distinct (and intersterile with it) to be recognized as a separate species, *G. bifida.*

Historically speaking most amphidiploids fall into one of two categories: those which arose under the observation of cytogeneticists; and those which were established species later discovered to be amphidiploids. The earliest known amphidiploids were of the first category; at first they were of accidental origin but later on they were synthesized artificially, a procedure now very commonly practised in plant breeding both to confirm the origin of suspected amphidi-ploids and to produce new ones.

Among the amphidiploids which arose while under more or less direct observation are: *Spartina "townsendii"* (from *S. alterniflora* x *S. stricta*), *Primula "kewensis"* (from *P. floribunda* x *P. verticillata*), *"Raphanobrassica"* (from *Brassica oleracea* x *Raphanus sativus*), *"Triticale"* (from *Secale cereale* x *Triticum* spp.), *Senecio cambrensis* (from *S. squalidus* x *S. vulgaris*) and *Saxifraga potternensis* (from *S. granulata* x *S. rosacea*). Nomenclaturally the F_1 hybrid must be treated as a hybrid, and the amphidiploid as a species, and separate names are needed for each. In the past arguments have often taken place concerning the correct application of some well-known binomials, and in other cases the available names have not been validly published; for this reason some of the above names are put in inverted commas. *"Spartina townsendii"*, for example, has been shown to refer to the F₁ sterile hybrid, and so a multiplication sign should be inserted. A valid name for the amphidiploid is not yet available; *S. anglica*, suggested by Hubbard (1968), is a *nomen nudum*.

Confirmation of the amphidiploid nature and of the ancestry of such plants comes from their genome analysis, followed by their experimental resynthesis. Genome analysis consists of a series of hybridization experiments designed to pinpoint the origin of the two genomes (in a tetraploid; three in hexaploid, and so on), principally involving backcrosses to suspected diploids. If a tetraploid amphidiploid (AABB) is backcrossed to one of the parental diploids (say AA) a sterile triploid is formed (AAB). During meiosis the two A genomes should pair fully to form bivalents, but the B genome remains as univalents. A similar but opposite situation is obtained by backcrossing to the other parental diploid, when ABB results. Crosses to non-parental diploids (CC, etc.) produce triploids (ABC) with no meiotic pairing. The interpretation of such experiments assumes that A, B, C, etc. are non-homologous, and that no heterogenetic pairing occurs. This point is taken up again below.

Established species are usually first suspected of being amphidiploids by the discovery that their chromosome number lies at a tetraploid or higher level when compared to that of closely related species. Possible parental diploids are those which resemble the polyploid closely in structural features, or which possess certain relatively infrequent characters in common with it. Such diploids are then crossed with the polyploid in a programme aimed at genome analysis. In this way the probable ancestry of a great many established amphidiploid species has been worked out either partially or fully, e.g. *Poa annua* (Tutin, 1957), *Dryopteris dilatata, D. cristata* and *D. carthusiana* (Walker, 1955), *Diplotaxis muralis* (Harberd and MacArthur, 1972), *Rorippa microphylla* (Manton, 1935), as well as that of many economically important plants such as wheat, cotton, tobacco and potato.

There are, however, many difficulties and pitfalls associated with genome analysis. In the first place it might not be possible to hybridize the amphidiploid with its putative parents, the causes of this interspecific incompatibility being possibly any of those isolating mechanisms surveyed previously. A good example of this is *Galeopsis tetrahit*, known to be derived from *G. pubescens* x *G. speciosa*, which will not cross with either of its parents. In this case its ancestry was proven by resynthesis from diploids which appeared likely on morphological grounds (Müntzing, 1930).

De Wet and Harlan (1972) pointed out some shortcomings of genome analysis which are operative even when viable backcrosses are obtainable. It is clear that there are many factors which can complicate the notion that homogenetic and heterogenetic pairing can be differentiated by observations on a few backcrosses. There are now numerous examples which show that, in the absence of a strictly homologous partner (AA pairing), chromosomes will pair with a homoeologous (i.e. not quite homologous) one (AA′ pairing), and sometimes the relationship can be quite remote yet pairing still occur. Often the presence of a homologous partner suppresses the tendency to pair with a homoeologous one.

Thus the F_1 *Primula floribunda* x *P. verticillata* forms a full or nearly full complement of bivalents, but the allotetraploid also forms mostly bivalents and is fully fertile (Newton and Pellew, 1929). Sterility in the F_1 is due to cryptic structural hybridity. In this case the F and V genomes can be said to be homoeologous. In the tetraploid *Tripsacum dactyloides* (Gramineae), however, some plants form only bivalents at meiosis while others in the same population form trivalents or quadrivalents by homoeologous pairing (De Wet and Harlan, 1972), so that homologous pairing does not always suppress homoeologous pairing. In hybrids between the bivalent-forming autotetraploid species of *Erucastrum* (see above) and other species, the chromosomes of the two genomes from the former parent associate together to form bivalents, irrespective of the homologies of the genome(s) from the latter parent (D. J. Harberd, pers. comm. 1973). For these reasons genome analysis has often given equivocal results in studies of amphidiploid ancestries.

Nicotiana tabacum is an amphidiploid of which one parent is a progenitor of *N. sylvestris.* There have been at least three suggestions as to the other parent, all of which pair fully with one of the *N. tabacum* genomes and all of which are therefore equal contenders on the genome analysis evidence. It is possible, of course, that in fact the true second parent is none of those three modern, closely related species, but the common progenitor of all of them (the three having diverged since the origin of *N. tabacum*). Walker (1955) made a similar suggestion in relation to the origin of the tetraploids *Dryopteris dilatata* and *D. spinulosa (D. carthusiana)*, whose common parent he suggested "is now represented by diploid forms of '*D. dilatata*' in Europe and Madeira", the Madeiran and European diploids now being morphologically distinct.

Whether or not these are correct interpretations in these cases, not all anomalous pairing data can be explained in this way, and it is certain that not only homoeologous but markedly non-homologous pairing does on occasions take place. The most telling evidence for this comes from certain triploid hybrids, such as in *Fragaria* ($2n = 21$), where 10 bivalents and one univalent or nine bivalents and one trivalent form at meiosis. Also, in the fertile hybrid between *Papaver nudicaule* ($2n = 14$) and *P. striatocarpum* ($2n = 70$), the 42 chromosomes regularly form 21 bivalents (Ljungdahl, 1924). Recognizing that chromosomes on occasions satisfy a basic pairing tendency without regard to homology, Grell (1965) proposed two sorts of pairing: exchange (homologous) and distributive (non-homologous). The "secondary pairing" found in some polyploids, whereby a bivalent tends to form a very loose attraction for another bivalent, is probably at least sometimes explicable on the basis of homoeology. It seems that non-homologous pairing is genically controlled (Chheda and Harlan, 1962; Riley, 1966; Solbrig, 1968), but its frequency is as yet unknown; if it is very common then the use of genome analysis requires some reappraisal.

The best-known genetic system governing the degree of pairing is the multivalent-suppressor system of hexaploid wheats (AABBDD) (Riley and Chapman, 1958). For some time hexaploid wheat ($2n = 42$) was thought to be a true amphidiploid, as it forms only bivalents at meiosis. However, it was found that whenever one particular chromosome (in fact chromosome 5 of genome B) was missing (viable strains of wheat lacking in turn each of the 21 different chromosomes are known) substantial numbers of multivalents are formed by homoeologous, inter-genome pairing. In other words chromosome 5B carries a gene system which limits pairing to strict homologues. This discovery ranks as a major milestone in cytogenetic research.

The karyotype morphology of a given genome in a hybrid as opposed to that in a pure species may also sometimes alter. Navaschin (1934), for example, found in *Crepis* hybrids that the morphology of a genome depends upon the genome with which it is combined. He called these effects amphiplastic changes, which are of two main types. In *Crepis capillaris* x *C. tectorum* the single satellited chromosome of the latter genome, but not that of the former, loses its satellite (differential amphiplasty); while in *C. capillaris* x *C. neglecta* all the chromosomes of the latter genome become shorter and thicker than in *C. neglecta* itself (neutral amphiplasty). Elliot (1950) also demonstrated differential amphiplasty in *Leontodon hispidus* x *L. taraxacoides.* Ellengorn and Petrova (1948) claimed that in various intergeneric grass hybrids in the Hordeeae the chromosomes of one parent stain more intensely at meiosis than those of the other parent, thus enabling the two to be differentiated and pairing behaviour to be analysed more fully.

In order to discover whether allopolyploids are true amphidiploids (i.e. with no A–B homology) or segmental allopolyploids it is customary to observe meiosis in plants with each genome represented only once, i.e. in the case of an allotetraploid AABB, to use AB plants. These are obtained in two ways: either as F_1 hybrids of the two parents, or as polyhaploids. If the two parents are known and F_1 hybrids are obtainable the former method is obviously the easier; but where one or both of these conditions are not met the use of polyhaploids has proved valuable. Polyhaploids are effectively the reverse of polyploids in that they are obtained from the latter by a halving of the chromosome number, usually by the induction of parthenogenesis, whereby the reduced egg develops into an embryo. Such plants derived from tetraploids are in fact diploids (or triploids from hexaploids), and are thus called polyhaploids to distinguish them from true haploids or monohaploids. Thus Manton and Walker (1954) obtained polyhaploids of *Dryopteris dilatata* and *D. filix-mas* (both tetraploids, $2n = 164$); in the former case 82 univalents appeared at meiosis and in the second case mostly univalents with 2–5 bivalents. This may be interpreted as showing that these species are true amphidiploids, for 2–5 bivalents (often very loosely attached) out of a possible total of 41 is hardly indicative of close homology.

Polyhaploids are sometimes found in natural situations, in which case they may be difficult to distinguish from F_1 hybrids. In the case of *Spartina* "townsendii" apparent F_1 hybrids have been found sporadically on the coasts of England, Wales and Ireland, among amphidiploids, yet one of the parents *(S. alterniflora)* is confined to the vicinity of Southampton Water. The most likely explanation is that these sporadic apparent F_1 hybrids are in fact polyhaploids, although some of them might represent introductions of F_1 hybrids from Southampton (see below under *Spartina*).

One aspect concerning the ancestry of amphidiploids which is too frequently overlooked is that it might now be impossible to trace the parents, because of the consequences of evolution since the origin of the amphidiploid. This point was touched upon above in discussing the parentage of *Nicotiana tabacum* and *Dryopteris dilatata*. In many cases the genome in the diploid and the originally identical one now in the amphidiploid must have diverged considerably, so that they are no longer strictly homologous. The same is true of the morphological characters of the two taxa. This is obvious when no ancestors seem apparent, but tends to be forgotten when diploids which seem likely parents do still exist, especially in groups where searches for amphidiploid ancestors have been traditionally successful.

HYBRID SWARMS AND INTROGRESSION

Sterile interspecific hybrids are usually fairly easily recognized in the field, but fertile hybrids are often much less so, both because of the absence of the sterility criterion and because of the existence of a range of hybrids (multiform hybrid) rather than of one sort (uniform hybrid). The two most obvious effects of hybrid fertility are the production of hybrid swarms, and the occurrence of introgression. Any two species capable of interbreeding and producing fertile hybrids are, theoretically at least, able to show either or both of these phenomena, but, by strict definition of the terms, not both in the same population. Which of the two situations is the commoner is not known; there are probably more proven examples of hybrid swarms, but these are much easier to detect. According to Grant (1956) hybrid swarms and introgression are more common in outbreeding taxa than in inbreeding ones; in the latter case polyploidy is the more likely outcome of hybridization.

A fertile F_1 generation will produce an F_2 generation and is also likely to produce backcrosses to either or both parents; if the F_1 is self-sterile a single plant will be able to reproduce only by backcrossing, and even where there are several F_1 plants backcrossing is more likely than the production of an F_2 if the parents are more common than the hybrids and the breeding system is predominantly outcrossing. As a result of segregation and recombination in the F_2 and successive generations the range of hybrid variation becomes broader and broader at each generation, and not only approaches at its extremes the range of the parents but also produces recombinants markedly different from either parent or F_1 hybrid. Backcrossing soon produces plants at various stages of intermediacy between the parents and F_1 hybrid. Thus the result of fairly free interbreeding between two species and their hybrid is a hybrid swarm of plants showing every degree of variation from one parental extreme to the other, and in addition plants with various recombinations of characteristics of both the parents. Such a taxonomic situation corresponds with the homogamic complex in Grant's (1953) classification of hybrid complexes. The precise structure of a hybrid swarm, i.e. the proportion of individuals of each genotype, is largely dependent upon environmental conditions, as mentioned later in this chapter.

In the British Isles several good examples of hybrid swarms are found, probably the best-known being *Geum rivale* × *G. urbanum* and *Primula elatior* × *P. vulgaris*. Others are *Linaria repens* × *L. vulgaris*, *Calystegia sepium* × *C.*

silvatica, Senecio aquaticus x *S. jacobaea, Tragopogon porrifolius* x *T. pratensis, Pilosella aurantiaca* x *P. officinarum* and *Prunella laciniata* x *P. vulgaris*. It is important to note that hybrid swarms between these pairs of species are in no cases found in every locality where the two species coexist, and in some cases are relatively infrequent. The barrier to the formation of hybrid swarms is not the initial production nor the fertility of F_1 hybrids, but the difficulty in the establishment of every grade of intermediate.

Introgression can be looked upon as taking place when the conditions are not conducive to the establishment of hybrid swarms. Anderson and Hubricht (1938) defined introgression (introgressive hybridization) thus: "through repeated backcrossing of the hybrids to the parental species there is an infiltration of the germplasm of one species into that of another." In other words the fertile F_1 hybrids selectively interbreed with one or both parents rather than among themselves, or the products of the former crosses survive selectively in preference to the products of the latter. In practice introgression very rarely appears to occur to both parents in the same limited area, although many cases are known where introgression to one parent occurs in one area and to the other elsewhere, e.g. *Quercus petraea* x *Q. robur*. There are several good reviews of the subject of introgression, in particular those of Anderson (1949, 1953a),* and it is thus not necessary to survey the topic here in great detail.

A typically introgressed population is one which is clearly referable to a particular species, but which possesses an extreme of variability outside that normally shown by that species and attributable to the influence of a second species. Thus the range of variability of the first species is extended without affecting that of the second, and without the necessary existence of any appreciable numbers of F_1 hybrids. Indeed, even F_1 hybrids which are relatively rare, ill-adapted to their environment and partially sterile can act as effective bridges for gene-flow from one species to another. P. F. Yeo (see under *Euphrasia*) considers that in *Euphrasia* introgression "seems to occur mainly where there are strong barriers to interbreeding", and Moore (1959) found introgression between *Viola lactea* and *V. riviniana* despite the F_1 being of low fertility. Cases where the actual F_1 fertility has been measured include *Salvia apiana* x *S. mellifera* (2%) (Epling, 1947) and *Geum montanum* x *G. reptans* (0.7%) (Gajewski, 1957).

The selective survival of hybrids close to one or both parents in various characteristics is usually attributable to the pressures of the environment. If two species are well adapted to two distinct habitats, their F_1 hybrids are likely to be well adapted to neither. In the absence of intermediate habitats F_2 hybrids which fall closer to the parents than to the F_1 hybrids are likely to be at a selective advantage, and the results of backcrosses repeatedly made to the parents are in a similar position. Strictly intermediate plants remain rare because of the absence of suitable, intermediate habitats. Where, on the other hand, there is a greater range of intermediate habitats, one should expect a greater range of hybrids. In most cases such intermediate habitats owe their existence to the activities of man, who has cleared natural vegetation, managed

* See also Heiser (1973).

the land in a great variety of ways, and in many cases intermingled different sorts of habitat which were once quite distinct ("hybridization of the habitat"; Anderson, 1948).

As an example introducing the concept of introgression Anderson (1949) described the Lower Mississippi Irises, *I. hexagona* var. *giganti-caerulea, I. fulva* and their hybrids, the former species growing in tidal marshes and the latter on wet clay by drainage ditches, often in semi-shade and at a very slightly higher altitude. In localities where the area of the population was divided into small-holdings, each of which had been managed somewhat differently, it was possible to correlate the type of land-management of the farm with the range of plants to be found there. This is in many cases the key to the population structure where two species capable of producing fertile hybrids meet. The whole range of possibilities is not realized unless a similar range of suitable habitats is provided. Where this does occur one has a situation usually described as a hybrid swarm. Where the habitat caters only for near-parental-types introgression results. If over the area occupied by hybrid populations there is a gradient of environments, then the hybrid segregants might form a cline along it. Melville (1939) described such a situation for *Ulmus glabra* x *U. plotii* and coined the term nothocline to describe hybrid clines in general.

One can look upon the distinction between hybrid swarms and introgression as one of degree, or alternatively consider introgressed populations to be a special (extreme) form of a hybrid swarm. Unfortunately in recent years there has been a tendency to misuse the term introgression to cover hybridization to produce fertile hybrids in general; used in its strict (and more useful) sense it covers only one end of that spectrum. According to Heiser (1961) situations where there is no spread of genes away from the area of hybridization "hardly deserve to be designated as introgression", but this is a much narrower use of the term, and one which requires information not always available and which was not implicated in the original definition.

There is a number of cases in the British Isles where introgression has been demonstrated, though not nearly as many as in North America. As with examples of hybrid swarms given above, in no cases does introgression occur whenever the species meet; sometimes hybrid swarms are found, sometimes mainly F_1 hybrids, and sometimes no hybrids at all. The following examples may be cited: *Quercus petraea* x *Q. robur, Betula pendula* x *B. pubescens, Gentianella amarella* x *G. uliginosa, Silene alba* x *S. dioica, Euphrasia anglica* x *E. confusa, Senecio squalidus* x *S. vulgaris, Viola lactea* x *V. riviniana.*

Introgression does not always result solely from a shortage of habitats suitable for the establishment of intermediate plants. It may be due to obvious effects concerning the relative crossability of the species. For example hybrids between *Erica mackaiana* and *E. tetralix* in Eire mostly backcross to the latter parent only, presumably because *E. mackaiana* is, for some unknown reason, largely sterile in Ireland (Webb, 1951). The cross *Primula veris* female x *P. vulgaris* male is far more successful than the reciprocal, with the result that hybrids are more often found with the former species than with the latter, and therefore more often backcross with it (Valentine, 1955; Woodell, 1965).

The situations described above refer to sympatric introgression, but allopatric introgression (between species separated geographically) also occurs and the very separation of the parental species can contribute to the occurrence of introgression. For allopatric introgression to occur pollen has to be transported from one population to another over reasonable distances, a feat obviously more easily accomplished by anemophilous species. If this happens infrequently there will be very few F_1 hybrids established in the vicinity of either population, whether or not there are habitats suitable for them, and in an outbreeding situation these will tend to backcross to the adjacent parent. In British oakwoods of either *Q. petraea* or *Q. robur* the presence of trees containing some characteristics of the other species may be attributable to such an effect.

This sort of situation may be very difficult to distinguish from one where the alien genes in an otherwise pure population owe their existence to the historical presence of sympatric populations, i.e. they are the remnants of a once more extensive distribution of the introgressing species. Pritchard (1961) found that in parts of south-eastern England *Gentianella amarella* has been introgressed by *G. germanica*, although the latter species does not occur there now. In part of this area (Berkshire and Oxfordshire) there is historical evidence of the presence of *G. germanica*, but in the other part (Kent and Surrey) there is not, although *G. germanica* does occur still on the other side of the English Channel in the Pas de Calais, where, interestingly enough, *G. amarella* is rare and exclusively littoral (Rose and Géhu, 1960).

There are several other similar examples in the British Isles, e.g. *Saxifraga (hirsuta)* × *S. spathularis* in parts of Eire, *Viola (lactea)* × *V. riviniana* in parts of south-eastern England, *Nuphar lutea* × *(N. pumila)* and *Ranunculus flammula* × *(R. reptans)* in northern England, and *Carex aquatilis* × *(C. recta)* in north-eastern Scotland. In each case the extinct or near-extinct species is placed in parentheses.

A third process by which hybrids may occur in an area not occupied by one parent (or both) involves the independent dispersal of hybrids beyond the areas in which they originated. Such dispersal might be by F_1 or F_2 seed or, more usually, by vegetative propagation. Hybrids such as *Rorippa* × *sterilis*, *Circaea* × *intermedia* and several in the genus *Mentha* have certainly been distributed in this way, and fr⸱ ⸱ntly deliberate plantings of hybrids have obscured the original distributio⸱ pattern, e.g. in *Salix* and *Populus*. In the case of purely artificial hybrids such as *Crocosmia* × *crocosmiiflora* and *Platanus* × *hybrida* there is no natural distribution pattern, and such hybrids are not treated in this book. Many of them, although very common now, originate from a single hybridization, while others have been frequently resynthesized. The proportion of sterile natural hybrids which are represented by a single clone now widely dispersed as opposed to those of polytopic origin is unknown, but almost certainly both situations exist.

One often talks of the species now extinct having been "hybridized out of existence", or of the introgressed species having shown "aggressive hybridization". These terms ascribe a strongly active role to the remaining species, and might be appropriate in some cases, but in others it is likely that the extinction of the other species is due to some factor not directly connected with the former, which is now acting as a "museum", so to speak, preserving the remains of a species. Where one species is very much rarer than another, for instance when it has been recently introduced, hybridization is a likely fate, as has been mentioned previously in connexion with species of *Linaria* studied by Dillemann. *Pinguicula grandiflora* x *(P. vulgaris)* in south-western Eire is a further example. It should perhaps be pointed out here (cf. Allen, 1957) that Druce (1896, 1897) observed precisely these results in Oxford in 1891 onwards following the introduction in 1890 of *Linaria repens* into an area where *L. vulgaris* is native. His description (Druce, 1897) of introgression, made prior to the rediscovery of Mendelism, is as clear as any to be found before about 1935, and easily predates the statements of Ostenfeld (1928) which were said by Anderson (1949, 1953a) and Heiser (1949) to be the earliest in this field.

What has been described by Valentine (1970) as " a subtle and complicated form of introgression" can occur between related diploids and polyploids by the production of viable diploid gametes by the diploid parent. Thus tetraploid (and possibly fertile) hybrids are obtained in one step. Valentine was referring to autotetraploids which could gain genes from various diploids by this method; the occasional injection of genes from a diploid into the autotetraploid gene pool would constitute introgression. There are several examples known, e.g. *Centaurea jacea* (Gardou, 1967) and *Delphinium hansenii* (Lewis, 1967).

A parallel but even more striking example is the similar production of tetraploid hybrids between an allotetraploid and one of its constituent diploids. If the diploid is AA and the tetraploid AABB, the hybrid derivative will be AAAB. In Britain examples are afforded by *Betula pendula* (diploid) x *B. pubescens* (tetraploid), and *Holcus lanatus* (diploid) x *H. mollis* (tetraploid to heptaploid). In the latter case obvious wild hybrids are triploids, but some of the present range of variation of *H. mollis* is thought to be due to past hybridization with *H. lanatus* (see below under *Holcus*).

Yeo (1956) has found that in *Euphrasia* introgression from a tetraploid to a diploid may take place. This, he suggested, occurs because of the occasional production of haploid gametes by the triploid F_1 plants. Yeo (see below under *Euphrasia*) recognizes introgression, hybrid swarms and incipient speciation as distinct hybridization phenomena, the last resembling introgression in its mode of origin but differing in that intermediate populations with a distinctive ecological and geographical range have resulted (see also Chapter 2).

Whenever two interfertile species meet and hybrid swarms develop the question of the ultimate disappearance of one or both parents in their pure forms arises. While this does happen in some cases it usually does not, partly because of the selective advantage the parents appear to have in their own original habitats. If these habitats are destroyed by man's activities it becomes more likely that the parental species might become extinct. The habitat thus imposes a barrier to indiscriminate hybridization which may be considered a further (and final) isolating mechanism, best termed (by analogy with a previous one) "Non-fitness of F_2 and backcross hybrids". There are of course other barriers to the "swamping" of two species by fertile hybrids; for example, according to Anderson (1939), "Linkage by itself is a force strong enough to prevent the complete swamping of interfertile species".

11. Artificial hybridization

As has been mentioned in Chapter 4, many more hybrids can be made artificially than exist in nature, and encompassing species of much wider relationships. In the Orchidaceae, of over 45,000 hybrids known among the cultivated genera only about 150 are found in the wild. On the other hand there are numerous wild hybrids that have never, despite careful attempts, been resynthesized, e.g. *Agrimonia eupatoria* x *A. odorata, Glyceria fluitans* x *G. plicata, Juncus effusus* x *J. inflexus.* In the first place it should be remembered that even if artificial cross-pollinations have been carried out in their thousands this probably represents but a tiny fraction of the natural cross-pollinations that occur every year. Secondly natural cross-pollinations probably involve a great many more different genetic stocks of the parental species, and a far wider range of environmental and physiological conditions. Thus it is not surprising that in certain cases natural hybridization has been more successful than artificial.

That particular conditions might sometimes be especially conducive to hybridization is indicated by the distribution in Britain of hybrids in the genera *Equisetum* and *Juncus.* Page (1973) showed that, of five known British hybrids of *Equisetum* subgen. *Equisetum,* four are confined to the Hebrides. *Juncus balticus* x *J. inflexus* is known in three different sites on the Lancashire coast, and *J. balticus* x *J. effusus* in two sites in the same area, yet for these hybrids elsewhere in the world there is only one known site (for the latter hybrid) (Stace, 1972).

Put simply, attempts at artificial hybridization aim to overcome the various isolating mechanisms that act in nature, and the techniques employed are of necessity as diverse as the mechanisms themselves. No attempt is made here to cover all the methods which are involved, but some idea is given of the range of possibilities which are available. There are many more extensive reviews, e.g. Hayes *et al.* (1955), Poehlman (1959).

EXTERNAL FACTORS

External isolating mechanisms are of course much easier to overcome than internal ones. Geographical and ecological barriers may be broken by cultivation in adjacent glass-houses, plant-beds or plant-pots, with, if necessary, different soil or climate provided. Similarly ethological and mechanical isolation is circumvented by means of manual pollination. Temporal isolation, even if complete, offers few problems, as pollen usually remains viable for many hours and can be kept until the later species flowers, or overnight until anthesis of the earlier one occurs. Alternatively, by adjusting artificially the temperature or the day–night cycle, the two species can usually be brought into flower simultaneously. Species which are seasonally isolated by a few weeks can also be hybridized by broadly the same two methods, and chemicals such as ether have been used to hasten flowering. D. H. Valentine (pers. comm., 1973) has stored pollen of *Primula* and *Viola* in normal laboratory conditions for several weeks before using it in pollination experiments; in general it can be said that the viability of pollen varies tremendously from species to species, from a few hours to many months. Plants whose flowering is separated by several months need more careful attention, but the life-cycles can often be made to coincide by alteration of the day-length, and in many cases pollen can be stored successfully for long periods in a refrigerator or deep-freeze. Thus Santamour (1972) was able to store pollen through the summer or winter and perform reciprocal pollinations between vernal- and autumnal-flowering species of *Ulmus*.

EMASCULATION AND POLLINATION

The actual act of manual pollination is usually the simplest part of the whole operation, and even in minute flowers can usually be performed successfully by the use of large excesses of pollen. But in self-compatible species used as the female care has to be taken to prevent self-pollination, usually by removal of the anthers (emasculation) before they dehisce, followed by enclosure of flower in a pollen-proof bag. If the flowers are large and open before or at about the time of anthesis emasculation presents no difficulties, but in many plants which are habitually self-fertilized the flowers are very small and moreover anthesis and self-pollination occur before the flowers open. At Manchester we have been engaged in hybridization programmes with various species of *Juncus* and *Vulpia*, and with *Vicia sativa* agg., all of which present this problem to varying degrees. In all cases the young flower-buds are dissected open and emasculated, and cross-pollination is carried out between 1 and 3 days later, as the stigmas become receptive. Apart from the small size and delicate nature of the flowers at the time of emasculation the main difficulty lies in preventing the death of the gynaecium from desiccation of the opened flower. Among the various techniques employed to overcome this are: covering the opened flower with a small gelatin capsule; placing a piece of hollow grass-culm over the bared style; and closing the opened perianth around the gynaecium by means of sellotape or a cylindrical clamp consisting of a length of grass-culm slit along one edge. With larger flowers no such precautions may be needed, or a piece of damp cotton-wool placed in the pollen-proof bag may be sufficient. When hybridization is to be carried out in the open rather more resilient materials are needed. Johnson and Bradley (1946) used a double pollen-proof bag in their work on anemophilous forest trees. The outer, protective bag could be removed and pollen was shot on to the female flowers by means of a "pollen-gun" via a small hole punctured in the inner bag. Such bags should be made of unattractive, dull materials to escape the attention of inquisitive birds or children.

Details of various techniques employed by workers engaged in crossing grasses are given by Jenkin (1924, 1931). A method employed by some legume- and grass-hybridizers is to emasculate by means of a small suction-pipette which removes the anthers without damaging the rest of the flower. In some grasses the flowers can be induced to open prematurely, without the anthers dehiscing, by suitable temperature treatments, and the exserted anthers can then be removed with forceps (Jordan, 1957). Members of the Compositae are usually difficult to emasculate, but, in those species of the subfamily Tubuliflorae where the ligulate (ray-) florets are female only, the desired effect is achieved by simply removing all the disk-florets.

In some species which are known to be completely self-incompatible emasculation may not be necessary, but most self-incompatible species which have been thoroughly investigated have been found to exhibit some degree of seed-set from selfing, and it is dangerous to assume in any species that all seed obtained by cross-pollinating unemasculated flowers is hybridogenous. It has in fact been shown that compatible pollen from a second species can sometimes induce self-compatibility in a normally self-incompatible species, e.g. in *Galium mollugo* and *G. verum* (Fagerlind, 1937) and *Calamagrostis purpurea* (Nygren, 1949). The phenomenon of pseudo-self-fertility investigated and described first in *Nicotiana* by East and Park (1917) involves the change to self-fertility of a normally self-sterile plant at the end of the flowering period or in adverse environments (e.g. low temperatures), and clearly needs adequate consideration when carrying out crosses. In certain genera male-sterile mutants are known, and such plants are very convenient for use as female parents (e.g. *Triticum*).*

PROGENY TESTING

With both self-compatible and self-incompatible species, whether or not emasculation has been undertaken, it is necessary to grow the putative hybrid seed to ensure that the offspring are indeed hybrids. Gymer and Whittington (1973a), for example, when crossing *Festuca pratensis* and *Lolium perenne*, did not bother to emasculate as both species show a low degree of self-fertility. Hybrids are easily detected in the flowering state, and it was judged more expedient to carry out a large number of crosses and select the hybrids from the next generation than to undertake far fewer crosses with a higher success rate. This technique is, however, more satisfactory if the hybrids can be distinguished by a single conspicuous characteristic. This is achieved by the use of dominant markers on the male parent (or recessive markers on the female parent). In the *Vicia sativa* aggregate, for example, white-petalled recessive mutants are available, and in the cross *Vulpia membranacea* female x *Festuca rubra* male the hairy, reddish, lower leaf-sheaths of *F. rubra* are dominant to the glabrous, greenish ones of the female parent, a feature which can be detected in the early seedling stage.

In monoecious pteridophytes it is scarcely practicable to emasculate prothalli, and crosses are usually carried out by flooding selected prothalli with water containing antherozoids of the male parent. Usually the antheridia and archegonia develop sequentially on prothalli, the former earlier, so the selected

* See also Wit (1974).

female prothallus is generally an older one. Interesting accounts of nineteenth century methods of fern hybridization were given by Druery (1900, 1907). In *Equisetum* the nutrient status has differential effects on the proportion of male and female sex organs. In all these cases it is necessary to grow the progeny and examine individual plants for hybrid origin.

CROSSES IN VARIOUS COMBINATIONS

The success rate of artificial hybridization is improved if careful regard is given to a number of well-known phenomena, any of which might be of particular importance to the combination being attempted. Three situations are of widespread significance; these are the very different degrees of success sometimes obtained by the use of reciprocal crosses, of crosses at different ploidy levels, and of crosses between different races of the two parents. Several examples often with different physiological or genetical bases have been given previously of different results obtained from reciprocal hybridizations. In addition Valentine (1952, 1955) found that, in crosses between *Primula veris* on the one hand and *P. elatior* or *P. vulgaris* on the other, artificial hybrids were only obtained when *P. veris* was the female parent; Stace and Cotton (1974) have been able to obtain the hybrid *Festuca rubra* x *Vulpia membranacea* only when *V. membranacea* was used as female; and Ritchie (1955b) and others have similarly obtained *Vaccinium myrtillus* female x *V. vitis-idaea* male but not the reverse. There are a great many other examples. Usually the reasons for the differences are unknown, but sometimes they are quite obvious. In *Prunus* species the style-length varies considerably from species to species, and hybridization is much more successful where the shorter-styled species is used as female. Crosses between diploids and tetraploids can also be more successful in one direction than the other (see Chapter 8). Since it is usually impossible to predict the outcome of such experiments it is always desirable to perform reciprocal hybridizations in the first instance.

There are many genera known where it has been found easier to produce hybrids by crossing polyploids than by crossing diploids. Fagerlind (1937) found that both *Galium mollugo* and *G. verum* are either diploid or tetraploid; at the diploid level they are intersterile yet at the tetraploid level fertile hybrids are formed. The diploid and tetraploid races of each species are indistinguishable (cryptic polyploidy), and as the diploids are mainly southern in European distribution and the tetraploids mainly northern one finds hybrids over only part of the area occupied by both parents. Grant (1965) also found hybridization between tetraploid species or races of *Lotus* much easier than that between diploids. This has led to the technique of obtaining crosses impossible or difficult at the diploid level by using natural or artificially induced tetraploids. In some cases, however, the reverse is true; Nordenskiöld found crossing more difficult and the fertility of the hybrids lower in *Luzula* the higher the ploidy level (diploid, tetraploid and hexaploid).

It is a common experience that, after many unsuccessful attempts to hybridize two plants, hybrids have been produced relatively easily once a new race of one of the species is utilized. Blakeslee and Satino (1949) found marked differences in the crossability of different, morphologically indistinguishable strains of *Datura leichhardtii* from various parts of Australia with various non-Australian species, and Valentine (1970) reported similar results in crosses between Japanese *Viola grypoceras* and various European sources of *V. riviniana*. Baker (1951) discussed this topic under the heading "Variance of Fertility", and gave other examples.

INTERNAL FACTORS

Numerous techniques have been employed to overcome internal isolating mechanisms; an interesting survey with special reference to *Datura* was made by Blakeslee (1945). The seemingly naïve suggestion by Jørgensen (1928) that it might be possible to achieve normally incompatible crosses by the admixture of a small amount of pollen from the maternal parent with the foreign pollen has in fact several times been followed up with success. Beasley (1940) was able to obtain new cotton hybrids in this way, and Yenikeyev (1966) has used the method consistently in producing hybrids between relatively remotely related species of fruits, notably apple x pear and plum x cherry. Presumably the compatible pollen in some way triggers off an acceptance mechanism in the carpel, and the foreign pollen is able to take advantage of it.

Where the incompatibility mechanism resides in the stigma or style (gametophytic isolation) it is sometimes possible to obtain hybrids by excising the style and pollinating the cut stump, e.g. Davies (1957) in *Lathyrus*. Similarly Mangelsdorf and Reeves (1931) were able to pollinate successfully *Zea* with *Tripsacum* or *Euchlaena* pollen by greatly shortening the styles of the *Zea* female parent. It was also found (Anon., 1966) that stylar amputation in *Lathyrus* was made even more effective by smearing the stump with an artificial seed-setting hormone (1% α-naphthyl-acetamide in lanolin), which also promoted hybridization in the absence of surgery. Other techniques which have on a few occasions successfully negotiated stylar inhibition are the grafting of compatible styles from the pollen-parent on to the seed-parent (Buchholz et al., 1932), the use of periclinal chimeras, whereby an epidermal layer of a compatible plant covers the incompatible seed-parent (Satina, 1944), and the introduction of pollen directly into the loculus of the ovary (Bosio, 1940).

Seed-incompatibility has on many occasions been circumvented by artificial embryo-culture, and in many plant-breeding establishments (especially concerning cereals) it is a standard technique. The young embryo is dissected from the seed, before the stage at which degeneration commences, and is grown in sterile conditions on a nutrient medium until it can be transplanted to soil. The technique was apparently first devised by Laibach (1925) for *Linum*, but since then its efficiency, particularly with respect to the nutrients employed, has risen markedly, and nowadays very young embryos indeed can be successfully cultured. The work of Blakeslee and co-workers (e.g. Blakeslee, 1945) and of Brink and Cooper (e.g. Brink et al., 1944) was particularly important in these developments.

Weak F_1 hybrids can often be cultured successfully to the flowering stage, whereas in the wild they would succumb to competition. Williams (1959) reported that the apple x pear hybrids obtained at the John Innes Horticultural Institution could not be grown to maturity on their own roots, but survived if the shoots were grafted on to pear or (preferably) apple rootstocks. Smith (1948) grafted chlorophyll-deficient *Melilotus alba* x *M. dentata* on to stocks of commercial *Melilotus* in order to overcome F_1 seedling death.

INDIRECT METHODS

Where hybrids are not obtainable by direct means hybridization can sometimes be carried out indirectly. Andersson-Kotto and Gairdner (1931) were able to combine the genomes of several incompatible species of *Dianthus* by crossing one of the species with a hybrid between the other species and a third one, and sometimes quadruple hybrids (obtained by crossing two binary hybrids) were necessary. Kihara (1940) used similar principles in *Triticum* (Gramineae). It is sometimes possible to cross an amphidiploid with species with which neither of the progenitors of the amphidiploid will cross, e.g. *Raphanobrassica*. Such combinations are not the same as the combinations desired, but they may enable the interaction of the characteristics and of the chromosomes of the two species in question to be studied.

Stebbins (1950, p. 215) reported some Russian work which produced a new *Elymus* x *Triticum* hybrid, the two species concerned being normally incompatible. Embryos of the *Triticum* were removed from their caryopses and grafted on to the endosperm in ovules of a growing *Elymus* plant. The mature *Triticum* plants eventually obtained were normal in appearance, but were then able to be used successfully as seed parents in crosses with the *Elymus*.

The technique of inducing fertility in sterile hybrids by the artificial production of allotetraploids is too well known to require description here. Goodspeed and Bradley (1942) provided an early survey of the means available; nowadays the application of the drug colchicine is the most usually employed method, having been first introduced in 1937.

12. Structural features

The traditional concept of an interspecific hybrid is that it should be intermediate between its parents, and sterile. The latter criterion has been discussed in Chapter 10, the former is described here with respect to morphological and anatomical features.

DEGREES OF INTERMEDIACY

Most hybrids are indeed intermediate between their parents in diagnostic characters, but there are many grades of intermediacy and even half-way intermediacy can be attained in a variety of ways. Thus in two species differing mostly in quantitative characters the hybrids may fall half-way between them not only overall but also with respect to each individual variable. *Calystegia sepium* x *C. silvatica* illustrates this case well. Stace (1961) constructed a numerical scale running from 4 to 40, based upon four diagnostic quantitative variables, so arranged that plants of *C. sepium* cover the range 6–12 and those of *C. silvatica* 23–36; thus by definition plants falling into the range 13–22 are intermediate, and presumed to be hybrids. It would be possible, however, for plants from the upper part of each species range (say 12 and 36) to produce hybrids which in fact fall within the range of one of the parents (24 in the example). Since fertility is (at least often) not reduced in the hybrids, and qualitative characters are lacking, it is impossible to be sure of the exact limits of each species and the hybrids. Moreover F_1 hybrids alone can adequately account for all degrees of intermediacy between the parents, and so it is not possible simply by analysing morphological characters to determine whether some wild hybrids represent F_2 or later generations or backcrosses, or all are F_1

On the other hand a hybrid between two closely related species differing by small, qualitative features is likely to inherit the latter in a simple Mendelian fashion. Due to the chance distribution of dominance between the parents, the hybrid is likely to possess some characters of one parent, some of the other, and some (due to incomplete dominance or polygenic inheritance) intermediate; the net result may be no less intermediate than the *Calystegia* example above. In some instances, by chance, the characters of each parent are manifested on different parts of the hybrid, for example on vegetative and reproductive parts respectively, so that during the growth of the hybrid it may at one time resemble one parent closely and later the other. The hybrid between *Festuca rubra* (perennial) and *Vulpia membranacea* (annual) is virtually indistinguishable from some variants of the former parent in its vegetative organs, but clearly intermediate in floral characters. In *Ulmus* Melville (1955) described a specimen which resembled *U. plotii* early in development and in the epicormic shoots, but where *U. carpinifolia* was almost totally dominant in the adult phase. Somatic segregation, whereby different parts of a hybrid may appear to revert to one parental pure species or the other, may appear similar, but has different genetic bases. Examples of it are given by Allan (1937), and some of the causes discussed by Swanson (1958). It is much more easily brought about in fungi which have formed hybrid heterokaryons but before the unlike nuclei have fused.

It is not possible to lay down rules governing the outcome of matings involving particular characters. Hybrids between annuals and perennials are usually perennials, but that between *Senecio squalidus* (a short-lived perennial in most areas of Britain, despite many statements to the contrary) and *S. vulgaris* (annual) is annual. The artificial hybrid between *Fragaria vesca* (herbaceous perennial) and *Potentilla fruticosa* (shrub) is a herb (Ellis, 1962). Crosses between plants with epigeal and those with hypogeal cotyledons may bear either sort or intermediate sorts; in *Phaseolus coccineus* (runner bean, hypogeal) x *P. vulgaris* (dwarf bean, epigeal) different crosses can show any of these situations.

Hybrids between dioecious and bisexual species are only possible in those few cases where the parents are sufficiently closely related. Baker (1958) reviewed the situation in the campions (Caryophyllaceae: Silenoideae), where the F_1 progeny can be all male, all female, all bisexual or mixtures of any of these. In at least one case an F_1 hybrid produced bisexual flowers for two years but only male flowers the next. Other crosses produced noticeably differing offspring when made in successive seasons.

The production of conspicuously different offspring from repeat crossings of the same two parental species has been documented on many occasions. Often it results from the use of different races of the parents whose varying interactions give different results. Sometimes, however, it is caused by the production of unreduced gametes by one parent. Hybrids between *Festuca pratensis* (FF) and *Lolium perenne* (LL) may be diploids (FL) roughly half-way between the parents, festucoid triploids (FFL) more closely resembling *F. pratensis,* or lolioid triploids (FLL) more closely resembling *L. perenne* (Gymer and Whittington, 1973b). Similarly *Ammophila arenaria* (AAAA, tetraploid) and *Calamagrostis epigejos* (CCCC, tetraploid) produce three sorts of hybrids known as nm. *intermedia* (AACC), nm. *subarenaria* (AAAACC) and nm. *epigeoidea* (AACCCC), according to their morphological appearance, but in this case it has been suggested that the last nothomorph (AACCCC) in fact arises from reduced gametes of an octoploid race of *C. epigejos* rather than from unreduced gametes of a tetraploid race (Westergaard, 1943). Either way, nm. *epigeoidea* possesses twice as many *Calamagrostis* as *Ammophila* genomes, and thus resembles the former parent more closely. The existence of two nothomorphs of *Agropyron repens* x *Hordeum secalinum,* one more or less intermediate and the other much closer to *A. repens,* may also be explicable on the same basis, but the cytological details have not yet been sufficiently investigated.

Experimental hybridizations have often given rise to both diploids and triploids, and sometimes, by parthenogenesis, to maternal (non-hybrid) haploids as well. In *Primula veris* female x *P. elatior* male crosses the genomic constitution of the offspring can be designated V, VE and VEE, and their morphological appearance varies accordingly (Valentine, 1952).

Hybrids of different appearance resulting from reciprocal crosses are in most instances explicable on the basis of cytoplasmic inheritance, for almost all the cytoplasm in an angiosperm zygote is derived from the female parent. The best understood interspecific examples of this phenomenon are probably in *Epilobium,* which have been studied by several German geneticists (notably E. Lehmann and P. Michaelis), and the results have been summarized by Caspari (1948). In any species where conspicuous features of the morphology are determined by cytoplasmic genes, it is to be expected that hybrids will more closely resemble the female than the male parent (matrocliny). The hybrid *Digitalis lutea* x *D. purpurea,* for example, resembles its female parent more closely than its male parent in several characters, and thus exists as two distinct nothomorphs (Hill, 1929).

The marked matrocliny in hybrids involving species of *Rosa* section *Caninae* is explained more fully under that genus. Since the male parent contributes only one genome (7 chromosomes) and the female $2n - 1$ genomes ($2n = 4x$ or $5x$ in most species), the direction of the cross is shown by cytological as well as morphological features. A second special case, this time not involving matrocliny and therefore where reciprocal differences are again not due to cytoplasmic inheritance, is found in those species of *Oenothera* subgen. *Oenothera* which are permanently heterozygous for numerous chromosome translocations. These plants are mostly self-pollinated, and thus give rise to progeny all alike and like their parent. When different races are reciprocally crossed, however, different sorts of progeny usually result. This is because of the system of gametophytic balanced lethals possessed by many species, whereby one of the two sets of chromosomes (Renner complex) can pass only through the male gametes and the other Renner complex only through the female gametes. Other species, which do not show marked reciprocal differences, have only zygotic lethals, where both complexes can pass through both sorts of gametes but homozygous zygotes die while heterozygous ones survive. The species with gametophytic lethals can thus produce by reciprocal hybridization two different sorts of progeny, possibly both new, and moreover each of these can then give rise by selfing to whole populations of similar individuals (see Cleland, 1972).

True matrocliny is a totally different phenomenon from the much closer resemblance some hybrids have to one parent than to the other, whatever the direction of the cross, i.e. apparent matrocliny or patrocliny according to the direction. The latter effect is due to the dominance of most of the characters of one parent over those of the other. Baker (1951) suggested that in crosses between wild species and derived cultivated ones the hybrids might tend to resemble the former more closely because the cultivated plant may well have accumulated many recessive genes. He cited the case of *Lycopersicon esculentum* (cultivated) x *L. hirsutum* (wild) (Solanaceae) as an example of this, and also gave a list of hybrids which had been shown to resemble one parent much more than the other. Where the hybrid is very similar indeed to one parent it probably goes largely undetected in the wild, e.g. *Geranium purpureum* x *G. robertianum*, which is virtually indistinguishable (apart from its sterility) from the latter parent. Davis and Heywood (1963) quoted the case of *Digitalis purpurea* x *D. thapsi*, which is similarly close to the latter parent.

Millardet (1894) coined the term false hybrids ("faux hybrides") to account for the occurrence in some interspecific *Fragaria* crosses of offspring which were apparently exactly maternal or exactly paternal in appearance. Mangelsdorf and East (1927) showed that the former resulted from self-fertilization or pseudogamy in the female parent, while the latter were female diploid x male octoploid crosses where the characters of the octoploid male swamped those of the female. Dominance by characters of a polyploid over those of a diploid species is a common cause of the non-intermediacy of hybrids.

In crosses between dioecious species the sexes of the hybrid progeny are not always in the expected 1 : 1 ratio. In many *Salix* hybrids (both natural and artificial), for example, male plants are rare or absent; the reasons for this appear to be unknown. This phenomenon is not to be confused, of course, with the common situation in *Salix* and *Populus* whereby one strain (or tree) of a hybrid has been propagated and commercially distributed, and hence become far commoner than other nothomorphs. The occurrence of only one sex is known in several hybrid animals, where it has been suggested ("Haldane's Rule") that it is in general the heterozygotic sex that is missing or rare (see Dobzhansky, 1937). So far as is known *Salix* does not have an XY sex-determining mechanism.

Directly opposite to the production of hybrids of different appearance from the same interspecific cross is the identical appearance shown by some hybrids of different parentage. In the systematic part of this book C. D. K. Cook covers *Ranunculus aquatilis* x *R. fluitans* and *R. fluitans* x *R. trichophyllus* together because they cannot be distinguished on morphological characters, although the two hybrids usually differ in chromosome number because the non-common parents mostly do so. Similarly B. Matfield treats *Potentilla erecta* x *P. reptans* and *P. anglica* x *P. reptans* together. Although the non-common parents again differ in chromosome number (*P. erecta* $2n = 28$; *P. anglica* $2n = 56$; *P. reptans* $2n = 28$) the chromosome number of hybrids does not give an answer to their parentage because of the known occasional production of unreduced gametes by *P. reptans.* Indeed, since *P. anglica* is thought to represent the amphidiploid (EERR) of the other two species (EE and RR), hybrids between *P. anglica* and *P. reptans* (ERR) are thought to have the same genomic constitution as hybrids between *P. erecta* and *P. reptans* where the latter parent contributes unreduced gametes. The morphological similarity of the two hybrids is therefore wholly expected. *Festuca rubra* x *Vulpia membranacea* and *F. juncifolia* x *V. membranacea* are treated separately by A. J. Willis, but the latter hybrid is known certainly from one locality only, and the slender differences between it and the former might be lost if and when more plants with wider morphological ranges are discovered.

Such situations give rise to polyphyletic hybrid taxa and, if doubling of the chromosome number occurs, amphidiploids of a much greater degree of polyphylesis than those (such as *Tragopogon mirus*) mentioned in Chapter 10. It is likely, for example, that *Ranunculus penicillatus* arose from the two *Ranunculus* hybrids mentioned above, and *Dactylis glomerata* is probably a tetraploid complex which represents the amphidiploid (and perhaps some autopolyploid) derivatives of many different diploid hybrids, for at least 10 diploid species are known and the morphological range of *D. glomerata* appears to encompass most of them (Stebbins and Zohary, 1959; Borrill, 1961). Allan (1937) described similar examples in *Celmisia* (Compositae).

An amphidiploid usually resembles very closely the sterile hybrid from which it is derived, except for its fertility (which can be reflected in distinct morphological features of flowers or fruit), but it is often robust and typically possesses larger pollen and stomata. If it is of very ancient origin the original parents (and hence their hybrids) and/or the amphidiploid may have gained extra or different characteristics in the normal course of evolution, and hence be more easily distinguishable, but the size differences of pollen and stomata may have disappeared.

DISCORDANT VARIATION

Anderson (1951) introduced the concepts of concordant and discordant variation, and suggested that the latter was characteristic of hybrids. He asserted that the corresponding parts of related species differ in three ways: in magnitude, in proportion and in the trend in proportion. He illustrated the last by reference to two species each with a range of leaf-shape; leaves identical in magnitude and proportion (size and shape) could be gathered from the two species, but when the *range* in shapes was graphically represented it was found that these two leaves came from the crossover or convergence points of two differently placed, straight-line relationships. The fact that in each species the range of leaf-shape can be expressed as a straight line means the variation is concordant, and the same may be true for F_1 hybrids. Anderson's data on *Berberis* x *gladwynensis* (Anderson, 1953b) illustrate this beautifully, without specifically referring to the fact. In F_2 hybrids between such species variation becomes discordant, because the different variables which define the parental species become randomly re-assorted and hence to some extent independently variable. Anderson claimed that discordancy was subjectively discernible to the trained eye, and was a valuable aid to the recognition of hybrid populations. Melville (1955) made the same claim for certain *Ulmus* hybrids; one interesting discordant feature is the presence in some hybrids of symmetrical leaves, which appear to combine two short sides or two long sides rather than the long and short sides of normal asymmetrical leaves. The upsetting of the normal distribution of male and female flowers in certain *Carex* hybrids is perhaps a further discordant feature.

F_2 SEGREGATION

While an F_1 generation is usually fairly uniform in appearance, the F_2 is much more variable as a direct consequence of Mendelian segregation; discordancy is in part due to the resultant recombination. Given a sufficiently great sample size an F_2 will theoretically reproduce parental types and every grade of intermediate between them. An F_1 which is male-sterile or female-sterile can equally produce every degree of intermediacy by backcrossing to both parents, and on occasions (as above under *Calystegia*) an F_1 alone can apparently attain the same effect. In an F_2 generation, however, not only are parental and intermediate types produced but recombinant types also appear. If all the characters of one parent are represented at one corner of a cube, and all those of the other parent at the opposite corner, then the theoretical range of F_2 possibilities is represented by the volume of the cube. In fact the actual range observed is represented by a narrowly ellipsoidal spindle, termed the recombination spindle by Anderson (1939, 1949), which runs between the corners occupied by the two species. Figure 3 depicts Anderson's recombination spindle, and Fig. 4 the corollas of the *Nicotiana* hybrids which are involved. The reduction in the volume of variation from the cube to the ellipsoidal spindle is brought about by the genetic linkage between the characters of the two parents, which in turn is related to the

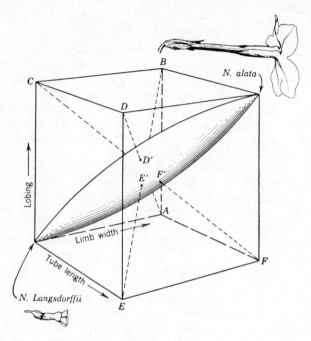

Fig. 3. The recombination spindle. The two species, *Nicotiana alata* and *N. langsdorffii*, occupy opposite corners of the cube, and representative flowers of them are shown. The theoretical range of F_2 hybrid segregants possible is represented by the volume of the cube, of which plants at positions A–F would be extremes. The actual F_2 range encountered is represented by the volume of the spindle, of which plants at positions A'–F' are extremes. From Anderson (1939).

number of loci involved compared to the number of chromosomes (loci on different chromosomes not being linked). Anderson (1939) outlined three taxonomic consequences of these considerations. Firstly, hybrids intermediate in one character will *tend* to be intermediate in others, or if like one parent in one character will *tend* to be like the same parent in others; secondly, there will *tend* to be less variation between individuals the closer they are to the parents; and thirdly, the novelty of hybrids will *tend* to be due not so much to the appearance of new characters but to new combinations of parental characters.

While these tendencies exist in theory and probably usually in practice too, there is no rigid adherence to them in the wild. Exceptions to the first have already been given in this chapter. Where the environment imposes a strong selection for parental types the number of half-way intermediates can be very low compared with the number of intermediates very close to the parents, so that often the range in near-parental sorts will be the greater. Thirdly, there are many cases known where "new" characters have appeared in hybrids, both F_1 and F_2. These result from the new and unpredictable interactions of the parental germplasms. The odd-looking hybrids in *Epilobium, Papaver* and *Brassica*, giving

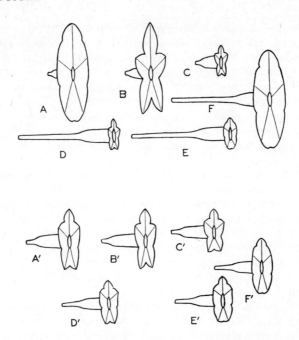

Fig. 4. Corollas of F_2 segregants of *N. alata* x *N. langsdorffii.* Corollas A–F represent theoretical extremes; corollas A′–F′ represent extremes actually encountered. In each case the lettering corresponds with positions on Fig. 3. From Anderson (1939).

the appearance of diseased plants, were mentioned in Chapter 9. Cockayne and Atkinson (1926) found bullate leaves and leaves with longitudinal ridges in *Nothofagus* (Fagaceae) hybrids, although these characters were not present in the parents, and, also in New Zealand, Cockayne (1929, *fide* Allan, 1937) reported that the hybrid *Plagianthus betulinus* (forest-tree) x *P. divaricatus* (divaricate shrub) (Malvaceae) is a small twiggy tree which represents a new life-form. Cook (1970) gave several examples of new characters (such as new leaf-shapes, yellow petals, two whorls of sepals) which emerged in F_2 hybrids of *Ranunculus* subgen. *Batrachium.* Marsden-Jones (1930) found that *Geum rivale* x *G. urbanum* possesses a type of hair not found in either parent. Other examples were given by Stebbins (1950). The presence in amphidiploids of characters not present in any of the possible parental diploids known may be further valid examples of new characters, but in other cases they probably indicate that the amphidiploid is of ancient origin, and that since then either the parental diploids carrying the character have died out or the amphidiploid has gained the character by mutation. A character of such unknown significance is the hairy apex of the ovaries of the tetraploid *Vulpia membranacea* and the hexaploid *Festuca arundinacea.* All related diploids known, in both cases, have glabrous ovaries.

A greater restriction in the F_2 than that due to the linkage effects shown by the recombination spindle has been noticed in crosses within the *Vicia sativa* aggregate (Yamamoto, 1966; E. Hollings and C. A. Stace, unpublished). Crosses between certain segregates produce an intermediate F_1, but the F_2 consists of three sorts resembling the two parents and the F_1 respectively. The parental sorts appear to breed true, but the F_1 sorts segregate again in the F_3 into the three distinct groups. Yamamoto suggested that this might be explained on the basis that gametes containing a set of chromosomes derived wholly from one parent are in some way far fitter than those containing chromosomes derived from both parents, but clearly more work needs to be carried out on this intriguing topic.

Cook (1970) has stressed that individual plants of the F_2 generation, while often showing greater genetic variation, greater plasticity and various new characters, etc. (i.e. greater versatility than their progenitors), sometimes exhibit markedly decreased versatility. For example, certain F_2 segregants of crosses between amphibious species of *Ranunculus* subgen. *Batrachium* (which bear both broad, floating and dissected, submerged leaves) produce only one sort of leaf, and thus have become only terrestrial or only aquatic instead of amphibious. It is possible that some of the terrestial and aquatic species of this subgenus arose in this way.

Allan (1931, *fide* Allan, 1937) coined three useful and frequently used terms describing the variability of a hybrid according to whether they showed no diversity, little diversity or great diversity: uniform, pauciform and multiform. In the New Zealand flora, as it was known in 1931, there were 4, 62 and 264 hybrids respectively of each sort.

HYBRID VIGOUR

One of the best-known characteristics of many hybrids, as well as their usual sterility, is the phenomenon of hybrid vigour, of which great use is made by plant and animal breeders. It may be considered in some ways the opposite of inbreeding depression, whose effects are equally widely appreciated. Although hybrid vigour is generally equated with the term heterosis there are in fact two opposite aspects of heterosis—positive and negative—as pointed out by Sveschnikova (1940), and hybrid vigour represents only one side of the coin. Nevertheless it is a side so well documented (see particularly Gowen, 1952) that prolonged discussion of it is not required here. Hybrid vigour is generally considered to represent the fortuitous coincidence in a hybrid of most of the favourable genetically-controlled growth factors of both parents. Viewed in this light the equal occurrence of negative heterosis becomes totally expected. From artificial *Primula elatior* female x *P. vulgaris* male crosses Valentine (1947) raised F_1 plants showing a wide range of vigour from marked positive to marked negative heterosis. Because of the chance association bringing about heterosis it is to be expected that its effects will be diminished in subsequent generations, and indeed it is generally found to be associated mainly with the F_1 generation.

From a taxonomic point of view three aspects of heterosis require emphasizing. Firstly, heterosis may be manifested in a great variety of ways, and it may be confined to certain aspects of the hybrid. For instance plants showing hybrid vigour may be taller, or stouter, or more branched than usual, or any combination of these. Similarly there may be more flowers and fruit, or larger ones, or both, or they may ripen more quickly, and so on. Secondly, hybrid vigour may not be maintained throughout the life of a plant. Passmore (1934) found that the initial vigour in *Cucurbita pepo* (Cucurbitaceae) hybrids was not always maintained, and in many British forestry plantations the same has been found regarding *Larix decidua* x *L. kaempferi*, which has been quite widely planted because of the hybrid vigour it exhibits as a sapling. Thirdly, different races of the same two species sometimes produce markedly different degrees of heterosis when crossed, e.g. *Juglans nigra* x *J. regia* (Schuster, 1937, *fide* Richens, 1945).

Hybrid vigour is apparent in many British wild hybrids, e.g. *Juncus acutiflorus* x *J. articulatus* and various *Epilobium* hybrids. The hybrid *Juncus balticus* x *J. inflexus,* which is known only as three separate large colonies, all on the Lancashire coast, has stems up to 2 m high (parents up to *c* 0.5 m and *c* 1.0 m respectively), stems and rhizomes over twice the diameter of those of either parent, and a rate of rhizome extension-growth far in excess of that of any other European *Juncus.*

ANATOMICAL CHARACTERS

Anatomical characters have often been used to very good effect in identifying hybrids and in studying their variation. This is not in any way unexpected, but far too frequently the opportunities offered by microscopic examination are completely overlooked. In those groups where anatomical features have been traditionally employed in species recognition they are of equal importance in the study of hybrids. In *Juncus* subgen. *Genuini,* for example, it is not always possible to distinguish between the hybrids *J. balticus* x *J. effusus* and *J. balticus* x *J. filiformis* on morphological characters, but the anatomy of the pith-cells makes determination easy, and in several other hybrids anatomical observations (particularly of the pith-cells, epidermal cells and sclerenchyma) are desirable for confirmation (Stace, 1970 and unpublished). In this genus the use of anatomical features is not new; F. Buchenau and P. Ascherson and P. Graebner were employing them to characterize *Juncus* hybrids almost a century ago. Other examples of their early use are given by Roberts (1929), and Cannon (1909) provided one of the earliest detailed studies of the inheritance of anatomical features of trichomes (in species and hybrids of *Juglans, Oenothera, Papaver* and *Solanum*). Examples of all the aspects of variation in structural features of hybrids that have been outlined above are doubtless to be equally found in anatomical characters. One will suffice. Cutler (1972), in his study of the leaf anatomy of *Gasteria* and *Aloe* (Liliaceae), found that several characters appeared in hybrids which were not found in either parent, and the inheritance of characters in the hybrids showed the same unpredictable variation patterns as have been mentioned previously.

XENIA

The phenomena known collectively as xenia deserve a brief mention. The term xenia was coined by Focke (1881) to cover simply effects of parental pollen on a hybrid offspring. As pointed out by Crane and Lawrence (1956), who gave a lucid discussion of the subject with numerous examples, this definition should be viewed in its "pre-genetic" context, for it includes such straightforward genetic phenomena as the effects of the paternal genes in heredity. These are reflected in the very nature of the F_1 hybrid and, owing to the process of double fertilization in angiosperms, of the endosperm. In fact the latter effects are still sometimes referred to as xenia. In addition the nature of the pollen can sometimes show effects in the maternal tissue itself, usually the fruit or the solely maternally-derived parts of the seed, and these latter effects, now usually described as metaxenia, are less easily explained. As pointed out by Bunyard (1907), as well as by Crane and Lawrence (1956), they are in fact the only category of true xenia. They are most commonly exhibited in the shape, size, colour or flavour of the fruit or seed, or in its time taken to reach maturity. It is probable that metaxenia result from growth substances or nutritive factors which are produced by the hybrid embryo (and hence in part determined by the pollen source) and passed out into the maternal tissue.

NON-HYBRID INTERMEDIATES

Finally, it is important to realize that plants intermediate between two species are by no means always hybrids. Many species with the epithets *"hybridus"* or *"intermedius"*, e.g. *Trifolium hybridum*, *Drosera intermedia*, were at one time thought to be hybrids, but in fact are quite normal species. In other cases the "intermediates" represent odd growth-forms or mutations, or the effects of attack by insects or fungi. Wingless-seeded variants of *Spergularia media* have often been misidentified as *S. marina* x *S. media*, and sterile specimens of *Juncus inflexus* (probably caused by environmental factors, and not uncommon) as *J. effusus* x *J. inflexus*. Often the "intermediacy" is apparent in only the most conspicuous features, and when the other characters are examined the true identity becomes clear.

Parsons and Kirkpatrick (1972) and Kirkpatrick *et al.* (1973) described what have been called "phantom hybrids" in *Eucalyptus**, where intermediate populations exist far from either putative parent species. Such populations are undoubtedly truly intermediate, but they now behave as distinct species which do not show sterility or progeny segregation to any greater degree than either parent. They represent either ancient hybrids which have now become stabilized, or they have arisen by means not involving hybridization. For example they might represent fragments of a cline once more or less continuous between the two species, or they might have evolved from one species or the other by strong selective pressures similar to those which caused the separation of the two species in the first place. A similar discussion has surrounded the origin of *Abies borisii-regis* in the Balkan Peninsula (Mattfeld, 1930). It should not be forgotten that hybrids can also form topoclines over quite large areas (Melville, 1939).

* (Myrtaceae)

Where two distinct taxa are separated by an area containing intermediate plants there are clearly several distinct explanations which can be forwarded. Turrill (1934) provided one (for the presence mainly in south-eastern Europe of intermediates between *Ajuga chia* and *A. chamaepitys*) different from any of those mentioned above. He postulated that the intermediates in fact represent the progenitors of both species, which had evolved from the pool of intermediates by adaptative radiation. This is one example of the "genetic pool hypothesis", which is essentially an alternative to the hybridization hypothesis. Despite the fact that there seem to be no reasons at all why both hypotheses should not be correct in different cases, several evolutionists have argued very strongly for one or the other. Zoologists, unused to the idea of widespread hybridization, have more often propounded the genetic pool hypothesis. E. Anderson, well known for his pioneer work on introgressive hybridization, has, on the other hand, investigated a number of actual cases with the aim of distinguishing between the two hypotheses. In all cases hybridization was found to be the cause of the variation pattern, and Anderson (1953a) concluded that the genetic pool concept "may be disregarded until definite evidence for such an hypothesis has been put on record." One can only protest, like Davis and Heywood (1963), that "there is plenty of room for both processes."

13. Chemical studies

Chemical plant taxonomy has become one of the more active fields of taxonomic research in the past 20 years, and, because of the precise nature and known genetic basis of many of the characters involved, hybrids have for some time been favourite plants for study. The best general review of the subject with regard to the higher plants is still that of Alston and Turner (1963a), and their conclusions are the same as those one would draw today. The number of examples one could now cite is, of course, vastly greater.

In many cases a study of chemical characters has helped to clarify the taxonomic position of certain taxa, or confirmed (or otherwise) opinions concerning hybrid swarms or complexes. Although the nature of the evidence and therefore the way in which it might aid the taxonomist varies a great deal, some general conclusions can be drawn. Most of the work has been carried out using either secondary plant products (alkaloids, flavonoids, terpenes, etc.), or proteins.

SECONDARY PLANT PRODUCTS

In hybrids between two taxa differing in a single gene governing the production of a secondary substance, the gene will either show a dominance–recessiveness relationship or incomplete dominance. For instance a red-flowered and a white-flowered plant differing by a single gene will usually be either red-flowered (where red is dominant; i.e. a single dose of the red-producing gene in the hybrid is sufficient to produce as much pigment as two doses in the red-flowered parent) or pink-flowered (where red is incompletely dominant; i.e. the single dose produces only about half as much colour as the double dose). Exactly the same is likely to be true of any other chemical characters differing by a single gene, whether or not the characters are visible ones.

Where two hybridizing taxa differ by more than one chemical character, e.g. species A has X but no Y, while species B has Y but no X, one might expect that the hybrid will produce both X and Y; but since the X- and Y-producing genes are each present only in a single dose there might in each case be a full amount or a reduced amount of the two substances. In general these expectations are fully realized; chemical characters of hybrids are mostly "additive", i.e. they represent the sum of the compounds present in both parents. Many examples of this can be quoted, of which the work by R. E. Alston, B. L. Turner and co-workers in *Baptisia* (Leguminosae) and by D. M. Smith and D. A. Levin in *Asplenium* is probably the best known. The additive effect in hybrids is not confined to chemical characters, but may be expected whenever the two parental species possess different, simply determined, non-homologous features. Heywood (1967) provided drawings of the sepal trichomes of *Digitalis purpurea*, *D. thapsi* and their hybrid, which show that the uniseriate hairs of the first species are eglandular, of the second glandular, and of the hybrid both glandular and eglandular.

Smith and Levin (1963), mostly using herbarium or cultivated material, investigated the distribution of flavonoids in a group of Appalachian *Asplenium* species (particularly the diploids *A. montanum*, *A. platyneuron* and *A. rhizophyllum*) between which hybrids, derived amphidiploids and secondary hybrids are known in many combinations. The genomic constitutions of most of these derived taxa had been previously worked out using morphological and cytological criteria by Wagner (1954) and others, and they thus provided a good opportunity to test the "additive" or "complementary" nature of chemical characters. In virtually all cases the expected complementation was observed. For instance the hybrid *A.* x *ebenoides* (*A. platyneuron* x *A. rhizophyllum* = PR) combined the constituents of its two parents, and its derived amphidiploid (PPRR) provided identical chromatograms. Moreover the triploid hybrid *A.* x *kentuckyense*, which combines all three genomes (MPR), is also completely complementary.

Alston, Turner and co-workers had previously produced similar evidence concerning the flavonoids of various hybrids in *Baptisia*, e.g. *B. leucophaea* x *B. sphaerocarpa* (see Alston and Turner, 1962). Later (Alston and Turner, 1963b) they analysed complex field populations consisting of the two above species as well as *B. nuttalliana* and *B. leucantha* and various hybrids between them. They were able to detect about 125 different compounds, which enabled precise species-characterization and the detection of hybridization even where its demonstration by cytological and morphological means would have been extremely difficult or impossible. Within one field they found all four species together with all six possible binary hybrids, and in several instances the presence of unexpected specific chemical compounds in plants otherwise typical of one particular species indicated the existence of introgression. Alston *et al.* (1962) also demonstrated that chemical studies may be used to differentiate between two hybrids which are indistinguishable on more normal criteria. Their example concerned *B. lanceolata* x *B. pendula* and *B. alba* x *B. lanceolata,* but similar studies might be as effectively employed in the case of the *Ranunculus* hybrids mentioned in Chapter 12. Pryor and Bryant (1958) studied variation in the oil composition of an F_2 generation of *Eucalyptus cinerea* x *E. macarthuri.* As might be expected the plants showed a great deal of variation between the parental extremes, and considerable evidence for recombination was apparent.

Results such as these established beyond all doubt the potential value of chemical analyses in hybridization studies. They mostly served to confirm or extend traditional taxonomic conclusions, but in other cases less expected results have been obtained. Mirov (1956) investigated populations of pines intermediate between *Pinus contorta* and *P. banksiana,* the former species containing β-phellandrine and the latter α- and β pinene in their terpenes; the artificial F_1 hybrid has (as expected) both, but in unequal amounts (75% pinenes). In the areas supporting naturally occurring intermediates the *P. banksiana* trees and the intermediate trees possessed the expected constituents, but trees apparently of *P. contorta* in fact possessed either the expected β-phellandrine only, or both β-phellandrine and pinenes (just as often), or even pinenes only (rarely). Thus the chemical and morphological data do not coincide; presumably the presence of terpenes of another species indicates a previous hybridization with that species, although morphological evidence for its participation is now lost. Adams and Turner (1970) investigated populations of *Juniperus ashei* in areas where previous workers had concluded that hybridization, principally with *J. virginiana,* was occurring or had occurred. Yet an analysis of their terpenes provided no evidence of hybridization, which led the authors to discuss various other possibilities reminiscent of the genetic pool versus introgression arguments outlined in Chapter 12. Finally, Hemingway *et al.* (1961) found that *Brassica juncea* (an amphidiploid derived from *B. rapa* x *B. nigra*) possesses either the volatile oil characteristic of *B. rapa,* or that of *B. nigra,* or a mixture of both, according to the geographical origin of the *B. juncea* utilized.

The existence of new chemical substances in hybrids is now as well established as that of new structural features. Up until about 1965 the presence of new hybrid compounds was in doubt because of their possible existence in the parents in very small quantities. For example Schwarze (1959) found that the leaf flavonoids in the hybrid *Phaseolus coccineus* x *P. vulgaris* represented the sum of those of the parents plus four additional substances, but he stated that these four did in fact occur in minute quantities in *P. coccineus.* Their presence in greater quantities in the hybrid would therefore be a kind of heterotic effect. In other cases "new" substances have been found in one or other parent but there restricted to one particular organ, whereas in the hybrid they occur more generally.

Alston *et al.* (1965) demonstrated the existence in various *Baptisia* hybrids of several flavonoids which were not present in the parents, at least in detectable quantities. They postulated two different methods of formation of new compounds, and suggested that both operated in the plants they studied. Firstly, the enzymes necessary for the formation of the new flavonoids are present but inoperative in one parental species, and are rendered operative only after the combination of the two parental germplasms; and secondly, the new compounds are formed by the separate but additive effects of both parental sets of enzymes on the same basic chemical skeleton. Harborne (1968) reported unpublished work by F. Nilsson and N. Nybom on hybrids between red and black currants *(Ribes nigrum* x *R. sativum)* which similarly possessed the fruit anthocyanins of both of the parents as well as two extra compounds.

PROTEINS

Work on proteins, both by means of serology and electrophoresis, has shown the same two basic phenomena of complementation and the formation of new compounds. The study of wheat (hexaploid *Triticum aestivum*) and its possible progenitors by Johnson and Hall (1965) forms a good example of the sort of results that have been obtained from electrophoretic analyses of seed proteins.

In many ways the study of enzymic proteins is more revealing, as enzymes clearly have a functional significance which may be closely related to evolutionary patterns. As long ago as 1960, Schwartz, working on maize endosperm esterases, found that hybrids showed both complementation and the presence of new enzymes which neither parent produced. H. N. Barber and co-workers (e.g. Barber, 1970) worked with the esterases of hexaploid wheat, diploid rye *(Secale cereale)* and their allooctoploid derivative *"Triticale"*. Within one group of esterases studied *Secale* has one, *Triticum* three and *"Triticale"* five iso-enzymes or isozymes (enzymes very closely related structurally, and perhaps catalysing the same reactions, but separable by electrophoretic means and probably becoming active in different conditions).

The five *"Triticale"* isozymes represent the four of the two parents plus one new one. Barber and co-workers explained this on the basis that active isozymes are oligomers of basic polypeptide units, one of which is produced by each genome (i.e. by each diploid species); new hybrid enzymes are thus isozymes which consist of more than one sort of basic unit produced by different (homoeologous) genes. If the active enzyme is a dimer (of two basic units), then in *Triticum* (with three different genomes) there would be six possible isozymes (three pure ones plus their three possible hybrid combinations) and in *"Triticale"* (with four different genomes) ten possible isozymes (four pure ones plus their six possible hybrid combinations). The lower numbers (three and five respectively) actually encountered can be explained on the basis that some are produced in very small quantity, that the techniques used do not effect complete separation, or that certain combinations of polypeptide units do not form viable isozymes. That an explanation of this sort (though not so obviously over-simplified) is the likely one is shown by studies by the same group on nullisomics of wheat. When the nullisomic involves the loss of the chromosome actually concerned with the production of the relevant polypeptide the number of isozymes produced is considerably reduced, showing that not only is a pure isozyme missing but so too are all the hybrid isozymes which incorporated that polypeptide unit. Multiple isozymes do, however, also occur in many undoubted diploids.

These results help to show the ways in which hybrids and polyploids can acquire new characters, including the ability to withstand conditions beyond the limits exploited by the parents, and similarly throw light on the phenomenon of hybrid vigour.

14. Recognition of hybrids

The broad criteria upon which the positive identification of hybrids exists have changed relatively little in the half-century since Cockayne (1923) listed three "principal conditions" which need to be fulfilled. His conditions are the morphological intermediacy of hybrids, their proximity in the field to their putative parents, and, if fertile, their segregating F_2 progeny. These, in turn, bear close similarity to suggestions previously made by Rolfe (1900) and Diels (1921). The main developments since 1923 have been to increase the means by which these three criteria might be applied; to add a fourth, that the hybrid should be artificially resynthesized; and to emphasize the point made by Cockayne himself that any or even all of these criteria might not be fulfilled even in undoubted hybrids.*

FERTILITY

From what has been said in Chapter 10 it will be clear that there are no tenable laws regarding the fertility or otherwise of hybrids, either absolutely or in relation to fertility in closely related taxa or hybrids. If detailed chromosomal studies are made much more evidence may be forthcoming, because the causes of sterility might then be ascertained, but the existence of cryptic structural hybridity, causing sterility where it might not be expected, and of homoeologous pairing, sometimes confounding the interpretations of genome analysis, among other phenomena, show that no absolute reliance can be placed on information of this sort either. Again, work such as that by Ellengorn and Petrova (1948), which suggests that in some hybrids the parental chromosomes might be not only morphologically distinct but also show heteropycnotic effects, is countered by the discovery by Navaschin (1934) of amphiplastic changes of karyotypes in hybrids, whereby parental chromosome differences disappear in hybrids.

Jeffrey (1914) investigated the extent to which misshapen or empty spores

* See also Gottlieb (1974).

may be used as a criterion of hybrid sterility, and Diels (1921) chose misshapen pollen or badly developed fruit as visual indications of such a condition. The diagnosis of sterility was taken a step further by Huskins (1929), who considered both pollen inviability and meiotic irregularities as characteristic features. He concluded that neither could be used as absolute criteria of hybridity. Dillemann (1954) provided a more extended discussion of the same conclusion.

FIELD EVIDENCE

The importance of field-studies has been emphasized by Allan (1937), Anderson (1949) and Baker (1951), among many others. The variable nature of the distribution of the hybrid in relation to that of the parents, whereby one or both parents may often be absent, for various reasons, from areas occupied by the hybrid, has also been described in Chapter 10. It is probably this sort of situation that Diels (1921) had in mind when he included irregular distribution as one of his hybrid criteria, and his suggestion that the hybrid should be frequently absent from areas where the parents are sympatric is still very useful.

Despite these difficulties there are many genera in which study in the field is the only sure way of confirming the identification of a hybrid by morphological analysis alone. In *Carex* and *Euphrasia,* for example, there are several hybrids of similar appearance whose parentage can only be ascertained with the knowledge of the possible parents growing in the vicinity. Sometimes the geographical locality is sufficient, as alternative parental candidates may be markedly allopatric, but frequently more detailed information is necessary.

In many cases the opinions of field-workers differ from those of bio-systematists or herbarium-based taxonomists, and, where there is not yet sufficient experimental evidence, either might be correct. Hence E. C. Wallace (see below under *Carex*) believes that the lowland *C. vesicaria* is not involved in the parentage of the upland *C.* x *grahamii,* although this is implied by herbarium studies. Similarly Nilsson (1928) concluded from progeny testing of natural and artificial hybrids that *Salix* x *laurina* is derived from *S. caprea* x *S. viminalis,* although most field- and herbarium-workers believe that *S. phylicifolia* is involved (see below under *Salix* for R. D. Meikle's comments).

The importance of field-work data is clearly greatest in those cases where the ecological or geographical distributions of the two putative parents differ conspicuously, so that those of any possible hybrids can be studied in relation to them. Studies of the floral biology and pollination of sympatric species can similarly be of great importance, particularly where introgression is involved. Field-work is also essential where extensive sampling is necessary. The gathering and preserving of mass collections in an attempt to bring field-work into the laboratory (Anderson and Turrill, 1935; Anderson, 1941) may be useful in some cases, but is hardly a satisfactory substitute.

PHENETIC INTERMEDIACY

The degree to which intermediacy is found in hybrids has been discussed in some detail in Chapter 12. Although most hybrids are more or less intermediate in most characters, many are not, and careful attention needs to be paid to the latter possibilities. For example hybrids may be very much closer in appearance to one parent than to the other, or exist as conspicuously different nothomorphs due to reciprocal hybridizations or other causes. Furthermore new characters often appear in hybrids, which may also sometimes resemble teratological or diseased specimens or show varying degrees of discordant variation, and in some situations hybrids display additive rather than intermediate conditions (especially so with chemical characters). Moreover intermediate plants are by no means always of hybrid origin. The necessity of guarding against the misinterpretation of such situations has been realized for many years (see Rolfe, 1900; Diels, 1921). Baker (1947) commented on these points in greater detail. In particular he was very critical of diagnosing hybrids without performing artificial crosses, and stressed the need to precede morphological studies with experimental ones. In a later paper (Baker, 1951) he enlarged upon these principles.

Although one may, with Baker, deplore the "decline of experiment", it is necessary to be realistic. Most taxonomic problems which are investigated present themselves first either in the field or in the herbarium, and where hybridization is possible it is surely expedient to present data for its suspected occurrence as a *first* step, to be followed by experimental investigation. It is regrettable that the latter step is often not taken (not always because of laziness or from preference), but it is likely that if experimental hybridizations were to be regarded as the essential first step then many fewer taxonomic problems would ever be solved, and that much time would be wasted in attempting to produce hybrids which a closer morphological study would have revealed as unlikely in the first place. Furthermore, there is much evidence to suggest that the criterion of experimental synthesis is no more absolute than any other.

The techniques employed in the demonstration of intermediacy in hybrids are largely those which are used to study the various morphological, anatomical and chemical characters themselves, but a number of special biometric devices are available in addition. These vary greatly in complexity; the simplest and by far the most widely employed are the hybrid indices and pictorialized scatter diagrams which were developed in the 1930s and 1940s by E. Anderson and co-workers in the Missouri Botanical Garden.

Pictorialized scatter diagrams are particularly useful when some of the characters are quantitative (to be plotted on the axes of the graph) and the others qualitative, or capable of being expressed qualitatively. These are added to the scattered points by furnishing the latter with rays, or by the use of differently shaped or shaded points. Such ornamented points are known as metroglyphs. If the individuals of the two parental species form clusters of points well separated on the scatter diagram intermediates are likely to fall somewhere in between, and by the careful choice of the system used to record the additional characters (ensuring that the presence of rays always refers to the characters of one species, and their absence to those of the other) a strong visual impact can be obtained. Figure 5 is a pictorialized scatter diagram of British herbarium specimens of *Luzula forsteri* (top left), *L. pilosa* (bottom right) and hybrids, taken from Ebinger (1962).

Anderson (1949), by the application of what he called "the method of

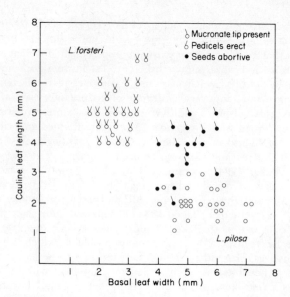

Fig. 5. Scatter diagram of herbarium specimens of *Luzula forsteri* (top left), *L. pilosa* (bottom right) and their hybrid. From Ebinger (1962).

extrapolated correlates" to pictorialized scatter diagrams, was able to predict the second parent of a hybrid in a population consisting only of plants of one parent and the hybrid. The method is based on the concept of the recombination spindle, which in turn relies upon the linkage of parental characters in the hybrids. A scatter diagram of a hybrid population where one parent is missing represents the centre and one end of the spindle; the other parent represents the other end of the spindle, whose characters are known by reading off the scatter diagram. Anderson claimed to have successfully predicted in this way previously unknown parents of several hybrid populations; the example shown in Fig. 6 is taken from Anderson (1953a), and concerns the introgression of *Oxytropis lambertii* into *O. albiflora*.

Several other methods of the pictorial representation of hybridizing populations have been employed, among them pie-diagrams, radiate indicators and polygonal graphs. The choice of the method to be employed depends entirely on the clarity of impression which can be visually gained from the various techniques, and it is usually necessary to make trial runs before the correct decision can be made. Figure 7 is of three polygonal graphs of *Viola lactea, V. riviniana* and their hybrid (Moore, 1959), which shows that this method is suited to situations where quantitative characters are involved.

The hybrid index is a numerical presentation of a plant on a scale ranging from the extreme shown by one parent to the opposite extreme shown by the other, usually presented as a histogram. Such histograms closely resemble those obtained by plotting a single character, but differ because they represent the

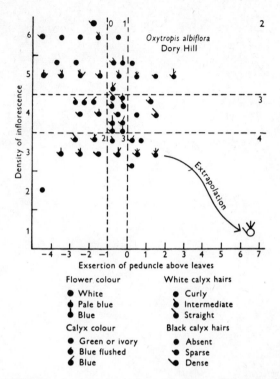

Fig. 6. The method of extrapolated correlates. The scatter diagram is of an introgressed colony of *Oxytropis albiflora*. The identity of the introgressing species (*O. lambertii*) is obtained by extrapolating to the bottom right of the Figure. From Anderson (1953a).

sum of several characters, some of which might be qualitative rather than quantitative. If we suppose species A and B differ by six characters each of which can be scored for A-ness (0), B-ness (2) or intermediacy (1), the hybrid index is obtained by adding together for each plant investigated the six scores of 0, 1 or 2. A plant typical of species B in all six characters will thus score 12, and one of species A will score 0. Intermediates will have intermediate scores; it is clear that plants which score 6 might have scores 2 for three characters and 0 for three, or 1 for all six, or other combinations. It is not necessary to limit the possible scores to 0, 1 or 2. If the characters are quantitative a wider range can be employed, say 0–10.

According to the *Primula* data presented by Woodell (1965) it is not even necessary to treat all the characters in the same way; Woodell scored five characters 0–4, one 0–7 and one 0–9, thus producing an additive scale of 0–36. This is effectively a method of weighting those characters with a wider range as being more significant, although Woodell did not state that this was his intention. Such a system of weighting had been previously suggested by Anderson (1936), when first proposing the hybrid index method: "when, for

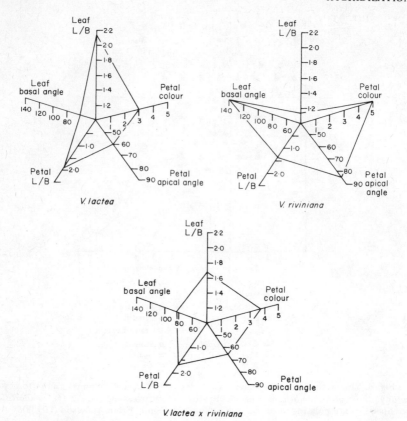

Fig. 7. Polygonal graphs of five quantitative characters of *Viola lactea*, *V. riviniana* and their hybrid. From Moore (1959).

instance, certain of the characters are thought to be more or less reliable than the others they can be appropriately weighted in combining the index". The notion that characters can be legitimately subjectively weighted in this way has often been strongly criticized, e.g. by Baker (1947), who described it as "particularly alarming", and by Davis and Heywood (1963), who called it a "hazardous procedure". Nevertheless Woodell's scoring system appears to have worked successfully, and it might be argued that subjective weighting is no more objectionable than the subjective enforcement of equality between characters. Moreover weighting might occur in less obvious ways, e.g. by the use of two strongly linked or even interdependent characters.

Figure 8 shows two histograms of hybrid indices of two mixed colonies of *Calystegia sepium* and *C. silvatica*, one without and one with putative hybrids. Because of the rampant, intertwining nature of these species, and because each branch bears only one flower open at any one time, each flower was of necessity treated as a separate unit. One could assume that only two and three clones

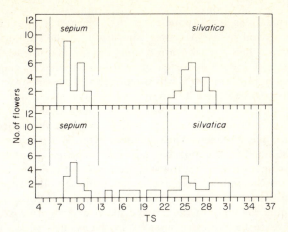

Fig. 8. Hybrid indices of individual flowers of two large, mixed colonies of *Calystegia sepium* and *C. silvatica*, the lower with and the upper without putative hybrids. From Stace (1961).

respectively were present in each colony. Figure 9, taken from the same source (Stace, 1961), shows mean hybrid indices for 70 separate colonies, each of which appeared uniform. In all these cases the additive scale runs from 4–40, being based on four quantitative characters each scored 1–10. Frequently the percentage rather than the number of plants is plotted on the y axis in order to mask differences in sample sizes.

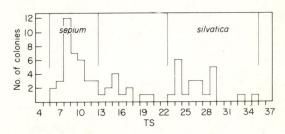

Fig. 9. Mean hybrid indices of 70 separate, uniform colonies of *C. sepium*, *C. silvatica* and putative hybrids. From Stace (1961).

Carlisle and Brown (1965) studied *Quercus petraea* x *Q. robur* by means of hybrid indices, but found that they could not score all the characters for every specimen. Rather than reject all the characters not scored for every plant (or all the plants not fully scored) they constructed a "percentage hybrid index", which was the hybrid index for each plant expressed as a percentage of the total score possible for that plant.

Gay (1960), investigating hybridizing populations of *Erica ciliaris* and *E. tetralix*, devised a means of comparing the degree of hybridization shown by different populations, each population first being scored in the form of a conventional hybrid index. A simple comparison of the arithmetical means of the hybrid index does not take into account the spread of the hybrid index, nor the proportion of the two parents, in each population. Gay therefore used in addition the "hybrid number", which represents the deviation in hybrid index shown by a specimen from the nearer of the two parental extremes. If the hybrid index runs from 0 to 12 (with six characters), the hybrid number runs from 0 to 6, and the higher the hybrid number the nearer to exactly intermediate is the specimen. When the mean hybrid index is plotted against the mean hybrid number the possible positions of populations on the scatter diagram are contained within a triangle, the corners of which would be occupied by populations containing solely one species, solely the other species, and solely exact intermediates, respectively. The actual position of populations along the sides and within the body of the triangle allows degrees of variation from these three extremes to be expressed visually. In the example shown (Fig. 10, from Gay, 1960), population 377 falls close to a situation with one species only,

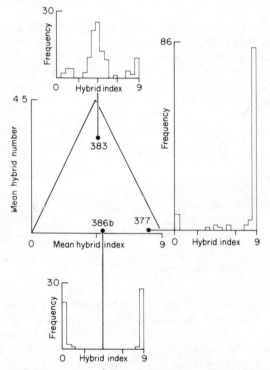

Fig. 10. Mean hybrid number plotted against mean hybrid index. The range of possibilities is contained within the triangle, in which the position of three populations (377, 383 and 386b) is marked. The hybrid index histogram of each population is also shown. From Gay (1960).

population 383 falls rather further from a situation with exact intermediates only, and population 386b falls close to a situation with equal numbers of the two parents but no intermediates.

Cousens (1965) investigated introgression in *Quercus petraea* and *Q. robur* by means of scatter diagrams where two primary characters (relative petiole-length and peduncle-length) formed the y and x axes. Four other secondary characters were scored for *robur*-ness versus *petraea*-ness (0, 1 and 2), and a hybrid index ("heterogeneity index") constructed on a scale 0–8. When the mean values of the primary characters of populations with each of the nine heterogeneity indices were plotted on the scatter diagram the nine classes formed a curve linking the two extremes of 0 and 8. This curve was called the "introgression path", and the parts of it occupied by populations from different geographical areas could be used to determine the extent to which introgression was occurring in each of those areas.

Clifford (1955) investigated putative hybrid populations of *Eucalyptus elaeophora* x *E. goniocalyx* by the construction of an index which combined two characters in such a way that the greatest possible degree of separation of the parents was obtained, although neither character used separately was an absolute discriminant. This index is essentially a variant of the hybrid index.

Anderson (1953a) described his "semi-graphical, semi-mathematical" methods for investigating hybridization and introgression as a "stop-gap" which should "ultimately be supplanted by more mathematically elegant methods". Some of these have since become available, but for many situations a simple hybrid index or pictorialized scatter diagram is perfectly adequate, and there is no point in spending time and money on more sophisticated techniques which produce exactly the same result. The "more elegant methods" are, however, of great value in cases where the weighting of characters appears justifiable, or where the data are otherwise not amenable to the simpler forms of processing. The statistical methods involved mostly come under the heading of multivariate analysis.

Whitehead (1954) illustrated the use of discriminant analysis in the separation of three annual species of *Cerastium*, whereby each of the characters is weighted to a degree commensurate with the extent of its discrimination between the taxa. No hybrids were involved in Whitehead's work, but Pritchard (1961) used the same technique to distinguish between *Gentianella amarella*, *G. germanica* and their hybrid, and to detect introgression of *G. germanica* into *G. amarella*.

Hathaway (1962) constructed what he termed a "weighted hybrid index" using the same criterion, viz. "The contribution of a character to an index should be in proportion to its usefulness in demonstrating a known or suspected relationship", but employing a different technique, i.e. canonical analysis. The weighted index was not expressed as a histogram, but on the x axis of a scatter diagram of which a further discriminating character formed the y axis. Hathaway warned that the use of multivariate analysis in biological research "should be advocated with caution", and "applied only after more general methods, especially metroglyphic analysis (i.e. pictorialized scatter diagrams), have thrown light on the nature of the problem." Goodman (1967), in fact, investigated a segregating population of *Gossypium barbadense* x *G. hirsutum* (Malvaceae) by means of several statistical methods, and found that "the simpler hybrid indices were as accurate as the more complex discriminant functions".

In recent years an enormous amount of work has gone into the use of computerized numerical methods in taxonomic discrimination ("numerical taxonomy"), but relatively little attention has been paid to situations involving hybridization. Sneath (1968) has, however, made an interesting numerical study of the graft-hybrid + *Laburnocytisus adamii*, in which separate branches of a single tree were used as OTUs and 102 characters studied. It was found that the chimera is about equally like either parent, but "lay well offset from a line joining the parents".

An excellent example of a modern multivariate approach to the analysis of suspected hybridizing populations has been provided by Adams and Turner (1970). They studied *Juniperus ashei* in Texas and searched for evidence of its hybridization with *J. virginiana* and *J. pinchotii*. After studying morphological and chemical characters by means of analysis of variance, SNK tests, contour mapping of various characters and numerical taxonomic methods they concluded that it had not taken part in past hybridization with these species, despite the contrary results previously obtained by Hall (1952) in what are often considered classic studies using Anderson's methods.

F_2 SEGREGATION

The detection of F_2 segregation is a valuable aid in hybrid diagnosis where hybridization experiments cannot easily be carried out, either because of difficulties in producing F_1 hybrids or (with trees) because of the long period required for maturation. The earliest use of this technique is apparently the analysis by MacDougal (1907) of the progeny of a tree of *Quercus* x *intermedia*; the F_2 included saplings almost indistinguishable from *Q. rubra* and *Q. phellos*, as well as intermediates, and the parentage of the hybrid was accordingly diagnosed. Baker (1947) criticized the concept of F_2 progeny testing as a clue to parentage, but as noted above this method is to be advocated where more direct ones are not available.

In growing an F_2 it is most important to attempt to nurture all available seeds (or an utterly random selection of them) to maturity. Brink and Cooper (1947), for example, found that certain *Triticum* F_2 caryopses were either plump or shrivelled owing to poor endosperm development. The latter contained unfavourable chromosome combinations, and it was essential to grow these to maturity (if necessary by special culture methods) in order to obtain the range of variation which they exhibited and which would otherwise be lost.

The chances of recovering exact parental types in the F_2 are very low, and are related to the numbers of chromosomes and genes and to the degree of crossing over; one should merely expect to obtain near-parental types. The subject has been discussed by Anderson (1949) and Baker (1951). Figure 11 shows the range of a garden-raised F_2 generation of *Juncus conglomeratus* x *J. effusus*, along with the wild F_1 and its parents. Fairly precise quantitative methods for

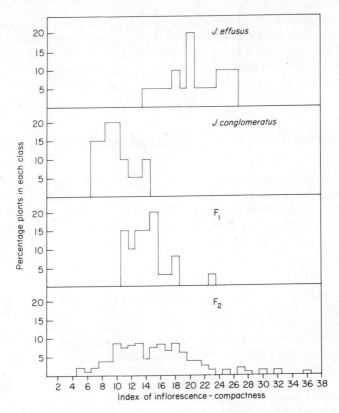

Fig. 11. F_2 segregation in a single character (inflorescence-compactness) of *J. conglomeratus* × *J. effusus*. The histograms shown the ranges, in descending order, of *J. effusus*, of *J. conglomeratus*, of the wild, putative F_1 of the above two populations, and of a garden-raised F_2 from the above F_1.

the analysis of F_2 variation in *Eucalyptus* hybrid swarms have been described by Clifford (1954) and Clifford and Binet (1954).

RESYNTHESIS OF HYBRIDS

Whenever possible the suspected parents of a hybrid should be crossed and the progeny compared with the wild hybrid plants. As pointed out by Baker (1947), even where the plants take a long time to reach maturity the initial cross can be made, and at least the possibility of the suggested origin be demonstrated. Long-term experiments have been carried out in genera such as *Betula* and *Quercus*, apart from many tree genera which are frequently crossed for commercial purposes. Many examples will be found in the Systematic Section of this book. It should, however, never be forgotten that the inability to synthesize a hybrid is not proof that it does not exist in the wild, and that the artificial production of a hybrid is no more conclusive evidence than is the demonstration of the existence of phenetic intermediates. The lines of evidence which can be obtained by the above methods are simply clues which can add up to a probable but never a definite solution to the problem.

15. Significance of hybridization

In the preceding chapters many aspects of hybrids and hybridization have been discussed, but it remains to consider to what extent hybridization may be considered an important biological force, and to what extent it may be utilized or manipulated by man within or beyond this framework. These aspects may be briefly discussed under three headings.

ROLE IN EVOLUTION

Lotsy (1916) was the first to claim a dominant role for hybridization in evolution, not simply because of the wider possibilities that it offers for genetic recombination but also because it was thought to provide radically new characters or combinations of characters in one or a few steps. Some agreed with the view, but others did not. We now know that interspecific hybridization *can* give rise to new fertile species in one step, and that new characters *can* arise from hybridization; but usually neither happens and the most important evolutionary contributions made by hybridization are manifested in a more subtle and long-term fashion. Even so there is still considerable disagreement concerning the importance of hybridization *vis-à-vis* other factors (e.g. mutation), and many eminent modern authorities have come down quite heavily on one side or the other.

That opinion is still so divided is evidence that the true answer will not come from further discussion or argument as much as from further research, particularly in areas of the world which have suffered least at the hands of man and in which biosystematic information is most lacking. Up until now most research has taken place in areas greatly altered by man, and it has long been realized (Wiegland, 1935) that in such places hybridization is of the greatest significance (see Chapter 9). The importance of hybridization varies also from taxon to taxon, which points to the need for the study of groups of plants that have hitherto been little investigated in the living state. Thirdly, hybridization has been of varying importance from one era to another, particularly in relation to the amount of environmental disturbance. Anderson and Stebbins (1954) held the view that in this way hybridization has contributed to bursts of evolutionary activity associated with such events as glaciation and the modern rapid exploitation of continents such as America and Africa.

Thus the differing views of modern workers are likely to be due in part to their various experiences, and for that reason no attempt is made here to draw a firm conclusion concerning the overall evolutionary significance of hybridization. Useful general discussions of this subject are given by Anderson (1949, 1953a), Heiser (1949), Stebbins (1950, 1959, 1969), Wagner (1968) and others.* One may, on the other hand, indicate ways in which hybridization *has* provided evolutionary stimuli, and as these are borne in mind during further studies a more complete picture will eventually emerge.

Stebbins (1959) outlined three main ways in which progeny resulting from hybridization may become stabilized and hence make some long-term contribution to evolution. These processes are stabilization by amphidiploidy, stabilization by introgression and the stabilization of hybrid segregates, but the last category is in fact a *pot pourri* consisting of several sorts of situations.

The regaining of fertility and hence the formation of an essentially new species by a doubling of the chromosome number (amphidiploidization) was first formulated as an evolutionary mechanism by Winge (1917), and has been discussed in some detail in Chapter 10. Its effects are now very well known, and in some groups it has undoubtedly been the dominant form of diversification, e.g. in many fern genera.

Similarly introgression, a concept introduced by Anderson and Hubricht (1938), has also been described in Chapter 10, and many examples (mainly American) have been discovered. Just as the onset of amphidiploidy is more likely the more sterile is the F_1 hybrid, introgression more often occurs where the F_1 is only very slightly fertile. In these situations the F_1 is more likely to be fertilized by the parental species than by its own, largely sterile pollen, and this will favour the production of near-parental types. Moreover the evolutionary significance of introgression is greater the more imperceptible it becomes (Anderson, 1949), i.e. the more extreme is the degree of backcrossing.

Amphidiploidy is often described as an escape from sterility, but other such escapes exist, notably apomixis—both vegetative apomixis (vegetative reproduction in the absence of reproduction by seeds) and agamospermy ("seed apomixis"). In many genera which display apomictic properties the sexual species or lines are diploid, or sometimes tetraploid, etc., but the apomictic representatives are polyploids (often triploids, pentaploids, etc.) or aneuploids. Examples are to be found in *Hieracium, Taraxacum, Rubus, Sorbus*, etc. Such apomicts are undoubtedly of hybrid origin, and apomixis represents for them a method of stabilization. Indeed, there is evidence that in certain cases hybridization itself may actually promote apomictic reproduction, e.g. in *Calamagrostis canescens* x *C. epigejos* (Nygren, 1948).

Polyploidy and apomixis have been called evolutionary dead-ends, in that they are only short-term solutions to sterility, but in fact there is no evidence which suggests that this must always be so. Estimates of polyploidy among angiosperms range from about 15% to about 40%, which scarcely suggests a lack of long-term success. In many genera polyploids are far more successful than diploids, for varying reasons. In cases like *Dactylis*, for instance, there is a kind of polyploid pillar complex in which a few widespread tetraploids (notably *D. glomerata*) encompass most of the range of variation of many much more

* See also Heiser (1973).

restricted diploids, from which the tetraploids have been derived by hybridization (Borrill, 1961). *Juncus bufonius,* a similarly widespread hexaploid apparently derived from a series of diploid and tetraploid Mediterranean taxa, is probably a further example (T. A. Cope and C. A. Stace, unpublished).

Apomictic genera are also very common, and, when apomixis is facultative, the group may be in a state of very active evolution. There is circumstantial evidence suggesting that such activity is related to the appearance of open or disturbed habitats (e.g. *Rubus, Taraxacum, Hieracium*); if so, facultative apomixis should be looked upon as a special method of hybridization giving rise to a great many genotypes, which at a later stage will become stabilized by selection. In *Rubus,* A. Newton (see below under *Rubus*) has recognized various groups of species which are essentially hybrids of different ages and therefore degrees of stability and success. The situation in *Rosa,* where the contribution to a hybrid by the female parent of four-fifths of the chromosomes may be considered a kind of partial apomixis, and where crypthybrids (of ancient origin) and phenhybrids (of recent origin) have been distinguished (Blackburn and Harrison, 1924), seems to be a parallel case.

Davis and Heywood (1963), while discussing Stebbins' three categories of stabilization, pointed out that stabilization is only one aspect of hybridization, the other of which is non-stabilization (or indeed de-stabilization). From what has been said above, it is possible to look upon de-stabilization as an essential phase in a cyclic pattern of hybridization followed by stabilization, so that the varying degrees of stabilization found in various taxa at the present day might in part reflect different stages of the cycle occupied at the moment. Rattenbury (1962) has discussed a concept of cyclic hybridization in a somewhat different sense in relation to the flora of New Zealand.

Conspicuous examples of non- or de-stabilization are shown by hybrid swarms and aggressive hybridization ("hybridization out of existence"), yet it is easy to visualize how each of these situations might revert to a stable condition, the first by the selection of F_2 segregates and the second by a balanced degree of introgression. Indeed, Stebbins (1959) was mainly considering such fertile F_2 hybrid segregates in his third category of stabilization, but other processes leading to new, stable, fertile hybrids are known and have been described in previous chapters. Among them are the production of new diploid species derived from occasional haploid gametes formed by sterile, triploid F_1 hybrids, as in *Euphrasia;* the origin of new fertile tetraploids from the fusion of non-reduced gametes, as in *Delphinium* and *Betula;* the genesis of fertile allotetraploids by crossing two autotetraploids, as in *Diplotaxis* (D. J. Harberd, pers. comm.); and the peculiar genetic system of *Oenothera* which can lead to new, stable biotypes by a single hybridization.

In other cases new, fertile, diploid species can arise by hybridization in one or a few steps. Thus Lewis and Epling (1959) described *Delphinium gypsophilum,* which is morphologically intermediate between the interfertile *D. recurvatum* and *D. hesperium,* and from hybrids between which it appears to have arisen without change in chromosome number and without the erection of any additional isolating mechanisms. The hybridization of *Penstemon centranthi-*

folius and *P. grinnellii* (Straw, 1955) mentioned in Chapter 9 is probably another example.

In Chapter 12 several examples were given of the appearance in hybrids of characters not found in either parent, or of ranges of variation outside that of the parents, and other cases are cited by Stebbins (1969) and Levin (1970). The *extent* to which this phenomenon leads to the rapid formation of new radically different species is unknown, although some workers (e.g. Lotsy, 1916; Cugnac, 1937; Levin, 1970) have argued that it is a powerful evolutionary mechanism. That it is a perfectly feasible mechanism at the DNA-protein level has been shown by the work of Barber (1970). In addition, it has been shown in several genera that hybridization can actually stimulate the rate of mutation (see Stebbins, 1969).

Thus one can logically look upon interspecific hybridization as an essential stage in evolution, providing, by recombination and the emergence of new characters (which might simply be due to recombination at the molecular level), a fresh complex of biotypes from which new successful lines will develop under the pressures of selection. These two phases are de-stabilization and stabilization respectively. The "fresh complex of biotypes" might be fertile from the first, or might be sterile, in which case polyploidy or apomixis might provide an escape mechanism. The re-establishment of stability, whether as a new non-hybridizing situation or as a new steady-state in a continuously hybridizing one, is governed by factors such as those which have been held by various workers (e.g. Heiser, 1949) to prevent the frequent swamping of species by hybridization (see Chapter 10).

VALUE IN GENETIC AND TAXONOMIC STUDIES

The artificial hybridization of organisms, whether of the same or of different species, is a routine genetic and taxonomic procedure which hardly requires further discussion here. The ability to interbreed two related species is an invaluable asset to the experimental biologist, who is only too well aware of the frustrations and consequent lack of information obtainable when such miscegenation is not possible. The data provided by the study of the immediate interactions of the germplasms of two distinct species are not available from any other sources, and indeed it is often considered that hybrids offer the only method of direct comparison between species. Certainly, in a group such as the angiosperms whose known fossils offer little in the way of phylogenetic information, hybridization experiments are the main source of our knowledge of evolutionary patterns, which are therefore much better understood at the lower end of the hierarchy than at the upper, despite the care needed in the interpretation of the results of genome analysis.

In terms of taxonomic classification the significance of hybridization is less certain, and opinions differ as to the extent to which such data (particularly compared to morphological characters) should be utilized (see Chapter 3). There is no doubt, however, that they *are* highly significant and *should* be utilized. Moreover hybrids, both natural and artificial, are extensively used as "test-cases" on which to try out ideas gained from other fields of taxonomic study, e.g. systematic anatomy, phytochemistry and numerical taxonomy. Stated simply, the value of hybrids in such situations is that together with their parents they represent three taxa whose inter-relationships are known with complete accuracy, i.e. one is a product of the other two. Thus the reliability of numerical techniques can be tested in practice better than by using species whose relationships are less well known. The complementation of chemical characters shown by many hybrids, including amphidiploids, has made them a favourite subject for chemical taxonomists, and anatomists are beginning to realize their potential as well.

In addition to its use as a means of direct comparison of two species, either for genetical or taxonomic purposes, hybridization is often employed indirectly as well. For example interspecific pollination, even when it fails to produce a hybrid zygote, may stimulate embryo development, thus in some cases effecting self-fertilization in a normally self-incompatible species, or in others initiating the production of non-fertilized (i.e. haploid; either monohaploid or poly-haploid) offspring (see Kimber and Riley, 1963). This method has been used for producing polyhaploids in *Nicotiana rustica* (Ivanov, 1928), which by sub-sequent doubling of the chromosome number have been used to secure completely homozygous individuals with the normal chromosome number (East, 1935). The pollen used by Ivanov was not always from a closely related species, but often came from a plant of another genus. A different use of remotely related pollen has been made by Yenikeyev (1966), in this case to obtain "wide hybrids" not obtainable by direct methods.

Hybrids have been much used in the investigation of breeding systems, as analysis of interspecific hybrids shows whether or not the incompatibility systems are identical in the parental species. Lewis and Crowe (1958) assessed the value and shortcomings of interspecific crosses in revealing such similarities and differences, and discussed related topics such as the occurrence of self-compatibility versus self-incompatibility and of unilateral interspecific incompatibility. Hybrids are frequently male-sterile, and have thus been used in plant breeding programmes to avoid the need for emasculation (see Edwardson, 1970).

Finally, the evolution of isolating mechanisms is clearly of great relevance to interspecific hybridization, and hence to the assessment of the latter as taxonomic information. The breeding barrier whose origin can most easily be visualized is geographical isolation, which can arise during geological history, and the isolated populations could then presumably diverge by chance mutations which would affect not only morphological and biochemical characteristics but also physiological features, such as further (internal or external) breeding barriers. Such methods are usually postulated in animal populations, and probably hold true in most plants too, but sympatric speciation (i.e. the initial formation of barriers other than geographical ones) can also take place. If hybrids between two interfertile species are considerably less well adapted than their parents, then selection may well favour factors which tend to reduce the frequency of hybridization. But in such cases (of which there are many possible examples), the two species have already diverged, and their hybridization probably occurs when the erstwhile allopatric populations meet once more. In many genera it has been found that sympatric species have strong, non-spatial isolating barriers, but allopatric ones often have not. To use this to suggest that in sympatric species selective pressures favour the origin of non-spatial barriers is a circular argument, for the only truly sympatric species which become differentiated are those between which such barriers *have* developed; what remains undocumented is the great number of cases where sympatric speciation has not occurred because non-spatial barriers did not arise. This obvious point has often been made, but is too frequently overlooked.

Nevertheless mechanisms by which sympatric speciation can arise are known, particularly in flowering plants, and include polyploidy, a high degree of inbreeding (or apomixis), selection by pollinators for mutations in flower morphology, different phenological spectra (especially vernal and autumnal flowering), and the origin by mutation of genic or chromosomal sterility (inter-sterile races of various species, such as *Galeopsis tetrahit*, are perhaps the first stages of this process).

Stebbins (1950) has discussed the idea that new internal breeding barriers might "be compounded from old ones through the segregation of fertile derivatives from partially sterile interspecific hybrids". In other words the results of an interspecific hybridization might be the formation of a third species from a fertile F_2 line which is at least partly intersterile with its parents, rather than the obliteration of the limits between the parental species. More recently Levin (1971) has surveyed possible modes of origin of isolating mechanisms.

IMPORTANCE IN AGRICULTURE AND HORTICULTURE

A great many plants of agricultural or horticultural value today are of hybrid origin, and hybridization figures prominently in the research programmes of experimental stations engaged in their study and improvement. According to Stebbins (1950) and Allard (1960) hybridization has been most important in the case of ornamentals, followed successively by fruit crops (both tree- and soft-fruits), forage crops, cereals and vegetables. Timber should probably be placed at the end of the list. The ways in which hybridization is utilized in crop improvement or breeding are varied.

In the case of a great many hybrid ornamental plants, such as irises, narcissi, dahlias, chrysanthemums, roses, gladioli, violas and the like, the main objective is to obtain a plant which is as visually attractive as possible, and other considerations are secondary. The keen gardener or astute nurseryman is prepared to take particular care to protect susceptible plants from frost, wind-damage or fungal attack if they are more prizeworthy than the more resistant cultivars. Moreover it is not very important to maintain fertility, and a great many ornamental species are in fact sterile F_1s which are repeatedly made anew or propagated clonally rather than multiplied by seed. In this line of work novelty, a common result of hybridization, is a considerable asset, and is a further reason for the importance of interspecific crossing. Where a standard quality is desired, particularly one which can be held constant over many generations and also propagated by seed, wholesale hybridization is less suitable, and more subtle methods need to be employed.

Many crops, such as wheat, tobacco, potato, swede, cotton and others, are themselves of hybrid origin, via amphidiploidization, and a study of their evolutionary pathways is of great value in their future improvement. This can be effected in two ways. In the first place their long evolutionary history, channelled by unconscious human selection, can be mimicked over the course of a few seasons, so that new versions of the crop are produced from specially selected parents which are superior to the wild plants which give rise to the usual version of the crop. Superiority can be measured in terms of yield, fertility, hardiness, disease resistance or other factors. Secondly, once the genomic constitution and homology of the crop is known, ways can be sought to replace one or more chromosomes or a genome by hybridization with plants containing desirable characteristics and by careful selection of the segregating progeny. When the individual chromosomes can be cytologically identified, selection of progeny by field trials can be checked by microscopic examination.

Work on the above lines has been particularly active, among temperate crops, in cereals and Brassicas, where resistance to fungal diseases is perhaps the most important consideration. The fact that the crops in use today are still basically the unconsciously selected amphidiploids, rather than their modern mimics, is a measure of the difficulties encountered in breeding crops anew. In particular it is often found that whether or not the desirable effect is produced, there are undesirable side-effects too, such as loss of hardiness or storage life, loss of flavour, or loss of fertility. Thus much of the crop improvement which has taken place in modern times has resulted from crosses between races of a single amphidiploid, rather than from the synthesis of new amphidiploids or crosses between different ones. In other cases the desirable characteristics of a crop are due not so much to particular genes but more to the interactions between the whole genotype, and any interference with the latter immediately creates an inferior race.

Nevertheless the potential for crop improvement by interspecific hybridization is enormous; in the case of *Brassica*, for example, crosses can not only be performed between most of the species, but between *Brassica* itself and 13 other genera (D. J. Harberd, pers. comm. 1973). The possibilities offered by somatic hybridization are too obvious to require further explanation; whether they can be realized is a topic to be explored in the next few years.

Section B

Systematic

1. Explanation of text

As explained previously, the main aim of this book is to present a synthesis of the knowledge which has been accumulated concerning hybrids and hybridization among the vascular plants native or naturalized in the British Isles (including the Channel Isles). I have attempted to include at least a mention of every hybrid combination that has been seriously claimed to have been found in the British Isles, even where it is now known that such claims were quite ill-founded. There are, however, certain casual notices of possible hybrids which have been deliberately ignored, because their inclusion would add nothing of interest or value to this book, and because their existence has never been championed by reliable authorities. For instance, in the *Report of the Winchester College Natural History Society,* **1909–11** (1911), there is, facing page 7, a photograph of the well-known peloric monstrosity of *Digitalis purpurea,* with the caption "Hybrid between foxglove and Canterbury Bell (?)". At various times a hybrid origin has been suggested for other abnormal plants, or for plants which fall more or less between two other species in some morphological characters; many of the latter bear the specific names *intermedius* or *hybridus.* Such plants are not treated in the systematic accounts. To ensure as complete a coverage as possible of seriously recorded hybrids the obvious source-books and reference works have been searched, including all the publications of the Botanical Society of the British Isles, all the Reports of the Botanical Exchange Club, the Watson Botanical Exchange Club and similar bodies, and all the volumes of several other British journals (such as the *Journal of Botany*) which contain taxonomic information. Although some hybrids will have probably escaped attention, they are hopefully very few in number.

The problems of deciding what constitutes a hybrid have also been discussed previously. In this book hybrids between recognized species, between taxa variously considered species or subspecies or varieties, and between taxa usually treated as subspecies or varieties but with different chromosome numbers, have been included. But fertile hybrids between geographical races which do not differ in chromosome number (and which constitute the traditional subspecies) have not been treated; one should expect hybrids between such races wherever they meet. Amphidiploids of known origin, e.g. *Spartina "anglica", Brassica napus,* are included so long as one of the parental species is also native or naturalized in the British Isles.

It is not possible to define exactly how well established a foreign species or hybrid must become before it can be said to be naturalized. For this reason the plants included in Dandy (1958) have in virtually every case been taken as the definitive list. Hybrids which appear in this list but which have not arisen naturally in the British Isles are only included if at least one of the parents also appears in the list. Thus many naturalized hybrids, such as *Crocosmia* x *crocosmiiflora* and hosts of composites, roses, narcissi and the like, have been excluded. Many of them, in fact, are of purely artificial origin. Artificial hybrids between native or naturalized species which do not hybridize in nature are similarly excluded; they must number hundreds, but it is not to be denied that a compilation of a list of them, together with the conditions of hybridization, would prove very valuable. In addition, mention is made of hybrids which have not been found in the British Isles but which have been reported from elsewhere and which involve two (or more) species listed by Dandy. I am fully aware of the imperfections of the list of hybrids in this category; many of those included are probably based on erroneous identifications, and there are doubtless many omissions, particularly in the list of countries from which they are recorded. But mention of such hybrids does point to possible genetic relationships which are not apparent from a study of the flora of the British Isles alone. In some cases the parents are not sympatric in this country; in others the hybrid should be sought here wherever the parents coexist. Wherever possible in this work it has been made clear whether the information provided concerns the situation in the British Isles or elsewhere.

It has not been possible to include any illustrations in the systematic part of this book. In an attempt to overcome this shortcoming a reference has been included, under each hybrid, to any available publications giving one or more authentic illustrations of the hybrid in question.

B. FORMAT

Accounts appear in one of three forms: a full account separated into paragraphs **a** to **f**, reserved for those hybrids which are reasonably well documented or even if not so are almost certainly authentic; a short account in a single paragraph, for doubtful or erroneous hybrids usually with rather little available information; and a brief sentence mentioning the occurrence of foreign hybrids. Plants in the third category are not described and no references are given, but such information is included, whenever appropriate, in the second category entries.

The author(s) of each generic (or inter-generic hybrid) account are given in parentheses after the name of the genus, although in a few genera separate subgenera have different authors. In most cases of joint authorship all the authors have taken responsibility for the whole account, but in some instances the hybrids were divided between the authors.

Where hybridization is extensive an introductory section is sometimes provided to cover all the hybrids of the genus or, in two cases (Orchidaceae, Gramineae), of the family. This serves as a general explanation and helps to avoid undue repetition. In *Epilobium, Rubus pro parte* and *Taraxacum* this opening section is more extensive and is the only descriptive information provided for those genera.

Genera, species and subspecies are numbered and lettered according to Dandy (1958). Hybrids are treated in strict numerical order, i.e. 1 x 2, 1 x 3, 1 x 4, etc., 2 x 3, 2 x 4, etc., but the species in a hybrid formula are given alphabetically, according to the *International Code of Botanical Nomenclature.* Except as noted under *Rosa,* the genetic convention (also permitted by the *Code*), whereby the female parent is the first-mentioned one in a formula, has been ignored, to avoid confusion. Where the direction of the cross is known it is specifically stated. Species not numbered in Dandy are provided with a code comprising the first three letters of the specific epithet; infraspecific taxa not lettered in Dandy are coded with the species number followed by the first letter of the infraspecific epithet.

In general the nomenclature follows that of Dandy's *List of British Vascular Plants* (1958) and his subsequent revision (Dandy, 1969a), and synonyms are only provided where of particular significance; when the accepted name is other than that listed by Dandy the latter is always given as well.

Paragraph **a** gives the valid binomial for the hybrid, if such a name exists, together with some important synonyms. Synonyms or invalid names are placed in parentheses, whether or not a valid name exists. Our state of knowledge concerning the nomenclature of hybrids is far less advanced than that concerning species, and a number of apparently valid hybrid-names not listed by Dandy have come to light. Others are of uncertain validity; because of the lack of time not all of them have been properly investigated and some of the hybrid names provided will undoubtedly have to be rejected or replaced. In particular many of the names which appeared in the Botanical Exchange Club Reports, and in the *British Plant List* (Druce, 1928), are *nomina nuda,* although many are not. In most cases the supplementary information supplied by Druce (1929a, b) is sufficient to determine which are which.

Paragraph **b** summarizes the general characteristics of the natural hybrid, including morphology, fertility and vigour, backcrossing and introgression, variation, etc. Enough details of the appearance of the hybrid are given to enable one to get a reasonable idea of how to recognize it, but it has not been possible to provide a full description, nor to include detailed differences between the two parents.

In paragraph **c** the ecological and geographical distribution of the hybrid is given. The degree of detail entered is dependent upon our very varied state of knowledge of these topics; in general wherever precise information is available it has been given or is referred to, e.g. by reference to the *Atlas of the British Flora*

(Perring and Walters, 1962) and its *Critical Supplement* (Perring and Sell, 1968). Obviously the author's knowledge of the distribution of a hybrid is often incomplete, but in many such cases a list of vice-counties is nevertheless provided. This will at least form a basis for the addition of further records by other workers. Items such as the degree of co-habitation with either parent, or the degree of persistence of the hybrid, are also provided when available. The geographical regions mentioned are in general those used in *Flora Europaea* (Tutin *et al.*, 1964), except that the Channel Isles are considered separate from France. The British Isles comprise Great Britain (including the Isle of Man), the Channel Isles and Ireland, and the adjective "British" refers to Great Britain alone. The term "the Continent" is used to indicate mainland Europe excluding the British Isles, Iceland, Faeröer and the Açores.

Paragraph d comprises a survey of experimental work which has been carried out. The results of artificial hybridizations and some details of the appearance and cytological and genetical behaviour of any artificial hybrids obtained form the most important contributions here. Observations on the pollination, fertilization and meiotic behaviour of natural hybrids are also usually given in this paragraph.

Paragraph e is a statement of the known chromosome number(s) of the two parents and their hybrids. Where the counts are on other than material of known wild origin in the British Isles, i.e. on cultivated or on wild foreign material, they are placed in parentheses. In almost all cases the chromosome number is given as the diploid ($2n$) number, whether it is based on mitotic or meiotic counts, but in a very few instances haploid (n) numbers are cited where they refer to mitotic counts of the gametophytic tissue of chromosomally variable hybrids.

The background literature is cited in paragraph f. The list has been kept as brief as possible by including key references only, but where no recent reviews have appeared the list is often necessarily longer. References to literature (in the form Smith, 1966) in paragraphs a to e are made only where essential; it has not been found practicable to cite the origin of every piece of information, much of which has in any case been gathered from multiple sources or is hitherto unpublished.

C. ABBREVIATIONS

Geographical Regions. Apart from the separation of the Channel Isles from France the geographical regions of Europe are taken from *Flora Europaea* (Tutin *et al.*, 1964):

Al Albania	Co Corse
Au Austria	Cr Kriti
Az Açores	Da Denmark
Be Belgium & Luxembourg	Fa Faeröer
Bl Islas Baleares	Fe Finland
Br Great Britain (excl. Channel Isles; incl. Isle of Man)	Ga France (excl. Corse)
	Ge Germany
Bu Bulgaria	Gr Greece (excl. Kriti)

Hb	Ireland (incl. Northern Ireland)	Po	Poland
He	Switzerland	Rm	Romania
Hs	Spain (excl. Islas Baleares)	Rs	Russia
Ho	Netherlands	Sa	Sardegna
Hu	Hungary	Si	Sicilia
Is	Iceland	Su	Sweden
It	Italy (excl. Sardegna & Sicilia)	Tu	Turkey
Ju	Jugoslavia	CI	Channel Isles
Lu	Portugal	BI	British Isles (i.e. CI, Br and Hb)
No	Norway		

v.c. Vice-county(ies). These are defined as in Dandy (1969b) and Praeger (1901).* The 112 vice-counties of Br and 40 of Hb are as follows:

1	W. Cornwall (incl. Scilly)	34	W. Gloucester
2	E. Cornwall	35	Monmouth
3	S. Devon	36	Hereford
4	N. Devon	37	Worcester
5	S. Somerset	38	Warwick
6	N. Somerset	39	Staffs.
7	N. Wilts.	40	Salop
8	S. Wilts.	41	Glamorgan
9	Dorset	42	Brecon
10	Isle of Wight	43	Radnor
11	S. Hants.	44	Carmarthen
12	N. Hants.	45	Pembroke
13	W. Sussex	46	Cardigan
14	E. Sussex	47	Montgomery
15	E. Kent	48	Merioneth
16	W. Kent	49	Caernarvon
17	Surrey	50	Denbigh
18	S. Essex	51	Flint
19	N. Essex	52	Anglesey
20	Herts.	53	S. Lincs.
21	Middlesex	54	N. Lincs.
22	Berks.	55	Leicester (incl. Rutland)
23	Oxford	56	Notts.
24	Bucks.	57	Derby
25	E. Suffolk	58	Cheshire
26	W. Suffolk	59	S. Lancs.
27	E. Norfolk	60	W. Lancs.
28	W. Norfolk	61	S. E. Yorks.
29	Cambs.	62	N. E. Yorks.
30	Beds.	63	S. W. Yorks.
31	Hunts.	64	Mid-W. Yorks.
32	Northants.	65	N. W. Yorks.
33	E. Gloucester	66	Durham

* *Proc. R. Ir. Acad.*, **23**.

67 S. Northumberland
68 Cheviot
69 Westmorland (incl. Furness)
70 Cumberland
71 Isle of Man
72 Dumfries
73 Kirkcudbright
74 Wigtown
75 Ayr
76 Renfrew
77 Lanark
78 Peebles
79 Selkirk
80 Roxburgh
81 Berwick
82 Haddington
83 Edinburgh
84 Linlithgow
85 Fife (incl. Kinross)
86 Stirling
87 W. Perth (incl. Clackmannan)
88 Mid Perth
89 E. Perth

90 Forfar
91 Kincardine
92 S. Aberdeen
93 N. Aberdeen
94 Banff
95 Elgin
96 Easterness (incl. Nairn)
97 Westerness
98 Argyll
99 Dunbarton
100 Clyde Isles
101 Kintyre
102 S. Ebudes
103 Mid Ebudes
104 N. Ebudes
105 W. Ross
106 E. Ross
107 E. Sutherland
108 W. Sutherland
109 Caithness
110 Outer Hebrides
111 Orkney
112 Zetland

H1   S. Kerry
H2   N. Kerry
H3   W. Cork
H4   Mid Cork
H5   E. Cork
H6   Waterford
H7   S. Tipperary
H8   Limerick
H9   Clare (incl. Aran Isles)
H10 N. Tipperary
H11 Kilkenny
H12 Wexford
H13 Carlow
H14 Leix
H15 S. E. Galway
H16 W. Galway
H17 N. E. Galway
H18 Offaly
H19 Kildare
H20 Wicklow

H21 Dublin
H22 Meath
H23 Westmeath
H24 Longford
H25 Roscommon
H26 E. Mayo
H27 W. Mayo
H28 Sligo
H29 Leitrim
H30 Cavan
H31 Louth
H32 Monaghan
H33 Fermanagh
H34 E. Donegal
H35 W. Donegal
H36 Tyrone
H37 Armagh
H38 Down
H39 Antrim
H40 Derry

Herbaria. The abbreviations employed are those in *Index Herbariorum, Part I* (5th ed.) (Lanjouw and Stafleu, 1964) and *British Herbaria* (Kent, 1958).

Periodicals. The abbreviations employed are those in the *World List of Scientific Periodicals*, 4th ed. (Brown and Stratton, 1963–65), and later additions, with the exception of:

> *Proc. B.S.B.I.*—Proceedings of the Botanical Society of the British Isles.
> *Rep. B.E.C.*—Report of the Botanical (Society and) Exchange Club of the British Isles.
> *Rep. Watson B.E.C.*—Report of the Watson Botanical Exchange Club.

Miscellaneous.

FIG. After a reference; implies that an illustration of the hybrid (or of diagnostic parts of it) is included there.

As above. After author and date; refers to that reference given previously under the same author and date in the *same generic* (or intergeneric) *account* (or, in the Orchidaceae and Gramineae, in the same family account). Such previous references appear either in the lists following previous hybrids or (occasionally) in the general list following an introductory generic or familial account. In most cases, however, the latter are not repeated after each individual hybrid account as they are relevant to all of them.

Op. cit. or *Loc. cit.* After author and date; refers to a reference given in the *General Bibliography* at the end of the book.

$2n$ Sporophytic chromosome number.

n Gametophytic chromosome number.

2. Systematic accounts

4. *Equisetum* L.
(by J. G. Duckett and C. N. Page)

1 x 3. *E. byemale* L. x *E. ramosissimum* Desf.

  a. *E.* x *moorei* Newm. (*E. occidentale* (Hy) Coste, *E.* x *samuelssonii* (W. Koch) Rothm.).

  b. This hybrid is intermediate between its putative parents in stem-diameter, ridge-number, sheath-length and appression to stem, sheath-tooth persistence, stem persistence in winter, stomatal length, and stem anatomy, but similar to *E. byemale* in stomatal width and to *E. ramosissimum* in sheath-width. Cones are produced from May to November but the spores are completely abortive and very variable in size.

  c. The hybrid is common on dunes and banks for about 30 miles by the sea in v.c. H12 and H20, and is known in v.c. 17 (introduced) and 88. It is also recorded from Au, Cz, Ga, Ge, He, Hu, It, Po, Rs, Su and Japan on sandy river-banks, lake-shores and railway-tracks, though the exact distribution is perhaps uncertain due to confusion with *E. byemale*. Over most of its range both parents are sympatric but its presence in Hb (and in Gotland) where *E. ramosissimum* is absent (and also by the Caspian Sea where *E. byemale* is absent) is highly problematical.

  d. Artificial hybrids are readily produced with either parent as female. Thus the main natural barriers to crossing are spatial and ecological. In the hybrid, chromosome pairing at meiosis is highly irregular.

  e. *E. byemale* and hybrid $2n = c\ 216$; *E. ramosissimum* ($2n = c\ 216$).

  f. BORG, P. (1967). Studies on *Equisetum* hybrids in Fennoscandia. *Ann. bot. fenn.*, **4**: 35-50. FIG.

    DUCKETT, J. G. (1968). *Developmental morphology and sex determination in prothalli of Equisetum.* Ph.D. Dissertation, University of Cambridge.

    DUCKETT, J. G. (1970). Spore size in the genus *Equisetum. New Phytol.*, **69**: 333-346.

    HAUKE, R. L. (1963). A taxonomic monograph of the genus *Equisetum* subgenus *Hippochaete. Beih. Nova Hedwigia*, **8**: 1-163. FIG.

    KUMMERLE, J. B. (1931). Equiseten-Bastarde als verkannte Artformen. *Magy. bot. Lap.*, **30**: 146-160.

    MANTON, I. (1950). *Problems of cytology and evolution in Pteridophyta.* Cambridge.

PERRING, F. H. and WALTERS, S. M. (1962). *Op. cit.*, p. 3.

1 × 4. ***E. hyemale*** **L.** × ***E. variegatum*** **Schleich. ex Weber & Mohr**
a. *E.* × *trachyodon* A. Braun.
b. The hybrid is intermediate between its putative parents in all characters except the length/width ratio of the leaf-sheath (exceeds both parents) and stomatal size (almost identical with *E. variegatum*). Confusion with *E. hyemale* is unlikely but small plants may be mistaken for *E. variegatum*. Separation from the latter is afforded by the long sheath and long teeth with narrow membranous margins in the hybrid. Cones are produced from June to October but usually fail to shed their completely abortive spores.
c. The plant is widespread but very local on shady, marshy stream-banks and lake-shores in Hb and Scotland in v.c. 88, 90–92, 110; H1, 2, 4, 9, 14–16, 21, 26–28, 33–36 and 38–40, often widely separated from one or both of the parents. It is also recorded from scattered localities in Cz, Fe, Ga, Ge, He, Hu, Is, No, Rs, Su, Greenland and temperate North America.
d. Artificial hybrids have not so far been produced with either parent as female, although attempts have been made. Chromosome pairing at meiosis is highly irregular in the hybrid.
e. Both parents and hybrid $2n = c\ 216$.
f. BIR, S. S. (1960). Chromosome numbers of some *Equisetum* species from the Netherlands. *Acta bot. neerl.*, 9: 224-234.
    BORG, P. (1967). As above. FIG.
    HAUKE, R. L. (1963). As above. FIG.
    JERMY, A. C. (1960). A revised preliminary census list of British pteriodophytes. *Br. Fern Gaz.*, 9 (Suppl.): 1-12.
    MANTON, I. (1950). As above.
    PERRING, F. H. and SELL, P. D. (1968). *Op. cit.*, p. 1.

3 × 4. ***E. ramosissimum*** **Desf.** × ***E. variegatum*** **Schleich. ex Weber & Mohr**
= *E.* × *meridionale* (Milde) Chiov. is reported from Cz and He.

6 × 4. ***E. palustre*** **L.** × ***E. variegatum*** **Schleich. ex Weber & Mohr**
is recorded from Su, but its existence seems doubtful.

5 × 6. ***E. fluviatile*** **L.** × ***E. palustre*** **L.**
a. None.
b. A plant tentatively assigned to this hybrid appears intermediate between the two supposed parents in the number and depth of ridges and furrows on the shoot, and in the ratio of the central hollow to the diameter of the stem. Shoots are semi-prostrate and simple or sparingly branched, and the rhizomes of the hybrid bear tubers as do those of *E. palustre*. Individual endodermises around each vascular bundle in both shoot and rhizome, and the presence of a central hollow in the rhizome, affirm that *E. fluviatile* is one parent.

c. This hybrid has been reported from the Isle of Harris, v.c. 110, where a small colony occurs in a roadside ditch. It was found within *c* 100m of some *E. palustre* but *E. fluviatile* was not close by. Gametophytes of the two species have been seen growing together on bare lake- and reservoir-mud and hybrids may perhaps arise in such situations.

d. Artificial hybrids have been produced with *E. fluviatile* as the female parent. In cultivation the hybrid plants lose their decumbent habit and become more abundantly and regularly branched.

e. Both parents $2n = c$ 216.

f. DUCKETT, J. G. (1968). As above.

   DUCKETT, J. G. and DUCKETT, A. R. (1974). The ecology of *Equisetum* gametophytes. *Am. J. Bot.,* **61** (Suppl.): 36.

   PAGE, C. N. (1963). A hybrid horsetail from the Hebrides. *Br. Fern Gaz.,* **9**: 117-119. FIG.

9 x 5. *E. arvense* L. x *E. fluviatile* L.

a. *E.* x *litorale* Kuhlew. ex Rupr.

b. Hybrids are usually intermediate between the parents in most characters but are also extremely variable and may sometimes very closely resemble either parent (e.g. in drier habitats they approach *E. arvense* and in wet ones *E. fluviatile*). This variability appears to be totally induced by environment. Hybrids can usually be distinguished from the parents by external features (e.g. length of first internodes of branches, shape of upper part of stem, number of ridges, number, length and direction of branches) but in some cases (e.g. separation from unbranched forms of *E. fluviatile*) anatomical characters (e.g. size of central hollow and vallecular canals) are also needed. Cones are produced from June to July but the spores are completely abortive.

c. This hybrid is widespread but uncommon in BI in wooded swamps, on the shores of lakes and rivers, and in gravel-pits, ditches, wet meadows and dune-slacks in v.c. 1–3, 5, 6, 8, 10–12, 14, 17, 21, 34, 41, 42, 45, 47–51, 53, 54, 57, 59, 65–67, 69, 71, 78, 88, 90, 104, 107, 110, 112; H1–3, 9–11, 14, 16, 17, 19, 21, 22, 24–31, 33, 35 and 37–39. It is not always in close proximity to either parent but is more often associated with *E. fluviatile* than with *E. arvense*. Gametophytes of the two parents have been seen growing together on bare lake- and reservoir-mud and hybrids perhaps often arise in such situations. The hybrid is also widespread in Fennoscandia, and probably throughout the North Temperate range of overlap of its parents.

d. Artificial hybrids have been produced with *E. fluviatile* as the female parent.

e. Both parents and hybrid $2n = c$ 216.

f. BIR, S. S. (1960). As above.

   BORG, P. (1967). As above. FIG.

   DUCKETT, J. G. (1968). As above.

   DUCKETT, J. G and DUCKETT, A. R. (1974). As above.

HAUKE, R. L. (1965). An analysis of a variable population of *Equisetum arvense* and *E.* x *litorale*. *Am. Fern J.*, 55: 123-135. FIG.

JERMY, A. C. (1960). As above.

MANTON, I. (1950). As above.

MARIE-VICTORIN (1927). Les Equisetinées du Québec. *Contr. Lab. Bot. Univ. Montréal*, 9. FIG.

PERRING, F. H. and SELL, P. D. (1968). *Op. cit.*, p. 1.

9 x 6. *E. arvense* L. x *E. palustre* L.

a. *E.* x *rothmaleri* C. N. Page.

b. Shoots of this hybrid are intermediate in morphology between those of the parents, and give the appearance of somewhat yellow-green shoots of *E. palustre* with a broader outline and more conspicuously angled branches. The monomorphic habit of *E. palustre* is inherited in the hybrid. Its spores are abortive.

c. A single small colony of this hybrid was discovered in the Trotternish Peninsula, Isle of Skye, v.c. 104, in 1971. It appears to have spread vegetatively from two roadside ditches into adjacent marshy fields. Previous reports of plants of this combination from Ge (as *E.* x *torgesianum* Rothm.) have been refuted.

d. Artificial hybrids have been produced with *E. palustre* as the female parent.

e. Both parents $2n = c$ 216.

f. DUCKETT, J. G. (1968). As above.

PAGE, C. N. (1973). Two hybrids in *Equisetum* new to the British flora. *Watsonia*, 9: 229-237. FIG.

6 x 10. *E. palustre* L. x *E. telmateia* Ehrh.

a. *E.* x *font-queri* Rothm.

b. Shoots of this hybrid are intermediate in morphology between those of the parents, and give the appearance of overgrown shoots of *E. palustre* with the conspicuous ivory-white internodes of *E. telmateia*. The hybrid has the shallowly biangulate branch-ridges and 2-ribbed teeth of *E. telmateia*, but inherits the monomorphic habit of *E. palustre*. It is vegetatively prolific. Cones are produced in abundance, and some of its spores appear to be well-formed.

c. This hybrid was discovered in 1968 in the Trotternish Peninsula, Isle of Skye, v.c. 104, where its shoots are abundant over an area of approximately two square miles. It occupies a wide range of habitats, becoming particularly abundant in damp depressions, irrigated slopes, flushes, seepage lines, scree banks, drainage channels, ditches, stream-banks, roadside verges and rubble. It appears to have displaced one of its parents, *E. palustre*, from most of its area. Records of this plant from Hs and Ga have been authenticated.

d. Artificial hybrids have been produced with *E. palustre* as the female parent. Plants remain distinct from parents and other hybrids in cultivation.

e. Both parents $2n = c$ 216.

f. DUCKÈTT, J. G. (1968). As above.

PAGE, C. N. (1972). An assessment of inter-specific relationships in *Equisetum* subgenus *Equisetum. New Phytol.*, 71: 355-369.

PAGE, C. N. (1973). As above. FIG.

7 x 8. *E. pratense* Ehrh. x *E. sylvaticum* L.

= *E.* x *mildeanum* Rothm. has been reported from Ge, Rs and Su.

9 x 8. *E. arvense* L. x *E. pratense* Ehrh.

= *E.* x *montellii* Hiitonen has been reported from Fe, Su and the Canadian Arctic.

9 x 10. *E. arvense* L. x *E. telmateia* Ehrh.

= *E.* x *dubium* Dostál has been reported from Cz.

7. *Hymenophyllum* Sm.
(by J. D. Lovis)

1 x 2. *H. tunbrigense* (L.) Sm. x *H. wilsonii* Hook.

has not been found anywhere, but a hybrid complex, which includes an amphidiploid derivative of *H. tunbrigense* x *H. wilsonii*, exists in Madeira.

25 x 8. *Polypodium* L. x *Pteridium* Scop.
(by C. A. Stace)

25/1 x 8/1. *Polypodium vulgare* L. x *Pteridium aquilinum* (L.) Kuhn

was at one time said to be the parentage of *Polypodium vulgare* var. *cambricum* (L.) Lightf.

LINTON, E. F. (1907). Hybrids among British phanerogams. *J. Bot., Lond.*, 45: 296-304.

15 × 14. *Asplenium* L. × *Phyllitis* Hill = × *Asplenophyllitis* Alston
(by J. D. Lovis)

Three hybrids of uncertain parentage are mentioned at the end of this account.

15/1a × 14/1. *Asplenium adiantum-nigrum* L. × *Phyllitis scolopendrium* (L.) Newm.

a. × *A. jacksonii* Alston (*Asplenium adiantum-nigrum* var. *microdon* T. Moore).

b. This plant is closer to *A. adiantum-nigrum* than to *P. scolopendrium*, but has simply pinnate fronds, and fewer, longer sori than in *A. adiantum-nigrum*, and some sori may be arranged in a scolopendrioid manner. It is distinguishable from × *A. microdon* by its frond shape, which is triangular instead of decrescent, and by the longer sori (4–6 mm) placed nearer to the midrib than to the margin of the pinnae. It is highly sterile.

c. The first records were made in 1856 near Hartland, v.c. 4, and in Guernsey, CI. Subsequently it was once found in Jersey and in two or three further localities in Devon between then and 1872. It is unknown elsewhere.

d. Two hybrids were synthesized in Leeds in 1967, using *A. adiantum-nigrum* from Cornwall and *P. scolopendrium* from Hu. They compared very closely indeed with the wild examples of × *A. jacksonii.*

e. *A. adiantum-nigrum* 2n = 144; *P. scolopendrium* 2n = 72.

f. ALSTON, A. H. G. (1940). Notes on the supposed hybrids in the genus *Asplenium* found in Britain. *Proc. Linn. Soc. Lond.,* **152**: 132-144.

   LOVIS, J. D. and VIDA, G. (1969). The resynthesis and cytogenetic investigation of × *Asplenophyllitis microdon* and × *A. jacksonii. Br. Fern Gaz.,* **10**: 53-67. FIG.

   MOORE, T. (1860). *The octavo nature-printed British ferns,* **2**: London.

15/2 × 14/1. *Asplenium billotii* F. W. Schultz (*A. obovatum* auct., *non* Viv.) × *Phyllitis scolopendrium* (L.) Newm.

a. × *A. microdon* (T. Moore) Alston (=*Asplenium lanceolatum* var. *microdon* T. Moore).

b. This hybrid is intermediate between its parents, but quite distinct from either. It can only be confused with × *A. jacksonii* but differs in that the robust, simply pinnate fronds are broadest just below the middle, with crinkled pinnae bearing small sori (2–3 mm) situated towards the edge of the pinnae. It is highly sterile.

c. This was also first discovered in Guernsey, CI, in 1855; it was found there again on two or three occasions in the subsequent thirty years and was rediscovered by P. J. Girard in 1965. Elsewhere it has been found only once, near Penzance, v.c. 1, in 1856. x *A. microdon* occurs in Guernsey on the face of roadside hedge-banks in well-shaded situations.

d. Hybrids were synthesized in 1967 in Leeds using *A. billotii* from He and *P. scolopendrium* from Hu, and again in 1968 using parental strains from Guernsey. These artificial hybrids are indistinguishable from the wild examples.

e. *A. billotii* $2n = 144$; *P. scolopendrium* $2n = 72$; hybrid $2n = 108$.

f. ALSTON, A. H. G. (1940). As above.

GIRARD, P. J. (1967). Discovery of a rare fern. *Rep. Trans. Soc. Guernes.*, **18**: 167-172.

GIRARD, P. J. and LOVIS, J. D. (1968). The rediscovery of x *Asplenophyllitis microdon*, with a report on its cytogenetics. *Br. Fern Gaz.*, **10**: 1-8.

LOVIS, J. D. and VIDA, G. (1969). As above. FIG.

McCLINTOCK, D. (1968). *Asplenium billotii* x *Phyllitis scolopendrium* = x *Asplenophyllitis microdon* (T. Moore) Alston. *Proc. B.S.B.I.*, **7**: 387-389.

MOORE, T. (1860). As above. FIG.

15/5q x 14/1. *Asplenium trichomanes* L. subsp. *quadrivalens* D. E. Meyer x *Phyllitis scolopendrium* (L.) Newm.

a. x *A. confluens* (T. Moore ex Lowe) Alston.

b. These plants are closer to *Asplenium trichomanes*, but more robust, with stouter rachides and thicker pinnae, and with the upper pinnae confluent; they are highly sterile.

c. There are three definite BI records: Levens Park, v.c. 69, in 1865; near Killarney, v.c. H2?, in 1875; and Whitby, v.c. 62, sometime during the same period. There are no records from BI in this century, and elsewhere only one in Ju.

d. Artificial hybrids were raised accidentally in cultivation by Hans (1916) in New York, and recently (1968) synthesized using tetraploid *A. trichomanes* from Japan and *P. scolopendrium* from Hu (J. D. Lovis, unpub.).

e. *A. trichomanes* subsp. *quadrivalens* $2n = 144$; *P. scolopendrium* $2n = 72$.

f. ALSTON, A. H. G. (1940). As above.

HANS, A. (1916). An interesting hybrid. *Am. Fern J.*, **6**: 37-39.

LOWE, E. J. (1867). *Our native ferns*, **2**: 207 and 560. London. FIG.

LOWE, E. J. (1891). *British ferns*, pp. 36-37. London.

MAYER, E. (1962). *Asplenophyllitis confluens* (Lowe) Alston − Prvi intergenerični hibrid praproti v flori Jugoslavije. *Biol. Věst.*, **10**: 3-5.

x *Asplenophyllitis claphamii* (T. Moore) Alston
is known only from a specimen of horticultural origin and unknown parentage.

x *Asplenophyllitis hendersonii* (Houlston & T. Moore) D. E. Meyer
is a plant of quite unknown parentage and origin (other than its cultivation

about 1850 at Wentworth House), and may indeed never have existed in nature.

x *Asplenophyllitis lobata* (Rouy) D. E. Meyer
> from Lu was originally considered to be *A. marinum* L. x *P. scolopendrium* but recently it has been suggested that the *Asplenium* parent is *A. onopteris* L.

15. *Asplenium* L.
(by J. D. Lovis)

One hybrid of uncertain parentage is discussed at the end of this account.

1a x 1b. *A. adiantum-nigrum* L. x *A. onopteris* L. (*A. adiantum-nigrum* subsp. *onopteris* (L.) Luerss.)
> = *A.* x *ticinense* D. E. Meyer is only known from He and It.

1a x 2. *A. adiantum-nigrum* L. x *A. billotii* F. W. Schultz (*A. obovatum* auct., *non* Viv.)
> a. *A.* x *sarniense* Sleep. (Guernsey Spleenwort).
> b. This hybrid is approximately intermediate between the parents, but is easily overlooked as a state of *A. adiantum-nigrum,* which it closely resembles in its triangular frond and conspicuous basal pinnae. The influence of *A. billotii* can best be detected in the middle of the lamina, where, even towards the distal end of the pinnae, the pinnules are oval in shape, distinctly stalked, and with broad, mucronate teeth. Meiosis is irregular and the spores are abortive.
> c. *A.* x *sarniense* was discovered by A. Sleep in 1971 in Guernsey, CI. It occurs, together with both parents, in sheltered hedge-banks in the south-west of the island. It is not known elsewhere.
> d. None.
> e. Both parents and hybrid $2n = 144$.
> f. SLEEP, A. (1971). A new hybrid fern from the Channel Islands. *Br. Fern Gaz.,* **10**: 209-211.
>   SLEEP, A. and RYAN, P. (1973). The Guernsey Spleenwort — a new fern hybrid. *Rep. Trans. Soc. Guernes.,* **19**: 212-213. FIG.

1a x cun. *A. adiantum-nigrum* L. x *A. cuneifolium* Viv.
> = *A.* x *centovallense* D. E. Meyer is known in He and It.

2 x 1b. *A. billotii* F. W. Schultz (*A. obovatum* auct., *non* Viv.) x *A. onopteris* L. (*A. adiantum-nigrum* L. subsp. *onopteris* (L.) Luerss.)
= *A.* x *joncheerei* D. E. Meyer is known only from the type collection from Madeira.

1 x 4. *A. adiantum-nigrum* L. *sensu lato* x *A. marinum* L.
has been recorded by Druce and Blackman but the specimens are states of *A. adiantum-nigrum.* This combination has also been suggested as the parentage of x *Asplenophyllitis jacksonii,* but this attribution is now known to be incorrect.
ALSTON, A. H. G. (1940). Notes on the supposed hybrids in the genus *Asplenium* found in Britain. *Proc. Linn. Soc. Lond.,* **152**: 132-144.

1 x 5. *A. adiantum-nigrum* L. *sensu lato* x *A. trichomanes* L.
= *A.* x *dolosum* Milde is recorded for Au, Be, Ga and It.

1 x 7. *A. adiantum-nigrum* L. *sensu lato* x *A. rutamuraria* L.
= *A.* x *perardii* Litard. is recorded from Ga, Ge, He, Ho, Hs, It and Po.

1a x 8. *A. adiantum-nigrum* L. x *A. septentrionale* (L.) Hoffm.
a. None.
b. This is quite distinct in appearance from all European species of *Asplenium,* but is extremely similar to the non-British *A. billotii* x *A. septentrionale.* In Br it is likely to be confused only with *A.* x *alternifolium,* from which it differs most obviously in the relatively much longer lowest pair of pinnae, which give the frond a broadly triangular outline like that of *A. adiantum-nigrum.* It is highly sterile.
c. A series of fronds distributed by T. Butler in the 1870s, all originating in cultivation from a single juvenile plant collected years earlier on Craig Ddu, Pass of Llanberis, v.c. 49, and the subject at that time of several opinions (all incorrect) regarding their true identity, constitute the only British record. Outside BI it is known only from two localities in Ga (T. Reichstein, pers. comm).
d. None.
e. *A. adiantum-nigrum* $2n = 144$; *A. septentrionale* $2n = (72), 144$; hybrid $2n = 144$.
f. ANONYMOUS (1968). *Asplenium adiantum-nigrum* x *septentrionale*: a hybrid new to Britain. *Br. Fern Gaz.,* **10**: 37.
BUTLER, T., BAKER, J. G. and SYME, J. T. B. (1881). *Asplenium germanicum,* Weiss. *Rep. B.E.C.,* **1**: 38-39.
BUTLER, T. and BABINGTON, C. C. (1882). *Asplenium 'Germanicum,* Weiss'. *Rep. B.E.C.,* **1**: 59.

2 x 4. *A. billotii* F. W. Schultz (*A. obovatum* auct., *non* Viv.) x *A. marinum* L.

has been erroneously regarded to be the parentage of x *Asplenophyllitis microdon.*

2 × 8. A. billotii F. W. Schultz (*A. obovatum* auct., *non* Viv.) × *A. septentrionale* (L.) Hoffm.
= *A.* x *souchei* Litard. has been found in Ga.

3 × 5. A. fontanum (L.) Bernh. × *A. trichomanes* L.
= *A.* x *corbariense* Rouy is reported from Ga and Hs.

3 × 6. A. fontanum (L.) Bernh. × *A. viride* Huds.
= *A.* x *gastonii-gautieri* Litard. is known from Ga and He.

4 × 5. A. marinum L. × *A. trichomanes* L.
has been erroneously considered to be the parentage of x *Asplenophyllitis confluens.*

5q × 5t. A. trichomanes L. subsp. *quadrivalens* D. E. Meyer × *A. trichomanes* L. subsp. *trichomanes*
a. *A.* x *lusaticum* D. E. Meyer.
b. This sterile hybrid is intermediate between the parents in morphology, but in practice distinguishable only by abortive spores, chromosome number and hybrid vigour.
c. So far it has been detected only in v.c. 48 and 89. Outside BI it has been found in Au, Be, Ga, Ge, He, Hu, No and U.S.A.
d. Hybrids have been synthesized in Leeds and Be.
e. *A. trichomanes* subsp. *trichomanes* $2n = 72$; subsp. *quadrivalens* $2n = 144$; hybrid $2n = 108$.
f. BENOIT, P. M. (1964). The two types of *Asplenium trichomanes. Nature Wales,* **9**: 75-79.
BOUHARMONT, J. (1968). Les formes chromosomiques d'*Asplenium trichomanes* L. *Bull. Jard. bot. État Brux.,* **38**: 103-114.
LOVIS, J. D. (1955). The problem of *Asplenium trichomanes,* in LOUSLEY, J. E., ed. *Species studies in the British flora,* pp. 99-103. London.
MEYER, D. E. (1958). Zur Zytologie der Asplenien Mitteleuropas, 16-20. *Ber. dt. bot. Ges.,* **71**: 11-20.
WAGNER, W. H. and WAGNER, F. S. (1966). Pteridophytes of the Mountain Lake area, Giles Co., Virginia; Biosystematic studies, 1964-65. *Castanea,* **31**: 121-140.

5 × 6. A. trichomanes L. × *A. viride* Huds.
The hybrid involving *A. trichomanes* subsp. *quadrivalens* D. E. Meyer, = *A.* x *bavaricum* D. E. Meyer, is known from Au and Ge, and that involving subsp. *trichomanes,* = *A.* x *protoadulterinum* Lovis & Reichstein, from Au

and He. A plant found in Levens Park, v.c. 69, described as intermediate between *A. trichomanes* and *A. viride*, with rachides black for only an inch from the caudex, was surely not this hybrid, but was very probably a variant of *A.* x *clermontiae*. It has been suggested that *A.* x *refractum* (q.v.) was *A. trichomanes* x *A. viride*, but this is certainly incorrect. *A. adulterinum* Milde ($2n = 144$), confined to serpentine and other ultrabasic rocks in Au, Cz, Fe, Ge, Gr, He, It, Ju, No, Po and Su, is the amphidiploid derivative of *A. trichomanes* subsp. *trichomanes* x *A. viride*. Hybrids between it and *A. viride*, = *A.* x *poscharskyanum* (Hoffm.) Dörfl., are known from Au, Ge, It, No and Su, and the other backcross, *A. adulterinum* x *A. trichomanes* subsp. *trichomanes*, = *A.* x *trichomaniforme* Woynar, has been detected in Au, It and No. Hybrids between *A. adulterinum* and *A. trichomanes* subsp. *quadrivalens*, *A. rutamuraria* and *A. adiantum-nigrum* have also been found on the Continent.

7 x 5q. *A. rutamuraria* L. subsp. *rutamuraria* x *A. trichomanes* L. subsp. *quadrivalens* D. E. Meyer

a. *A.* x *clermontiae* Syme.

b. This hybrid is rather closer to *A. trichomanes,* from which it is readily distinguished by the less numerous pinnae, the rachis green above, and the toothed indusium. Furthermore, the lowest pinnae are usually distinctly larger than those above, and in luxuriant fronds are tri-lobed. Small examples might be mistaken for a peculiar form of *A. viride.* Foreign examples may be either very sterile, or partially fertile.

c. In view of the abundance of both parents on mortared walls and comparable habitats in many parts of BI and Europe, this is an excessively rare hybrid. It was first found in 1863 at Newry, v.c. H38, and has not been recorded since in BI. A plant found in Levens Park, v.c. 69, and then tentatively attributed to *A. trichomanes* x *A. viride*, appears from the description to have been *A.* x *clermontiae*, but no specimen is known. Outside BI this hybrid has been found in Au, It and U.S.A.

d. None.

e. Both parents and hybrid $2n = 144$.

f. ALSTON, A. H. G. (1940). As above. FIG.

LOVIS, J. D., MELZER, H. and REICHSTEIN, T. (1966). *Asplenium* x *stiriacum* D. E. Meyer emend. und *A.* x *aprutianum hybr. nov.*, die zwei *Asplenium lepidum* x *trichomanes*-Bastarde. *Bauhinia*, **3**: 87-101. FIG.

SYME, J. T. B. (1886). *English Botany*, ed. 3, **12**: 132-134 and t.1879. London. FIG.

8 x 5q. *A. septentrionale* (L.) Hoffm. x *A. trichomanes* L. subsp. *quadrivalens* D. E. Meyer

= *A.* x *heufleri* Reichardt has been found in Au, Be, Ge, He, Hu, It and Rm.

8 × 5t. *A. septentrionale* (L.) Hoffm. × *A. trichomanes* L. subsp. *trichomanes*

a. *A.* × *alternifolium* Wulfen (*A.* × *breynii* auct., *non* Retz.; *A.* × *germanicum* auct., *non* Weiss.).

b. The frond is narrowly triangular in outline with alternate pinnae, 2–5 on either side of the rachis, the lowest deeply incised or trifid and the middle ones undivided and narrowly cuneiform. This can be confused only with other hybrids, e.g. *A. adiantum-nigrum* × *A. septentrionale, A.* × *beufleri, A.* × *murbeckii* and *A.* × *souchei.* It is highly sterile, probably completely so in the wild.

c. *A.* × *alternifolium* is restricted to lime-free rocks. It is known in v.c. 48 and 49 and has been seen relatively recently in v.c. 70. Older records supported by herbarium specimens exist for v.c. 5, 68, 80, 83 and 89, but others require confirmation. It is frequent in central Europe, certainly known from Au, Be, Cz, Fe, Ga, Ge, He, Hu, It, No, Po, Rm and Su, and eastwards to Kashmir.

d. Synthesized hybrids of comparable morphology to wild examples have been produced in Leeds and in Berlin. There is at least one record, dating from the last century, of progeny being obtained in cultivation from the wild hybrid.

e. *A. trichomanes* subsp. *trichomanes* $2n = 72$; *A. septentrionale* $2n = (72)$, 144; hybrid $2n = 108$.

f. ALSTON, A. H. G. (1940). As above.

BOUHARMONT, J. (1966). Note sur *Asplenium* × *alternifolium* Wulfen. *Bull. Jard. bot. État Brux.,* **36**: 383-391.

GUÉTROT, M. (1926). Histoire de l'*Asplenium (germanicum) Breynii. Bull. Soc. bot. Deux-Sèvres* **1926**: 15-31.

HYDE, H. A., WADE, A. E. and HARRISON, S. G. (1969). *Welsh ferns,* ed. 4. Cardiff. FIG.

LOVIS, J. D. and SHIVAS, M. G. (1954). The synthesis of *Asplenium* × *breynii. Proc. B.S.B.I.,* **1**: 97.

MEYER, D. E. (1952). Untersuchungen über Bastardierung in der Gattung *Asplenium. Biblthca bot.,* **30 (123)**: 1-34.

PERRING, F. H. and SELL, P. D. (1968). *Op. cit.,* p. 2.

7 × 6. *A. rutamuraria* L. × *A. viride* Huds.

= *A.* × *meyeri* Rothmaler is reported from Bavaria, Ge, but requires confirmation.

cun × 6. *A. cuneifolium* Viv. × *A. viride* Huds.

= *A.* × *woynarianum* Aschers. & Graebn. is known in Au and probably also in Cz, Ge and Po.

7 × **8.** *A. rutamuraria* L. × *A. septentrionale* (L.) Hoffm.
   a. *A.* × *murbeckii* Dörfler.
   b. This hybrid is intermediate between the parents. Small forms can be confused with small states of *A.* × *alternifolium*, from which it can be distinguished by the more conspicuous but narrower teeth at the tips of the pinnae and the irregular edge of the indusium. The only British example available for study in recent years is highly sterile but some Continental examples are partially fertile. Very recently, a single wild backcross hybrid with *A. septentrionale* has been detected in Hu (G. Vida, pers. comm.).
   c. It grows only on rocks of neutral or complex composition. It has been found in BI in recent years only in v.c. 70, but there are authentic old records from v.c. 83 and 89, and it is also known from Au, Ga, Ge, He, Hu, Hs, Rm and Su.
   d. Synthesized examples have been produced by A. Heilbronn, D. E. Meyer & G. Vida (pers. comm.). Progeny have been readily obtained in cultivation in both Basel and Leeds from spores gathered in nature from a German hybrid. Examples of this progeny have been crossed with examples of the parents originating from the same locality. These artificial backcross hybrids are highly sterile (J. D. Lovis & T. Reichstein, unpub.).
   e. Both parents $2n = (72)$, 144; hybrid $2n = 144$.
   f. ALSTON, A. H. G. (1940). As above.
      HEILBRONN, A. (1910). Apogamie, Bastardierung und Erblichkeitsverhältnisse bei einigen Farnen. *Flora*, **101**: 1-42.
      LOVIS, J. D. (1963). Meiosis in *Asplenium* × *murbeckii* from Borrowdale. *Br. Fern Gaz.*, **9**: 110-113.
      LOVIS, J. D. (1964). Autopolyploidy in *Asplenium*. *Nature, Lond.*, **203**: 324-325.
      MEYER, D. E. (1952). As above.
      MEYER, D. E. (1959). Zur Zytologie der Asplenien Mitteleuropas, 21-23. *Ber. dt. bot. Ges.*, **72**: 37-48. FIG.

Asplenium × *refractum* T. Moore
   is clearly a hybrid but the parentage is doubtful. Alston (1940) suggested *A. obovatum* (= *A. billotii*) × *A. trichomanes*, which is a very plausible attribution on morphological grounds. In general appearance *A.* × *refractum* is closer to *A. trichomanes*, but it has the fronds bipinnatifid below and the pinnae deeply serrate above, and the rachis is green to halfway. The plant was described from cultivated material said to have been found in Scotland, where *A. billotii* is, significantly, almost absent. It has never been refound in the wild.
   ALSTON, A. H. G. (1940). As above. FIG.
   MOORE, T. (1855). t.35A, in LINDLEY, J., ed. *The ferns of Great Britain and Ireland*. London. FIG.

15 × 16. *Asplenium* L. × *Ceterach* DC.
= × *Asplenoceterach* D. E. Meyer
(by J. D. Lovis)

15/7 × 16/1. *Asplenium rutamuraria* L. × *Ceterach officinarum* DC.
= × *A. badense* D. E. Meyer is only known from the type plant discovered on the Kaiserstuhl, Ge.

19. *Cystopteris* Bernh.
(by A. C. Jermy)

1. *C. fragilis* (L.) Bernh.
exists as tetraploid, hexaploid and octoploid races, $2n = 168, 252$ and 336, on the Continent, and as the first two in Br. In Co, He and It sterile pentaploids ($2n = 210$), presumably hybrids between the tetraploid and hexaploid races, have been found.

2 × 1. *C. dickieana* Sim × *C. fragilis* (L.) Bernh.
has been found in He and No; investigated plants were tetraploids ($2n = 168$). A specimen in **BM** collected from Glengarry, v.c. 88, about 1870 may be this hybrid. It has abortive spores, a few of which show an intermediate wall sculpturing pattern.

1 × 3. *C. fragilis* (L.) Bernh. × *C. montana* (Lam.) Desv.
has been recorded from He.

20. *Woodsia* R.Br.
(by A. C. Jermy)

2 × 1. *W. alpina* (Bolton) Gray × *W. ilvensis* (L.) R.Br.
= *W.* × *gracilis* (Lawson) Butters (*W.* × *pilosella sensu* C. Hartm., ?*non* Rupr.) has been reported from Fe, No, Su, U.S.A. and Canada.

21. *Dryopteris* Adans.
(by A. C. Jermy and S. Walker)

All hybrids involving *D. pseudomas* can reproduce apogamously and could rightly be regarded as species. They are included in this account as in most cases they have been recorded in the literature as hybrids.

1 x 2. *D. filix-mas* (L.) Schott x *D. pseudomas* (Woll.) Holub & Pouzar (*D. borreri* Newm.)
  a. *D.* x *tavelii* Rothm.
  b. Hybrids are intermediate between the parents in leaf-segment shape and texture and in the amount of ramenta, but the dark pigment spot at the bases of the pinnae is characteristic of *D. pseudomas*. Indusia characteristic of both parents (i.e. tucked under and spread out flat on the lamina) are found on the same pinna. Plants are usually much larger than either parent; they reproduce by apogamy.
  c. Hybrids occur in the vicinity of both parents, and in similar habitats, in v.c. 2, 3, 8, 9, 14, 35, 36, 49, 69, 94, 97, 98, 103 and most likely throughout the range of the parents in Br, and in Au, Be, Ga, Ge, He, Ho, Hs, Hu, Ju, Lu and No.
  d. Hybrids can be synthesized using *D. pseudomas* as the male parent — being apogamous the latter cannot be used as female. These hybrids are comparable in fertility with wild hybrids, though they make take up to five years to become fertile. Any isolation appears to be spatial only.
  e. *D. filix-mas* $2n = 164$; *D. pseudomas* $2n = 82, 123$; hybrid $2n = 164, 205$.
  f. DÖPP, W. (1939). Cytologische und genetische Untersuchungen innerhalb der Gattung *Dryopteris*. *Planta*, **29**: 481-533. FIG.
     DÖPP, W. (1955). Experimentell erzeugte Bastarde zwischen *Dryopteris filix-mas* (L.) Schott und *D. paleacea* (Sw.) C. Chr. *Planta*, **46**: 70-91. FIG.
     MANTON, I. (1950). *Problems of cytology and evolution of the Pteridophyta*. Cambridge.
     REICHLING, L. (1953). *Dryopteris paleacea* (Sw.) Handel-Mazzetti et *Dryopteris Tavelii* Rothmaler au grand-duché de Luxembourg et en Belgique. *Bull. Soc. r. bot. Belg.*, **86**: 39-57. FIG.
     SEGAL, S. (1963). Pteridologische aantekeningen, 2. *Dryopteris tavelii* in Nederland. *Gorteria*, **1**: 121-128. FIG.

3 x 1. *D. abbreviata* (DC.) Newm. x *D. filix-mas* (L.) Schott
  a. *D.* x *mantoniae* Fraser-Jenkins & Corley.
  b. This hybrid is intermediate between its parents though difficult to distinguish from *D. filix-mas*. The lower basiscopic pinnule is often upturned as in *D. abbreviata* and both types of indusia (inrolled and adpressed margins) are found on a single pinna. Abortive spores lead to sterility.

c. It is found in the vicinity of the parents on marginal land, rocky gullies, screes and in scrub wherever protected from grazing. Recorded from v.c. 48, 49, 68, 69, 94 and probably frequent throughout northern and north-western Br, it is also known in Ge and He.

d. Hybrids have been synthesized using *D. abbreviata* as the male parent. Such hybrids are sterile.

e. *D. abbreviata* $2n = 82$; *D. filix-mas* $2n = 164$; hybrid $2n = 123$.

f. FRASER-JENKINS, C. R. and CORLEY, H. V. (1972). *Dryopteris caucasica* — an ancestral diploid of the male fern aggregate. *Br. Fern Gaz.*, **10**: 221-231.
   MANTON, I. (1950). As above. FIG.

6 x 1. *D. carthusiana* (Vill.) H. P. Fuchs (*D. lanceolatocristata* (Hoffm.) Alston) x *D. filix-mas* (L.) Schott

a. (*D.* x *remota sensu* Druce, *Lastraea remota sensu auct. brit., non Aspidium remotum* (A. Br.) A. Br.).

b. This little-known plant is intermediate between the parents in the frond characters of blade-width, degree of pinnation and teeth, but the habit is as in *D. filix-mas.* It is sterile.

c. It has been found once only in wet woodland by Windermere Lake, v.c. 69. Hybrids of this parentage have been reported from Scotland and Hb in error and also from the Continent. Of the latter those from Zastlertal, Ge, and Licstal, He, may be correctly identified.

d. None.

e. Both parents and hybrid $2n = 164$.

f. MANTON, I. (1950). As above. FIG.

7 x 1. *D. dilatata* (Hoffm.) A. Gray x *D. filix-mas* (L.) Schott

a. *D.* x *subaustriaca* Rothm.

b. This unexpected hybrid is intermediate between the parents and resembles *D.* x *woynarii,* but the rachis is less scaly. The spores are abortive and sterile.

c. It has been recorded from v.c. 3 but the plant requires location and cytological verification. The plants recorded from Loch Lomond, v.c. 99, and from H15, proved to be apogamous triploids and therefore are not of this parentage. Continental records exist for Au, Ga, He, Ju and Su.

d. None.

e. Both parents $2n = 164$.

f. WALTER, E. (1929). *Nephrodium (subalpinum) Borbasio*, in GUÉTROT, M., ed. *Plantes hybrides de France*, 3 and 4: 104-105. Paris.

ass x 1. *D. assimilis* S. Walker x *D. filix-mas* (L.) Schott

might occur in Br amongst the putative *D. pseudomas* x *D. assimilis* hybrids (q.v.).

3 × 2. *D. abbreviata* (DC.) Newm. × *D. pseudomas* (Woll.) Holub & Pouzar (*D. borreri* Newm.)

may be the parentage of variants of *D. pseudomas* with more distinctly toothed pinnules with a rounded, not truncate, apex and with the lower basiscopic pinnules inflexed forward. The spores are intermediate in sculpturing between the putative parents and show signs of reproducing apogamously. Such plants may well be the origin of triploid *D. pseudomas*.

7 × 2. *D. dilatata* (Hoffm.) A. Gray × *D. pseudomas* (Woll.) Holub & Pouzar (*D. borreri* Newm.)

a. *D.* × *woynarii* Rothm. (*D. remota sensu* Lowe *et auct. al.*).
b. This resembles *D. dilatata* but is more scaly on the rachis and has a lesser degree of frond-dissection, with distinctly square ends to the pinnule-segments. The sporangia possess spores of two sizes and the plant is most likely apogamous.
c. It has been recorded from amongst its parents in shady scree at Lochinver, v.c. 108, but it needs cytological confirmation. The record from Loch Lomond, v.c. 99, has since proved to be based on an apogamous triploid and therefore most likely *D.* × *remota* (A. Br.) Druce. It is also recorded for Au and Ge but has no doubt been confused with *D.* × *remota*.
d. None.
e. *D. dilatata* 2*n* = 164; *D. pseudomas* 2*n* = 82, 123.
f. None.

ass × 2. *D. assimilis* S. Walker × *D. pseudomas* (Woll.) Holub & Pouzar (*D. borreri* Newm.)

a. *D.* × *remota* (A. Br.) Druce, *excl. specim.* (*Lastrea remota* (A. Br.) Moore, *non auct. brit.*).
b. This is more or less intermediate in frond-shape and -dissection between the parents. The rachis and stipe are paleaceous with golden to grey-brown scales with a dark line in the centre. This plant reproduces apogamously.
c. Plants from Loch Lomond, v.c. 99, originally reported as *Lastrea remota* or "*L. boydii*" with the putative parentage of *D. dilatata* × *D. filix-mas*, are likely to be this hybrid. A similar but paler-scaled, apogamous, triploid plant (originally named *D. carthusiana* × *D. filix-mas*) has been collected in Dalystown, v.c. H15, although *D. assimilis* has yet to be found in Hb. The other records from L. Cowey, v.c. H38, and elsewhere, attributed by Praeger (1951) to R. D. Meikle, are typographical errors for *D. carthusiana* × *D. dilatata* (*fide* R. D. Meikle pers. comm. 1972). In Europe this hybrid is often confused with *D. carthusiana* × *D. filix-mas* (or × *D. pseudomas*), but Ge, He and Hu records have been cytologically confirmed. On geographical and morphological grounds it seems likely that A. Braun's *Aspidium remotum* was this hybrid.
d. None.
e. *D. assimilis* 2*n* = 82; *D. pseudomas* 2*n* = 82, 123; hybrid 2*n* = 123.

f. BENL, G. and ESCHELMÜLLER, A. (1973). Über *"Dryopteris remota"* und ihr Vorkommen in Bayern. *Ber. bayer. bot. Ges.*, **44**: 101-141. FIG.
   MANTON, I. (1950). As above. FIG.
   PRAEGER, R. L. (1951). *Loc. cit.*

3 x 8. *D. abbreviata* (DC.) Newm. x *D. aemula* (Ait.) Kuntze

a. *D.* x *pseudoabbreviata* Jermy.
b. This is intermediate between its parents in the shape of its frond and the texture of the lamina. The base of the stipe has the characteristic purple-red colour of *D. aemula*; the lamina is glandular and hay-scented and the costae are dark. The spores are abortive.
c. It has been seen only once in the Isle of Mull, v.c. 103, in oak-birch scrub on a steep, west-facing hillside.
d. None.
e. Both parents $2n = 82$.
f. JERMY, A. C. (1968). Two new hybrids involving *Dryopteris aemula*. *Br. Fern Gaz.*, **10**: 9-12. FIG.

6 x 5. *D. carthusiana* (Vill.) H. P. Fuchs (*D. lanceolatocristata* (Hoffm.) Alston) x *D. cristata* (L.) A. Gray

a. *D.* x *uliginosa* (Newm.) Kuntze ex Druce.
b. The hybrid is intermediate in the characteristics of its parents, particularly in the shape of the frond and its degree of dissection. Fertile fronds are similar to those of *D. carthusiana* but the lowest pinnae at least are usually sterile and show distinctly less dissection. The rhizome-apex is flat and very young croziers translucent, as in *D. cristata*. Plants are characteristically larger than either parent, often forming clumps 75 cm across; they are sterile with abortive spores.
c. It occurs in *Phragmites* mires or soliflugenous bogs of pH 5.5–6.5, usually with *Sphagnum* species. It is rare now in Br, but has been recorded from v.c. 27, 28, 40, 56, 58 and 64, and also from Be, Da, Ga, Ge, Hu, No, Su, Siberia and North America.
d. None.
e. Both parents and hybrid $2n = 164$.
f. MANTON, I. (1950). As above. FIG.
   SORSA, V. and WIDÉN, C.-J. (1968). The *Dryopteris spinulosa* complex in Finland. A cytological and chromatographic study of some hybrids. *Hereditas*, **60**: 273-293. FIG.
   WALKER, S. (1955). Cytogenetic studies in the *Dryopteris spinulosa* complex, 1. *Watsonia*, **3**: 193-208. FIG.
   WALKER, S. (1961). Cytogenetic studies in the *Dryopteris spinulosa* complex, 2. *Am. J. Bot.*, **48**: 607-614. FIG.

6 × 7. *D. carthusiana* (Vill.) H. P. Fuchs (*D. lanceolatocristata* (Hoffm.) Alston) × *D. dilatata* (Hoffm.) A. Gray

   a. *D.* × *deweveri* (Jansen) Jansen & Wachter (*D.* × *ambigua* Druce, *nom. nud.*).

   b. This is again intermediate between its parents in morphology, including the shortly creeping, semi-prostrate rhizome, ramenta with a diffuse darker central portion, and glandular indusia. The spores are usually abortive but occasionally a few are fully formed. Plants often form large clumps over 50 cm across.

   c. It is found regularly in areas where both parents are present, usually in damp, shady woodland, in v.c. 2, 3, 6, 7, 11, 12, 14, 15, 17, 18, 22, 27, 34–36, 40, 41, 49, 61, 63, 64, 67, 69, 77, 85, 88, 92, 94, 95, 98, 100, H15, H34 and H38, and also in Be, Da, Ga, Ge, Ho and Hs.

   d. The hybrid has been produced artificially and was in agreement with wild specimens.

   e. Both parents and hybrid $2n = 164$.

   f. MANTON, I. (1950). As above. FIG.

      SORSA, V. and WIDÉN, C.-J. (1968). As above. FIG.

      WALKER, S. (1955). As above. FIG.

      WIDÉN, C.-J., SARVELA, J. and AHTI, T. (1967). The *Dryopteris spinulosa* complex in Finland. *Acta bot. fenn.*, 77: 1-24. FIG.

ass × 6. *D. assimilis* S. Walker × *D. carthusiana* (Vill.) H. P. Fuchs (*D. lanceolatocristata* (Hoffm.) Alston)

   has been found in Fe.

8 × 7. *D. aemula* (Ait.) Kuntze × *D. dilatata* (Hoffm.) A. Gray

   a. None.

   b. This hybrid is morphologically intermediate between its parents but could easily be taken for *D. dilatata.* The lamina is more finely cut, glandular, with a coumarin smell, and the stipe is purple-red at the very base. Spores are usually abortive but occasionally some are fully-formed. It still requires cytological confirmation; hybrids between *D. aemula* and *D. dilatata* and the very similar *D. assimilis* would be very difficult to separate on morphological characters.

   c. It is found in mossy woodland in v.c. 98, 101, 103, 104 and H2, and has been reported from v.c. 69 and H26, possibly in error. It is not known elsewhere.

   d. None.

   e. *D. aemula* $2n = 82$; *D. dilatata* $2n = 164$.

   f. JERMY, A. C. (1968). As above. FIG.

ass × 7. *D. assimilis* S. Walker × *D. dilatata* (Hoffm.) A. Gray

   a. None.

   b. The hybrid is intermediate between the two parents but very difficult to distinguish owing to the variation within each of the latter. In wild, mixed populations of the parents it is readily detected by its sterility, indicated by abortive spore production.

c. It is always found in the vicinity of the parents, particularly among rocks and boulders in mountain areas from sea level to 3000 ft (900 m), in v.c. 92, 96, 98, 101, 105 and 111, and in Au, Ge, He and Su within the range of the parents.

d. Hybrids are readily synthesized but sterile. Crosses can be made with either parent as female. The only apparent barrier to crossing is spatial.

e. *D. assimilis* $2n = 82$; *D. dilatata* $2n = 164$; hybrid $2n = 123$.

f. CRABBE, J. A., JERMY, A. C. and WALKER, S. (1970). The distribution of *Dryopteris assimilis* S. Walker in Britain. *Watsonia*, 8: 3-15.

DÖPP, W. and GÄTZI, W. (1964). Der Bastard zwischen tetraploider und diploider *Dryopteris dilatata. Ber. schweiz. bot. Ges.*, 74: 45-53. FIG.

SORSA, V. and WIDÉN, C.-J. (1968). As above. FIG.

WALKER, S. (1955). As above. FIG.

WIDÉN, C.-J., SARVELA, J. and AHTI, T. (1967). As above. FIG.

8 × ass. *D. aemula* (Ait.) Kuntze × *D. assimilis* S. Walker
is possibly the parentage of sterile plants from v.c. 98 and 103. They are very similar to *D. assimilis* but the presence of glands, the purple-red stipe-base and the upturned lamina-teeth suggest the *D. aemula* parentage.

22. *Polystichum* Roth
(by A. Sleep)

2 × 1. *P. aculeatum* (L.) Roth × *P. setiferum* (Forsk.) Woynar

a. *P. × bicknellii* (Christ) Hahne.

b. Hybrids are generally intermediate between the parents, although often somewhat difficult to recognize owing to the great variation shown by the latter. The fronds are usually bipinnate, resembling the *P. setiferum* parent, and, although the pinnules may be either adnate or stalked, the angle at the base of the pinnule is invariably acute. Typically, the frond is narrower than in *P. setiferum*, and is usually contracted towards the base as in *P. aculeatum*. The spores are abortive. Hybrids are usually sterile and of the F_1 type, although in some cases there may be partial fertility. The possibility of occasional backcrossing to either parent cannot be completely excluded, and introgression may occur in some localities.

c. There has been much confusion in this country in the past between the two parent species, and as a result the distribution of the hybrid is only imperfectly known, there being several erroneous records. There are as yet no cytologically confirmed specimens from BI. Specimens have been seen from v.c. 1, (2), 3–5, 7, 8, (9), 10, 12, 17, (18), 23, 34, 36, 39, (40), (42), 47, (57), 63, 65, 70, 73, (87), 98, 99, 101, H1, H14, H19, H26 and H30; there is still some doubt attached to the records in parentheses. Although there is some geographical and ecological separation of the parents the ranges of these two species do overlap, and they may occasionally be found growing together in habitats such as wooded ravines and disused quarries. This hybrid is also known from Au, Ga, Ge, He, Hu, It, Ju, Rm and Tu.

d. Hybrids have been synthesized with ease using either parent as the female. These hybrids are triploid, and the meiotic pairing observed strongly suggests that *P. setiferum* is part-parental to *P. aculeatum*.

e. *P. aculeatum* $2n = 164$; *P. setiferum* $2n = 82$; hybrid ($2n = 123$).

f. MANTON, I. (1950). *Problems of cytology and evolution in the Pteridophyta.* Cambridge.

MANTON, I. and REICHSTEIN, T. (1961). Zur Cytologie von *Polystichum braunii* (Spenner) Fée und seiner Hybriden. *Ber. schweiz. bot. Ges.*, **71**: 370-383. FIG.

SLEEP, A. (1966). *Some cytotaxonomic problems in the fern genera Asplenium and Polystichum.* Ph.D. Thesis, University of Leeds.

SLEEP, A. (1971). *Polystichum* hybrids in Britain. *Br. Fern Gaz.*, **10**: 208-209.

VIDA, G. (1966). Cytology of *Polystichum* in Hungary. *Bot. Közl.*, **53**: 137-144. FIG.

3 × 1. *P. lonchitis* (L.) Roth × *P. setiferum* (Forsk.) Woynar

= *P.* × *lonchitiforme* (Halácsy) Becherer has been reported from southern Gr.

2 × 3. *P. aculeatum* (L.) Roth × *P. lonchitis* (L.) Roth

a. *P.* × *illyricum* (Borbás) Hahne.

b. Hybrids are generally intermediate between the parents; the fronds are narrowly lanceolate, rather stiff and simply pinnate, but with deeply pinnatisect pinnae which, in the middle third of the frond, are again divided into tiny, acute, obliquely inserted pinnules, at least in the proximal part. Although confusion with stunted or juvenile *P. aculeatum* can easily occur, the hybrid can always be recognized by its shrivelled, abortive spores. There is growing evidence of partial fertility in Continental populations, although this hybrid is generally regarded as being highly sterile and of the F_1 type.

c. This hybrid has only recently been reported from BI. A single herbarium specimen (**DBN**) was detected in 1972, collected in 1932 from Glenade, v.c. H29, where it was rediscovered *in situ* in 1973. It was also found in 1973 on limestone scree near Inchnadamph, v.c. 108. It is frequent in the mountains of Europe, being known from Al, Au, Bu, Cz, Ga, Ge, He, Hs, Hu, Ju and Rm.

d. Hybrids can be synthesized with ease using either parent as the female. The resultant hybrids are triploid, and the meiotic pairing observed strongly suggests that *P. lonchitis* is the second parent of *P. aculeatum*.

e. *P. aculeatum* $2n = 164$; *P. lonchitis* $2n = 82$; hybrid ($2n = 123$).

f. MANTON, I. (1950). As above. FIG.

SLEEP, A. (1966). As above.

SLEEP, A. and REICHSTEIN, T. (1967). Der Farnbastard *Polystichum* x *meyeri* hybr. nov. = *Polystichum braunii* (Spenner) Fée x *P. lonchitis* (L.) Roth und seine Cytologie. *Bauhinia*, 3: 299-374. FIG.

SLEEP, A. and SYNNOTT, D. (1972). *Polystichum* x *illyricum*: a hybrid new to the British Isles. *Br. Fern. Gaz.*, 10: 281-282. FIG.

STIRLING, A. McG. (1974). A fern new to Scotland — *Polystichum* x *illyricum* in West Sutherland. *Watsonia*, 10: 231.

VIDA, G. (1966). As above. FIG.

25 × 22. *Polypodium* L.
× *Polystichum* Roth
(by C. A. Stace)

25/1 x 22/2. *Polypodium vulgare* L. x *Polystichum aculeatum* (L.) Roth

was at one time said to be the parentage of *Polypodium vulgare* var. *serratum* Willd.

LINTON, E. F. (1907). Hybrids among British phanerogams. *J. Bot., Lond.*, 45: 296-304.

24G. *Gymnocarpium* Newm.
(*Thelypteris* Schmidel *pro parte*)
(by A. C. Jermy)

4 x 5. *G. dryopteris* (L.) Newm. (*T. dryopteris* (L.) Slosson) x *G. robertianum* (Hoffm.) Newm. (*T. robertiana* (Hoffm.) Slosson)

is known in the U.S.A. and has been reported from Su.

25. *Polypodium* L.
(by J. A. Crabbe and M. G. Shivas)

1a × 1v. *P. australe* Fée × *P. vulgare* L.

a. *P.* x *font-queri* Rothm.

b. Hybrids are intermediate between the parents in frond-shape, the number of indurated annular cells, and the position and shape of the sori; they are completely sterile. Abnormally large and translucent sporangia and spores are interspersed between the shrivelled sporangia and spores in the hybrids. The paraphyses characteristic of *P. australe* have not been found in any of the wild hybrids examined.

c. The ecology of the natural hybrid has not been studied but the distribution in BI is limited by the range of *P. australe*. Hybrids have so far been recorded for v.c. 6, 14, 49, 52 and 72 and also occur in Ga, He and It.

d. Two artificial hybrids have been synthesized using *P. australe* as the female parent, but none from the reciprocal cross. The hybrid shows 111 univalents at meiosis.

e. *P. australe* $2n = 74$; *P. vulgare* $2n = 148$; hybrid $2n = 111$ (triploid).

f. MANTON, I. (1950). *Problems of cytology and evolution in the Pteridophyta.* Cambridge.
ROTHMALER, W. (1937). In CADEVALL, J. *Flora de Catalunya*, 6: 353. Barcelona.

1i × 1v. *P. interjectum* Shivas × *P. vulgare* L.

a. *P.* x *mantoniae* (Rothm.) Shivas.

b. The hybrid is intermediate between the parents in frond-shape and in the number of indurated annular cells. The mature sori are more or less circular (as in *P. vulgare*) but smaller. The sporangia and spores are generally shrivelled but some sporangia may be abnormally large and pale and contain large, pale spores. Thus the hybrid may not be completely sterile. One backcross hybrid from Teesdale has been recorded and cytologically confirmed.

c. The hybrid occurs in at least 29 vice-counties of Br and Hb and is of frequent occurrence where the parents coexist. Its detailed distribution is not known in Europe but cytologically confirmed material has been collected from several areas in He and in northern It.

d. Five synthesized hybrids were produced using *P. interjectum* as the female parent, but so far the reciprocal cross has not been successful. No prothalli grew from the spores of the hybrid. The hybrid shows 74 bivalents and 37 univalents at meiosis.

e. *P. interjectum* $2n = 222$; *P. vulgare* $2n = 148$; hybrid $2n = 185$ (pentaploid).

f. MANTON, I. (1950). As above.
ROTHMALER, W. (1962). In ROTHMALER, W. and SCHNEIDER, U. Die Gattung *Polypodium* in Europa. *Kulturpflanze, Beih.* 3: 234-248.

SHIVAS, M. G. (1961). Contributions to the cytology and taxonomy of species of *Polypodium* in Europe and America, 1. Cytology; 2. Taxonomy. *J. Linn. Soc., Bot.*, **58**: 13-38.
SHIVAS, M. G. (1970). Names of hybrids in the *Polypodium vulgare* complex. *Br. Fern Gaz.*, **10**: 152.

1a × 1i. *P. australe* Fée × *P. interjectum* Shivas

a. *P.* × *shivasiae* Rothm. (*P* × *rothmaleri* Shivas).
b. Hybrids are intermediate between the parents in the shape of the frond and sori and in the number of indurated annular cells. Paraphyses typical of *P. australe* were observed in the hybrids by Roberts (1970) but not by Shivas (1961). The sporangia and spores are generally shrivelled and misshapen, although some sporangia are abnormally large and translucent. The hybrids are highly sterile.
c. This hybrid is recorded from v.c. 14, 35, 52 and 73. Cytologically-examined hybrids have also been collected in It and Tu. The range of the hybrid will probably be found to coincide with the range of overlap of the two parents.
d. Hybrids have been synthesized so far using only *P. interjectum* as the female parent; they appeared to be completely sterile. The hybrid shows 37 bivalents and 74 univalents at meiosis.
e. *P. australe* $2n = 74$; *P. interjectum* $2n = 222$; hybrid $2n = 148$ (tetraploid).
f. ROBERTS, R. H. (1970). A revision of some of the taxonomic characters of *Polypodium australe* Fée. *Watsonia*, **8**: 121-134.
ROTHMALER, W. (1962). As above.
SHIVAS, M. G. (1961). As above.
SHIVAS, M. G. (1970). As above.

32. *Larix* Mill.
(by C. A. Stace)

1 × kae. *L. decidua* Mill. × *L. kaempferi* (Lamb.) Carrière (*L. leptolepis* (Sieb. & Zucc.) Endl.)

a. *L.* × *eurolepis* Henry.
b. Hybrids are more rapidly-growing trees (at least in early years) and they show more resistance to insect and fungal attack than either parent. The cone-scales are slightly recurved (not recurved in *L. decidua*; conspicuously so in *L. kaempferi*) and the leaves are intermediate in length, breadth, colour and the number of abaxial rows of stomata. The hybrid shows a

high degree of fertility and the expected segregation of characters in F_2 and later generations.

c. This hybrid first arose around or soon after 1900 at Dunkeld, Perthshire, from seed borne on *L. kaempferi* trees of Japanese origin by natural pollination from *L. decidua*. In recent years it has been much grown in Br on a forestry scale, as well as for ornament (its rapid early growth favouring its choice in preference to its parents), the seed originating from interplanted parent trees. The hybrid may be found, like its parents, in semi-natural conditions, but there seem to be no records of seedlings outside forest areas (where they are sometimes common).

d. The hybrid has been reciprocally synthesized on a number of occasions, the characters of the hybrid being variable within the expected limits.

e. Both parents (2n = 24).

f. DALLIMORE, W. and JACKSON, A. B. (1966). *A handbook of Coniferae and Ginkgoaceae.* 4th ed. by HARRISON, S. G., pp. 299-300. London.

HENRY, A. and FLOOD, M. G. (1919). The history of the Dunkeld hybrid larch, *Larix eurolepis,* with notes on other hybrid conifers. *Proc. R. Ir. Acad.,* Ser. B, **35**: 55-66. FIG.

RICHENS, R. H. (1945). Forest tree breeding and genetics. *Imperial Agricultural Bureaux Joint Publication,* **8**: 17-19.

36. *Caltha* L.

(by C. A. Stace)

1 × rad. *C. palustris* L. × *C. radicans* T. F. Forst.

Fertile plants showing all stages of intermediacy between these two taxa exist widely in Hb and northern and western Br. Since the only clear-cut difference between them is the nodal rooting of *C. radicans* (most of the other characters are greatly affected by the environment), and the different chromosome races found within the genus are not correlated with a different morphology, the above two taxa seem best treated as varieties or forms. There are no sterility barriers within each chromosome number, but hybrids between plants of different ploidy level are sterile.

KOOTIN-SANWU, M. and WOODELL, S. R. J. (1971). The cytology of *Caltha palustris*: Cytogenetic relationships. *Heredity,* **26**: 121-135.

PANIGRAHI, G. (1955). *Caltha* in the British flora, in LOUSLEY, J. E., ed. *Species studies in the British flora,* pp. 107-110. London.

PRAEGER, R. L. (1934). *Caltha radicans* in Ireland. *Ir. Nat. J.,* **5**: 98-102.

PRAEGER, R. L. (1951). *Op. cit.,* p. 5-6.

WCISŁO, H. (1969). Further studies on experimental hybrids in *Caltha palustris* L. s.l. *Acta biol. cracov., Bot.,* **11**: 87-103.
WOODELL, S. R. J. and KOOTIN-SANWU, M. (1971). Intraspecific variation in *Caltha palustris. New Phytol.,* **70**: 173-186.

43. *Anemone* L.
(by C. A. Stace)

1 × 2. *A. nemorosa* L. × *A. ranunculoides* L.
= *A.* × *lipsiensis* Beck has been recorded from Au, Cz and He.

46. *Ranunculus* L.
Subg. *Ranunculus*
(by P. A. Candlish)

1 × 2. *R. acris* L. × *R. repens* L.
= *R.* × *roblenae* Domin is reported from Au and Cz, but is doubtful.

1 × 3. *R. acris* L. × *R. bulbosus* L.
= *R.* × *goldei* Meinsh. is reported from Be, Cz and Rm, but requires confirmation.

12 × 13. *R. flammula* L. × *R. reptans* L.
  a. (*R.* × *levenensis* Druce, *nom. nud.*).
  b. Hybrids (often misidentified as *R. reptans*) have a creeping habit with nodal rooting and extremely slender, arching internodes. They have a high degree (up to 90%) of pollen sterility, contrasting with the fully fertile pollen of *R. reptans* from Is and Su. Between this sort of hybrid and normal, upright *R. flammula* there may occur, in any one area, a wide range of plants varying in their degree of robustness, nodal rooting and pollen sterility. It is thought that these are hybrid swarms resulting from introgressive hybridization between *R. reptans* (which is now no longer found pure in Br) and *R. flammula*.

c. The hybrid is only known with certainty from v.c. 70 and 85, on stony lake-shores, although it probably occurs elsewhere in similar habitats. It has not been reported elsewhere in Europe.

d. The postulated hybrid when brought into cultivation retains its distinctive morphology, in contrast to *R. flammula* var. *tenuifolius* Wallr. (a lake-shore variant mimicking *R. reptans*) which in garden conditions becomes upright and robust.

e. *R. flammula* $2n = 32$; *R. reptans* ($2n = 32$).

f. MOSS, C. E. (1920). *Ranunculus flammula*, in *The Cambridge British Flora*, 3: 128-129 and pl. 132. Cambridge. FIG.

PADMORE, P. A. (1957). The varieties of *Ranunculus flammula* L. and the status of *R. scoticus* E. S. Marshall and of *R. reptans* L. *Watsonia*, 4: 19-27. (P. A. Padmore = Mrs P. A. Candlish.)

Subg. *Batrachium* (DC.) A. Gray
(by C. D. K. Cook)

17 × 18. *R. omiophyllus* Ten. (*R. lenormandii* F. W. Schultz) × *R. tripartitus* DC.

a. None.

b. The hybrid resembles *R. tripartitus* in general appearance except that the petals are larger (up to 6 mm long) and the entire leaves are frequently 5-lobed with shallower sinuses and curved lobes. A curious feature is that the pedicels often remain erect at maturity. The fertility is variable from plant to plant. Hybrid populations are also morphologically variable.

c. Hybrids occur in and around pools in the New Forest, v.c. 11, and to a large extent replace the parent species in this region. They have also been reported from v.c. 41; elsewhere they are known only in Ga.

d. This hybrid and numerous backcrosses have been artificially produced. The F_1 is only slightly fertile but highly fertile strains can be selected in the F_2 and backcrosses. It has also been possible to make tri-hybrid *R. (omiophyllus* × *R. tripartitus*) × (*R. peltatus*).

e. *R. omiophyllus* $2n = (16)$, 32; *R. tripartitus* $2n = 48$; hybrid $2n = 40$. Experimentally produced F_2 plants have shown some aneuploidy, $2n = 36–42$.

f. COOK, C. D. K. (1966). A monographic study of *Ranunculus* subgenus *Batrachium*. *Mitt. bot. StSamml. Münch.*, 6: 47-237. FIG.

17 × **22b. *R. omiophyllus*** Ten. (*R. lenormandii* F. W. Schultz) × *R.*
peltatus Schrank (*R. aquatilis* L. subsp. *peltatus* (Schrank) Syme)
  a. *R.* × *hiltonii* H. & J. Groves.
  b. This hybrid resembles *R. peltatus* in vegetative characteristics but lacks the
     very finely divided leaves of *R. peltatus* and roots regularly at the
     internodes. In floral characteristics the hybrid is intermediate between both
     parents but has blue sepals, nectar pits that are lunate or lacking and a
     hairy receptacle. It is highly fertile.
  c. The only confirmed locality is Copthorne Common, v.c. 14. The colony
     was discovered in 1896 and persisted until at least 1926, but is now
     probably extinct. It has also been reported from near Truro, v.c. 1, but the
     record has not been confirmed.
  d. In spite of repeated attempts this hybrid has not been successfully
     resynthesized.
  e. *R. omiophyllus* $2n = (16), 32$: *R. peltatus* $2n = (16, 32), 48$.
  f. COOK, C. D. K. (1966). As above.
     GROVES, H. and GROVES, J. (1901). A new hybrid water *Ranunculus. J.*
       Bot., Lond., **39**: 121-122. FIG.
     MOSS, C. E. (1920). *Ranunculus,* in *The Cambridge British Flora,* **3**:
       125-151, pl. 146. Cambridge. FIG.
     WILLIAMS, J. A. (1926). *Ranunculus Hiltoni* Groves. *J. Bot., Lond.,* **64**:
       250.

22a × **18. *R. aquatilis*** L. × *R. tripartitus* DC.
  a. None.
  b. In habit the hybrid resembles *R. aquatilis* but the leaf-form is highly
     unstable and the majority of the leaves are morphologically intermediate
     between divided and entire. In size and shape the flowers are intermediate
     between the two parents but the sepals are blue and the nectar pits lunate.
     The hybrid is sterile.
  c. It is found in v.c. 1, usually near both parents. The hybrid, being unstable
     in leaf-form, is weak and rarely persists for more than a few months; it is
     not known elsewhere.
  d. None.
  e. Both parents $2n = 48$.
  f. None.

20 × **19. *R. circinatus*** Sibth. × *R. fluitans* Lam.
  a. None.
  b. The hybrid is intermediate between its parents but has a hairy receptacle.
  c. Specimens have recently been found with *R. circinatus* in the River
     Blackadder, v.c. 81; they were first reported in 1972 from Ge.
  d. None.
  e. *R. circinatus* $2n = 16$; *R. fluitans* $2n = 16, 32$.
  f. VOLLRATH, H. and KOHLER, A. (1972). *Batrachium*-Fundorte aus
     bayerischen Naturräumen. *Ber. bayer. bot. Ges.,* **43**: 63-75.

19 × **21.** *R. fluitans* Lam. × *R. trichophyllus* Chaix
22a × **19.** *R. aquatilis* L. × *R. fluitans* Lam.
   a. *R.* × *bachii* Wirtgen. (probably **19** x **21**) (*R. pseudofluitans* auct. angl., *pro parte*, is also referable to these hybrids).
   b. It is not possible to distinguish between these two hybrids on morphological characters. All three parent species are genetically very variable and show a high degree of phenotypic plasticity. The hybrids are also genotypically and phenotypically variable. Morphologically the hybrids resemble *R. penicillatus* (Dumort.) Bab. (*R. aquatilis* subsp. *pseudofluitans* (Syme) Clapham) but are highly sterile.
   c. These hybrids are very robust and are effectively spread vegetatively. They are almost confined to flowing water and have replaced *R. fluitans* in large areas of Br. They are known in several British river systems in v.c. 9, 34, 38, 39, 50, 57, 67 and 83, and also in Au, Be, Da, Ga, Ge, He and Lu. The v.c. 57 record refers to cytologically-confirmed *R. aquatilis* x *R. fluitans*.
   d. The *R. penicillatus* group ($2n$ = 32, 48) is believed to represent amphidiploids derived from these hybrids. Under experimental conditions a sterile triploid *R. fluitans* x *R. trichophyllus* from southern Ge produced some hexaploid offspring that were indistinguishable from *R. penicillatus*. Attempts to resynthesize both hybrids artificially have not so far been successful, probably due to technical difficulties.
   e. *R. fluitans* $2n$ = 16, 32; *R. trichophyllus* $2n$ = (16), 32, (48); *R. aquatilis* $2n$ = 48; *R. fluitans* x *R. trichophyllus* ($2n$ = 24); *R. aquatilis* x *R. fluitans* $2n$ = 40.
   f. COOK, C. D. K. (1966). As above.

19 × **22b.** *R. fluitans* Lam. × *R. peltatus* Schrank (*R. aquatilis* subsp. *peltatus* (Schrank) Syme)
   a. None.
   b. This hybrid is morphologically intermediate between both parents except that the receptacle is hairy and that it develops some leaves that are morphologically intermediate between divided and entire. In summer it develops some entire leaves that resemble those of *R. peltatus*. It is very robust (usually rather larger than either parent) and is sterile.
   c. The hybrid is known from between Market Deeping and Crowland in the River Welland and some of its tributaries in v.c. 32, 53 and 55. It spreads vegetatively and where it grows it has largely replaced its parent species. It is also known in Ga and Lu.
   d. It is considered possible that some races of *R. penicillatus* (*R. aquatilis* subsp. *pseudofluitans* (Syme) Clapham) ($2n$ = 32, 48) may be amphidiploid derivatives from this hybrid.
   e. *R. fluitans* $2n$ = 16, 32; *R. peltatus* $2n$ = (16, 32), 48; hybrid $2n$ = 40.
   f. COOK, C. D. K. (1966). As above.

20 × **21.** *R. circinatus* Sibth. × *R. trichophyllus* Chaix
   a. *R.* × *glueckii* A. Félix (*Batrachium* x *cookii* Soó).

b. This hybrid is morphologically intermediate between both parents with
the exception that the mature pedicels remain erect and that the
internodes and pedicels are usually much longer than those of either
parent. It is sterile.
c. It is known in BI only from some artificial pools near Mildenhall, v.c. 26;
it also occurs in Au, Ga, Ge and Hu.
d. None.
e. *R. trichophyllus* 2n = (16), 32, (48); *R. circinatus* 2n = 16.
f. COOK, C. D. K. (1966). As above.

22a × 21. *R. aquatilis* L. × *R. trichophyllus* Chaix

a. *R.* × *lutzii* A. Félix *pro parte* (*Batrachium* × *lutzii* (A. Félix) Soó) (Much
of Félix's material is referable to either one or the other parent.)
b. The hybrid is morphologically intermediate between both of its parents
with the exception that it develops some leaves that are intermediate
between capillary and entire. It is sterile.
c. The hybrid is found where both parents grow together; although it is
perennial it has not been found to persist away from them. It is sporadic in
occurrence: v.c. 16, 20, 29, 30, 32, 55 and 57. It is also known in Au, Be,
Da, Ge, Ho, Hu, Po and Su.
d. None.
e. *R. trichophyllus* 2n = (16), 32, (48); *R. aquatilis* 2n = 48; hybrid
2n = (32), 40.
f. COOK, C. D. K. (1966). As above.
TURAŁA, K. and WOŁEK, J. (1971). A natural tetraploid hybrid of
Ranunculus subgenus *Batrachium* from the Nowy Targ Basin (Poland).
Acta biol. cracov., Bot., **14**: 154-157.

22b × 21. *R. peltatus* Schrank (*R. aquatilis* L. subsp. *peltatus* (Schrank) Syme) × *R. trichophyllus* Chaix

(= *R.* × *grovesiana* Druce, *nom. nud.*) was recorded from v.c. 17 and 34, in
both cases growing with its alleged parents and reputedly quite intermediate in appearance between them, but no specimens have been seen.
Putative hybrids have also been found in Ge. (Note added in proof: I have
recently seen convincing material from v.c. 38.)
GROVES, H. and GROVES, J. (1901). *R. peltatus* × *trichophyllus*. *Rep.
B.E.C.*, **2**: 4.
WHITE, J. W., HIERN, W. P. and WHELDON, J. A. (1916). *Ranunculus
peltatus* × *trichophyllus*. *Rep. B.E.C.*, **4**: 310.

23 × 21. *R. baudotii* Godron × *R. trichophyllus* Chaix

a. *R.* × *segretii* A. Félix (*R.* × *durandii* A. Félix, *R.* × *grovesii* Druce, *nom.
nud.*).
b. In shape and size of the flowers this hybrid is intermediate between both
parents except that the achenes are hairy. Vegetatively the hybrid

resembles *R. trichophyllus* in autumn, winter and early spring but in late spring and summer it develops intermediate leaves that are not deeply divided but also not entire. The hybrid is sterile.

c. It is found in coastal regions where both parents grow together. It is not known to persist and is sporadic in occurrence: v.c. 10, 16, 17, 28, 51 and 53. It is known in Ge and Su.

d. This hybrid has been experimentally synthesized. The synthesized hybrid resembles the natural one and is a sterile tetraploid.

e. *R. baudotii* $2n = 32$; *R. trichophyllus* $2n$ (16), 32, (48); experimental hybrid ($2n = 32$).

f. COOK, C. D. K. (1966). As above.

FÉLIX, A. (1927). *Batrachium Durandi*, in GUÉTROT, M. *Plantes hybrides de France*, 1 and 2: 24-25. Lille.

22a × 22b. *R. aquatilis* L. × *R. peltatus* Schrank (*R. aquatilis* subsp. *peltatus* (Schrank) Syme)

= *R.* × *virzionensis* A. Félix is known in Ga and Ge. There are very doubtful records for v.c. 19 and 69.

22a × 23. *R. aquatilis* L. × *R. baudotii* Godron

a. *R.* × *lambertii* A. Félix (*Batrachium* × *lambertii* (A. Félix) Soó).

b. This sterile hybrid is florally and vegetatively intermediate between both the parents with the exception that the sepals are blue, the achenes hairy, the mature pedicels uncurved and some leaves are morphologically intermediate between divided and entire.

c. The hybrid is found in coastal regions where both parents grow together in v.c. 1. It is not known to persist and is sporadic in occurrence. It is also known in Da, Ga, and Su.

d. None.

e. *R. baudotii* $2n = 32$; *R. aquatilis* $2n = 48$; hybrid ($2n = 40$).

f. ERIKSON, J. (1905). Några växtfund från Blekinge *Batrachium baudotii* × *peltatum. Bot. Notiser*, **1905**: 319-320.

SØRENSEN, T. (1955). Hybriden *Ranunculus baudotii* × *R. radians. Bot. Tidsskr.*, **52**: 113-124.

Subg. *Ficaria* (Huds.) L. Benson
(by B. M. G. Jones)

24b × 24f. *R. ficaria* L. subsp. *bulbifer* Lawalrée × *R. ficaria* L. subsp. *fertilis* (Clapham) Lawalrée

a. None. It is probable that subsp. *bulbifer* is the type subspecies (subsp. *ficaria*).
b. Several populations containing triploid plants have been found in BI but as yet their status is uncertain. Morphologically, many of them fall within the range of the diploid subsp. *fertilis* (i.e. they lack axillary bulbils) and are cytologically autopolyploid, but it is possible that some may have arisen from hybridization between diploid plants and the tetraploid subsp. *bulbifer*. Triploid plants are always completely sterile.
c. The triploids so far detected are widely scattered with one locality each in v.c. 3, 17, 28, 64, 92, H3, H9, H16 and H20. In some instances they occur alone, in others with one or both of the other cytotypes close by. They appear to be relatively frequent in Hb.
d. Fruits are formed when diploids are pollinated by tetraploids; the reciprocal cross is ineffective.
e. *R. ficaria* subsp. *fertilis* $2n = 16$ (often with 1–7 B chromosomes); *R. ficaria* subsp. *bulbifer* $2n = 32$ (with no B chromosomes); triploids $2n = 24$ (with no B chromosomes).
f. GILL, J. J. B., JONES, B. M. G., MARCHANT, C. J., McLEISH, J. and OCKENDON, D. J. (1972). The distribution of chromosome races of *Ranunculus ficaria* L. in the British Isles. *Ann. Bot.*, n.s., 36: 31-47.

53 × 54. *Berberis* L. × *Mahonia* Nutt. = × *Mahoberberis* Schneid.
(by C. A. Stace)

53/1 × 54/1. *Berberis vulgaris* L. × *Mahonia aquifolium* (Pursh) Nutt.
= × *M. neubertii* (Baum.) Schneid. was recorded in v.c. 5 in 1907, but the specimens were subsequently identified as *B. aristata* DC. The true hybrid is supposed to have arisen in a garden in Alsace, Ga, in 1850, and is now a recognized garden plant.

AHRENDT, L. W. A. (1961). *Berberis* and *Mahonia*. A taxonomic revision. *J. Linn. Soc., Bot.*, 57: 1-410.
MARSHALL, E. S. (1907). A natural *Berberis*-hybrid in England. *J. Bot., Lond.*, 45: 393-394.
MARSHALL, E. S. (1909). *Berberis aquifolium* × *vulgaris*. *Rep. B.E.C.*, 2: 358-359.

55. *Nymphaea* L.
(by Y. Heslop-Harrison)

1 × occ. *N. alba* L. × *N. occidentalis* (Ostenf.) Moss
has been recorded from northern Scotland. Small plants of *N. alba* erroneously identified by G. C. Druce as *N. candida* C. Presl in 1911 were finally raised to the rank of species by C. E. Moss in 1920 as *N. occidentalis* (Ostenf.) Moss. Their status is in dispute, however, since *N. alba* is inherently very variable, and small plants are frequently to be found in nutrient-poor waters. It is possible, however, that a cline exists in Europe, and that reports of hybrids refer to intermediate variants from this cline.

HESLOP-HARRISON, Y. (1955). *Nymphaea* L. em. Sm. (*nom. conserv.*), in Biological Flora of the British Isles. *J. Ecol.,* **43**: 719-734.

HESLOP-HARRISON, Y. (1955). British water lilies. *New Biology,* **18**: 111-120. Harmondsworth.

56. *Nuphar* Sm.
(by Y. Heslop-Harrison)

1 × 2. *N. lutea* (L.) Sm. × *N. pumila* (Timm) DC.
a. *N.* × *spennerana* Gaudin (*N.* × *intermedia* Ledeb.).
b. Putative natural hybrids are intermediate between the parents in meristic details, size and shape of the floral parts, leaf-size and -venation, and various anatomical characteristics of the peduncles and petioles. Pollen and seed fertility is poor, c 15% and 20% respectively. Backcrossing and introgression are rarely encountered in Br because the two parents are nowadays widely separated geographically. One colony of *N. pumila* in Br shows a much reduced pollen fertility, and is suspected of having suffered from introgression from *N. lutea.* Introgression has probably also occurred in several Continental stations, where populations show a considerable range of morphological variation and reduced pollen fertility.
c. The presumed hybrid occurs in England only at Chartners Lough, v.c. 67. It is assumed to have arisen here when the parents coexisted in the vicinity, and to have been perpetuated vegetatively, probably since quite early post-glacial times. The hybrid has been recorded from several localities in Scotland. Records from v.c. 72, 73, 85, 89, 97, 99 and 101 are

probably authentic; those from v.c. 48, 92, 109, H1 and H2 are probably erroneous. The two parents are now reasonably well isolated in Europe as well, both geographically and ecologically. F_1 hybrids and/or introgressants are recorded from Lappland, Ga, Ge, Po and Rs.

d. No artificial hybrids have been made with British plants, but the Continental parents have been successfully crossed reciprocally, and the F_1 hybrids brought to maturity. Pollen fertility of the two parents is rarely less than 95%; in the hybrids it is c 15%, so that genetical isolation obviously strengthens the ecological and geographical isolation.

e. Both parents and hybrid $2n = 34$.

f. FOCKE, W. O. (1881). *Op. cit.*, pp. 22-23.

HESLOP-HARRISON, Y. (1953). *Nuphar intermedia* Ledeb., a presumed relict hybrid, in Britain. *Watsonia*, 3: 7-25. FIG.

HESLOP-HARRISON, Y. (1955). *Nuphar* Sm., in Biological Flora of the British Isles. *J. Ecol.*, 43: 342-364. FIG.

PERRING, F. H. and SELL, P. D. (1968). *Op. cit.*, p. 6.

58. *Papaver* L.

(by S. Rogers)

2 x 1. *P. dubium* L. x *P. rhoeas* L.

a. (*P.* x *nicholsonii* Druce, *nom. nud.*, *P. strigosum* auct.).

b. Seedlings of this hybrid often exhibit a highly aberrant morphology including foliar distortions and chlorosis, presumably resulting from genetic imbalance. Established plants become more normal in appearance but are highly variable in foliage and floral characters. The flower buds are similar in shape to those of *P. rhoeas* but the flowers are generally smaller with petals intermediate in colour and reduced in extreme forms to a strap-like shape. The ovary is intermediate in shape but tapers to the base like *P. dubium*. The hybrid is highly sterile.

c. The hybrid nature of many reported hybrids and herbarium specimens is open to question due to widespread confusion with plants of *P. rhoeas* which have adpressed pedicel hairs or which are sterile as a result of heterozygosity for one or more chromosome segments. Wherever both species cohabit in Br hybrids may be expected, but they are nonpersistant annuals. Genuine records exist for v.c. 17 (Salmon, 1919) and at Cothill Quarry and in a beet-field near Fyfield, v.c. 22 (I. H. McNaughton *in litt.* 1972). Hybrids are also reported from Ge.

d. *P. rhoeas* is almost completely self-sterile, while *P. dubium* is partially self-fertile. Under experimental conditions about 25% of *P. rhoeas* or *P.*

dubium capsules pollinated with pollen of the other species set seed and there is no obvious reciprocal difference in behaviour. The hybrid embryos are poorly developed and give a low percentage germination. Pollen sterility appears complete and no successful backcrosses have been made to either parental species.

e. *P. rhoeas* 2*n* = 14; *P. dubium* 2*n* = (28), 42; hybrid 2*n* = 28.

f. FOCKE, W. O. (1881). *Op. cit.*, pp. 30-31.

KOOPMANS, A. (1970). Preliminary notes on crosses between *Papaver dubium* L. (2*n* = 42) and *P. rhoeas* L. (2*n* = 14). *Acta bot. neerl.*, **19**: 533-534.

McNAUGHTON, I. H. and HARPER, J. L. (1960). The comparative biology of closely related species living in the same area, 1. External breeding-barriers between *Papaver* species. *New Phytol.*, **59**: 15-26.

McNAUGHTON, I. H. and HARPER, J. L. (1960). The comparative biology of closely related species living in the same area, 2. Aberrant morphology and a virus-like syndrome in hybrids between *Papaver rhoeas* and *P. dubium*. *New Phytol.*, **59**: 27-41. FIG.

ROGERS, S. (1969). Studies on British poppies, 1. Some observations on the reproductive biology of the British species of *Papaver*. *Watsonia*, **7**: 55-63.

ROGERS, S. (1969). Studies on British poppies, 2. Some observations on hybrids between *Papaver rhoeas* L. and *P. dubium* L. *Watsonia*, **7**: 64-67. FIG.

ROGERS, S. (1969). Studies on British poppies, 3. A note on sterility in *Papaver rhoeas* L. *Watsonia*, **7**: 128-129.

ROGERS, S. (1971). Studies on British poppies, 4. Some aspects of variability in the British species of *Papaver* and their relation to breeding mechanisms and ecology. *Watsonia*, **8**: 263-276.

SALMON, C. E. (1919). *Papaver rhoeas, P. dubium* and the hybrid between them. *New Phytol.*, **18**: 111-117. FIG.

WOODRUFFE-PEACOCK, E. A. (1913). Poppy hybrids. *J. Bot., Lond.*, **51**: 48-50.

3 × 1. *P. lecoqii* Lamotte × *P. rhoeas* L.

(=? *P. rhoeas* var. *chelidonioides* O. Kuntze) has been recorded doubtfully from v.c. 53 and 54 but has never been confirmed.

REYNOLDS, B. (1912). *Papaver rhoeas* var. *chelidonioides* O. Kuntze. *J. Bot., Lond.*, **50**: 348.

WOODRUFFE-PEACOCK, E. A. (1913). As above.

1 × 6. *P. rhoeas* L. × *P. somniferum* L.

(=? *P. rupifragum* Boiss. & Reut., =? *P. trilobum* Wallr.) was said to have been reported from Ashford, v.c. 15, in 1929, but the original report was far less certain of the identity. It is recorded from Ge.

ANONYMOUS (1930). *Proc. R. hort. Soc., Scientific Committee, September 19, 1929, p. cxviii.*

DRUCE, G. C. (1931). *Papaver somniferum* x *Rhoeas* = *P. rupivagum* [sic] . *Rep. B.E.C.*, 9: 258.
FOCKE, W. O. (1881). *Op. cit.*, p. 30.

2 x 3. *P. dubium* L. x *P. lecoqii* Lamotte

has been recorded from v.c. 54. It was said to have an intermediate number of stigmatic rays and a pale yellow sap, taking a little time to attain this colour in air. Artificial hybrids, however, have white sap.
McNAUGHTON, I. H. and HARPER, J. L. (1960). The comparative biology of closely related species living in the same area, 3. The nature of barriers isolating sympatric populations of *Papaver dubium* and *P. lecoqii*. *New Phytol.*, 59: 129-137.
WOODRUFFE-PEACOCK, E. A. (1912). As above.
WOODRUFFE-PEACOCK, E. A. (1914). *Papaver dubium* x *lecoqii*. *Rep. B.E.C.*, 3: 307.

2 x 6. *P. dubium* L. x *P. somniferum* L.

= *P.* x *godronii* Rouy has been recorded from Ga.

66. *Fumaria* L.
(by M. G. Daker)*

4 x 2. *F. bastardii* Bor. x *F. capreolata* L.

is probably the parentage of *F. occidentalis* Pugsl. An artificial hybrid ($2n = c$ 56) between the two parental species ($2n = 48$ and 64 respectively), when treated with colchicine, produced branches which were morphologically similar to wild *F. occidentalis*, and, by doubling, had the same chromosome number, $2n = c$ 112.

4 x 6a. *F. bastardii* Bor. x *F. muralis* Sond. ex Koch subsp. *boraei* (Jord.) Pugsl.

a. None.

b. These two species have been claimed as the parents of sterile material thought to be of hybrid origin. The flowers (9 mm long) are described as more or less intermediate in character. Features to examine critically are sepal and fruit characteristics, and raceme- and peduncle-lengths. The existence of such hybrids requires confirmation.

* With assistance from P. M. Benoit.

c. There is an isolated record for Guernsey.

d. Using *F. bastardii* as the male parent, successful artificial hybrids have been made, and these produced approximately 6% of well-developed fruits.

e. Both parents $2n = 48$.

f. DAKER, M. G. (1964). *Cytotaxonomic studies on European Fumaria L.* Ph.D. Thesis, University of Wales.

PUGSLEY, H. W. (1919). A revision of the genera *Fumaria* and *Rupicapnos. J. Linn. Soc., Bot.*, **44**: 233-355.

6a × 8. *F. muralis* Sond. ex Koch subsp. *boraei* (Jord.) Pugsl. × *F. officinalis* L.

a. (?*F. painteri* Pugsl.).

b. The hybrid from Bryncrug was described by P. M. Benoit (*in litt.* 1973) as morphologically intermediate between the parents. The best developed racemes were 15–19-flowered. The sepals were broadly ovate and subacute as in *F. muralis* subsp. *boraei*, but smaller (3.5 x 2.5 mm), and the flowers were 10 mm. Pollen was mostly (wholly?) imperfect, while out of a possible *c* 950 fruits only nine were developed; these were also intermediate in shape. *F. painteri* was likewise intermediate between *F. muralis* subsp. *boraei* and *F. officinalis* (sepals ovate-lanceolate, laciniate or irregularly dentate, 3–3.5 x 1.5 mm; corolla 10–11 mm; fruits truncate or emarginate). However, although *F. painteri* is generally considered to have been a hybrid, it was anomalous in being fertile and there is some doubt as to its true nature.

c. A single plant of the sterile hybrid involving *F. officinalis* subsp. *officinalis* was discovered by P. M. Benoit in 1972 growing with both parents at Pont Dysynni, Bryncrug, v.c. 48. There are also records of sterile putative hybrids from v.c. 4, 9 and 15 and from Guernsey. *F. painteri* was recorded from two places in v.c. 40, but there are no recent reports of it.

d. Viable seeds were obtained in artificial crosses using *F. officinalis* subsp. *wirtgenii* (Koch) Arcang. as the male parent, but these failed to survive beyond the seedling stage.

e. *F. muralis* subsp. *boraei* $2n = 48$; *F. officinalis* subsp. *officinalis* $2n = 32$; *F. officinalis* subsp. *wirtgenii* $2n = 48$.

f. PERRING, F. H. and SELL, P. D. (1968). *Op. cit.*, p. 7.

PUGSLEY, H. W. (1912). The genus *Fumaria* L. in Britain. *J. Bot., Lond.*, **50** (Suppl.): 1-76, t. 519. FIG.

PUGSLEY, H. W. (1920). *Fumaria*, in MOSS, C. E. *The Cambridge British Flora*, **3**: 171-190. Cambridge.

7 × 8. *F. densiflora* DC. (*F. micrantha* Lag.) × *F. officinalis* L.

a. (*F.* × *salmonii* Druce, *nom. nud.*).

b. The Dunstable hybrid was described by P. M. Benoit (*in litt.* 1973) as intermediate between the parent species and highly sterile. The flowers were 7–8 mm; the sepals were broadly ovate and subacute as in *F. densiflora* but smaller, 2.5 x 1.8 mm. Only two fruits were developed on this plant. The specimen grown from seed collected near Therfield was similar but had larger sepals.

c. A single plant of the hybrid involving *F. officinalis* subsp. *officinalis* was found by P. M. Benoit in 1972 with both parents at Dunstable, v.c. 20. There are also older, somewhat doubtful records from v.c. 17, 24 and 29. Another hybrid plant appeared among *F. densiflora* grown in 1965 by P. M. Benoit from seed collected near Therfield, v.c. 20, where *F. officinalis* was growing with *F. densiflora*.

d. None.

e. *F. densiflora* 2*n* = 32; *F. officinalis* 2*n* = 32, 48.

f. PUGSLEY, H. W. (1912). As above.

SALMON, C. E. (1907). A *Fumaria* hybrid. *J. Bot., Lond.*, 45: 120.

7 × 9. *F. densiflora* DC. (*F. micrantha* Lag.) × *F. vaillantii* Lois.

was said to have been the parentage of a plant collected in Kent, *teste* Pugsley, but Pugsley later stated that all hybrids he had seen involved *F. officinalis*.

DRUCE, G. C. (1910). *Fumaria densiflora* × *Vaillantii. Rep. B.E.C.*, 2: 412.

PUGSLEY, H. W. (1912). As above.

8 × 9. *F. officinalis* L. × *F. vaillantii* Lois.

= *F.* × *albertii* Rouy & Fouc. has been recorded from Ga, Ge and Gr. In Ga hybrids involving both *F. officinalis* subsp. *officinalis* and subsp. *wirtgenii* have been claimed. Puglsey considered that subsp. *wirtgenii* of *F. officinalis* "may have originated" from the cross between the typical subspecies and *F. vaillantii*, and that a fertile plant from v.c. 20 suspected by Marshall of being this hybrid was in fact *F. officinalis* subsp. *wirtgenii*. This is apparently the sole mention of this hybrid in BI.

BOUCHARD, J. (1949). Observations sur un Fumeterie de nature hybride voisin du × *F. alberti* (Rouy et Foucaud). *Bull. Soc. fr. Éch. Pl. vasc.*, 3: 17.

DAKER, M. G. (1964). As above.

LITTLE, J. E., MARSHALL, E. S. and PUGSLEY, H. W. (1915). *F. vaillantii* Lois? *Rep. Watson B.E.C.*, 2: 482-483.

PUGSLEY, H. W. (1912). As above.

ROUY, G. and FOUCAUD, J. (1893). *Flore de France*, 1: 178-179. Tours.

8 × 10. *F. officinalis* L. × *F. parviflora* Lam.

a. None.

b. A single plant was thought to be a hybrid between these two species. Although large, with several hundred racemes, there was no fruit development. The foliage was described as *F. officinalis*-like but small, and the flowers were 6 mm long and pale pink, with sepals not exceeding 1.5 mm.

c. The single record was from Mickleham, v.c. 17.

d. None.

e. *F. parviflora* 2n = 32; *F. officinalis* 2n = 32, 48.

f. PUGSLEY, H. W. (1912). As above.

PUGSLEY, H. W. (1920). As above.

67. *Brassica* L.

(by D. J. Harberd)

1 x 3. *B. oleracea* L. x *B. rapa* L.

a. *B. napus* L. is the fertile allotetraploid; the primary hybrid is unknown in the wild. (Swede, etc.).

b. *B. napus* is most easily distinguished from its parental diploids in the vegetative state: by the glaucous leaves (a *B. oleracea* character) and the presence of setose hairs (from *B. rapa*) on the veins of the underside of the young leaf. These characters are of little value in the flowering state since the upper stem-leaves of all three species may be glaucous and glabrous. In most other characters *B. napus* lies between the parents, with overlapping ranges.

c. The allotetraploid species is commonly cultivated (swede, rape, cole, etc.) and frequently occurs as an escape.

d. The parents are extremely difficult to cross, but the hybrid has been produced experimentally. It resembles a weak *B. napus*, but is completely sterile with irregular chromosome pairing. By chromosome doubling (or by using tetraploid forms of the parental species for crossing) fertile *B. napus* can be synthesized, and this procedure features in several plant breeding programmes.

e. *B. oleracea* 2n = 18; *B. rapa* 2n = 20; artificial F_1 2n = 19; *B. napus* 2n = 38.

f. FRANDSEN, K. J. (1947). The experimental formation of *Brassica napus* L. var *oleifera* DC. and *Brassica carinata* Braun. *Dansk bot. Ark.*, 12: 1-16.

OLSSON, G. (1960). Species crosses within the genus *Brassica*, 2. Artificial *Brassica napus* L. *Hereditas*, 46: 351-386.

PERRING, F. H. and WALTERS, S. M. (1962). *Op. cit.*, p. 33.

ROSS-CRAIG, S. (1949). *Drawings of British plants*, 3: 48. London. FIG.

U, N. (1935). Genome-analysis in *Brassica* with special reference to the experimental formation of *B. napus* and peculiar mode of fertilisation. *Jap. J. Bot.*, 7: 389-452.

YARNELL, S. H. (1956). Cytogenetics of the vegetable crops, 2. Crucifers. *Bot. Rev.*, **22**: 81-166.

4 × 1. *B. nigra* (L.) Koch × *B. oleracea* L.

The fertile allotetraploid, *B. carinata* Al. Braun. ($2n$ = 34), is found in Ethiopia and neighbouring states.

2 × 3. *B. napus* L. × *B. rapa* L.

a. *B.* × *harmsiana* O. E. Schulz.
b. The hybrid is intermediate between the parents, though nearer to *B. napus*, and only attracts attention because of its sterility. Its true identity can be checked solely by cytological examination.
c. Hybrids occur sporadically in crops of swede when seed stocks have been exposed to pollination by *B. rapa*.
d. Artificial hybrids are easily made by pollinating *B. napus* stigmas with pollen of *B. rapa*. The hybrids frequently have malformations on the roots, referred to as "hybridization nodules", resembling club-root infections but without the pathogen. The hybrid has much reduced fertility though it does set some seed, and progenies have been raised for several successive generations. In the F_1 the chromosome association at meiosis forms a regular pattern of 10 bivalents and 9 univalents. Later generations are frequently more fertile and approximate in chromosome number and behaviour to either of the parents or to the hexaploid derivative "*B. napocampestris*".
e. *B. napus* $2n$ = 38; *B. rapa* $2n$ = 20; artificial hybrid $2n$ = 29.
f. DAVEY, V. McM. (1939). Hybridization in Brassicae and the occasional contamination of seed stocks. *Ann. appl. Biol.*, **26**: 634-636.
  U, N. and NAGAMATSU, T. (1933). On the difference between *Brassica campestris* L. and *B. napus* L. in regard to fertility and natural crossing. *Journ. imp. agric. Exp. Stn Nishigahara*, **2**: 113-128.
  YARNELL, S. H. (1956). As above.

4 × 3. *B. nigra* (L.) Koch × *B. rapa* L.

a. *B. juncea* (L.) Czern. & Coss. is the fertile allotetraploid; the primary hybrid is unknown in the wild. (Indian or Chinese Mustard).
b. The allotetraploid, though intermediate in its characters, is quite distinct from both parents. The leaves resemble those of *B. nigra* in shape, particularly in the absence of auricles from the stem-leaves, but the fruits are much closer to those of *B. rapa*, being long, with a long beak, round in section and spreading. It is cultivated for its seed-oil.
c. The allotetraploid is a casual in cornfields and near ports, etc. It is native to Asia, but occurs as a casual or established weed throughout Europe, particularly in the east and south-east.
d. The parents are not easily crossed, though the hybrid has been prepared and examined several times. It resembles *B. juncea* in all but its lack of

vigour and its sterility. There is very little chromosome pairing at meiosis.
e. *B. nigra* (2*n* = 16); *B. rapa* 2*n* = 20; artificial F₁ 2*n* = 18; *B. juncea* (2*n* = 36).
f. FRANDSEN, K. J. (1943). The experimental formation of *Brassica juncea* Czern. et Coss. *Dansk bot. Ark.*, **11**: 1-7.
MIZUSHIMA, U. (1950). Karyogenetic studies of species and genus hybrids in the tribe Brassiceae of Cruciferae. *Tohoku J. agric. Res.*, **1**: 1-14.
OLSSON, G. (1960). Species crosses within the genus *Brassica*, 1. Artificial *Brassica juncea* Coss. *Hereditas*, **46**: 171-223.
PRAKASH, S. (1973). Artificial synthesis of *Brassica juncea. Genetica*, **44**: 249-263.
YARNELL, S. H. (1956). As above.

jun × 3. *B. juncea* (L.) Czern. & Coss. × *B. rapa* L.
a. *B.* × *turicensis* O. E. Schulz & Thellung.
b. The hybrid closely resembles *B. juncea* but it is sterile and the upper stem-leaves are minutely auriculate.
c. The hybrid apparently occurs sporadically where the parents have flowered together, since seed of *"B. juncea"* from botanic gardens and research stations collected from open-pollinated plants frequently includes the hybrid. There appear to be no wild records for BI.
d. The parents cross readily when *B. juncea* is used as the female parent. At meiosis the chromosomes characteristically form 10 bivalents and 8 univalents.
e. *B. rapa* 2*n* = 20; *B. juncea* (2*n* = 36); hybrid (2*n* = 28).
f. MORINAGA, T. (1929). Interspecific hybridisation in *Brassica*, 2. The cytology of F₁ hybrids of *B. cernua* and various other species with 10 chromosomes. *Jap. J. Bot.*, **4**: 227-289.
SINSKAIA, E. (1927). Genetic systematic studies on cultivated *Brassica. Trudȳ prikl. Bot. Genet. Selek.*, **17**: 3-166.
YARNELL, S. H. (1956). As above.

72. *Diplotaxis* DC.
(by D. J. Harberd)

1 × 2. *D. muralis* (L.) DC. × *D. tenuifolia* (L.) DC.
= *D.* × *wirtgenii* Rouy & Fouc. is reported from Ge, Ju and Su.

2 × vim. *D. tenuifolia* (L.) DC. × *D. viminea* (L.) DC.

a. *D. muralis* (L.) DC. is the fertile allotetraploid; the primary hybrid is unknown in the wild. (Wall Rocket, Stinkweed).

b. *D. muralis* is quite distinct from both of its parental diploids. It differs from *D. tenuifolia* in having its leaves mainly radical and green (not glaucous), smaller flowers, and fruits sessile on the receptacle. From *D. viminea* it differs in its much greater size overall, perennial (though short-lived) duration, and in the anthers of the short stamens being rarely much smaller than those of the long stamens.

c. The species is thoroughly naturalized in many parts of southern Br on limestone cuttings, etc.

d. The parents can be crossed with difficulty and the hybrids resemble *D. muralis* apart from their sterility. There is an almost complete lack of chromosome pairing at meiosis in the F_1 hybrid.

e. *D. tenuifolia* $2n = 22$; *D. viminea* ($2n = 20$); *D. muralis* $2n = 42$.

f. HARBERD, D. J. and McARTHUR, E. D. (1972). The chromosome constitution of *Diplotaxis muralis* (L.) DC. *Watsonia*, 9: 131-135.

PERRING, F. H. and WALTERS, S. M. (1962). *Op. cit.*, p. 35.

ROSS-CRAIG, S. (1949). *Drawings of British plants*, 3: 57. London. FIG.

74. *Raphanus* L.

(by D. J. Harberd and Q. O. N. Kay)

2 × 1. *R. maritimus* Sm. × *R. raphanistrum* L.

a. None.

b. Artificial F_1 hybrids are intermediate between the parents in habit and in the characters of the leaves and mature fruits. They are 50–60% pollen- and seed-fertile under good conditions, but less fertile under stress when they tend to produce abortive flowers.

c. Some populations of *R. maritimus* in south-western Br, especially those growing in disturbed habitats, vary in the direction of *R. raphanistrum* and appear to have been affected by introgression from the latter. Wild, apparently F_1, hybrids have been seen with the parents in north-western Ga.

d. Both species are fairly strongly self-incompatible; they form hybrid seeds readily when crossed. However, differences in ecology and time of flowering form a partial barrier between the species; in BI the perennial *R. maritimus* typically grows in base-rich drift-line and sea-cliff habitats and flowers in June and early July, but the annual *R. raphanistrum* typically grows as a weed of light, acid, arable land and flowers from July onwards.

e. Both parents and artificial F_1 hybrids $2n = 18$.

f. None.

1 × 3. *R. raphanistrum* L. × *R. sativus* L.

a. *R.* × *micranthus* (Uechtr.) O. E. Schulz.

b. Artificial F$_1$ hybrids are intermediate between the parents in habit and in the characters of the leaves, but the fruit is usually lomentaceous and the plants lack tubers; they are vigorous and are usually 50–70% pollen- and seed-fertile. The white petal-colour of *R. sativus* is dominant to the yellow colour often found in *R. raphanistrum,* and the presence of anthocyanin venation in the petals is dominant to its absence. F$_2$ plants and backcrosses show various combinations of the parental characters, but tuber formation is rare.

c. Morphologically intermediate and probably hybrid plants are not infrequent in BI, usually occurring in sites where *R. sativus* has flowered and *R. raphanistrum* grows as a weed, or as "bolters" in crops of *R. sativus.* Hybrids have been reported from Ga, Ge and U.S.A. Populations of hybrid origin sometimes persist for several years.

d. The parents are fairly strongly self-incompatible, and form hybrid seeds readily when crossed in either direction. Gene-flow between the species is probably fairly free; a small proportion of yellow-flowered but otherwise morphologically normal plants (*c* 0.1%) has been observed to occur in large populations of fodder radish *(R. sativus),* and, conversely, the white petal-colour of many *R. raphanistrum* plants may have originated from *R. sativus.* The differences between the species are probably maintained by selection and the rarity of intermediate habitats, at least in BI.

e. Both parents and artificial hybrids $2n = 18$.

f. FROST, H. B. (1923). Heterosis and dominance of size factors in *Raphanus. Genetics, Princeton,* **8**: 116-153.

PANETSOS, C. A. and BAKER, H. G. (1968). The origin of variation in "wild" *Raphanus sativus* (Cruciferae) in California. *Genetica,* **38**: 243-274.

TROUARD-RIOLLE, Y. (1914). Recherches morphologiques et biologiques sur les radis cultivés. *Ann. Sci. agron. fr.,* **4**: 295-322, 346-550. FIG.

TROUARD-RIOLLE, Y. (1920). Les hybrides de *Raphanus. Revue gén. Bot.,* **43**: 438-447.

77. *Cakile* Mill.

(by C. A. Stace)

1b × 1a. *C. edentula* (Bigelow) Hooker (*C. maritima* subsp. *integrifolia* (Hornem.) Hyland.) × *C. maritima* Scop.

Plants from northern and western Scotland variously referred to *C.*

edentula (Allen, 1952) or to *C. maritima* subsp. *integrifolia* (Dandy, 1958) are in fact *C. maritima* subsp. *maritima* (Ball, 1964), and thus sterile intermediates between them and typical subsp. *maritima* recorded from the Isle of Harris, v.c. 110 (Harrison, 1953) are not interspecific hybrids.

ALLEN, D. E. (1952). *Cakile edentula* (Bigelow) Hooker in Britain. *Watsonia*, 2: 282-283.

BALL, P. W. (1964). A revision of *Cakile* in Europe. *Reprium nov. Spec. Regni veg.*, 69: 35-40.

DANDY, J. E. (1958). *Op. cit.*, p. 15.

HARRISON, J. W. H. (1953). The new British Sea Rocket, *Cakile edentula* (Bigelow) Hooker. *Vasculum*, 38: 30.

86. *Capsella* Medic.

(by C. A. Stace)

1 × 2. *C. bursa-pastoris* (L.) Medic. × *C. rubella* Reut.

a. *C.* × *gracilis* Gren.

b. *C. rubella* differs from the common plant in its petals (only a little longer than the sepals, not twice as long), sepals (reddish, not green), silicula (with concave, not straight, sides and a deeper apical notch) and its pedicels (ascending, not spreading). *"C. gracilis"* is generally considered to be the hybrid between them. It is intermediate in the petal, sepal and pedicel characters. The siliculae are small and often abortive (even when fully formed they may produce shrivelled seeds); they have slightly concave sides but differ from those of either parent in the shallower apical notch with protruding style. *Capsella* plants are frequently partially or wholly seedless due to male sterility, particularly towards the end of the season, and such plants have been recorded as hybrids, sometimes between various segregates of *C. bursa-pastoris*.

c. *C. rubella* and the hybrid are rarely-established aliens in Br. The latter has been recorded from v.c. 17, 55, 63 and most probably elsewhere, but the study of *Capsella* has been neglected in Br in recent years. Elsewhere it is known across Europe from Ga and It to Tu, but is nowhere common.

d. Almquist cultivated many sorts of *Capsella* and treated as distinct taxa those which bred true from seed. From a long series of hybridizations in America Shull (1929) found that there were two breeding groups (diploids and tetraploids) between which hybrids could be made but were sterile. Hybrids between taxa within each group were fertile. Among the many crosses made was *C. bursa-pastoris* × *C. rubella*, which was sterile and resembled closely specimens of *C.* x *gracilis* collected in the wild. The cross was apparently effected with ease.

e. *C. bursa-pastoris* (2n = 32), *C. rubella* (2n = 16).

f. ALMQUIST, E. (1923). Studien über *Capsella bursa-pastoris* (L.), 2. *Acta Horti Bergiani*, **7**: 41-95.

ALMQUIST, E. (1930). The stability of forms in our floras. *Rep. B.E.C.*, **9**: 199-200.

ALMQUIST, E. and DRUCE, G. C. (1921). *Bursa pastoris* Weber. *Rep. B.E.C.*, **6**: 179-207.

MOSSERAY, R. (1935). Matériaux pour une Flore de Belgique, 3. *Capsella rubella* Reut. et *Capsella Bursa-pastoris* (L.) Medic. *Bull. Soc. r. Bot. Belg.*, **17**: 180-192.

MURR, J. (1899). Beiträge zur Kenntnis der Gattung *Capsella*. *Öst. bot. Z.*, **49**: 168-172, 277-279. FIG.

ROCHER, E. (1927). *Capsella [gracilis] gracili*, in GUÉTROT, M., ed. *Plantes hybrides de France*, **1** and **2**: 55. Lille.

SHULL, G. H. (1929). Species hybridisations among old and new species of Shepherd's Purse, in DUGGAR, B. M., ed. *Proc. International Congress of Plant Sciences, Ithaca, New York, August 16–23, 1926*, **1**: 832-888. Menasha, Wisconsin.

88. *Cochlearia* L.

(by J. J. B. Gill)

2 x 1. *C. alpina* (Bab.) M. C. Wats. (*C. pyrenaica* DC. *pro parte*) x *C. officinalis.* **L.**

a. None.

b. The putative hybrid is intermediate in all morphological characters between the two parents.

c. A single possible record exists from Ben Cruachan, v.c. 98; it is also recorded from He.

d. Artificial hybrids resemble the Ben Cruachan plant. The pollen in material grown by the author varies in fertility between 15 and 75%. The hybrids set a small amount of apparently normal seed but are largely sterile. Meiosis is highly irregular with the formation of many trivalents at metaphase I.

e. *C. officinalis* 2n = 24; *C. alpina* 2n = 12; artificial hybrid 2n = 18.

f. BEEBY, W. H. (1898). *Cochlearia officinalis*, L. ♂ x *micacea*, Marsh. *Rep. B.E.C.*, **1**: 510-511.

GILL, J. J. B. (1973). Cytogenetic studies in *Cochlearia* L. (Cruciferae). The origins of *C. officinalis* L. and *C. micacea* Marshall. *Genetica*, **44**: 217-234.

ROHNER, P. (1954). Zytologische Untersuchungen an einigen schweizerischen Hemi-Oreophyten. *Mitt. naturf. Ges. Bern*, **11**: 43-107.

3 × 1. *C. micacea* E. S. Marshall (*C. pyrenaica* DC. *pro parte*) × *C. officinalis* L.

a. None.

b. The hybrid is intermediate in size between the two parents but shows the perennial habit of *C. micacea*. The leaves are intermediate in size between the two parents and are darker green than those of *C. officinalis*. The fruits approach those of *C. micacea* in shape but like those of *C. officinalis* are distinctly reticulate-veined. The plant is more or less fully fertile and sets apparently normal seed. The pollen is about 95% fertile.

c. Spontaneous hybrids were said to have arisen when the two species were grown together at Walton-on-Thames, v.c. 17.

d. The two parental species may be crossed with ease; the F_1 is as described above. It is likely that in the wild isolation between the two species is spatial and ecological; *C. micacea* is not found below about 1000 m and *C. officinalis* is essentially maritime.

e. *C. officinalis* $2n = 24$; *C. micacea* $2n = 26$; hybrid $2n = 25$.

f. BEEBY, W. H. (1898). As above.
   GILL, J. J. B. (1973). As above.

1 × 4. *C. officinalis* L. × *C. scotica* Druce

has been reported from Hb in v.c. H1, 21, 35 and 38–40. Plants larger than *C. scotica* in leaf and flower measurements and found growing with it have been identified as this hybrid. However, some doubt must be cast on the validity of *C. scotica* as a distinct species and therefore on its putative hybrids. All the plants phenotypically identifiable with *C. scotica* and collected from Scotland by the author are cytologically identical with *C. officinalis* ($2n = 24$), and the single reported count of $2n = 14$ for *C. scotica* (Gairdner, 1939) has not been confirmed. Hybrids between plants of the two phenotypes, both with $2n = 24$, are like many other interpopulational hybrids within *C. officinalis*, and are completely fertile.

GAIRDNER, A. E. (1939). In MAUDE, P. F. The Merton Catalogue. A list of the chromosome numerals of British flowering plants. *New Phytol.*, **38**: 1-31.

GILL, J. J. B. (1973). As above.

PRAEGER, R. L. (1932). Some noteworthy plants found in or reported from Ireland. *Proc. R. Ir. Acad.*, **41**, Ser. B: 95-104.

PRAEGER, R. L. (1951). *Loc. cit.*

5 × 1. *C. danica* L. × *C. officinalis* L.

a. None.

b. Plants intermediate between these two species in growth-habit and flowering time have been recognized as hybrids. These plants usually have the pale mauve flower-buds and "ivy-shaped" stem leaves of *C. danica*, but the flowers are somewhat larger. All combinations of the two parental phenotypes occur. It is possible that many plants identified as *C. scotica* but having a chromosome number of $2n = 24 + B$ chromosomes are in fact segregants from these hybrids. The positive identification of these as such has not, however, been achieved. Putative hybrids set a large amount of seed.

c. Hybrids have been reported from v.c. 48, 49, H21, H27, H35, H38, H39 and H40, but may be more common than this if some of the *C. officinalis* specimens containing B chromosomes are segregants from such hybrids.

d. The two species can be crossed artificially with ease. The F_1 hybrid shows about 70% pollen fertility but has highly irregular meiosis with the formation of many univalents and multivalents at metaphase I. Backcrosses can, however, be made to either parent. Isolation in nature is probably both spatial and ecological.

e. *C. officinalis* $2n = 24$; *C. danica* $2n = 42$; F_1 hybrid $2n = 33$.

f. BENOIT, P. M. (1971). *In litt.*

CRANE, M. B. and GAIRDNER, A. E. (1923). Species-crosses in *Cochlearia*, with a preliminary account of their cytology. *J. Genet.,* **13**: 187-220.

PRAEGER, R. L. (1932). As above.

PRAEGER, R. L. (1951). *Loc. cit.*

SAUNTE, L. H. (1955). Cytogenetical studies in the *Cochlearia officinalis* complex. *Hereditas,* **41**: 499-515.

6 × 1. *C. anglica* L. × *C. officinalis* L.

a. *C. × hollandica* Henrard (*C. anglica* var. *hortii* Syme, *C. × briggsii* Druce).

b. Hybrids are intermediate between the parents in most characters. The basal leaves vary from cuneate to cordate at the base, but are oblong rather than orbicular in overall shape. In this character they approach *C. anglica*. The flowering shoots are less robust than in *C. officinalis* and the fruiting head less dense. The flowers and fruits are intermediate in size between the two parents. The F_1 hybrid is as fertile as either species but the fertility drops in backcrosses. A large amount of introgression occurs.

c. Hybrids occur more or less wherever the two species meet, but particularly on salt-marshes. *C. anglica* is commonest on the wet, muddy areas of the marsh, with *C. officinalis* being most common on the drier areas. The intermediate habitats are frequently occupied by extensive hybrid swarms. Hybrids are also known in Da and Su.

d. The two species can be artificially crossed with ease. The F_1 hybrid shows little or no reduction in fertility compared with the parents. Backcrosses can also be easily made but backcross progeny show reduced fertility. The isolating mechanism in nature is thus ecological coupled with low fertility in backcrosses.

e. *C. officinalis* 2n = 24; *C. anglica* 2n = 48; F₁ hybrid 2n = 36.
f. CRANE, M. B. and GAIRDNER, A. E. (1923). As above.
  LÖVKVIST, B. (1963). Något om de skånska *Cochlearia*-arterna. *Bot. Notiser*, 116: 326-330.
  LÖVKVIST, B. (1963). Taxonomic problems in aneuploid complexes. *Regnum Vegetabile*, 27: 51-57.
  PRAEGER, R. L. (1932). As above.
  PRAEGER, R. L. (1951). *Loc. cit.*
  SAUNTE, L. H. (1955). As above.
  VAN DER MAAREL, E. (1962a). Aantekeningen over *Cochlearia officinalis* L. s.l., 1. Herbariumonderzoek van *Cochlearia officinalis* L. en *C. anglica* L. *Gorteria*, 1: 75-79.
  VAN DER MAAREL, E. (1962b). Aantekeningen over *Cochlearia officinalis* L. s.l., 2. Populatieonderzoek aan *Cochlearia officinalis* L. en *C. anglica* L. *Gorteria*, 1: 86-90.

5 × 4. *C. danica* L. × *C. scotica* Druce
  has been reported from Dooagh on Achill Island, v.c. H27, and from Giant's Causeway, v.c. H39; plants so identified had the habit of *C. scotica* but the larger, pale mauve flowers of *C. danica*. As noted above, under *C. officinalis* × *C. scotica*, the identity of *C. scotica* and therefore of its hybrids remains uncertain.
  PRAEGER, R. L. (1939). A further contribution to the Flora of Ireland. *Proc. R. Ir. Acad.*, 45, Ser. B: 231-254.
  PRAEGER, R. L. (1951). *Loc. cit.*

95. *Erophila* DC.
(by C. A. Stace)

2 × 1. *E. spathulata* Láng × *E. verna* (L.) Chevall.
  = *E.* × *fauconnetii* O. E. Schultz was reported from Br by Clapham, but no specimens have been seen and nothing further is known of its occurrence here. It has been recorded from Be, Ga, He and Hu; it is said to have very small, abortive, seedless fruits.
  CLAPHAM, A. R. (1962). *Erophila*, in CLAPHAM, A. R., TUTIN, T. G. and WARBURG, E. F. *Op. cit.*, pp. 159-160.
  SCHULZ, O. E. (1927). *Erophila*, in ENGLER, A., ed. *Das Pflanzenreich*, 89 (IV, 105): 343-372. Leipzig.

3 × 1. *E. praecox* (Stev.) DC. × *E. verna* (L.) Chevall.
= *E.* × *chavinii* Favrat is recorded from Ga, Ge and He.

3 × 2. *E. praecox* (Stev.) DC. × *E. spathulata* Láng
= *E.* × *vincentii* O. E. Schulz is recorded from Ga.

97. *Cardamine* L.
(by B. M. G. Jones)

1pa × 1pr. *C. palustris* (Wimmer & Grab.) Peterm. × *C. pratensis* L. *sensu stricto.*

At least two polyplotypes of *C. pratensis* L. agg. are present in BI: *C. pratensis sensu stricto, 2n = 30*, and *C. palustris, 2n = 56*. These frequently hybridize on the Continent where they occur together and plants of hybrid origin, with reported chromosome numbers between 38 and 48, persist. These hybrid plants at times may comprise whole populations, owing to extensive vegetative reproduction. The extent to which these cytotypes are morphologically recognizable in BI is not yet fully worked out.

2 × 1. *C. amara* L. × *C. pratensis* L.

= *C.* × *ambigua* O. E. Schulz (*C.* × *mixta* Druce) has been recorded from v.c. 23. Experimentation has revealed a well-developed incompatibility between these species; the little hybrid seed produced from many pollinations contained inviable embryos. The reputed wild hybrids involving *C. pratensis sensu stricto* which have been reported both in BI and on the Continent therefore require verification, but hybrids involving *C. rivularis* Schur, a non-British, diploid cytodeme of the *C. pratensis* agg. have been reliably reported from He.

URBANSKA-WORYTKIEWICZ, K. and LANDOLT, E. (1972). Natürliche Bastarde zwischen *Cardamine amara* L. und *C. rivularis* Schur aus den Schweizer Alpen. *Ber. geobot. ForschInst. Rübel*, **41**: 88-101. FIG.

4 × 1. *C. flexuosa* With. × *C. pratensis* L.

a. *C.* × *haussknechtiana* O. E. Schulz (*C. hayneana auct. angl., non* Welw.).

b. Hybrids are intermediate between the parent species and can be readily recognized by their stems being usually branched below and by their short and narrow (10 x 4 mm), lilac petals. Their pollen is sterile and fruit does not develop. They are perennial and reproduce vegetatively by adventitious buds and roots on the margins of leaflets shed in autumn. For this reason, hybrid clones are occasionally locally abundant.

c. Hybrids are rare and scattered through BI, usually in open meadows. Authentic specimens have been seen from v.c. 5, 9, 14, 17, 22, 35 and 99. They are also recorded from Da, Ge and He.

d. The hybrid has been synthesized, using *C. pratensis* with low chromosome numbers, by the bud-pollination technique. This circumvents the incompatibility which has developed by anthesis.

e. *C. pratensis* $2n = 30$, 56 (also 38, 48, 58, 72 and many more outside BI); *C. flexuosa* $2n = 32$; wild hybrid $2n = c$ 30.

f. ALLEN, D. E. (1971). *In litt.*
   LÖVKVIST, B. (1956). The *Cardamine pratensis* complex—outlines of its cytogenetics and taxonomy. *Symb. bot. upsal.,* 14: 1-131.
   SCHULZ, O. E. (1903). Monographie der Gattung *Cardamine. Bot. Jb.,* 32: 280-623.

5 × 1. *C. hirsuta* L. × *C. pratensis* L.

was listed by Druce with the addition "=?*Hayneana* Welw.", so that the plants to which he was referring were probably *C. pratensis* x *C. flexuosa.* DRUCE, G. C. (1928). *Op. cit.,* p. 7.

7 × 1. *C. latifolia* Vahl (*C. raphanifolia* Pourr.) × *C. pratensis* L.

= *C.* x *undulata* Larambergue has been found in Ga.

2 × 4. *C. amara* L. × *C. flexuosa* With.

= *C.* x *keckii* Kerner has been reported on several occasions in Au, Ga, Ge and He.

2 × 5. *C. amara* L. × *C. hirsuta* L.

has been reported from Au.

4 × 5. *C. flexuosa* With. × *C. hirsuta* L.

a. *C.* x *zahlbrucknerana* O. E. Schulz.

b. The hybrid resembles *C. flexuosa* rather than being strictly intermediate. Its pollen is highly sterile and fruits are only occasionally developed. They contain only one or two seeds.

c. In BI specimens have only been seen from v.c. 47; they are also recorded from Au, Ga and He.

d. The hybrid is readily synthesized when *C. flexuosa* is the female parent. Meiosis in the synthetic F_1 is irregular, with univalents at metaphase I and chromosome bridges at anaphase I. The seeds occasionally produced by F_1 plants give an F_2 which is variable in morphology, fertility and vigour. The more viable plants are close to *C. flexuosa* in appearance and more or less fertile.

e. *C. flexuosa* 2*n* = 32; *C. hirsuta* 2*n* = 16; synthetic hybrid (2*n* = 24).

f. BENOIT, P. M. (1957). Synthesised *Cardamine flexuosa* x *hirsuta. Proc. B.S.B.I.*, 3: 86.

ELLIS, R. P. (1969). *Taxonomy of the Arabideae (Cruciferae) with special reference to the genus Cardamine.* Ph.D. Thesis, University of London.

ELLIS, R. P. and JONES, B. M. G. (1969). The origin of *Cardamine flexuosa* With.: evidence from morphology and geographical distribution. *Watsonia.* 7: 92-103 (Corrected title).

98. *Barbarea* R.Br.
(by C. A. Stace)

2 x 1. *B. stricta* Andrz. x *B. vulgaris* R.Br.

= *B.* x *schulzeana* Hausskn. has been recorded from Cz and Ge.

3 x 1. *B. intermedia* Bor. x *B. vulgaris* R.Br.

= *B.* x *gradlii* J. Murr has been recorded from Au and Ge.

100. *Arabis* L.
(by B. M. G. Jones)

2 x 4. *A. alpina* L. x *A. hirsuta* (L.) Scop.

= *A.* x *palezieuxii* Beauv. has been recorded from Ga and He.

5 x 4. *A. brownii* Jord. x *A. hirsuta* (L.) Scop.

a. None.

b. Plants more or less intermediate between these species and occurring in western Hb, where the endemic *A. brownii* grows on calcareous dunes, have been interpreted as hybrids. However, populations of plants with intermediate morphology (e.g. *A. hirsuta* var. *glabra* (L.) Hartman and *A. hirsuta* subsp. *ciliata* var. *hispida* Syme) are not confined either to Hb or to dunes and they commonly occur in the absence of both of the putative parents. The situation in nature indicates microspecies formation between populations of a single, variable, inbreeding species in the classic Jordanian manner, rather than hybridization between distinct species.

c. Intermediates are found in scattered localities in BI.

d. Both "species" are normally self-pollinated. They can be hybridized experimentally without difficulty to give progeny which are usually highly fertile. Isolation in nature is primarily topographical, reinforced by facultative autogamy.

e. Both "species" and intermediates $2n = 32$.

f. JONES, B. M. G. (1963). *Experimental taxonomy of the genus Arabis.* Ph.D. Thesis, University of Leicester.

102. *Rorippa* Scop.

(by C. A. Stace)

2 × 1. *R. microphylla* (Boenn.) Hyland. (*Nasturtium microphyllum* (Boenn.) Reichb.) × *R. nasturtium-aquaticum* (L.) Hayek (*Nasturtium officinale* R.Br.)

a. *R.* × *sterilis* Airy Shaw.

b. The parents can be distinguished with certainty by the chromosome number, stomatal index (c 10–12% in the tetraploid and c 15–18% in the diploid), the fruits (longer and thinner and borne on a longer and thinner pedicel in the tetraploid), and the seeds (borne in one row and with c 100 cell outlines on each face in the tetraploid; in two rows and with c 25 cell outlines on each face in the diploid). The pollen grains and fruits of the hybrid are largely abortive. There is in each siliqua usually about one good seed, which is intermediate between that of the two species in anatomy. The stomatal index of the hybrid is c 13–15%. Despite the apparent fertility of the triploid hybrid no evidence of backcrossing in the wild has been found. In any case such plants would be difficult to identify without cytological study.

c. The triploid hybrid is scattered throughout virtually the whole of BI, from CI to Orkney (Perring and Sell, 1968). Cultivated watercresses are either diploid *R. nasturtium-aquaticum* or triploid hybrids, both reproducing vigorously by vegetative means, and thus the hybrid is very frequently found in the absence of one or both parents and in all sorts of wet habitats. Records exist for Au, Be, Ga and Ge.

d. The triploid has been synthesized by the cross female tetraploid × male diploid, but not by the reciprocal combination. Artificial autotetraploid *R.*

nasturtium-aquaticum retains the characters of the diploid plant, except that it is much less fertile, and crosses between this and wild tetraploid *R. microphylla* have produced tetraploid hybrids which, however, are no more fertile than the triploids. Meiosis in the triploid hybrid shows 16 bivalents and 16 univalents, indicating that *R. microphylla* is an allotetraploid of which *R. nasturtium-aquaticum* is one parent. The other parent is unknown; species of *Cardamine*, in which one-rowed seeds are usual, have been suggested. Plants raised from selfed triploids possess chromosome numbers from near diploid to near tetraploid.

e. *R. nasturtium-aquaticum* $2n = 32$; *R. microphylla* $2n = 64$; hybrid $2n = 48$.

f. HOWARD, H. W. and LYON, A. G. (1950). The identification and distribution of the British watercress species. *Watsonia*, 1: 228-233.

HOWARD, H. W. and LYON, A. G. (1951). Distribution of the British watercress species. *Watsonia*, 2: 91-92.

HOWARD, H. W. and LYON, A. G. (1952). *Nasturtium* R.Br., in Biological Flora of the British Isles. *J. Ecol.*, 40: 228-245.

HOWARD, H. W. and MANTON, I. (1946). Autopolyploid and allopolyploid watercress with the description of a new species. *Ann. Bot.*, n.s. 10: 1-13. FIG.

MANTON, I. (1935). The cytological history of watercress (*Nasturtium officinale* R.Br.). *Z. indukt. Abstamm.- u. VererbLehre*, 69: 132-157. FIG.

PERRING, F. H. and SELL, P. D. (1968). *Op. cit.*, p. 9.

SHAW, H. K. A. (1951). A binary name for the hybrid watercress. *Watsonia*, 2: 73-75.

4×3. *R. palustris* (L.) Bess. (*R. islandica auct. mult., non* Borbás) $\times$ *R. sylvestris* (L.) Bess.

a. (*R.* x *barbaraeoides* (Tausch) Čelak. *pro parte*).

b. The hybrid is perennial with ascending stems, as in *R. sylvestris*, but the petal-length and pod-size and -shape are intermediate. The fertility is variable, but backcrossing has not been observed.

c. There are some records from Br (v.c. 88 and 89) but none has been confirmed. On the Continent the only material considered authentic is Swedish, but the hybrid now seems extinct in Su. Unconfirmed records exist for many countries, including Au, Da, Fe, Ge, Ho, No, Po, Rm and Rs but at least some are errors. Jonsell (1968) says that in BI "some specimens exist that may be *R. palustris* x *sylvestris* but they do not permit any safe decision".

d. None of the crosses between tetraploid, pentaploid or hexaploid *R. sylvestris* (female) and *R. palustris* produced F_1 plants.

e. *R. sylvestris* $2n = 32$, (40), 48; *R. palustris* $2n = 32$.

f. JONSELL, B. (1968). Studies in the north-west European species of *Rorippa s. str. Symb. bot. upsal.*, 19(2): 1-222.

NYÁRÁDY, E. I. (1955). *Rorippa*, in SĂVULESCU, T. *Flora Republicii populare Române*, 3: 215-250. Bucharest. FIG.

5 × 3. *R. amphibia* (L.) Bess. × *R. sylvestris* (L.) Bess.

a. *R.* × *anceps* (Wahlenb.) Reichenb. (*R.* × *barbaraeoides* (Tausch) Čelak. *pro parte*).

b. This hybrid possesses the habit of *R. sylvestris* but the leaves (particularly the upper stem-leaves) are intermediate in serration between those of the two parents, the styles at fruiting are as long as those of *R. amphibia* (1.0–2.5 mm; 0.5–1.0 mm in *R. sylvestris*) and the fruiting pedicels also are usually deflexed as in *R. amphibia.* The pods are intermediate in shape and size (3–10 × 1.2–2.5 mm) but are often abortive. The degree of fertility of both parents is very variable, since they are highly self-incompatible, and thus sterility is not a reliable indication of hybridity. The fertility of hybrids is also variable, partly because they may be tetraploid or pentaploid. The fertile ones backcross to both parents in mixed populations, and specimens of *R. sylvestris* showing introgression from *R. amphibia* have been found in several parts of BI. The opposite situation appears much less common.

c. Records have been confirmed from v.c. 6, 13, 17, 23, 29, 31, 33–37, 40, 61/65, H6, H9, H25, H29 and H33. It is recorded also in Al, Au, Be, Cz, Da, Fe, Ga, Ge, Gr, He, Ho, Hu, It, Ju, Lu, No, Po, Rm, Rs and Su. Hybrid plants are found in similar wet places to the parents, but one or both of the latter may be absent from the vicinity due to the sexual and vegetative reproductive capacities of the hybrid. This is the most frequent hybrid between yellow-flowered Rorippas.

d. Crosses involving the parents of various ploidy levels have given differing results. F_1 plants were raised from reciprocal crosses between tetraploid parents, and from the cross hexaploid *sylvestris* × male tetraploid *amphibia*, but other combinations either gave inviable seed or no seed was formed. Backcrosses between tetraploid wild hybrids and tetraploid *amphibia* on one hand and tetraploid, pentaploid and hexaploid *sylvestris* on the other all gave viable offspring. The above crosses all involved Scandinavian material; the degree of success was always lower than in intraspecific crosses. Genetic barriers are greater between different ploidy levels than within one level. Artificially-formed hybrids resemble some of the natural ones, but none of them could be backcrossed to either parent.

e. *R. amphibia* $2n = 16, 32$; *R. sylvestris* $2n = 32, (40), 48$; hybrid $2n = 32, 40$.

f. JAVŮRKOVÁ-KRATOCHVÍLOVA, V. and TOMŠOVIC, P. (1972). Chromosome study of the genus *Rorippa* Scop. em. Reichenb. in Czechoslovakia. *Preslia*, 44: 140-156.

JONSELL, B. (1968). As above. FIG.

LAWALRÉE, A. (1971). L'hybride *Rorippa amphibia* × *R. sylvestris* en Belgique. *Gorteria*, 5: 170-171.

TOMŠOVIC, P. (1969). Nejdůležitější výsledky revise československých rukví (*Rorippa* Scop. em. Reichenb.). *Preslia*, 41: 21-38. FIG.

6 × 3. *R. austriaca* (Crantz) Bess. × *R. sylvestris* (L.) Bess.
= *R.* × *armoracioides* (Tausch) Fuss occurs in Au, Bu, Cz, Da, Fe, Ga, Ge, Ju, No, Po, Rm, Rs and Su.

5 × 4. *R. amphibia* (L.) Bess. × *R. palustris* (L.) Bess. (*R. islandica auct. mult., non* Borbás)

a. *R.* × *erythrocaulis* Borbás.

b. Hybrid plants are perennial and vigorous like *R. amphibia*, but the upper stem-leaves are intermediate and have well-developed auricles as in *R. palustris*. The petals are intermediate in size between those of the parents (2.5–3.5 mm, about 1.5 times as long as sepals). The fruiting pedicels are usually deflexed and frequently bear abortive pods, as in *R. amphibia*, but the shape of fertile pods is intermediate. The smaller petals and larger auricles are the best characters distinguishing this hybrid from *R. amphibia* × *R. sylvestris*. The triploid hybrid is completely sterile but the tetraploid hybrid is fertile, although backcrossing has not been detected.

c. The triploid hybrid is known from the Thames near Putney (v.c. 17 and 21) along with both parents. The tetraploid hybrid was found in a roadside ditch not far from the Avon at Tewkesbury (v.c. 34) with neither parent close by. Neither of the above has been confirmed from any other locality in Br or elsewhere, although there are other reports from Br and numerous records from Scandinavia and eastern and central Europe. The two parents are often found close together without hybrids.

d. Due to the small, autogamous flowers of *R. palustris* it has been used in crosses only as the male parent. Hybrids were obtained with tetraploid but not with diploid *R. amphibia*. They resembled the natural hybrids very closely and were largely sterile.

e. *R. amphibia* $2n = 16, 32$; *R. palustris* $2n = 32$; hybrid $2n = 24, 32$.

f. BRITTON, C. E. (1909). A *Radicula* hybrid. *J. Bot., Lond.*, **47**: 430.

   BRITTON, C. E. (1910). *Radicula amphibia* × *R. palustris*. *Rep. B.E.C.*, **2**: 436-437.

   HOWARD, H. W. (1947). Chromosome numbers of British species of the genus *Rorippa* Scop. (Part of the Genus *Nasturtium* R.Br.). *Nature, Lond.*, **159**: 66.

   JONSELL, B. (1968). As above. FIG.

4 × 6. *R. austriaca* (Crantz) Bess. × *R. palustris* (L.) Bess. (*R. islandica auct. mult., non* Borbás)
= *R.* × *neogradensis* Borbás is recorded from Rm and Rs.

5 × 6. *R. amphibia* (L.) Bess. × *R. austriaca* (Crantz) Bess.
= *R.* × *hungarica* Borbás is reported from Au, Cz, Ge, Hu, Rm and Su, but erroneously so in the last and perhaps always so.

113. *Viola* L.
(by D. H. Valentine)

The violets (Section *Viola*) set seed from both open and cleistogamous flowers. All the hybrids in this Section are more or less infertile, and in summer it is often easy to see the withered cleistogamous flowers which have failed to set seed, or an occasional small, few-seeded capsule. The infertility is an important criterion of hybridity. The pansies (Section *Melanium*) do not have cleistogamous flowers.

General References
BECKER, W. (1910). *Violae Europaeae.* Dresden.
FOTHERGILL, P. G. (1944). The somatic cytology and taxonomy of our British species of the genus *Viola. New Phytol.,* **43**: 23-35.
GAMS, H. (1965). *Viola,* in HEGI, G., ed. *Illustrierte Flora von Mittel-Europa,* 2nd ed., **5**: 586-656. Munich.
GREGORY, E. S. (1912). *British Violets.* Cambridge.
HARVEY, M. J. (1962). *The cytotaxonomy of some rostrate violets.* Ph.D. Thesis, University of Durham.
SCHMIDT, A. (1961). Zytotaxonomische Untersuchungen an europäischen *Viola*-Arten der Sektion *Nomimium. Öst. bot. Z.,* **108**: 20-88.
VALENTINE, D. H. (1962). Variation and evolution in the genus *Viola. Preslia,* **34**: 190-206.

2 × 1. *V. hirta* L. × *V. odorata* L.
  a. *V.* × *permixta* Jord. (*V. sepincola* auct., *non* Jord).
  b. Hybrids are partially fertile and intermediate in most characters; stolons are usually present but shorter than in *V. odorata* and the flowers are slightly fragrant. The hybrids may vary because of variability in the flower-colour and indumentum of the *V. odorata* parent, but also because backcrossing probably occurs, producing plants closer to *V. hirta* (often distinguished as *V. permixta* Jord.) and closer to *V. odorata* (often distinguished as *V. sepincola* auct.).
  c. Hybrids typically occur at the ecotone between the more shady habitats of *V. odorata* and the more sunny habitats of *V. hirta,* usually on base-rich soil. They are recorded from v.c. 2–4, 6–9, 11–14, 16, 17, 20, 22–24, 26, 29, 30, 32–38, 41, 46, 55, 56, 62, 64 and 65, and from Au, Cz, Ge and He.
  d. Reciprocal crosses give a good yield of seed and vigorous hybrids which are partially fertile, producing 2–3 seeds in a cleistogamous capsule. From the hybrid *V. hirta* female × *V. odorata* male an F_2 generation was raised which showed segregation for flower-colour, stolon-length and fertility; some of the plants were weak. The vigorous F_1 hybrids could be matched with hybrids from natural habitats.
  e. Both parents 2*n* = 20; hybrid (2*n* = 20, 22, 24).

f. DIZERBO, A. H. (1967–68). Observations sur *Viola odorata* L., *Viola hirta* L. et leur hybride *Viola permixta* Jord. dans le Massif Armoricain. *Bull. Soc. sci. Bretagne*, **42**: 113-124, 215-223; **43**: 71-79, 225-236. FIG.

GREGORY, E. S. (1912). As above. FIG.

SCHÖFER, G. (1954). Untersuchungen über die Polymorphie einheimischer Veilchen. *Planta*, **43**: 537-565.

SNOW, R. and CHATTAWAY, M. M. (1930). An artificial cross between *Viola hirta* and *V. odorata. J. Bot., Lond.*, **68**: 115-116.

WALTERS, S. M. (1946). Observations on varieties of *Viola odorata* L. *Rep. B.E.C.*, **12**: 834-839.

6 × 2 × 1. *V. canina* L. × *V. hirta* L. × *V. odorata* L.

(as *V. canina* x *V. sepincola*) was recorded in 1917 from v.c. 12 by Miss Palmer. I have not seen the specimen, but the existence of such a hybrid is very unlikely.

1 × 5. *V. odorata* L. × *V. reichenbachiana* Jord.

= *V. x olimpia* Beggiato has been recorded from Hu but must be regarded as very doubtful.

2 × 3. *V. hirta* L. × *V. rupestris* Schmidt

= *V. x wilczekiana* Beauv. is recorded from Au, but must be considered very doubtful.

2 × 5. *V. hirta* L. × *V. reichenbachiana* Jord.

was recorded in 1908 from v.c. 36 by A. Ley, but was thought by E. S. Gregory to be *V. riviniana* forma *villosa* N.W. & M. The occurrence of this hybrid is very unlikely.

4 × 3. *V. riviniana* Reichb. × *V. rupestris* Schmidt.

a. *V. x burnatii* Gremli (*V. x leesii* Druce, *nom. nud.*).

b. The hybrid is intermediate between the parents in leaf and flower characters and also in indumentum; it has the fine pubescence of peduncle and petiole characteristic of British populations of *V. rupestris*, though the pubescence is even finer. No seeds have ever been observed, from either open or cleistogamous flowers, but hybrids are capable of vegetative reproduction by root soboles; this character is derived from the *V. riviniana* parent.

c. Two populations only are known in Britain, both from Widdy Bank Fell, Upper Teesdale, v.c. 66; one is now covered by the water of the new reservoir, though many of the plants are in cultivation at Durham and Manchester. The hybrid occurs in grazed grassland on sugar-limestone and may form clones up to a metre in diameter. The hybrid is not recorded from the neighbourhood of the two other British populations of *V.*

rupestris (on Long Fell and Arnside Knott). In the literature up to *c* 1930, it was recorded by G. C. Druce and E. S. Gregory for v.c. 20, 22, 23 and 52, but these were errors, based on the mistaken impression that subglabrous varieties of *V. rupestris* were widespread in lowland England (see also *V. canina* x *V. rupestris*). The hybrid is recorded from Au, Cz, Ga, Ge, Ho, Hu, Rs and Su.

  d. Artificial hybrids have been made from the cross *V. riviniana* female x *V. rupestris* male. They resemble natural hybrids in appearance and, in cultivation for 2 years, no seeds were produced. Meiosis in an artificial hybrid with 2*n* = 32 showed three bivalents.

  e. *V. rupestris* 2*n* = 20; *V. riviniana* 2*n* = 37–46, (35–47); hybrid (v.c. 66) 2*n* = 33, 34.

  f. DUNBAR, S. C. (1971). *The cytotaxonomy of Viola riviniana.* M.Sc. Thesis, University of Manchester.

  GADELLA, T. W. J. (1963). A cytotaxonomic study of *Viola* in the Netherlands. *Acta bot. neerl.,* 12: 17–39.

  VALENTINE, D. H. (1949). Vegetative and cytological variation in *Viola riviniana* Rchb., in WILMOTT, A. J., ed. *British flowering plants and modern systematic methods.* Arbroath. FIG.

5 x 3. *V. reichenbachiana* Jord. x *V. rupestris* Schmidt

= *V.* x *iselensis* W. Becker is recorded from Au.

6a x 3. *V. canina* L. subsp. *canina* x *V. rupestris* Schmidt

= *V.* x *braunii* Borb. (*V.* x *heslopii* Druce, *nom. nud.*) was recorded by G. C. Druce from v.c. 105 and 112 and by E. S. Gregory from v.c. 65. The specimens are probably *V. canina* x *V. riviniana.* Druce and Gregory identified many dwarf plants of *V. riviniana* as glabrous or subglabrous forms of *V. rupestris,* but inspection of numerous herbarium specimens both by P. M. Hall and by myself leaves no doubt that these identifications are all erroneous. *V.* x *braunii* is recorded from Ge, He and Su.

6b x 3. *V. canina* L. subsp. *montana* (L.) Hartm. x *V. rupestris* Schmidt

= *V.* x *villaquensis* Benz is recorded from Au, Ga, He and Su.

8 x 3. *V. persicifolia* Schreber (*V. stagnina* Kit.) x *V. rupestris* Schmidt

= *V.* x *vilnaensis* W. Becker is recorded from Rs.

5 x 4. *V. reichenbachiana* Jord. x *V. riviniana* Reichb.

  a. (*V. intermedia* Reichb., *non* Krock).

  b. The hybrid is intermediate between the parents; it more closely resembles the *V. riviniana* parent in the breadth of the petals and length of the calcycine appendages, but it has a rather dark spur, like the *V.*

reichenbachiana parent. Unless compared with the parents in the field it is often difficult to identify, but its high infertility is a useful character. Few good pollen grains are produced and very little good seed is set from either open or cleistogamous flowers, though quantitative observations on hybrid fertility in the field have not been made. Introgression has not been studied in British populations, but Schöfer (1954) obtained some evidence of backcrossing in populations near München, Ge.

c. Hybrids are widely distributed in England, usually occurring wherever the parents meet, though not in large numbers. *V. reichenbachiana* occurs mainly in woodland on calcareous soils, and, though it flowers 2–3 weeks earlier than *V. riviniana*, the flowering times overlap. Hybrids are recorded from v.c. 3–5, 8, 9, 13, 17, 24, 27, 29, 33, 34, 36, 37, 41, 57, H8 and H39, and from Au, Cz, Ga, Ge, He, It and Su.

d. Good yields of seed and good germination are obtained in reciprocal experimental crosses. The hybrids are vigorous and long-lived. Some seed can be obtained both by backcrossing and, more easily, by automatic selfing from the cleistogamous flowers. Observations on artificial hybrids in an experimental garden have shown that the fertility of the hybrids is about 1% of that of the parents; at meiosis they showed ten bivalents.

e. *V. reichenbachiana* $2n = 20$; *V. riviniana* $2n = 37–46$; artificial hybrids $2n = 30–34$; natural hybrids from Ge ($2n = 30, 40$). F_2 plants of one family had $2n = 18, 20, 21$ and of another family $2n = 41 \pm 1$.

f. DUNBAR, S. C. (1971). As above.
GADELLA, T. W. J. (1963). As above.
SCHÖFER, G. (1954). As above.
VALENTINE, D. H. (1949). As above.
VALENTINE, D. H. (1950). The experimental taxonomy of two species of *Viola. New Phytol.*, **49**: 193-212. FIG.

6b × 5 × 4. *V. canina* L. subsp. *montana* (L.) Hartm. × *V. reichenbachiana* Jord. × *V. riviniana* Reichb.

= *V.* × *mixta* Kerner is recorded from Hu.

6a × 4. *V. canina* L. subsp. *canina* × *V. riviniana* Reichb.

a. (*V.* × *berkleyi* Druce, *nom. nud.*).

b. The hybrid is intermediate between the parents; using morphological characters, and the fact that it is sterile, it is fairly easy to recognize in the field. Quantitative estimates on hybrid fertility in the field have not been made. There is no evidence of backcrossing or introgression.

c. Subsp. *canina* occurs often on dry soils (e.g. dunes), sometimes in moist meadows, but nearly always in well-illuminated habitats. It thus meets *V. riviniana* where this species occurs outside woodland, often on heaths and dunes where the soil is not too acid. Hybrids are recorded from v.c. 1, 3, 4, 6–9, 14, 16, 17, 19, 20, 24, 26, 29, 32–37, 41, 46, 55–57, 59, 103, H2, H8 and H15, but are nowhere common, and from Au, Bu, Cz, Ga, Ge, He, Hu, It, Rs and Su.

d. Experimental crosses gave a good yield of viable seed with *V. canina* as seed-parent. The hybrids are vigorous and highly infertile. When cultivated in an experimental garden seed-yields from cleistogamous flowers in a 2-week period of September were two good seeds from 20 plants, and neither seed germinated. Meiosis in artificial hybrids showed ten bivalents.

e. *V. canina* $2n = 40$; *V. riviniana* $2n = 37-46$; artificial hybrids $2n = 43$. In Da Clausen (1931) recorded *V. canina* ($2n = 43$) and *V. canina* x *V. riviniana* ($2n = 47$).

f. CLAUSEN, J. (1931). *Viola canina* L., a cytologically irregular species. *Hereditas*, **15**: 67-88.

DUNBAR, S. C. (1971). As above.

GADELLA, T. W. J. (1963). As above.

GREGORY, E. S. (1912). As above. FIG.

MOORE, D. M. and HARVEY, M. J. (1961). Cytogenetic relationships of *Viola lactea* Sm. and other West European arosulate violets. *New Phytol.*, **60**: 85-95.

SCHÖFER, G. (1954). As above.

VALENTINE, D. H. (1949). As above.

6 x 7 x 4. *V. canina* L. x *V. lactea* Sm. x *V. riviniana* Reichb.

= *V.* x *lambertii* Léveillé has been recorded from Kynance Downs, v.c. 1, but no conclusive *in situ* studies have been made.

7 x 4. *V. lactea* Sm. x *V. riviniana* Reichb.

a. (*V.* x *curnowii* Druce, *nom. nud.*).

b. The hybrid is intermediate in leaf-shape and in the shape and colour of the petals between the parents; it often forms large, floriferous clumps which may extend vegetatively (by means of soboles) to cover areas of several m^2. The plants are highly sterile (six kept 2 years in a greenhouse produced only a single seed) but population studies indicate that some introgression occurs, mainly in the direction of *V. lactea*.

c. *V. riviniana* occurs throughout the range of *V. lactea* (mainly southern England). The two species show a wide ecological overlap on heaths and downs where the pH is between 5 and 6, the humus content of the soil fairly high and the exchangeable calcium low. Hybrids have been recorded from v.c. 1-5, 8, 9, 14, 17, 34, 41, 52 and H3, and from Ga and Lu.

d. Reciprocal crosses gave a good yield of seed, but viable hybrids were obtained only from the cross with *V. lactea* as seed parent, and germination was low (7%). The hybrids were vigorous and sterile, showing ten bivalents at meiosis.

e. *V. lactea* $2n = 58$; *V. riviniana* $2n = 37-46$; hybrid $2n = 49 + 1$ supern.

f. DUNBAR, S. C. (1971). As above.

GADELLA, T. W. J. (1963). As above.

GREGORY, E. S. (1912). As above. FIG.

MOORE, D. M. (1959). Population studies on *Viola lactea* Sm. and its wild hybrids. *Evolution, Lancaster, Pa.*, **13**: 318-332.

MOORE, D. M. and HARVEY, M. J. (1961). As above.
VALENTINE, D. H. (1949). As above.

8 × 4. *V. persicifolia* Schreber (*V. stagnina* Kit.) × *V. riviniana* Reichb.
= *V.* × *najadum* Wein is recorded from Ge.

6 × 5. *V. canina* L. × *V. reichenbachiana* Jord.
a. *V.* × *borussica* W. Becker.
b. The putative hybrids are presumably intermediate between the parents, but I have not seen the specimens.
c. They were recorded from v.c. 24 by G. C. Druce, from v.c. 34 by H. J. Riddelsdell, and from v.c. 37 by W. J. Rendall. In view of the ecological differences one would not expect these species to meet frequently but, if they did, hybridization would be possible (see below), and hybrids are recorded from Cz, Ga, Ge and Hu. Artificial hybrids resemble closely *V. canina* × *V. riviniana* and natural hybrids could only be identified with certainty in the field, or by a chromosome count.
d. Experimental crosses gave a good yield of viable seed with both subsp. *canina* and subsp. *montana* (L.) Hartm. as female parent. The hybrids, which were vigorous and intermediate in appearance between the parents, were infertile, but less so than the *V. canina* × *V. riviniana* hybrids. In the experimental garden, during 2 weeks in September, 25 hybrid plants produced 20 seeds, one of which germinated. At meiosis there were about three bivalents.
e. *V. canina* (both subspp.) $2n = 40$; *V. reichenbachiana* $2n = 20$; artificial hybrids $2n = 30$.
f. None.

7 × 5. *V. lactea* Sm. × *V. reichenbachiana* Jord.
(= *V.* × *fouilladei* W. Becker, *nom. nud.*.) has been recorded from Ga.

8 × 5. *V. persicifolia* Schreber (*V. stagnina* Kit.) × *V. reichenbachiana* Jord.
has been recorded from Ge.

6a × 6b. *V. canina* L. subsp. *canina* × *V. canina* subsp. *montana* (L.) Hartm. (*V. montana* L.)
(= *V.* × *fryeri* Druce, *nom. nud.*) has been recorded from Woodwalton Fen, v.c. 31, but these two taxa, which are interfertile and have the same chromosome number, are no longer considered distinct species. They both behave similarly when hybridized with other species.

6a × 7. *V. canina* L. subsp. *canina* × *V. lactea* Sm.
a. *V.* × *militaris* Savouré.

b. Hybrids are intermediate between the parents in diagnostic characters of leaf-shape, flower-colour and petal-breadth, but are vigorous, with longer internodes than the parents, and very floriferous. Fertility is estimated at 10% of that of the parents, but there is no evidence of backcrossing or introgression.

c. Though both occur on soils ranging in pH from 5.0 to 6.2, *V. canina* subsp. *canina* is found on soils with much lower humus content and much higher exchangeable calcium than *V. lactea.* Also, *V. lactea* is restricted geographically to southern England and scattered localities in Wales and Scotland. These isolating factors are effective, and the hybrid is rare; it occurs in rather open, heathy habitats, e.g. with *Calluna* and *Ulex gallii* or *U. minor.* It is recorded from v.c. 1–4, 8–12, 14, 16, 17, 22, 24, 34, 41, 52, 62, H3 and H16, but according to Moore (1959) some of these records may refer to *V. lactea* x *V. riviniana.* It is also recorded from Ga.

d. Reciprocal experimental crosses, using *V. canina* subsp. *canina,* gave a good yield of seed; germination was higher from the cross with *V. canina* as female parent. The hybrids were vigorous. The fertility of the plants, estimated by the production of viable seed from cleistogamous flowers in autumn, is about 5% of that of parents. At meiosis the artificial hybrid showed about 20 bivalents.

e. *V. canina* $2n = 40$; *V. lactea* $2n = 58$; artificial hybrid $2n = 49$.

f. GREGORY, E. S. (1912). As above. FIG.

MOORE, D. M. (1958). *Viola lactea* Sm., in Biological Flora of the British Isles. *J. Ecol.,* 46: 527-535.

MOORE, D. M. (1959). As above.

MOORE, D. M. and HARVEY, M. J. (1961). As above.

6 x 8. *V. canina* L. x *V. persicifolia* Schreber (*V. stagnina* Kit.)

a. *V.* x *ritschliana* W. Becker (*V.* x *gregoriae* Druce, *nom. nud., V.* x *stagninoides* Druce, *nom. nud.*).

b. This hybrid is intermediate between the parents in leaf characters, but the flowers resemble more the *V. canina* parent and it is soboliferous like the *V. persicifolia* parent; it is very vigorous, sometimes taller than either parent and forming large plants. It is quite sterile, so far as is known.

c. Hybrids are known from v.c. 23, 27, 29 (extinct), 31, H13, H26 and H33. The English localities are all in fenland on peaty soil which is flooded in winter and probably of high base status; in v.c. 23 and H26 *V. persicifolia* occupies the wetter and *V. canina* the drier habitats, with the hybrids between. In v.c. 31 *V. canina* subsp. *montana* is the parent of the hybrid; elsewhere it is probably *V. canina* subsp. *canina.* Druce (1928) recorded *V. canina* subsp. *canina* x *V. canina* subsp. *montana* x *V. persicifolia* from v.c. 31. The hybrid with subsp. *canina* is known from Da, Ge, Rs and Su and with subsp. *montana* from He and Rs.

d. An experimental cross with *V. persicifolia* as seed-parent and subsp. *canina* as male gave good seed and good germination; the hybrids were highly infertile, with about ten bivalents at meiosis. A reciprocal cross also gave viable hybrids.

e. *V. canina* $2n = 40$; *V. persicifolia* $2n = 20$; hybrid $2n = 30$.

f. DRUCE, G. C. (1928). *Op. cit.*, p. 13.

MOORE, D. M. and HARVEY, M. J. (1961). As above.

WOODELL, S. R. J. (1965). *Viola stagnina* in Oxfordshire. *Proc. B.S.B.I.*, 6: 32-36.

9b × 9a. *V. palustris* subsp. *juressi* (Link ex K. Wein) Coutinho × *Viola palustris* L. subsp. *palustris*

was recorded from v.c. 1, 10–12 and 41 at a time when subsp. *juressi* was though to be *V. epipsila* Ledeb., a species of Continental Europe. Subsp. *palustris* and subsp. *juressi* differ mainly in indumentum, shape of leaf and length of spur, and it is likely that intermediates occur in a number of places. The two subspecies do not differ in habitat, but subsp. *juressi* is much the more local and is found mainly in southern England and Wales. Its chromosome number is unknown.

GREGORY, E. S. (1912). As above.

11 × 12. *V. lutea* Huds. × *V. tricolor* L.

a. None.

b. In the natural hybrid population described by Fothergill (1938), about half the plants were intermediate and about half resembled *V. lutea* more or less closely; some showed hybrid vigour and many were fertile to some extent and produced viable seeds by open pollination. The variability in phenotype was matched by a variability in chromosome number, and possibly the hybrids were both intercrossing and backcrossing to the parents. *V.* × *wittrockiana*, the garden pansy, is thought to have arisen about 1830 by hybridization between *V. tricolor*, *V. lutea*, and perhaps other non-British species. It is commonly found as a garden escape.

c. Hybrids are known only from v.c. 57 and 67 (banks of the Tyne), and possibly also from v.c. 59 (Lancashire coast dunes) and 90. *V. tricolor* subsp. *tricolor* is mainly lowland, *V. lutea* montane; seeds of the latter may have travelled down the Tyne and produced the hybrid population in v.c. 67 at an altitude of 100 m. Elsewhere, the hybrid is recorded from Cz and Ga.

d. Clausen crossed *V. tricolor* "*hortensis*" with *V. lutea* reciprocally, and obtained a good yield of viable seeds; the hybrids on selfing had low seed-set and low germination (10%), but a viable F_2 was obtained. Meiosis in hybrids from v.c. 67 was usually irregular, e.g. 1–3 quadrivalents, 2 trivalents, 15–19 bivalents and 5 univalents.

e. *V. tricolor* $2n = 26$; *V. lutea* $2n = 48$; hybrid plants (v.c. 67) $2n = 26$, c 42, c 46, 47, 48, 50, 51, c 52, 53, 54.

f. CLAUSEN, J. (1931). Cytogenetic and taxonomic investigations on *Melanium* violets. *Hereditas*, 15: 219-308.

FOTHERGILL, P. G. (1938). Studies in *Viola*, 1. The cytology of a naturally occurring population of hybrids between *Viola tricolor* L. and *V. lutea* Huds. *Genetica*, **20**: 159-185.

13 × 12. *V. arvensis* Murr. × *V. tricolor* L.

a. (?*V. contempta* Jord.).

b. Intermediates between the parents in floral characters, including corolla-colour, shape of stylar flap and pollen-type, are found in wild populations. Some have $2n = 30$, are partially fertile and are presumed to be F_1 hybrids; others have $2n = 34$, are fully fertile and are thought to be true-breeding variants of *V. arvensis* (Pettet, 1964). There is no evidence of introgression.

c. The F_1 hybrid is rare; it is recorded from v.c. 7–9, 11, 21, 27, 29, 46, 57, 66, 85, 105, H33 and H38, often as a single plant in a population of *V. arvensis*. *V. tricolor* occurs mainly on acidic and *V. arvensis* mainly on basic soils and the species rarely meet. Hybrids are recorded also from Da, where Clausen (1951) stated that introgression occurs, and Ge.

d. Reciprocal hybridization gives a good yield of seed and of viable hybrids. These have about 70% good pollen and set good seed on selfing to yield a variable F_2. The F_2s can be selfed and bred for several generations and more or less pure lines established. The F_1 hybrids usually show 13–14 bivalents at meiosis.

e. *V. arvensis* $2n = 34$; *V. tricolor* $2n = 26$; hybrid $2n = 30$. F_2 plants usually have $2n = 28$, and in further breeding the number tends to approach $2n = 26$, but Clausen bred lines with ($2n = 28$, 32 and 42).

f. CLAUSEN, J. (1931). As above.

CLAUSEN, J. (1951). *Stages in the evolution of plant species*. Ithaca, New York.

DRABBLE, E. (1909). The British pansies. *J. Bot., Lond.*, **47** (Suppl. 2): 1-32.

PETTET, A. (1958). Delimitation of *Viola tricolor* and *V. arvensis*. *Proc. B.S.B.I.*, **3**: 97-98.

PETTET, A. (1964). Studies on British pansies, 2. The status of some intermediates between *Viola tricolor* L. and *V. arvensis* Murr. *Watsonia*, **6**: 51-69.

13 × 12 × wit. *V. arvensis* Murr. × *V. tricolor* L. × *V.* × *wittrockiana* Gams (*V. hortensis* auct.)

was recorded (as *V. contempta* Jord. × garden pansy) from v.c. 20 in 1931. It is presumably distinguishable from *V. contempta* on the basis of the same characters as given under *V. tricolor* × *V.* × *wittrockiana*.

DRABBLE, E. and LITTLE, J. E. (1931). *Viola contempta* Jord. *Rep Watson B.E.C.*, **4**: 59-60.

12 × wit. *V. tricolor* L. × *V.* × *wittrockiana* Gams (*V. hortensis* auct.)

occurs sporadically in a few places where *V. tricolor* occurs as a weed;

there are records from v.c. 17 and 57. It differs from *V. tricolor* in its larger flowers and in the stipules, which are very broad at the base and have short lateral lobes. Since *V. tricolor* and *V. lutea* have both entered into the percentage of *V.* x *wittrockiana* this hybrid may be considered a backcross.

DRABBLE, E. (1909). As above.

FOCKE, W. O. (1881). *Op. cit.*, pp. 48-49.

13 x **wit.** *V. arvensis* Murr. x *V.* x *wittrockiana* Gams (*V. hortensis* auct.)

has similarly been recorded from v.c. 57 and 58 and is distinguishable from *V. arvensis* on the basis of the same characters as given under *V. tricolor* x *V.* x *wittrockiana.*

DRABBLE, E. (1909). As above.

114. *Polygala* L.

(by I. C. Trueman)

3 x **1.** *P. calcarea* F. W. Schultz x *P. vulgaris* L.

a. None.

b. The hybrids are like *P. vulgaris* in general appearance, but are intermediate in habit, degree of lateral wing vein-anastomosis and length of corolla-tube. Most hybrids are highly sterile but similar characteristics to the above are found in fertile and partially fertile plants of the same population.

c. Hybrids appear to occur rarely throughout the range of *P. calcarea* in close proximity to the two species in calcareous grassland. The hybrids are found both in the denser swards occupied by *P. vulgaris* and the more open areas favoured by *P. calcarea*. There are confirmed records from near Sevenoaks, v.c. 16, near Dorking, v.c. 17, and near Wallingford, v.c. 23.

d. At meiosis the sterile hybrids generally show 23 or 24 bivalents and 3 or 5 univalents.

e. *P. vulgaris* $2n = 68$; *P. calcarea* $2n = 34$; more or less fertile hybrids $2n = 34$ or 68; sterile hybrids $2n = 51$.

f. YEO, P. F. (1952). A possible hybrid between *Polygala vulgaris* L. and *P. calcarea* F. Schultz. *Year Book B.S.B.I.*, **1952:** 56.

4/2 x **1.** *P. amarella* Crantz (*P. austriaca* Crantz) x *P. vulgaris* L.

a. (*P.* x *amara-vulgaris* Brögg.).

b. Putative hybrids are vigorous, variable, bitter-tasting plants, usually with enlarged, spathulate, lower leaves and the corolla-tube shorter than the free petals. They often resemble *P. alpestris* in that they have enlarged, crowded leaves beneath the inflorescence. Seeds are set but the pollen fertility is reduced.

c. Such plants have been found on a sloping area of almost bare chalk in association with *P. vulgaris* and *P. amarella* near Wye, v.c. 15. Hybrids between these two species are also recorded from Cz.

d. None.

e. *P. vulgaris* 2n = 68; *P. amarella* 2n = 34.

f. ASCHERSON, P. and GRAEBNER, P. (1916). *Polygala*, in *Synopsis der Mitteleuropäischen Flora*, 7: 387. Leipzig.

115. *Hypericum* L.

(by N. K. B. Robson)

1 × 3. *H. androsaemum* L. × *H. hircinum* L.

a. *H.* × *inodorum* Mill. (*H. elatum* Ait.). (Tall Tutsan.)

b. Circumstantial evidence suggests that *H.* × *inodorum* is the result of the above cross. It is intermediate between the suspected parents in characters of flower and fruit, while in other characters it resembles one or other of them. Its variability in the size of all parts parallels that of *H. hircinum*, but the odour usually present in that species is absent in most plants of *H.* × *inodorum*. The cultivar 'Elstead' may have resulted from backcrossing with *H. androsaemum* and, if it has such an origin, this implies that *H.* × *inodorum* is at least partially fertile.

c. None of the records of *H.* × *inodorum* from BI (Perring and Walters, 1962) suggests that hybridization has occurred in natural habitats. The cross is thought to have taken place in gardens and the hybrid to have escaped subsequently. Records of *H.* × *inodorum* from areas of Co, Ga, It and Hs where the natural areas of the suspected parents overlap may be the result of natural hybridization; but its apparent occurrence in natural habitats in Madeira (outside the areas of both suspected parents) remains to be explained. It is found as an escape in Ga and He.

d. No records of artificial hybrids are known but one garden specimen alleged to be of this parentage agreed with *H.* × *inodorum* in appearance.

e. Both parents and hybrid (2n = 40).

f. DUPONT, P. and DUPONT, S. (1958). Un hybride d'*Androsaemum*. *Monde des Plantes*, 53: 3.

PERRING, F. H. and WALTERS, S. M. (1962). *Op. cit.*, p. 57.

ROBSON, N. K. B. (1967). Materials for a Flora of Turkey, 11. Notes on Turkish species of *Hypericum. Notes R.bot. Gdn, Edinb.*, **27**: 185-204.

SCHNEIDNER, F. (1966). *Hypericum. Meded. Inst. Vered. TuinbGewass.*, **252**: 18-22, 37-39. Partly reprinted in *Dendroflora*, **2**: 18-22 (1967).

6 × 5. *H. maculatum* Crantz × *H. perforatum* L.

a. *H.* × *desetangsii* Lamotte.

b. Hybrids occur in two nothomorphs, corresponding to the two subspecies of *H. maculatum*. Only nm. *desetangsii* (*H. maculatum* subsp. *obtusiusculum* (Tourlet) Hayek × *H. perforatum*) has been recorded in BI. It is intermediate between the parents in the F_1, the most reliable characters being the sepals of intermediate width with the apex both erose-denticulate and apiculate. The hybrid is at least partly fertile, and backcrossing produces a complete series of intermediates between the parents.

c. Records are from scattered localities (28 vice-counties) throughout Br south of the Highland Line, mainly in south-eastern and northern England and southern Scotland, but not from Hb. They are doubtless mostly records of F_1 hybrids, as the backcrosses are less easily recognized. The hybrid tends to occupy habitats intermediate in wetness between those of the parents which, however, have overlapping habitat requirements. *H.* × *desetangsii* nm. *desetangsii* sometimes occurs in the absence of one or both parents, presumably owing partly to a capacity for limited vegetative propagation and partly to its limited fertility. It also occurs in Au, Be, Ga, He, Hu and Ju.

d. Tetraploid *H. perforatum* has been crossed with diploid *H. maculatum* subsp. *maculatum* (= *H.* × *desetangsii* nm. *carinthiacum* (Fröhlich) N. Robson) to give triploid hybrids if *H. perforatum* is the pollen-parent but usually pentaploid hybrids if it is the seed-parent (ovules 97% apomictic and tetraploid) (Noack, 1939, using non-British material). The former hybrids are intermediate morphologically, the latter indistinguishable from *H. perforatum*; both are apparently sterile. *H. maculatum* subsp. *maculatum*, when treated with colchicine, has given rise to tetraploid plants identical with subsp. *obtusiusculum* (Robson, 1957). Pugsley crossed subsp. *obtusiusculum* with *H. perforatum* to give plants resembling *H.* × *desetangsii* nm. *desetangsii*.

e. *H. perforatum* 2*n* = 32, (48 from botanic garden material); *H. maculatum* subsp. *maculatum* 2*n* = 16; *H. maculatum* subsp. *obtusiusculum* 2*n* = 32; *H.* × *desetangsii* nm. *desetangsii* 2*n* = 32 (40, 48 from botanic garden material).

f. BUTCHER, R. W. and STRUDWICK, F. E. (1930). *Further illustrations of British plants*, p. 96. Ashford. FIG.

FRÖHLICH, A. (1913). Über *Hypericum maculatum* Cr. × *perforatum* L. und *H. desetangsii* Lamotte. *Öst. bot. Z.*, **63**: 13-19.

FRÖHLICH, A. (1915). Über zwei der Steiermark eigentümliche Formen aus dem Verwandschaftkreis des *Hypericum maculatum* Cr. *Mitt. naturw. Ver. Steierm.*, **51**: 216-245.

KAASA, J. (1966). Characters which separate *Hypericum maculatum* Cr. and *H. perforatum* L. *Blyttia*, **1966**: 247-250.

NOACK, K. S. (1939). Über *Hypericum* Kreuzungen, 6. Fortpflanzungsverhältnisse und Bastarde von *H. perforatum* L. *Z. indukt. Abstamm.- u. VererbLehre*, **76**: 569-601.

PUGSLEY, H. W. (1940). On *Hypericum quadrangulum* L. *J. Bot., Lond.*, **78**: 25-36.

ROBSON, N. K. B. (1957). *Hypericum maculatum* Crantz. *Proc. B.S.B.I.*, **2**: 237-238.

SALMON, C. E. (1913). *Hypericum desetangsii* Lamotte in Britain. *J. Bot., Lond.*, **51**: 317-319. FIG.

5 × 8. *H. perforatum* L. × *H. tetrapterum* Fr.

= *H.* × *medium* Peterm. (*H.* × *dubioides* Druce, *nom. nud.*) is recorded from Au and Ge. There are several British records before about 1940, but all of them result from the nomenclatural confusion which surrounded *H.* × *desetangsii* and in fact refer to the latter hybrid.

9 × 5. *H. humifusum* L. × *H. perforatum* L.

= *H.* × *assurgens* Peterm. has been doubtfully recorded from Ge.

6 × 8. *H. maculatum* Crantz × *H. tetrapterum* Fr.

= *H.* × *laschii* Fröhlich is recorded from Au, Cz, Ge, He, Po and Su.

9 × 10. *H. humifusum* L. × *H. linarifolium* Vahl

a. (*H.* × *caesariense* Druce, *nom. nud.*).

b. Suspected hybrids have been recorded from CI. The supposed parents are part of a closely related group of three species which form apparent intermediates (? hybrids) where any two of them are sympatric. Decumbent variants of both British species that occur in CI and used to grow at Cape Cornwall, v.c. 1, have been wrongly recorded as hybrids, but true hybrids may occur in CI.

c. Possible hybrids are known only in Jersey, Guernsey and Alderney. Decumbent plants of unknown status also grow on exposed cliffs near the sea in Ga.

d. A suspected hybrid (morphologically intermediate) arose spontaneously at the Royal Botanic Garden, Edinburgh, in 1956, but attempts to make the cross artificially failed.

e. Both parents $2n = 16$.

f. DRUCE, G. C. (1929). Notes on the second edition of the "British Plant List". *Rep. B.E.C.*, **8**: 867-877.

PUGSLEY, H. W. (1915). British forms of *Hypericum humifusum* and *H. linarifolium. J. Bot., Lond.*, **53**: 162-170.

118. *Helianthemum* Mill.

(by M. C. F. Proctor)

2 × 1. *H. apenninum* (L.) Mill. × *H. nummularium* (L.) Mill. (*H. chamaecistus* Mill.)

a. *H.* × *sulphureum* Willd.

b. This hybrid is intermediate between the parents in most diagnostic characters, including the pale yellow flower-colour. It is usually very highly, perhaps fully, fertile, but in some cases fertility is relatively low.

c. It occurs with the parent species at Purn Hill, Bleadon, v.c. 6, where it grows in thin limestone turf around the margins of the rocky areas occupied by *H. apenninum*. The hybrid occurred for a number of years at Fishcombe Point, Brixham, v.c. 3, following the introduction of *H. apenninum*, and persisted after that species had died out again. Hybrids have been recorded from various localities where the parents meet in southern and western Europe.

d. Hybrids are readily produced in cultivation. In Ga the wild hybrid has been found to possess pollen varying from 0 to 97% fertility, and at meiosis to have from three to ten bivalents.

e. Both parents $2n = 20$; hybrid ($2n = 20$).

f. BOUHARMONT, J. (1968). Observations sur la cytologie et la fertilité chez *Helianthemum* × *sulphureum* Willd. (Cistaceae). *Bull. Jard. bot. nat. Belg.*, **38**: 415-420.

GROSSER, W. (1903). *Helianthemum*, in ENGLER, A., ed. *Das Pflanzenreich*, **14** (**IV, 193**): 61-123. Leipzig.

PROCTOR, M. C. F. (1956). *Helianthemum apenninum* (L.) Mill., in Biological Flora of the British Isles. *J. Ecol.*, **44**: 688-692.

3 × 1. *H. canum* (L.) Baumg. × *H. nummularium* (L.) Mill. (*H. chamaecistus* Mill.)

= *H.* × *bickhamii* E. S. Marshall was reported from Great Orme's Head, v.c. 49, by Marshall (1913). The specimens seen are small, rather hairy examples of *H. nummularium*; they are fertile.

MARSHALL, E. S. (1913). A new hybrid rockrose. *J. Bot., Lond.*, **51**: 182-183.

PROCTOR, M. C. F. (1956). *Helianthemum canum* (L.) Baumg., in Biological Flora of the British Isles. *J. Ecol.*, **44**: 677-682.

123. *Silene* L.
(by S. M. Walters)

2 × 1. *S. maritima* With. (*S. vulgaris* subsp. *maritima* (With.) A. & D. Löve) × *S. vulgaris* (Moench) Garcke

a. (*S.* × *intermedia* Druce, *nom. nud.*).

b. Recognition of natural hybrid populations in Br is difficult because of the absence of precise characters which can be said to separate unambiguously the two taxa. The ecological separation of the normally maritime *S. maritima* from the inland, often ruderal, *S. vulgaris* seems to be very effective, and typical *S. maritima*, with its dwarf habit and large, often single flowers, contrasts strongly with *S. vulgaris* with its straggly habit and lax, cymose inflorescences. Such characters are, however, ecotypic in nature, and inland mountain populations in Britain and elsewhere share some of the "*maritima*" characters.

c. Marsden-Jones and Turrill said that hybrids were "relatively rare". They gave records for v.c. 1-3, 6, 9, 11, 28, 46 and 49 and for Scandinavia, but this hybrid also occurs in northern Br.

d. Very extensive crossing experiments were carried out by Marsden-Jones and Turrill (1957). F_1 plants of artificial crosses between British *S. maritima* and *S. vulgaris* of various origins showed no significant sterility; the two species in fact behaved as classical ecotypes with no genetic barriers to gene exchange.

e. Both parents and hybrid $2n = 24$.

f. MARSDEN-JONES, E. F. and TURRILL, W. B. (1957). *The Bladder Campions*. London. FIG.

SAVIDGE, J. P. (1969). Campion enquiry. *Proc. B.S.B.I.*, **7**: 557-559.

6a × 6q. *S. anglica* L. × *S. quinquevulnera* L.

has been reported from Jersey, but these two taxa are now known to be completely interfertile minor variants of *S. gallica* L., with which they also hybridize.

MELVILL, M. C. (1880). *Silene eu-gallica* in Jersey. *J. Bot., Lond.*, **18**: 146.

11 × 10. *S. italica* (L.) Pers. × *S. nutans* L. = *S.* × *pseudonutans* Panč.
has been reported from Rm, but requires careful checking.

14 × 13. *S. alba* (Mill.) E. H. L. Krause × *S. dioica* (L.) Clairv.

a. (*Lychnis* × *intermedia* (Schur) Druce).
b. The F_1 hybrid is intermediate in most differential characters, the most obvious of which is the pink petal-colour. It is quite highly fertile, showing up to 30% bad pollen. Detailed criteria for diagnosis of hybridity are given by Baker (1947). In English counties where *S. dioica* is very local (mainly v.c. 29 and 31) occasional pink-flowered plants within populations of *S. alba* are plausibly of hybrid origin through occasional long-distance cross-pollination. White- or pale-flowered variants of *S. dioica* occur, however, right outside the range of *S. alba* (as in Shetland), and do not necessarily indicate hybridization.
c. Hybrid swarms of plants ranging in flower-colour from reddish-purple, through pink, to white are quite frequent in areas of lowland England where *S. alba* is a common arable and ruderal weed, especially where recent disturbance (such as hedgerow removal or road widening) has brought the two parental species close together (Perring and Sell, 1968). In Scotland, Wales and Hb they are much more sparsely scattered. There is evidence to support the view that *S. alba* came into Br with early agriculture (probably Neolithic), and that it is still spreading in northern and western Br where previously only *S. dioica* occurred. In such areas new hybrid populations arise. The hybrid is also widespread on the Continent.
d. Artificial hybrids can be made easily, and the main natural barrier to hybridization in England is the ecological separation of the two species. Artificial F_2s have variable percentages of bad pollen, but are usually quite fertile.
e. *S. alba* and hybrid $2n = 24$; *S. dioica* $2n = 24$ (24, 48).
f. BAKER, H. G. (1947). *Melandrium,* in Biological Flora of the British Isles. *J. Ecol.,* **35:** 271-292.
   COMPTON, R. H. (1920). *Melandryum,* in MOSS, C. E. *The Cambridge British Flora,* 3: 70-74. Cambridge.
   FOCKE, W. O. (1881). *Op. cit.,* pp. 65-69.
   GAGNEPAIN, F. (1929). *Melandryum,* in GUÉTROT, M., ed. *Plantes hybrides de France,* 3 and 4: 102-103. Paris, FIG.
   PERRING, F. H. and SELL, P. D. (1968). *Op. cit.,* p. 11.

13 × pre. *S. dioica* (L.) Clairv. × *S. preslii* (Sekera) ined. (*Lychnis preslii* Sekera)

(= *Lychnis* × *troweriae* Druce, *nom. nud.*) was reported to have arisen spontaneously in gardens where the two parents were cultivated, but *L. preslii* is now considered only a glabrous variant of *S. dioica* (*Melandrium rubrum* var. *glaberrimum* Rohrbach).
DRUCE, G. C. (1924). *Lychnis dioica* × *Preslii. Rep. B.E.C.,* 7: 377.

124 × 123. *Lychnis* L. × *Silene* L.
= × *Lychnisilene* Ciferri & Giacomini
(by C. A. Stace)

124/3 × 123/13. *Lychnis flos-cuculi* L. x *Silene dioica* (L.) Clairv.
was recorded between Birtley and Fatfield, v.c. 66, by J. W. H. Harrison in
1920. Although such a combination seems unlikely, and no specimens
have been seen, Gärtner claimed last century to have performed this cross
successfully using female *S. dioica,* and reciprocal crosses have been made
more recently.
FOCKE, W. O. (1881). *Op. cit.*, pp. 69-70.
GRAHAM, G. G., SAYERS, C. D. and GAMAN, J. H. (1972). *A check-list
of the vascular plants of County Durham*, p. 11. Durham.

123 × 124V. *Silene* L. × *Viscaria* Bernh.
(*Lychnis* L. *pro parte*) = × *Vislene* Dost.
(by C. A. Stace)

**123/10 x 124V/2. *Silene nutans* L. x *Viscaria vulgaris* Bernh.
(*Lychnis viscaria* L.)**
= x *V. hybrida* Dost. is recorded from Cz, but needs careful checking.

124V. *Viscaria* Bernh. (*Lychnis* L. *pro parte*)
(by C. A. Stace)

**1 x 2. *V. alpina* (L.) G. Don (*L. alpina* L.) x *V. vulgaris* Bernh. (*L.
viscaria* L.)**
is recorded from Fe and Su.

127. *Dianthus* L.
(by C. A. Stace)

1 × 3. *D. armeria* L. × *D. carthusianorum* L.
= *D.* × *javorkae* Kárp. is recorded from Ge and Rm.

1 × 8. *D. armeria* L. × *D. deltoides* L.
= *D.* × *hellwigii* Borbás is known in much of central Europe from Be to the Ukraine, Rs, and is said to be the commonest natural hybrid in the genus.

3 × 8. *D. carthusianorum* L. × *D. deltoides* L.
= *D.* × *dufftii* Hausskn. is recorded from Au, Cz and Ge.

8 × 7. *D. deltoides* L. × *D. gratianopolitanus* Vill.
is recorded from Ga.

The following hybrids, and probably others, are grown in gardens and may escape. Several have been reported in the wild in Europe, but apparently none in BI: *D. armeria* × *D. barbatus* L., *D. barbatus* × *D. carthusianorum*, *D. barbatus* × *D. plumarius* L., *D. barbatus* × *D. caryophyllus* L., *D. barbatus* × *D. deltoides*, *D. carthusianorum* × *D. plumarius*, *D. caryophyllus* × *D. plumarius* and *D. gratianopolitanus* × *D. plumarius*.

131. *Cerastium* L.
(by J. K. Morton)

2 × 3. *C. arvense* L. × *C. tomentosum* L.
= *C.* × *maureri* M. Schulze has been recorded from Ge and He, where it has arisen in Botanic Gardens, and it has recently been found in Canada in derelict gardens, on rubbish dumps and along roadsides.

4 × 2. *C. alpinum* L. × *C. arvense* L.
= *C.* × *brueggeranum* Dalla Torre & Sarnthein has been reported from Au and He.

2 × 7. *C. arvense* L. × *C. fontanum* Baumg. (*C. holosteoides* Fr.)
a. *C.* × *pseudoalpinum* Murr.

b. The F_1 hybrid more closely resembles *C. arvense*, differing from it in its
slightly smaller flowers, abortive anthers, better developed inflorescences
with more flowers, somewhat broader leaves, and a short, velutinous
pubescence over the whole plant which gives it a slightly greyer coloration.
It is highly sterile with considerable meiotic irregularities and largely sterile
pollen. Occasional capsules are formed but these are short (equalling the
calyx) and contain small, shrivelled, sterile seeds. Apparent backcrosses to
C. fontanum have been encountered at Lambton Castle, v.c. 66. These
resemble large plants of that parent with somewhat larger petals than is
usual, and the whole plant is velutinous and grey-green in colour.
Occasional capsules are formed but these only slightly exceed the calyx
and although they sometimes contain well-formed seed no germination has
been obtained.

c. Only three British localities are known for this hybrid: between Stamford
and Grantham, v.c. 53 (discovered in 1951 and flourishing for several years
until destroyed by road widening operations); banks of the River Wear just
below Sunderland Bridge and at Lambton Castle, v.c. 66, whence they
largely disappeared with the decline of rabbits following the introduction
of myxomatosis. The hybrids compete vigorously in rough grassland where
both parents are growing. There is also a doubtful record for Au.

d. Hybrid seeds were produced artificially for the first time in 1972.

e. *C. arvense* $2n = 72$ (36, 72); *C. fontanum* $2n = 144$ in most cases but a
range of numbers from 72 to 180 reported by Blackburn and Morton
(1957); F_1 hybrid $2n = 108$; backcross to *C. fontanum* $2n = c\ 126$.

f. ASCHERSON, P., GRAEBNER, P. and CORRENS, C. (1917). *Cerastium*,
in ASCHERSON, P. and GRAEBNER, P. *Synopsis der Mitteleuropäischen Flora*, 5(1): 571-690. Leipzig.

BLACKBURN, K. B. and MORTON, J. K. (1957). The incidence of
polyploidy in the Caryophyllaceae of Britain and Portugal. *New
Phytol.*, 56: 344-351.

4 × 5. *C. alpinum* L. × *C. arcticum* Lange

a. (*C.* × *blytii* auct,? Baenitz; *C. alpinum* var. *pubescens* Syme, *pro parte*).

b. Hybrids are intermediate between the parents and often (? always in Br)
sterile. Hultén (1956) considers that these two species, together with
several other arctic species, form hybrid swarms and have undergone
world-wide introgressive hybridization.

c. Sterile intermediates are frequent with the parents on the higher
mountains of North Wales (e.g. Snowdon, v.c. 49) and the Scottish
mainland (e.g. Ben Lawers and Ben Heasgarnich, v.c. 88). Hultén records it
from Scandinavia (including Is).

d. None.

e. *C. alpinum* $2n = 72$; *C. arcticum* $2n = 108$.

f. DRUCE, G. C. (1910). *Cerastium alpinum* × *nigrescens* = *C. Blytii* Baenitz.
Rep. B.E.C., 2: 498.

HULTÉN, E. (1956). The *Cerastium alpinum* complex. A case of world-wide introgressive hybridisation. *Svensk bot. Tidskr.,* **50**: 411-495.

4 × 7. *C. alpinum* L. × *C. fontanum* Baumg. (*C. holosteoides* Fr.)

  a. *C.* × *symei* Druce (*C. alpinum* var. *pubescens* Syme, *pro parte*).
  b. Hybrids are intermediate between the parents and are rather variable in leaf-shape and in the number of flowers in the inflorescence. Some plants always have solitary flowers and resemble *C. arcticum*; others have several-flowered inflorescences. Frequently the stems are pigmented with anthocyanin which is not found in British *C. alpinum.* The anthers are very small and abortive. Occasional small capsules (equalling the calyx) are formed but the seeds are small and sterile.
  c. This hybrid is frequent in the vicinity of the parents on rills and moist turf of the Scottish mountains in v.c. 88, 94 and 96–98. It displays considerable heterosis and spreads widely through the grass. It is also known in No and Su.
  d. Experimental pollinations using several hybrid clones from Ben Lawers and both parents failed to produce fertile seed.
  e. *C. alpinum* $2n = 72$; *C. fontanum* $2n = 144$ in most cases; hybrids (Ben Lawers) $2n = 108$.
  f. ASCHERSON, P., GRAEBNER, P. and CORRENS, C. (1917). As above.
     DRUCE, G. C. (1910). *C. alpinum* × *vulgatum* = *C.* × *symei* Druce. *Rep. B.E.C.,* **2**: 498.
     DRUCE, G. C. (1911). The alpine Cerastia of Britain. *Ann. Scot. nat. Hist.,* **77**: 38-44.
     DRUCE, G. C. (1920). *Cerastium,* in MOSS, C. E. *The Cambridge British Flora,* **3**: 43-56. Cambridge. FIG.
     HULTÉN, E. (1956). As above.
     MURBECK, S. (1898). Studier ofver kritiska karlvaxtformer, 3. De Nordeuropeiska formerna af slägtet *Cerastium. Bot. Notiser,* **1898**: 241-268.

5 × 7. *C. arcticum* Lange × *C. fontanum* Baumg. (*C. holosteoides* Fr.)

  a. *C.* × *richardsonii* Druce.
  b. This hybrid is intermediate between the parents but has large flowers.
  c. It has been found with the parents on Snowdon, v.c. 49, Ben Nevis, v.c. 97, and Ben Enaiglair, Ben Dearg and Eididh nan Clach Geala, v.c. 105.
  d. None.
  e. *C. arcticum* $2n = 108$; *C. fontanum* $2n = 144$ in most cases; probable hybrid $2n =$ more than 108 (Brett, 1953).
  f. BRETT, O. E. (1953). *Cerastium arcticum* Lange. *Nature, Lond.,* **171**: 527-528.
     DRUCE, G. C. (1910). *Cerastium nigrescens* × *vulgatum* = × *C. Richardsonii. Rep. B.E.C.,* **2**: 498.

7 × 8. *C. fontanum* Baumg. (*C. holosteoides* Fr.) × *C. glomeratum* Thuill.

was recorded from v.c. 53/54 along with several other very unlikely hybrids, and there is also a record from Ge, but both are very doubtful.
WOODRUFFE-PEACOCK, E. A. (1910). *A check list of Lincolnshire plants,* p. 20. Louth.

7 × 11. *C. fontanum* Baumg. (*C. holosteoides* Fr.) × *C. pumilum* Curt.

has been recorded from Su, but must be regarded as very doubtful.

7 × 12. *C. fontanum* Baumg. (*C. holosteoides* Fr.) × *C. semidecandrum* L.

= *C.* × *oxoniense* Druce was recorded by Druce from Headington Quarries and Stow Wood, v.c. 23. The plants were fertile and from the description appeared close to typical *C. semidecandrum,* but had been identified variously as *C. vulgatum, C. pumilum* and *C. tetrandrum* by other authorities.
DRUCE, G. C. (1927). *The flora of Oxfordshire,* 2nd ed., pp. 71-72. Oxford.

9 × 8. *C. brachypetalum* Pers. × *C. glomeratum* Thuill.

= *C.* × *triculinum* Nyár. & Prodan has been recorded from Rm, but is doubtful.

133. *Stellaria* L.
(by C. A. Stace)

2 × 4. *S. media* (L.) Vill. × *S. neglecta* Weihe

has been reported from a few parts of Br and the Continent, but no published records have been traced in Br and no specimens have been verified from anywhere in Europe. The hybrid (highly sterile) has been synthesized in Br and Su.
PETERSEN, D. (1936). *Stellaria*-Studien. *Bot. Notiser,* **1936:** 281-419.

7 × 6. *S. graminea* L. × *S. palustris* Retz.

= *S.* × *glauciformis* Bouvet has been reported from Ga, Ge, Rs and Su.

8 × 6. *S. alsine* Grimm × *S. palustris* Retz.
= *S.* × *hybrida* Rouy & Fouc. has been reported from Ge.

8 × 7. *S. alsine* Grimm × *S. graminea* L.
= *S.* × *adulterina* Focke has been reported from Ge.

136. *Sagina* L.
(by F. N. Hepper)

1 × 2. *S. apetala* Ard. (*S. apetala* subsp. *erecta* (Hornem.) F. Hermann) × *S. ciliata* (*S. apetala* subsp. *apetala, S. reuteri* auct.)
was reported with the parents at Malvern, v.c. 37, by R. F. Towndrow and S. H. Bickham in 1909. Nothing else is known of this plant.
DRUCE, G. C. (1910). *Sagina apetala* × *Reuteri. Rep. B.E.C.*, 2: 413.

4 × 6. *S. procumbens* L. × *S. saginoides* (L.) Karst.
a. *S.* × *normaniana* Lagerh. (*S. scotica* (Druce) Druce; *S.* × *hybrida* Kerner ex Dalla Torre & Sarnth.). (Scottish Pearlwort).
b. This hybrid is more similar to *S. saginoides* than to *S. procumbens*, although its stems are longer, more slender and rooting, and the capsules, if developed at all, are smaller.
c. In BI it is only recorded from the mountains of Scotland: v.c. 88–90, 92, 98 and 104. It is also found in Au, No and Su.
d. Specimens cultivated, presumably from seeds gathered in the wild, at Potterne Biological Station and preserved at K show well-developed capsules.
e. *S. procumbens* ($2n = 22$); *S. saginoides* $2n = 22$.
f. CLAPHAM, A. R. and JARDINE, N. (1964). *Sagina*, in TUTIN, T. G. *et al.*, eds *Flora Europaea*, 1: 146-148. Cambridge.
DALLA TORRE, K. W. von and SARNTHEIN, L. G. (1909). *Flora von Tirol*, 6(2): 155. Innsbruck.
DRUCE, G. C. (1913). *Sagina scotica* Druce. *J. Bot., Lond.*, 51: 89-91.
LAGERHEIM, G. (1898). *K. norske Vidensk. Selsk. Skr. (Trondheim)* 1898(1) (not seen).

4 × 9. *S. procumbens* L. × *S. subulata* (Sw.) C. Presl
a. *S.* × *micrantha* Boreau ex É. Martin.
b. This hybrid has the habit and rooting branches of *S. procumbens,* and the pubescence of *S. subulata.* Recent examination of G. C. Druce's Shetland

material by J. A. Richardson has shown that 64 out of 100 pollen grains were apparently viable. Seed-set is variable with many empty capsules and a mean of 4.1 and a maximum of 12 seeds in a capsule. The Merioneth plant had apparently normal pollen and it was morphologically nearer *S. procumbens*; Benoit suggests it was probably a backcross.

c. The hybrid was found in v.c. 48 by P. M. Benoit in 1961, and in v.c. 112 by G. C. Druce in 1920. It is also recorded from v.c. 101 and from Ga.

d. An experimental hybrid was produced in 1962 by Benoit with *S. procumbens* as the female parent. The F_1 had long but subglabrous pedicels, mostly 4-merous flowers, very sparsely glandular sepals appressed to the capsule in fruit, petals equalling the sepals, 5–7 stamens, and *c* 68% sterile pollen. Selfing produced normally-developed capsules and seeds.

e. *S. procumbens* (2n = 22); *S. subulata* 2n = 22, (18).

f. BENOIT, P. M. and RICHARDS, M. (1963). *A contribution to a Flora of Merioneth*, 2nd ed., p. 15. Haverfordwest.

MARTIN, É. (1875). *Catalogue des plantes vasculaires et spontanées des environs de Romorantin*, p. 64. Romorantin.

8 × 6. *S. intermedia* Fenzl × *S. saginoides* (L.) Karst.

was reported by Druce from Ben Lawers, v.c. 88, where "Dr Ostenfeld considered" certain plants to be of this parentage. No specimens have been traced.

DRUCE, G. C. (1911). The International Phytogeographical Excursion in the British Isles, 3. The Floristic Results. *New Phytol.*, **10**: 306-328.

143. *Spergularia* (Pers.) J. & C. Presl

(by J. A. Ratter)

2 × 5. *S. bocconii* (Scheele) Aschers. & Graebn. × *S. marina* (L.) Griseb.

a. None.

b. The putative wild hybrid differs from *S. bocconii* in being larger, with fewer-flowered, less-crowded inflorescences and with much longer, deflexed fruiting peduncles (about equal to the calyx in length), and in having longer, stouter leaves. Its fruits are similar to those of *S. bocconii* but larger.

c. Lousley (1935, 1936) recorded a single plant from Par, v.c. 2, and he also assigned a specimen from Jersey to this hybrid.

d. Artificial hybrids can be produced relatively easily with either parent as the female. Unlike the putative wild hybrids they are completely sterile, with totally abortive pollen, and produce long, straggling, sterile cymes; after examination of the specimens of the putative wild hybrid I have concluded that they are large, robust specimens of *S. bocconii.*

e. Both parents and artificial hybrids $2n = 36$.

f. LOUSLEY, J. E. (1935). Short notes on some interesting British plants. *J. Bot., Lond.,* **73**: 256-260.

LOUSLEY, J. E. (1936). *Spergularia Bocconei* (Soleir.) Steudel x *S. salina* Presl. *Rep. B.E.C.,* **11**: 25.

RATTER, J. A. (1965). Cytogenetic studies in *Spergularia,* 3. Some interspecific hybrids involving *S. marina* (L.) Griseb. *Notes R. bot. Gdn, Edinb.,* **26**: 224-236.

5 x 3. *S. marina* (L.) Griseb. x *S. rupicola* Lebel ex Le Jolis

a. (*S.* x *symeana* Druce, *nom. nud.*).

b. Hybrids are vigorous prostrate perennials intermediate in morphology between the parental species. They are very floriferous and produce long, straggling, many-flowered inflorescences which never bear any capsules. The flowers are notable for the very small stamens which lie in a ring around the base of the ovary. The pollen is completely abortive and the hybrid is absolutely sterile.

c. Hybrids have only been found in three localities: three plants on the west breakwater of Par harbour, v.c. 2; one plant on the harbour pier at Lyme Regis, v.c. 9; and nine plants on the top of a sea-cliff at Stackpole Head, v.c. 47. In all cases the hybrids occurred in mixed populations of the parental species.

d. Artificial hybrids can be made relatively easily with either parent as the female. Like the natural hybrids they are completely sterile. Isolation of the parental species is therefore genetical. The artificial hybrid normally exhibits *c* 9 bivalents and 18 univalents at meiotic metaphase I and shows further irregularities in the later stages of meiosis.

e. Both parents and artificial hybrid $2n = 36$.

f. PUGSLEY, H. W. (1911). Lyme Regis plants. *J. Bot., Lond.,* **49**: 365-366.

RATTER, J. A. (1963). *Spergularia marina* x *rupicola. Proc. B.S.B.I.,* **5**: 25.

RATTER, J. A. (1965). As above.

RATTER, J. A. (1973a). Plant Records. *Watsonia,* **9**: 378.

5 x 4. *S. marina* (L.) Griseb. x *S. media* (L.) C. Presl.

a. (*S.* x *morei* Druce, *nom. nud.*).

b. Plants assigned to this hybrid have either been individuals of *S. marina* with dimorphic (winged and wingless) seeds, or specimens of the wingless-seeded race (or races) of *S. media* (*Lepigonum marinum* var. *apterum* E. S. Marshall).

c. The hybrid has been reported from salt-marshes and spray-washed rocks in
v.c. 5/6, 10 and 16 and from western parts of the Continent from
Scandinavia to North Africa, including Jersey.

d. Experimental investigations have shown that hybridization of the two
species is completely blocked by a barrier of seed incompatibility, the
young embryos aborting at an early stage of development.

e. *S. media* 2*n* = 18; *S. marina* 2*n* = 36.

f. MOSS, C. E. (1920). *Spergularia,* in *The Cambridge British Flora,* 3:
17-23. Cambridge. FIG.

RATTER, J. A. (1959). *A cytogenetic study in Spergularia.* Ph.D. Thesis,
University of Liverpool.

RATTER, J. A. (1973b). Cytogenetic studies in *Spergularia,* 8. Barriers to
the production of viable interspecific hybrids. *Notes R. bot. Gdn,
Edinb.,* **32:** 297-301.

148. *Scleranthus* L.

(by C. A. Stace)

1 × 2. *S. annuus* L. × *S. perennis* L.

= *S.* × *intermedius* Kittel has been recorded from Ga, He, Scandinavia and
parts of central Europe.

153. *Amaranthus* L.

(by C. A. Stace)

2 × 1. *A. hybridus* L. × *A. retroflexus* L.

a. *A.* × *adulterinus* Thell.

b. The hybrid is characterized by the female flowers having oblong perianth
segments with obtuse to acute apices and a more or less excurrent midrib.
In habit and inflorescence form it is intermediate between the parents. It
has been described as producing "almost no fruits".

c. This has been seen in Br only recently and as an impermanent casual, e.g. v.c. 34 (1959), v.c. 37 (1961) and v.c. 21 (1969). On the Continent it is recorded from Au, Cz, Da, Ga, Ge, He, Hu, It and Rm. These two New World species are probably not sympatric in their native ranges, but widespread introduction to most parts of the world has given rise to sporadic hybrids, e.g. in the U.S.A.

d. Both species are monoecious and protogynous, but usually selfed by wind pollination. Reciprocal crosses using American material were easily carried out and F_1 plants raised. These were "highly sterile" but were induced to form fertile amphidiploids by colchicine treatment. Artificial F_1 hybrids showed intermediate morphological characters.

e. *A. hybridus* ($2n = 32$); *A. retroflexus* ($2n = 34$).

f. AELLEN, P. (1959). *Amaranthus*, in HEGI, G. *Illustrierte Flora von Mitteleuropa*, 2nd ed., 3: 465-516. München.

   BRENAN, J. P. M. (1965). *Amaranthus hybridus* × *retroflexus* = *A.* × *adulterinus* Thell. *Proc. B.S.B.I.*, 6: 122-123.

   MURRAY, M. J. (1940). The genetics of sex determination in the family Amaranthaceae. *Genetics, Princeton*, 25: 409-431.

   PRISZTER, S. (1958). Über die bisher bekannten Bastarde der Gattung *Amaranthus*. *Bauhinia*, 1: 126-135.

   THELLUNG, A. (1914). *Amaranthus*, in ASCHERSON, P. and GRAEBNER, P. *Synopsis der Mitteleuropäischen Flora*, 5(1): 225-356. Leipzig.

154. *Chenopodium* L.
(by C. A. Stace)*

The genus *Chenopodium* poses many taxonomic problems, which are heightened by the extreme plasticity shown by many morphological characters, especially the leaf-shape. In the first half of this century plants which were intermediate in such features were almost invariably determined as hybrids, sometimes between more than two species. These identifications were often reinforced by the apparent sterility of the plants, although many of them were casual aliens which in any case seldom fruit in northern Europe. Even when fruits were formed, examination of the seeds to observe the critical testa-markings was rarely carried out. Most of the European hybrid identifications of the past are undoubtedly erroneous, and only three of the following (given fuller treatment) can be considered as possibly genuine in BI.

---

* With assistance from M. J. Cole.

13 × 2. C. hybridum L. × **C. polyspermum** L.

= C. × perhybridum Ponert has recently been described from Cz.

4 × 3. C. album L. × **C. vulvaria** L.

has been very doubtfully reported from Ge.

4 × 5. C. album L. × **C. suecicum** J. Murr (*C. viride* auct.)

= C. × fursajewii Aellen &Iljin has been recorded from v.c. 9, 14, 25, 41, 63 and 77. *C. viride* L. has been widely used as a synonym of *C. suecicum* and has very frequently been mistaken for cymose variants of *C. album*. Records of hybrids probably refer to similar but less extreme variants of *C. album*; those specimens seen by Cole were determined by him as *C. album*. This hybrid has also been recorded from Ge and Rs.

COLE, M. J. (1957). *Variation and interspecific relationships of C. album L. in Britain.* Ph.D. thesis, University of Southampton.

UOTILA, P. (1972). Chromosome counts on the *Chenopodium album* aggregate in Finland and N.E. Sweden. *Ann. bot. Fenn.*, 9: 29-32.

4 × 6. C. album L. × **C. berlandieri** Moq. (incl. *C. zschackei* J. Murr)

a. *C.* × variabile Aellen.

b. This variable putative hybrid is intermediate in most characters but usually resembles *C. berlandieri* more closely than *C. album*. The leaves are only indistinctly three-lobed and are obtuse and mucronate at the apex. The stems are green with conspicuous purple axillary patches. The inflorescence is variable in form, as in *C. album*, but usually consists of larger glomerules, and the perianth segments are more conspicuously keeled than in *C. album*. The testa is marked both with furrows as in *C. album* and with irregular shallow pocks as in *C. berlandieri*. The plant appears to be fertile, but in Europe is usually only a casual.

c. Most British records are of casual plants on rubbish tips, etc., where it is one of the only two frequent putative hybrids in the genus and commoner than *C. berlandieri* itself. It has been recorded from scattered localities throughout most of Br, especially in southern England, and from Au, Ge and Ho.

d. Attempts to hybridize the two species and *C.* × variabile with *C. album* have all been unsuccessful (Cole, 1957, 1961) and have led to doubts concerning the hybrid nature of this taxon. These doubts are re-inforced by the chromosome numbers.

e. *C. album* $2n = 54$; *C. berlandieri* $2n = 36$; *C.* × variabile $2n = 36$.

f. BRENAN, J. P. M. and LOUSLEY, J. E. (1947). *Chenopodium album* L. × *Berlandieri* Moq. subsp. *Zschackei* (Murr) Zobel var. *typicum* (Ludwig) Aellen. *Rep. B.E.C.*, 13: 165-166.

COLE, M. J. (1957). As above.

COLE, M. J. (1961). Interspecific relationships and intraspecific variation of *Chenopodium album* L. in Britain, 1. The taxonomic delimitation of the species. *Watsonia*, 5: 47-58.

4 × 6 × 10. *C. album* L. × *C. berlandieri* Moq. × *C. pratericola* Rydb.
(*C. leptophyllum* auct.)
has been recorded from He.

4 × 6 × str. *C. album* L. × *C. berlandieri* Moq. × *C. striatum* (Krasan)
J. Murr
= *C.* × *drucei* J. Murr has been recorded from Au.

4 × 7. *C. album* L. × *C. opulifolium* Schrad. ex Koch & Ziz
a. *C.* × *preissmannii* J. Murr.
b. A great number of variants has been assigned this parentage. They show an
   enormous range in leaf, stem and inflorescence characters, paralleling those
   in *C. album* itself, and it is impossible to construct a worthwhile overall
   description. The leaves of *C. opulifolium* are usually more glaucous and
   shorter and broader than those of *C. album*, and the putative hybrids
   usually show more or less intermediate characters. The testa markings of
   the two species are markedly different, but past descriptions of the
   putative hybrids have ignored these characters. Some of the intermediate
   plants have been reported to be sterile.
c. The comments under *C. album* × *C. berlandieri* apply equally here, except
   that *C. opulifolium* is a much more frequent plant in Br than is *C.
   berlandieri* or either putative hybrid. There are records of *C. album* × *C.
   opulifolium* as a casual from a number of localities in England, Scotland
   and Wales. Elsewhere records exist for many parts of Europe, northern
   Africa and western Asia.
d. Attempts to hybridize the two species have been unsuccessful (Cole, 1957,
   1961) and have caused Cole to doubt the hybrid origin of intermediate
   plants.
e. Both parents $2n = 54$.
f. ASCHERSON, P. and GRAEBNER, P. (1913). *Synopsis der Mitteleuropäischen Flora*, 5(1): 70-78. Leipzig.
   COLE, M. J. (1957). As above.
   COLE, M. J. (1961). As above.
   MURR, J. (1922). *C. album* L. *Rep. B.E.C.*, 6: 302-305.

4 × 8. *C. album* L. × *C. hircinum* Schrad.
   has been recorded from v.c. 64 as a casual. It is said to resemble *C.
   hircinum* in its habit, dense mealiness and smell, and *C. album* in its
   inflorescence, with leaves variable but more lobed than in *C. hircinum*.
   Whether or not such plants are in fact hybrids is unknown, but specimens
   so-labelled seen by Cole were identified by him as *C. album*. They are also
   reported from Ge and He.
   ASCHERSON, P. and GRAEBNER, P. (1913). As above, pp. 88-89.
   COLE, M. J. (1957). As above.

4 × 9. *C. album* L. × *C. ficifolium* Sm.

a. *C.* × *zahnii* J. Murr.

b. Plants reported as hybrids have intermediate leaf and inflorescence characters, although the relatively distinctive 3-lobed leaves with a long, narrow central lobe and the small, well-separated glomerules of flowers of *C. ficifolium* are closely approached by some variants of *C. album*. The seeds of the putative hybrids, which appear to be fertile, are said to be either smooth (cf. *C. album*) or with elongate pits (cf. *C. ficifolium*), often on one plant. Specimens so-labelled seen by Cole were identified by him as *C. album*.

c. This plant is a rather rare casual in Br. Records exist for v.c. 32, 39 and 61. It is also recorded from Au, Cz, Ga, Ge and He.

d. Attempts to hybridize the two species have been unsuccessful (Cole, 1957, 1961) and have caused Cole to doubt the hybrid origin of intermediates.

e. *C. album* $2n = 54$; *C. ficifolium* $2n = 18$.

f. ASCHERSON, P. and GRAEBNER, P. (1913). As above, pp. 87-88.
COLE, M. J. (1957). As above.
COLE, M. J. (1961). As above.

4 × 10. *C. album* L. × *C. pratericola* Rydb. (*C. leptophyllum* auct.)

= *C.* × *leptophylliforme* Aellen has been recorded from He.

4a × 4r. *C. album* L. × *C. reticulatum* Aellen

has been recorded from v.c. 23, 30 and H39, but *C. reticulatum* itself has been shown by Cole to be a minor genetic variant of *C. album*, differing only in testa characters. Intermediate plants have sseds of both sorts and/or of an intermediate nature. The two taxa have the same chromosome number, and can be easily crossed in either direction to give rise to intermediates.
COLE, M. J. (1957). As above.
COLE, M. J. (1961). As above.

4 × str. *C. album* L. × *C. striatum* (Krasan) J. Murr

has been recorded under a variety of binomials (e.g. *C. interjectum* J. Murr, *C. pseudostriatum* Zschacke, *C. substriatum* J. Murr) as casual plants in widely separated parts of Br, in v.c. 6, 9, 23, 34, 41 and 79. Little is known of the true nature of these plants or of the specific distinctness of the two species; *C. striatum* is often considered an infraspecific variant of *C. album*. Specimens identified as this hybrid were determined by Cole as *C. album*. Hybrids have also been recorded in Au, Ga, Ge, He and Rm.
COLE, M. J. (1957). As above.
MURR, J. (1922). As above.

7 × 5. *C. opulifolium* Schrad. ex Koch & Ziz × *C. suecicum* J. Murr

(= *C.* × *aellenianum* Blom, *nom. nud.*) has been recorded from Su.

str × 5. *C. striatum* (Krasan) J. Murr × *C. suecicum* J. Murr (*C. viride* auct.)

has been mentioned as occurring in Br and elsewhere, but both supposed parents have been greatly confused with *C. album* and old records of such hybrids are best ignored.

6 × 7. *C. berlandieri* Moq. × *C. opulifolium* Schrad. ex Koch & Ziz

has been recorded from Ge and He.

6 × 8. *C. berlandieri* Moq. × *C. hircinum* Schrad

has been reported from Ga and Ge.

6 × 9 × 8. *C. berlandieri* Moq. × *C. ficifolium* Sm. × *C. hircinum* Schrad.

has been very doubtedly reported from Ge.

6 × 9. *C. berlandieri* Moq. × *C. ficifolium* Sm.

has been very doubtfully reported from Ge.

6 × 10. *C. berlandieri* Moq. × *C. pratericola* Rydb. (*C. leptophyllum* auct.)

= *C.* × *binzianum* Aellen & Thellung has been recorded from He.

6 × str. *C. berlandieri* Moq. × *C. striatum* (Krasan) J. Murr

has been recorded from v.c. 48, but much more detailed work would be needed to confirm that the plant was not one of the variants of the putative *C. album* × *C. berlandieri* group.

9 × 7. *C. ficifolium* Sm. × *C. opulifolium* Schrad. ex Koch & Ziz.

Single plants, one on a canal bank and one in a field, were reported by J. A. Wheldon in 1901 and 1912 in the Liverpool district, v.c. 59, each growing with both parents. The second was considered by J. Murr and E. S. Marshall to be *C. opulifolium,* and the above combination has not been confirmed from any other localities in BI or abroad.

MARSHALL, E. S., MURR, J. & WHELDON, J. A. (1913). *Chenopodium opulifolium* × *C. serotinum? Rep. B.E.C.,* 3: 279.

7 × str. *C. opulifolium* Schrad. ex Koch & Ziz × *C. striatum* (Krasan) J. Murr

(= *C.* × *vachelliae* Druce, *nom. nud.*) has been recorded from Barry, v.c. 41, and from Ga and central Europe under other binomials, but the parentage of all of these plants, or if they are indeed hybrids, is very uncertain.

MURR, J. (1922). As above.

9 × 8. *C. ficifolium* Sm. × *C. hircinum* Schrad.

has been very doubtfully recorded from Ge.

8 × 10. *C. hircinum* Schrad. × *C. pratericola* Rydb. (*C. leptophyllum* auct.)

= *C.* × *pseudoleptophyllum* Aellen has been recorded from He.

8 × str. *C. hircinum* Schrad. × *C. striatum* (Krasan) J. Murr

= *C.* × *haywardiae* J. Murr was reported from v.c. 61, 64 and 79 as a casual. It apparently differs from putative *C. album* × *C. hircinum* in the purple striping on the stems, and its status as a hybrid is equally doubtful.

MURR, J. (1914). Weiteres zur Adventivflora von Grossbritannien. *Allg. bot. Z.*, **20**: 25-26.

MURR, J. (1915). × *Chenopodium Haywardiae* Murr. *Rep. B.E.C.*, **4**: 19, pl. 2 and 3. FIG.

15 × 14. *C. botryodes* Sm. (*C. chenopodioides* (L.) Aellen) × *C. rubrum* L.

has been recorded from Ge.

16 × 14. *C. glaucum* L. × *C. rubrum* L.

= *C.* × *schulzeanum* J. Murr has been reported from Ge and Rm.

155. *Beta* L.

(by C. A. Stace)

1m × 1v. *B. maritima* L. (*B. vulgaris* subsp. *maritima* (L.) Thell.) × *B. vulgaris* L.

occurs readily wherever the parents occur close together, with apparently no loss of fertility, though intermediates have scarcely ever been reported in BI. The interfertility is good evidence in favour of the subspecific treatment for these two taxa.

TJEBBES, K. (1933). The wild beets of the North Sea region. *Bot. Notiser*, **1933**: 305-315. FIG.

156. *Atriplex* L.

(by E. M. Jones)

4 × 1. *A. glabriuscula* Edmondst. × *A. littoralis* L.
   a. None.
   b. Plants intermediate between *A. littoralis* and *A. glabriuscula* in leaf-shape, bracteoles, seeds and habit are thought to be of hybrid origin. The fertility is variable; some plants are highly sterile.
   c. Intermediates have only been found at Gibraltar Point, v.c. 54. Their appearance followed the breakdown of their normal ecological separation by the destruction of sand-bars by storms.
   d. None. Attempts to synthesize hybrids are made difficult by the high degree of self-fertility in both parents and by the problems of emasculation.
   e. Both parents $2n = 18$.
   f. JONES, E. M. (1971). *Taxonomic and ecological studies in the genus Atriplex.* D. Phil. Thesis, University of Oxford.

3 × 1. *A. hastata* L. × *A. littoralis* L.
   has been recorded as rare in Da.

3 × 2. *A. hastata* L. × *A. patula* L.
   has been very doubtfully recorded from v.c. 3, 10 and 14 and from Ge on the basis of apparently intermediate specimens. Artificial hybrids have been synthesized (Hulme, 1958); they are sterile triploids which are intermediate in morphology, but they do not resemble any wild plants seen by Hulme.
   HULME, B. A. (1957). *Studies on some British species of Atriplex.* Ph.D. thesis, University of Edinburgh.
   HULME, B. A. (1958). Artificial hybrids in the genus *Atriplex. Proc. B.S.B.I.,* 3: 94.
   WOLLEY-DOD, A. H. (1937). *Flora of Sussex,* p. 372. Hastings.

4 × 2. *A. glabriuscula* Edmondst. × *A. patula* L.
   has been suggested as the parentage of straggly, prostrate plants with small, slightly hastate and widely-spaced leaves, but there is little to substantiate this. A. J. Wilmott identified specimens from v.c. 3 and 6 (**BM**).
   THOMPSON, H. S. and WILMOTT, A. J. (1932). *Atriplex glabriuscula* Edmondston? var. *Babingtonii* Moss and Wilmott. *Rep. Watson B.E.C.,* 4: 138.

4 × 3. *A. glabriuscula* Edmondst. × *A. hastata* L.
   a. None.
   b. Plants intermediate between *A. hastata* and *A. glabriuscula* in leaf-shape, bracteoles, seeds and habit are thought to be of hybrid origin. They are nearly fully fertile.

c. The intermediates appear to be common in BI where the two species are in contact. This occurs when weedy habitats are introduced into maritime areas by landslides or the building of sea-defences etc. Intermediates have so far been found in v.c. 6, 10, 11, 14, 28, 54 and CI, but Wilmott identified specimens (BM) from coastal areas from Cornwall and Kent to Shetland. They have also been recorded in Da, Ho and Su, but their hybrid status has been questioned (Meijden, 1970). Gustafsson (1973b) mentioned three records in Da and five in Su.

d. The seeds of intermediate plants germinate more rapidly than the seeds of either species. Gustafsson (1973a) has successfully synthesized this hybrid from Scandinavian parental material. Germination of batches of hybrid seed varied from 0 to 40%, but those seeds which did germinate gave rise to well developed plants which showed a high degree of fertility and nine bivalents at meiosis. The F_2 generation was equally fertile.

e. Both parents $2n = 18$; artificial hybrid ($2n = 18$).

f. BARTON, W. C. (1917). *Atriplex hastata* L. x *A. Babingtonii* Woods. *Rep. B.E.C.*, **4**: 585-586.

GUSTAFSSON, M (1973a). Evolutionary trends in the *Atriplex triangularis* group of Scandinavia, 1. Hybrid sterility and chromosomal differentiation. *Bot. Notiser*, **126**: 345-392.

GUSTAFSSON, M. (1973b). Evolutionary trends in the *Atriplex triangularis* group of Scandinavia, 2. Spontaneous hybridization in relation to reproductive isolation. *Bot. Notiser*, **126**: 398-416.

JONES, E. M. (1971). As above.

MEIJDEN, R. van der (1970). Biosystematic notes on *Atriplex patula* L., *A. hastata* L. and *A. littoralis* L. (Chenopodiaceae). *Blumea*, **18**: 53-63.

MOSS, C. E. and WILMOTT, A. J. (1914). *Atriplex*, in MOSS, C. E. *The Cambridge British Flora*, **2**: 168-182. Cambridge.

PEDERSEN, A. (1968). Nogle kritiske, danske *Atriplex*-arte. *Bot. Tidsskr.*, **63**: 289-302.

TASCHEREAU, P. M. (1972). Taxonomy and distribution of *Atriplex* species in Nova Scotia. *Canad. J. Bot.*, **50**: 1571-1594.

TURESSON, G. (1922). The genotypical response of the plant species to the habitat. *Hereditas*, **3**: 211-260.

160. *Salicornia* L.
(by D. H. Dalby)

There are records from Br for five of the six possible hybrid combinations involving the four annual species of this genus recognized by Dandy (1958):

2 × 3. *S. dolichostachya* Moss × *S. europaea* L.
2 × 4. *S. dolichostachya* Moss × *S. ramosissima* Woods
3 × 4. *S. europaea* L. × *S. ramosissima* Woods
    (= *S.* × *salisburii* Druce, *nom. nud.*)
3 × 5. *S. europaea* L. × *S. pusilla* Woods
    (= *S.* × *intermedia* Woods, *pro parte*)
5 × 4. *S. pusilla* Woods × *S. ramosissima* Woods
    (= *S.* × *marshallii* Druce, *nom. nud.*, *S.* × *townsendii* Druce, *nom. nud.*, *S.*
    × *bartonii* Druce, *nom. nud.*)
    In addition hybrids have been recorded between many of the segregates of
these species, notably of *S. ramosissima*. *S. pusilla* is a diploid ($2n = 18$), *S.
dolichostachya* is a tetraploid ($2n = 36$), and *S. ramosissima* and *S.
europaea* probably exist as both diploid and tetraploid cytodemes. No
triploid plants have been encountered, and in view of this, and of the
uncertainty regarding the delimitation of most of the annual species, only
one hybrid seems worthy of further consideration.

General References

BALL, P. W. (1964). A taxonomic review of *Salicornia* in Europe.
    Reprium nov. Spec. Regni veg., **69**: 1-8.
BALL, P. W. and TUTIN, T. G. (1959). Notes on annual species of
    Salicornia in Britain. *Watsonia*, 4: 193-205.
DALBY, D. H. (1962). Chromosome number, morphology and breeding
    behaviour in the British Salicorniae. *Watsonia*, 5: 150-162.
MOSS, C. E. (1911). Some species of *Salicornia*. *J. Bot., Lond.*, **49**:
    177-185.
MOSS, C. E. (1912). The International Phytogeographical Excursion in the
    British Isles, 12. Remarks on the characters and nomenclature of some
    critical plants noticed on the excursion. *New Phytol.*, 11: 398-414.
MOSS, C. E. and SALISBURY, E. J. (1914). *Salicornia*, in MOSS, C. E.
    The Cambridge British Flora, 2: 187-196. Cambridge. FIG.

5 × 4. *S. pusilla* Woods × *S. ramosissima* Woods

a. (Three *nomina nuda* listed above).
b. Putative hybrids are characterized by having cymules with different
    numbers of flowers (1, 2 or 3) on the same plant. The size and colour of
    the mature fruiting segments may be rather nearer typical *S. pusilla*. When
    open-pollinated in the field putative hybrids set fertile seed.
c. Hybrids have been reliably recorded only from v.c. 3, 9–11, 13–15, 19, 25
    and 28 on the southern and south-eastern coasts of England. In several
    sites the hybrids occur on the upper part of a salt-marsh closely associated
    with almost unbranched competition-forms of the two parent species.
    Close proximity to the parents is probably significant, as species of
    Salicornia are predominantly or perhaps sometimes wholly self-pollinated
    in nature.
d. Progeny from plants open-pollinated in the field are rather variable in

morphology, but most plants show variation in the number of flowers per cymule. Most are nearer to *S. pusilla* in colour, segment-shape, etc.

e. Both parents $2n = 18$.

f. BARTON, W. C. (1918). *Salicornia. Rep. B.E.C.*, 5: 51, 125.
   BARTON, W. C. (1918). *Salicornia. Rep. Watson B.E.C.*, 3: 71-72.
   LITTLE, J. E. and MARSHALL, E. S. (1917). *S. disarticulata* x *ramosissima. Rep. Watson B.E.C.*, 3: 30.
   LITTLE, J. E., MARSHALL, E. S. and SALMON, C. E. (1920). *S. disarticulata* Moss x *ramosissima* Woods. *Rep. Watson B.E.C.*, 3: 113.
   MARSHALL, E. S. (1915). A new *Salicornia* variety and hybrid. *J. Bot., Lond.*, 53: 362-363.
   MOSS, C. E. and SALISBURY, E. J. (1914). As above.

162. *Tilia* L.
(by C. D. Pigott)

2 x 1. *T. cordata* Mill. x *T. platyphyllos* Scop.

a. *T.* x *vulgaris* Hayne (*T.* x *europaea* auct., *non* L.). (Common Lime).

b. Hybrids are characterized by an abundance of epicormic shoots and they show higher rates of elongation and growth in diameter of shoots than the parental species. The distribution of simple and stellate hairs on the shoots and leaves, the shape and size of the leaves, and the prominence of the veins on the lower surface of the leaves are intermediate between the parental species, but the inflorescence is almost always pendulous in hybrids, as in *T. platyphyllos*, although the number, shape and structure of the fruits are intermediate. A low proportion of the fruits are fertile and give rise to seedlings showing segregation in many vegetative characters. The name *T.* x *vulgaris* refers to the commonly planted tree which is probably an F_1 hybrid but is of unknown origin. Trees which are morphologically similar occur in natural populations but are usually associated with individuals intermediate between them and both parental species. It is not known if these are F_2 and subsequent generation recombinants or the products of backcrossing.

c. Hybrids are found on cliffs and screes of limestone or other calcareous rock, and are associated with both or one parental species, in the Wye valley, v.c. 34–36, and on limestone in v.c. 57 and 63. They are possibly also native in v.c. 37, but in scattered localities elsewhere they are more likely to have been planted; they are not uncommon within the range of *T. platyphyllos* in central Europe.

d. In most but not all years the flowering periods of the parental species do not overlap. *T. platyphyllos* flowers in late June before *T.* x *vulgaris* and *T. cordata* which flower in July.
e. Both parents and hybrid (2*n* = 82).
f. PIGOTT, C. D. (1969). The status of *Tilia cordata* and *T. platyphyllos* on the Derbyshire limestone. *J. Ecol.*, **57**: 491-504.
   SYME, J. T. B. (1864). *Tilia*, in *English Botany*, 2nd ed., **2**: 171-177, t. 286. London. FIG.

163. *Malva* L.
(by D. H. Dalby)

1 x 2. *M. moschata* L. x *M. sylvestris* L.
   = *M.* x *inodora* Poñert has been recently described from Cz.

4 x 2. *M. neglecta* Wallr. x *M. sylvestris* L.
   = *M.* x *zoernigii* Fleischer has been recorded from Au, Cz, Ga and Ge.

4 x 5. *M. neglecta* Wallr. x *M. pusilla* Sm
   = *M.* x *adulterina* Wallr. has been recorded from Au, Cz, Ge, Po, Rm and Su.

4 x 6. *M. neglecta* Wallr. x *M. parviflora* L. (*M. oxyloba* Boiss.)
   a. None.
   b. Artificial hybrids are intermediate between the parents in the patterning on the dorsal faces of the ripe mericarps and in the development of a marginal wing, while the fruiting pedicels (the upper very short, the lower much longer and deflexed) combine the parental expressions. The petals are similar in size to those of *M. neglecta*.
   c. A plant determined as this hybrid by Schinz was recorded from a waste heap at Welwyn, v.c. 20, in 1932. No other records have been encountered.
   d. Kristofferson found hybrids easy to produce. They had pollen and seed fertilities of about 95%, and a segregating F_2 was produced by selfing.
   e. *M. neglecta* (2*n* = 42).
   f. KRISTOFFERSON, K. B. (1926). Species crossings in *Malva*. *Hereditas*, **7**: 233-354. FIG.
      PHILLIPS, H. (1933). *Malva neglecta* x *parviflora*. *Rep. B.E.C.*, **10**: 21.

165. *Althaea* L.

(by C. A. Stace)

fic x ros. *A. ficifolia* (L.) Cavanilles x *A. rosea* (L.) Cavanilles
   a. *A.* x *cultorum* Bergmans.
   b. All three taxa are cultivated hollyhocks, of which *A. ficifolia* is the least common. The leaves of *A. rosea* are 5-7-lobed or -angled, whereas in *A. ficifolia* they are deeply lobed and in the hybrid intermediate. Many other parental characteristics (especially of the flowers) are very variable, and the hybrid is thus similarly variable. It seems fully fertile. *A. ficifolia* is often regarded as a variety of *A. rosea*, or as a hybrid of it with some other species, and *A. rosea* itself might be of hybrid origin since it is not known as a wild plant.
   c. The hybrid is considered by some to be the commonest garden hollyhock and hence the one most commonly found on rubbish dumps, etc.
   d. Hybrids have been synthesized, probably on numerous occasions, the characters being inherited in the expected fashion.
   e. *A. rosea* ($2n$ = 26, 42, 56); *A. ficifolia* ($2n$ = 42, 44).
   f. CLAPHAM, A. R. (1962). *Althaea*, in CLAPHAM, A. R., TUTIN, T. G. and WARBURG, E. F. *Op. cit.*, pp. 296-297.
   FOCKE, W. O. (1881). *Op. cit.*, p. 74.
   WEBB, D. A. (1968). *Alcea*, in TUTIN, T. G. *et al.*, eds *Flora Europaea*, 2: 253-254. Cambridge.

168. *Geranium* L.

(by D. McClintock)

1 x 16. *G. pratense* L. x *G. robertianum* L.
   was reported from Lathkilldale, v.c. 57, in 1908, but no specimen has been traced. The single plant was sterile and described as obviously a hybrid, and intermediate between the putative parents in many floral characters, particularly the petals, but in habit it resembled a small plant of *G. pratense*. The existence of such a hybrid, however, seems very unlikely and it appears not to have been recorded elsewhere.
   DRABBLE, E. and DRABBLE, H. (1908). *Geranium pratense* x *Robertianum*. *J. Bot., Lond.*, **46**: 301.

3 × 4. *G. endressii* Gay × *G. versicolor* L. (*G. striatum* L.)

a. (Cultivar names only).

b. This hybrid is intermediate between the parents in varying degree, for it is fertile and backcrosses easily. The veining of the petals comes through from *G. versicolor* and their pinkness from *G. endressii*; and the leaf-shape is also intermediate. Various clones are grown in gardens, e.g. 'Arthur Johnson', 'George Claridge Druce', 'Lady Moore's Variety' and 'Somerset Huish', and there are others unnamed.

c. Neither parent is native in BI, so the hybrid is likely to arise in the wild only where one or both parents happen to have become established there. It seems probable that most wild hybrid plants originated as such from gardens, where they were planted or arose naturally. In wild gardens where both species are grown the hybrid often predominates. Because it is tainted as a garden plant it has relatively rarely been reported or collected wild, but it has been recorded from scattered localities throughout southern England and northwards to at least v.c. 80, and in Hb from v.c. H22.

d. Var. *thurstonianum* Turrill was described as a variety of *G. endressii* but almost certainly belongs to this hybrid. It occurs only where the putative parents grow, and Thurston's field-notes (Turrill, 1928) indicate that the original plant was in a hybrid swarm. Sansome (1936) suggested that a factor carried by *G. versicolor*, when homozygous in cytoplasm derived from *G. endressii*, causes the contabescence and petalloidy of the anthers which characterize this variety. The petals of this male-sterile plant are smaller in size, very narrow and deep in colour. Despite its sterility, seeds obtained from this taxon in my garden were sown and one plant, which appears identical with its parent, raised. The earliest reference to it seems to be in Bowles (1914).

e. Both parents and var. *thurstonianum* (garden specimens) $2n = 26$ (G. E. Marks pers. comm. 1971).

f. BOWLES, E. A. (1914). *My garden in summer,* pp. 102-103. London.

LEY, A. and MARSHALL, E. S. (1911). *Geranium endressi* × *striatum. Rep. B.E.C.*, 2: 551.

SANSOME, F. W. (1936). Experiments with *Geranium* species. *J. Genetics,* 33: 359-363.

TURRILL, W. B. (1928). *Geranium endressi* J. Gay in Cornwall. *J. Bot., Lond.,* 66: 44-46.

13 × 9. *G. molle* L. × *G. pyrenaicum* Burm. f.

= *G. × luganense* Chenevard has been recorded from He.

14 × 9. *G. pusillum* L. × *G. pyrenaicum* Burm. f..

= *G. × hybridum* Hausskn. has been recorded from Ge.

13 × 14. *G. molle* L. × *G. pusillum* L.

= *G.* × *oenense* J. Murr has been recorded from Au.

15 × 16. *G. lucidum* L. × *G. robertianum* L.

(= *G.* × *hybridum* F. A. Lees, *non* L., *nec* Hausskn., *nec* C. Bailey) has been recorded from Aysgarth, v.c. 65, hill districts in western Yorkshire, and Bridgnorth, v.c. 40, but no specimens have been traced. The Aysgarth plants were said to be "handsome but ephemeral (because sterile) individuals" halfway between the putative parents in morphology. The parent species occur in immediate proximity in many parts of the country, but there is no other evidence of hybridization in Br or elsewhere.

LEES, F. A. (1887). *G. purpureum, Angl. Auct. G. hybridum mihi. Bot. Loc. Rec. Club,* 3: 118.

LEES, F. A. (1888). *The flora of West Yorkshire,* p. 182. London.

17 × 16. *G. purpureum* Vill. × *G. robertianum* L.

a. None.

b. Some people find difficulty in distinguishing certain variants of these two species; consequently it has been suggested some plants are hybrids. Baker (1955, 1957) wrote that certain populations of *G. purpureum* have suffered introgression from *G. robertianum,* and that a combination of the features of *G. purpureum* subsp. *purpureum* and prostrate *G. robertianum* could produce the characters of *G. purpureum* subsp. *forsteri.* Moreover *G. purpureum* subsp. *forsteri* always grows in sites near which the other two taxa are known to have occurred.

c. The distribution of subsp. *forsteri* is given by Perring and Sell (1968). Baker quoted specimens of *G. purpureum* showing evidence of hybridization from v.c. 2, 3, 34, 44 and H6.

d. P. M. Benoit (pers. comm. 1972) has synthesized the hybrid in both directions, and notes that except for their complete sterility these hybrids seem indistinguishable from *G. robertianum.* Baker (1951) found the same to be true of his artificial hybrids.

e. *G. robertianum* $2n = 64$; *G. purpureum* (both subspecies) $2n = 32$.

f. BAKER, H. G. (1951). Hybridisation and natural gene-flow between higher plants. *Biol. Rev.,* 26: 302-337.

BAKER, H. G. (1955). *Geranium purpureum* Vill. and *G. robertianum* L. in the British Flora, 1. *Geranium purpureum. Watsonia,* 3: 160-167.

BAKER, H. G. (1957). Genecological studies in *Geranium* (Section *Robertiana*). General considerations and the races of *G. purpureum* Vill. *New Phytol.,* 56: 172-192.

PERRING, F. H. and SELL, P. D. (1968). *Op. cit.,* p. 17.

169. *Erodium* L'Hérit.
(by P. M. Benoit)

3 × 4. *E. cicutarium* (L.) L'Hérit. *sensu stricto* × *E. glutinosum* Dumort. (*E. cicutarium* subsp. *bipinnatum* Tourlet)

a. *E.* × *anaristatum* Andreas.

b. This hybrid can be distinguished by its showy inflorescences of large flowers, few or no developed fruits, and pollen grains very variable in size and colour (20–75 μm diam., red to yellow when fresh) in the same anther. Petal-colour and other characters are intermediate between those of the parents. The hybrid occurs as single plants with the parents, is not variable, and does not form complex populations, suggesting that most plants are F_1 s. Hybrids should not be confused with stunted, sterile plants having small, crumpled petals but otherwise the characters of *E. cicutarium* or *E. glutinosum*; they are apparently diseased. A hexaploid ($2n = 60$) found in Da, *E. danicum* K. Larsen, has been interpreted as the amphidiploid derived from this hybrid.

c. The hybrid is known from coastal duneland at Newton Burrows, v.c. 41, Towyn Burrows, v.c. 44, Morfa Dyffryn and Morfa Harlech, v.c. 48, Newborough Warren, v.c. 52, and Formby, v.c. 59; and from Ho.

d. Hybrids are readily obtained by pollinating *E. glutinosum* with pollen of *E. cicutarium*, but the reverse cross has not been successful. Synthesized hybrids are identical with the wild hybrid in morphology and fertility. They do not develop fruit from pollination with hybrid or *E. glutinosum* pollen, but do so when backcrossed to *E. cicutarium*, though few seeds are produced. The viability of this seed has not been tested.

e. *E. cicutarium* (subsp. *cicutarium* and subsp. *dunense*) $2n = 40$; *E. glutinosum* $2n = 20$; hybrid ($2n = 30$).

f. ANDREAS, C. M. (1947). De inheemsche Erodia van Nederland. *Ned. kruidk. Archf,* **54**: 138-231.

BENOIT, P. M. (1967). *Erodium cicutarium* agg. *Proc. B.S.B.I.,* **6**: 364-366.

LARSEN, K. (1958). Cytological and experimental studies on the genus *Erodium* with special references to the collective species *E. cicutarium* (L.) L'Hérit. *K. danske Vidensk. Selsk. Skr., Biol.,* **23(6)**.

170. *Oxalis* L.
(by C. A. Stace)

2 × 4. *O. corniculata* L. × *O. europaea* Jord. has been recorded from He.

187. *Ulex* L.
(by M. C. F. Proctor)

1 × 2. *U. europaeus* L. × *U. gallii* Planch.
 a. (*U.* × *douieae* Druce, *nom. nud.*).
 b. Hybrid bushes are intermediate between the parents, the most useful
    characters being the pedicel-thickness, the size of the bracteoles and the
    pod, the number of ovules (mostly 7–10), the indumentum of the calyx
    and the flowering time (August to March, maximum September and
    October). The hybrid seems generally to be highly fertile, and extensive
    hybrid swarms occur at least locally.
 c. The hybrid has been found in v.c. 1, 3, 6, 48, 49 and 52, and R. D. Meikle
    (pers. comm.) has reported it "common" in Hb, in heath and acid hill
    grassland. Plants from v.c. 10 referred to this parentage by Druce (1921)
    were *U. europaeus,* but those reported by Wilson *et al.* (1924) from v.c. 71
    might have been correctly identified (Benoit *in litt.,* 1971). Corillion
    (1951) and Lambinon (1962) have reported it from the coast of Brittany,
    Ga, though the latter author stated that the hybrids flower in summer so
    the records may refer at least in part to *U. europaeus.*
 d. None.
 e. *U. europaeus* (2*n* = 64, 96); *U. gallii* (2*n* = 80).
 f. BENOIT, P. M. (1962). *Ulex europaeus* × *gallii. Proc. B.S.B.I.,* 4: 414-415.
    CORILLION, R. (1951). Sur l'existence et la répartition des *Ulex* hybrides
      des landes bretonnes. *C.r. hebd. Séanc. Acad. Sci., Paris,* 232: 344-346.
    DRUCE, G. C. (1921). *U. europaeus* × *Gallii* Planch., nov. hybr. *Rep.
      B.E.C.,* 6: 17-18.
    LAMBINON, J. (1962). Note sur les *Ulex* du Massif Armoricain. *Lejeunia,*
      9: 64-66.
    MILLENER, L. H. (1952). *Experimental studies on the growth forms of
      the British species of Ulex L.* Ph.D. thesis, University of Cambridge.
    WILSON, A., WHELDON, J. A., DRUCE, G. C. and FRASER, J. (1924).
      Ulex europaeus × *U. Gallii? Rep. B.E.C.,* 7: 379.

1 × 3. *U. europaeus* L. × *U. minor* Roth
 has occasionally been considered in error to be the parentage of *U. gallii*
 Planch.

2 × 3. *U. gallii* Planch. × *U. minor* Roth
 has been reported from Ga, but requires confirmation.

189. *Ononis* L.
(by J. K. Morton)

1 × 2. *O. repens* L. × *O. spinosa* L.
a. None.
b. Hybrids are intermediate between the parents in habit, leaf-shape, stem-pubescence, and the relative lengths of the capsule and calyx. The possession of spines is of no value in determining hybridity as both spineless forms of *O. spinosa* and spiny forms of *O. repens* occur. Introgression has been demonstrated by biometric means in the Durham coast colonies (Morton, 1956), but Morisset (1964) considers that the structure of these populations, and of similar ones which he studied in v.c. 29, is due to abnormal variation in the parent species, coupled with the occasional occurrence of F_1 hybrids.
c. The parents are usually ecologically separated and only rarely occur together. *O. repens* favours rough grassland by roadsides, on sand-dunes and overlying chalk or limestone. *O. spinosa* usually favours rough grassland on neutral to calcareous clays, especially when these overlay chalk or limestone. Hybrids are only rarely encountered and there is little mention of them in the literature. They occur on limestone on the coast at Marsden and Blackhall Rocks, v.c. 66, and on clays associated with the chalk at Orwell, v.c. 29. On the Continent they appear to be recorded only from north-eastern Ge.
d. Artificial hybrids are very difficult to produce. Hybrid pollinations frequently result in the formation of apparently fairly normal, though sometimes rather small, fruit, but the seeds are either shrivelled and immature or, if well-formed, fail to germinate. A single, fertile seed has been obtained from the hundred or so pollinations made by the writer. This has produced a young hybrid plant intermediate in vegetative characters and chromosome number. Both parents are pollinated by bees, etc., and do not usually self-pollinate, although selfing usually leads to fertile seed production. The resulting fruit usually has only a single seed which is often smaller and which grows into a less vigorous plant than usual.
e. *O. repens* $2n = 60$ (30, 60), *O. spinosa* $2n = 30$. The presence of one pair of chromosomes in *O. spinosa* and two pairs in *O. repens* with a long secondary constriction probably accounts for the many reports of $2n = 32$ and 64 which occur in the literature. Hybrids (presumed F_1) $2n = c$ 44, 45, 48.
f. ERDTMANN, J. (1964). Zur Verbreitung und Taxonomie der Gattung *Ononis* in Nordöst Deutschland. *Reprium nov. Spec. Regni veg.*, **69**: 103-131. FIG.
MORISSET, P. (1964). Hybridization in *Ononis. Proc. B.S.B.I.*, **5**: 378-379.

MORTON, J. K. (1956). Studies on *Ononis* in Britain, 1. Hybridity in the Durham coast colonies. *Watsonia*, **3**: 307-316.

190. *Medicago* L.

(by P. L. Curran)

1 × 2. *M. falcata* L. (*M. sativa* L. subsp. *falcata* (L.) Arcangeli) × *M. sativa* L.

a. *M.* × *varia* Martyn. (Sand Lucerne).
b. The hybrids are variously intermediate between the parents but tend to diverge into two sorts: one with relatively large leaves, long internodes and slightly spiral fruits resembling *M. sativa*; the other with small leaves, short internodes and lunate pods resembling *M. falcata*. Chimaeras are frequent. Pollen-germinability and seed-set are variable but generally low. Most seedlings are abnormal and very few survive.
c. Sand Lucerne occurs in scattered localities in Br and CI as far north as v.c. 83. It is commonest in parts of East Anglia and the East Midlands where *M. falcata* is native; elsewhere it is mainly coastal in distribution and has arisen following the agricultural introduction of the parents or of the hybrid itself. In Hb it is known only on maritime sands in v.c. H21. It is widespread throughout Europe and, as a result of introductions, elsewhere in temperate regions.
d. Gilmour and others have reported a high fertility in the F_1, but many hybrids are largely sterile. Hybrids are cross-fertile with both parent species but the frequency of successful backcrosses is very low.
e. *M. sativa* $2n = 32$ (most), 16, 64; *M. falcata* $2n = 16$ (most), 32; hybrid $2n = 28, 30, 32$.
f. BUTCHER, R. W. and STRUDWICK, F. E. (1930). *Further illustrations of British plants*, t. 124, Ashford. FIG.
   FARRAGHER, M. A. and CURRAN, P. L. (1962). Observations on native sand-lucerne (*Medicago* × *varia* Martyn). *Scient. Proc. R. Dubl. Soc.*, Ser. B, **1**: 59-66.
   GILMOUR, J. S. L. (1933). The taxonomy of plants intermediate between *Medicago sativa* L. and *M. falcata* L. and their history in East Anglia. *Rep. B.E.C.*, **10**: 393-395.
   HO, K. M. and KASHA, K. J. (1972). Chromosome homology at pachytene in diploid *Medicago sativa*, *M. falcata* and their hybrids. *Can. J. Genet. Cytol.*, **14**: 829-838.

LEDINGHAM, G. F. (1940). Cytological and developmental studies of hybrids between *Medicago sativa* and a diploid form of *M. falcata*. *Genetics, Princeton,* **25**: 1-15.

PERRING, F. H. and SELL, P. D. (1968). *Op. cit.,* p. 20.

191. *Melilotus* Mill.
(by C. A. Stace)

1 x 2. *M. altissima* Thuill. x *M. officinalis* (L.) Pall.
= *M.* x *haussknechtii* O. E. Schulz has been recorded from Rm, but is doubtful.

3 x 2. *M. alba* Medic. x *M. officinalis* (L.) Pall.
= *M.* x *schoenheitianus* Hausskn. has been recorded from Rm, but is doubtful.

192. *Trifolium* L.
(by D. McClintock)

4 x 2. *T. medium* L. x *T. pratense* L.
= *T.* x *permixtum* Neumann has been recorded from Ge and Rm.

7 x 2. *T. incarnatum* L. x *T. pratense* L.
was suggested by Marquand (1901) to be the parentage of a plant from Guernsey described under *T. pratense* as "resembling the type generally but having flower-heads shaped like those of *T. incarnatum* and on very long pedicels". No specimen has been found and, in view of the strong breeding barriers between virtually all species of *Trifolium*, the existence of such a hybrid seems very unlikely.

EVANS, A. M. (1962). Species hybridization in *Trifolium*, 1. Methods of overcoming species incompatibility. *Euphytica*, **11**: 164-176.

MARQUAND, E. D. (1901). *The flora of Guernsey*, p. 76. London.
TAYLOR, N. L., STROUBE, W. H., COLLINS, G. B. and KENDALL, W. A. (1963). Interspecific hybridization of red clover (*Trifolium pratense* L.). *Crop Sci.*, 3: 549-552.

195. *Lotus* L.
(by C. A. Stace)

5 × 4. *L. angustissimus* L. × *L. subbiflorus* Lag. (*L. hispidus* Desf. ex DC.)

= *L.* × *davyae* Druce was recorded from Start Point, v.c. 3, along with both the species in 1915. It was said to be closer to *L. hispidus* in vegetative and to *L. angustissimus* in fruiting characters. The type specimen in herb. Druce (**OXF**) is fertile and appears to be a fairly typical example of *L. angustissimus*; it is similarly annotated by E. F. Warburg.

DRUCE, G. C. (1916). *Lotus angustissimus* L. × *hispidus* Desf. = × *L. Davyae* Druce. *Rep. B.E.C.*, 4: 194.

206. *Vicia* L.
(by C. A. Stace)

15 × 14. *V. angustifolia* L. (*V. sativa* L. subsp. *nigra* (L.) Ehrh.) × *V. sativa* L.

Hybrids between these two taxa are sometimes encountered, and similarly between the two varieties of *V. angustifolia*, var. *angustifolia* (*V. sativa* subsp. *nigra sensu stricto*) and var. *segetalis* (Thuill.) Koch (*V.sativa* subsp. *segetalis* (Thuill.) Gaud.). Such hybrids are intermediate in appearance and vary in fertility according to the cytological characteristics of the parents. Hybrids are easily made experimentally, and their rarity in nature is due to the almost total self-pollination of the parents. The three taxa are probably best treated as subspecies of *V. sativa*.

HOLLINGS, E. (1971). *Experimental taxonomic studies of the pattern of variation within the Vicia sativa aggregate.* Ph.D. thesis, University of Manchester.

KILLICK, H. J. (1974). Two suspected natural hybrids. *Watsonia,* **10**: 228.

15 x 16. V. angustifolia L. (V. sativa L. subsp. nigra (L.) Ehrh.) xV. lathyroides L.

is often noted from sand-dune areas, and sometimes published references are made (e.g. Thompson and Barton, 1924, and several foreign sources). All specimens seen are referable to one species or the other (usually small *V. sativa* subsp. *nigra*), and artificial hybrids have all died before flowering, often as seedlings.

HOLLINGS, E. (1971). As above.

THOMPSON, H. S. and BARTON, W. C. (1924). *V. lathyroides* L. *Rep. Watson B.E.C.,* **3**: 250.

209. *Spiraea* L.
(by C. A. Stace)

2 x 1. S. douglasii Hook. x S. salicifolia L.

a. *S.* x *billiardii* Hérincq.

b. These three taxa are all grown in gardens and have all at various times been found as more or less naturalized escapes. *S.* x *billiardii* is intermediate between its parents in the important diagnostic leaf character (tomentose beneath in *S. douglasii*, glabrous in *S. salicifolia*). The hybrid appears to be highly fertile.

c. Both parents and hybrid are found in a naturalized or semi-naturalized state in shrubberies, neglected woodland and hedgerows, etc. There are definite records of the hybrid from v.c. 17, 54 and 92 (J. E. Lousley, *in litt.* 1974). It is recorded for Ge and No but probably occurs elsewhere.

d. According to Focke (1881) hybrid seedlings may be found where the two species are grown close together. The hybrid was apparently first raised in France in *c* 1850 and has presumably been synthesized many times since by horticulturalists.

e. Both parents ($2n = 36$); *S.* x *billiardii* ($2n = 45, 54$); (but all correctly identified?).

f. CHITTENDEN, F. J., ed. (1951). *Dictionary of gardening,* **4**: 2001. Oxford.

FOCKE, W. O. (1881). *Op. cit.*, p. 115.

JØRGENSEN, P. M. (1973). Forvillede arter og hybrider av slekten *Spiraea* i Norge. *Blyttia*, 31: 29-33.

McCLINTOCK, D. (1959). *Supplement to the Pocket guide to wild flowers*, p. 52. Platt, Kent.

211. *Rubus* L.

(by A. Newton)

In the classification followed here the native and naturalized species of *Rubus* found in BI belong to six subgenera. Since five of these each possess three or fewer species in this country, and all of them are sexual, hybridization within and between them is relatively straightforward; these cases are dealt with first.

Subg. *CHAMAEMORUS* (Focke) Focke x Subg. *CYLACTIS* (Raf.) Focke

1 x 2. *R. chamaemorus* L. x *R. saxatilis* L.

= *R.* x *tranzschelii* Juz. has been recorded from northern Fe and Rs.

3 x 1. *R. arcticus* L. x *R. chamaemorus* L.

= *R.* x *neogardicus* Juz. has been recorded from northern Fe and Rs.

Subg. *CHAMAEMORUS* (Focke) Focke x Subg. *IDAEOBATUS* Focke

1 x 6. *R. chamaemorus* L. x *R. idaeus* L.

has been recorded from northern Fe.

Subg. *CYLACTIS* (Raf.) Focke

3 x 2. *R. arcticus* L. x *R. saxatilis* L.

= *R.* x *castoreus* Laestad. is recorded from the northern parts of Fe, No, Rs and Su.

Subg. *CYLACTIS* (Raf.) Focke x Subg. *IDAEOBATUS* Focke

6 x 2. *R. idaeus* L. x *R. saxatilis* L.

has been recorded from Su.

Subg. *CYLACTIS* (Raf.) Focke × Subg. *GLAUCOBATUS* Dumort.
9 × 2. *R. caesius* L. × *R. saxatilis* L.

= *R.* × *areschougii* A. Blytt has been recorded from Ge, Hu, No and Su.

Subg. *ANOPLOBATUS* Focke × Subg. *IDAEOBATUS* Focke
6 × 4. *R. idaeus* L. × *R. odoratus* L.

= *R.* × *nobilis* Regel is grown in gardens and might be found as an escape. It is doubtful whether it has ever arisen spontaneously either in the wild or in cultivation.

Subg. *IDAEOBATUS* Focke
6 × 7. *R. idaeus* L. × *R. phoenicolasius* Maxim.

= *R.* × *paxii* Focke was collected in 1930 by J. A. Wheldon in Hall Road, Liverpool, v.c. 59, and determined by W. C. R. Watson.

WHELDON, J. A. (1931). *Rubus idaeus* × *phenicolasius*. *Rep. B.E.C.*, **9**: 260.

Subg. *IDAEOBATUS* Focke × Subg. *GLAUCOBATUS* Dumort.
9 × 6. *R. caesius* L. × *R. idaeus* L.

= *R.* × *idaeoides* Ruthe *sec.* Focke appears to be rare in BI. It has been reported only from v.c. 11, 15, 19, 57, 62 and 66. It is apparently frequent in No and Su, and is also recorded from Au, Be, Ga, Ge and He. The genetics of the F_1 hybrid (which is largely infertile) and of later generations and backcrosses have been investigated by Focke, Lidforss (1905-1907) and Rosanova (1934). Heslop-Harrison found a British plant to be pentaploid and European material varies from triploid to hexaploid; *R. idaeus* is a sexual diploid, and *R. caesius* a tetraploid or pentraploid apomict. The hybrid resembles *R. caesius* in habit and stem characters; the leaves resemble those of *R. idaeus* but are rarely 7-nate (usually 3- or 5-nate), and the inflorescence is short and corymbose.

Subg. *IDAEOBATUS* Focke × Subg. *RUBUS*
6 × 11. *R. idaeus* L. × *R. fruticosus* L. agg.

Recent hybrids between raspberries and blackberries are apparently rare. *R. idaeus* × *R. ulmifolius* Schott was reported by G. C. Druce (det. W. C. R. Watson) from Hungerford Park, v.c. 22, in 1930. The fruits were dull red, easily separated from the receptacle and tasted of raspberries; the stems rooted at the tip and the leaves were 7-nate. Two sterile hybrids, *R. idaeus* × *R. plicatus* Erikson = *R.* × *eriksonii* Sudre and *R. idaeus* × *R. bifrons* Vest. = *R.* × *albinitus* Sudre, have been reported from Su and Ge respectively. It is possible that some members of subgenus *Rubus* section *Suberecti* (see below) are ancient hybrids between *R. idaeus* and *R. fruticosus* agg.

R. *loganobaccus* L. H. Bail. is derived from a hybrid between *R. idaeus* and *R. vitifolius* Cham. & Schlect., an American dioecious species in subgenus *Rubus* section *Ursini* Focke. It is frequently grown in gardens and allotments but is only rarely found as an established outcast.

CRANE, M. B. (1940). The origin of new forms in *Rubus*, 2. The Loganberry, *R. loganobaccus* Bailey. *J. Genet.*, **40**: 129-140. FIG.

DRUCE, G. C. and WATSON, W. C. R. (1930). *Rubus idaeus* x *ulmifolius*. *Rep. B.E.C.*, **9**: 20-21.

The Problem of Taxonomy in Subg. Rubus

The main monographs (Sudre, 1908–13; Focke, 1914), the results of bramble study on a continental scale, classified the genus into groups of species to each of which was subordinated a hierarchy of taxa, including in the former case sterile hybrids. As the number of different Rubi considered on this scale is so vast, and since they are nearly all originally of assumed hybrid origin, it is difficult to decide on a treatment which does justice to the incidence of the plants in the field and at the same time constitutes a viable system. Because of the differing views taken in the past when describing brambles *de novo* the ascription of names has been a haphazard process, in extreme cases single bushes unlike any other having received names while widespread taxa have remained undescribed. The situation in large measure has depended on the attention given by assiduous batologists, on the range of their activities, on their concept of "species" differentiation, and on their view of the propriety of publishing descriptions.

It is suggested that if a variant can be shown to be distinct, and forms a significant part of the vegetation over a substantial portion of an area of say 100 km^2, consideration should be given to naming it as a species. While many of these taxa have obvious affinities to one another and may in fact be derived from putative parents, it is perhaps unwise to adopt a hierarchical system, with its implication that certain taxa are more "primordial" than others. It is considered more practicable to categorize Rubi on distributional criteria and it is proposed that the following categories should be recognized:

1. Taxa occurring widely in a part of Europe and also widely in Br, e.g. *R. vestitus* Weihe & Nees, *R. rufescens* Lef. & Muell., *R. radula* Weihe ex Boenn.

2. Taxa occurring in restricted areas of Br but widespread or local in Europe, e.g. *R. questieri* Lef. & Muell., *R. cavatifolius* P. J. Muell.

3. Taxa occurring widely in Br in more than one geographical unit, e.g. Dartmoor and the Forest of Dean, and perhaps also very locally in Europe, e.g. *R. bloxamii* Lees, *R. murrayi* Sudre, *R. dasyphyllus* (Rogers) E. S. Marshall.

4. Taxa locally common in Br but confined to one geographical unit, perhaps with odd outliers but otherwise unknown elsewhere, e.g. *R. glareosus* E. S. Marshall ex Rogers, *R. mucronatiformis* Sudre.

5. Taxa confined to a small area within a single geographical unit, e.g. *R. regillus* A. Ley, *R. hirsutissimus* Sudre & A. Ley.

6. Isolated bushes or clumps unlike any other growing in the vicinity or included in the previous categories.

7. Good fruiting isolated bushes or clumps of intermediate character between others occurring in the vicinity.

8. Poorly fruiting isolated bushes or clumps of intermediate character between others occurring in the vicinity.

Categories 1–5 can be regarded as established "ancient hybrids" which have achieved range and stability and deserve unique names, category 6 as taxa which may or may not turn out to be deserving of subsequent description, and categories 7 and 8 as putative hybrids. The ascription by Watson (1958) of hybrid notation to validly described species of wide occurrence (in south-western England particularly) appears unjustifiable; it is thought better to retain the names given by the original authors in these cases, reserving hybrid formulae for categories 7 and 8.

General references

CRANE, M. B. (1940). Reproductive versatility in *Rubus*, 1. Morphology and inheritance. *J. Genet.*, **40**: 109-118.

CRANE, M. B. and DARLINGTON, C. D. (1927). The origin of new forms in *Rubus*, 1. *Genetica*, **9**: 241-278.

FOCKE, W. O. (1877). *Synopsis Ruborum Germaniae*. Bremen.

FOCKE, W. O. (1881). *Loc. cit.*

FOCKE, W. O. (1902–1903). *Rubus*, in ASCHERSON, P. and GRAEBNER, P. *Synopsis der Mitteleuropäischen Flora*, 6(1): 440-648. Leipzig.

FOCKE, W. O. (1911–1914). *Species Ruborum*. Stuttgart.

GUSTAFSSON, A. (1943). Genesis of the European blackberry flora. *Acta Univ. lund.*, **39**: 1-200.

HASKELL, G. (1954). The genetic detection of natural crossing in blackberry. *Genetica*, **27**: 162-172.

HASKELL, G. (1960). Role of the male parent in crosses involving apomictic *Rubus* species. *Heredity*, **14**: 101-113.

HESLOP-HARRISON, Y. (1953). Cytological studies in the genus *Rubus*, 1. Chromosome numbers in the British *Pubus* flora. *New Phytol.*, **52**: 22-39.

LIDFORSS, B. (1905–1907). Studier över artbildningen inom släktet *Rubus*, 1-2. *Ark. Bot.*, **4, 6**.

ROSANOVA, M. A. (1934). Origin of new forms in the genus *Rubus*. *Bot. Zhurnal*, **19**.

SUDRE, H. (1908–1913). *Rubi Europae*. Paris.

THOMAS, P. T. (1940). Reproductive versatility in *Rubus*, 2. The chromosomes and development. *J. Genet.*, **40**: 119-128.

VAARAMA, A. (1939). Cytological studies on some Finnish species and hybrids of the genus *Rubus* L. *J. scient. agric. Soc., Finland*, **11**.

WATSON, W. C. R. (1958). *Handbook of the Rubi of Great Britain and Ireland*. Cambridge.

Subg. *GLAUCOBATUS* Dumort. × Subg. *RUBUS*

9 × 11. *R. caesius* L. × *R. fruticosus* L. agg. = *Rubus* sect. *Triviales* P. J. Muell.

R. caesius, in addition to its status as a Linnaean species, is a very distinctive plant throughout its wide European range. Tetraploid and pentaploid races are known. According to Gustafsson (1943) the pollen is thoroughly good and it is a prolific producer of hybrids. As a polyploid, however, it presumably is itself of hybrid origin, the product of extinct parents.

The *Triviales* are tetraploid, pentaploid and hexaploid hybrids between *R. caesius* and the true blackberries, but despite frequent crossing the boundaries of *R. caesius* are never obliterated. This is partly due to the fact that it is entirely apomictic (Gustafsson, 1943) and also that the *Triviales* formed with *R. caesius* as a pollen parent have poor pollen themselves and exhibit reduced fruit-setting ability. The following features distinguish the *Triviales* from subgenus *Rubus*: they are still in vigorous development, as exemplified by the large number of "recent hybrids" encountered; they tend to inhabit ground inhospitable to the true blackberries, e.g. low lying clay and damp soils; throughout their range they are the most active colonizers of open habitats, e.g. cinder-heaps, railway- and river-banks, sand-pits and sub-littoral zones.

Despite this behaviour there are many "ancient hybrids" within the section which are now stable and constant throughout wide areas both in BI and north-western Europe, and it is not sufficient to deal with all in the manner adopted by Sudre (1908-1913) (by a hybrid formula). Examples of species in categories 3 and 4 above are *R. sublustris* Lees, *R. balfourianus* Blox. ex Bab., *R. eboracensis* W. Wats., and *R. latifolius* Bab., and of category 5 *R. tenuiarmatus* Lees and *R. rubriflorus* Purchas. Other taxa equally deserving of names still await description. On the other hand, there are large numbers of plants of doubtful parentage in category 8 which tend to confuse the delimitation of further *Triviales* species.

The *Triviales* exhibit the complete range of stem characters by which subg. *Rubus* is divided into sections; unnamed plants which are most similar to *Sylvatici* and *Discolores* are often known for convenience as "*Corylifolii*" and to *Appendiculati* and *Glandulosi* as "*R. dumetorum agg.*" Comprehensive study is necessary to establish more precise taxonomic groups. Hybrids of *R. caesius* with species of the *Suberecti* appear to be very scarce, particularly in BI, but in northern Europe the term "*sub-idaei*" has been applied to supposed derivatives of *R. caesius* x *idaeus* with the *Triviales; R. sublustris* is the British representative closest to this group.

Subg. *RUBUS* L. (11. *R. fruticosus* L. agg.)

R. ulmifolius Schott is a sexual diploid species; all other investigated British species are polyploids (triploids to heptaploids, x = 7) and as far as is known they are all apomictic. The following sections can be usefully recognized: 1. *Suberecti* P. J. Muell. 2. *Sylvatici* P. J. Muell. (incl. *Sprengeliani* Focke). 3. *Discolores* P. J. Muell. (= Section *Rubus*). 4. *Appendiculati* Genev. 5. *Glandulosi* P. J. Muell.

The *Suberecti* are a small group of species chiefly inhabiting northern and western Europe on heath and moor borders on peaty soils; they rarely form "recent hybrids". The category 5 species *R. daltrii* Edees & Rilst. is of suberect habit but exhibits glandular development and was on this account included in section *Appendiculati* by Watson (1958). The *Suberecti* at the present time show no colonizing propensities in BI; owing to diminution of suitable habitats they form a decreasing remnant of primitive, post-glacial vegetation. They may be triploid, tetraploid or hexaploid. A few of the *Suberecti*, e.g. *R. nessensis* and *R. scissus* W. Wats., could perhaps be described as subg. *Idaeobatus* x subg. *Rubus* on account of their strongly suckering habit, erect canes and reddish fruit. On balance, however, it seems preferable to include these species with the greater number of other suberect *Rubus* species which do not exhibit *idaeus*-like characteristics to such a marked degree.

The *Sylvatici* include three groups of which one *(Grati)* stands close to the *Suberecti* and another *(Discoloroides)* close to the *Discolores*. The taxa in these two constitute "ancient hybrids" which may have originated by hybridization between the third group *(Virescentes)* and species of the *Suberecti* and *Discolores* respectively. Recent hybrids involving this section are rarely encountered, in spite of the fact that four of them *(R. selmeri* Lindeb., *R. lindleianus* Lees, *R. sprengelii* Weihe and *R. polyanthemus* Lindeb.) are common species of wide distribution in BI. Category 7 plants of this section are most likely to be found, if at all, among ancient bramble populations as single individuals exhibiting intermediate characters. As an example, a single plant exactly intermediate between *R. mollissimus* Rogers and *R. milfordensis* Edees was found on Yateley Common, v.c. 12, where both species are frequent. Species of this section may be triploid, tetraploid or hexaploid.

The *Discolores* include *R. ulmifolius*, the only sexually-breeding and only diploid species of subg. *Rubus* to be found in BI. Other species of the section may be from triploids to pentaploids (or heptaploids). Ecologically *R. ulmifolius* has much in common in BI with *R. caesius* and the *Triviales* (e.g. preference for clay and calcareous soils) and is often completely absent from the bramble communities of heaths, moors and woods. *R. ulmifolius* has a wide distribution from Hb through southern and eastern England to Ju and North Africa. It is a very variable species. Sudre described 20 microspecies and 92 varieties; it is likely that these are all diploid recombinations of cross-fertilized populations. Large numbers of hybrids between it and other members of all the groups were also described by him, mostly sterile or poorly fruiting. In BI *R. ulmifolius* hybridizes most actively with *R. caesius* and the *Triviales* and also with *R. vestitus* (section *Appendiculati*) (all species with which it is very frequently in contact). Hybrids with other species occur occasionally, and are usually distinguished by vigorous vegetative capacity and high sterility. Evidently *R. ulmifolius* was an equally prolific hybridizer in past times; a large proportion of constant and uniform *Discolores* species are very likely to be ancient hybrids originally derived from it.

The *Appendiculati* include groups (e.g. *Vestiti, Radulae, Apiculati* and *Hystrices*) which might with advantage be elevated to higher status. The species vary from triploids to hexaploids. They form a very significant portion of the British *Rubus* flora and include several species of categories 1 and 3, e.g. *R. vestitus, R. dasyphyllus, R. longithyrsiger* Lees ex Bak., *R. hebecaulis* Sudre, *R.*

radula, R. rudis Weihe & Nees and *R. echinatus* Lindl. The influence of
R.vestitus may be seen in many category 7 plants, and also appears likely to have
played a part in the formation of many category 5 species in the past. *R.
hebecaulis* and *R. longithyrsiger* exhibit similar propensities in south-western Br
(from which *R. vestitus* is absent) while the widespread and often abundant *R.
dasyphyllus, R. radula, R. rudis* and *R. echinatus* seem (like the common
Sylvatici) to remain as discrete entities. The likelihood is that all these species
are now virtually obligate apomicts but may in some cases have exhibited a
different breeding behaviour in the past.

The *Glandulosi* form a very distinct section of tetraploids (and pentaploids)
which are relatively scarce in BI. *R. bellardii* Weihe & Nees and *R. hylonomus*
Lef. & Muell. are thinly but widely distributed, and there are a number of
category 5 species with distinct affinities to the *R. serpens* group (*sensu* Focke &
Sudre) in the Severn Valley, and to *R. hirtus* agg. in eastern and south-eastern
England. A few of the British representatives may be themselves hybrids of
recent formation but the majority appear to be good-fruiting and of ancient
origin; like the *Suberecti* at the other end of the *Rubus* spectrum, they behave as
remnants of former communities without the vigour to expand from their
woodland refugia.

212. *Potentilla* L.

(by B. Matfield and S. M. Walters)

6 × 7. *P. argentea* L. × *P. recta* L.
   = *P. × pseudocanescens* Błocki has been reported from several parts of
   central and southern Europe. *P. inclinata* Vill. may also be this hybrid.

6 × 11. *P. argentea* L. × *P. tabernaemontani* Aschers.
   has been recorded from the Continent.

12 × 11. *P. crantzii* (Cr.) Beck ex Fritsch × *P. tabernaemontani*
Aschers.
   a. (*P. × cryeri* Druce, *nom. nud.*; ?*P. × beckii* J. Murr).
   b. Both these species are pseudogamous apomicts mostly at the hexaploid or
      heptaploid level in Br, although sexual tetraploid *P. crantzii* occurs on the
      Continent. The possibilities of hybridization are therefore limited. It seems
      clear, however, that high polyploid apomictic plants of hybrid origin can
      arise as rare, aberrant progeny when an unreduced egg-cell is fertilized by

normal (reduced) pollen. Such hybrid plants are intermediate between the parents in habit, flower-size and stipule-shape.

c. Two topodemes which have been much discussed in the British literature because of their puzzling intermediate morphology have almost certainly had such an origin. These are near Grassington, v.c. 64, and on Little Craigandal, v.c. 92. A similar plant has been found in v.c. 68. High polyploid intermediate plants are also known in Su; *P.* x *beckii* was described from Au.

d. Artificial cross-pollinations have yielded occasional hybrid offspring in predominantly maternal (apomictic) progenies.

e. *P. tabernaemontani* $2n = 42, 49, 56$; *P. crantzii* $2n = 42, 49$ ($28, 42, 49, 64$); hybrids $2n = 63–64, 61–63$ (from the Grassington and Craigandal populations respectively).

f. CRYER, J. *et al.* (1911). *Potentilla Crantzii*, G. Beck. *Rep. B.E.C.*, 2: 555.

  SMITH, G. L. (1963a). Studies in *Potentilla* L., 1. Embryological investigations into the mechanism of agamospermy in British *P. tabernaemontani* Aschers. *New. Phytol.*, 62: 264-282.

  SMITH, G. L. (1963b). Studies in *Potentilla* L., 2. Cytological aspects of apomixis in *P. crantzii* (Cr.) Beck ex Fritsch. *New Phytol.*, 62: 283-300.

  SMITH, G. L., BOZMAN, V. G. and WALTERS, S. M. (1971). Studies in *Potentilla* L., 3. Variation in British *P. tabernaemontani* Aschers. and *P. crantzii* (Cr.) Beck ex Fritsch. *New Phytol.*, 70: 607-618. FIG.

  SWAN, G. A. and SWAN, M. (1962). A Northumbrian plant thought to be extinct. *Vasculum*, 47: 29-30.

14 x 13. *P. anglica* Laich. x *P. erecta* (L.) Räusch.

a. *P.* x *suberecta* Zimm.

b. Hybrids are intermediate between the parents in many characters including leaflet- and petal-number, petiole-length and flower-diameter. The suberect to decumbent stems of the hybrid resemble *P. erecta* in habit, but they may root at the nodes in late summer as in *P. anglica*. The percentage of stainable pollen grains varies from 0 to 40% and a few achenes are usually formed (rarely more than 10 per flower). Backcrossing does occur, but its frequency is unknown as the morphological variability of the parents and the F_1 hybrid makes backcross derivatives very difficult to distinguish without cytological examination.

c. Hybrids are widely distributed in Br and Hb wherever *P. erecta* and *P. anglica* meet. They are also known to occur in Au, Da, Ga, Ge, Po, Rs and Su.

d. Artificial cross-pollinations have been highly successful using either species as the female parent. Natural isolation is therefore mainly ecological. Experimental backcrosses have been continued for two generations with *P. erecta* and for four with *P. anglica*. The backcross derivatives are vigorous and their fertility increases with each successive generation, showing that introgression is possible.

e. *P. erecta* $2n = 28$; *P. anglica* $2n = 56$; natural hybrids $2n = (35), 42, 45$ (the first and last probably backcrosses); artificial F_1 hybrid $2n = 42$. Experimental backcross derivatives are intermediate in chromosome number between the F_1 hybrid and the recurrent parent.

   f. MATFIELD, B., JONES, J. K. and ELLIS, J. R. (1970). Natural and experimental hybridization in *Potentilla. New Phytol.*, **69**: 171-186. FIG.

   MURBECK, S. (1890). Studies on *Potentilla* hybrids. *Bot. Notiser*, **1890**: 193-235.

   SKALINSKA, M. and CZAPIK, R. (1958). Studies in the cytology of the genus *Potentilla* L. *Acta biol. cracov.*, Sér. bot., **1**: 137-149.

13 × 15. *P. erecta* (L.) Räusch. × *P. reptans* L.
14 × 15. *P. anglica* Laich. × *P. reptans* L.

   a. *P. × mixta* Nolte ex Reichb. (*P. × italica* Lehm.).
   b. Hybrids with a strong resemblance to *P. reptans* have variously been identified as *P. reptans* × *P. anglica* and *P. reptans* × *P. erecta*. They have the same ability to form runners as *P. reptans* and are most easily distinguished from the latter by the presence of 3- and 4- as well as 5-nate leaves, 4- as well as 5-merous flowers, and by their sterility; the pollen stainability is generally less than 10% and achene formation is rare. The hybrids are very variable, particularly in vegetative characters such as leaf-size, and petiole- and internode-length, but this variation is continuous so that it is not possible to recognize more than one hybrid taxon. There is no evidence for natural backcrossing.
   c. *P. × mixta* is widely distributed in Br and Hb and has also been recorded in Au, Be, Bu, Cz, Ga, Ge, He, Ho, Hu, It, Ju, Po, Rs and Su. It commonly occurs beside roads and footpaths, where the runners provide a highly efficient means of dispersal, and it is frequently isolated from its parents.
   d. Large numbers of reciprocal crosses between *P. reptans* and *P. anglica* have been attempted but only three experimental hybrids have been obtained. The experimental hybrids resembled natural *P. × mixta* in morphology, fertility and chromosome number. Incompatibility barriers also exist between *P. erecta* and *P. reptans*. Many attempted cross-pollinations have failed but one viable hybrid, with $2n = 28$, was raised in 1970. This was intermediate between the parents in most characters but was smaller than either of them, weak and highly sterile. It differed strikingly from the vigorous natural hybrid, *P. × mixta*. Five hybrids have also been obtained from crosses between Swiss specimens of *P. erecta* and *P. reptans* (Schwendener, 1969). Three of these hybrids were tetraploids with $2n = 28$, one was an aneuploid with $2n = 39–40$, and one was a hexaploid with $2n = 42$. The latter was thought to have arisen from an unreduced gamete in *P. reptans* and was very similar to *P. × mixta* in morphology and fertility. Experimental hybridizations have therefore revealed two possible origins for *P. × mixta*: hybridization between *P. reptans* and *P. anglica* or between *P. reptans* and *P. erecta* with *P. reptans* contributing an unreduced gamete. Whether both or only one of these crosses occurs

naturally is not known. The similarity between hybrids arising in different ways is explained by the fact that *P. anglica* itself almost certainly arose as the allopolyploid hybrid between *P. erecta* and *P. reptans*. Experimental backcrosses between *P. x mixta* and *P. anglica* have produced vigorous aneuploid progenies whose fertility and chromosome numbers increased with each successive generation. Sterile, morphologically aberrant off-spring have been obtained from crosses between *P. x mixta* and *P. erecta* but all crosses attempted between *P. x mixta* and *P. reptans* have failed.

e. *P. reptans* and *P. erecta* $2n = 28$ (tetraploid); *P. anglica* $2n = 56$ (octoploid); *P. x mixta* $2n = 42$ (hexaploid).

f. CZAPIK, R. (1968). Karyological studies on *Potentilla reptans* L. and *P. mixta* Nolte. *Acta biol. cracov.*, Sér. bot., **11**: 187-197.

MATFIELD, B. (1972). *Potentilla reptans* L.—Identification of its hybrids. *Watsonia*, **9**: 137-139. FIG.

MATFIELD, B. and ELLIS, J. R. (1972). The allopolyploid origin and genomic constitution of *Potentilla anglica*. *Heredity*, **29**: 315-327. FIG.

MATFIELD, B., JONES, J. K. and ELLIS, J. R. (1970). As above. FIG.

MURBECK, S. (1890). As above.

SCHWENDENER, J. (1969). Experimente zur Evolution von *Potentilla procumbens* Sibth. *Ber. schweiz. bot. Ges.*, **79**: 49-92.

215. *Fragaria* L.
(by C. A. Stace)

2 x 1. *F. moschata* Duchesne x *F. vesca* L.

= *F. x intermedia* Bach is known from Au, Cz, Ga, Ge and Rm, mostly as a garden escape.

3 x 1. *F. x ananassa* Duchesne (*F. chiloensis* auct.) x *F. vesca* L.

was reported from v.c. 22 in 1894, but W. R. Linton considered it a variant of *F. vesca*. It has apparently not been recorded elsewhere and experimental hybrids, which are variable in appearance but all pentaploids, are completely sterile, often dying well before flowering. *F. x ananassa* (octoploid) is itself derived from *F. chiloensis* (L.) Duchesne x *F. virginiana* Duchesne, and is commonly more or less naturalized. It is not treated here as neither parent is found wild in BI.

DRUCE, G. C. and LINTON, W. R. (1895). *Fragaria*. *Rep. B.E.C.*, **1**: 445-446.

YARNELL, S. H. (1931). Genetic and cytological studies on *Fragaria*. *Genetics, Princeton*, 16: 422-454.

216. *Geum* L.
(by A. E. S. Pike)

3 × 1. *G. rivale* L. × *G. urbanum* L.
a. *G.* × *intermedium* Ehrh. (Intermediate Avens).
b. A wide range of hybrids is common in many localities but those more or less exactly intermediate between the parents in petal-colour are most readily noticed. It is not possible in the field to distinguish the F_1 hybrid from later generations or backcrosses, and thus the extent of backcrossing and introgression which occurs is uncertain. Hybrid plants are often very bushy; they show a high degree of fertility and frequently completely link the two parents by a swarm of intermediates. The flowers, fruits and leaves each provide several diagnostic features.
c. Hybrids are usually found in fairly close association with both species, but seem more often to resemble *G. rivale* than *G. urbanum* in habitat preference. Hybrids are found throughout BI where both parents occur, except in southern Hb where the habitats of the two species rarely meet. The hybrid occurs throughout Europe and eastwards to the Altai in western Siberia and Tian-Shan in central Asia.
d. Artificial hybrids can be made relatively easily when *G. urbanum* is used as the female parent, but with greater difficulty in the opposite direction. In BI the main natural barriers to crossing are the differing habitats and the partially asynchronous flowering times of the parent species. A range of F_2 hybrids and both types of backcross plants are very easily raised. F_1 hybrids have about 90% fertility of the parents.
e. Both parents and hybrid ($2n = 42$).
f. GAJEWSKI, W. (1957). A cytogenetic study of the genus *Geum* L. *Monographiae bot.*, 4.
   MARSDEN-JONES, E. M. (1930). The genetics of *Geum intermedium* Willd. haud Ehrh., and its backcrosses. *J. Genet.*, 23: 377-395. FIG.
   PERRING, F. H. and SELL, P. D. (1968). *Op. cit.*, p. 28.
   PRYWER, C. (1932). Genetische Studien über die Bastarde zwischen *Geum urbanum* L. und *G. rivale* L. *Acta Soc. bot. Pol.*, 9: 87-114.
   ROSEN, D. (1916). Kreuzungsversuche *Geum urbanum* L. ♀ × *rivale* L. ♂. *Bot. Notiser*, 1916: 163-172.

SYME, J. T. B., ed. (1864). *English Botany*, 3rd ed. 3: 199, t. 458. London. FIG.

WEISS, F. E. (1912). *Geum intermedium* Ehr. and its segregates. *Rep. Br. Ass. Advmt Sci., Dundee*, Ser. K, pp. 675-676.

218. *Agrimonia* L.

(by C. A. Stace)

1 × 2. *A. eupatoria* L. × *A. odorata* (Gouan) Mill.

a. *A.* × *wirtgenii* Aschers. & Graebn.
b. Hybrid plants are intermediate between the parents in the indumentum, the petal-length, and the toothing of the leaflet-margins. No fruits are formed in the hybrids.
c. One (or few) plant(s) were found by N. H. Brittan in the late 1940s in a grassy verge in Hulne Park, v.c. 67, with both parents in the immediate vicinity. Hybrids have also been reported from Au, Cz, Ga, Ge and Po.
d. Artificial crosses carried out by C. M. Medd produced some hybrid progeny, but all died after the seedling stage. The natural hybrid showed a somewhat irregular meiosis. The barrier to hybridization is thus genetic and operates at more than one stage.
e. *A. eupatoria* $2n = 28$; *A. odorata* $2n = 56$; hybrid $2n = 42$.
f. BRITTAN, N. H. (1950). The *cytology, taxonomy and geographical distribution of some species of Agrimonia, with special reference to the British species*. Ph.D. thesis, University of Newcastle.

BRITTAN, N. H. (1953). Cytotaxonomy of some species of *Agrimonia* L., in OSVALD, H. and ÅBERG, E., eds *Proceedings of the Seventh International Botanical Congress, Stockholm*, p. 278.

MEDD, C. M. (1955). Meiotic behaviour in *Agrimonia* pollen mother cells. *Proc. B.S.B.I.*, 1: 376.

MEDD, C. M. (1956). *Meiotic behaviour in the species and interspecific hybrids of the genus Agrimonia L., with special reference to a haploid plant of Agrimonia eupatoria* L. Ph.D. thesis, University of Newcastle.

SKALICKÝ, V. (1962). Ein Beitrag zur Erkenntnis der europäischen Arten der Gattung *Agrimonia* L. *Acta Horti bot. prag.*, **1962**: 87-108.

220. *Alchemilla* L.
(by S. M. Walters)

1 × 2. *A. alpina* L. × *A. conjuncta* Bab.
= *A.* × *bakeri* Druce was said to have appeared in a rock-garden at Kew in
1869, along with the two parents, but no specimen has been traced. It is
possible that the plant was *A. plicatula* Gand., which is somewhat
intermediate in appearance and is grown in gardens. Hybrids involving
these species are not to be expected because so far as is known they are
obligate apomicts. *A. conjuncta* has occasionally been considered to be a
hybrid between *A. alpina* and *A. vulgaris* agg.; if so, it is obviously of
ancient origin.
DRUCE, G. C. (1918). *Alchemilla argentea* G. Don. *Rep. B.E.C.*, **5**: 20-26.
WALTERS, S. M. (1969). *Alchemilla*, in SYNGE, P. M., ed. *Supplement to
the Dictionary of gardening*, 2nd ed., pp. 167-168. Oxford.

225. *Rosa* L.
(by R. Melville)

The breeding-system in the majority of the British roses is very unusual and is
the major cause of the great taxonomic confusion which exists in the genus. Two
of the native species (as well as the four non-natives given by Dandy (1958)) are
normally-behaving sexual taxa, either diploid or tetraploid. All the others (of
which 13 are recognized in this account) are what may be termed unbalanced
polyploids, mostly pentaploids. In a pentaploid such as *R. canina* two of the five
sets of chromosomes pair at meiosis while the other three sets remain unpaired.
During the formation of pollen the latter sets are lost and each pollen grain
contains only one of the two paired sets (7 chromosomes). In the formation of
the embryo-sac the three unpaired sets are incorporated with one of the two
paired sets, so that the mature embryo-sac possesses four sets (28 chromosomes).
The fusion of haploid and tetraploid gametes reconstitutes the pentaploid state
(35 chromosomes) characteristic of these plants. One species, *R. villosa*, is an
unbalanced tetraploid, with haploid pollen and triploid embryo-sacs, and some
of the other British species may exist as unbalanced tetraploids, hexaploids or
heptaploids, as well as pentaploids. At one time it was thought that the
unbalanced polyploid roses were apomictic. While this may be true for some it is

now considered not to be the case generally, and many workers are of the opinion that apomixis is not of normal occurrence in British species.

This remarkable method of reproduction has several important consequences. 1. The haploid pollen of the whole group may be shared between its members with tremendous potential for hybridization. 2. Hybrid progeny are largely matroclinal, having in the pentaploids four sets of chromosomes from the female parent and only one set from the male. As a result of this, it is often possible to judge with reasonable certainty in which direction a cross has been made. 3. When the hybrids are between parents of different degrees of polyploidy, the chromosome number of the progeny will differ according to the direction in which the cross is made. For example female *R. arvensis* ($2n = 14$) x male *R. canina* ($2n = 35$) should be diploid, and the reciprocal hybrid pentaploid; female *R. pimpinellifolia* ($2n = 28$) x male *R. canina* should be triploid and the reciprocal hybrid hexaploid. Blackburn and Harrison (1921, 1924) found, however, that some of the hybrids between *R. pimpinellifolia* (female) and species of the *Villosae (R. tomentosa, R. sherardii* and *R. villosa)* were not triploids but fertile hexaploids ($2n = 42$), and they suggested that these had arisen by doubling of the chromosome number.

Rhodologists in the past have been unwilling to admit a plant as a hybrid unless there was obvious evidence of infertility. In fact, the hybrids range from completely infertile to relatively highly fertile. The more fertile hybrids in consequence were assigned to species to which they bore some resemblance or were recognized as distinct species. This not only obscured the origion of the hybrids, but effectively confused the limits of the true species. The "groups" *Andegavenses* and *Scabratae* under *R. canina* and *Deseglisei* and *Mercicae* under *R. dumetorum* are probably all hybrid; so too are at least some of the plants placed by Wolley-Dod (1930-1931) under "group" *Subcaninae* of *R. afzeliana* and "group" *Subcollinae* of *R. coriifolia.* Many of these have been identified as hybrids in this account, but more study is necessary, especially of the northern taxa.

The following taxonomic treatment differs from that given by Dandy (1958) and Warburg (1962) in that *R. dumetorum* is recognized in addition to *R. canina,* and *R. coriifolia* in addition to *R. afzeliana.* The last name is used instead of *R. dumalis* Bechst., the identity of which is extremely dubious. Under a the name given in bold-face is in the majority of cases that used by Dandy. Other names are added only if additional to the information given by Dandy, or if it is possible to typify each of the reciprocal crosses. No attempt has been made to mention every hybrid combination that has been recorded in BI, since many plants have been variously and wrongly identified, but some account is given of all those that appear to have been reliably recorded. Very few triple hybrids are given; probably more exist, but so far have been unrecognized due to the inconspicuousness of the characters of the third parent. Moreover no crosses between *R. tomentosa, R. villosa* and *R. sherardii,* between *R. canina* and *R. dumetorum,* or between *R. afzeliana* and *R. coriifolia* are described; it would be very difficult to prove the existence of these, and similarly difficult to distinguish two hybrids involving either of two species within one of these groups with a third species.

To save space paragraphs **d, e** and **f** are added only where there is relevant information available. Under **a, b** and **c** the expressions 1 x 7 and 7 x 1, for example, refer to reciprocal hybrids in which the female parent is given first, but in the main heading strictly alphabetical order is observed without regard to the direction of the cross. The detailed distributions refer to specimens seen by me, unless otherwise stated; no reference is made here to foreign records.

Chromosome Numbers

1. *R. arvensis* Huds. $2n = 14$
3. *R. multiflora* Thunb. $(2n = 14)$
4. *R. pimpinellifolia* L. $2n = 28$.
5. *R. rugosa* Thunb. $(2n = 14)$
7. *R. stylosa* Desv. $2n = 35, 42$ (unbalanced)
8. *R. canina* L. $2n = 35$ (unbalanced)
8d. *R. dumetorum* Thuill. $2n = 35$ (unbalanced)
9. *R. afzeliana* Fr. $2n = 35$ (unbalanced)
9c. *R. coriifolia* Fr. $2n = 35$ (unbalanced)
10. *R. obtusifolia* Desv. $2n = 35$ (unbalanced)
11. *R. tomentosa* Sm. $2n = 35$ (unbalanced)
12. *R. sherardii* Davies $2n = 28, 35, 42$ (unbalanced)
13. *R. villosa* L. $2n = 28, 56$ (unbalanced)
14. *R. rubiginosa* L. $2n = 35$ (unbalanced)
15. *R. micrantha* Borrer ex Sm. $2n = 35, 42$ (unbalanced)
17. *R. agrestis* Savi $2n = 35, 42$ (unbalanced)
gal. *R. gallica* L. $(2n = 28)$
pen. *R. pendulina* L. $(2n = 28)$

General References

BLACKBURN, K. B. (1949). Chromosomes and classification in *Rosa*, in WILMOTT, A. J., ed. *British flowering plants and modern systematic methods*, pp. 53-57. London.

CRÉPIN, F. (1894). Rosae hybridae. Études sur les roses hybrides. *Bull. Soc. r. bot. Belg.*, Mem. **33**: 7-149.

DANDY, J. E. (1958). *Op. cit.*, pp. 56-58.

MATTHEWS, J. R. (1920). Hybridism and classification in the genus *Rosa. New Phytol.*, **19**: 153-171.

MELVILLE, R. (1967). The problem of classification in the genus *Rosa. Bull. Jard. bot. nat. Belg.*, **37**: 39-44.

WOLLEY-DOD, A. H. (1930-1931). A revision of the British roses. *J. Bot., Lond.*, **68** and **69**, Suppl.

1 x 7. *R. arvensis* Huds. x *R. stylosa* Desv.

a. *R.* x *bibracteoides* W.-Dod (= 1 x 7) (*R. stylosa* var. *pseudo-rusticana* Crép. ex Rogers (= 7 x 1)).

b. 1 x 7 differs from *R. arvensis* f. *major* Coste in the larger, often pink petals, more or less regularly-pinnated sepals, and larger leaflets with more numerous teeth and an intermediate shape. 7 x 1 differs from *R. stylosa* in the long, arching branches with few prickles, larger, deeply serrate, usually glabrous leaflets, white petals, long-exserted styles, and rather densely glandular-hispid pedicels.

   c. 1 x 7 is recorded from v.c. 1/2, 3/4, 5/6, 18/19 and 37, and 7 x 1 from v.c. 3/4, 9 and 17.

1 x 8. *R. arvensis* Huds. x *R. canina* L.

   a. *R.* x *wheldonii* W.-Dod (*R. canina* var. *flexilis* (Déségl.) Rouy, var. *rousselii* (Rip.) Rouy, var. *schottiana* Ser., and var. *pouzinii* (Tratt.) W.-Dod f. *anglica* Dingl. and f. *wolley-dodii* Sudre) (= 8 x 1); (*R.* x *debilis* W.-Dod (= 1 x 8)).

   b. 1 x 8 is a usually small, rather slender bush with dark-pigmented branches, small intermediate-shaped, irregularly biserrate leaflets, often pink petals, more or less pinnated sepals, intermediate-shaped hips, and pedicels which are usually non-glandular and shorter than in *R. arvensis.* 8 x 1 is a variable complex of plants resembling *R. canina* but separated from it by various characters of *R. arvensis.*

   c. 1 x 8 is recorded from v.c. 17 and 34, and 8 x 1 from v.c. 3/4, 15/16, 17, 33/34, 36, 47 and 59/60.

   f. WOLLEY-DOD, A. H. (1936). Some rose notes. *Rep. B.E.C.*, **11**: 68-81.

1 x 8 x 10. *R. arvensis* Huds. x *R. canina* L. x *R. obtusifolia* Desv.
see under *R. canina* x *R. obtusifolia.*

1 x 8d. *R. arvensis* Huds. x *R. dumetorum* Thuill.

   a. (*R. dumetorum* var. *deseglisei* (Bor.) Chr. and var. *incerta* (Déségl.) W.-Dod *pro parte*) (= 8d x 1).

   b. 8d x 1 has the general appearance of *R. dumetorum*. Var. *deseglisei* differs from the type in the long, glandular-setose pedicels and small leaflets, and var. *incerta* differs from the latter in the shorter pedicels with rather fewer glandular setae and subglobose to ovoid hips. 1 x 8d resembles *R. arvensis* but has elliptic to suborbicular, more or less acutely serrate leaflets which are pubescent beneath, pink petals, free, protruding styles, and long, glandular pedicels.

   c. 1 x 8d is recorded from Beenham, v.c. 22, and 8d x 1 from v.c. 3/4, 5/6, 17, 21, 33/34 and 35.

1 x 10. *R. arvensis* Huds. x *R. obtusifolia* Desv.
see under *R. canina* x *R. obtusifolia.*

1 x 14. *R. arvensis* Huds. x *R. rubiginosa* L.

   a. (*R. arvensis* var. *gallicoides* (Déségl.) Crép.) (= 1 x 14).

   b. 1 x 14 differs from *R. arvensis* in the presence of acicles on the pedicels and branches, the glandular biserration of the leaflets, the often sessile glands, and the intermediate shape of the leaflets. 14 x 1 is a small, weak bush differing from *R. rubiginosa* in the small leaflets with an intermediate shape and less compound serration, pale pink flowers, and sparsely glandular-setose to aciculate pedicels. The hips are not developed.

c. 1 x 14 is recorded from v.c. 3/4, 23, 29, 33/34, 36, 38, 55 and 57, and 14 x 1 from v.c. 44.

1 x 15. *R. arvensis* Huds. x *R. micrantha* Borrer ex Sm.

a. *R.* x *inelegans* W.-Dod (= 15 x 1).

b. 15 x 1 differs from *R. micrantha* in the larger leaflets which are pale beneath, the frequently abortive hips which are not contracted below the disk, and in the glandular-hispid pedicels.

c. 15 x 1 is recorded from v.c. 32; 1 x 15 has not been recorded.

1 x gal. *R. arvensis* Huds. x *R. gallica* L.

has been suggested as the parentage of *R. gallicoides* Déségl. (= *R. arvensis* var. *gallicoides* (Déségl.) Crép.)—see under *R. arvensis* x *R. rubiginosa*.

3 x 14. *R. multiflora* Thunb. x *R. rubiginosa* L.

a. None.

b. The 5–7 leaflets are 15–25 mm long, broadly ovate, strongly biserrate, sparsely pubescent above, and pubescent and dotted with *R. rubiginosa*-type glands beneath; the slender, glandular-setose pedicels are 12–17 mm long and bear panicles of flowers and narrowly to broadly ellipsoid hips which are 6–10 mm long. These characters suggest that *R. multiflora* was the female parent.

c. This hybrid is recorded only from Herm, C.I.

8 x 4. *R. canina* L. x *R. pimpinellifolia* L.

a. (*R. hibernica* Templeton var. *grovesii* Baker) (= 4 x 8).

b. 4 x 8 resembles the better-known *R.* x *hibernica* (q.v.) but differs in the glabrous leaves and in the main stems with mainly stout, straight or arching, *R. canina*-type prickles and few slender prickles.

c. 4 x 8 is known from Barnes, Wimbledon and Ham Commons, v.c. 17, and Luddesdown, v.c. 16. Northern records are errors for *R. afzeliana* x *R. pimpinellifolia*; 8 x 4 has not been recorded.

e. Hybrid ($2n$ = 42 (hexaploid)).

f. TACKHOLM, G. (1922). Zytologische Studien über die Gattung *Rosa*. *Acta Horti Bergiani*, 7: 97-381.

8d x 4. *R. dumetorum* Thuill. x *R. pimpinellifolia* L.

a. *R.* x *hibernica* Templeton (= 4 x 8d).

b. 4 x 8d differs from *R. pimpinellifolia* in its usually larger, suborbicular to elliptic leaflets with sharp serration and occasional secondary teeth, the hairs on the rachis and at least the leaflet veins beneath, the intermediate armature with many rather stout prickles with elongated bases, and the red suborbicular to urn-shaped hips on smooth pedicels 8–15 mm long.

c. 4 x 8d is recorded from v.c. 58, 70, H33 and H38; 8d x 4 has not been recorded.

d. *R. pimpinellifolia* is apparently self-sterile (Harrison, 1955). The artificial cross *R. dumetorum* female x *R. pimpinellifolia* produced a plant "very close" to *R. x hibernica*, but with a more dwarf *R. pimpinellifolia*-like habit (Williams, 1959).

f. HARRISON, J. W. H. (1921). The genus *Rosa*, its hybridology and other genetical problems. *Trans. nat. Hist. Soc. Northumb.*, n.s., **5**: 244-325. FIG.

HARRISON, J. W. H. (1955). Durham wild roses. *Proc. B.S.B.I.*, **1**: 369-371, 373-374.

SYME, J. T. B., ed. (1864). *English Botany*, 3rd ed., **3**: 205-206, t. 463. FIG.

WILLIAMS, W. (1959). Department of Plant Breeding. *Rep. John Innes hort. Instn*, **49**: 7-13.

9 x 4. *R. afzeliana* Fr. x *R. pimpinellifolia* L.

a. (*R. hibernica* var. *glabra* Bak., *R. glabra* (Bak.) W.-Dod, *non* Andr.) (= 4 x 9).

b. 4 x 9 has glabrous, rounded to elliptic, sharply serrate or biserrate leaflets, a few strongly arching prickles on the rachis, usually red stems with slender and rather stout prickles, and smooth, globular to ovoid hips. 9 x 4 is a small, erect shrub with small, elliptic, acute leaflets similar to those of *R. afzeliana*, main stems with numerous, long-based, moderately stout, slightly arching prickles, branches with intermediate, slightly arching prickles, smooth pedicels, smooth, ellipsoid hips, and pink petals. It is reported as completely sterile (Harrison, 1955).

c. 4 x 9 is recorded from v.c. 61/65, and 9 x 4 from v.c. 66.

f. HARRISON, J. W. H. (1955). As above.

9c x 4. *R. coriifolia* Fr. x *R. pimpinellifolia* L.

a. *R. x setonensis* W.-Dod (*R. x margerisonii* W.-Dod) (= 4 x 9c).

b. 4 x 9c has leaves resembling those of *R. pimpinellifolia* but the leaflets are elliptic to obovate with nearly simply serrate or slightly biserrate margins. The stem bears fine, erect prickles interspersed with stouter, strongly arched ones, the hips are 8–10 mm long, oblate-spheroid and smooth with erect, simple sepals, and the pedicels are smooth. This hybrid is fertile.

c. 4 x 9c is recorded from v.c. 61/65, 67 and 82; 9c x 4 has not been recorded.

d. Harrison (1955) reared an F_2 generation of *R. pimpinellifolia* x *R. caesia* which showed even greater fertility than the wild F_1. He also synthesized the hybrid, using *R. pimpinellifolia* as the female parent, and reared it to the F_2 generation, which closely resembled *R. pimpinellifolia*. The chromosome numbers of the F_1 plants varied very considerably, both euploid and aneuploid, but in the F_2 were mostly $2n = 28$. A plant found on the coast of v.c. 67 was considered by Blackburn and Harrison (1921) to be *R. pimpinellifolia* female x (*R. pimpinellifolia* x *R. coriifolia*); it had $2n = 28$ with rather variable but often high fertility.

f. BLACKBURN, K. B. and HARRISON, J. W. H. (1921). The status of the British rose forms as determined by their cytological behaviour. *Ann. Bot.*, **35**: 159-188.
  HARRISON, J. W. H. (1921). As above. FIG.
  HARRISON, J. W. H. (1955). As above.
  WOLLEY-DOD, A. H. (1936). As above.

4 x 11. *R. pimpinellifolia* L. x *R. tomentosa* Sm.

a. (*R.* x *wilsonii* Borrer, *pro parte?*).
b. The leaves have usually seven leaflets which are suborbicular to broadly ovate or elliptic, rather acutely biserrate, thinly hairy above and more or less densely tomentose beneath, and a pubescent rachis with small often subsessile glands and occasional pricklets. The pedicels are 15–25 mm long and thinly to rather densely glandular-setose to glandular-aciculate, and the hips are *c* 7 mm diameter, broadly ovoid or obovoid and narrowed below the sepals which are spreading to reflexed and rather thinly glandular dorsally.
c. This hybrid is recorded from v.c. 13/14, 15/16, 37 and 38.
d. An artificial hybrid between *R. tomentosa* Sm. var. *dumosa* (Pug.) Rouy (female) and *R. pimpinellifolia* was said to resemble the holotype of *R.* x *wilsonii* in most characters, but to approach the male parent more in foliage and fruit colour (Williams, 1959). The specimen of *R.* x *wilsonii* in herb. Borrer is, however, *R. pimpinellifolia* x *R. sherardii*, as is the figure in Blackburn and Harrison (1924). The conflicting results of Wylie (Darlington, 1951) and Blackburn and Harrison (1924) concerning the cytology, fertility and direction of the cross are probably explained by the fact that different taxa were being investigated. Hybrids involving both directions of crossing probably occur.
f. BLACKBURN, K. B. and HARRISON, J. W. H. (1924). Genetical and cytological studies in hybrid roses, 1. The origin of a fertile hexaploid form in the Pimpinellifoliae-Villosae crosses. *Br. J. exp. Biol.*, **1**: 557-570 (Figure of *R.* x *wilsonii* probably represents *R. pimpinellifolia* x *R. sherardii*).
  DARLINGTON, C. D. (1951). Cytology Department. *Rep. John Innes hort. Instn*, **41**: 19-24.
  HARRISON, J. W. H. (1921). As above. FIG.
  SYME, J. T. B., ed. (1864). *English Botany*, 3rd ed., **3**: 206, t. 464. London. FIG.
  WILLIAMS, W. (1959). As above.

4 x 12. *R. pimpinellifolia* L. x *R. sherardii* Davies

a. *R.* x *involuta* Sm. (*R.* x *gracilis* Woods, *R.* x *rubella* Sm., *R.* x *wilsonii* Borrer, *pro parte?*) (= 4 x 12).

b. In 4 x 12 the leaves have usually seven leaflets which are suborbicular to broadly ovate or elliptic and biserrate, and a thinly and closely pubescent to subglabrous rachis with some glandular-setose hairs and small, straight or slightly arching pricklets. The pedicels are 10–15 (–20) mm long and usually rather densely glandular-aciculate, the hips are subglobose, 7–10 mm diameter and usually rather densely covered with stout setae or acicles, and the sepals are reflexed to erect and glandular dorsally. This hybrid is fertile, and R. x rubella is probably a segregate or a backcross to R. pimpinellifolia. 12 x 4 is a tall bush with large leaves with often seven leaflets up to 30 mm long and having a R. sherardii-type indumentum but a suborbicular to elliptic shape and an intermediate serrature. The pedicels and hips approach those of R. sherardii.

c. 4 x 12 is recorded from v.c. 49, 50, 55, 58, 61/65, 66, 67, 69, 79, 87/89, 100, 104 and H33, and 12 x 4 from v.c. 67, 98, 104 and 107.

d. Harrison (1955) reared 4 x 12 up to the F_4 generation, which was even more fertile than the wild F_1. Artificial hybrids were also made and found to be fertile; the F_2 inclined more to R. pimpinellifolia. The chromosome number of the artificial F_1 plants varied considerably, both euploid and aneuploid, but in the F_2 were mostly $2n = 28$ and some plants of this generation resembled R. x rubella. Some of the results obtained by Blackburn and Harrison (1924), and referred to R. wilsonii and R. sabinii, almost certainly apply to this hybrid.

f. BLACKBURN, K. B. and HARRISON, J. W. H. (1924). As above. FIG (as R. sabinii and R. wilsonii).

HARRISON, J. W. H. (1921). As above. FIG.

HARRISION, J. W. H. (1955). As above.

SYME, J. T. B., ed. (1864). English Botany, 3rd ed., 3: 204-205, t. 462. London. FIG.

4 x 14 x 12. R. pimpinellifolia L. x R. rubiginosa L. x R. sherardii Davies

a. R. x perthensis Rouy.

b. This is a bush to 2 m high having a complete mixture of characters of the three putative parents. The often seven leaflets and scattered slender prickles on the stem are R. pimpinellifolia characters; the leaflet indumentum is predominantly of the R. sherardii type but also includes glands of the R. rubiginosa type, and the acicles on the hips, pedicels and stems below the pedicels are also characters of the latter species. It is fertile.

c. The type locality of R. x perthensis is at Auchterarder, v.c. 88. Matthews also noted it from near Forres, v.c. 95.

d. Matthews (1934) grew c 75 "seeds" of this plant and found that all the offspring were alike; he considered that R. x perthensis was not a recent hybrid but an ancient "crypthybrid" close to R. sherardii.

f. BLACKBURN, K. B. and HARRISON, J. W. H. (1924). As above. FIG (as R. omissa x R. pimpinellifolia).

HARRISON, J. W. H. (1921). As above. FIG (as *R. omissa* x *R. pimpinellifolia* from Auchterarder).

MATTHEWS, J. R. (1934). *Rosa perthensis* Rouy and its history as a British plant. *J. Bot., Lond.*, **72**: 167-171.

4 x 13. *R. pimpinellifolia* L. x *R. villosa* L.

a. *R.* x *sabinii* Woods (= 4 x 13); (*R.* x *mayoensis* W.-Dod, *R.* x *marshallii* W.-Dod (= 13 x 4)).

b. 4 x 13 has leaves with 5–7(9), broadly ovate to broadly elliptic, biserrate leaflets, and a pubescent rachis with scattered short, glandular hairs and occasional small prickles. The pedicels are 10–20 mm long and thinly glandular-aciculate, the hips are 15 x 10–15 mm, ovoid-urceolate and have scattered setae or acicles, and the sepals are erect and often somewhat fleshy at the base. 13 x 4 has rather large leaves with 7–9, elliptic to orbicular, softly pubescent leaflets, stems with numerous, erect prickles of varying size, pedicels and hips which are thinly aciculate, and subglobose hips with erect sepals. It was reported as completely sterile in v.c. 66 (Harrison, 1955).

c. 4 x 13 is recorded from v.c. 67, 87/89, 90, 92/93 and 94, and 13 x 4 from v.c. 87/89.

d. Blackburn and Harrison (1921) found that "*R.* x *sabinii*" was a sterile hexaploid forming 14 bivalents and 14 univalents at meiosis, but their drawing of this plant apparently represents *R. pimpinellifolia* x *R. sherardii*.

f. BLACKBURN, K. B. and HARRISON, J. W. H. (1921). As above.

BLACKBURN, K. B. and HARRISON, J. W. H. (1924). As above. FIG (as *R. pimpinellifolia* x *R. mollis*).

HARRISON, J. W. H. (1921). As above. FIG.

HARRISON, J. W. H. (1955). As above.

SYME, J. T. B., ed. (1864). *English Botany*, 3rd ed., **3**: 206-207, t. 465. London. FIG.

4 x 14. *R. pimpinellifolia* L. x *R. rubiginosa* L.

a. *R.* x *cantiana* (W.-Dod) W.-Dod (= 4 x 14).

b. In 4 x 14 the armature of the stems is intermediate in nature. The leaves resemble those of *R. pimpinellifolia* but have suborbicular to elliptic leaflets with triangular-dentate to serrate and glandular-serrulate margins and lower surfaces sparsely pubescent and dotted with glands of the *R. rubiginosa* sort. The pedicels are 7–12 mm long and the hips subglobose to ovoid, glandular-setose to -aciculate, and with erect, nearly simple sepals. This hybrid is fertile.

c. 4 x 14 is recorded from v.c. 15/16, 66, 80, 82, 87/89 and 90; 14 x 4 has not been recorded.

d. Harrison (1955) reared plants up to the F_3 generation, which was more fertile than the wild F_1. Using *R. pimpinellifolia* as the female parent he made artificial hybrids which showed some fertility and a range of euploid

and aneuploid chromosome numbers. Many of the F_2 plants had $2n = 28$ and closely resembled *R. pimpinellifolia,* and an F_3 generation was raised.
f. HARRISON, J. W. H. (1921). As above. FIG.
   HARRISON, J. W. H. (1955). As above.

pen × 4. *R. pendulina* L. (*R. alpina* L.) × *R. pimpinellifolia* L.

was one of the parentages wrongly attributed to *R. x rubella* (see under *R. pimpinellifolia* x *R. sherardii*) by some authors.

8 × 5. *R. canina* L. × *R. rugosa* Thunb.

a. *R. x praegeri* W.-Dod.
b. This hybrid has stems with slender, erect prickles and a few acicles and leaves with (5)7–9, elliptic leaflets which are dull and slightly bullate on the upper surface and pubescent on the veins on the lower surface. The pedicels are 20–25 mm long and smooth or with scattered acicles. The petals are bright crimson, the sepals long and simple with an expanded tip and glandular-setose dorsally, and the hips are globose to ovoid.
c. *R. x praegeri* is recorded from only one of the many localities in which *R. rugosa* is naturalized, viz. Cushendun, v.c. H39.
d. This hybrid has been synthesized by Gustafsson (1944) with *R. canina* as the female parent. The F_1 plants had $2n = 35$ and showed a meiosis with up to 14 bivalents and 7 univalents.
f. GUSTAFSSON, A. (1944). The constitution of the *Rosa canina* complex. *Hereditas,* **30:** 405-428. FIG.
   PRAEGER, R. L. (1928). A new hybrid rose. *J. Bot., Lond.,* **66:** 87-88.

8 × 7. *R. canina* L. × *R. stylosa* Desv.

a. (*R. stylosa* var. *garroutei* (Pug. & Rip.) Rouy) (= 7 x 8); (*R. canina* var. *andegavensis* (Bast.) Desf., *pro parte* (= 8 x 7)).
b. 7 x 8 differs from *R. stylosa* in that the leaflets are weakly biserrate and pubescent only on the midrib below, and the styles are less exserted and not fused. 8 x 7 differs from *R. canina* var. *canina* and parts of var. *andegavensis* in having leaflets of the *R. stylosa* shape, more or less glandular pedicels, and exserted though not fused styles.
c. 7 x 8 is recorded from v.c. 5/6, 7/8, 17, 31 and 33/34, and 8 x 7 from v.c. 3 and 5/6.

8d × 7. *R. dumetorum* Thuill. × *R. stylosa* Desv.

a. *R. x rufescens* W.-Dod (= 8d x 7).
b. 8d x 7 is similar to *R. dumetorum* with the leaflets pubescent below and glabrous above, but has styles in a column on a conical disk, narrowly ovoid, often abortive hips, and sparsely glandular-setose pedicels.
c. 8d x 7 is recorded from v.c. 15/16; 7 x 8d has not been recorded.
f. WOLLEY-DOD, A. H. (1936). As above.

17 × 7. R. agrestis Savi × **R. stylosa** Desv.

a. (*R. stylosa* var. *belnensis* (Ozan) Rouy) (= both 7 x 17 and 17 x 7).

b. 7 x 17 is a vigorous bush with elliptic to narrowly elliptic leaflets which are acute at the apex, rounded at the base, much less glandular than in *R. agrestis*, and have an intermediate serrature. The pedicel length and number of flowers is variable, and the hips subglobose with exserted, usually fused styles borne on a slightly conical disk. 17 x 7 is less vigorous and has very glandular, small leaflets resembling those of *R. agrestis* except for the intermediate serrature. In the other characters it resembles 7 x 17.

c. 7 x 17 is recorded from v.c. 5/6 and 17, and 17 x 7 from v.c. 17.

9 × 8. R. afzeliana Fr. × **R. canina** L.

a. (*R. afzeliana* var. *subcanina* (Chr.) W.-Dod and var. *denticulata* (Kell.) W.-Dod) (= 9 x 8).

b. 9 x 8 has glabrous leaflets intermediate between those of the parents in shape but simply serrate in var. *subcanina* and biserrate in var. *denticulata*. The hips are subglobose and bear reflexed sepals and woolly styles which are fewer and looser than those of var. *afzeliana*.

c. 9 x 8 is recorded from v.c. 66, 89 and 111; 8 x 9 has not been recorded.

8 × 9c. R. canina L. × **R. coriifolia** Fr.

a. (*R. coriifolia* var. *subcollina* (Chr.) W.-Dod, var. *caesia* (Sm.) W.-Dod and probably most of the other varieties in "group" *Subcollinae* of *R. coriifolia*.) (= 9c x 8).

b. 9c x 8 has ellipsoid to subglobose hips with reflexed sepals. In var. *subcollina* the leaflets are sparsely pubescent below and simply serrate; in var. *caesia* they are often more densely pubescent and biserrate.

c. 9c x 8 is recorded from v.c. 66, 85, 88 and 93; 8 x 9c has not been recorded.

e. Blackburn and Harrison (1921) found a plant they determined as *R. coriifolia* var. *lintonii* x *R. lutetiana* (the latter a variant of *R. canina* in the present sense) to have $2n = 35$.

f. BLACKBURN, K. B. and HARRISON, J. W. H. (1921). As above.

8 × 10. R. canina L. × **R. obtusifolia** Desv.

a. *R. x concinnoides* W.-Dod (*R. x subobtusifolia* W.-Dod, *R. x surreyana* W.-Dod) (= 10 x 8).

b. 10 x 8 has the general aspect of *R. obtusifolia* but has less pubescent leaflets with intermediate serrature. It is possible that *R. dumetorum* rather than *R. canina* is the pollen parent in some cases. In nothomorphs *surreyana* and *concinnoides* the acutely biserrate leaflets suggest *R. canina*, but the flat disk and long styles might be due to introgression from *R. arvensis* as well; Dandy (1958) in fact gave the parentage of these as *R. arvensis* x *R. obtusifolia*.

c. 10 x 8 is recorded from v.c. 9, 15/16, 17 and 58; 8 x 10 has not been recorded.

f. DANDY, J. E. (1958). *Op. cit.*, p. 56.
WOLLEY-DOD, A. H. (1936). As above.

8 × 11. *R. canina* L. × *R. tomentosa* Sm.

a. *R.* × *curvispina* W.-Dod (= 8 × 11); (*R. sherardii* var. *woodsiana* (Groves) W.-Dod (= 11 × 8)).

b. 8 × 11 is intermediate in leaflet-shape and -serrature; the leaflets are glabrous or thinly pubecent below, the pedicels are 12–20 mm long and glandular-hispid, and the hips are subglobose. 11 × 8 has 5–7 narrowly elliptic to elliptic, acute leaflets with deep, narrow, glandular, double serrations and a glandular-pubescent rachis with prickles. The pedicels are 10–15 mm and glandular-hispid, the hips broadly ovoid, and the sepals erect and glandular-hispid to nearly glabrous.

c. 8 × 11 is recorded from v.c. 40, and 11 × 8 from v.c. 18/19, 33/34 and 40.

8 × 12. *R. canina* L. × *R. sherardii* Davies

was recorded by Harrison from v.c. 66. He said it was fertile and that he had raised an F_2 generation.
HARRISON, J. W. H. (1955). As above.

8 × 14. *R. canina* L. × *R. rubiginosa* L.

a. *R.* × *latens* W.-Dod (*R. canina* var. *aspernata* (Déségl.) Briggs, var. *blondeana* (Rip.) Rouy, *pro parte*, var. *latebrosa* (Déségl.) N. E. Br and var. *verticillacantha* (Mér.) Baker, *R. dumetorum* var. *mercica* W.-Dod and var. *seticaulis* W.-Dod, *R. obtusifolia* var. *rothschildii* (Druce) W.-Dod) (= 8 × 14).

b. Nothomorph *latens* has glabrous leaflets with scattered *R. rubiginosa*-type glands. In most cases the leaflet-shapes are more or less intermediate, and the *R. rubiginosa*-type acicles are generally manifested to some extent. Some of the nothomorphs described as varieties of *R. canina* may be backcrosses to that species; those described as varieties of *R. dumetorum* and *R. obtusifolia* have pubescence on the lower leaf-surface. Occasional nothomorphs with leaflet-shape, indumentum and armature more like those of *R. rubiginosa* are probably 14 × 8. Some of these have narrowly ellipsoidal hips with few "seeds", whereas in 8 × 14 fertility appears to be higher.

c. 8 × 14 is recorded from southern England and the Midlands, and 14 × 8 from v.c. 67/68.

d. Gustafsson (1944) synthesized this hybrid reciprocally, the F_1 generation in general quite closely resembling the female parent. With *R. canina* as female the F_1 was very heterogeneous and rather highly sterile; with *R. rubiginosa* as female it was homogeneous and nearly as fertile as the parents. The F_1 plants had $2n = 35$ with up to seven bivalents at meiosis.

e. *R. canina* var. *blondeana* (*sec.* Gustafsson) ($2n = 42$).

f. GUSTAFSSON, A. (1944). As above. FIG.

8 × 15. R. canina L. × R. micrantha Borrer ex Sm.
  a. *R.* × *toddii* W.-Dod (= 15 × 8).
  b. 15 × 8 has leaflets of intermediate shape, glabrous above and sparsely pubescent below with some *R. micrantha*-type glands; the glands on the biserrations are less developed than in *R. micrantha*. The stems have rather sparse, stout, arching prickles, the pedicels are glabrous to glandular-hispid, and the sepals glandular and pinnated.
  c. 15 × 8 is recorded from v.c. 23 and 61/65; 8 × 15 has not been recorded.

8 × gal. R. canina L. × R. gallica L.
  = *R.* × *alba* L. has been recorded, presumably as an escape from or relic of cultivation.

8d × 10. R. dumetorum Thuill. × R. obtusifolia Desv.
  see under *R. canina* × *R. obtusifolia*.

8d × 11. R. dumetorum Thuill. × R. tomentosa Sm.
  = *R.* × *aberrans* W.-Dod was recorded by Wolley-Dod (1930–31) from v.c. 58, but it would be extremely difficult to distinguish this hybrid from *R. canina* × *R. tomentosa*.

8d × 13. R. dumetorum Thuill. × R. villosa L.
  was recorded by Harrison from v.c. 66. He said it was fertile and that he raised an F$_2$ generation.
  HARRISON, J. W. H. (1955). As above.

8d × gal. R. dumetorum Thuill. × R. gallica L.
  = *R.* × *collina* Jacq., *non* Woods has been recorded, presumably as an escape from or relic of cultivation.

9 × 11. R. afzeliana Fr. × R. tomentosa Sm.
  a. *R.* × *rogersii* W.-Dod (= 11 ×9).
  b. 11 × 9 has often seven leaflets which are elliptic to ovate, rather strongly biserrate and glabrous or nearly so except for glands. The rachis is glandular-hispid and has numerous prickles. The hips are ellipsoidal and often infertile and the pedicels are slender and glandular-setose. Wolley-Dod (1930–31) attributed *R.* × *rogersii* to *R. canina* × *R. tomentosa*.
  c. 11 × 9 is recorded from v.c. 67/68, 69 and 105; 9 × 11 has not been recorded.

9 × 12. R. afzeliana Fr. × R. sherardii Davies
  a. (*R. sherardii* f. *glabrata* Ley, *R. afzeliana* var. *pseudohaberiana* (Kell.) W.-Dod) (= 9 × 12).
  b. 9 × 12 has elliptic, biserrate leaflets which are glabrous but sparsely glandular below. The stem-prickles are mostly rather slender and arching,

but with some stouter ones of the *R. afzeliana* type. The hips are subglobose to obovoid, usually smooth, and borne on pedicels which are 8–10 mm long and thinly glandular-setose. 12 x 9 differs mainly in the leaflets which are uniformly pubescent as well as glandular below, and in the pedicels which are more hispid and usually 10–12 mm long.

c. 9 x 12 is recorded from v.c. 67/68, 97, 103, 105 and 111, and 12 x 9 from v.c. 59/60, 64 and 96.

9 x 13. *R. afzeliana* Fr. x *R. villosa* L.

= *R.* x *glaucoides* W.-Dod was recorded by Wolley-Dod (1930–31), but Dandy (1958) placed *R. glaucoides* under *R. sherardii*.

9 x 14. *R. afzeliana* Fr. x *R. rubiginosa* L.

a. (*R. dumetorum* var. *fanasensis* R. Kell.) (= 9 x 14).

b. In 9 x 14 the leaflets are nearest to those of *R. afzeliana* in shape, but strongly glandular-biserrate and on some of the lower leaves broader as in *R. rubiginosa*. The lower leaf-surface is pubescent on the veins and has a few glands. The prickles on the stem are like those of *R. afzeliana* but more slender and suberect on the flowering branchlets. The hips are ellipsoid with woolly styles and reflexed sepals.

c. 9 x 14 is recorded from v.c. 66; 14 x 9 has not been recorded.

9c x 14. *R. coriifolia* Fr. x *R. rubiginosa* L.

a. (*R. coriifolia* var. *obovata* (Baker) W.-Dod) (= 9c x 14).

b. In 9c x 14 the leaflets are obovate with a cuneate base to elliptic with a rounded base, rather sharply glandular-biserrate, glabrous above and pubescent beneath with numerous glands of the *R. rubiginosa* type. The stout stem-prickles are like those of *R. coriifolia* but mixed with more slender ones on the flowering branchlets. The hips are subglobose to obovoid with reflexed to spreading sepals.

c. 9c x 14 is recorded from v.c. 66; 14 x 9c has not been recorded.

d. Gustafsson (1937) produced ·artificial hybrid seed but failed to obtain full-grown F$_1$ plants.

f. GUSTAFSSON, A. (1937). Experimentella undersökringer över fort-plantningssätt och formbildung hos de apomiktiska rosorna, 1. *Bot. Notiser*, **1937**: 323-331.

10 x 14. *R. obtusifolia* Desv. x *R. rubiginosa* L.

a. *R.* x *tomentelliformis* W.-Dod (= 10 x 14).

b. 10 x 14 resembles *R. obtusifolia* but has acicles on the branchlets below the flowers and *R. rubiginosa*-type glands on the leaves.

c. 10 x 14 is recorded from v.c. 58; 14 x 10 has not been recorded.

12 x 11. *R. sherardii* Davies x *R. tomentosa* Sm.

= *R.* x *suberectiformis* W.-Dod was recorded from v.c. 70 by Wolley-Dod (1930–31). The *R. sherardii* parent was stated to be *R. sherardii* var.

suberecta (Ley) W.-Dod, which is itself in my opinion *R. rubiginosa* x *R. sherardii.*

14 x 11. *R. rubiginosa* L. x *R. tomentosa* Sm.

a. None.

b. 11 x 14 has the general aspect of *R. tomentosa* with rather slender, straight or declining prickles and rare acicles on the flowering branches. The leaflets are mostly ovate or elliptic, sparsely pubescent on both surfaces and with a few stout glands below. The pedicels are glandular-hispid or more or less glandular-aciculate, and the hips narrowly ovoid and rather infertile. 14 x 11 resembles *R. rubiginosa* more closely in the stouter, arching stem-prickles with some acicles below the inflorescence, and in the suborbicular to elliptic leaflets which are glabrous above and sparsely pubescent with numerous, stout glands below. The hips are 5–8 mm long and subglobose, and the pedicels glandular-hispid with some acicles.

c. 11 x 14 is recorded from v.c. 7/8, 12 and 44, and 14 x 11 from v.c. 45 and H20.

12 x 13. *R. sherardii* Davies x *R. villosa* L.

= *R.* x *shoolbredii* W.-Dod was recorded by Wolley-Dod (1930–31) from northern Br, but included by Dandy (1958) under *R. sherardii.*

14 x 12. *R. rubiginosa* L. x *R. sherardii* Davies

a. *R.* x *burdonii* W.-Dod (*R. sherardii* var. *suberecta* (Ley) W.-Dod, *pro parte*) (= 12 x 14); (*R. sherardii* var. *suberecta* f. *glabrata* (Ley) W.-Dod (= 14 x 12)).

b. 12 x 14 is intermediate or closer to one parent or the other in leaflet-shape and -serrature, and the stem-armature is largely irregular. The pedicels are more or less glandular-hispid and often with a few acicles, and the hips subglobose to obovoid and glabrous or occasionally with a few acicles. 14 x 12 is closer to *R. rubiginosa* in leaflet-shape, -serrature and -indumentum, and there are often stout, arching prickles of the *R. rubiginosa* type on the flowering branches. The pedicels are glandular-aciculate. Wolley-Dod (1936) gave the parentage of *R.* x *burdonii* as *R. rubiginosa* x *R. tomentosa.*

c. 12 x 14 is recorded from v.c. 67/68, 79, 104 and 106, and 14 x 12 from v.c. 68 and 85.

f. WOLLEY-DOD, A. H. (1936). As above.

17 x 12. *R. agrestis* Savi x *R. sherardii* Davies

a. None.

b. 12 x 17 is a shrub with rather slender, zig-zag branches and foliage similar to those of *R. sherardii*, but with the leaves at the base of the flowering shoots more like those of *R. agrestis*. The pedicels are *c* 20 mm long with glandular setae and the petals pink. 17 × 12 is closer to *R. agrestis* in habit and armature but the leaflets are mostly elliptic and acute but more broadly elliptic at the base of the flowering shoots. The leaflets are glabrous to sparingly pilose above and sparsely pubescent with numerous *R. agrestis*-type glands below. The pedicels are 15–20 mm long and glabrous and the petals flushed pink but turning white.

   c. 12 × 17 is recorded only from the shores of Lough Derg, v.c. H9/10/15, and 17 × 12 only from Ballyvaughan, v.c. H9.

14 × 13. *R. rubiginosa* L. × *R. villosa* L.

   a. *R.* × *molliformis* W.-Dod (= 14 × 13).

   b. 14 × 13 has suborbicular leaflets rather densely pubescent on both surfaces and with numerous glands. The pedicels are rather short and glandular-aciculate, the hips subglobose and sparsely acicular, and the sepals suberect and glandular.

   c. 14 × 13 is recorded from v.c. 96/97; 13 × 14 has not been recorded.

   f. WOLLEY-DOD, A. H. (1936). As above.

15 × 14. *R. micrantha* Borrer ex Sm. × *R. rubiginosa* L.

   a. (*R.* × *dubia* W.-Dod, *non* Wibel, *R. micrantha* f. *trichostyla* R. Kell.) (= 15 × 14).

   b. 15 × 14 resembles *R. micrantha* but has a few acicles on the stems below the flowers and has styles which are glabrous in *R.* × *dubia* or slightly hispid in *R. micrantha* f. *trichostyla*.

   c. 15 × 14 is recorded from v.c. 17; 14 × 15 has not been recorded.

17 × 15. *R. agrestis* Savi × *R. micrantha* Borrer ex Sm.

   a. *R.* × *bishopii* W.-Dod (= 17 × 15).

   b. 17 × 15 has the habit of *R. agrestis* but small leaflets with fewer glands and intermediate serrature. The pedicels are *c* 10 mm long and glandular-hispid.

   c. 17 × 15 is recorded from v.c. 9, 13 and 17; 15 × 17 has not been recorded.

226. *Prunus* L.

(by C. A. Stace)

2 × 1. *P. domestica* L. *sensu lato* × *P. spinosa* L.

   a. *P.* × *fruticans* Weihe.

   b. *P. spinosa* appears to hybridize with both subsp. *domestica* and subsp.

insititia to form a range of plants varying between the two parental extremes. They are intermediate in habit, thorniness, pubescence, size of stomata, flowers, pollen, leaves and fruit, and stone-shape. The flowers generally appear with the leaves. The plants are apparently fertile but also spread quite vigorously by means of suckers. Some plants differ from *P. spinosa* only in the slightly greater size and pubescence of their parts, and are known as *P. spinosa* var. *macrocarpa* Wallr. Since *P. spinosa* is thought to be one of the ancestral parents of *P. domestica*, the hybrid may be considered a sort of backcross.

c. Intermediates are fairly frequent over the southern part of Br, where *P. domestica* is cultivated, and have been found at least as far north as Yorkshire. They are reported as common in Da and Su and are scattered across Europe as far as Rm.

d. *P. spinosa* has been hybridized with various plum cultivars on many occasions, particularly in Rs. The F_1 is intermediate in appearance (though very variable according to the *P. domestica* parent used) and usually pentaploid and thus largely sterile, though some fruits are set and some fully fertile hexaploids are obtained. The cross has been made reciprocally, being easier when the shorter-styled *P. spinosa* is used as female. Backcrosses have been made, especially to *P. domestica* cultivars, the progeny having chromosome numbers varying between the tetraploid, pentaploid and hexaploid levels.

e. *P. domestica* $2n = 48$; *P. spinosa* $2n = 32$; hybrid $2n = 40$.

f. LITTLE, J. E. and FRASER, J. (1920). *Prunus spinosa* L., var. *macrocarpa* Wallr. *Rep. Watson B.E.C.*, 3: 97-98.

MELVILLE, R. and RILSTONE, F. (1947). x *Prunus fruticans* Weihe. *Rep. B.E.C.*, 13: 156.

RYBIN, V. A. (1962). Wide hybridisation as a method of studying origin and improvement through breeding of fruit-crop plants—illustrated by the plum *Prunus domestica* L., in TSITSIN, N. V., ed. *Wide hybridisation in plants*, pp. 77-82. Jerusalem.

SALESSES, G. (1967). Connaissances cytogénétiques et hybridation interspécifique dans le sous-genre *Prunophora*, section *Euprunus*. *Annls Amél. Pl.*, 17: 397-408.

WEIMARCK, H. (1943). Om pollenkorn och klyvöppningar hos *Prunus Insititia*, *P. spinosa* och hybriden dem emellan. *Bot. Notiser*, 1943: 389-398.

YENIKEYEV, K. H. (= ENIKEEV, H. H.) (1965). The method of pollination with a pollen mixture to obtain interspecific hybrids of plum and cherry. *Genetica*, 36: 301-306.

YENIKEYEV, K. H. (1968). Breeding new plum cultivars for the central areas of the non-chernozem zone. *Pl. Breed. Abstr.*, 39: 165-166 (1969).

3 x 1. *P. cerasifera* Ehrh. x *P. spinosa* L.

is known from various parts of Europe and from the Caucasus, and is thought, by chromosome doubling, to have given rise to *P. domestica* agg. In the northern Caucasus region both triploid and hexaploid hybrids have been found, the latter fertile and resembling certain variants of *P. domestica*. (The two parents are tetraploid and diploid respectively). *P. cerasifera* var. *pissardii* (Carrière) L. H. Bailey x *P. spinosa* has been found in the wild in Ga, among *P. spinosa*, and named *P.* x *simmleri* Palézieux.

2a x 2b. *P. domestica* L. subsp. *domestica* x *P. domestica* subsp. *insititia* (L.) C. K. Scheid. (*P. insititia* L.)

a. *P.* x *italica* Borkh. em. Kárp. (*P. domestica* L. subsp. *italica* (Borkh.) Hegi). (Greengage).

b. These two taxa (plum and bullace) are often treated as separate species, but the differences between them are small and mostly quantitative in nature. Intermediacy is best marked in the pubescence of the twigs, pedicels and sepals, and in the shape of the fruit and "stone" and the degree to which the latter is free from the "flesh". The plants apparently do not show diminished fertility.

c. Records of intermediates exist for many parts of southern Br where plums, bullaces and damsons are grown commercially. It is likely that many intermediates encountered arose in cultivation whence they escaped. In the wild British plants have been studied in most detail by J. E. Little in v.c. 20. Semi-wild hybrids are widespread in Europe and elsewhere.

d. Cultivars of *P. domestica* and *P. insititia* are either self-compatible or -incompatible, and many inter-cultivar pollinations are incompatible. The two taxa have been interbred many times for commercial purposes and the progeny have given rise to new cultivars. The genetic behaviour of such crosses is strong evidence for including *P. domestica* and *P. insititia* in a single species.

e. Both parents and hybrid $2n = 48$ (hexaploid).

f. KÁRPÁTI, Z. E. (1967). Taxonomische Betrachtungen am Genus *Prunus*. *Reprium nov. Spec. Regni Veg.*, 75: 47-53.

LITTLE, J. E. and FRASER, J. (1920). *P. insititia* L. *Rep. Watson B.E.C.*, 3: 98-99.

LITTLE, J. E. and FRASER, J. (1931). *Prunus domestica* x *insititia. Rep. Watson B.E.C.*, 4: 67.

MELVILLE, R. and WARBURG, E. F. (1950). *Prunus domestica* L. x *insititia* L. *Year Book B.S.B.I.*, 1950: 90-91.

2 x 3. *P. cerasifera* Ehrh. x *P. domestica* L.

= *P.* x *syriaca* Borkh. em Kárp. is known as a semi-wild plant in several areas of Europe, e.g. Au.

4 x 5. *P. avium* (L.) L. x *P. cerasus* L.

= *P.* x *gondouinii* (Poiteau & Turpin) Rehder (Duke Cherry) is cultivated in Europe and is sometimes found outside gardens (e.g. Au, Cz, Ga, Ge), but it has apparently not been recorded wild in BI.

227 × 232. *Cotoneaster* Medic. × *Sorbus* L. = × *Sorbocotoneaster* Pojark.

(by C. A. Stace)

A natural hybrid between two non-British species of these genera was reported from eastern Siberia in 1953.

229. *Crataegus* L.

(by A. D. Bradshaw)

1 × 2. *C. laevigata* (Poiret) DC. (*C. oxyacanthoides* Thuill.) × *C. monogyna* Jacq.

a. *C.* × *media* Bechst. (*C.* × *intermixta* Beck.).

b. Hybrids are intermediate between the parents in all characters, of which leaf-shape, particularly the amount of indentation, is the most diagnostic. But there is a complete gradation of characteristics in mixed populations, and strictly intermediate plants seem no commoner than plants closer to the two parents. There is no indication of hybrid sterility from either pollen fertility or fruit production. This suggests that most populations are composed of a mixture of F_1 s and subsequent generations. There appear also to be parental populations suffering from small amounts of introgression.

c. Hybrids occur wherever *C. laevigata* has occurred, in woodlands on heavy clay in central and south-eastern England north to v.c. 57 and widely in north-western Europe as far as Po. They are most frequent where there has been disturbance of dense woodlands, allowing the invasion of *C. monogyna* and subsequent hybridization. In undisturbed woodland hybrids occur at the edges. The hybrid grows in a variety of habitats in between those of the parents; they appear to be at a slight disadvantage in dense woodland where they do not flower as freely as *C. laevigata,* and in exposed open conditions where they grow less well than *C. monogyna.* But they appear to be able to persist for a long time in hedgerows made of natural hybrids taken from woodlands, probably even from the Saxon period.

d. Crosses between the two parents, which are self-sterile, are as successful as crosses made within either parent species. Normal seeds are produced, but their ability to grow into adults has not been tested. *C. laevigata* flowers

about eight days before *C. monogyna* but there is about 40% overlap. The main barrier to crossing therefore appears to be spatial due to the ecological separation of the two species.
e. Both parents and hybrid $2n = 34$.
f. ASCHERSON, P. and GRAEBNER, P. (1906). *Mespilus*, in *Synopsis der Mitteleuropäischen Flora*, 6(2): 12-47. Leipzig.
   BRADSHAW, A. D. (1953). Human influence on hybridisation in *Crataegus*, in LOUSLEY, J. E., ed. *The changing flora of Britain*, pp. 181-183. Oxford. FIG.
   BRADSHAW, A. D. (1971). The significance of hawthorns, in *Hedges and local history, Standing conference for local history*, pp. 20-29. London.
   BYATT, J. I. (1975). Hybridization between *Crataegus monogyna* Jacq. and *C. laevigata* (Poiret) DC. in south-eastern England. *Watsonia*, 10: 253-264.

229 × 230. *Crataegus* L. × *Mespilus* L. = × *Crataemespilus* Camus

(by A. D. Bradshaw)

229/2 × 230/1. *Crataegus monogyna* Jacq. × *Mespilus germanica* L.
a. × *C. grandiflora* (Sm.) Camus (× *Crataegomespilus grandiflora* (Sm.) Bean).
b. Hybrids are intermediate between the parents in leaf and flower characters. The leaves are *c* 7.5 × 4 cm, ovate and have finely toothed margins and downy surfaces; on the long shoots they are more elongated and have several large indentations. The flowers are *c* 2.5 cm diameter and borne on short, woolly stalks in groups of about three; the fruits are *c* 2 cm diameter, yellowish-brown, and have two stones. The hybrid is said to be fairly sterile, although some fruits are produced: the ratio of fruit to flowers suggests a fertility of about 20%. It must be distinguished from the two well-known cultivated graft-hybrids of the same parentage: + *Crataegomespilus dardarii* Simon-Louis ex G. Ballair 'Dardarii' differs in remarkable instability in shoot morphology; and + *C. dardarii* 'Jules d'Asnières' has smaller, rounder leaves, and flowers similar to those of *Crataegus monogyna*.
c. Naturally occurring hybrids, although very distinct, are extremely rare and have been reported as isolated trees in open places in only a few places in Br; there are records from v.c. 3, 8, 17, 22, 48, 55 and 83. In part this is because *M. germanica* is itself an uncommon cultivated plant in Br. The

hybrid is reported from Ga, but seems to be planted elsewhere. It is quite widely planted as an ornamental shrub, which confuses the natural distribution.

d. None.

e. *C. monogyna* $2n = 34$; *M. germanica* ($2n = 34$).

f. ASCHERSON, P. and GRAEBNER, P. (1906). *Mespilus*, in *Synopsis der Mitteleuropäischen Flora*, 6(2): 12-47. Leipzig.

BEAN, W. J. (1914). *Trees and shrubs hardy in the British Isles*, 8th ed., 1: 760-761 and 790. London.

CHITTENDEN, F. J. (1951). *Dictionary of gardening*, 2: 567-568. Oxford. FIG.

HOOKER, W. J. (1835). x *Crataego-mespilus grandiflora. Curtis bot. Mag.*, 62: 3442. FIG.

SANDWITH, N. Y. (1948). x *Crataegomespilus grandiflora* (Sm.) Bean. *Rep. B.E.C.*, 13: 260.

229 × 232. *Crataegus* L. × *Sorbus* L. = × *Crataegosorbus* Makino ex Koidz.

(by C. A. Stace)

Two hybrids of this parentage have been reported, one involving *Sorbus aucuparia.*

231 × 232. *Amelanchier* Medic. × *Sorbus* L. = × *Amelasorbus* Rehder

(by C. A. Stace)

A natural hybrid between two non-British species of these genera was reported from the U.S.A. in 1925.

232. *Sorbus* L.
(by A. J. Richards)

Of the 20 British species recognized by Warburg (1962) three (*S. aucuparia, S. aria* and *S. torminalis*) are sexual diploids, and their genomes are represented in the following account by B, A and T respectively. The other species have not all been investigated cytologically but those which have are polyploids and at least partially apomictic; it is possible that this is true of all of them. In that case hybrids are likely to arise at the present time only between sexual species or between a sexual and an apomictic species where the latter is the male parent. The 17 presumed polyploid apomicts fall into four groups: those similar to *S. aria* (8 species); those intermediate between *S. aria* and *S. aucuparia* (5 species in *S. intermedia* agg.); those intermediate between *S. aria* and *S. torminalis* (3 species in *S. latifolia* agg.); and *S. pseudofennica*, which is intermediate between *S. aucuparia* and *S. arranensis* (in *S. intermedia* agg.). It is thus likely that many of the agamospecies recognized today are of hybrid origin involving the three sexual diploids and their derivatives or ancestors.

General References
WARBURG, E. F. (1962). *Sorbus*, in CLAPHAM, A. R., TUTIN, T. G. and
  WARBURG, E. F. *Op. cit.*, pp. 423-437. FIG.
WARBURG, E. F. and KÁRPÁTI, Z. E. (1968). *Sorbus*, in TUTIN, T. G.
  et al., eds *Flora Europaea*, 2: 67-71. Cambridge.

1 x 4/1. *S. aucuparia* L. x *S. intermedia* (Ehrh.) Pers.
  a. (*S.* x *pinnatifida* auct.).
  b. The leaves are intermediate in appearance with 0–5 (usually 1–2) free, obtuse pinnae of which the basal pair are more or less distant; they are grey-green tomentose beneath. The fruits are 9–10 mm long, globose and scarlet. This plant is very similar to some trees of *S.* x *thuringiaca*, from which it can usually be separated by the more ovate, narrower leaves, distant basal pinnae, smaller scarlet fruits and an absence of pollen. It is rather invariable and probably always pollen sterile and apomictic. Good fruit is set, but there is no evidence of second-generation plants occurring.
  c. It is found as occasional, spontaneous trees in the company of both parents in v.c. 10, 11, 13, 16–18, 23, 33, 34, 36, 39, 40 and 48. It is infrequently planted elsewhere. It is probably spontaneous in western and central Europe, but is only definitely recorded as such from Cz.
  d. The anthers lack pollen, the pollen mother cells usually degenerating at meiosis when they may be replaced by plasmodia, which do not form pollen but undergo disturbed mitotic divisions and become highly polyploid. In cases where male meiosis has been observed, 1 trivalent, 12–16 bivalents and 16–20 univalents are characteristic. A similar pattern is observed in the female meiosis, but apparently viable fruits are set,

although there is no record of germination. Fruit-set has been observed in the absence of cross-pollination by Hedlund (1901) and by Liljefors (1953, 1955) and apomixis presumed, although one case of apparent sexuality is recorded. The hybrid has apparently been synthesized on a number of occasions, usually for horticultural purposes, but plants of experimental hybrid origin are still grown in Uppsala, Su. It is likely that *S. intermedia* itself is of hybrid origin (see under *S. aucuparia* x *S. rupicola*).

e. *S. aucuparia* ($2n = 34$) (BB); *S. intermedia* $2n = 68$ (AABB); hybrid $2n = 51$ (ABB).

f. HEDLUND, T. (1901). Monographie der Gattung Sorbus. *K. svenska Vetensk-Akad. Handl.*, 35: 1-147.

LILJEFORS, A. (1953). Studies on propagation, embryology and pollination in *Sorbus. Acta Horti Bergiani*, 16: 277-329.

LILJEFORS, A. (1955). Cytological studies in *Sorbus. Acta Horti Bergiani*, 17: 47-113. FIG.

SALMON, C. E. (1930). Notes on *Sorbus. J. Bot., Lond.*, 68: 172-177.

WILMOTT, A. J. (1934). Some interesting British Sorbi. *Proc. Linn. Soc. Lond.*, 146: 73-79.

4/2 x 1. *S. arranensis* Hedl. x *S. aucuparia* L.

a. *S. pseudofennica* E. F. Warburg. (Bastard Mountain-ash).

b. The leaves are intermediate in appearance, with 0–2 free pinnae which are not distant, shortly and broadly triangular in outline and thinly grey-green tomentose beneath. The fruits are 8–10 mm long, scarlet and ovoid. *S. pseudofennica* is very similar to *S.* x *thuringiaca*, but has 7–8 rather than 10–12 pairs of veins, and to the Scandinavian *S. hybrida* L., which probably had a similar origin. *S. pseudofennica* is invariable and probably an apomict, although there seems to be no published information on this point. It is likely that *S. arranensis* itself is of hybrid origin (see under *S. aucuparia* x *S. rupicola*).

c. It is known only on steep, rocky stream-banks in Glen Catacol, Arran, v.c. 100, with both presumed parents.

d. None.

e. *S. aucuparia* ($2n = 34$) (BB); *S. arranensis* $2n = 51$ (AAB).

f. LILJEFORS, A. (1953). As above.

LILJEFORS, A. (1955). As above.

PERRING, F. H. and SELL, P. D. (1968). *Op. cit.*, p 34.

WILMOTT, A. J. (1934). As above.

5/1 x 1. *S. aria* (L.) Crantz x *S. aucuparia* L.

a. *S.* x *thuringiaca* (Ilse) Fritsch (*S.* x *semipinnata* Hedl., *non* Borbás).

b. The leaves are intermediate in appearance, with 1–4 free, more or less acute pinnae of which the basal pair are not distant; they are triangular in outline, more or less acute at the apex and grey-green tomentose beneath.

The fruits are 9–11 mm long, brownish-red and ovoid. *S.* x *thuringiaca* is very similar to *S. aucuparia* x *S. intermedia* (see above) and to *S. pseudofennica*, from which it differs in the 10–12 pairs of veins and brownish fruits. It is very variable in habit and leaf-shape since it is sexual and seed-fertile, although with reduced pollen-fertility. No F_2, plants have been discovered in BI, although these may well be very difficult to detect.
  c. The northern range of this hybrid, which contrasts markedly with that of *S. aucuparia* x *S. intermedia,* perhaps reflects the fact that the parents are usually ecologically separated within their natural range and only meet when *S. aria* is planted in northern Br. A few planted trees are known, but they are thought to be spontaneous in v.c. 6, 15, 55, 56, 65, 75, 78, 80, 82, 88, 90 and 94. They are also probably native in western and central Europe, but are only recorded as spontaneous in Cz, Ga and Asian Tu.
  d. Male meiosis is rather regular in the hybrid, with 16 or 17 bivalents. Tetrads are usually normal in appearance, although cases of tetrads with two aborted cells, and of diads, have been reported. However, the pollen is irregular in size, not more than 50% of the grains having contents. The hybrids are fertile and sexual, as are both parents. Hybrids of experimental origin are grown in Uppsala, Su, but this hybrid does not seem to have been synthesized for horticultural purposes as frequently as *S. aucuparia* x *S. intermedia.*
  e. *S. aucuparia* (2*n* = 34) (BB); *S. aria* 2*n* = 34 (AA); hybrid (2*n* = 34) (AB).
  f. DE POUCQUES, M.-L. (1953). La différenciation de certains *Sorbus* par le pollen. *Bull. Séanc. Soc. Sci. Nancy,* **12**(4).
  GABRIELAN, E. (1961). The genus *Sorbus* in Turkey. *Notes R. bot. Gdn, Edinb.,* **23**: 483-495.
  HEDLUND, T. (1901). As above. FIG.
  LILJEFORS, A. (1953). As above.
  LILJEFORS, A. (1955). As above. FIG.

1 x 5/7. *S. aucuparia* L. x *S. rupicola* (Syme) Hedl.

  a. None.
  b. A number of apomictic triploids may have arisen as a result of this cross. These include the British triploid species *S. minima* (A. Ley) Hedl. (AAB), *S. leyana* Wilmott (AAB) and *S. arranensis* (AAB). In addition *S. intermedia* and *S. anglica* Hedl. may be derived from backcrosses to *S. aucuparia.*
  c. Three of these species are very localized endemics, *S. minima* and *S. leyana* being restricted to v.c. 42, and *S. arranensis* to v.c. 100. *S. anglica* is an endemic more widespread in Wales and south-western England, and *S. intermedia* is an introduced plant found over much of Br.
  d. There is no indication that the primary hybrid has ever been discovered in Br or abroad, but these species, together with the Scandinavian *S. lancifolia* Hedl., *S. subpinnata* Hedl. and *S. neglecta* Hedl., probably represent the direct apomictic progeny of primary hybrids between

tetraploid *S. aria sensu lato* (of which *S. rupicola* is the commonest) and diploid *S. aucuparia*.

e. *S. aucuparia* (2*n* = 34) (BB); *S. rupicola* 2*n* = 68 (AAAA); *S. minima* 2*n* = 51 (AAB); *S. arranensis* 2*n* = 51 (AAB); *S. intermedia* 2*n* = 68 (AABB); *S. anglica* 2*n* = 68 (AABB).

f. LILJEFORS, A. (1953). As above.
LILJEFORS, A. (1955). As above.
PERRING, F. H. and SELL, P. D. (1968). *Op. cit.*, pp. 34-35.
WILMOTT, A. J. (1934). As above. FIG.

6 × 4/4. *S. latifolia* (Lam.) Pers. agg. × *S. minima* (A. Ley) Hedl.

was mentioned by Marshall (1916), but is unlikely to exist as these two species are thought to be always apomictic.

MARSHALL, E. S. (1916). Notes on *Sorbus. J. Bot., Lond.*, **54**: 10-14.

4/5 × 6. *S. anglica* Hedl. × *S. latifolia* (Lam.) Pers. agg.

was recorded by Druce (1928), but is unlikely to be correct as these two species are thought to be always apomictic.

DRUCE, G. C. (1928). *Op. cit.*, p. 41.

5/1 × 5/7. *S. aria* (L.) Crantz × *S. rupicola* (Syme) Hedl.

a. None.

b. This hybrid is intermediate between the parents; the leaves are broadest at the middle and have 9–11 pairs of veins, and the fruits are *c* 12 mm long, crimson and ovoid. It is very similar to, and probably indistinguishable from, *S. vexans* E. F. Warb. and *S. lancastriensis* E. F. Warb. which probably originated in the same way. Plants of this affinity are usually referred to the hybrid when found away from the very limited ranges of these two species. Fruit-set in the hybrid is good; it is probably a triploid apomict (AAA).

c. The hybrid is recorded from the Avon Gorge, v.c. 6, and from the Wye Valley, v.c. 34, in company with both parents. *S. lancastriensis* is restricted to a small area of v.c. 60 and 69, and *S. vexans* to v.c. 4 and 5.

d. None.

e. *S. aria* 2*n* = 34 (AA); *S. rupicola* 2*n* = 68 (AAAA).

f. PERRING, F. H. and SELL, P. D. (1968). *Op. cit.*, p. 37.
WILMOTT, A. J. (1934). As above.

5/1 × 7. *S. aria* (L.) Crantz × *S. torminalis* (L.) Crantz

a. *S.* × *vagensis* Wilmott.

b. This is intermediate between its parents; the leaves are usually very shallowly lobed in the distal half and closely yellowish- or grey-green tomentose beneath, and the fruits are 7–12 mm long and brownish-orange.

S. *subcuneata* Wilmott and S. *bristoliensis* Wilmott have smaller, narrower leaves and orange fruits, but S. *devoniensis* E. F. Warb. is dubiously distinguishable from the hybrid, only the fruit-shape being slightly different. All three species probably arose as a result of this cross, but differ from the hybrid in being triploid and apparently apomictic. A number of central European species also share the same presumed parents. The diploid hybrid, S. x *vagensis*, is very variable, perhaps as a result of the variability of the sexual parents, and can be sterile or fertile and sexual.

c. S. x *vagensis* is known from a limited area of v.c. 34, 35 and 36, where it seems to be fairly frequent in areas where the parents meet. The three British agamospecies of this presumed parentage are restricted to localized areas of south-western England: S. *devoniensis* in v c. 2–4; S. *subcuneata* in v.c. 4 and 5; and S. *bristoliensis* in v c. 6 and 34. S. x *vagensis* is probably found in a number of areas in western and central Europe where the parents meet, but is only definitely recorded from Cz, Ga and Hu.

d. There is some indication (De Poucques, 1951) that hybrids with S. *aria* as the female parent produce fertile fruits, whereas those with S. *torminalis* as the female parent are sterile.

e. S. *aria* $2n = 34$ (AA); S. *torminalis* $2n = 34$ (TT); S. x *vagensis* $2n = 34$ (AT); S. *bristoliensis* $2n = 51$ (AAT).

f. DE POUCQUES, M.-L. (1951). Étude chromosomique des *Sorbus latifolia* Pers. et *Sorbus confusa* Gremli. *Bull. Soc. bot. Fr.*, **98**: 89-92.

DE POUCQUES, M.-L. (1953). As above.

DILLEMANN, G. and DE POUCQUES, M.-L. (1954). Le pollen du *Sorbus latifolia* Pers. et son origine hybride. *Bull. Soc. bot. Fr.*, **101**: 239-240.

HEDLUND, T. (1901). As above.

HENSEN, K. J. W. (1967). In Nederland geweekle tussenvormen tussen *Sorbus aria* en S. *torminalis*. *Dendroflora*, **4**: 51-60.

KOVANDA, M. (1961). Spontaneous hybrids of *Sorbus* in Czechoslovakia. *Acta Univ. Carol., Biol.*, **1**: 43-83.

LILJEFORS, A. (1955). As above. FIG.

PERRING, F. H. and SELL, P. D. (1968). *Op. cit.*, p. 38.

WILMOTT, A. J. (1934). As above. FIG.

5/2 × 5/7. S. *leptophylla* E. F. Warb. × S. *rupicola* (Syme) Hedl.

was the parentage assigned by A. J. Wilmott and E. F. Warburg to specimens collected from Mynydd Llangattwg, v.c. 42. They are inter-mediate between their putative parents in characters of leaves and fruits. The leaves are broadest at the middle, 12–14 cm long (exceeding those of either parent), and have 9–10 pairs of veins; the fruits are 18–19 mm long and carmine. However, both' alleged parents are tetraploids and are probably obligate apomicts, so that it is more likely that these specimens represent abnormal S. *aria* x S. *rupicola*, although S. *aria* is not now known in this area.

WILMOTT, A. J. (1934). As above.

5/5 × 7. *Sorbus porrigentiformis* E. F. Warb. × *S. torminalis* (L.) Crantz

a. None.
b. This hybrid is very similar to *S.* x *vagensis,* but has smaller leaves with an obtuse apex and fewer pairs of veins. Fertile plants assigned here by H. J. Riddelsdell seem to be *S. porrigentiformis,* but a sterile tree on which fruit has never been known to set, collected and determined as this hybrid by A. Ley, seems to be correct.
c. A single tree in company with both parents was found at Symonds Yat, v.c. 34.
d. None.
e. *S. porrigentiformis* 2n = 68 (AAAA); *S. torminalis* 2n = 34 (TT). It is possible that a specimen from Symonds Yat that was found to be triploid, 2n = 51, and was assigned to *S. porrigentiformis* by Warburg, was either this hybrid, or *S. aria* x *S. porrigentiformis.*
f. DILLEMANN, G. and DE POUCQUES, M.-L. (1954). As above.

5/7 × 7. *S. rupicola* (Syme) Hedl. × *S. torminalis* (L.) Crantz

a. None.
b. This hybrid is again very similar to *S.* x *vagensis,* but has more obtuse leaves with fewer veins, fruits broader than long, and a smaller, stiffer habit. It is apparently fertile and probably apomictic.
c. It has been collected from v.c. 41 and 42 in company with the presumed parents; specimens were determined by A. J. Wilmott, apparently correctly so.
d. None.
e. *S. rupicola* 2n = 68 (AAAA); *S. torminalis* 2n = 34 (TT).
f. DILLEMANN, G. and DE POUCQUES, M.-L. (1954). As above.
WILMOTT, A. J. (1934). As above.

233 × 232. *Pyrus* L. × *Sorbus* L. = × *Sorbopyrus* C. K. Schn.

(by C. A. Stace)

233/1 × 232/5/1. *Pyrus communis* L. × *Sorbus aria* (L.) Crantz = ×*S. auricularis* (Knoop) C. K. Schn.

originated in Alsace, Ga, before 1619, and is found wild in parts of central Europe.

234 × 232. *Malus* Mill. × *Sorbus* L.
= × *Malosorbus* Browicz
(by C. A. Stace)

234/1 × 232/7. *Malus sylvestris* Mill. × *Sorbus torminalis* (L.) Crantz = × *M. florentina* (Zuccagni) Browicz occurs in Al, Gr, It, Ju and Tu. Some authors consider *Malus pumila* Mill. or *M.dasyphylla* Borkh. rather than *M. sylvestris* to be one of the parents, and others believe the plant is not a hybrid but a species of *Malus* or some other genus.

235. *Sedum* L.
(by C. A. Stace)

8 × 9. *S. acre* L. × *S. sexangulare* L.
has been recorded from He, but is doubtful.

239. *Saxifraga* L.
(by D. A. Webb)

1 × 2. *S. nivalis* L. × *S. stellaris* L.
   a. *S.* × *crawfordii* Marshall.
   b. Although the parent species are usually considered to belong to the same section, the description of the hybrid, as has been noted by Engler and Irmscher (1919), is much more suggestive of *S. nivalis* than of *S. stellaris*, there being no character which unambiguously points to the latter. The existence of this hybrid must, therefore, be considered very doubtful.
   c. This hybrid was described from a small colony of plants collected in the Cairngorms, v.c. 96, in 1902 by F. C. Crawford; they were not seen *in vivo* by Marshall.
   d. None.

240 HYBRIDIZATION

e. *S. nivalis* (2n = 60); *S. stellaris* (2n = 28).
f̂. ENGLER, A. and IRMSCHER, E. (1919). *Saxifraga* I, in ENGLER, A., ed.
   Das Pflanzenreich, **69(IV, 117)**: 658. Leipzig.
   MARSHALL, E. S. (1909). A new hybrid saxifrage from Scotland. *J. Bot.,*
   Lond., **47**: 98-99.

5 × 4. *S. spathularis* Brot. × *S. umbrosa* L.

a. *S.* × *urbium* D. A. Webb (*S. umbrosa* var. *crenatoserrata* Bab., *sec.* Pugsl.).
   (London Pride).
b. This plant is intermediate between the presumed parents in the length and
   hairiness of the petiole and in the conspicuousness of the cartilaginous
   border to the leaf. In the form of the leaf-margin it is, though
   intermediate, closer to *S. umbrosa,* the margin being always crenate, not
   dentate, and the terminal crenation is distinctly wider (and usually slightly
   lower) than its neighbours; the crenations are, however, bolder than in *S.
   umbrosa.* In the size of the ovary (and of the capsule when developed) it
   comes closer to *S. spathularis.* It is nearly always sterile, for reasons which
   are not clear, but a few of the populations in BI (notably in Orkney) are
   fertile; they show no morphological differences from the sterile ones.
c. It is unknown except as an escape from gardens, the parent species being
   nowhere sympatric. It is apparently of 17th or 18th century garden origin,
   and is locally naturalized, usually in woods or by streams, in Br, Hb and
   Ga. As a garden plant it is very common in BI, but is seldom seen in
   Continental gardens, where it is replaced by *S.* × *geum.*
d. A hybrid of this parentage has been synthesized; it was identical in
   appearance with the garden plant. There is no record as to its fertility. If
   the hybrid is used as a pollen-parent and either of the parent species as a
   seed-parent a small amount of seed is produced, but such backcross plants
   have not been reared to maturity.
e. Both parents and hybrid 2n = 28.
f. MARTIN, W. K. (1969). *Concise British flora in colour,* 2nd ed., t. 32.
   London. FIG.
   PUGSLEY, H. W. (1936). The British Robertsonian saxifrages. *J. Linn.
   Soc., Bot.,* **50**: 267-289.
   WEBB, D. A. (1950a). Hybridization and variation in the Robertsonian
   saxifrages. *Proc. R. Ir. Acad.,* **53B**: 85-97.
   WEBB, D. A. (1963). *Saxifraga* × *urbium* D. A. Webb, *hybr. nov. Reprium
   nov. Spec. Regni veg.,* **68**: 199-200.

6 × 4. *S. hirsuta* L. × *S. umbrosa* L.

= *S.* × *geum* L. (but not of earlier British authors) occurs in the western
and central Pyrenees where the parents are sympatric. It is commonly
grown in gardens on the Continent and is naturalized in Au, Be, Ga, Ge
and It; in BI it is rarely grown and not naturalized.

6 × 5. *S. hirsuta* L. × *S. spathularis* Brot.
  a. *S.* × *polita* (Haw.) Link (*S. hirsuta* auct., *non* L.).
  b. This binomial is applicable to a very wide range of variants. In plants
     which approximate to the F_1 hybrid the lamina is suborbicular to very
     broadly elliptical and crenate-dentate, and the petiole narrow, but
     perceptibly flattened and fairly hairy. The hybrid is fully fertile and
     backcrosses with both parents, so that a full range of intermediates can be
     found. It is an arbitrary decision as to where one draws the line between
     hybrids and slightly introgressed variants of either species. *S. punctata*
     auct., *non* L., and *S. elegans* Mackay constitute such variants of *S.
     spathularis* and *S. hirsuta* respectively
  c. Throughout the range of *S. hirsuta* in Hb this hybrid is very common, and
     in many localities commoner than *S. hirsuta* itself. *S. hirsuta* grows only in
     very damp and sheltered situations (mainly by mountain streams or in
     lowland woods), but *S.* × *polita* inherits from *S. spathularis* a greater
     tolerance of exposure. There is reason to believe that, at least in some
     regions, *S. hirsuta* is being swamped by introgression from the more
     tolerant, widespread and aggressive *S. spathularis*. Corroboration for this
     idea is given by the fact that in at least three places in v.c. H16 and 27
     remote from the present range of *S. hirsuta* small colonies of *S.* × *polita*
     have been recorded. In the area of north-western Hs where the two parent
     species are sympatric *S.* × *polita* is occasionally found, but it is much rarer
     than in Hb, and there is little evidence of introgression. The hybrid is
     found as an escape from cultivation in a few areas of Br.
  d. Very full experimental work on the synthesis of this hybrid and on its
     segregation following self-pollination is reported by Scully (1916).
  e. Both parents and hybrid $2n = 28$.
  f. PERRING, F. H. and SELL, P. D. (1968). *Op. cit.,* p. 40.
     PUGSLEY, H. W. (1936). As above.
     REICHENBACH, H. G. L. (1829). *Iconographia Botanica seu Plantae
        Criticae,* 7: t. 622. Leipzig.
     SCULLY, R. W. (1916). *Flora of County Kerry,* pp. 96-106. Dublin.
     WEBB, D. A. (1950a). As above.

pan × 5. *S. paniculata* Miller (*S. aizoon* Jacq.) × *S. spathularis* Brot.
  = *S.* × *andrewsii* Harvey was reported as having been collected in 1845 in
  Glencar, v.c. H1, by W. Andrews; it was described from a cultivated
  specimen in 1848 but it is not known whether plants in cultivation today
  are clonally derived from the original hybrid or from a subsequent
  hybridization. There is no doubt that *S.* × *andrewsii* is a hybrid between *S.
  paniculata* and a member of the section *Gymnopera* D. Don (*S. umbrosa*
  and its allies). As the former does not occur wild in BI the occurrence of
  such a hybrid in a remote part of Hb far from gardens is extremely
  unlikely; it is generally assumed, therefore, that it arose in Andrews'
  garden and was mislabelled (as were several others of his plants).

HARVEY, W. H. (1848). Account of a new British saxifrage. *Hooker's London J. Bot.*, 7: 569-571. FIG.
SCULLY, R. W. (1916). As above, p. 106.
SYME, J. T. B., ed. (1865). *English Botany*, 4: 71-72, t. 549. London. FIG.

15 × 8. *S. hypnoides* L. × *S. tridactylites* L.

a. (*S.* × *farreri* Druce, *nom. nud.*).
b. Judging from Farrer's scanty description which, as far as it goes, agrees with the description by Marsden-Jones and Turrill of the artificial hybrid of this parentage, the hybrid is more like *S. hypnoides* in general appearance, being perennial, with axillary buds on the short, ascending side-shoots, and with petals only slightly smaller. The leaves, however, are simple or three-lobed, and the habit is dwarf and compact. The hybrid appears to be completely sterile.
c. This plant was found originally in very small quantity on bare limestone rocks on Ingleborough, v.c. 64, together with both parents, by Farrer in 1906. It has not been recorded elsewhere in the wild state.
d. The parent species were crossed by Marsden-Jones and Turrill, who gave a full description of the hybrids.
e. *S. tridactylites* ($2n = 22$); *S. hypnoides* $2n = 30$–64.
f. DRUCE, G. C. (1908). *Saxifraga hypnoides* × *tridactylites* = × *S. Farreri*, Druce. *Rep. B.E.C.*, 2: 256-257.
  FARRER, R. (1919). *The English rock garden*, 2: 273. London and Edinburgh.
  MARSDEN-JONES, E. M. and TURRILL, W. B. (1938). Further interspecific *Saxifraga* hybrids. *J. Genet.*, 36: 431-445. FIG.

12 × 9. *S. cespitosa* L. × *S. granulata* L.

has been recorded from Su.

9 × 14. *S. granulata* L. × *S. rosacea* Moench

= *S.* × *freiburgii* Ruppert (*S. potternensis* Marsden-Jones & Turrill) is recorded from Ge where it occurs both in the wild and spontaneously in botanic gardens, and it arose similarly in a garden in Br about 1927.
MARSDEN-JONES, E. M. and TURRILL, W. B. (1930). The history of a tetraploid saxifrage. *J. Genet.*, 23: 83-92. FIG.

10 × 11. *S. cernua* L. × *S. rivularis* L.

= *S.* × *opdalensis* Blytt is locally frequent in No.

15 × 14. *S. hypnoides* L. × *S. rosacea* Moench

a. None.
b. The great variability of *S. rosacea* (and to a lesser extent of *S. hypnoides*) makes the morphology of the hybrid very difficult to characterize. It

differs from *S. hypnoides* in its coarser, less apiculate leaf-segments, and in the absence of conspicuous axillary buds. It differs from *S. rosacea* in its diffuse habit, with prostrate non-flowering shoots, and in its distinctly mucronate or shortly apiculate leaf-segments. Both these characters can be found in variants of *S. rosacea*, but not in combination. Hybrids have also been recorded between various segregates of each of these two species.

   c. It has been recorded in small quantity from the Galtee mountains, v.c. H7, and the Burren district of v.c. H9, the only regions in BI (and virtually in Europe) where the parent species grow together. A large number of garden plants have also been ascribed to this parentage. It seems fairly certain that both species have contributed to the "mossy saxifrages" of rock-gardens, and the finer-leaved, relatively small-flowered clones among these probably do not have any other species in their parentage.

   d. The hybrid has been synthesized by Marsden-Jones and Turrill, and Webb has obtained some seed by crossing several different variants of the parent species.

   e. *S. rosacea* $2n = 56–64$; *S. hypnoides* $2n = 30–64$.

   f. MARSDEN-JONES, E. M. and TURRILL, W. B. (1956). Additional breeding experiments on *Saxifraga. J. Genet.*, 54: 186-193. FIG.

   WEBB, D. A. (1950b). The Dactyloid saxifrages of north-west Europe. *Proc. R. Ir. Acad.*, 53B: 207-240.

   WEBB, D. A. (1950c). *Saxifraga* L. (Section *Dactyloides* Tausch), in Biological Flora of the British Isles. *J. Ecol.*, 38: 185-213.

   WEBB, D. A. (1951). The mossy saxifrages of the British Isles. *Watsonia*, 2: 22-29.

16 × 17. *S. aizoides* L. × *S. oppositifolia* L.

has never been found, but *S. nathorstii* (Dusen) Hayek, which is endemic to Greenland, is usually regarded as an allotetraploid hybrid between the above species.

242. *Chrysosplenium* L.

(by C. A. Stace)

2 × 1. *C. alternifolium* L. × *C. oppositifolium* L.

has been recorded from the Continent, but is doubtful.

246. *Ribes* L.

(by C. A. Stace)

1 × 2. *R. rubrum* L. (*R. sylvestre* (Lam.) Mert. & Koch) × *R. spicatum* Robson

is cultivated and according to Warburg (1962) it "might occur as an escape", but no records of hybrids found in the wild have been traced from Br or elsewhere.

WARBURG, E. F. (1962). *Ribes*, in CLAPHAM, A. R., TUTIN, T. G. and WARBURG, E. F. *Op. cit.* pp. 460-463.

247. *Drosera* L.

(by D. A. Webb)

2 × 1. *D. anglica* Huds. × *D. rotundifolia* L.

a. *D.* × *obovata* Mert. & Koch.

b. In most characters this hybrid is intermediate between the parents, and not very variable. The leaf-lamina is narrowly obovate, usually 2½–3 times as long as wide. It can best be distinguished from *D. intermedia* by its lack of stoloniferous growth, its pseudoterminal inflorescence and its small capsule and empty seeds. It is completely sterile.

c. In BI the range of the hybrid covers most of the joint range of the parent species, with the greatest recorded frequency in north-western Scotland. The hybrid has also been recorded from most of the regions of northern and central Europe where both parent species occur, and there are also records from northern Asia and North America.

d. None.

e. *D. rotundifolia* (2n = 20); *D. anglica* (2n = 40).

f. PERRING, F. H. and SELL, P. D. (1968). *Op. cit.*, p. 40.

3 × 1. *D. intermedia* Hayne × *D. rotundifolia* L.

a. *D.* × *beleziana* Camus.

b. The leaf-shape of this hybrid is closer to that of *D. rotundifolia*, though some laminae are slightly obovate, but the scape is ascending and arises from the axil of one of the lower leaves, as in *D. intermedia*. The seeds are abortive, with a finely tuberculate testa.

c. It has been recorded from boggy places with the parents in the New

Forest, v.c. 11, Dersingham, v.c. 28, and Borth Bog, v.c. 46, and also in Au, Ga and Ge.

d. None.
e. Both parents (2n = 20).
f. CAMUS, E. G. (1891). Note sur les *Drosera. J. Bot., Paris*, **5**: 198. FIG.

DRUCE, G. C. (1912). x *Drosera longifolia* x *rotundifolia. Rep. B.E.C.*, **3**: 20.

ROSENBERG, O. (1909). Cytologische und morphologische Studien an *Drosera longifolia* x *rotundifolia. K. svenska Vetensk-Akad. Handl.*, **43**: 1-64.

SALTER, J. H. (1941). Hybrid sundews. *NWest. Nat.*, **16**: 92.

SCHUSTER, J. (1907). Über *Drosera beleziana* Camus. *Allg. bot. Z.*, **13**: 180-183.

2 x 3. *D. anglica* Huds. x *D. intermedia* Hayne
has been reported from Au and Ge, but needs checking.

251. *Daphne* L.
(by I. K. Ferguson)

2 x 1. *D. laureola* L. x *D. mezereum* L.
a. *D.* x *houtteana* Lindl. & Paxt.
b. This hybrid is intermediate between its parents in most characters. The leaves are deciduous (appearing after the flowers) or sometimes evergreen; they are thicker in texture and more shining than in *D. mezereum*. The flowers are mostly in threes and less markedly terminal than in *D. laureola*. The hypanthium and sepals are whitish-green, though often tinged red outside, and glabrous as in *D. laureola*. There appears to be no information on pollen-fertility or seed-set.
c. This hybrid is apparently very rare. Specimens have been seen from v.c. 6 (1907), 13 (1902) and from Fiezar Wood, v.c. 64 (1954). No additional literature records with reference to Br or elsewhere have been found.
d. Both self-fertile and self-sterile forms of *D. mezereum* have been observed.
e. *D. mezereum* 2n = 18; *D. laureola* (2n = 18).
f. BLAISE, S. (1959). Contribution a l'étude caryologique et palynologique de quelques Thyméléacées. *Revue gén. Bot.*, **66**: 109-161.

LUDWIG, F. (1900). On self-sterility. *Jl R. hort. Soc.*, **1900**: 214-217.

MARSHALL, E. S. (1903). West Sussex plant notes for 1902. *J. Bot., Lond.*, **41**: 227-232.

MARSHALL, E. S. (1910). *Daphne Laureola* x *Mezereum* in N. Somerset. *J. Bot., Lond.*, **48**: 79.

254. *Epilobium* L.

(by C. A. Stace)

The widespread and prodigious occurrence of hybrids in this genus has been well known for a long time, and a considerable number of experimental hybridizations have been carried out. Hybrids are, however, rarely of more than sporadic occurrence in *Epilobium,* and cannot be said to contribute greatly to the taxonomic difficulties which are often encountered or professed in this genus. These difficulties arise mainly from the considerable phenotypic plasticity of many characters, particularly the leaves and the *quantity* of the indumentum, and the failure of some botanists to examine critically the proven diagnostic characters, notably the *quality* of the indumentum.

Real taxonomic problems among the British species are largely confined to the generic delimitation of *Epilobium* and to species delimitation within *E. tetragonum* agg. The first need not concern us here, for species of *Chamerion* (Raf.) Raf. (*Chamaenerion* auct., *non* Ség.) do not hybridize with those of *Epilobium sensu stricto*, and experimental hybridizations have not produced offspring either. The same is true of the dwarf introductions from New Zealand, *E. brunnescens* (Cockayne) Raven & Engelhorn (*E. nerteroides* auct., *E. pedunculare* auct.) and its allies, *vis-à-vis* the rest of *Epilobium*. *E. tetragonum* L. (*E. adnatum* Griseb.) and *E. lamyi* F. W. Schultz have traditionally been regarded as separate species in Br, but recently British botanists have more often reduced the latter to subspecific level, and P. H. Raven (1970 *in litt.*) has informed me that he now considers the two names to represent a single taxon. They are kept as separate species in this account merely for the convenience of preserving separately the information which has been recorded under each name.

Interest in *Epilobium* hybrids in Br was first kindled by Marshall (1890, 1891, 1895), who started observations in 1888, having read the authoritative monograph of Haussknecht (1884). Unfortunately neither Marshall nor G. M. Ash, the leading British authority from about 1930 to 1959, seem to have attempted artificial hybridizations. Marshall's opinions were not accepted by all British botanists, several of whom denied the widespread occurrence of hybrids. Brown (1892) said that the specimens of supposed hybrids identified by Haussknecht and Marshall "appear to me at the utmost but trifling variations of one or other of their supposed parents, the differences between the supposed hybrid and the species it most resembles being no greater and sometimes not as great as may often be found between individuals in a bed of seedlings from one plant." Similar opinions are still voiced by some botanists at the present day, but no-one who has carefully studied Epilobia in the field over several seasons could possibly doubt the existence of hybrids, albeit infrequently, in many combinations.

Of the hybrids possible between the 12 British species of *Epilobium* (*sensu stricto* and excluding *E. brunnescens* and allies), all but 12 have been recorded in Europe. The 12 latter combinations all involve either *E. anagallidifolium* or *E. alsinifolium* (the two upland species) as one parent and, since several of them have been synthesized artificially, they probably remain unrecorded in the wild because the two potential parents never grow close together. Thus there is no evidence of anything but weak genetic incompatibility between all these species. The relative rarity of hybrids in the wild is probably a measure of the prevalence of inbreeding in all species, except perhaps in *E. hirsutum* L., and perhaps also of the lack in many cases of habitats suitable for colonization by hybrid offspring. In support of the latter idea is the fact that hybrids are most often encountered in disturbed habitats, typically quarries and waste-land (including, formerly, bombed sites) and, in my experience, old neglected gardens and allotments. In such places hybrids are more likely to find a suitable habitat among the wide spectrum available, and there are some reports of spectacular occurrences. For example G. M. Ash and J. H. G. Chapple found at least five hybrids in 1936 in a stone-quarry at Groby, v.c. 55 (specimens in **BM**); G. M. Ash (Dony, 1948) identified five species and seven hybrids in a gravel-pit at Eaton Socon, v.c. 30; Shaw (1951) reported five species and four hybrids in a neglected garden in south-eastern London, v.c. 16; Townsend (1953) reported eight species and seven hybrids in a wood near Forthampton, v.c. 34; and Ash (Jewell and Polunin, 1960) discovered at least five species and four hybrids in clay-pits at Brook, v.c. 17.

All 45 possible combinations between the 10 British lowland species of *Epilobium* have been recorded from Europe, and all but two of these (albeit some perhaps erroneously so) from Br. As might be expected some combinations are more frequently encountered than others. Thakur (1965) listed five hybrids as being particularly common: *E. montanum* x *E. obscurum*, x *E. parviflorum*, and x *E. roseum*; *E. obscurum* x *E. palustre*; and *E. parviflorum* x *E. roseum*. In my own experience in southern England the first two of these plus *E. adenocaulon* x *E. montanum* and x *E. obscurum* have been most frequently encountered, and other field-workers would probably wish to make other modifications to these lists. While several hybrids are rare or unknown because their would-be parents are rare or rarely grow together, it seems fairly clear that hybrids involving *E. hirsutum* are relatively uncommon even when this species grows in mixed populations with others. Whether this is because of genetic incompatibility or to differing habits of pollinating insects, or to both, has not been established for certain, but Thakur (1965) found a lower degree of success in artificial pollinations involving *E. hirsutum* than in those involving other species.

Artificial hybridizations have been carried out by at least a dozen different workers, the first dating from 1842 when Salter (1852) crossed *E. tetragonum* with *E. montanum*. Most of the earlier experiments were done on a small scale and were aimed merely at confirming the origin of putative hybrids collected in the wild. Two very extensive programmes of crosses have been carried out: by Geith (1924), who synthesized about 18 of the British hybrids, and by Thakur (1965), who synthesized almost twice as many but whose work is unfortunately not yet published. The results of the early crosses not only helped to interpret wild hybrids, and to understand the breeding relationships and barriers within *Epilobium sensu lato*, but also opened a new phase in the study of cytoplasmic inheritance. It was found by many workers, notably by Lehmann, at first in 1918, that there were often big differences in reciprocal crosses, e.g. the hybrid *E. parviflorum* female x *E. roseum* male differed in appearance from the reciprocal *E. roseum* female x *E. parviflorum* male. It was soon realized that these differences resulted from cytoplasmic genes (which were collectively called the plasmon), for in the former cross the cytoplasm of the hybrid is almost entirely derived from *E. parviflorum*, and in the latter from *E. roseum*. Moreover crosses using strains of the female parent from different sources also produced different F_1 offspring, and this was interpreted as a measure of the differing interaction between nuclear genes and cytoplasmic genes of different origins. Furthermore, by repeatedly backcrossing an F_1 hybrid to its male parent over many generations, it is possible to obtain a plant with a nuclear complement almost entirely of one plant (the original male), and a cytoplasmic complement almost entirely of the other (the original female), thus allowing a direct comparison of the expression of the nuclear genes of one species in an environment of its own cytoplasm with that in the cytoplasm of a second species. This work has been summarized by Lehmann (1925, 1941), Lehmann and Schwemmle (1927), Michaelis (1940, 1954), Caspari (1948), Lehmann and Düppel (1950) and others.

These results are also of value in understanding the taxonomic problems posed by the hybrids. In practical terms, the great variability found in the artificial hybrids shows that it is not possible to define precisely the characters of a hybrid of any one species-combination. One can say that the characters of a hybrid will be intermediate between those of the parents, but the extent of the intermediacy, both quantitatively and qualitatively, is unpredictably variable, and sometimes new, unexpected characters arise. In the wild there is probably considerable selection of F_1 plants, because most wild putative hybrids seem to approximate to the expected intermediate condition. But it may well be that these are the only hybrids which have been recognized as such. Hybrid variability extends to the degree of fertility. Hybrids are rarely fully fertile; most of them are largely sterile but produce a few viable seeds. Many experimental hybrids are completely sterile and may never flower, the plants being stunted and often deformed. In the wild they would go unnoticed, or in any case probably soon die out. The considerable fertility of some wild hybrids is shown by the occasional discovery of apparent backcrosses or even hybrid swarms, and a few instances of triple hybrids, e.g. *E. montanum* x *E. parviflorum* x *E.*

roseum. Moreover meiosis in the hybrids is uniformly highly regular, with no univalents or multivalents in most pollen mother cells. Sterility appears to be a function of the early abortion of the developing F_2 embryos.

Because of this variability, descriptions of the individual hybrids are not given here, but general hints in the determination of hybrids are provided. It is, of course, necessary to have a good knowledge of the diagnostic characters of putative parents, and a great help to know which species were growing in the vicinity of putative hybrids. Ash (1953) mentioned that the following general characteristics were good indications of the hybrid origin of a plant: taller and more branched habit; flowers unusually large or small in size; petals markedly deeper in colour at the tips; and fruits shortened and undeveloped with mostly abortive seeds. Many field-workers, including myself, have found these clues of value, but it must be remarked that they only cover one part of the spectrum of hybrid variability.

For present purposes we may divide the 12 species with which we are concerned into the following 5 Groups:

1. Stigmas cruciform:
   2. Both long, spreading, eglandular hairs and shorter, glandular hairs present
      Group A (*E. hirsutum E. parviflorum*)
   2. Long spreading, eglandular hairs and glandular hairs both absent
      Group B (*E. montanum, E. lanceolatum*)
1. Stigmas clavate:
   3. Spreading glandular hairs abundant
      Group C (*E. adenocaulon, E. roseum*)
   3. Spreading glandular hairs absent or scarce:
      4. Plants ± erect, lowland
         Group D (*E. lamyi, E. obscurum, E. palustre, E. tetragonum*)
      4. Plants ± decumbent, upland
         Group E (*E. alsinifolium, E. anagallidifolium*)

In identifying Epilobia it is in my experience best to decide first whether the plant is one of these 12 species or is a hybrid, and in the latter case to decide whether the parents belong to the same or to different Groups. It is naturally easier to assess the parentage when it involves species from two different Groups. In hybrids between Groups A or B on one hand and C, D or E on the other the stigmas are intermediate in morphology, although sometimes clavate and/or cruciform stigmas may occur along with them on a single plant. "Intermediate" stigmas may be obscurely cruciform or have 2–4 variously and irregularly developed lobes. Hybrids involving species of Group A usually show the long spreading hairs, though these are often deflexed apically to some degree, and in the case of *E. hirsutum* the large flowers. The different leaf-toothing of the two species in this Group is also an important character in diagnosing their parentage. The most useful features of the species in Group B, apart from the cruciform stigmas and lack of both spreading and glandular hairs, are the distinctive leaves, whose characters are usually clearly apparent in hybrids. The combination of clavate stigmas and abundant spreading glandular hairs is usually sufficient to detect the species of Group C in hybrids, but the presence of spreading glandular hairs in species of Group A also can be confusing. The very different leaf-shapes, particularly the leaf-bases and petioles, of *E. roseum* and *E. adenocaulon* normally provide conclusive evidence of parentage. The most important characters of species of Group D, all of which have clavate stigmas but have few or no glandular hairs, involve the leaves and the organs of perennation: *E. tetragonum* and *E. lamyi* bear narrow leaves (more or less decurrent in the former) and subsessile basal rosettes; *E. obscurum* has broader leaves and elongated leafy stolons; and *E. palustre* has narrow, sessile yet non-decurrent leaves and very long, largely naked stolons terminated by a small, compact bud. All these characters appear in hybrids involving these species, but it is probably true to say that the greatest difficulties in detecting the parentage of hybrids are met with in cases involving species of this Group. The involvement of species of Group E is usually apparent from the habitat and locality and the presence of one or other of the two species, which are themselves best separated on leaf-shape and -toothing.

As stated above no individual descriptions (b) are provided in the following accounts, and similarly under d the data given are sufficient only to follow-up the information available in the literature. The distributions given under c are very incomplete. They are based on published records made or confirmed by C. Haussknecht, E. S. Marshall and G. M. Ash, on specimens identified by these workers in **BM** (including Hb. Ash), and on a few other reliable sources. Thus, whereas few of these records are errors, many other records exist, but it is thought worth listing the above records as a starting point for future additions. Chromosome numbers (e) are not given. In all the British species $2n = 36$ (Raven and Moore, 1964), and the same number has been recorded in all the hybrids that have been examined (about a quarter of the total). Specific synonyms are not cited in the separate accounts. *E. tetragonum* L. is more often given in the British literature as *E. adnatum* Griseb.; *E. lamyi* F. W. Schultz is often known as *E. tetragonum* subsp. *lamyi* (F. W. Schultz) Leveillé; and *F. alpinum* auct. has often been used to cover *E. anagallidifolium* Lam. (much more rarely *E. alsinifolium* Vill.).

General References

ASH, G. M. (1953). *Epilobium adenocaulon* in Britain, in LOUSLEY, J. E., ed. *The changing flora of Britain*, pp. 168-170. Oxford.

BROWN, N. E. (1892). *Epilobium* in SYME, J. T. B., ed. *English Botany*, 3rd ed., Suppl., pp. 172-178. London.

CASPARI, E. (1948). Cytoplasmic inheritance. *Adv. Genet.*, 2: 1-66.

DONY, J. G. (1948). Excursions 1948. July 19–22, Bedford and district. *Rep. B.E.C.*, 13: 220-223.

GEITH, K. (1924). Experimentell-systematische Untersuchungen an der Gattung *Epilobium* L. *Bot. Arch.*, 6: 123-186.

HAUSSKNECHT, C. (1884). *Monographie der Gattung Epilobium.* Jena.

JEWELL, A. L. and POLUNIN, O. (1960). Obituaries: Gerald Mortimer Ash. *Proc. B.S.B.I.*, 4: 106-107.

LEHMANN, E. (1925). Die Gattung *Epilobium. Biblphia genet.,* 1: 363-416.

LEHMANN, E. (1941). Zur Genetik der Entwicklung in der Gattung *Epilobium,* 4. Das Plasmon in der Gattung *Epilobium. Jb. wiss. Bot.,* 89: 687-753; 90: 49-98.

LEHMANN, E. and DÜPPEL, W. (1950). Plasmonbegriff und Störungs- systeme in der Gattung *Epilobium. Züchter,* 20: 103-125.

LEHMANN, E. and SCHWEMMLE, J. (1927). Genetische Untersuchungen in der Gattung *Epilobium. Biblthca bot.,* 95: 1-156.

MARSHALL, E. S. (1890). *Epilobium* notes for 1889. *J. Bot., Lond.,* 28: 2-10.

MARSHALL, E. S. (1891). *Epilobium* notes for 1890. *J. Bot., Lond.,* 29: 6-9.

MARSHALL, E. S. (1895). Two hybrid Epilobia new to Britain. *J. Bot., Lond.,* 33: 106-108.

MICHAELIS, P. (1940). Über reziprok verschiedene Sippen-Bastarde bei *Epilobium hirsutum. Z. indukt. Abstamm.- u. VererbLehre,* 78: 187-222, 223-237, 295-337.

MICHAELIS, P. (1954). Cytoplasmic inheritance in *Epilobium* and its theoretical significance. *Adv. Genet.,* 6: 287-401.

RAVEN, P. H. and MOORE, D. M. (1964). Chromosome numbers of *Epilobium* in Britain. *Watsonia,* 6: 36-38.

SALTER, T. B. (1852). On the fertility of certain hybrids. *Phytol.,* 4: 737-742.

SHAW, H. K. A. (1951). An interesting *Epilobium* population. *Year Book B.S.B.I.,* 1951: 78.

THAKUR, V. (1965). *Biosystematics of some species of Epilobium.* Ph.D. thesis, University of Durham.

TOWNSEND, C. C. (1953). *Epilobium. Watsonia,* 2: 412.

1 x 2. *E. hirsutum* L. x *E. parviflorum* Schreb.

  a. (*E.* x *intermedium* Ruhmer, *non* Mérat).

  c. There are records from v.c. 1, 3, 6, 7, 10, 13–15, 17, 23, 24, 28–30, 32–34, 37, 41, 48, 55, 65, 69 and H39, and from Au, Be, Cz, Da, Ga, Ge, Ho, Hu, Po, Rm, Rs and Su.

  d. Crosses (mostly reciprocal) have been made by Compton (1913), Geith (1924), Michaelis (1944) and Thakur (1965), who also carried out backcrosses.

  f. BENNETT, A. and LITTLE, J. E. (1924). *Epilobium hirsutum* x *parviflorum. Rep. Watson B.E.C.,* 3: 252.

COMPTON, R. H. (1913). Further notes on *Epilobium* hybrids. *J. Bot., Lond.,* 51: 79-85.

COMPTON, R. H. and MARSHALL, E. S. (1913a). *Epilobium hirsutum,* L., ♀ x *E. parviflorum,* Schreb., ♂. *Rep B.E.C.,* 3: 254.

GEITH, K. (1924). As above, p. 135. FIG.

HAUSSKNECHT, C. (1884). As above, p. 64.

MICHAELIS, P. (1944). Untersuchungen an reziprok verschiedenen
Artbastarden bei *Epilobium*, 1. Über die Bastarde verschiedener Sippen
der Arten *E. hirsutum* mit *E. parviflorum*, resp. *E. montanum*. *Flora*,
137: 1-23. FIG.
MICHAELIS, P. and ROSS, H. (1944). Untersuchungen an reziprok
verschiedenen Artbastarden bei *Epilobium*, 2. Über Abänderungen an
reziprok verschiedenen und reziprok gleichen *Epilobium*-Artbastarden.
Flora, 137: 24-56. FIG.
THAKUR, V. (1965). As above.

1 × 3. *E. hirsutum* L. × *E. montanum* L.
a. *E.* × *erroneum* Hausskn.
c. There are records from v.c. 3, 6, 9, 16, 17, 23, 24, 31, 33, 37, 41, 55, 57,
64, 66 and 88/89, and from Au, Da, Ga, Ge, Ho and Rm.
d. Reciprocal crosses have been made by Compton (1913), Åkerman (1921),
Geith (1924), Michaelis (1944) and Thakur (1965).
f. ÅKERMAN, Å. (1921). Untersuchungen über Bastarde zwischen
Epilobium hirsutum und *Epilobium montanum*. *Hereditas*, 2: 99-112.
FIG.
COMPTON, R. H. (1913). As above.
COMPTON, R. H. (1914). *Epilobium hirsutum* ♂ × *montanum* ♀. *Rep.
B.E.C.*, 3: 469.
COMPTON, R. H. and MARSHALL, E. S. (1936). *Epilobium hirsutum*, L.,
♀ × *E. montanum*, L., ♂. *Rep. B.E.C.*, 3: 254.
GEITH, K. (1924). As above, p. 137. FIG.
HÅKANSSON, A. (1924). Beiträge zur Zytologie eines *Epilobium*-
Bastardes. *Bot. Notiser*, 1924: 269-278.
HAUSSKNECHT, C. (1884). As above, pp. 62-63.
MICHAELIS, P. (1944). As above. FIG.
MICHAELIS, P. and ROSS, H. (1944). As above. FIG.
MICHAELIS, P. and WERTZ, E. (1935). Entwicklungsgeschich-
tlichgenetische Untersuchungen an *Epilobium*, 6. Vergleichende Unter-
suchungen über das Plasmon von *Epilobium hirsutum*, *E. luteum*,
E. montanum und *E. roseum*. *Z. indukt. Abstamm - u. VererbLehre*,
70: 138-159. FIG.
THAKUR, V. (1965). As above.

1 × 4. *E. hirsutum* L. × *E. lanceolatum* Seb. & Mauri
= *E.* × *surreyanum* E. S. Marshall was recorded by Marshall from near
Worplesdon, v.c. 17, in 1889 (**BM, CGE**), but there is doubt attached to
the identity of the specimens. This hybrid combination has not been
found elsewhere.
MARSHALL, E. S. (1890). As above.
THAKUR, V. (1965). As above.

1 × 5. *E. hirsutum* L. × *E. roseum* Schreb.
a. *E.* × *goerzii* Rubner.
c. There are records from v.c. 17, 34 and 57, and from Be, Cz, Ge, Ho and
  Po.
d. Reciprocal crosses have been made by Lehmann and Schwemmle (1927)
  and Thakur (1965).
f. HAUSSKNECHT, C. (1884). As above, p. 65.
  LEHMANN, E. and SCHWEMMLE, J. (1927). As above.
  MARSHALL, E. S. (1918). *Epilobium hirsutum* x *roseum* in Surrey. *J.
  Bot., Lond.*, 56: 332-333.
  THAKUR, V. (1965). As above.

6 × 1. *E. adenocaulon* Hausskn. × *E. hirsutum* L.
a. None.
c. There are records from v.c. 12, 17, 24, 28, 30 and 85, the first from v.c.
  12 in 1936, collected by P. M. Hall (**BM**).
d. None.
f. HALL, P. M. and SLEDGE, W. A. (1937). *Epilobium adenocaulon*
  Hausskn. x *hirsutum* L. *Rep. B.E.C.*, 11: 223.

1 × 7. *E. hirsutum* L. × *E. tetragonum* L.
a. *E.* × *brevipilum* Hausskn.
c. There are records from v.c. 17, 30, 55 and 64, and from Ga, Ge and Ho.
d. The hybrid was synthesized by Compton (1911, 1913) using *E. hirsutum*
  as female, and reciprocal crosses were made by Geith (1924).
f. COMPTON, R. H. (1911). Notes on *Epilobium* hybrids. *J. Bot., Lond.*, 49:
  158-163.
  COMPTON, R. H. (1913). As above.
  COMPTON, R. H. and MOSS, C. E. (1911). *Epilobium hirsutum* ♀ x *E.
  tetragonum* ♂. *Rep. B.E.C.*, 2: 562.
  GEITH, K. (1924). As above, p. 143. FIG.
  HAUSSKNECHT, C. (1884). As above, p 103.

1 × 8. *E. hirsutum* L. × *E. lamyi* F. W. Schultz
a. None.
c. There are records from v.c. 17 and 30, and from Ge.
d. Thakur (1965) synthesized the hybrid using *E. lamyi* as female, and made
  backcrosses to both parents.
f. THAKUR, V. (1965). As above.

1 × 9. *E. hirsutum* L. × *E. obscurum* Schreb.
  = *E.* × *anglicum* E. S. Marshall was recorded by Marshall from v.c. 17, but
  the specimens in **BM** have been redetermined by G. M. Ash as *E.
  parviflorum* and *E. obscurum* x *E. parviflorum*. There is also a record from
  v.c. 58 which was determined by E. S. Marshall.

MARSHALL, E. S. (1890). As above.
WOLLEY-DOD, A. H. (1893). *Epilobium hirsutum* x *obscurum* in Cheshire. *J. Bot., Lond.,* **31**: 372.

1 x 10. *E. hirsutum* L. x *E. palustre* L.
a. *E.* x *waterfallii* E. S. Marshall.
c. There are records from v.c. 3, 15 and 58, and from Cz and Rs.
d. Reciprocal crosses have been made by Geith (1924) and Thakur (1965), who backcrossed the F₁ to *E. hirsutum.*
f. GEITH, K. (1924). As above, p 141. FIG.
   HAUSSKNECHT, C. (1884). As above, p 63.
   MARSHALL, E. S. (1916a). A new hybrid willow-herb. *J. Bot., Lond.,* **54**: 75-76.
   MARSHALL, E. S. (1916b). *Epilobium hirsutum* x *palustre* and *E. palustre* x *tetragonum* in E. Kent. *J. Bot., Lond.,* **54**: 114-115.
   THAKUR, V. (1965). As above.

3 x 2. *E. montanum* L. x *E. parviflorum* Schreb.
a. *E.* x *limosum* Schur.
c. There are records from v.c. 3, 4, 6, 11, 14–17, 20, 23, 24, 28, 30, 32–34, 37, 41, 55, 69, H30, H33 and H39, and from Au, Cz, Da, Ga, Ge and Ho.
d. Crosses, mostly reciprocal, have been made by Compton (1913), Lehmann (1919), Geith (1924) and Thakur (1965), who also carried out back-crosses.
f. COMPTON, R. H. (1913). As above.
   COMPTON, R. H. and MARSHALL, E. S. (1913c). *Epilobium montanum,* L., ♀ x *E. parviflorum,* Schreb., ♂. *Rep. B.E.C.,* **3**: 254-255.
   GEITH, K. (1924). As above, p. 144. FIG.
   HAUSSKNECHT, C. (1884). As above, pp. 79-80.
   LEHMANN, E. (1919). Weitere *Epilobium*-Kreuzungen. *Ber. dt. bot. Ges.,* **37**: 347-357. FIG.
   LEHMANN, E. (1924). Über Sterilitätserscheinungen bei reziprok verschiedenen Epilobiumbastarden. *Biol. Zbl.,* **44**: 243-254.
   LEHMANN, E. and SCHWEMMLE, J. (1927). As above.
   THAKUR, V. (1965). As above.

3 x 2 x 5. *E. montanum* L. x *E. parviflorum* Schreb. x *E. roseum* Schreb.
   was recorded by E. S. Marshall from near Worplesdon, v.c. 17, in 1889. The specimens (CGE) were confirmed by C. Haussknecht and Thakur (1965) also agreed with the determination.
   MARSHALL, E. S. (1890). As above.
   THAKUR, V. (1965). As above.

3 x 9 x 2. *E. montanum* L. x *E. obscurum* Schreb. x *E. parviflorum* Schreb.

was said by Thakur (1965) to exist in various herbaria among specimens of *E. obscurum* x *E. parviflorum*.

4 x 2. *E. lanceolatum* Seb. & Mauri x *E. parviflorum* Schreb.

a. *E.* x *aschersonianum* Hausskn.

c. There are old records from Plymouth, v.c. 3 (J. R. A. Briggs 1867 (**BM**) det. C. Haussknecht; H. Trimen 1877 (**BM**) det. G. M. Ash). Specimens from v.c. 13 originally determined as this hybrid by E. S. Marshall were redetermined by G. M. Ash as *E. parviflorum* x *E. roseum.* The hybrid is also known from Ho.

d. Thakur (1965) made artificial hybrids using *E. lanceolatum* as female, and he also performed backcrosses.

f. ASH, G. M. (1937a). *Epilobium parviflorum* Schreb. x *roseum* Schreb. *Rep. B.E.C.,* **11**: 254.

DRUCE, G. C. (1919). *Epilobium lanceolatum* x *parviflorum. Rep. B.E.C.,* **5**: 284.

HAUSSKNECHT, C. (1884). As above, p 95.

THAKUR, V. (1965). As above.

2 x 5. *E. parviflorum* Schreb. x *E. roseum* Schreb.

a. *E.* x *persicinum* Reichb.

c. There are records from v.c. 8, 13, 15–17, 23, 24, 30, 32–34, 36, 37, 43, 55, 57, 62 and 65, and from Au, Be, Cz, Da, Ga, Ge, Gr, He, Ho, Hs, Hu, Ju, Po, Rm, Rs and Su. Backcrosses to *E. parviflorum* were recorded from v.c. 16 by E. S. Marshall.

d. Reciprocal crosses have been made by Lehmann (1918), Geith (1924), Schwemmle (1924) and Thakur (1965).

f. GEITH, K. (1924). As above, p. 145. FIG.

HAUSSKNECHT, C. (1884). As above, pp. 72-73.

LEHMANN, E. (1918). Über reziproke Bastarde zwischen *Epilobium roseum* und *parviflorum. Z. Bot.,* **10**: 497-511. FIG.

LEHMANN, E. (1924). As above.

LEHMANN, E. and SCHWEMMLE, J. (1927). As above.

SCHWEMMLE, J. (1924). Zur Kenntnis der reziproken Bastarde zwischen *Epilobium parviflorum* und *roseum. Z. indukt. Abstamm. - u. VererbLehre,* **34**: 145-185. FIG.

THAKUR, V. (1965). As above.

6 x 2. *E. adenocaulon* Hausskn. x *E. parviflorum* Schreb.

a. None.

c. There are records from v.c. 9, 13, 16, 17, 19, 21, 22, 24, 27, 28, 30 and 32–34 and from Ho. The first British records were from v.c. 17 in 1934.

d. Thakur (1965) made artificial hybrids and backcrosses to both parents.

f. ASH, G. M. and SANDWITH, N. Y. (1935). *Epilobium adenocaulon* Hausskn. in Britain. *J. Bot., Lond.*, **73**: 177-184.
THAKUR, V. (1965). As above.

2 × 7. *E. parviflorum* Schreb. × *E. tetragonum* L.

a. *E.* × *weissenburgense* F. W. Schultz.
c. There are records from v.c. 3, 14, 16, 17, 22–24, 27, 34, 37 and 55, and from Au, Be, Cz, Ga, Ge, He, Ho, Hu, Po and Rm. The material at **BM** collected by E. S. Marshall from v.c. 23 has been redetermined as *E. obscurum* × *E. parviflorum*.
d. Thakur (1965) synthesized hybrids using *E. parviflorum* as female, and Geith (1924) made reciprocal hybrids.
f. GEITH, K. (1924). As above, p. 149. FIG.
HAUSSKNECHT, C. (1884). As above, p. 105.
THAKUR, V. (1965). As above.

8 × 2. *E. lamyi* F. W. Schultz × *E. parviflorum* Schreb.

a. *E.* × *palatinum* F. W. Schultz.
c. Hybrids have been recorded from v.c. 6, 15–17, 24, 28, 30 and 33, and from Au, Be, Ga, Ge and Po.
d. Reciprocal crosses were made by Thakur (1965).
f. HAUSSKNECHT, C. (1884). As above, pp. 111-112.
MARSHALL, E. S. (1890). As above.
THAKUR, V. (1965). As above.

8 × 9 × 2. *E. lamyi* F. W. Schultz × *E. obscurum* Schreb. × *E. parviflorum* Schreb.

was said by Thakur (1965) to be represented by a specimen at **CGE**, determined by E. S. Marshall, among specimens of *E. obscurum* × *E. parviflorum*.
THAKUR, V. (1965). As above.

9 × 2. *E. obscurum* Schreb. × *E. parviflorum* Schreb.

a *E.* × *dacicum* Borbás.
c. There are records from v.c. 3, 4, 11, 13–17, 22–24, 27, 32–34, 37, 39, 41, 48, 50, 51, 55, 57, 58, 66, 69, 89, 90, H16 and H39, and from Au, Be, Cz, Ga, Ge, He, Ho, Po and Rm.
d. Reciprocal crosses and backcrosses were carried out by Thakur (1965).
f. ASH, G. M. (1937b). *Epilobium obscurum* Schreb. × *parviflorum* Schreb. *Rep. B.E.C.*, **11**: 402.
HAUSSKNECHT, C. (1884). As above, pp. 122-123.
THAKUR, V. (1965). As above.

10 × 2. *E. palustre* L. × *E. parviflorum* Schreb.

a. *E.* × *rivulare* Wahlenb.

c. There are records from v.c. 4, 14, 17, 20, 27, 34, 37, 48, 58, 70, 76, 88, 101, 111, 112, H33, H39 and H40, and from Au, Be, Cz, Da, Ga, Ge, He, Ho, Po, Rm and Su.

d. Reciprocal crosses have been made by Lehmann (1919), Geith (1924) and Thakur (1965), who also made backcrosses.

f. GEITH, K. (1924). As above, p. 147. FIG.

HAUSSKNECHT, C. (1884). As above, pp 138-140.

LEHMANN, E. (1919). As above.

LEHMANN, E. (1924). As above. FIG.

LITTLE, J. E. and PUGSLEY, H. W. (1923). *Epilobium palustre* x *parviflorum? Rep. Watson B.E.C.*, 3: 216-217.

THAKUR, V. (1965). As above.

12 x 2. *E. alsinifolium* Vill. x *E. parviflorum* Schreb.

= *E.* x *gerstlaueri* Rubner has been recorded from Au and Ge.

4 x 3. *E. lanceolatum* Seb. & Mauri x *E. montanum* L.

a. *E.* x *neogradense* Borbás.

c. There are records from v.c. 1–4, 16, 17, 24, 34 and 35, and from CI, Au, Ga, Ge and Ho.

d. Reciprocal crosses were made by Thakur (1965).

f. HAUSSKNECHT, C. (1884). As above, pp. 93-94.

MARSHALL, E. S. and RIDDELSDELL, H. J. (1912). *E. lanceolatum* x *montanum. Rep. B.E.C.*, 3: 94.

THAKUR, V. (1965). As above.

3 x 5. *E. montanum* L. x *E. roseum* Schreb.

a. *E.* x *mutabile* Boiss. & Reut.

c. There are records from v.c. 1, 3, 6, 11, 16, 17, 21, 24, 33–35, 37, 38, 41, 55, 62, 64, 85, 90 and H21, and from Au, Be, Cz, Ga, Ge, He, Ho, Hu, Po, Rm and Su. Marshall (1890) reported a backcross to *E. roseum* from near Worplesdon, v.c. 17.

d. Crosses have been made using *E. montanum* as female by Focke (1881), and reciprocally by Geith (1924) and Thakur (1965); Thakur also made backcrosses.

f. FOCKE, W. O. (1881). *Op. cit.*, p. 528.

GEITH, K. (1924). As above, p. 154. FIG.

HAUSSKNECHT, C. (1884). As above, pp. 80-81.

MARSHALL, E. S. (1890). As above.

THAKUR, V. (1965). As above.

6 x 3. *E. adenocaulon* Haussk. x *E. montanum* L.

a. None.

c. There are records from v.c. 8, 11–17, 20, 24, 30, 31, 33, 34, 48, 71 and 85; the first three were from v.c. 17 in 1930, 1933 and 1934.

d. Hybrids have been synthesized by Raven and Moore (1964) (direction not
   stated), Thakur (1965) (reciprocally), and Brockie (1970) (using *E.
   adenocaulon* as female).
f. ASH, G. M. and SANDWITH, N. Y. (1935). As above.
   BROCKIE, W. B. (1970). Artificial hybridisation in *Epilobium* involving
   New Zealand, European and North American species. *N.Z.J.Bot.*, **8**:
   94-97.
   LOUSLEY, J. E. (1935). *Epilobium adenocaulon* Hausskn. *J. Bot., Lond.*,
   73: 257.
   RAVEN, P. H. and MOORE, D. M. (1964). As above.
   THAKUR, V. (1965). As above.

3 × 7. *E. montanum* L. × *E. tetragonum* L.
a. *E.* × *beckhausii* Hausskn.
c. There are records from v.c. 17, 24 and 33, and from Cz, Ga and Ge.
d. Hybrids were made by Compton (1911), using *E. tetragonum* as the
   female, and the two parents were crossed reciprocally by Salter (1852) and
   Geith (1924).
f. COMPTON, R. H. (1911). As above.
   COMPTON, R. H. and MOSS, C. E. (1911). *Epilobium montanum*♂ ×
   tetragonum♀. *Rep. B.E.C.*, **2**: 563.
   GEITH, K. (1924). As above, p. 157. FIG.
   HAUSSKNECHT, C. (1884). As above, p. 104.
   SALTER, T. B. (1852). As above.

8 × 3. *E. lamyi* F. W. Schultz × *E. montanum* L.
a. *E.* × *haussknechtianum* Borbás.
c. There are records from v.c. 6, 17 (**BM**) and 23, and from Au, Cz, Ga and
   Ge.
d. Hybrids were made artificially by Haussknecht (1884), and reciprocal
   crosses and backcrosses were made by Thakur (1965).
f. HAUSSKNECHT, C. (1884). As above, p. 27, 110-111.
   THAKUR, V. (1965). As above.

3 × 9. *E. montanum* L. × *E. obscurum* Schreb.
a. *E.* × *aggregatum* Čelak.
c. There are records from v.c. 2–7, 12–17, 23, 24, 30, 31, 33, 34, 36, 38, 40,
   41, 47–49, 57, 58, 80, 88, 101, H27, H29 and H40, and from Au, Cz, Ga,
   Ge, Ho and Ju.
d. Crosses have been made using *E. montanum* as female by Focke (1881),
   and reciprocally by Geith (1924) and Thakur (1965), who also made
   backcrosses to both parents, studied segregation in the F_2 generation and
   made backcrosses from various F_2 segregants to both original parents.
f. FOCKE, W. O. (1881). *Op. cit.*, p. 528.
   GEITH, K. (1924). As above, p. 158. FIG.

HAUSSKNECHT, C. (1884). As above, pp. 78-79.
THAKUR, V. (1965). As above.

3 × 10. *E. montanum* L. × *E. palustre* L.

= *E.* × *montaniforme* Knaf ex Čelak. has been recorded from Au, Cz, Ga, Ge, He, Ho and Rm. It was recorded for BI by Dandy (1958), but I have seen no published records. There are specimens in **BM** determined as this hybrid by E. S. Marshall from v.c. 70 and 96, but G. M. Ash has redetermined the latter as *E. palustre*. Artificial hybrids have been made reciprocally by Geith (1924).

DANDY, J. E. (1958). *Op. cit.*, p. 65.
GEITH, K. (1924). As above, p. 156. FIG.
HAUSSKNECHT, C. (1884). As above, p. 79.

12 × 3. *E. alsinifolium* Vill. × *E. montanum* L.
  a. *E.* × *grenieri* Rouy & Camus (*E.* × *salicifolium* Facch., *non* Stokes).
  c. There are records from v.c. 65 (Sledge, 1945) 87/88 (Marshall, 1891), 90 (**BM**) and 94 (**BM**), and from Au, Cz, Ga, Ge, He and Rm.
  d. Reciprocal crosses were made by Thakur (1965). A spontaneous hybrid which arose in cultivation at Durham was backcrossed to both its parents by Thakur.
  f. HAUSSKNECHT, C. (1884). As above, pp. 168-169.
     MARSHALL, E. S. (1891). As above.
     SLEDGE, W. A. (1945). *E. alsinifolium* × *montanum. Naturalist, Hull,* 812: 24.
     THAKUR, V. (1965). As above.

4 × 5. *E. lanceolatum* Seb. & Mauri × *E. roseum* Schreb.
  a. *E.* × *abortivum* Hausskn.
  c. There are records from v.c. 16 and 40 and from Ge, and the hybrid also arose spontaneously in E. S. Marshall's garden in v.c. 17.
  d. Thakur (1965) made artificial hybrids using *E. roseum* as female.
  f. HAUSSKNECHT, C. (1884). As above, p. 95.
     MARSHALL, E. S. (1895). As above.
     THAKUR, V. (1965). As above.

6 × 4. *E. adenocaulon* Hausskn. × *E. lanceolatum* Seb. & Mauri
  a. None.
  c. There are records from v.c. 17 (1949, G. M. Ash, **BM**) and 24 (1961, A. F. Wood).
  d. Artificial hybrids were made by Brockie (1970), using *E. lanceolatum* as female.
  f. BROCKIE, W. B. (1970). As above.

4 × 7. *E. lanceolatum* Seb. & Mauri × *E. tetragonum* L.
= *E.* × *fallacinum* Hausskn. has been recorded from Ga, Ge and Ho. There
are specimens at **BM** and **CGE** identified as this by E. S. Marshall, found
by him in his garden in v.c. 17, where they arose spontaneously from the
cultivated parents, but those in **BM** have been redetermined by G. M. Ash
as *E. tetragonum.*

8 × 4. *E. lamyi* F. W. Schultz × *E. lanceolatum* Seb. & Mauri
a. *E.* × *ambigens* Hausskn.
c. There are records from v.c. 6, 16, 17 and 23, and from Ga and Ge. Further
   study of these is, however, necessary. Of the British specimens in **BM** the
   one from v.c. 16 (det. E. S. Marshall) has been redetermined by G. M. Ash
   as *E. lanceolatum* × *E. obscurum,* the one from v.c. 17 (det. C.
   Haussknecht) is named by Ash *E. lamyi,* and one collected by Marshall in
   his garden in v.c. 17 is named by Ash *E. lamyi* × *E. tetragonum.*
d. Thakur (1965) synthesized the hybrid using *E. lanceolatum* as female.
f. HAUSSKNECHT, C. (1884). As above, p 110.
   MARSHALL, E. S. (1890). As above.
   MARSHALL, E. S. (1895). As above.
   THAKUR, V. (1965). As above.

4 × 9. *E. lanceolatum* Seb. & Mauri × *E. obscurum* Schreb.
a. *E.* × *lamotteanum* Hausskn.
c. There are records from v.c. 3 and 15–17, and from Ga, Ge and Ho.
   Specimens from v.c. 17 recorded by E. S. Marshall were both wild hybrids
   and spontaneous ones which had arisen in his garden.
d. None.
f. HAUSSKNECHT, C. (1884). As above, p. 94.
   MARSHALL, E. S. (1895). As above.

4 × 10. *E. lanceolatum* Seb. & Mauri × *E. palustre* L.
= *E.* × *langeanum* Hausskn. arose spontaneously in E. S. Marshall's garden
in v.c. 17 (**BM**), and similarly in a botanic garden in Da. Thakur (1965) has
synthesized the hybrid using *E. lanceolatum* as female.
HAUSSKNECHT, C. (1884). As above, pp 94-95.
MARSHALL, E. S. (1895). As above.
THAKUR, V. (1965). As above.

12 × 4. *E. alsinifolium* Vill. × *E. lanceolatum* Seb. & Mauri
is represented in **BM** by a specimen which arose spontaneously in 1895 in
E. S. Marshall's garden in v.c. 17, but Thakur (1965), who synthesized the
hybrid reciprocally, said that all specimens he had seen were highly
doubtful.
THAKUR, V. (1965). As above.

6 × 5. *E. adenocaulon* Hausskn. × *E. roseum* Schreb.

a. None.

c. There are records from v.c. 16–18, 24, 33 and 85, the earliest being from v c. 16 (1947, H. K. A. Shaw).

d. Thakur (1965) carried out reciprocal crosses and obtained F_1 seed, whose viability was not tested. Brockie (1970) obtained F_1 hybrids using *E. roseum* as female, and also grew an F_2 generation.

f. BROCKIE, W. B. (1970). As above.

SHAW, H. K. A. (1951). As above.

THAKUR, V. (1965). As above.

5 × 7. *E. roseum* Schreb. × *E. tetragonum* L.

a. *E.* × *borbasianum* Hausskn.

c. There are records from v.c. 16, 17 and 24, and from Au, Be, Cz, Ga, Ge, Ho and Po.

d. Reciprocal crosses were made by Geith (1924).

f. GEITH, K. (1924). As above, p. 172. FIG.

HAUSSKNECHT, C. (1884). As above, pp. 105-106.

8 × 5. *E. lamyi* F. W. Schultz × *E. roseum* Schreb.

= *E.* × *dufftii* Hausskn. has been recorded from Cz, Ga and Ge.

9 × 5. *E. obscurum* Schreb. × *E. roseum* Schreb.

a. *E.* × *brachiatum* Čelak.

c. There are records from v.c. 14, 17, 22, 24, 37, 55 and 57, and from Au, Cz, Ga, Ge, Ho, Hu and Rm, but the specimens from v.c. 37 and 55 at BM (det. E. S. Marshall) have been redetermined by G. M. Ash as *E. adenocaulon.*

d. Reciprocal crosses have been made by Thakur (1965).

f. HAUSSKNECHT, C. (1884). As above, pp. 123-124.

MARSHALL, E. S. (1890). As above.

THAKUR, V. (1965). As above.

10 × 5. *E. palustre* L. × *E. roseum* Schreb.

a. *E.* × *purpureum* Fr.

c. The only British specimen was collected in 1889 by E. S. Marshall near Worplesdon, v.c. 17. The hybrid has also been recorded from Au, Cz, Ge, Ho, Rm and Rs.

d. Reciprocal crosses were made by Geith (1924).

f. GEITH, K. (1924). As above, p. 173. FIG.

HAUSSKNECHT, C. (1884). As above, p. 140.

MARSHALL, E. S. (1890). As above.

12 × 5. *E. alsinifolium* Vill. × *E. roseum* Schreb.

= *E.* × *gemmiferum* Boreau has been recorded from Au, Ga, Ge, He and Rm.

6 × 7. *E. adenocaulon* Hausskn. × *E. tetragonum* L.
a. None.
c. There are records from v.c. 17 (1938, G. M. Ash, **BM**) and 24 (1954, A. F. Wood, **BM**, det. G. M. Ash), and from Ho (before 1962).
d. None.
f. None.

6 × 8. *E. adenocaulon* Hausskn. × *E. lamyi* F. W. Schultz
a. None.
c. There are records from v.c. 17 (1936 and 1938, G. M. Ash, **BM**) and 24 (1954, A. F. Wood, **BM**, det. G. M. Ash).
d. F_1 seed was obtained by Thakur (1965) using *E. adenocaulon* as female, but the viability of the seed was not tested.
f. THAKUR, V. (1965). As above.

6 × 9. *E. adenocaulon* Hausskn. × *E. obscurum* Schreb.
a. None.
c. There are records from v.c. 3, 4, 8, 12, 14–17, 24, 31–33, 48 and 85, the earliest being from v.c. 17 (1934, two localities) and v.c. 12 (1935).
d. Reciprocal crosses were made by Thakur (1965). Brockie (1970) obtained artificial hybrids using *E. obscurum* as female, and he also grew a vigorous F_2 generation.
f. ASH, G. M. (1947). *Epilobium adenocaulon* Hausskn. × *obscurum* Schreb. *Rep. B.E.C.*, **13**: 160.
ASH, G. M. and SANDWITH, N. Y. (1935). As above.
BROCKIE, W. B. (1970). As above.
THAKUR, V. (1965). As above.

6 × 10. *E. adenocaulon* Hausskn. × *E. palustre* L.
a. None.
c. There are records from v.c. 12 (1934–38, at edge of Fleet Pond, G. M. Ash, **BM**) and from Ho.
d. Thakur (1965) obtained F_1 seed from crosses using *E. adenocaulon* as female, but did not test the viability of the seed.
f. ASH, G. M. and SANDWITH, N. Y. (1935). As above.
THAKUR, V. (1965). As above.

8 × 7. *E. lamyi* F. W. Schultz × *E. tetragonum* L.
a. *E.* × *semiadnatum* Borbás.
c. There are records from v.c. 6, 17, 24, 30 and 33, and from Au, Cz, Ge, He and Hu.

   d. Crosses were made by Thakur (1965) using *E. lamyi* as female.
   f. ASH, G. M. (1937c). *Epilobium lamyi* Schultz x *tetragonum* L., em.
Curtis. *Rep. B.E.C.*, **11**: 402.
HAUSSKNECHT, C. (1884). As above, p. 103.
MARSHALL, E. S. (1890). As above.
MARSHALL, E. S. (1895). As above.

9 x 7. *E. obscurum* Schreb. x *E. tetragonum* L.

   a. *E.* x *thuringiacum* Hausskn.
   c. There are records from v.c. 4, 6, 9, 17, 28, 33, 36, 41, 61 and 70, and
from Au, Ge and Ho.
   d. Artificial hybrids have been made by Thakur (1965), using *E. obscurum* as
female.
   f. HAUSSKNECHT, C. (1884). As above, p. 104.
MARSHALL, E. S. (1889). Notes on Epilobia. *J. Bot., Lond.*, **27**:
143-147.
THAKUR, V. (1965). As above.

10 x 7. *E. palustre* L. x *E. tetragonum* L.

   a. *E.* x *laschianum* Hausskn.
   c. This hybrid was recorded from v.c. 15 (1913, near Dungeness, R. H.
Compton), and as a spontaneous weed in E. S. Marshall's garden in v.c. 17.
There are also records from Cz, Ga, Ge and Ho.
   d. Reciprocal hybrids were made by Geith (1924).
   f. GEITH, K. (1924). As above, p. 175. FIG.
HAUSSKNECHT, C. (1884). As above, pp. 104-105.
MARSHALL, E. S. (1895). As above.
MARSHALL, E. S. (1916b). As above.

8 x 9. *E. lamyi* F. W. Schultz x *E. obscurum* Schreb.

   a. *E.* x *semiobscurum* Borbás.
   c. There are records from v.c. 3, 6 and 17, and from Au and Ge. A further
specimen in **BM** from E. S. Marshall's garden in v.c. 17 has been
redetermined by G. M. Ash as *E. lamyi*.
   d. Reciprocal hybrids have been made by Thakur (1965).
   f. HAUSSKNECHT, C. (1884). As above, p. 111.
MARSHALL, E. S. (1890). As above.
THAKUR, V. (1965). As above.

8 x 10. *E. lamyi* F. W. Schultz x *E. palustre* L.

   = *E.* x *probstii* Léveillé has been recorded from Cz and Ge.

9 x 10. *E. obscurum* Schreb. x *E. palustre* L.

   a. *E.* x *schmidtianum* Rostk.
   c. There are records from v.c. 4, 9, 15–17, 24, 27, 32, 41, 48, 57, 58, 62,

88–90, 94, 98, 108, 110, H3, H16, H21, H23, H30 and H39, and from
Au, Be, Cz, Fe, Ga, Ge, He, Ho, Po, Rm and Su. Specimens in BM from
v.c. 17 determined by Marshall as backcrosses to the two parents have
been redetermined by G. M. Ash as *E. obscurum* and *E. palustre*
respectively.
d. Reciprocal crosses were made by Thakur (1965), but the seed obtained
was not tested for viability. Geith (1924) obtained reciprocal hybrids.
f. ASH, G. M. (1937d). *Epilobium obscurum* Schreb. x *palustre* L. *Rep.
B.E.C.*, **11**: 402.
GEITH, K. (1924). As above, p. 175. FIG.
HAUSSKNECHT, C. (1884). As above, pp. 121-122.
MARSHALL, E. S. (1890). As above.
MARSHALL, E. S. (1891). As above.
THAKUR, V. (1965). As above.

11 × 9. *E. anagallidifolium* Lam. × *E. obscurum* Schreb.
a. *E.* x *marshallianum* Hausskn.
c. There are records from v.c. 86 and 108, and from Ga and Ge. Specimens
collected from *c* 1100–1600 ft on Ben More of Assynt, v.c. 108, in 1887
and 1890 by E. S. Marshall were grown in his garden in v.c. 17 between
1891 and 1895. They increased greatly in vigour, but remained
"constantly sterile" (BM).
d. None.
f. MARSHALL, E. S. (1891). As above.

12 × 9. *E. alsinifolium* Vill. × *E. obscurum* Schreb.
a. *E.* x *rivulicola* Hausskn.
c. There are records from v.c. 87 and 94, and from Cz, Ga, Ge and Hs.
Specimens in BM from v.c. 94 were collected by E. S. Marshall in 1905 at
1000–1100 ft, by Couglass Water, near Tomintoul.
d. Reciprocal crosses were made by Thakur (1965), but the viability of the
F_1 seeds obtained was not tested.
f. HAUSSKNECHT, C. (1884). As above, pp. 169-170.
MARSHALL, E. S. (1895). As above.

11 × 10. *E. anagallidifolium* Lam. × *E. palustre* L.
a. (*E.* x *dasycarpum* auct.).
c. There are records from v.c 96 and 108, and from Au, Fe, Hs and No.
Specimens in BM from v.c. 108 were collected by E. S. Marshall in 1890 at
1500–2000 ft, on Ben More of Assynt.
d. None.
f. HAUSSKNECHT, C. (1884). As above, pp. 157-158.
MARSHALL, E. S. (1891). As above.

B. SYSTEMATIC

B. SYSTEMATIC

12 × 10. *E. alsinifolium* Vill. × *E. palustre* L.

a. *E.* × *haynaldianum* Hausskn.
c. There are records from v.c. 66 (Cauldron Snout), 68 (coll. Sowerby, **BM**), 69 (1894, High Cup Nick, **BM**), 70 (1912, Fisher Ghyll, **BM**), 86 (Balglass Corrie, J. Mitchell *in litt.* 1971), 94 (Tomintoul, **BM**) and 108 (1890, Ben More of Assynt, **BM**), and from Au, Cz, Ga, Ge, He, Hs, Ju and Rm. The altitude from which British specimens in **BM** were collected is 1000–2000 ft; specimens grown in E. S. Marshall's garden in v.c. 17 became very vigorous.
d. Reciprocal hybrids were made by Thakur (1965).
f. HAUSSKNECHT, C. (1884). As above, pp. 170-171.
MARSHALL, E. S. (1891). As above.
THAKUR, V. (1965). As above.

12 × 11. *E. alsinifolium* Vill. × *E. anagallidifolium* Lam.

a. *E.* × *boissieri* Hausskn.
c. There are records from v.c. 89 (1892, **BM**), 90 (1892, **BM**), 92 (1892, **BM**), and 108 (1890, **BM**), and from Au, Cz, Ga, Ge, He, Hs, and Rm. The altitude from which British specimens in **BM** were collected is 1500–3000 ft, mostly over 2500 ft; specimens grown in E. S. Marshall's garden in v.c. 17 became very vigorous.
d. None.
f. HAUSSKNECHT, C. (1884). As above, pp. 166-167.
MARSHALL, E. S. (1889). As above.
MARSHALL, E. S. (1891). As above.

256. *Oenothera* L.

(by C. A. Stace)

1 × 2. *O. biennis* L. × *O. erythrosepala* Borbás (*O. lamarckiana* de Vries)

a. None.
b. *O. erythrosepala* differs from *O. biennis* in the red-based hairs on the stems (absent in *O. biennis*), reddish or red-striped calyx (green in *O. ·biennis*), relatively longer style and stigmas and considerably larger flowers. Hybrids are apparently completely fertile and can give rise to extensive hybrid swarms exhibiting all degrees of variation between the parental extremes and to introgressed variants of both species. In parts of v.c. 59 all plants which resemble *O. biennis* in fact have many red-based hairs on the stems.

c. Hybrids are widespread on the coastal dunes of v.c. 58, 59 and 60 with both parents, where they were apparently first reported in 1914, and Davis (1926) cited specimens from v.c. 6 (1883) and Jersey (1871). They have also been occasionally reported from inland localities in Br and from eastern, southern and western Ga.

d. Both species are self-compatible. Hybrids have been synthesized on a great many occasions utilizing a wide range of parental races. The chromosomal configurations (Renner complexes) and the system of balanced lethals found in these species, together with variable amounts of in- and out-breeding, have resulted in a complex pattern of variability which has been the subject of research since the 1880s. *O. biennis* is a native of North America, but *O. erythrosepala* is of European origin involving hybridization between introduced American taxa, probably *O. biennis* and *O. hookeri* or close relatives. Apparently no experimental work aimed at investigating natural hybridization between *O. biennis* and *O. erythrosepala* has been undertaken.

e. Both parents and artificial hybrid (2*n* = 14).

f. CLELAND, R. E. (1972). *Oenothera. Cytogenetics and evolution.* London and New York.

DAVIS, B. M. (1926). The history of *Oenothera biennis* Linnaeus, *Oenothera grandiflora* Solander, and *Oenothera lamarckiana* of de Vries in England. *Proc. Am. phil. Soc.,* 65: 349-378.

GATES, R. R. (1914). Some Oenotheras from Cheshire and Lancashire. *Ann. Mo. bot. Gard.,* 1: 383-400.

LINDER, R. (1957a). Les *Oenothera* récemment reconnus en France. *Bull. Soc. bot. Fr.,* 104: 515-525.

LINDER, R. (1957b). Aperçu des Oenotheres rencontrées dans le Sud-Ouest en 1957. *Bull. Cent. Étud. Rech. scient., Biarritz,* 1: 575-576.

WHELDON, J. A. (1913). The *Oenothera* of the South Lancashire coast. *Lancs. Nat.,* 6: 205-210.

1 × 4. *O. biennis* L. × *O. parviflora* L. (*O. muricata* L.)

= *O.* × *braunii* Doell has been reported from Au, Cz, Ga, Ge and He.

258. *Circaea* L.
(by P. M. Benoit)

3 × 1. *C. alpina* L. × *C. lutetiana* L.

a. *C.* × *intermedia* Ehrh. (Intermediate or Upland Enchanter's Nightshade).

b. *C.* × *intermedia* is presumed to be the hybrid of this parentage because it is highly sterile and, though variable, is more or less intermediate in its characters between the only two European species of *Circaea*. It can be distinguished from *C. lutetiana* by its weaker habit, more cordate and dentate leaves, tardily or not dehiscent anthers containing sterile pollen, and fruits not developing but falling soon after the flowers. *C. alpina* is quite different in its completely glabrous petioles and stem (except for stalked glands in the inflorescence) and in its small flowers (petals *c* 1 mm) crowded together at the top of the stem, and it has normal fertility. *C.* x *intermedia* may produce some fruits, typically with a distinctive, narrow shape, from backcrossing to *C. lutetiana*. The viability of such seed has not been tested, but partially fertile plants intermediate between *C.* x *intermedia* and *C. lutetiana* probably originate from this backcross.

c. *C.* x *intermedia* is recorded from Au, Be, Cz, Da, Ga, Ge, He, Ho, Hs, Hu, It, Ju, No, Po, Rm, Rs and Su. It is frequent, often in the absence of one or both of the species, in shady places with rich humus in the mountainous districts of north-western Br and northern Hb. It occurs over a large area of BI in which *C. alpina* is unknown, probably owing partly to the loss of *C. alpina* with the post-glacial advance of temperate vegetation since hybridization took place, and partly to accidental introduction of *C.* x *intermedia*. In parts of Wales *C.* x *intermedia* is predominantly a weed of gardens and roadsides.

d. Hybrids are readily obtained from pollinating *C. alpina* with the pollen of *C. lutetiana*. The reverse cross has not been successful. Morphologically the synthesized hybrid is similar to wild *C.* x *intermedia* though rather nearer than usual to *C. alpina*. The synthesized hybrid is male-sterile and produces fruit only from backcrossing to *C. lutetiana*.

e. Both parents and hybrid $2n = 22$.

f. BENOIT, P. M. (1966). Synthesised *Circaea alpina* x *lutetiana*. *Proc. B.S.B.I.*, 6: 271.

PERRING, F. H. and WALTERS, S. M. (1962). *Op. cit.*, p. 149.

RAVEN, P. H. (1963). *Circaea* in the British Isles. *Watsonia*, 5: 262-272.

RAVEN, P. H. (1968). *Circaea*, in TUTIN, T. G. *et al.*, eds *Flora Europaea*, 2: 305-306. Cambridge.

ROSS-CRAIG, S. (1958). *Drawings of British plants*, 11: 36. London. FIG.

262. *Callitriche* L.

(by C. A. Stace)

4 x 1. *C. intermedia* Hoffm. x *C. stagnalis* Scop.
4 x 3. *C. intermedia* Hoffm. x *C. obtusangula* Le Gall
Hybrids of these two suggested parentages have been reported, the latter

only doubtfully, but it is very unlikely that either has in fact occurred. Specimens so-labelled have always been identifiable with one or other species, and no experimental hybridizations have succeeded in producing mature plants. Hybrids are extremely scarce elsewhere in Europe; all those confirmed involve at least one non-British species.

SAVIDGE, J. P. (1959). The experimental taxonomy of European *Callitriche. Proc. Linn. Soc. Lond.,* **170**: 128-130.

272. *Eryngium* L.
(by C. A. Stace)

2 × 1. *E. campestre* L. × *E. maritimum* L.
= *E.* × *rocheri* Corb. ex Guétrot is recorded from Ga.

274 × 273. *Anthriscus* Pers. × *Chaerophyllum* L. = × *Anthrichaerophyllum* P. Fourn.
(by C. A. Stace)

274/2 × 273/2. *Anthriscus sylvestris* (L.) Hoffm. × *Chaerophyllum aureum* L.
= × *A. loretii* (Rouy & Camus) P. Fourn. has been recorded from Ga.

285. *Apium* L.
(by T. G. Tutin)

2 × 3. *A. nodiflorum* (L.) Lag. × *A. repens* (Jacq.) Lag.
(= *A.* × *riddelsdellii* Druce, *nom. nud.*) was reported doubtfully from

Binsey Common and Port Meadow, v.c. 23, in 1917, but all the specimens
seen appear to be variants of *A. nodiflorum.*
RIDDELSDELL, H. J. (1917a). *Helosciadium* in Britain. *Rep. B.E.C.*, 4:
409-412.
RIDDELSDELL, H. J. (1917b). *H. repens* x *nodiflorum? Rep. B.E.C.*, 4:
570.

4 x 2. *A. inundatum* (L.) Reichb. f. x *A. nodiflorum* (L.) Lag.
a. *A.* x *moorei* (Syme) Druce.
b. This hybrid resembles *A. inundatum* but is usually larger and the segments
of the lower leaves are linear or ligulate instead of filiform. It flowers
much less freely than the parents and appears to be completely sterile.
c. The hybrid occurs with the parents in shallow water or on damp mud. It is
very local in eastern central England but fairly frequent and widespread in
Hb: v.c. 32, 53, 54, 57; H3, 8, 9, 14–17, 19, 21–28, 30, 32–34, 37–39
(records of Perring and Sell (1968) updated). Riddelsdell (1914) also gave
v.c. H36 and H40. It is apparently unknown in any other country.
d. None.
e. Both parents $2n = 22$.
f. DRUCE, G. C. (1912 and 1914). *Apium moorei* (Syme) mihi. *Rep. B.E.C.*,
3: 20, 324-325, 470.
PERRING, F. H. and SELL, P. D. (1968). *Op. cit.*, p. 42.
RIDDELSDELL, H. J. (1914). *Helosciadium moorei. Ir. Nat.*, 23: 1-11.

294. *Pimpinella* L.
(by C. A. Stace)

2 x 1. *P. major* (L.) Huds. x *P. saxifraga* L.
= *P.* x *intermedia* Figert is recorded from Au, Cz and Ge.

307. *Angelica* L.
(by C. A. Stace)

2 x 1. *A. archangelica* L. x *A. sylvestris* L.
= *A.* x *mixta* Nyár. has been recorded from Rm.

311. *Heracleum* L.
(by D. McClintock)

2 × 1. *H. mantegazzianum* Somm. & Levier × *H. sphondylium* L.
 a. None.
 b. The hybrid is intermediate in size of stem, leaf and umbel, in leaf-outline, in the shape and length of the fruit and vittae, in the hairiness of the stem and sheath, and in the smell when bruised. Hybrids have reduced or no fertility.
 c. Praeger (1951) recorded the hybrid from v.c. H21, but no voucher seems to exist. Authentic specimens were collected in Scotland in 1970 and shown me by Miss C. Muirhead, and nearly all records are from that country. These are from v.c. 16, 21, 80–83, 96, 106, 109 and H13, but hybridization is doubtless much more widespread. R. K. Brummitt has been studying a colony near Heathrow, v.c. 21, since 1969, and M. G. Collett had recognized the hybrid in 1962 near the Brent, Ealing, also in v.c. 21. No foreign records have been traced.
 d. The whole umbel of *H. sphondylium* is protandrous, the stigma becoming mature only after all the anthers have opened and most have fallen off (Proctor and Yeo, 1973).
 e. *H. sphondylium* $2n = 22$.
 f. McCLINTOCK, D. (1973). *Heracleum sphondylium* x *H. mantegazzianum. Watsonia*, **9**: 429-430.
    PRAEGER, R. L. (1951). *Op. cit.*, p. 17.
    PROCTOR, M. C. F. and YEO, P. F. (1973). *The pollination of flowers*, pp. 62-63. London.

314. *Daucus* L.
(by C. A. Stace)

1a × 1b. *D. carota* L. subsp *carota* × *D. carota* subsp. *gummifer* Hook. f.
   has been reported from Brittany, Ga, but apparently not from BI.

319. *Euphorbia* L.
(by P. M. Benoit and C. A. Stace)

17 × 3. *E. amygdaloides* L. × *E. pilosa* L.

= *E.* × *turneri* Druce was recorded from Prior Park, Bath, v.c. 6, in 1916, close to the only known site of *E. pilosa* and among *E. amygdaloides*. It was said to differ from the latter in having hairy capsules and more hairy leaves, but E. F. Warburg examined the type specimen (**OXF**) and concluded that it was "merely *E. amygdaloides* and not very unusual" (*in litt. ad* N. Y. Sandwith 1955, *fide* A. J. Willis *in litt.* 1972).

DRUCE, G. C. (1917). *Euphorbia Amygdaloides* × *pilosa* = × *E. Turneri*. *Rep. B.E.C.*, 4: 428.

13 × 12. *E. paralias* L. × *E. portlandica* L.

a. None.

b. This hybrid is a partially sterile plant with more or less intermediate characters. It has a low, open, branched habit and diffuse inflorescences like *E. portlandica*. But it is more robust and longer-lived, often producing barren overwintering shoots from below as well as above ground-level; it is very glaucous; and it has acute primary bracts (though without the wide bases of *E. paralias*) and concave floral bracts. The inflorescences have a conspicuous "prickly" appearance owing to the strongly bicuspidate nectar glands, 1.1–1.8 mm wide, larger than in *E. portlandica* and quite different from the even larger irregularly laciniate glands of *E. paralias*. The pollen is partially sterile and the fruits have only 0–1, seldom more, developed cells containing intermediate seeds (*c* 2.2 mm long, somewhat rugose, but with a caducous caruncle as in *E. paralias*). Despite a degree of fertility, the hybrid does not form complex populations but occurs as single plants with the parents and is not normally variable. A more fertile plant morphologically nearer *E. portlandica*, seen at Morfa Harlech in 1970, was possibly a backcross.

c. *E. paralias* × *E. portlandica* is so far known from old fixed dunes or sandy or gravelly banks near the shore at Kenfig Burrows, v.c. 41, Morfa Dyffryn and Morfa Harlech, v.c. 48, Newborough Warren, v.c. 52, and Curracloe, v.c. H12. It is apparently not known elsewhere.

d. None.

e. Both parents ($2n = 16$).

f. PRAEGER, R. L. (1951). *Op. cit.*, p. 11.

RIDDELSDELL, H. J. (1907). A flora of Glamorganshire, p. 56. *J. Bot., Lond.*, 45 (Suppl.).

16 × 15. *E. cyparissias* L. × *E. esula* L.

= *E.* × *pseudoesula* Schur (*E.* × *figertii* Dörfl.) is known in Au, Be, Cz, Ga, Ge, Ho, Hu, Rm and Canada.

16 × 14. *E. cyparissias* L. × *E. uralensis* Fisch. ex Link (*E. virgata* Waldst. & Kit., *non* Desf.; *E. esula* subsp. *tommasiniana* (Bertol.) Nyman)

= *E.* × *gayeri* Bor. & Soó ex Soó was the parentage somewhat doubtfully assigned to specimens collected from Hulne Park, near Alnwick, v.c. 68, in 1908 (Lady M. Percy), 1925 (Miss Vachell) and 1929 (Mrs A. Leith). The first two were determined by Thellung (all **OXF**). These three specimens obviously represent the same taxon, which in leaf-shape and -crowding appears intermediate between *E. cyparissias* and *E. uralensis*, but it is extremely difficult to make accurate determinations of herbarium material in this critical group. This hybrid is recorded from Au, Cz, Ge, He and Hu.

DRUCE, G. C. (1920). *Euphorbia Cyparissias* × *virgata. Rep. B.E.C.*, 5: 575.

DRUCE, G. C. (1928). *E. Cyparissias* × *virgata? Rep. B.E.C.*, 8: 417.

15 × 14. *E. esula* L. × *E. uralensis* Fisch. ex Link (*E. virgata* Waldst. & Kit., *non* Desf.; *E. esula* subsp. *tommasiniana* (Bertol.) Nyman)

a. *E.* × *pseudovirgata* (Schur) Soó (*E.* × *podperae* Croizat, *E.* × *intercedens* Podp., *non* Pax).

b. According to Moore (1958) the two species are best separated by leaf characters: linear-lanceolate or obovate-linear, broader above the middle, tapering to the base and sessile in *E. esula*; lanceolate to ovate, broader below the middle, some widening at the base and with a short petiole in *E. uralensis*. Hybrids are intermediate in these features, but are difficult to recognize and have been confused with both parents and with hybrids between them and *E. cyparissias* and *E. lucida* Waldst. & Kit. All hybrids in this group are at least partly fertile; the type specimen of *E.* × *podperae* has 30% fertile pollen.

c. The common plant belonging to this group of spurges in Br has been considered as *E. esula* (by T. Pritchard, in Moore, 1958), as *E. uralensis* (by Warburg, 1962), and as the hybrid between them (by A.R. Smith, *in litt.* 1971). It occurs over much of Br, but mainly in south-eastern England (Perring and Walters, 1962). It has also been widely recorded in central Europe and in U.S.A. and Canada.

d. None.

e. "*E. virgata*" (2*n* = 20); "*E. esula*" (2*n* = 64); the common plant in Br and Canada 2*n* = 60.

f. CROIZAT, L. (1945). "*Euphorbia Esula*" in North America. *Am. Midl. Nat.*, 33: 231-243.

KUZMANOV, B. (1964). On the origin of *Euphorbia* subg. *Esula* in Europe (Euphorbiaceae). *Blumea*, 12: 369-379.

MOORE, R. J. (1958). Cytotaxonomy of *Euphorbia esula* in Canada and its hybrid with *Euphorbia cyparissias*. *Canad. J. Bot.*, 36: 547-559. FIG.

PERRING, F. H. and WALTERS, S. M. (1962). *Op. cit.*, p. 172.

B. SYSTEMATIC   273

WARBURG, E. F. (1962). *Euphorbia,* in CLAPHAM, A. R., TUTIN, T. G. and WARBURG, E. F. *Op. cit.,* pp. 535-542.
ZIMMERMANN, W. (1965). *Euphorbia,* in HEGI, G., *Illustrierte Flora von Mittel-Europa,* ed. 2, 5: 134-189. Munich.

320. *Polygonum* L.
(by C. A. Stace, B. T. Styles and J. Timson)

A large proportion of the possible hybrids in sections *Polygonum* and *Persicaria* have been recorded, but all appear to be rare and scattered; many specimens reported to be hybrids are in fact referable to one of the putative parent species. The pronounced autogamy characteristic of species in these sections must be a reason for the low frequency of hybrids even in mixed populations. Extremely few hybrids have been studied in the field, and information on variability and fertility is mostly lacking; they are probably all very largely sterile.

Section *Polygonum*
(by B. T. Styles)

1/4 × 1/1. *P. arenastrum* Bor. (incl. *P. calcatum* Lindm.) × *P. aviculare* L.
has been reported from Su and Canada; it is said to be closer morphologically to *P. aviculare* but sterile. It was recorded for v.c. 23 by Druce (det. Lindman) in 1912, but the specimen in herb. Druce (**OXF**) is *P. aviculare.* There are also doubtful records from elsewhere in Europe.
DRUCE, G. C. (1913). *Polygonum aviculare* L. *Rep. B.E.C.,* 3: 176-179.
STYLES, B. T. (1962). The taxonomy of *Polygonum aviculare* in Britain. *Watsonia,* 5: 177-214.

1/1 × 1/6. *P. aviculare* L. × *P. boreale* (Lange) Small
has been reported from boreal parts of North America.

1/4 × 1/5. *P. arenastrum* Bor. (*P. aequale* Lindm.) × *P. calcatum* Lindm.

(= *P.* × *lindmanii* Druce, *nom. nud.*) was recorded from v.c. 2 and 24 by Druce (det. Lindman) but *P. calcatum* is not now considered a species distinct from *P. arenastrum,* and the voucher specimens of these two records (**OXF**) are of the latter species

DRUCE, G. C. (1929). × *Polygonum lindmandii [sic]* Dr. *Rep. B.E.C.,* **8:** 873.

STYLES, B. T. (1962). As above.

Section *Persicaria* (Mill.) DC.
(by J. Timson)

10 × **9.** *P. lapathifolium* L. (incl. *P. nodosum* Pers.) × *P. persicaria* L.
a. *P.* × *lenticulare* Hy (*P.* × *danseri* Druce, *nom nud.*).
b. Hybrids are intermediate between the parent species, especially in fruit morphology and ochrea cilia-length. Most herbarium specimens are referable to one of the putative parents, both of which are very variable.
c. There are records for v.c. 3, 11, 17, 34, 35 and 41 and for Ga, Ge, He and Ho.
d. Attempts to produce artificial hybrids were unsuccessful.
e. *P. persicaria* $2n = 44$; *P. lapathifolium* $2n = 22$.
f. BRITTON, C. E. (1933). British Polygona, Section *Persicaria. J. Bot., Lond.,* **71:** 90-98.
BRITTON, C. E. and PEARSALL, W. H. (1931). *Polygonum. Rep. B.E.C.,* **9:** 522.
BRITTON, C. E. and THOMPSON, H. S. (1935). *Polygonum nodosum* Pers., var. *erectum* Rouy × *P. Persicaria* L. ? *Rep. B.E.C.,* **10:** 983-984.
TIMSON, J. (1965). A study of hybridization in *Polygonum* Section *Persicaria. J. Linn. Soc., Bot.,* **59:** 155-161.

12 × **9.** *P. hydropiper* L. × *P. persicaria* L.
a. *P.* × *intercedens* G. Beck.
b. Hybrids are intermediate between the two parents, especially in ochrea characters; there are long cilia as in *P. persicaria* mixed with short cilia like those of *P. hydropiper. P. mite* has sometimes been mistaken for this hybrid.
c. This hybrid is recorded from v.c. 17, 22, 23 and 57, and from Ga, Ge and He.

d. Attempts to produce artificial hybrids were not successful.
e. *P. persicaria* 2*n* = 44; *P. hydropiper* 2*n* = 20.
f. BRITTON, C. E. (1933). As above.
TIMSON, J. (1965). As above.
MOSS, C. E. (1914). *Polygonum,* in *The Cambridge British Flora,* 2: 109-127. Cambridge.

13 x 9. *P. mite* Schrank x *P. persicaria* L.

a. *P.* x *condensatum* (F. W. Schultz) F. W. Schultz (*P.* x *axillare* Rigo).
b. Hybrids are intermediate between the two parents, esp florescence denseness. The leaves may be faintly blotched above. Many specimens referred to *P.* x *axillare* are *P. hydropiper* f. *obtusifolium* A. Braun.
c. Hybrids have been recorded from v.c. 17, 22, 23 and 35, and from Au, Ga, Ge, He, Ho and It.
d. None.
e. *P. persicaria* 2*n* = 44; *P. mite* 2*n* = 40.
f. BRITTON, C. E. (1930). *Polygonum mite* x *Persicaria. Rep. B.E.C.,* 9: 237-238.
BRITTON, C. E. (1933). As above.
MOSS, C. E. (1914). As above.
RECHINGER, K.-H. (1958). *Polygonum,* in HEGI, G. *Illustrierte Flora von Mittel-Europa,* 2nd ed., 3: 403-434. Munich.
ROUY, G. and FOUCAUD, J. (1910). *Polygonum,* in *Flore de France,* 12: 90-115. Asnières, Paris and Rochefort.
SCHOTSMAN, H. D. (1950). De Bouw der klieren van enige *Polygonum*-soorten en -bastaarden. *Ned. kruidk. Archf,* 57: 262-275. FIG.
TIMSON, J. (1965). As above.

14 x 9. *P. minus* Huds. x *P. persicaria* L.

a. *P.* x *braunianum* F. W. Schultz.
b. Hybrids are intermediate between the two species, especially in inflorescence structure which is often slender and interrupted at the base.
c. They have been recorded from v.c. 11–14, 22 and 23, and from Au, Be, Da, Ga, Ge, He, Ho and It.
d. None.
e. *P. persicaria* 2*n* = 44; *P. minus* 2*n* = 40.
f. BRITTON, C. E. (1933). As above.
GRENIER, C. and GODRON, D. A. (1855). *Polygonum,* in *Flore de France,* 3: 45-56. Paris.
RECHINGER, K.-H. (1958). As above.
MOSS, C. E. (1914). As above.
SCHOTSMAN, H. D. (1950). As above. FIG.
TIMSON, J. (1965). As above.

10 × 11. *P. lapathifolium* L. ×*P. nodosum* Pers.

has been recorded on several occasions, but *P. nodosum* is best regarded as one of several variants of the former species.

TIMSON, J. (1963). The taxonomy of *Polygonum lapathifolium* L., *P. nodosum* Pers., and *P. tomentosum* Schrank. *Watsonia,* 5: 386-395.

12 × 10. *P. hydropiper* L. × *P. lapathifolium* L. (incl. *P. nodosum* Pers.)

a. *P.* x *metschii* G. Beck.
b. Hybrids are intermediate between the parent species with leaves which are eglandular beneath and blotched above. The hybrids found in Br involved variants of *P. lapathifolium* identified as *P. nodosum.*
c. They have been recorded from v.c. 17, 29 and 31 and from Ga, Ge, Ho and Su.
d. Attempts to form artificial hybrids were not successful.
e. *P. lapathifolium* (incl. *P. nodosum*) 2n = 22; *P. hydropiper* 2n = 20.
f. BRITTON, C. E. (1933). As above.
   MOSS, C. E. (1914). As above.
   TIMSON, J. (1965). As above.

10 × 13. *P. lapathifolium* L. × *P. mite* Schrank

= *P.* x *bicolor* Borbás has been recorded from Be, Ga, Ge, Hu and Su.

10 × 14. *P. lapathifolium* L. × *P. minus* Huds.

= *P.* x *langeanum* Rouy (*P.* x*hervieri* Beck) has been recorded from Au, Ga and Ge.

12 × 13. *P. hydropiper* L. × *P. mite* Schrank

a. *P.* x *oleraceum* Schur.
b. The hybrid is said to be intermediate between the two parent species with an almost eglandular perianth and glandular ochreae. Variants of *P. persicaria* have sometimes been mistaken for this hybrid.
c. It has been recorded from v.c. 17, 34 and H30, and from Ga, Ge and Ho.
d. None.
e. *P. hydropiper* 2n = 20; *P. mite* 2n = 40.
f. BRITTON, C. E. (1933). As above.
   GRENIER, C. and GODRON, D. A. (1855). As above.
   RECHINGER, K.-H. (1958). As above.
   TIMSON, J. (1965). As above.

12 × 14. *P. hydropiper* L. × *P. minus* Huds.

a. *P.* x *subglandulosum* Borbás.
b. The hybrid is intermediate between the two species with slightly glandular perianth segments and with leaves which are suddenly attenuate at the base.

c. It has been recorded from v.c. 13, 22, 37 and 64, and from Ga and Ge.
d. None.
e. *P. hydropiper* $2n = 20$; *P. minus* $2n = 40$.
f. BRITTON, C. E. (1933). As above.
MOSS, C. E. (1914). As above.
SCHOTSMAN, H. D. (1950). As above. FIG.
ROUY, G. and FOUCAUD, J. (1893). As above.
TIMSON, J. (1965). As above.

14 × **13.** *P. minus* Huds. × *P. mite* Schrank
a. *P.* × *wilmsii* G. Beck.
b. The hybrid is said to be intermediate between the two parents with leaves
like those of *P. minus* but with inflorescences like those of *P. mite*. The
exact degree of fertility has not been established, but fruiting hybrids have
been recorded; since the parents have the same chromosome number this is
perhaps not unlikely.
c. It has been recorded from v.c. 22 and 23, and from Ga, Ge and Ho.
d. None.
e. Both parents $2n = 40$.
f. BRITTON, C. E. (1933). As above.
MOSS, C. E. (1914). As above.
ROUY, G. and FOUCAUD, J. (1893). As above.
TIMSON, J. (1965). As above.

Section *Tiniaria* Meisn.

(by C. A. Stace)

15 × **16.** *P. convolvulus* L. × *P. dumetorum* L.
= *P.* × *convolvuloides* Brügg. (*P.* × *heterocarpum* Beck) has been recorded
in Au, Be, Ga, Scandinavia and North America. Ascherson and Graebner
also recorded it (with doubt) from England, but no other notice of this has
been found.
ASCHERSON, P. and GRAEBNER, P. (1913). *Synopsis der
Mitteleuropäischen Flora*, **4**: 872-873. Leipzig.

320 × 325. *Polygonum* L. × *Rumex* L. = × *Polygonorumex* J. Weill
(by C. A. Stace)

320/12 × 325/12. *Polygonum hydropiper* L. × *Rumex obtusifolius* L.
= × *P. guinetii* J. Weill has been reported from Ga, but is very doubtful.

325. *Rumex* L.
(by J. E. Lousley and J. T. Williams)

In contrast to the situation in subgenera *Acetosella* and *Acetosa,* the species of subgenus *Rumex* cross freely and hybrids are frequent in nature. They are usually readily recognized by the irregular enlargement of the tepals, of which only a small proportion mature. The remainder drop off only partially developed. The panicle is often flushed with red. Partial sterility shows also in the pollen with a high proportion of shrivelled grains. Flowering continues late into the autumn beyond the normal period of the parents.

Most *Rumex* hybrids produce little or no viable seed. Even in the most fertile crosses (e.g. *R. crispus* × *obtusifolius*) production of seed is variable and sometimes low. Backcrosses to the parents occur but hybrid swarms have not been observed. The long persistence of dock hybrids in certain areas suggests that, once formed, a hybrid plant may persist for many years.

Only binary hybrids have been recorded in Britain, but crosses involving three species have been produced experimentally, and are known in Europe. All references to *R. obtusifolius* refer to *R. obtusifolius* L. subsp. *obtusifolius* (subsp. *agrestis* (Fries) Danser), and all references to *R. pulcher* refer to *R. pulcher* L. subsp. *pulcher*. No hybrids of the alien subspecies of these two species are known in BI, but some occur abroad. In view of the many errors of determination perpetrated in the literature the British distributions given here are based on records we have personally vetted. The only foreign hybrids not recorded in Br which are mentioned here are those which are likely to be correctly identified; several other dubious and erroneous records exist.

General References
  BECK von MANNAGETTA, G. (1904). *Rumex,* in REICHENBACH,
    H. G. L. and REICHENBACH, H. G. *Icones Florae Germaniae et
    Helvetiae,* **24**: 26-87, t. 158-201. Leipzig and Gera.

CAMUS, E. G. (1904). Renseignements bibliographiques sur les hybrides du genre *Rumex. Bull. Herb. Boissier, sér.* 2, 4: 1232-1240.

DANSER, B. H. (1924). Über einige Aussaatversuche mit *Rumex* Bastarden. *Genetica*, 6: 145-220.

HAUSSKNECHT, C. (1885). Beitrag zur Kenntniss der einheimischen Rumices. *Mitt. geogr. Ges. Thüringen*, 3: 56-79.

LAMBINON, J. (1956). Notes sur quelques *Rumex* hybrides de la flore Belge. *Bull. Soc. r. Bot. Belg.*, 88: 29-32.

LOUSLEY, J. E. (1939). Notes on British Rumices, 1. *Rep. B.E.C.*, 12: 118-157.

LOUSLEY, J. E. (1944). Notes on British Rumices, 2. *Rep. B.E.C.*, 12: 547-585.

LÖVE, Á. and KAPOOR, B. M. (1968). A chromosome atlas of the collective genus *Rumex. Cytologia*, 32: 328-342.

MOSS, C. E. (1914). *Rumex*, in *The Cambridge British Flora*, 2: 130-149. Cambridge.

RECHINGER, K. H. (1932). Vorarbeiten zur einer Monographie der Gattung *Rumex*, 1. *Beih. bot. Zbl.*, 49(2): 1-132.

RECHINGER, K. H. (1933). Vorarbeiten zur einer Monographie der Gattung *Rumex*, 2. *Reprium nov. Spec. Regni veg.*, 31: 225-283.

RECHINGER, K. H. (1957). *Rumex*, in HEGI, G. *Illustrierte Flora von Mitteleuropa*, 2nd ed., 3: 352-400. Munich.

11 × 4. *R. crispus* L. × *R. hydrolapathum* Huds.

a. *R. × schreberi* Hausskn.
b. This is a tall plant, sometimes even exceeding *R. hydrolapathum* in height, with several stems from the root. The cauline leaves are rather narrow and wavy on the margin. The tepals are variable, sometimes closely resembling those of *R. hydrolapathum* but smaller, at other times more like those of *R. crispus,* but always with red elongate tubercles.
c. It occurs rarely by marsh ditches with *R. hydrolapathum* in v.c. 25, 56 and H38, and in Ge, Ho and Su.
d. None.
e. *R. crispus* $2n = 60$. *R. hydrolapathum* ($2n = c\ 200$).
f. BECK, G. (1904). As above, t. 174. FIG.

   DANSER, B. H. (1922). Die Nederlandsche *Rumex*-Bastarden (Eerste deel). *Ned. kruidk. Archf,* 1921: 229-271. FIG.

   PRAEGER, R. L. (1942). *Rumex crispus* × *Hydrolapathum*—Down. *Ir. Nat. J.,* 8: 35-36.

4 × 12. *R. hydrolapathum* Huds. × *R. obtusifolius* L.

a. *R. × weberi* Fisch.-Benz. (*R. maximus* auct., *R. heterophyllus* auct.).
b. This is a stout plant resembling *R. hydrolapathum* in size and habit but with more open panicles and with only irregular production of fruit. The radical and lower cauline leaves are almost as large as those of *R. hydrolapathum,* but they have a thinner texture and a broad, cordate or

subcordate base. The tepals when developed resemble those of *R. hydrolapathum* but usually have a few distinct teeth towards the base.

c. Correctly-named material has been seen from water-meadows in v.c. 2, 6, 9, 10, 11, 13, 22, 26 and 36. It is also recorded from Au, Ge, Ho and Su.

d. Both Syme and Lousley have been unsuccessful in germinating hybrid fruit.

e. *R. hydrolapathum* (2*n* = *c* 200); *R. obtusifolius* 2*n* = 40.

f. DANSER, B. H. (1922). As above. FIG.

SYME, J. T. B. (1875). *Rumex Hydrolapathum*, Huds. *J. Bot., Lond.*, **13**: 374.

TRIMEN, H. (1874). On the Great Water-Dock of England. *J. Bot., Lond.*, **12**: 33-36. FIG.

15 × 4. *R. conglomeratus* Murr. × *R. hydrolapathum* Huds.

a. *R.* × *digeneus* G. Beck (*R.* × *hybridus* Hausskn., *non* Kindb.).

b. This is a tall plant (over 1 m) with a rather dense panicle with ascending branches recalling those of *R. hydrolapathum*. The flowers are reddish, only a few maturing. The tepals are mostly lingulate but a few are broader with large elongate tubercles showing more resemblance to those of *R. hydrolapathum* at maturity.

c. This has been found by dykes and canals in the company of the parents in v.c. 6, 14 and 15, and also in Ge.

d. None.

e. *R. conglomeratus* 2*n* = 20; *R. hydrolapathum* (2*n* = *c* 200).

f. BECK, G. (1904). As above, t. 176. FIG.

5 × 12. *R. alpinus* L. × *R. obtusifolius* L.

= *R.* × *mezei* Hausskn. has been reported from Au.

6 × 11. *R. confertus* Willd. × *R. crispus* L.

a. *R.* × *skofitzii* Błocki.

b. This is a tall robust plant resembling *R. confertus* but with the upper leaves narrower and with crisped margins, the panicle lax and with lanceolate, crisped bracts, and the tepals matured irregularly and with a less rounded apex.

c. *R.* × *skofitzii* was found in 1954 and 1955 in neglected pasture on chalk at Old Coulsdon, v.c. 17, associated with *R. confertus* (as an alien) and *R. crispus*. Elsewhere it has been found in Au and Fe.

d. None.

e. *R. confertus* (2*n* = 40); *R. crispus* 2*n* = 60.

f. ERKAMO, V. (1949). *Rumex confertus* × *crispus* ja *R. confertus* × *domesticus* loydetty Suomesta. *Archvm Soc. zool.-bot. fenn. 'Vanamo'*, **2**(1947): 128-130.

LOUSLEY, J. E. (1955). Botanical records for 1954. *Lond. Nat.*, **34**: 2-6.

6 × 12. *R. confertus* Willd. × *R. obtusifolius* L.

a. *R.* × *borbasii* Błocki.

b. This is intermediate in height between the two parents. The radical and lower cauline leaves resemble those of *R. confertus* but are thinner in texture; the panicle is lax with irregularly maturing, subacute mature tepals with sharply toothed margins.

c. *R.* × *borbasii* was found in 1954 in neglected pasture on chalk at Old Coulsdon, Surrey, v.c. 17, growing with *R. confertus* (as an alien) and *R. obtusifolius.* It was described from Au, and in 1973 was reported from Fe.

d. None.

e. *R. confertus* ($2n = 40$); *R. obtusifolius* $2n = 40$.

f. LOUSLEY, J. E. (1955). As above.

7 × 8. *R. aquaticus* L. × *R longifolius* DC.

= *R.* × *armoraciifolius* L. M. Neum. has been reported from Su.

7 × 11. *R. aquaticus* L. × *R. crispus* L.

= *R.* × *conspersus* Hartm. has been reported from Au and Ge.

7 × 11 × 12. *R. aquaticus* L. × *R. crispus* L. × *R. obtusifolius* L.

has been reported from the Continent.

7 × 12. *R. aquaticus* L. × *R. obtusifolius* L.

a. *R.* × *platyphyllos* Aresch. (*R.* × *schmidtii* Hausskn.).

b. This resembles *R. aquaticus* in general habit, but the radical and lower cauline leaves are more ovate and less triangular in outline, and are subobtuse. The panicle is lax with irregularly maturing, more or less acute tepals which are often reddish in colour and distinctly toothed near the base.

c. *R.* × *platyphyllos* is restricted in Br to the small area between Balmaha and Gartocharn on, or near, the shore of Loch Lomond, v.c. 86 and 99, occupied by *R. aquaticus,* where it is common. The hybrid replaces *R. aquaticus* away from the mouth of the Endrick Water. It has also been reported from Au, Fe, Ge, It and Su.

d. Danser grew a seed from the hybrid as far as the production of tepals.

e. *R. aquaticus* ($2n = c$ 200); *R. obtusifolius* $2n = 40$.

f. BECK, G. (1904). As above. t. 170. FIG.

DANSER, B. H. (1924). As above.

IDLE, E. T. (1968). *R. aquaticus* L. at Loch Lomondside. *Trans. Proc. bot. Soc. Edinb.,* **40**: 445-449.

JARETZKY, R. (1928). Histologische und karyologische Studien an Polygonaceen. *Ber. dt. bot. Ges.,* **45**: 48-54.

LÓVE, A. (1942). Cytogenetic studies in *Rumex,* 3. Some notes on the Scandinavian species of the genus. *Hereditas,* **28**: 289-296.

7 × 15 × 12. *R. aquaticus* L. × *R. conglomeratus* Murr. × *R. obtusifolius* L.

has been reported from the Continent.

7 × 15. *R. aquaticus* L. × *R. conglomeratus* Murr.

= *R.* × *ambigens* Hausskn. has been reported from Ge.

11 × 8. *R. crispus* L. × *R. longifolius* DC.

a. *R.* × *propinquus* Aresch.
b. This resembles *R. longifolius* in habit but has a less dense panicle which is sometimes strict and little-branched. The cauline leaves are usually broadly lanceolate and finely crisped on the margin, and the tepals are mostly well-developed, slightly less rounded than in *R. longifolius*, and one of them bears a distinct tubercle. This hybrid is easily detected when the panicle is lax; dense-panicled plants can be separated from *R. longifolius* by the presence of tubercles. It is distinguished from *R. longifolius* × *R. obtusifolius* by the leaves being more or less glabrous beneath and the tepals being entire. The parents can be separated on the stomatal index (9.2 in *R. longifolius*, 5.2 in *R. crispus*).
c. *R.* × *propinquus* is widely distributed in Scotland wherever the parents grow together, but it is probably often overlooked. Specimens have been seen from v.c. 81, 85, 90, 93, 94, 96, 97 and 104. Elsewhere it is known in Ge and Su.
d. None.
e. *R. crispus* 2n = 60; *R. longifolius* (2n = 60).
f. None.

8 × 12. *R. longifolius* DC. × *R. obtusifolius* L.

a. *R.* × *arnottii* Druce (*R. hybridus* Kindb., *R. conspersus* auct., *non* Hartm.).
b. The leaves are similar to those of *R. longifolius* but are usually broadly cordate at the base and slightly asperous below. The panicle is less dense than in *R. longifolius* and reddish; the tepals are of about the same size but less obtuse, with occasional tubercles, and often with a few teeth at the base. This is a variable hybrid which often appears fertile as most of the tepals mature. Syme claimed that the hybrid is fertile and comes true from seed. The parents can be separated on the stomatal index (9.2 in *R. longifolius*, 17.1 in *R. obtusifolius*), but the hybrid (9.1) resembles the former.
c. *R.* × *arnottii* arises freely in Scotland wherever the two parents grow together; convincing material has been seen from v.c. 80, 85, 88–92, 95, 96 and 104. It is also recorded from Da, Fe, Is, No and Su.
d. None.
e. *R. longifolius* (2n = 60); *R. obtusifolius* 2n = 40.
f. SYME, J. T. B., ed. (1868). *English Botany*, 3rd ed., 8: 48-49, t. 1217. London. FIG.

SYME, J. T. B. (1875a). *Rumex conspersus* Hartm. *Botl Exch. Club, Rep. Curators*, 1872-1874: 36.

8 × 14. *R. longifolius* DC. × *R. sanguineus* L.
has been reported from Su.

9 × 10. *R. cristatus* DC. × *R. patientia* L.
= *R.* × *xenogenus* Reichb. has been reported from the Continent.

11 × 9. *R. crispus* L. × *R. cristatus* DC.
a. (*R.* × *dimidiatus* Hausskn., *nom. nud.*).
b. This is a tall, robust plant resembling *R. cristatus* but with smaller tepals with small teeth and with cauline leaves narrower and with crisped margins.
c. Specimens have been seen only from v.c. 16 and 18. It is also known in south-eastern Europe.
d. None.
e. *R. crispus* 2n = 60.
f. None.

11 × 10. *R. crispus* L. × *R. patientia* L.
a. *R.* × *confusus* Simonk.
b. This is a tall, robust dock closely resembling *R. patientia* but with much smaller tepals and narrower upper cauline leaves which are usually crisped.
c. Specimens have been seen only from v.c. 18 and 34; on the Continent it is frequent within the area of *R. patientia*, especially in Au.
d. None.
e. *R. crispus* 2n = 60; *R. patientia* (2n = 60).
f. BECK, G. (1904). As above, t. 178. FIG.

12 × 10. *R. obtusifolius* L. × *R. patientia* L.
a. *R.* × *erubescens* Simonk.
b. This is a tall, stout plant resembling *R. patientia* generally but with leaves much broader in relation to their length and asperous on the midrib below, and with several, short teeth on the margin of the tepals.
c. *R.* × *erubescens* has been associated with *R. patientia* in v.c. 17, 18, 21, 34 and 39. It is recorded from Au.
d. None.
e. *R. obtusifolius* 2n = 40; *R. patientia* (2n = 60).
f. BECK, G. (1904). As above, t. 177. FIG.

17 × 10. *R. palustris* Sm. × *R. patientia* L.
= *R.* × *peisonis* Rech. pat. has been reported from Au.

11 × 12. *R. crispus* L. × *R. obtusifolius* L.

a. *R.* × *pratensis* Mert. & Koch (*R.* × *acutus* auct., *R.* × *khekii* Rech. pat., *R.* × *brittonii* Druce, *nom. nud.). (Meadow Dock).*

b. Hybrids tend to be intermediate between the parents but demarcated on the length/breadth ratio, hairiness and number of main lateral veins of the basal leaves; hairiness and number of veins of the cauline leaves; shape of the stipules and tepals; and on the tubercle number. Paper chromatography also separates the parents and hybrids.

c. This hybrid is common throughout BI wherever the parents grow in proximity, mainly in disturbed ground but also in neutral grassland. It is the commonest of all *Rumex* hybrids over most of Europe, involving at least two of the subspecies of *R. obtusifolius* and several varieties of *R. crispus.*

d. Genotypes of both parents may differ in their degree of protandry or protogyny (Cavers and Harper, 1964). Parents from a Warwickshire population where hybrids were present produced an artificial F_1 generation showing *c* 20% sterility, but pollen was 80% viable. Selfing produced some fertile fruits. Some F_2 seedlings from the Warwickshire crosses approached *R. crispus,* and backcrosses to *R. crispus* resembled that parent. Morphological evidence (Williams, 1971) suggests that backcrossing to *R. obtusifolius* as well as to *R. crispus* may occur in the field, but this has not been verified experimentally. Jensen (1936) described 15 bivalents and 10 univalents at meiosis.

e. *R. crispus* $2n = 60$; *R. obtusifolius* $2n = 40$; hybrid $2n = 44–56$.

f. BECK, G. (1904). As above, t. 175. FIG.

CAVERS, P. B. and HARPER, J. L. (1964). *Rumex crispus* L. and *Rumex obtusifolius* L., in Biological Flora of the British Isles. *J. Ecol.,* **52**: 737-766.

JENSEN, H. W. (1936). Meiosis in *Rumex,* 1. Polyploidy and the origin of new species. *Cytologia,* **7**: 1-22.

SYME, J. T. B., ed. (1868). *English Botany,* 3rd ed., **8**: 47-48, t. 1216. London. FIG.

WILLIAMS, J. T. (1971). Seed polymorphism and germination, 2. The role of hybridization in the germination polymorphism of *Rumex crispus* and *R. obtusifolius. Weed Res.,* **11**: 12-21.

11 × 13. *R. crispus* L. × *R. pulcher* L.

a. *R.* × *pseudopulcher* Hausskn.

b. In the more frequent nothomorph the branches are ascending so that the upright habit resembles *R. crispus* but it is readily distinguished by the raised reticulations and warty tubercles on the tepals, which are broad as in *R. pulcher.* Rarely the habit is bushy with the branches spreading at angles of about 90° with the stem, as in *R. pulcher,* but with shorter teeth on the irregularly-produced tepals.

c. *R.* × *pseudopulcher* occurs in warm, dry places with *R. pulcher*; specimens have been seen from v.c. 2, 3, 17 and 45. Elsewhere it has been recorded from Gr and Hs.

d. None.

e. *R. crispus* 2*n* = 60; *R. pulcher* (2*n* = 20).

f. BECK, G. (1904). As above, t. 191. FIG.

11 × 14. *R. crispus* L. × *R. sanguineus* L.

a. *R.* × *sagorskii* Hausskn.

b. In habit this hybrid is intermediate between its parents. The leaves are narrowly acuminate and narrowed to the base with margins conspicuously crisped; the panicle is lax, markedly infertile and with broadly lingulate tepals which usually bear tubercles.

c. *R.* × *sagorskii* is common on roadsides, wood-borders, and waste ground; specimens have been seen from v.c. 3, 14, 17, 19, 20, 24 and 33. It is also recorded from Au, Ga, Ge and Su.

d. None.

e. *R. crispus* 2*n* = 60; *R. sanguineus* 2*n* = 20.

f. BECK, G. (1904). As above, t. 172. FIG.

DANSER, B. H. (1922). As above. FIG.

LAMBERT, L. (1927). *Rumex sagorskii*, in GUÉTROT, M. *Plantes hybrides de France*, 1 and 2: 66. Lille.

15 × 11. *R. conglomeratus* Murr. × *R. crispus* L.

a. *R.* × *schulzei* Hausskn.

b. This is an erect plant with a panicle with ascending branches and leafy bracts. The lower leaves are lanceolate and only slightly crisped. The tepals are intermediate in outline between those of the parents, i.e. more elongate than those of *R. crispus* and with an occasional slightly elongate tubercle. It is highly infertile, although Danser found it not completely so. Hybrids involving *R. conglomeratus* rarely show evidence of widely spreading branches, which suggests that this character may be recessive.

c. *R.* × *schulzei* is common and widely distributed in England and Wales, and also on the Continent. Authentic specimens have been seen from v.c. 10, 16, 17, 25, 29, 34, 38, 48 and 49.

d. None.

e. *R. conglomeratus* 2*n* = 20; *R. crispus* 2*n* = 60.

f. BECK, G. (1904). As above, t. 172. FIG.

DANSER, B. H. (1922). As above. FIG.

DANSER, B. H. (1924). As above.

LAMBERT, L. (1927). *Rumex Schulzei*, in GUÉTROT, M. *Plantes hybrides de France*, 1 and 2: 67. Lille.

11 × 16. *R. crispus* L. × *R. rupestris* Le Gall

a. None.

b. This very rare hybrid is best distinguished by the mature tepals (when developed), which exhibit the broad outline of those of *R. crispus* combined with the more lingulate apex and very large, elongate tubercles of those of *R. rupestris*. Some plants suspected of this parentage are so infertile that no mature tepals develop. Where *R. crispus* and *R. rupestris* grow together in v.c. 41 and the Isles of Scilly plants referred to *R. crispus* occur with exceptionally large, elongate tubercles recalling those of *R. rupestris* and suggesting possible introgression.

c. It is known only from the Isles of Scilly, v.c. 1, and from Kenfig, v.c. 41, where it grows with the parents on sand near the sea.

d. None.

e. *R. crispus* $2n = 60$.

f. None.

11 × 17. *R. crispus* L. × *R. palustris* Sm.

a. *R.* × *areschougii* G. Beck.

b. The few specimens seen combine the characters of the parents in different combinations. The crisped leaf margins of *R. crispus* are often evident but the panicle usually shows the leafy bracts of *R. palustris*. The tepals combine the broader outline of those of *R. crispus* with the teeth (sometimes long) of those of *R. palustris*.

c. *R.* × *areschougii* occurs on the margins of lakes and reservoirs in v.c. 19 and 55. It is widespread almost throughout Europe.

d. None.

e. *R. crispus* $2n = 60$; *R. palustris* ($2n = 40$).

f. BECK, G. (1904). As above, t. 189. FIG.

LOUSLEY, J. E. (1962). *Rumex crispus* × *palustris* (*R.* × *areschougii* Beck, 1904). *Proc. B.S.B.I.*, **4**: 415-416.

11 × 18. *R. crispus* L. × *R. maritimus* L.

= *R.* × *fallacinus* Hausskn. has been reported from Au and Ge.

11 × obo. *R. crispus* L. × *R. obovatus* Danser

= *R.* × *bontei* Danser was reported as a casual at Avonmouth Docks, v.c. 34, in 1928 by C. I. and N. Y. Sandwith.

DANSER, B. H. (1926). Beitrag zur Kenntnis der Gattung *Rumex*. *Ned. kruidk. Archf*, **1925**: 466. FIG.

SANDWITH, C. I. and N. Y. (1936). *Rumex crispus* L. × *obovatus* Danser. *Rep. B.E.C.*, **11**: 40.

12 × 13. *R. obtusifolius* L. × *R. pulcher* L.

a. *R.* × *ogulinensis* Borbás (*R* × *mornetii* Lambert).

b. This is a very variable hybrid, sometimes closely resembling one parent and sometimes the other. It usually has a more open panicle than *R. obtusifolius*, but with branches ascending rather than making an angle of about 90° with the main stem as in *R. pulcher*. The influence of *R. obtusifolius* is usually clearly shown in the asperous midrib of the leaves, and of *R. pulcher* in the raised reticulations and warted tubercles of the tepals.

c. *R. × ogulinensis* occurs in the company of *R. pulcher* in dry habitats. Specimens have been seen from v.c. 1–3, 10, 17 and 34. It is also recorded from Ga and It.

d. None.

e. *R. obtusifolius* 2n = 40; *R. pulcher* (2n = 20).

f. LAMBERT, L. (1927). *Rumex morneti*, in GUÉTROT, M. *Plantes hybrides de France*, 1 and 2: 40. Lille.

LITTLE, J. E. (1924). *Rumex obtusifolius* L. × *R. pulcher* L. *J. Bot., Lond.*, 62: 330-331.

LOUSLEY, J. E. (1935). Short notes on some interesting British plants. *J. Bot., Lond.*, 73: 256-260.

12 × 14. *R. obtusifolius* L. × *R. sanguineus* L.

a. *R. × dufftii* Hausskn.

b. This is usually clearly intermediate between the parents, with ascending branches making an angle of about 45° with the main stem. The lower leaves resemble those of *R. obtusifolius* in outline, with a cordate base, but are more pointed. The tepals are intermediate, usually showing the influence of *R. sanguineus* in the more elongate, often lingulate, outline, and of *R. obtusifolius* in the presence of at least one elongate tubercle and occasional short teeth at the base.

c. *R. × dufftii* occurs in woodland rides and margins, and sometimes on waste ground. It is rather common in southern England; specimens have been seen from v.c. 3, 6, 9, 17, 23, 24, 28, 29, 32, 37, 38 and 48.

d. None.

e. *R. obtusifolius* 2n = 40; *R. sanguineus* 2n = 20.

f. BECK, G. (1904). As above, t. 173. FIG.

15 × 12. *R. conglomeratus* Murr. × *R. obtusifolius* L.

a. *R. × abortivus* Ruhmer.

b. In habit this resembles *R. obtusifolius*. The lower leaves are broad, subcordate at the base and obtuse or acute at the apex; the inflorescence is lax, with distant whorls and with leafy bracts below, and the tepals are very variable even on one plant, but they usually have several conspicuous teeth.

c. *R. × abortivus* is frequent and widespread in Br; specimens have been seen from v.c. 13, 14, 16, 17, 21, 22, 29, 38, 48, 86 and 88. It is widely distributed throughout most of Europe.

d. None.

e. *R. conglomeratus* 2n = 20; *R. obtusifolius* 2n = 40.

 f. BECK, G. (1904). As above, t. 173. FIG.
 LAMBINON, J. (1956). As above.
 LAMBERT, L. (1927). *Rumex (abortivus) abortivi,* in GUÉTROT, M.
 Plantes hybrides de France, 1 and 2: 65-66. Lille.

12 × 17. *R. obtusifolius* L. × *R. palustris* Sm.

a. *R.* × *steinii* Becker.
b. This is a rather tall, much-branched, leafy plant resembling *R. palustris* in
 the candelabra-like arrangement of the ascending branches, thin texture of
 the leaves, and long teeth on the tepals. The influence of *R. obtusifolius* is
 apparent in the broader, less acute, leaves which are asperous on the
 midrib below, and in the broader outline of the tepals.
c. *R.* × *steinii* occurs with *R. palustris* in marshes in v.c. 17, 21 and 29. It is
 also known in Ge, Ho and Su.
d. None.
e. *R. obtusifolius* $2n = 40$; *R. palustris* $(2n = 40)$.
f. None.

18 × 12. *R. maritimus* L. × *R. obtusifolius* L.

a. *R.* × *callianthemus* Danser.
b. This is a luxuriant plant with ascending branches and lanceolate leaves,
 resembling *R. maritimus* but larger and less spreading. The tepals are
 broad, resembling those of *R. obtusifolius* in outline, but with numerous
 long to very long teeth along both margins. It does not turn golden at
 maturity.
c. *R.* × *callianthemis* occurs with *R. maritimus* on the sides of ponds in v.c.
 20 and 29, and there is an authentic specimen in **BIRM** collected in 1892
 in v.c. 17. Elsewhere it is recorded from ?Au and Ho.
d. None.
e. *R. maritimus* $(2n = 40)$; *R. obtusifolius* $2n = 40$.
f. None.

13 × 14. *R. pulcher* L. × *R. sanguineus* L.

a. *R.* × *mixtus* Lambert (*R.* × *lambertii* Guétrot; *R.* × *warrenii* Druce, *nom.
 nud., non* Trim.).
b. In habit this resembles *R. pulcher* in the widely spreading branches, but it
 is more leafy. The tepals generally resemble those of *R. pulcher* but they
 are entire or with fewer, shorter teeth, and are less strongly reticulate and
 with less-warted tubercles. In the garden of the South London Botanical
 Institute a plant of the hybrid *R. pulcher* × *R. sanguineus* var. *sanguineus*
 appeared in 1941. The tepals were as described above, but the leaves had
 conspicuous red veins similar to those of the second parent.
c. This hybrid is associated with *R. pulcher* on dry soils, but is very rare.
 Specimens have been seen from v.c. 2, 3, 14, 23 and 29. It has also been
 reported from Ga.

d. None.

e. *R. pulcher* (2*n* = 20); *R. sanguineus* 2*n* = 20.

f. LAMBERT, L. (1927). *Rumex (mixtus) Lamberti,* in GUÉTROT, M. *Plantes hybrides de France,* 1 and 2: 40. Lille.

LAMBERT, L. (1929). *Rumex (mixtus) Lamberti,* in GUÉTROT, M. *Plantes hybrides de France,* 3 and 4: 115. Paris.

15 × 13. *R. conglomeratus* Murr. × *R. pulcher* L.

a. *R. × muretii* Hausskn.

b. This usually recalls *R. pulcher* in the wide branching, but it is taller and sometimes more leafy. The tepals have the strong reticulation of *R. pulcher* but are entire, or with only a few small teeth, and the tubercles are less warty.

c. *R. × muretii* is probably frequent in places where the two parents come together in southern England and Wales, but *R. pulcher* is a plant of dry, sunny places while *R. conglomeratus* grows in wetter habitats. Authentic specimens have been seen from v.c. 1, 3, 6, 9, 13–15, 17 and 35. It is rare on the Continent; there are records for Ga, Gr and He.

d. None.

e. *R. conglomeratus* 2*n* = 20; *R. pulcher* (2*n* = 20).

f. None.

13 × 16. *R. pulcher* L. × *R. rupestris* Le Gall

a. *R. × trimenii* Camus.

b. The influence of *R. pulcher* is seen in the wide branching, in the reticulation of the tepals, and in the warted tubercles. The influence of *R. rupestris* is apparent in the stout habit and leathery, broadly lanceolate leaves.

c. *R. × trimenii* has occurred very rarely in the company of *R. rupestris* on the shore of Samson, Isles of Scilly, v.c. 1, and at Whitesand Bay, v.c. 2. It also arose spontaneously in the garden of the South London Botanical Institute.

d. None.

e. *R. pulcher* (2*n* = 20).

f. BRIGGS, T. R. A. (1880). *Flora of Plymouth, an account of the flowering plants and ferns, etc.,* p. 295. London.

RECHINGER, K. H. (1948). Beiträge zur Kenntnis von *Rumex,* 9 *Candollea,* **11**: 229-241.

SYME, J. T. B. (1875b). *Rumex,* hybrid between *pulcher* and *nemorosus? Botl Exch. Club, Rep. Curators,* **1872-1874**: 34-35.

SYME, J. T. B. (1878). *Rumex* (hybrid between *pulcher* and *rupestris*). *Botl Exch. Club, Rep. Curators,* **1876**: 31.

15 × 14. *R. conglomeratus* Murr. × *R. sanguineus* L.

a. *R. × ruhmeri* Hausskn. (*R. × varians* Druce).

b. Hybrids are intermediate between the parents in the leaf-shape, the arrangement of veins arising from the midrib, the leafiness and denseness of the inflorescence, and on the number of tubercles (1, 2 or 3), but the tubercles are usually elongate (not spherical) and the branches ascending (not widely spreading). The variation pattern of the hybrid tends to overlap that of each parent.

c. *R. conglomeratus* occurs in wet, unshaded places and typically *R. sanguineus* grows in shade, though sometimes also in exposed situations. The intermediates occur mainly where the ecological ranges of the two species overlap (Rackham, 1961). In Br the hybrid may be frequent; authentic material has been seen from v.c. 3, 23, 37 and 38. It is also reported from Au, Ge and Ho.

d. Crosses using Warwickshire parental material have been made with a high degree of success utilizing emasculated *R. sanguineus* var. *viridis*; F_1 plants showed remarkable uniformity and tended to more closely resemble *R. conglomeratus*. They showed a high degree of sterility with 38% infertile pollen and many seeds inviable; those that were viable showed strong dormancy. Attempts to cross *R. conglomeratus* with *R. sanguineus* var. *sanguineus* failed, suggesting that the supposed mutant origin and subsequent selection by man of this variety has caused a degree of isolation. The F_1 selfed easily but 45% of the seedlings produced were unthrifty.

e. Both parents $2n = 20$; wild hybrid $2n = 19, 20, 21$ and 25.

f. BECK, G. (1904). As above, t. 171. FIG.
RACKHAM, O. (1961). Ecological significance of hybridisation between *Rumex sanguineus* and *R. conglomeratus*. *Proc. B.S.B.I.*, 4: 332.

15 × 17. *R. conglomeratus* Murr. × *R. palustris* Sm.

a. *R.* × *wirtgenii* Beck.

b. This hybrid resembles a bushy *R. palustris* with (usually) somewhat broader leaves. The tepals have a few long teeth, a lingulate apex, and three elongate tubercles.

c. *R.* × *wirtgenii* is found rarely with the parents by ponds, reservoirs and canals. Specimens have been seen from v.c. 17, 21, 30, 55 and 56. It is also recorded from Au, Ge, Ho and Su.

d. None.

e. *R. conglomeratus* $2n = 20$; *R. palustris* ($2n = 40$).

f. BECK, G. (1904). As above, t. 188. FIG.
LAMBINON, J. (1956). As above.

15 × 18. *R. conglomeratus* Murr. × *R. maritimus* L.

a. *R.* × *knafii* Čelak. (*R.* × *warrenii* Trim., *non* Druce).

b. This resembles *R. maritimus* in habit but the leaves are somewhat broader and the panicle has many very leafy bracts not turning gold at maturity. The tepals usually have 1–2 teeth on each side and an elongate tubercle.

c. *R.* x *knafii* occurs on the margins of ponds with the parents. Material has been seen from v.c. 13, 16–18, 22, 23, 28, 34 and 61–63. It is also recorded from Au and Ge.

d. None.

e. *R. conglomeratus* $2n = 20$; *R. maritimus* ($2n = 40$).

f. LOUSLEY, J. E. (1935). As above.

TRIMEN, H. (1874). On a *Rumex* from the south of England. *J. Bot., Lond.,* **12**: 161-163. FIG.

15 × 20. *R. conglomeratus* Murr. × *R. frutescens* Thou.

a. *R.* x *wrightii* Lousley.

b. This rhizomatous hybrid has coriaceous leaves as in *R. frutescens*, but the influence of *R. conglomeratus* is seen in the greater height (35–40 cm), the leaves being often subcordate or truncate at the base, the panicle having long branches with remote whorls subtended by bracts below, and in the lingulate outline of the fruiting tepals.

c. *R.* x *wrightii* has been found only at Braunton Burrows, v.c. 4, where it grew in a dune-slack with *R. frutescens, R. crispus* var. *littoreus* and formerly *R. conglomeratus.*

d. The hybrid has been grown for 20 years in the London garden of J. E. Lousley and the characters of the hybrid have remained constant.

e. *R. conglomeratus* $2n = 20$.

f. LOUSLEY, J. E. (1953). *Rumex cuneifolius* and a new hybrid. *Watsonia,* **2**: 394-397. FIG.

18 × 17. *R. maritimus* L. × *R. palustris* Sm.

= *R.* x *benrardii* Danser was determined by Danser on a specimen collected by Druce in 1929 in v.c. 24. Intermediates between these two closely related species do sometimes occur on the rare occasions when they grow together, and are probably hybrids, but the latter is very difficult to prove. While some of the specimens so-named are likely to be correct, further study of growing plants is desirable before the hybrid can be accepted with confidence for Br. It was described from Ho.

DANSER, B. H. (1915). Bijlage 3. Mededeelingen gehouden door den heer. *Ned. kruidk. Archf,* **1915**: 111-116.

DRUCE, G. C. (1930). *R. maritimus* x *palustris* = x *R. benrardii* Danser. *Rep. B.E.C.,* **9**: 36.

den × 18. *R. dentatus* Campd. × *R. maritimus* L.

= *R.* x *kloosii* Danser was raised by Danser in Amsterdam in 1921 from *R. dentatus* collected as an adventive by A. W. Kloos. It is fully described and illustrated by Danser (1922). It is said to have been found on Blackheath, v.c. 16, by Mrs [should be Miss] Gertrude Bacon (= Mrs T. F. Foggitt) in

1920. This is claimed as the first occurrence anywhere in the wild, but if correctly named the cross almost certainly originated out of Br.

DANSER, B. H. (1922). Fünf neue *Rumex*-Bastarde. *Receuil des travaux botaniques néerlandais*, **19**: 293-295. FIG.

DRUCE, G. C. (1925). x *Rumex klosii* [sic] Danser. *Rep. B.E.C.*, **7**: 452.

DRUCE, G. C. (1926). x *R. kloosii* Danser. *Rep. B.E.C.*, **7**: 781.

328. *Urtica* L.
(by C. A. Stace)

2 x 1. *U. dioica* L. x *U. urens* L.

    = *U.* x *oblongata* Koch has been recorded from Au and Cz, but is doubtful.

330. *Ulmus* L.
(by R. Melville)

1 x 2. *U. glabra* Huds. x *U. procera* Salisb.

    has been reported on a number of different occasions but these records are errors partly of determination but largely due to the nomenclatural complexities of the names *"U. montana"* and *"U. glabra"*. For instance the statement that the Huntingdon Elm, *U.* x *vegeta*, is of this parentage stems from a concept of *"U. glabra* Mill." which included both *U. procera* and *U. carpinifolia*. Richens (1967) recorded plants in Essex and Lincolnshire which appeared to be of this parentage, but further investigation of these is needed.

HENRY, A. (1910). On elm-seedlings showing Mendelian results. *J. Linn. Soc., Bot.*, **39**: 290-300.

MELVILLE, R. (1944). The British elm-flora. *Nature, Lond.*, **153**: 198-199.

RICHENS, R. H. (1967). Studies on *Ulmus*, 7. Essex Elms. *Forestry*, **40**: 185-206.

3 × 1. *U. angustifolia* (Weston) Weston × *U. glabra* Huds.

a. None.

b. This is a spreading tree sometimes reaching a large size in which the habit of *U. glabra* tends to be dominant. The short-shoots are often nearly at right angles to the branches as in *U. glabra,* but are more slender and less hairy, becoming sub-glabrous when old. Leaf-size and -shape are intermediate and the upper surface may be nearly smooth when mature. The shape of distal leaves of the short-shoots is often close to those of *U. angustifolia* but the leaves are large, the serrature is intermediate and a somewhat abbreviated apical cusp of *U. glabra* is present. The fruits are intermediate in size and tend to be orbicular as in *U. angustifolia.* Much less variation has been observed in hybrids of this parentage than in those of other combinations, suggesting a low degree of fertility.

c. This hybrid is restricted to areas in which *U. angustifolia* is native, and has been recorded from v.c. 1–3, 7 and 9–11.

d. None.

e. *U. glabra* ($2n = 28$).

f. None.

4 × 1. *U. coritana* Melville × *U. glabra* Huds.

a. None. Wrongly given as the parentage of *U.* × *hollandica* Mill. by Dandy (1958).

b. The spreading habit and stout branches of *U. glabra* are often dominant in this hybrid and are then associated with rather large, broad, coriaceous leaves intermediate in shape and with the serrature tending towards that of *U. coritana.* Alternatively the leaves may be rather narrow if *U. coritana* var. *angustifolia* has been involved. Other nothomorphs have a habit tending more towards that of *U. coritana* with more slender, twiggy branching and these may have rather pale green, coriaceous leaves tending towards those of *U. coritana* in shape but with the sharp serrature and an abbreviated apical cusp of *U. glabra.* Hybrids of this parentage are probably moderately fertile, but no direct observations are available.

c. As *U. glabra* occurs throughout the range of *U. coritana,* hybrids may occur anywhere within its area. There are records from v.c. 18, 20, 29, 30, 38 and 55.

d. None.

e. *U. glabra* ($2n = 28$).

f. DANDY, J. E. (1958). *Op. cit.,* p. 78.

4 × 1 × 6. *U. coritana* Melville × *U. glabra* Huds. × *U. plotii* Druce

a. *U.* × *diversifolia* Melville.

b. The habit is intermediate, but sometimes tends towards the broad shape of *U. coritana.* This hybrid is notable for its variations in leaf-shape. The leaves of normal short-shoots of the crown are intermediate in shape between those of the two primary parents—*U. coritana* and *U. plotii*—but tend towards *U. plotii* in the rather deep green colour and to *U. coritana* in the coriaceous texture and large axillary tufts of hairs on the lower leaf surface. The uniformly dispersed short pubescence on the lower surface and the rather tapering apical cusp of the distal leaves cannot be attributed to either of the major parents and are probably due to introgression from *U. glabra* Huds. which may occur in the field by crossing with other hybrids to which the latter has contributed. In addition to the normal kind of short-shoot, others occur in which one or more of the leaves are approximately symmetrical. The equal-based leaves generally have longer petioles in proportion to the lamina-length than usual; the leaf in effect consists of two short sides. Accordingly, the leaf base may be rounded with two short sides from *U. plotii* or cuneate with two from *U. coritana* or slightly asymmetrical with one side from each parent. Similar hybrid interactions have been observed in some other elm hybrids and this phenomenon may be compared with the deletion of parts of the lamina in some oak hybrids (Melville, 1960).

c. This hybrid occurs mainly in East Anglia and the eastern Midlands and has been reported from v.c. 18–20, 25–31 and 53.

d. None.

e. None.

f. MELVILLE, R. (1939). Contributions to the study of British elms, 2. The East Anglian Elm. *J. Bot., Lond.,* 77: 138-145. FIG.

MELVILLE, R. (1960a). A metrical study of leaf-shape in hybrids, 2. Some hybrid oaks and their bearing on Turing's theory of morphogenesis. *Kew Bull.,* 14: 161-177.

TUTIN, T. G. (1952). *Ulmus,* in CLAPHAM, A. R., TUTIN, T. G. and WARBURG, E. F. *Op. cit.,* pp. 715-724, t. 48. FIG. (NOTE typographical errors in Fig. 47 in this account: throughout account, *for* 47A *read* 47C, *for* 47B *read* 47A, *for* 47C *read* 47D, *for* 47D *read* 47B).

5 × 1. *U. carpinifolia* Gled. × *U. glabra* Huds.

a. *U. × vegeta* (Loud.) A. Ley (*U. × smithii* Henry). (Huntingdon Elm). Wrongly given by Tutin (1952) as the parentage of *U. × hollandica* also.

b. This hybrid includes medium to large trees generally with a spreading habit, but the typical nothomorph of *U. × vegeta* has a fountain-like habit with a number of ascending branches from a rather short trunk. The leaves of this have the general shape, size and serrature of those of *U. glabra,* but are distinguished by the almost smooth upper surface and the shape of the lower half of the long side of distal and subdistal leaves which closely resemble the corresponding parts in *U. carpinifolia.* The basal lobe is almost rectangular with a nearly straight margin joining the petiole, or often in subdistal leaves joining the lowest lateral vein, shortly before it unites with the midrib. Hybrids of this parentage may exhibit any combination of characters of the parents. The leaves may be as large as in *U. glabra* or as small as in *U. carpinifolia,* and rough or smooth above, but they commonly show the rectangular basal lobe to the long side as in the latter and the short-shoots make an angle of 80–90° with the branches as in the former.

c. As an apparently natural hybrid it occurs in v.c. 7, 17–26, 29–41 and 53–57. Elsewhere in BI it is probably planted or feral from seedlings of planted trees. Hybrids are also known from Au, Be, Ga, He and Ho.

d. None.

e. Both parents ($2n = 28$).

f. ELWES, H. J. and HENRY, A. (1913). *Trees of Great Britain and Ireland,* pp. 1879-1882, t. 395. Edinburgh. FIG.

MOSS, C. E. (1914). *Ulmus,* in *The Cambridge British Flora,* 2: 88-96, pl. 94 and 95. Cambridge. FIG.

TUTIN, T. G. (1952). As above.

5 × 1 × 6. *U. carpinifolia* Gled. × *U. glabra* Huds. × *U. plotii* Druce

a. *U. × hollandica* Mill. (Dutch Elm).

b. The nothomorph to which the binomial name strictly applies is a large, spreading tree suckering freely and producing large, corky flanges on the bark of suckers and epicormic shoots. In the normal short-shoots the subdistal leaves are rather large, broadly ovate and acute with the basal lobe of the long side rounded or showing flattening of its curvature as in *U. carpinifolia.* The distal leaves often have shapes recalling those of *U. plotii* but are larger and with sharper serrature and more numerous lateral veins. This typical nothomorph is widely planted, but innumerable others occur naturally, which present in their kaleidoscopic variation every possible combination of characters from the three parents. In order to recognize such hybrids it is necessary to scrutinize the trees for characteristic features of the parent species and to note how the corresponding characters of the species intergrade. Features to look for indicative of *U. glabra* are the broad spreading habit, the stout branchlets with the short-shoots set nearly at right angles, rufous hairs on the bud scales, large leaves with more than 12 vein pairs, a tapering apical cusp and a scabrid upper leaf surface. Features of *U. carpinifolia* are less easy to recognize. In the habit an ascending, fountain-like arrangement of the upper branches of the crown may be seen and the more slender, glabrous branchlet character modifies the stoutness and hairiness of *U. glabra.* The leaf characters most evident are the wedge-shaped leaf-base, particularly of the distal leaves, with the basal lobe of the long side forming nearly a right angle. Although the broad habit of *U. glabra* tends to be dominant, the tall, erect stem of *U. plotii* with its widely spaced branches and open crown does appear in some of the segregants. The influence of *U. plotii* is commonly shown in the dark green colour and smaller size of the leaves and by the blunting and broadening of the serrature of the other two species.

c. Hybrids of this parentage occur naturally in the Midlands and East Anglia. Elsewhere they are so frequently planted that it is difficult to determine the boundaries of their natural occurrence. In Hb they are introduced, *U.*

x *hollandica sensu stricto* being planted frequently around towns and villages. On the Continent the elm populations of Be, Ga, Ge and Ho consist almost exclusively of trees of this parentage.

d. Much of the breeding work carried out in Ho aimed at developing forms resistant to the Dutch elm disease has been carried out on selections from this hybrid. Examination of progenies has given additional evidence for the triple parentage.

e. *U. glabra, U. carpinifolia* and hybrid ($2n = 28$).

f. KRIJTHE, N. (1939). Verslag over de werkzaamheden voor het Iepen-ziekte-comité. *Tijdschr. PlZiekt.*, **45**: 63-70.

MOSS, C. E. (1914). As above, **2**: 88-96, pl. 96 and 97. FIG.

TUTIN, T. G. (1952). As above.

(5 × 1 × 6) × 3. (*U. carpinifolia* Gled. × *U. glabra* Huds. × *U. plotii* Druce) × *U. angustifolia* (Weston) Weston (= *U. angustifolia* × *U. × hollandica* Mill.)

a. *U.* × *sarniensis* (Loud.) Bancroft. (Jersey or Wheatley Elm).

b. The tree to which the binomial name strictly applies is of pyramidal habit with the lower branches widely spreading or slightly ascending and the upper branches steeply ascending. The foliage characters are closest to those of *U. angustifolia* in the rather small size, coriaceous texture and near glabrous surfaces but the serration is sharper and more complex than in that species. Nothomorphs of this parentage may occur wherever *U. angustifolia* comes into contact with any of the numerous nothomorphs of *U.* × *hollandica*. In the south-west confusion may occur with *U. angustifolia* × *U. glabra* and it is necessary to scrutinize such hybrids for characters of *U. plotii* and *U. carpinifolia*. The effect of *U. plotii* is manifested by the sporadic appearance of distal leaves of the *U. plotii* shape and of occasional semi-long-shoots and proliferating short-shoots. The Kentish population consists of trees verging towards *U. angustifolia* and may be the result of occasional introgression with that species. In the Midlands nothomorphs tending more towards *U. plotii* appear to have arisen by the crossing of a local population of the latter with avenue trees of the Jersey Elm. The Jersey Elm itself produces a fair proportion of good seed and this is probably true also of other nothomorphs.

c. Despite the association of some nothomorphs of the hybrid with the ragstone escarpment in v.c. 15, there do not appear to be any special habitat requirements elsewhere. It is reported from v.c. 1–3, 15 and CI, and in adjacent parts of Ga; elsewhere in BI the Jersey Elm is planted as an amenity tree.

d. None.

e. *U. carpinifolia, U. glabra*, the Dutch Elm and the Jersey Elm ($2n = 28$).

f. ELWES, H. J. and HENRY, A. (1913). As above, t. 398. FIG.

KRIJTHE, N. (1939). As above.

MELVILLE, R. (1960b). The names of the Cornish and the Jersey Elm. *Kew Bull.*, **14**: 216-218.

1 × 6. *U. glabra* Huds. × *U. plotii* Druce

a. *U.* × *elegantissima* Horwood.

b. Hybrids are intermediate in leaf-shape, leaf-serration, indumentum, and branchlet and habit characters. With a moderate degree of fertility and ready backcrossing and segregation almost any combination of characters of the species may occur. The nothomorphs can be arranged as a nothocline on leaf-shape, but other characters cut across this arrangement. Determination of the parentage depends upon the recognition of characteristic features of the parents in different combinations including spreading habit, stout branchlets, rufous bud-scale hairs and large, scabrid leaves with acuminate apical cusps of *U. glabra* and erect habit, slender glabrous branchlets, small dark green leaves with blunt serrature and proliferating shoots of *U. plotii*. The proliferating shoot character is not always to be seen on crown branches, but can usually be found on epicormic shoots. The change from more typical foliage on these shoots to successively rounder leaves with fewer lateral veins and blunter serrature is a good indication of the influence of genes from *U. plotii*. The pendulous branch characters of fully mature *U. plotii* may also be inherited in some nothomorphs and gives rise to some very handsome trees.

c. Hybrids of this parentage are frequent from the Trent valley southwards to Hertfordshire and Essex and have been recorded from v.c. 7, 8, 18–24, 26, 28–34, 36–40 and 53–57. Planted trees have been seen occasionally elsewhere in BI.

d. The abundance, frequency and variability of the hybrids suggest a high level of fertility. Open-pollinated seeds from an isolated hybrid tree near Banbury, v.c. 23, germinated readily and gave progeny segregating towards both parents.

e. *U. glabra* ($2n = 28$).

f. MELVILLE, R. (1955). Morphological characters in the discrimination of species and hybrids, in LOUSLEY, J. E., ed. *Species studies in the British flora*, pp. 55-64. London. FIG.
   TUTIN, T. G. (1952). As above.

5 × 2. *U. carpinifolia* Gled. × *U. procera* Salisb.

has been recorded as widely distributed in Essex by Richens (1967), but he expressed some degree of doubt about the identification and further work is necessary.
RICHENS, R. H. (1967). As above.

4 × 6. *U. coritana* Melville × *U. plotii* Druce

a. None.

b. These are usually rather spreading trees with the upper branches ascending and the crown rather open, but nothomorphs also occur with an erect habit tending more towards *U. plotii*. The leaf shapes are commonly intermediate, but often on one branch some short shoots bear leaves

tending towards those of *U. coritana* in shape, while others more closely
resemble those of *U. plotii*. The dark green of *U. plotii* leaves is usually
combined with a somewhat coriaceous texture and the rather large woolly
axillary tufts of the lower surface of *U. coritana.* Some proliferating
short-shoots of the *U. plotii* type are usually to be found, though the
branchlets of some trees may consist almost exclusively of semi-long-
shoots tailing off into small rounded leaves as in those of *U. plotii.* Such
trees may also have very small leaves, no more than 2 cm long, and appear
to be examples of negative heterosis. In parts of the Midlands where *U.
plotii* is frequent, nothomorphs with habit and foliage types tending
towards *U. plotii* are not uncommon, but in East Anglia the trees are more
often intermediate or resemble *U. coritana* more closely in their
characters. The wide range of nothomorphs occurring naturally suggests
that the hybrids are moderately fertile and segregate and backcross
without much difficulty.

c. The hybrid occurs through the range of its two parents, namely through
East Anglia and the Midlands as far west as Shropshire. It has been
recorded from v.c. 18, 19, 25–29, 40, 53, 54 and 56. Whereas *U. plotii* is
essentially a species of valley-bottom alluvium and *U. coritana* favours low
hills, though generally not above 400 ft, the hybrids extend over the
ecological range of both parents. They are found on sands and boulder
clay in East Anglia and on alluvium, clays and loams in the Midlands.

d. None.

e. None.

f. None.

5 × 6. *U. carpinifolia* Gled. × *U. plotii* Druce

a. *U. × viminalis* Lodd.

b. This combination consists of small to medium-sized trees with small, dark
green leaves and irregular habit, sometimes with tortuous branches, but
erect individuals occur tending towards *U. plotii* in habit. The leaf-shapes
are commonly intermediate with neat, rather fine and sharp serration and
small, rather inconspicuous axillary tufts to the veins below. The distal leaf
of the short-shoots frequently displays the rectangular basal lobe of the
long side of *U. carpinifolia.* At the same time, other short-shoots may have
leaf-shapes approaching those of *U. plotii*. The effect of *U. plotii* genes
generally is also shown by the presence of proliferating shoots continuing
their growth with smaller more rounded leaves. In some nothomorphs the
number of lateral veins in the leaves approximates to those of *U. plotii* and
the serrature may then be sharp and rather deep. Sharp, deep serrature is
also a feature of typical *U. × viminalis,* which has tortuous branches and
occasional leaves with bifid midribs. In wild trees, the latter character is
sometimes manifested and it may also be induced by hormone sprays.
Occasional nothomorphs tending towards *U. plotii* may have a preponderance of semi-long-shoots over normal short-shoots and the accompanying leaf characters of *U. plotii* are then emphasized.

c. This hybrid occurs in a band of country extending from Essex to

Oxfordshire and has been recorded from v.c. 18–20, 23, 29, 30 and 32. The rather extreme nothomorphs to which the binominal *U.* x *viminalis* strictly applies have been in cultivation for about a century and a half.

d. None.

e. *U. carpinifolia* ($2n = 28$).

f. MELVILLE, R. (1955). As above. FIG.

335. *Betula* L.
(by S. M. Walters)*

1 x 2. *B. pendula* Roth x *B. pubescens* Ehrh.

a. *B.* x *aurata* Borkh.

b. Trees variously intermediate between the parents in such characters as leaf-shape, pubescence, bark of trunk and twigs, habit, and fruiting catkins are usually considered hybrids, but their status needs further investigation (particularly cytogenetically). Most intermediates seem fully fertile and appear to backcross (and to obscure the species delimitation), but a few are highly sterile.

c. Intermediates are common and widespread in BI, not necessarily in close proximity to both or either species. Their common occurrence in BI (in contrast to their relative rarity in, e.g., Fe) seems to be correlated with the secondary status of much birch woodland in BI, and the breakdown of originally more effective ecological isolation between the two species. The hybrid is recorded from much of central and northern Europe.

d. Crosses with Swedish parental material could be made only with a very low level of success, but F_1 plants were raised. Isolation thus appears to be both genetical and ecological.

e. *B. pendula* $2n = 26–28$ (42 in sterile autotriploids); *B. pubescens* $2n = 56$; intermediates $2n = 41–43$ in sterile plants ($2n = 56$ in fertile plants).

f. BROWN, I. R. and TULEY, G. (1971). A study of a population of birches in Glen Gairn. *Trans. Proc. bot. Soc. Edinb.*, **41**: 231-245.

JOHNSSON, H. (1945). Interspecific hybridisation within the genus *Betula. Hereditas*, **31**: 163-176.

MARSHALL, E. S. (1914). *Betula*, in MOSS, C. E. *The Cambridge British Flora*, **2**: 80-86., t. 84. Cambridge. FIG.

NATHO, C. (1959). Variationsbreite und Bastardbildung bei Mitteleuropäischen Birkensippen. *Reprium nov. Spec. Regni veg.*, **61**: 211-273.

RICHENS, R. H. (1945). Forest tree breeding and genetics. *Imperial Agricultural Bureaux Joint Publication*, **8**: 29-31.

WALTERS, S. M. (1968). *Betula* in Britain. *Proc. B.S.B.I.*, **7**: 179-180.

* With assistance from D. Aston.

3 × 1. *B. nana* L. × *B. pendula* Roth
= *B.* × *plettkei* Junge is recorded from Ge and Su.

3 × 2. *B. nana* L. × *B. pubescens* Ehrh.

a. *B.* × *intermedia* Thomas ex Gaud. (*B.* × *alpestris* Fries).

b. Two nothomorphs occur: one (*B.* × *intermedia*) closer to *B. pubescens* and the other (*B.* × *alpestris*) to *B. nana*. Thus hybrid plants vary considerably in the main characters of the leaves, female catkins and achenes. The plants are rarely as much as 4 m and often under 1 m high. D. Aston (*in litt.* 1973) has found only one catkin (male) on the hybrid plants on Ben Loyal, v.c. 108; it showed 0% pollen viability, but *B. nana* from the same area showed only 14–18%.

c. This hybrid occurs within the range of *B. nana* in Scotland; there are old but probably reliable records from v.c. 88, 90, 92, 98, 106 and 108, and its occurrence on Ben Loyal, v.c. 108, has recently been confirmed. It is also known in He, Is, No, Rs, Su, Greenland and Siberia. Godwin (1956) documented the Late-glacial and early Post-glacial finds of female cone-scales of this hybrid in England, Scotland and Hb.

d. In Is introgression, mainly to *B. pubescens*, has been detected. Experimental reciprocal crosses (using whole plants of *B. nana* and grafts of *B. pubescens*) made by D. Aston (*in litt.* 1973) have failed to produce viable fruit, although some swelling of the female catkins occurred.

e. *B. pubescens* $2n = 56$; *B. nana* $2n = 28$; hybrid $2n = 37–42$.

f. ELKINGTON, T. T. (1968). Introgressive hybridization between *Betula nana* L. and *B. pubescens* Ehrh. in North-West Iceland. *New Phytol.*, **67**: 109-118.

GODWIN, H. (1956). *The history of the British flora*, p. 192. Cambridge.

KENWORTHY, J. B., ASTON, D. and BUCKNALL, S. A. (1972). A study of hybrids between *Betula pubescens* Ehrh. and *Betula nana* L. from Sutherland—an integrated approach. *Trans. bot. Soc. Edinb.*, **41**: 517-539.

MARSHALL, E. S. (1901). *B. alpestris* Fr. *J. Bot., Lond.*, **39**: 271.

MARSHALL, E. S. (1914). As above, t. 87. FIG.

NATHO, C. (1959). As above.

336. *Alnus* Mill.
(by C. A. Stace)

1 × 2. *A. glutinosa* (L.) Gaertn. × *A. incana* (L.) Moench

a. *A.* × *pubescens* Tausch.

b. Hybrid trees are somewhat variable in appearance, especially leaf-shape, though generally intermediate. They may be distinguished by the somewhat pubescent young twigs and undersurfaces of the leaves, the obtuse to shortly acuminate leaf-apex, and the shortly pedunculate female catkins. Focke (1881) claimed that the achenes were small and abortive, but Ehrenberg (1946) found that putative hybrids were equally as fertile as the parents.

c. Hybrids are quite frequently planted in Br, but have apparently been recorded only once in natural conditions (at Horley, v.c. 17, in 1949), in a wood with both parents. It is not clear whether they were planted or had arisen naturally. On the Continent hybrids are found across Europe from Scandinavia and Ga to Rs.

d. Focke reported that Klotzsch made the cross in 1846, and that after eight years the hybrid saplings were larger than those of the parents of the same age. Hybrids have been synthesized by foresters in more recent years, but no published details have been encountered. Swedish wild hybrids have a regular meiosis.

e. *A. glutinosa* $2n = 28$; *A. incana* and hybrid ($2n = 28$).

f. BRENAN, J. P. M. (1950). *Alnus incana*, *A. glutinosa* and the hybrid between them from a wood near Horley, Surrey. *Year Book B.S.B.I.*, **1950**: 56.

EHRENBERG, C. E. (1946). Till frågen: existerar *Alnus glutinosa* x *incana* i naturen? *Bot. Notiser*, **1946**: 529-535.

FOCKE, W. O. (1881). *Op. cit.*, p. 355.

McVEAN, D. N. (1953). *Alnus* Mill., in Biological Flora of the British Isles. *J. Ecol.*, **41**: 447-466.

REICHENBACH, L. and REICHENBACH, H. G. (1850). *Icones Florae Germanicae et Helveticae*, t. 1292. Leipzig. FIG.

WINKLER, H. (1904). Betulaceae, in ENGLER, A., ed. *Das Pflanzenreich*, **19(IV, 61)**: 128-129. Leipzig.

341. *Quercus* L.
(by J. E. Cousens)

4 x 3. *Q. petraea* (Mattuschka) Liebl. x *Q. robur* L.

a. *Q.* x *rosacea* Bechst. (*Q.* x *intermedia* Boenn. ex Reichb., *Q. robur* subsp. *puberula* (Lasch) Weim.).

b. Trees exist with intermediate characters and also with most possible combinations of parental characters. Species-delimitation is still a matter of opinion and with it the extent of hybridization and introgression that have occurred in BI. There is agreement that *Q. robur* is the more variable species of the two. It seems likely that the F_1 commonly occurs as a variant which superficially resembles *Q. robur*; however, both the slender peduncles and the abaxial leaf surface bear a well-developed stellate pubescence and the indented, sharply-reflexed base of the leaf does not produce auricles which overlap the petiole. There is a little evidence from Scotland that intermediates are less fertile than the parent species, but *Q. robur* subsp. *puberula* is described as fully fertile in Su.

c. Intermediates are found over most of BI but are less frequent in southern than in northern England and are most abundant in Scotland. On the Continent intermediates have been reported in greatest abundance from Scandinavia. The explanation that the flowering periods overlap more completely as the growing season becomes shorter is not convincing because it would appear that there is always at least a 50% overlap. Ecological interpretation is made difficult by the extent of planting of stock of *Q. robur*-affinity in nearly all oakwoods in Br, the acorns in some instances coming from the Continent. In Scotland intermediates are most abundant in oakwoods with a long history of management as coppice-with-standards.

d. Crosses in Po and Ge suggested low interfertility (*c* 2%), the maximum reported being 15% for *Q. petraea* male x *Q. robur* female.

e. Both parents (2*n* = 24).

f. CARLISLE, A. and BROWN, A. H. F. (1965). The assessment of the taxonomic status of mixed oak (*Quercus* spp.) populations. *Watsonia*, **6**: 120-127.

COUSENS, J. E. (1963). Variation in some diagnostic characters of the Sessile and Pedunculate Oaks and their hybrids in Scotland. *Watsonia*, **6**: 273-286.

COUSENS, J. E. (1965). The status of the pedunculate and sessile oaks in Britain. *Watsonia*, **6**: 161-176.

DENGLER, A. (1941). Bericht über Kreuzungversuche zwischen Trauben-und Stiel-eiche (*Q. sessiliflora* Smith u. *Q. pedunculata* Ehrh.) *Mitt. H.-Göring-Akad. dt. Forstwiss.*, **1**: 87-109.

GARDINER, A. S. (1970). Pedunculate and Sessile Oak (*Quercus robur* L. and *Quercus petraea* (Mattuschka) Liebl.). A review of the hybrid controversy. *Forestry*, **43**: 151-160.

HØEG, E. (1929). Om mellemformerne mellem *Q. robur* L. og *Q. sessiliflora* Mart. *Bot. Tidsskr.*, **40**: 411-427.

JONES, E. W. (1959). *Quercus* L., in Biological Flora of British Isles. *J. Ecol.*, **47**: 169-222.

JONES, E. W. (1968). The taxonomy of the British species of *Quercus*. *Proc. B.S.B.I.*, **7**: 183-184.

MOSS, C. E. (1914). *Quercus*, in *The Cambridge British Flora*, **2**: 71-76, t. 77. Cambridge. FIG.

WEIMARCK, H. (1947). De nordiska ekarna. *Bot. Notiser*, **1947**: 61-78, 105-134.

342. *Populus* L.
(by R. D. Meikle)

1 × 3. *P. alba* L. × *P. tremula* L.

a. *P.* × *canescens* (Ait.) Sm. (*P.* × *hybrida* Bieb.). (Grey Poplar).

b. The Grey Poplar is now generally regarded as a hybrid by Continental authors, and with justification, for *P.* × *canescens* is intermediate in almost every respect between the presumed parents. The material labelled *P.* × *hybrida* Bieb. by British authors, and supposed to be *P. canescens* × *P. tremula*, is most probably a nothomorph of *P. alba* × *P. tremula*, but could conceivably be a backcross of *P.* × *canescens* to *P. tremula*. *P.* × *canescens* is a tall tree with greyish, often conspicuously lenticellate bark. The leaves are extremely variable. Those of the long-shoots and suckers are broadly ovate-deltoid, usually shallowly lobed and irregularly toothed, soon becoming subglabrous above but often persistently whitish- or greyish-tomentose below, and with rather stout, plano-convex, often persistently tomentose petioles. Those of the short-shoots are broadly ovate-deltoid or oblate-deltoid, coarsely but bluntly toothed or sinuate-toothed, thinly tomentellose and glabrescent (or subglabrous from the start) on both surfaces, and often with long and strongly compressed petioles.

c. *P. tremula* is a native tree found throughout most of BI, but *P. alba* is introduced. *P.* × *canescens* has possibly arisen in BI but most specimens are planted in woodland and on roadsides, etc. The male tree is much commoner than the female in BI, and has almost certainly spread vegetatively. *P.* × *canescens* has been recorded from about 70 vice-counties in Br north to v.c. 90 and 98, from about 26 vice-counties in Hb, and from CI. It is widespread in central, southern and western Europe.

d. Experimental work has been carried out by Bartkowiak and Bialobok (1966).

e. Both parents and hybrid ($2n = 38$, but autotriploids, $2n = 57$, are also known).

f. BARTKOWIAK, S. and BIALOBOK, S. (1966). Morphological variability in artificial hybrids—*P.* × *canescens* Smith. *Arboretum kórn.*, **11**: 105-151.

BUTCHER, R. W. and STRUDWICK, F. E. (1930). *Further illustrations of British plants*, t. 338. Ashford. FIG.

MARCET, E. (1961). Taxonomische Untersuchungen in der Sektion *Leuce* Duby der Gattung *Populus* L. *Mitt. Schweiz. Anst. Forstl. Versuchsw.*, **37**: 269-321.

MOSS, C. E. (1914). *Populus*, in *The Cambridge British Flora*, **2**: 4-13, t. 3-5. Cambridge. FIG.

PERRING, F. H. and WALTERS, S. M. (1962). *Op. cit.*, p. 187.

SYME, J. T. B., ed. (1868). *English Botany*, 3rd ed., **8**: 194-196, t. 1300. London. FIG.

el × 4. *P. deltoidea* Marsh. × *P. nigra* L.

a. *P.* × *canadensis* Moench. (Black Italian Poplar).

b. Several nothomorphs (or cultivars) of this hybrid are extensively cultivated, and some, particularly nm. *serotina*, occur subspontaneously. *P.* × *canadensis* is commonly confused with *P. nigra*, a more stocky tree, usually with downward curving branches and a burred trunk; the leaves of *P. nigra* are generally cuneate at the base, and the catkins are distinctly smaller than those of *P.* × *canadensis*. *P.* × *canadensis* is a tall tree with erect branches, a spreading, often fan-shaped crown, and a trunk without burrs. The leaves are broadly deltoid, acuminate-cuspidate at the apex, very broadly cuneate to shallowly cordate at the base, regularly and bluntly serrate except near the base, with a narrow translucent border, and with very long, glabrous, compressed petioles. Nm. *serotina* (Hartig) Rehd. is a male variant in which each flower has 20–25 stamens with crimson anthers; nm. *marilandica* (Poir.) Rehd. is a female variant with longer and more slender catkins lengthening greatly in fruit and with almost spherical, glabrous ovaries and 2–4 thick, short stigmas with conspicuous basal lobes.

c. This hybrid occurs in woodland and parkland and by roadsides, etc., almost always planted, throughout BI. *P. nigra* is possibly native in parts of England, but *P. deltoidea* is only seen as a planted tree.

d. Many different hybrids between these two species have been raised for commercial purposes.

e. Both parents and hybrid ($2n = 38$).

f. BUTCHER, R. W. and STRUDWICK, F. E. (1946). As above, t. 340. FIG.

CANSDALE, C. S. (1938). *The Black Poplars and their hybrids cultivated in Britain.* Oxford.

HOUTZAGERS, G. (1937). *Het Geslacht Populus in Verband met zijn Beteekenis voor de Houtteelt*, pp. 83-146. Wageningen.

MOSS, C. E. (1914). As above, t. 15-16. FIG.

REHDER, A. (1940). *Manual of cultivated trees and shrubs hardy in North America*, 2nd ed., p. 80. New York.

SMITH, E. C. (1943). A study of cytology and speciation in the genus *Populus* L. *J. Arnold Arbor.*, **24**: 275-305.

343. *Salix* L.

(by R. D. Meikle)

The willows are nowadays so notorious for their hybrids that it is hard to understand how botanists, until the latter part of the last century, could ignore, or indeed positively refute, the possibility of such miscegenation. Smith (1828) vehemently set aside "the gratuitous suppositions of the mixture of species, or the production of new, or hybrid ones, of which, no more than of any change in established species, I have never met with an instance". Forty years later Syme (1868) was saying much the same thing, though with noticeably less vehemence.

It was this refusal to admit the possibility of hybridization that was responsible, to a very considerable extent, for the excessive number of *Salix* names in the literature of the late 18th and early 19th centuries. As Wimmer (1866) put it: "Salices hybridas esse nemo cogitabat, aut si cogitaverat praeter Laschium pronuntiare non audebat." The "Laschium" was Wilhelm Lasch (1786–1863), whose field-studies in eastern Prussia had led him to conclude that hybrids (and not only *Salix* hybrids) were more common than was generally supposed. His conclusions (Lasch, 1857) were at first taken lightly, but their acceptance by such eminent authorities as Wimmer (1866) and Andersson (1867) gave them weight, and within 20 years of the publication of Lasch's paper the prevalence of hybrids in the *Salicaceae* was generally conceded. Now the tendency is to move too far in this direction, and to assume that hybrids are so frequent that identification of the parent species is virtually impossible. This is not so. In theory, it should be possible for all the British species to cross either directly or indirectly. In fact, with the exception of *S. triandra* x *S. viminalis* (and the questionable *S. purpurea* x *S. triandra*), one does not find hybrids between species belonging to subg. *Salix* (true willows) and those belonging to either subg. *Chamaetia* Dumort. (dwarf, alpine willows) or subg. *Caprisalix* Dumort. (sallows and osiers). Furthermore, for ecological and phenological reasons, it is unlikely for hybrids to occur between lowland and highland species, or between species whose flowering periods do not overlap.

Willows seldom reproduce themselves from seed except as colonists of open, competition-free sites—sand-spits by rivers, gravel-pits, bombed sites, land-slides and the like—and it is in such sites that the majority of spontaneous hybrids and hybrid complexes are found. In closed communities reproduction from seed is unusual (Leefe, 1871), and most hybrids found in such communities have been introduced, either deliberately or unintentionally, by man, and propagated vegetatively. Some hybrids (e.g. *S.* x *calodendron*) are almost certainly single clones, and depend almost wholly upon man for their spread. Others (e.g. *S.* x *smithiana, S.* x *rubra, S.* x *mollissima*) are generally clonal, and man-distributed, but may occasionally arise spontaneously. Only three hybrids, *S. caprea* x *S. cinerea, S. aurita* x *S. cinerea* and *S. nigricans* x *S. phylicifolia,* are both common and spontaneous, and these frequently form complexes which defy precise identification. For the remainder, hybrids are either rare, and must be searched for in areas where both parents are abundant, or are so uniform (because of vegetative propagation) that they are readily distinguished and determined.

While many willow hybrids are fertile and can be backcrossed or re-crossed (Wichura, 1865; Nilsson, 1954) some (e.g. *S.* x *calodendron*) are normally sterile, even when a male plant of a related species is growing nearby. A considerable number of hybridization experiments have been undertaken in the past, particularly in Scandinavia, where hybrid plants involving as many as 13 species have been raised. In this country experimental work has not been carried out in such a systematic fashion, but several hybrids have been synthesized, mostly by E. F. Linton around the start of this century, and some were distributed as exsiccata.

In this account the taxonomic treatment given by Dandy (1958) is followed, except that *S. calodendron* is here treated as a hybrid (*S. caprea* x *S. cinerea* x *S. viminalis*), and no attention is paid to the different subspecies that are recognized. Hybrids of doubtful origin are discussed under their most likely parentage, often with cross-references under other combinations which have been suggested. To save space sections d, e and f are only included where there is relevant information available.

Chromosome Numbers

1. *S. pentandra* L. (2*n* = 76)
2. *S. alba* L. (2*n* = 76)
3. *S. babylonica* L. (2*n* = 76)
4. *S. fragilis* L. (2*n* = 76, 114)
5. *S. triandra* L. 2*n* = 38 (38, 44, 88)
6. *S. purpurea* L. (2*n* = 38)
9. *S. viminalis* L. (2*n* = 38)
11. *S. caprea* L. 2*n* = 38, 76
12. *S. cinerea* L. (2*n* = 76)
13. *S. aurita* L. 2*n* = 38, 76
14. *S. nigricans* Sm. (2*n* = 114)
15. *S. phylicifolia* L. (2*n* = 88, 114)
16. *S. repens* L. 2*n* = 38
17. *S. lapponum* L. (2*n* = 38, 76)
18. *S. lanata* L. (2*n* = 38)
19. *S. arbuscula* L. (2*n* = 38)
20. *S. myrsinites* L. (2*n* = 38, 152, 190)
21. *S. herbacea* L. (2*n* = 38)
22. *S. reticulata* L. (2*n* = 38)

General References

ANDERSSON, N. J. (1867). Monographia Salicum hucusque cognitarum. *K. svenska Vetensk-Akad. Handl.*, **6**.

BLACKBURN, K. B. and HARRISON, J. W. H. (1924). A preliminary account of the chromosomes and chromosome behaviour in the Salicales. *Ann. Bot.*, **38**: 361-378.

CAMUS, A. and CAMUS, E.-G. (1904–1905). *Classification des saules d'Europe et monographie des saules de France*. Paris.

HÅKANSSON, A. (1955). Chromosome numbers and meiosis in certain Salices. *Hereditas*, **41**: 454-483.

HEGI, G. (1912). *Illustrierte Flora von Mittel-Europa*, **3**: 13-57. Munich.

LASCH, W. (1857). Aufzählung der in der Provinz Brandenburg, besonders in der Gegend um Driesen, wildwachsenden Bastard-Pflanzen, nebst kurzen Notizen zur Erkennung solcher Gewächse. *Bot. Zeit.*, **15**: 505-517.

LEEFE, J. E. (1871). On hybridity in *Salix,* and the growth of willows from seed. *J. Bot., Lond.,* **9**: 225-227.

LINTON, E. F. (1913). A monograph of the British willows. *J. Bot., Lond.,* **51** (Suppl.).

MOSS, C. E. (1914). *Salix,* in *The Cambridge British Flora,* **2**: 13-68. Cambridge. FIG.

NILSSON, N. H. (1954). Über Hochkomplexe Bastardverbindungen in der Gattung *Salix. Hereditas,* **40**: 517-522.

RECHINGER, K. H. (1957). *Salix,* in HEGI, G. *Illustrierte Flora von Mittel-Europa,* 2nd ed.. **3**: 44-135. Munich.

REICHENBACH, L. (1849). *Icones Florae Germanicae et Helveticae,* t. 557-619. Leipzig. FIG.

SEEMEN, O. von. (1908–1910). *Salicaceae,* in ASCHERSON, P. and GRAEBNER, P. *Synopsis der Mitteleuropäischen Flora,* **4**: 54-350. Leipzig.

SMITH, J. E. (1828). *The English flora,* **4**: 164. London.

SYME, J. T. B., ed. (1868). *English Botany,* 3rd ed., **8**: 200-261, t. 1303-1379. London. FIG.

WHITE, F. B. (1890). A revision of the British willows. *J. Linn. Soc., Bot.,* **27**: 333-457.

WICHURA, M. (1865). *Die Bastardbefruchtung im Pflanzenreich erläutert an den Bastarden der Weiden.* Breslau.

WIMMER, F. (1866). *Salices Europaeae.* Bratislava.

2 × 1. *S. alba* L. × *S. pentandra* L.

a. *S.* × *ehrhartiana* Sm.

b. This is a graceful, ornamental tree with lustrous, but not highly polished, twigs. The leaves are lanceolate or narrowly elliptic with a long slender acumen, bright green above, glaucescent and thinly appressed-pubescent below, and with minutely and regularly serrate margins. The catkins appear with the leaves. There are (2)3–4(5) stamens per flower. Female catkins have not been seen.

c. In BI *S.* × *ehrhartiana* is probably always planted. It has been found on stream-banks in v.c. 17, 19, 20, 29, 69, 70 and H8, and in Au, Da, Ge, It and Su.

2 × 4 × 1. *S. alba* L. × *S. fragilis* L. × *S. pentandra* L.

= *S.* × *pentandroides* Rouy has been recorded from Ge and Su. Specimens from Hitchin, v.c. 20, determined as this by R. Görz, are in my opinion *S. alba* × *S. pentandra.*

GÖRZ, R. (1930). *Salix. Rep. Watson B.E.C.,* **4**: 12.

LITTLE, J. E. (1928). Salices. *Rep. Watson B.E.C.,* **3**: 417-419.

4 × 1. *S. fragilis* L. × *S. pentandra* L.

a. *S.* × *meyerana* Rostk. ex Willd. (Pointed-leaved Willow).
b. This is a tall shrub or small tree with highly polished twigs. The leaves are ovate or ovate-oblong, generally with a slender acumen, bright shining green above, slightly glaucescent below, glabrous, and with minutely and regularly serrate margins. The catkins appear with the leaves. The female catkins are cylindrical and stipitate. The male catkins are shortly cylindrical, stipitate, and have (2)3–4(5) stamens per flower, with the filaments hairy towards the base. Many specimens so named are *S. alba* × *S. pentandra* or narrow-leaved forms of *S. pentandra*. *S.* × *meyerana* varies considerably, and may resemble either parent. It is most readily distinguished from *S. pentandra* by its relatively narrow, cylindrical catkins.
c. *S.* × *meyerana* has been found on river-banks and in moist places, often planted, in v.c. 7, 21, 26, 29, 57, 59, 69, H16, H19 and H26, and in Au, Cz, Da, Ga, Ge, He, No, Rs and Su.
f. SPROTT, W. A. P. (1936). New light on a *Pentandra* hybrid willow. *J. Bot., Lond.,* 74: 230-233.

1 × 5. *S. pentandra* L. × *S. triandra* L.

= *S.* × *schumanniana* Seem. has been recorded from Cz, Ge and Po.

13 × 1. *S. aurita* L. × *S. pentandra* L.

= *S.* × *basaltica* Coste has been reported from Ga. It was also once reported by C. G. Trapnell from Sedbergh, v.c. 65, but (from the citation of *S.* × *ludificans*) was obviously a typographical error for *S. aurita* × *S. phylicifolia*.
TRAPNELL, C. G. (1926). *S. aurita* × *pentandra = ludificans* And. *Rep. B.E.C.,* 7: 895.

2 × 3. *S. alba* L. × *S. babylonica* L.

a. *S.* × *sepulcralis* Simonk.
b. This tree closely resembles *S. babylonica*, but is distinguished by its longer, distinctly stipitate catkins, shorter leaves with minutely and regularly serrate margins, and longer petioles. It may be questioned if any of the *S. babylonica* growing in Europe is identical with the willow, so named, which is indigenous in the Far East. Chinese specimens of *S. babylonica* usually have very short, sub-sessile or very shortly stipitate catkins.
c. *S.* × *sepulcralis* is widely cultivated in Br but can scarcely be considered as naturalized anywhere. It is also widely distributed on the Continent and in North America.
f. FRASER, J. (1932). Some planted or cultivated willows. *Rep. B.E.C.,* 9: 720-721.
REHDER, A. (1940). *Manual of cultivated trees and shrubs,* 2nd ed., p. 95. New York.

2 × 4. *S. alba* L. × *S. fragilis* L.

a. *S.* × *rubens* Schrank.

b. This is a tree with a rather spreading crown and glabrous, brownish, terete twigs. The leaves are lanceolate, generally with a slender acumen, glabrescent and dull or rather bright green above, often permanently appressed-pilose below, and with minutely, but frequently rather irregularly, serrate margins; the young leaves are often whitish-sericeous. Continental authors regard almost all British *S. fragilis* as hybrids between true *S. fragilis* (*S. decipiens* Hoffman), with wholly glabrous shoots and leaves, and *S. alba*. It is certainly true that *S. fragilis* L. *sec.* Sm. is completely connected to *S. alba* by a series of intermediates which I regard as hybrids. At least some hybrids show univalents and multivalents at meiosis.

c. *S.* × *rubens*, which is often confused with *S. fragilis*, is found by the sides of rivers and streams and at field-margins, often planted. It is recorded from 26 vice-counties in Br and 6 in Hb, and is widespread on the Continent from No to It and Bu.

e. Hybrid ($2n = 76$).

f. WILKINSON, J. (1941). The cytology of the Cricket-Bat Willow. *Ann. Bot.*, n.s., 5: 150-165.

2 × 5. *S. alba* L. × *S. triandra* L.

is the parentage which sometimes has wrongly been attributed to *S.* × *undulata* Ehrh.—see under *S. triandra* × *S. viminalis*. The true hybrid *S. alba* × *S. triandra* (? = *S.* × *erythroclados* Simonk.) has been recorded from Au, Ge and Hu.

FRASER, J. (1933). Revised nomenclature of *Salix*. *Rep. B.E.C.*, **10**: 367-371.

2 × 6. *S. alba* L. × *S. purpurea* L.

was recorded by Horwood and Noel (1933) from near Croft, v.c. 55. It is almost certainly an error, but I have not examined the specimen.

CLAIRE, C. (1927). *Salix Clairei*, in GUÉTROT, M. *Plantes hybrides de France*, 1 and 2: 41. Lille.

HORWOOD, A. R. and NOEL, C. W. F. (1933). *The flora of Leicestershire and Rutland*, p. 502. Oxford.

3 × 4. *S. babylonica* L. × *S. fragilis* L.

a. *S.* × *blanda* Anderss.

b. This is a tree with a rounded crown and slender, pendulous, glabrous, olive-brown branches and twigs, closely resembling *S. alba* × *S. babylonica* but differing in its glabrous petioles and more coarsely serrate leaves.

c. Planted or occasionally subspontaneous specimens are widely scattered in Br, but are less frequent than *S. alba* × *S. babylonica*; the precise distribution is impossible to ascertain from published records. The hybrid is widely distributed on the Continent and in North America.

f. FRASER, J. (1932). As above.
   REHDER, A. (1940). As above, p. 96.

4 × 5. *S. fragilis* L. × *S. triandra* L.

a. *S.* × *speciosa* Host.
b. This is a shrub or tree with brown, glabrous, terete, or more or less angled twigs. The leaves are lanceolate-acuminate, bright green above, paler below, glabrous or glabrescent on both surfaces, and with coarsely and rather irregularly serrate margins, and ovate-acuminate, not very persistent stipules. The male catkins are more or less erect, narrowly elongate-cylindrical, and with the rachis and catkin-scales densely pilose; the 2–3 stamens have filaments which are hairy towards the base. Female catkins have not been seen.
c. There are well-known records from v.c. 3 and 9, and others unconfirmed from v.c. 7, 8, 17, 24, 29, 37, 38, 55, 57, 93, H8 and H32. Most records are errors for *S. fragilis* var. *furcata* Seringe ex Gaud. The two well-known specimens are unsatisfactory: Linton's plant (v.c. 3) is said to be uniformly diandrous, and closely resembles forms of *S. fragilis*, though distinguished by its more or less angular twigs and suberect catkins; Briggs' plant (v.c. 9) is triandrous, and resembles Continental material of *S.* × *speciosa*, but it lacks mature foliage. The hybrid is reported from Au, Be, Bu, Cz, Da, Ga, Ge, It, Ju, No, Rm and Su.
f. FRASER, J. (1933). As above.

4 × 6. *S. fragilis* L. × *S. purpurea* L.

= *S.* × *margaretae* Seem. has been recorded from Ge.

4 × 9. *S. fragilis* L. × *S. viminalis* L.

= *S.* × *boulayi* F. Gérard has been recorded from Ga and Ge.

6 × 5. *S. purpurea* L. × *S. triandra* L.

a. *S.* × *leiophylla* G. & A. Camus.
b. The British plant is a sprawling shrub with twigs which are at first finely pubescent but soon glabrous and rather lustrous mahogany-brown. The leaves are lanceolate and acute, at first finely pubescent along the midrib and veins, and with regularly serrate margins, rather short petioles, and rather broadly auriculate, acute, persistent stipules. The small, cylindrical, erect catkins develop in advance of the leaves; the catkin-scales are shortly and bluntly ovate-lingulate, blackish above, and densely villose. The two stamens are connate below and with reddish or purplish anthers. Female catkins have not been seen. The identity of the British plant, which differs from Continental examples of *S. purpurea* × *S. triandra*, is puzzling. The pubescence on the midrib and veins of the young leaves suggests the

influence of a third contributor—perhaps *S. cinerea* or *S. viminalis*—unless this is an unidentified exotic.

c. *S.* × *leiophylla* is known from an osier-bed in v.c. 56. The hybrid also occurs in Po.

f. HOWITT, R. C. L. and HOWITT, B. M. (1969). Lowland willows. *Gdnrs' Chron.*, **166**(24): 12-13.

5 × 9. *S. triandra* L. × *S. viminalis* L.

a. *S.* × *mollissima* Hoffm. ex Elwert. (Sharp-stipuled Triandrous Willow).

b. This hybrid is represented in BI by two nothomorphs. Nm. *undulata* (Ehrh.) Wimmer is a tall shrub with erect or spreading, glabrous, olive-brown twigs and flaking bark (as in *S. triandra*). The leaves are lanceolate or linear-lanceolate, at first thinly pubescent, soon becoming glabrous, bright green above, paler below, and with closely glandular-serrate margins, rather short petioles, and ovate-acuminate, rather persistent stipules. The female catkins are erect, narrowly cylindrical, shortly stipitate, and appear with the leaves; the rachis is densely villose, the catkin-scales yellowish, elongate-lingulate, and thinly and softly villose, and the ovaries glabrous; I have not seen male catkins. Nm. *hippophaeifolia* (Thuill.) Wimmer differs in its linear-lanceolate or almost linear leaves which are dark green above, slightly paler below, and have subentire or minutely and distantly glandular-serrulate margins and rather blunt, generally caducous stipules. The male catkins are rather shortly cylindrical, with short, rounded, yellowish, densely villose catkin-scales and usually three stamens per flower. The female catkins have elongate-lingulate, villose catkin-scales, and ovaries which are at first densely appressed-pubescent and later glabrescent but never quite glabrous.

c. These plants occur on river-banks and roadsides, and in moist thickets and osier-beds, where they are often planted, chiefly in southern Br. Nm. *undulata* occurs in about 32 vice-counties in Br, in v.c. H39 and H40, and in CI, Au, Be, Bu, Da, Ga, Ge, Ho, Hs, No, Po, Rm, Rs and Su. Nm. *hippophaeifolia* is recorded from v.c. 17, 34, 36, 39, 55, 59 and 71, and from Be, Cz, Da, Ga, Ge and Rm.

12 × 5. *S. cinerea* L. × *S. triandra* L.

= *S.* × *krausei* Anderss. has been recorded from Ga and Ge.

13 × 5. *S. aurita* L. × *S. triandra* L.

= *S.* × *litigiosa* G. & A. Camus has been recorded from Ge.

7 × 6. *S. daphnoides* Vill. × *S. purpurea* L.

= *S.* × *calliantha* Kerner has been recorded from Au, Ge, He and Hu.

11 × 7 × 6. *S. caprea* L. × *S. daphnoides* Vill. × *S. purpurea* L.

= *S.* × *neuburgensis* Erder has been recorded from Ge.

7 × 6 × 16. *S. daphnoides* Vill. × *S. purpurea* L. × *S. repens* L.

= *S.* × *boettcheri* Seem. has been recorded from Ge.

8 × 11 × 6. *S. acutifolia* Willd. × *S. caprea* L. × *S. purpurea* L.

= *S.* × *scholzii* Rouy has been recorded from Ge.

6 × 9. *S. purpurea* L. × *S. viminalis* L.

a. *S.* × *rubra* Huds. (Green-leaved Osier).

b. This is a spreading shrub or small tree with branches which are at first often appressed-pubescent but soon glabrous or subglabrous. The leaves are linear or linear-lanceolate, acuminate (or rarely acute), dark green and usually rather lustrous above, slightly paler below and sometimes with a thin, appressed, sericeous indumentum. The shortly cylindrical catkins develop before the leaves. The two stamens are free, partly free or united; the filaments are glabrous; and the anthers reddish or yellow. Female catkins are densely white-tomentose; the ovaries are broadly and shortly flask-shaped; the style is distinct; and the stigmas are linear and entire. This hybrid varies considerably towards either parent, with which it may possibly backcross. Though variation has probably been extended artificially by osier-growers, since this is an important basket-willow, the greenish under-surface of the leaves generally distinguishes it from *S. viminalis* and the pubescence of the young growths from *S. purpurea*.

c. *S.* × *rubra* occurs on roadsides and in osier-beds and thickets, scattered in about 40 vice-counties in Br and 9 in Hb. Druce (1932) recorded *S.* × *rubra* from 69 vice-counties, but he almost certainly included records for *S. cinerea* × *S. purpurea* × *S. viminalis*. *S.* × *rubra* is also recorded from Au, Be, Bu, Da, Fe, Ga, Ge, He, Ho, Hs, Hu, It, Ju, Rm, Rs and Su.

e. *S.* × *rubra* ($2n = 38, 57$).

f. DRUCE, G. C. (1932). *The comital flora of the British Isles*, p. 272. Arbroath.

WILKINSON, J. (1944). The cytology of *Salix* in relation to its taxonomy. *Ann. Bot.*, n.s., **8**: 269-284.

11 × 6 × 9. *S. caprea* L. × *S. purpurea* L. × *S. viminalis* L.

= *S.* × *rubriformis* Tourlet has been recorded from Ge.

11 × 12 × 6 × 9. *S. caprea* L. × *S. cinerea* L. × *S. purpurea* L. × *S. viminalis* L.

= *S.* × *taylorii* Reching. f. was described by K. H. Rechinger f. as *S. dasyclados* × *S. purpurea* from foliage material collected on Barry Links near Carnoustie, v.c. 90. I have not succeeded in tracing the type specimen, nor have I seen flowering specimens of this very interesting record. Both parents occur in the area.

RECHINGER, K. H. (1950). Observations on some Scottish willows. *Watsonia*, **1**: 271-275.

12 × 6 × 9. S. *cinerea* L. × S. *purpurea* L. × S. *viminalis* L.

a. *S.* × *forbyana* Sm. (Fine Basket Osier).
b. This is an erect, vigorous shrub or small tree with yellowish twigs which are at first thinly pubescent, but soon glabrous or subglabrous. The leaves are narrowly oblong, acute, at first shortly floccose-tomentose but soon glabrous, and with subentire or remotely and very obscurely toothed margins; they turn black on drying. The catkins are erect or spreading, cylindrical, sessile, and develop in advance of the leaves. The filaments are united towards the base, and usually free above. The ovaries are shortly flask-shaped; the style is distinct; and the stigmas are shortly linear and entire. This plant is puzzling in the field, looking superficially like a *S. triandra* hybrid on account of its glossy leaves. It may be no more than a nothomorph of *S.* × *rubra* and the presence of *S. cinerea* has not been satisfactorily established. Until recently only the female plant was known, but the male has been submitted by R. C. L. Howitt from specimens originally found in v.c. 56.
c. *S.* × *forbyana* occurs on river-banks and moist thickets in v.c. 6, 9, 11, 14, 17, 20, 22, 28, 29, 38, 56, 57, 59, 68, 70, 71, H28, H29 and H33; it is also recorded from Au, Ga, Ge and He.

13 × 6 × 9. S. *aurita* L. × S. *purpurea* L. × S. *viminalis* L.

was said to have arisen spontaneously in E. F. Linton's garden in Bournemouth, v.c. 11. I have not seen specimens, nor can the hybrid, which has been recorded from Ge, be properly regarded as part of the natural British flora.
LINTON, E. F. (1903a). *A. aurita-viminalis* × *purpurea. Rep. Watson B.E.C.*, **1902-1903**: 20.

6 × 16 × 9. S. *purpurea* L. × S. *repens* L. × S. *viminalis* L.

has been recorded from Ge.

11 × 6. S. *caprea* L. × S. *purpurea* L.

= *S.* × *wimmerana* Gren. & Godr. has been recorded from Au, Ga, Ge, He and Hu.

12 × 6. S. *cinerea* L. × S. *purpurea* L.

a. *S.* × *sordida* A. Kerner.
b. This is an erect or spreading shrub with twigs which are at first thinly pubescent but soon glabrous. The leaves are oblong or narrowly obovate, shortly acute, glabrous and dark, rather lustrous green above, thinly pubescent or glabrous below, conspicuously glaucous, and with serrate margins with obscure and often rather remote teeth. The catkins develop in advance of the leaves; the catkin-scales are blackish and densely villose. The ovaries are subsessile, narrowly flask-shaped, distinctly longer than the catkin-scales and densely white-tomentose. The two stamens are free or

united for all or part of their length, with purplish or reddish anthers. According to F. B. White (letter at **K** to J. G. Baker, 21st April 1887) this hybrid is "not uncommon and quite wild" in Perthshire; he adds that it is more variable than the books suggest and runs quite into *S. cinerea*.

  c. *S.* × *sordida* occurs on river-banks and in moist thickets in v.c. 32, 33, 69, 72, 88 and 89; it is also recorded from Au, Cz, Da, Ga, Ge, He, Hu, It, Rm and Su.

13 × 12 × 6. *S. aurita* L. × *S. cinerea* L. × *S. purpurea* L.

  a. *S.* × *confinis* G. & A. Camus.
  b. This is an erect shrub. The leaves are narrowly oblong or oblanceolate, acute or obtuse, dark green and subglabrous above, greyish and sub-glabrous or thinly pubescent below, and with broadly ovate-acute, toothed, persistent stipules. Determination of the British plants as this hybrid is questionable, but the large, persistent stipules appear to indicate *S. aurita*; Druce's sterile specimen from Tyndrum, v.c. 88, was determined as *S. cinerea* × *S. purpurea* by J. Fraser.
  c. Specimens have been recorded from swampy ground and river-banks in v.c. 87 and 88, and also in Ge and He.

13 × 6. *S. aurita* L. × *S. purpurea* L.

  a. *S.* × *dichroa* Doell.
  b. This is an erect or sprawling shrub with slender twigs which are at first pubescent but soon glabrous or subglabrous. The leaves are oblong, obovate or oblanceolate, dark green and glabrous or subglabrous and often rather lustrous above, glaucous and at first thinly pubescent but later often glabrous below, and with narrowly ovate-acuminate or broadly auriculate, persistent or caducous stipules. The short, cylindrical male catkins develop in advance of the leaves; the catkin-scales are blackish and villose. The two stamens are free or united for the whole or part of their length and have purplish or reddish anthers. I have not seen female catkins. This plant is not satisfactorily distinguished from *S. cinerea* × *S. purpurea*, to which female specimens cited by Linton (1913) probably belong.
  c. In BI *S.* × *dichroa* is only known from v.c. 68. It is also recorded from Au, Be, Ga, Ge, He, Ho and Su.
  e. Hybrid ($2n = 38$).
  f. WILKINSON, J. (1944). As above.

13 × 15 × 6. *S. aurita* L. × *S. phylicifolia* L. × *S. purpurea* L.

  a. *S.* × *sesquitertia* F. B. White.
  b. This is an erect shrub with twigs which are at first very sparsely pubescent but soon glabrous and lustrous red-brown. The leaves are glabrous or subglabrous and dark, shining green above, glaucous and glabrous or glabrescent below, and with a rather prominent venation. The ovaries are a little longer than the catkin-scales and densely tomentose; the style is short

but distinct; and the stigmas are oblong and usually undivided. Male catkins have not been seen. B. Floderus has identified authentic material of this hybrid in **K** as "*S. atrocinerea* (x ?*phylicifolia*)".

c. This hybrid was found on a river-bank in v.c. 72; it is apparently endemic.

f. WHITE, F. B. (1892). Notes on Scottish willows. *Ann. Scot. nat. Hist.*, 1892: 66.

13 x 6 x 16. *S. aurita* L. x *S. purpurea* L. x *S. repens* L.

= *S.* x *pseudo-doniana* Rouy has been recorded from Ge.

14 x 6. *S. nigricans* Sm. x *S. purpurea* L.

a. *S.* x *beckiana* Beck.

b. This is a spreading shrub with twigs which are at first thinly pubescent but soon glabrous. The leaves are oblanceolate, acute, at first thinly pubescent on both surfaces, soon glabrous and slightly glaucous below, and with subentire margins; they become blackish on drying. The ovaries are shortly pedicellate and densely greyish-pubescent; the style is distinct; and the stigmas are shortly oblong and deeply bifid. Male catkins have not been seen.

c. *S.* x *beckiana* is known from river-banks and moist thickets in v.c. 64 and 67, and also from Au, Ge and Po.

f. HARRISON, J. W. H. (1949). Notes on local willows, with a record of a hybrid new to the district. *Vasculum*, **34**: 15.

15 x 6. *S. phylicifolia* L. x *S. purpurea* L.

a. *S.* x *secerneta* F. B. White.

b. This is a spreading shrub with twigs which are glabrous or subglabrous and dark, shining red-brown. The leaves are oblanceolate-elliptic, dark, shining green and rather lustrous above, glaucous and glabrous below, and with minutely and rather regularly serrate margins. The catkins develop with the leaves. The two free or variously connate stamens have yellow anthers. Female catkins have not been seen.

c. *S.* x *secerneta* is known from river-banks and moist places in v.c. 67, 72 and 99; it is apparently endemic.

f. WHITE, F. B. (1892). As above.

6 x 16. *S. purpurea* L. x *S. repens* L.

a. *S.* x *doniana* Sm. (Donian Willow).

b. This is a sprawling shrub 1–1.5 m high with reddish-brown twigs which are soon glabrous. The leaves are narrowly oblanceolate, acute, at first densely sericeous but soon glabrous and shining green above, and glaucescent and thinly appressed-pilose or glabrous below; they turn black on drying. The catkins develop in advance of the leaves; the catkin-scales are densely villose, yellowish at the base, reddish above, and almost black at the apex. The ovaries are densely white-tomentose; and the style is very short or

absent. Male plants are not known in the wild, but occurred spontaneously
in E. F. Linton's garden in v.c. 11; they have broader leaves, long filaments
pubescent towards the base and united almost to the apex, and reddish
anthers.

  c. *S.* x *doniana* occurs on stabilized sand-dunes and moist heathy ground in
v.c. 59 and 88, and also in Au, Cz, Da, Ga, Ge, He, Po, Rs and Su.

  d. Artificial hybrids were made at Bournemouth, v.c. 11, by E. F. Linton,
and distributed as exsiccata.

  f. LINTON, E. F. (1910d). *S. Doniana* Sm. *Rep. Watson B.E.C.,* **2**: 257.

17 x 6. *S. lapponum* L. x *S. purpurea* L.

= *S.* x *schatilowii* Schroeder has been recorded from Rs.

19 x 6. *S. arbuscula* L. x *S. purpurea* L.

= *S.* x *buseri* Favrat has been recorded from He.

7 x 9. *S. daphnoides* Vill. x *S. viminalis* L.

= *S.* x *digenea* Kerner has been recorded from Au, Be, Cz, Ge and Su.

11 x 7. *S. caprea* L. x *S. daphnoides* Vill.

= *S.* x *hungarica* Kerner has been recorded from Au, Cz, Ge, He, Hu and
Rs.

12 x 7. *S. cinerea* L. x *S. daphnoides* Vill.

= *S.* x *mariana* Wolosz. has been recorded from Au.

7 x 14. *S. daphnoides* Vill. x *S. nigricans* Sm.

= *S.* x*inticensis* Huter has been recorded from Au and He.

7 x 16. *S. daphnoides* Vill. x *S. repens* L.

= *S.* x *maritima* Hartig has been recorded from Ge.

8 x 11. *S. acutifolia* Willd. x *S. caprea* L.

= *S.* x *propinqua* G. & A. Camus has been recorded from Ge.

8 x 12. *S. acutifolia* Willd. x *S. cinerea* L.

has been recorded from Ge.

11 x 9. *S. caprea* L. x *S. viminalis* L.

  a. *S.* x *sericans* Tausch ex A. Kerner (*S.* x *laurina* Sm. *sec.* Nilsson, Dandy).

  b. This is an erect shrub or small tree rarely up to 30 ft high, with branches
which are at first rather densely pubescent but soon glabrescent, and
which have wood without striae. The leaves are narrowly oblong-ovate or
broadly lanceolate, dull green above, greyish and thinly tomentose below,

and with prominent reticulate primary and secondary venation. The catkins appear before the leaves and are cylindrical or narrowly ovoid. The style is distinct and often pubescent; and the stigmas are linear and undivided or rarely cleft almost to the base. This is a variable hybrid, not always easy to distinguish from *S. cinerea* x *S. viminalis* but generally with much longer, broader leaves, the undersides of which are more or less tomentose, especially when young; the bark of the twigs is commonly yellowish, and often rather glossy with age.

c. *S.* x *sericans* is a very common plant of hedgerows, thickets and waste ground throughout most of BI; it is also recorded from Au, Be, Cz, Ga, Ge, Gr, He, Ho, It, Rm, Rs and Su.

d. Nilsson has made detailed genetical studies on *S. caprea* x *S. viminalis*; the idea that *S.* x *laurina* is one of its nothomorphs is argued, but not convincingly proven.

e. Natural hybrid ($2n = 41$); artificial F_1 ($2n = 38$); artificial F_2 ($2n = 38, 57, 76$).

f. NILSSON, N. H. (1928). *Salix laurina. Acta Univ. lund.*, Avd. 2, **24(6)**: 1-88. FIG.

NILSSON, N. H. (1931). Über das Entstehen eines ganz *cinerea*-ähnlichen Typus aus dem Bastard *Salix viminalis* x *caprea. Hereditas*, **15**: 309-319. FIG.

NILSSON, N. H. (1935). Die Analyse der synthetisch hergestellten *Salix laurina. Hereditas*, **20**: 339-353. FIG.

WILKINSON, J. (1944). As above.

11 x 12 x 9. *S. caprea* L. x *S. cinerea* L. x *S. viminalis* L.

a. *S.* x *calodendron* Wimm. (*S.* x *acuminata* auct., *S.* x *dasyclados* auct.). (Long-leaved Sallow).

b. This is an erect shrub or small tree with one-year-old twigs and current year's shoots which are densely greyish-velutinous and which have sparsely striate wood. The leaves are oblong-elliptic, dull green and glabrescent or sparsely pubescent above, ashy-grey and thinly pubescent below, and with conspicuous (often brownish) venation. The catkins are erect, rather crowded, elongate-cylindrical, and develop in advance of the leaves. The style is long; and the stigmas are rather shortly linear, thickish, and undivided or slightly notched at apex. The parentage of *S.* x *calodendron* is dubious. It was thought to be *S. caprea* x *S. dasyclados* by Wimmer, but the status of *S. dasyclados* is itself questionable; some authors regard it as one of the nothomorphs of *S. caprea* x *viminalis*. The distinct striations on the wood of *S.* x *calodendron* suggest the presence of *S. cinerea* or *S. aurita*.

c. *S.* x *calodendron* is scattered over Br in hedgerows and waste ground, often planted, and has also been recorded from v.c. H3, H33 and H38, and from Da, Ge and Su.

e. *S. calodendron* ($2n = 76$); *S. dasyclados* ($2n = 38, 57, 76$); artificial (*S. viminalis* x *S. caprea*) x *S. cinerea* ($2n = 57$).

f. MEIKLE, R. D. (1952). *Salix calodendron* in Britain. *Watsonia*, 2: 243-248.

SWANN, E. L. (1957). West Norfolk willows. *Proc. B.S.B.I.*, 2: 337-345. FIG.

13 × 11 × 9. *S. aurita* L. × *S. viminalis* L. × *S. caprea* L.

a. *S.* × *stipularis* Sm. (Auricled Osier).

b. This is an erect shrub or small tree with one-year-old twigs and current year's shoots which are densely greyish-velutinous. The leaves are narrowly lanceolate or linear-lanceolate, dull green and sparsely pubescent above, densely whitish- or ashy-velutinous below, and with narrowly ovate or lanceolate, acuminate, persistent stipules which are velutinous below. The catkins develop in advance of the leaves and are erect and narrowly elongate-cylindrical; the catkin-scales are mid-brown or slightly darker towards the apex. The style is distinct; and the stigmas linear and undivided. Male catkins have not been seen. The parentage of this plant is dubious. Wimmer considered *S.* × *stipularis* a hybrid between *S. dasyclados* and *S. viminalis*, but *S. dasyclados* is now reckoned to be a hybrid itself. *S.* × *stipularis* comes nearest to *S. viminalis*, but is readily distinguished by its broad, persistent stipules, velutinous (not appressed-sericeous) leaf under-surfaces, and by its relatively large, villose catkins.

c. *S.* × *stipularis* is found in hedgerows and scrubland, probably often planted, throughout much of northern Br and in v.c. H33. It has also been recorded from CI, Ge, Rm, Rs and Su.

e. Hybrid ($2n = 114$).

f. MEIKLE, R. D. (1952). As above.

11 × 15 × 9. *S. caprea* L. × *S. phylicifolia* L. × *S. viminalis* L.

= *S.* × *tomentella* G. & A. Camus has been recorded from Rs.

11 × 16 × 9. *S. caprea* L. × *S. repens* L. × *S. viminalis* L.

= *S.* × *turfosa* G. & A. Camus has been recorded from Ge.

12 × 9. *S. cinerea* L. × *S. viminalis* L.

a. *S.* × *smithiana* Willd. (Silky-leaved Osier).

b. This is an erect shrub or small tree in which the shoots of the current year are pubescent, the one-year-old twigs are glabrescent or glabrous and the wood has a few or sometimes no longitudinal striae. The leaves are narrowly lanceolate, dark green and thinly pubescent above, ashy-grey and sometimes distinctly pubescent or thinly velutinous below, and with the midrib and primary lateral veins fairly prominent, but the secondary veins obscure. The catkin-scales are elongate-lingulate, subacute, and dark brown, sometimes with a darker tip. The style is distinct; and the stigmas shortly linear and undivided or sometimes deeply bifid. This hybrid intergrades with *S. caprea* × *S. viminalis* but is generally easily distinguished by its striate wood, and narrow, lanceolate leaves, the under-surfaces of which have a thin, short, crisped pubescence. The venation of *S.* x *smithiana* is normally much less prominently reticulate than that of *S. caprea* x *S. viminalis*, and the bark of the twigs is generally reddish-brown. Male specimens of *S.* x *smithiana* seem to be very uncommon; the uniformity of local populations suggests that the hybrid may frequently have been spread vegetatively.

c. *S.* x *smithiana* occurs in hedgerows, moist thickets, osier-beds and waste ground, etc., throughout most of BI. It has also been recorded from Au, Be, Bu, Cz, Ga, Ge, He, Ho, Hu, It, Ju, Rm, Rs and Su.

d. Artificial hybrids have been made by Håkansson; only rarely did male plants occur in the F_1. The hybrids exhibited a very retarded development of the embryo-sac.

e. Artificial hybrid ($2n = 57$).

13 x 12 x 9. *S. aurita* L. x *S. cinerea* L. x *S. viminalis* L.

= *S.* x *hirtei* Strachler has been recorded from Ge.

12 x 15 x 9. *S. cinerea* L. x *S. phylicifolia* L. x *S. viminalis* L.

= *S.* x *hirsutophylla* G. & A. Camus has been recorded from Rs.

12 x 16 x 9. *S. cinerea* L. x *S. repens* L. x *S. viminalis* L.

a. *S.* x *angusensis* Reching. f.

b. This is a spreading shrub with twigs which are at first rather densely pubescent but later glabrous and dark reddish-brown, and which have wood apparently without striae. The leaves are lanceolate, green and thinly appressed-pubescent above, and densely silvery and sericeous-tomentose below; the venation is not very prominent. I have not seen catkins.

c. *S.* x *angusensis* occurs on stabilized sand-dunes in v.c. 90; it is apparently endemic.

f. RECHINGER, K. H. (1950). As above.

13 x 9. *S. aurita* L. x *S. viminalis* L.

a. *S.* x *fruticosa* Doell.

b. This is an erect shrub or small tree with twigs which are at first densely grey-pubescent but later subglabrous (or even glabrous), and which have wood with scattered longitudinal striae. The leaves are linear-lanceolate, acuminate, dull green above, silver-grey with a dense indumentum of appressed or crispate hairs below, prominently veined, and have margins which are often conspicuously rugose-undulate and irregularly serrate; the stipules are large, ovate-acuminate, serrate and persistent. The style is short but distinct; and the stigmas linear. I have not seen male catkins.

c. *S.* x *fruticosa* occurs in hedgerows and thickets; it is scattered over Br and has also been recorded from v.c. H11, H33, H39 and H40, and from Au, Be, Da, Fe, Ga, Ge, No and Su.

d. Artificial hybrids have been made by Håkansson.
e. Artificial hybrid ($2n = 38$).

13 × 16 × 9. *S. aurita* L. × *S. repens* L. × *S. viminalis* L.

= *S.* × *aberrans* G. & A. Camus has been recorded from Ge and Rs.

16 × 9. *S. repens* L. × *S. viminalis* L.

a. *S.* × *friesiana* Anderss.
b. This is a sprawling shrub with twigs which are at first densely appressed-pubescent but later glabrous or subglabrous. The leaves are narrowly lanceolate, and densely appressed silver-sericeous below. The shortly cylindrical catkins develop a little in advance of the leaves; the catkin-scales are shortly lingulate and dark brown. The style is rather long; and the stigmas are cleft into two linear lobes. The stamens are free with yellow anthers.
c. *S.* × *friesiana* occurs in stabilized sand-dunes and heathy ground in v.c. 59, 107 and 108. It is also recorded from Au, Da, Ge, Ju, No, Rs and Su.
d. Artificial hybrids were made by Håkansson; at Bournemouth, v.c. 11, by E. F. Linton, who distributed them as exsiccata; and later at Edmondsham, v.c. 9, by E. F. Linton.
e. Artificial hybrid ($2n = 38$).
f. LINTON, E. F. (1898a). *S. viminalis* × *repens*, Lasch. *Rep. B.E.C.*, **1**: 565. LINTON, E. F. (1910c). *S. repens* × *viminalis*. *Rep. Watson B.E.C.*, **2**: 258.

17 × 9. *S. lapponum* L. × *S. viminalis* L.

has been recorded from Rs.

11 × 12. *S. caprea* L. × *S. cinerea* L.

a. *S.* × *reichardtii* A. Kerner.
b. This is an erect shrub or small tree with twigs which are at first pubescent but later glabrous and a rather lustrous reddish-brown, and which have wood with a few scattered striae, or without striae. The leaves are variable; they are generally obovate or broadly elliptic, dark green above, ashy-grey and glabrescent or more or less densely pubescent to tomentose below, and with a usually prominent and conspicuously reticulate venation, and commonly undulate margins. It is one of the commoner willow hybrids in Br, and backcrosses to such an extent as to make a complete gradation between the two parent species. "Pure" *S. caprea* has stouter twigs, larger (often closely sessile) catkins and broader, ovate or almost orbicular leaves which are densely tomentose below and often with the margins strongly undulate or crisped. *S. cinerea* has narrower, obovate or oblong leaves which are thinly pubescent below with a much less prominent venation and with flat, or at most slightly undulate, margins. Most of the British material must be *S. caprea* L. × *S. cinerea* L. subsp. *oleifolia* Macreight (*S.*

atrocinerea Brot.), but it would be well-nigh impossible to distinguish between hybrids of *S. caprea* and the two subspecies of *S. cinerea*.

  c. *S.* x *reichardtii* has been recorded from woodland margins, scrubland, hedgerows, waste ground, etc., over much of Br and from v.c. H21 and H39. It is also known in Au, Be, Da, Fe, Ga, Ge, He, Ho, It, No, Rm, Rs and Su.

  d. Artificial hybrids have been made by Håkansson—see under *S. cinerea* x *S. viminalis*.

  e. Natural hybrid ($2n = 76$); artificial hybrid ($2n = 57$).

  f. WILKINSON, J. (1944). As above.

13 x 11 x 12. *S. aurita* L. x *S. caprea* L. x *S. cinerea* L.

= *S.* x *woloszcakii* Zalewski was recorded by Druce (1926) from Ufton, v.c. 31. In the absence of detailed genetical data the identification of a hybrid with such a parentage must be regarded as questionable, but there are also records from Ge and Po.

DRUCE, G. C. (1926). Huntingdonshire plants. *Rep. B.E.C.*, **7**: 949-957.

13 x 11 x 12 x 16. *S. aurita* L. x *S. caprea* L. x *S. cinerea* L. x *S. repens* L.

= *S.* x *aschersoniana* Seem. has been recorded from Ge.

11 x 12 x 15. *S. caprea* L. x *S. cinerea* L. x *S. phylicifolia* L.

= *S.* x *ludibunda* G. & A. Camus (*S.* x *tephrocarpa* F. B. White, *non* Wimm.) was recorded by White (1890) from "Banks of the Tay, above Dunkeld, *C. McIntosh*". I have not seen specimens and cannot comment on the identification, though the hybrid is quite likely to occur in Br. It has been recorded from Ga and Ge.

13 x 11. *S. aurita* L. x *S. caprea* L.

  a. *S.* x *capreola* A. Kerner ex Anderss.

  b. This is an erect shrub or small tree with twigs which are at first grey-pubescent but soon glabrous or subglabrous and dark reddish-brown, and which have wood with scattered longitudinal striae. The leaves are ovate, obovate or broadly oblong-elliptic, commonly with an obliquely twisted, cuspidate apex, glaucous and more or less densely pubescent below, and with a usually prominent venation, and often strongly rugose-undulate margin; the stipules are broadly ovate-acuminate, serrate, and persistent or sometimes caducous. One suspects that some plants named as this hybrid are either small-leaved variants of *S. caprea* or hybrids between *S. aurita* and *S. cinerea* or between *S. caprea* and *S. cinerea*.

  c. *S.* x *capreola* occurs in woodland margins, scrubland and hedgerows; it is scattered over Br and also recorded from v.c. H23 and H30, and Au, Be, Cz, Da, Fe, Ga, Ge, He, Ho, It, No, Rm and Su.

d. Artificial hybrids were made by W. R. Linton at Shirley, v.c. 57, and distributed as exsiccata.

f. LINTON, W. R. (1900a). *S. aurita* x *Caprea. Rep. B.E.C.*, 1: 587.

13 x 11 x 15. *S. aurita* L. x *S. caprea* L. x *S. phylicifolia* L.

= *S.* x *schatzii* Sagorski has been recorded from Ge.

11 x 14. *S. caprea* x *S. nigricans* Sm.

a. *S.* x *latifolia* Forbes.

b. This is a spreading shrub or small tree with twigs which are at first thinly pubescent but later glabrous or subglabrous and a lustrous dark brownish-red. The leaves are obovate, broadly elliptic or oblong, acute, dark green and glabrous or subglabrous above, greyish-green and thinly pubescent below, and with bluntly and rather remotely serrate margins; they turn blackish on drying. The catkins, which develop with the leaves, are subsessile or shortly stipitate. The ovaries are narrowly flask-shaped and densely white-tomentose; the style is long; and the stigmas are shortly oblong and undivided.

c. *S.* x *latifolia* is found on banks of rivers and lake-shores, and in moist thickets, chiefly in northern Br, in v.c. 38, 69, 72, 80, 83, 88, 90 and 96; and also in Au, Fe, Ge, It and Su.

11 x 14 x 15. *S. caprea* L. x *S. nigricans* Sm. x *S. phylicifolia* L.

(= *S.* x *phylicioides* Druce, *nom. nud.*) was tentatively recorded from near Clova, v.c. 90, by Linton (1913). I have not seen any material of this hybrid, though it is likely to occur in northern Br. It is apparently unrecorded elsewhere.

11 x 15. *S. caprea* L. x *S. phylicifolia* L.

a. None.

b. This is an erect shrub with twigs which are at first very thinly pubescent but soon glabrous and reddish-brown. The leaves are obovate or obovate-elliptic, dark green above, and persistently pubescent or thinly tomentose below, and with a prominent venation. I have not seen catkins.

c. This hybrid occurs on river-banks and lake-shores and in moist thickets in v.c. 88–90 and 104, and in Fe, Ga, Ge, Rs and Su.

11 x 16. *S. caprea* L. x *S. repens* L.

a. *S.* x *laschiana* Zahn.

b. This is a suberect or spreading shrub, resembling *S. repens,* with slender twigs which are at first thinly tomentellous but soon glabrous and dark reddish-brown. The leaves are obovate, sometimes broadly so, at first thinly greyish appressed-pubescent but later dark green and glabrous or subglabrous above, at first whitish and densely sericeous-tomentose but later calvescent and thinly pubescent below, and acute or often obliquely apiculate at the apex. The catkins are rather sparse and develop with or a little in advance of the leaves; the catkin-scales are blackish, subacute and densely villose. The ovaries are densely greyish-tomentose; the style is short but distinct; and the stigmas are shortly oblong and undivided. The two stamens are free. In the past this hybrid has been confused with *S. caprea* x *S. lapponum* and with *S. lapponum* x *S. repens*. Continental material is more robust than the British, resembling *S. caprea*, whereas British specimens all come much nearer to *S. repens*, with sprawling branches and small leaves.

c. *S.* x *laschiana* occurs on mountain-sides and rocky moorland in v.c. 81, 89, 92 and 108; it is also recorded from Au, Cz, Fe, Ge, No and Su.

11 × 17. *S. caprea* L. × *S. lapponum* L.

a. *S.* x *laestadiana* Hartm.

b. This is a sprawling shrub or small tree with twigs which are at first rather densely pubescent but later subglabrous and dark brown. The leaves are obovate-elliptic, rather densely greyish-tomentellous below, and have entire margins. The catkins appear with the leaves; the catkin-scales are bluntly oblong-lingulate, rather dark brown and villose. The ovaries are narrowly flask-shaped and pedicellate; the style is short but distinct; and the stigmas are shortly oblong and undivided. Male catkins have not been seen.

c. *S.* x *laestadiana* has been recorded from cliffs and rocky banks at high altitudes in v.c. 89 and 90; and also in Fe, No, Rs and Su.

19 × 11 × 17. *S. arbuscula* L. × *S. caprea* L. × *S. lapponum* L.

has been recorded from No and Su.

11 × 18. *S. caprea* L. × *S. lanata* L.

= *S.* x *balfourii* E. F. Linton was artificially produced in E. F. Linton's gardens in v.c. 9 and 11, and distributed as exsiccata. Linton at one time supposed that a specimen from Glen Isla, v.c. 90, collected by J. H. Balfour in 1837 (**K**), was *S. caprea* x *S. lanata* but subsequently (Linton, 1898) he decided that it was simply a form of *S. lanata*, which, in my own opinion, it undoubtedly is. He seems to have forgotton his second thoughts, for the supposed Glen Isla hybrid is reinstated in his monograph (Linton, 1913).

LINTON, E. F. (1898b). Experiments in cross-fertilization of Salices. *J. Bot., Lond.*, **36**: 122-124.

LINTON, E. F. (1910a). *S. caprea* x *lanata*. *Rep. Watson B.E.C.*, **2**: 257.

11 × 20. *S. caprea* L. × *S. myrsinites* L.

a. *S.* x *lintonii* G. & A. Camus (*S.* x *scotica* Druce, *nom. nud.*).

b. This is a small, spreading shrub with twigs which are at first thinly pubescent but soon glabrous or subglabrous and dark brown. The leaves are broadly ovate or obovate, thinly pubescent above, greyish and rather more densely and persistently pubescent below, and with a prominent venation. The catkins develop with the leaves and are cylindrical and erect. The ovaries are pedicellate, narrowly flask-shaped, much longer than the catkin-scales and tomentose with somewhat iridescent hairs; the style is short but distinct; and the stigmas are very short and emarginate or bifid. Male catkins have not been collected in the wild, but have been distributed by E. F. Linton from artificial hybrids cultivated in his garden at Bournemouth, v.c. 11; these cultivated plants have rather large, cylindrical catkins, with subacute, reddish, long-villose scales and two free stamens with long, glabrous filaments and reddish anthers.

c. *S.* × *lintonii* is apparently endemic to rocky mountain slopes in v.c. 88–90.

d. Artificial hybrids were made using *S. caprea* as the female parent by E. F. Linton at Bournemouth, v.c. 11, and distributed as exsiccata.

f. LINTON, E. F. (1909). *S. caprea* × *myrsinites. Rep. Watson B.E.C.,* **2**: 204.

LINTON, E. F. (1910b). *S. caprea* × *myrsinites. Rep. Watson B.E.C.,* **2**: 257.

13 × 12. *S. aurita* L. × *S. cinerea* L.

a. *S.* × *multinervis* Doell.

b. This is an erect shrub with slender twigs which are at first densely and closely grey-pubescent but later glabrous or subglabrous and dark reddish-brown, and which have the wood generally marked with numerous prominent striae. The leaves are obovate, often with an obliquely twisted apex, dark green above, ashy-grey and often rather densely and softly pubescent below, and with usually prominent venation and often strongly rugose-undulate margins; the stipules are broadly auriculate, acute and coarsely dentate. Much of the material named as this hybrid is *S. cinerea* subsp. *cinerea,* which frequently has large, persistent stipules, like those of *S. aurita* × *S. cinerea.*.

c. *S.* × *multinervis* is recorded from hedgerows, heathy scrubland and woodland margins on acid soils in scattered localities over much of BI, and also in Au, Be, Cz, Ga, Ge, He, Ho and Rm.

d. Artificial hybrids have been made by Håkansson.

e. Natural hybrid ($2n = 76$); artificial hybrid ($2n = 57$).

f. WILKINSON, J. (1944). As above.

13 × 12 × 14. *S. aurita* L. × *S. cinerea* L. × *S. nigricans* Sm.

a. *S.* × *forbesiana* Druce.

b. This is an erect shrub with twigs which are at first pubescent but soon glabrous or subglabrous, and which have obscurely and sparsely striate wood. The leaves are rather narrowly oblong-obovate, dark green and subglabrous above, greyish-green and very thinly pubescent below, and with rather prominent venation; they turn blackish on drying. The stipules are broadly auriculate, acute, toothed and persistent. The catkins develop a little in advance of the leaves. The ovaries are narrowly flask-shaped, distinctly pedicellate, longer than the catkin-scales and densely tomentose; the style is rather long; and the stigmas are linear and often cleft to base.

c. There are records from v.c. 72 and 87, and from Fe, No, Rs and Su.

13 × 12 × 15. *S. aurita* L. × *S. cinerea* L. × *S. phylicifolia* L.

was recorded and described by Linton (1913) from material collected by E. Fingland near Thornhill, v.c. 72, and by F. B. White from Woody Island in the River Tay, v.c. 88. Its occurrence in northern Br is not at all unlikely, but I have not seen specimens. There are also records from Fe and Su.

13 × 12 × 16. *S. aurita* L. × *S. cinerea* L. × *S. repens* L.

a. *S.* × *straehleri* Seem.
b. This is a small, suberect shrub with twigs which are at first densely ashy-pubescent but later glabrous or subglabrous. The leaves are oblong-lanceolate, dark green and thinly pubescent above, ashy-grey and rather more densely pubescent below, and with prominent venation. The catkins develop with the leaves and are subsessile and cylindrical; the catkin-scales are blunt, blackish and densely villose. The styles are short or almost absent; and the stigmas are shortly oblong, erect, more or less connivent, and entire or bifid.
c. I have seen no British material, but this hybrid is recorded by Dandy (1958). It is also recorded from Fe, Ge and Su, and Swedish material was used for the above description.
f. DANDY, J. E. (1958). *Op. cit.*, p. 82.
   FLODERUS, B. (1931). *S. aurita* × *cinerea* × *repens*, in HOLMBERG, O. R. *Skandinaviens Flora*, **1b(1)**: 61. Stockholm.

13 × 12 × 17. *S. aurita* L. × *S. cinerea* L. × *S. lapponum* L.

has been recorded from Fe and Su.

12 × 14. *S. cinerea* L. × *S. nigricans* Sm.

a. *S.* × *strepida* Forbes.
b. This is an erect shrub or small tree with twigs which are dark reddish-brown and at first pubescent but later subglabrous and often somewhat lustrous. The leaves are oblong or obovate, glaucescent and at first rather densely pubescent but later subglabrous or thinly pubescent below, and with rather prominent venation; they turn blackish on drying. The catkins appear with or a little in advance of the leaves. The style is distinct; and the stigmas are shortly linear and frequently cleft to the base. This hybrid is easily confused with *S. nigricans* and overlooked.
c. *S.* × *strepida* occurs on river-banks and in moist thickets, chiefly in northern Br; it has been recorded from v.c. 13, 16, 65, 69, 72, 85, 87, 88,

90, 102, H19 and H30, and also from Au, Da, Fe, Ga, Ge, He, It, No, Rs and Su.

12 × 14 × 15. *S. cinerea* L. × *S. nigricans* Sm. × *S. phylicifolia* L.

(= *S.* × *cinerioides* Druce, *nom. nud.*) has been recorded from Fe, No, Rs and Su—see also under *S. cinerea* × *S. phylicifolia.*

12 × 15. *S. cinerea* L. × *S. phylicifolia* L.

a. *S.* × *laurina* Sm. (*S.* × *wardiana* Leefe ex F. B. White).

b. This is an erect shrub with twigs which are at first thinly pubescent but soon glabrous and lustrous. The leaves are oblong or obovate, acute, lustrous dark green and thinly pubescent at first but soon glabrescent above, ashy-grey and at first pubescent but soon glabrous or pubescent only along the midrib and lateral veins below, and with a not very prominent venation. The catkins develop with or a little in advance of the leaves. The ovaries are densely white-tomentose; the style is short but distinct; and the stigmas are shortly oblong and entire or deeply cleft. Nilsson (1928) concluded that *S.* × *laurina* is one of the variants of *S. caprea* × *S. viminalis,* a conclusion which is scarcely tenable in view of the many differences between the two hybrids. The notion that *S.* × *laurina* is *S. cinerea* × *S. nigricans* × *S. phylicifolia* may possibly explain why the leaves of *S.* × *laurina* turn blackish on drying. However, were *S. nigricans* involved, one would expect the leaves to be more persistently pubescent, and the ovaries less tomentose. I have seen only one old male specimen of *S.* × *laurina* (**K**) and am by no means satisfied that this flowering portion is correctly identified, or that it "belongs" to the adjacent foliage-bearing twig.

c. *S.* × *laurina* is recorded from thickets, hedgerows and river-banks, sometimes (or perhaps usually) planted, chiefly in northern Br; it has been recorded from v.c. 14, 38, 39, 64–66, 69, 72, 82, 83, 85, 88, 90, 99, 112, H4, H39 and H40, and also from Fe, Ga, Rs and Su.

d. Nilsson has made detailed genetical studies on *S.* × *laurina.*

e. *S.* × *laurina* (2*n* = 95).

f. NILSSON, N. H. (1928). As above.

NILSSON, N. H. (1931). As above.

NILSSON, N. H. (1935). As above.

12 × 16. *S. cinerea* L. × *S. repens* L.

a. *S.* × *subsericea* Doell.

b. This is a small, sprawling shrub with twigs which are at first densely pubescent but later subglabrous and reddish-brown. The leaves are obovate, entire or subentire, dark green and thinly appressed-pubescent above, silvery-grey and at first densely appressed-sericeous but later rather thinly pubescent below, and with small, ovate-acute, caducous stipules. The catkins develop with the leaves. The ovaries are narrowly flask-shaped and distinctly pedicellate; the style is very short or almost absent; and the stigmas are oblong and undivided or variously cleft. This hybrid is difficult to distinguish from the much commoner *S. aurita* x *S. repens*, and, on this account, perhaps overlooked. Dwarfed, starved specimens of *S. cinerea* subsp. *oleifolia* are also not infrequently mistaken for *S. cinerea* x *S. repens*.

c. *S.* x *subsericea* occurs on heaths and stabilized dunes in v.c. 27–29, 48, 58, 90, 101, 104, 108 and 110, and also in Au, Be, Cz, Ga, Ge, Ho and Su.

d. Artificial hybrids were made by Håkansson; only rarely did male plants appear in the F_1, which showed meiotic abnormalities and was sterile. W. R. Linton also performed this cross at Shirley, v.c. 57, and distributed exsiccata.

e. Artificial hybrid ($2n = 57$).

f. LINTON, W. R. (1900b). *Salix cinerea* x *repens*. *Rep. B.E.C.*, **1**: 587.

12 x 17. *S. cinerea* L. x *S. lapponum* L.

= *S.* x *canescens* Fr. has been recorded from Ge and Su. White (1890) identified as this hybrid a specimen (E) said by him to be labelled "*Salix cinerea*, Carlowrie, 1838" in J. H. Balfour's handwriting. Carlowrie is about 11 miles west of Edinburgh, and quite outside the normal range of *S. lapponum*, so either the specimen is mislabelled, or White's identification incorrect, for, as White himself remarked, it is most unlikely that such a rare hybrid should have been introduced.

13 x 14. *S. aurita* L. x *S. nigricans* Sm.

a. *S.* x *coriacea* Forbes.

b. This is an erect shrub or small tree with twigs which are at first canescent but later subglabrous or glabrous and dull reddish-brown. The leaves are ovate, often with an obliquely twisted apex, dull green and thinly pubescent above, and ashy-grey and pubescent below; they turn blackish on drying. The stipules are broadly ovate, acute and dentate. The catkins develop with the leaves. The ovaries are narrowly flask-shaped, distinctly pedicellate, much longer than the catkin-scales and densely tomentose; the style is very short; and the stigmas are oblong and deeply bifid. Male catkins have not been seen.

c. *S.* x *coriacea* occurs on river-banks and in moist thickets in v.c. 62/65, 69, 72, 81, 88, 90, 107, 110 and H28, and also in Au, ?Fe, Ge, He, Po, ?Rs and ?Su.

13 x 14 x 15. *S. aurita* L. x *S. nigricans* Sm. x *S. phylicifolia* L.

a. *S.* x *saxetana* F. B. White.

b. This is an erect shrub or small tree with twigs which are at first thinly pubescent but soon glabrous and lustrous reddish-brown, and which have wood which is very sparsely striate. The leaves are obovate or elliptic, subglabrous and shining green above, and glaucous and very thinly pubescent below; they turn a little blackish on drying. The catkins develop with the leaves. The ovaries are densely tomentose; the style is rather long; and the stigmas are narrowly oblong, erect, more or less connivent and cleft almost to the base. Male catkins have not been seen.

c. *S.* x *saxetana* has been found on rocky moorland and mountain-sides in v.c. 72, 88 and 89, and also in Fe, Rs and Su.

13 x 20 x 14. *S. aurita* L. x *S. myrsinites* L. x *S. nigricans* Sm.

(=*S.* x *whitei* Druce, *nom. nud.*) was recorded by Linton (1913) from v.c. 88 (as *S. andersoniana* x *S. aurita* x *S. myrsinites*), but the material distributed is simply *S. nigricans,* and B. Floderus has redetermined the material at K as such. The identity of the sterile specimen cited by Linton is dubious, and in the absence of satisfactory evidence the hybrid should not be admitted to the British list.

13 x 15. *S. aurita* L. x *S. phylicifolia* L.

a. *S.* x *ludificans* F. B. White.

b. This is an erect shrub or small tree with twigs which are at first thinly pubescent but soon dark reddish-brown and lustrous. The leaves are oblong or obovate, occasionally twisted obliquely at the apex, dark green and glabrous or subglabrous above, and rather glaucous and at first thinly pubescent but becoming glabrous or subglabrous below. The catkins develop with the leaves. The ovaries are narrowly flask-shaped, shortly pedicellate, much longer than the catkin-scales and densely tomentose; the style is distinct; and the stigmas are oblong and deeply bifid. Male catkins have not been seen.

c. *S.* x *ludificans* has been recorded from rocky mountain-sides, lake-shores, and river-banks in mountainous areas in v.c. 64, 69, 72, 79, 87, 88 and 112, and from Fe, No, Rs and Su.

e. Hybrid ($2n = 63, 82$).

f. WILKINSON, J. (1944). As above.

13 x 16. *S. aurita* L. x *S. repens* L.

a. *S.* x *ambigua* Ehrh. (Ambiguous Sallow).

b. This is a low, sprawling shrub with twigs which are at first rather densely pubescent but later glabrous and dark reddish-brown, and which have obscurely striate wood. The leaves are narrowly oblong or obovate, at first sericeous-pubescent on both surfaces, later dark green above and silvery-grey below; the stipules are usually conspicuous, ovate-acute and persistent. The catkins develop in advance of the leaves; the catkin-scales are pale brown with a darker tip and villose. The ovaries are narrowly flask-shaped, distinctly pedicellate and densely tomentose; the style is short but distinct; and the stigmas are shortly oblong, often erect or connivent and undivided or bifid. Small, starved forms of *S. aurita* and *S. cinerea* are sometimes confused with this hybrid, but it is usually recognizable by the appressed sericeous pubescence of the under-surface of the leaves.

c. *S.* x *ambigua* occurs on heaths and moors throughout much of BI, and also in Au, Be, Cz, Da, Fe, Ga, Ge, He, Ho, Hu, It, No, Rm, Rs and Su.

13 x 17 x 16. *S. aurita* L. x *S. lapponum* L. x *S. repens* L.

has been recorded from Ge.

13 x 21 x 16. *S. aurita* L. x *S. herbacea* L. x *S. repens* L.

a. *S.* x *grahamii* Borrer ex Baker (incl. *S.* x *moorei* F. B. White). (Graham's Willow).

b. This is a dwarf, trailing shrub with twigs which are at first rather densely pubescent but soon glabrous and dark, shining brown. The leaves are broadly ovate, obovate or suborbicular, and green and persistently but very thinly pubescent on both surfaces; the stipules are sometimes conspicuous, ovate, subentire or serrate and persistent on robust shoots. The catkins develop with the leaves; the catkin-scales are shortly suborbicular-lingulate (nm. *grahamii*) or narrowly oblong-lingulate (nm. *moorei*), pale brown or tinged reddish towards the apex and sparsely villose. The ovaries are narrowly flask-shaped and glabrous or sometimes thinly villose, especially towards the apex (nm. *moorei*); the style is rather long; and the stigmas are rather short and deeply bifid. Male catkins have not been seen. Apart from the difference in the shape of the catkin-scales nm. *grahamii* and nm. *moorei* are virtually indistinguishable, and must, presumably, have the same parentage. Exactly what the parentage is remains an unsolved mystery. The presence of *S. herbacea* is agreed by all those who have examined the plant; the other parent has been variously identified as *S. lapponum, S. myrsinites, S. phylicifolia, S. nigricans* and *S. repens.* Since *S.* x *grahamii* nm. *moorei* is an Irish plant one may conclude that the participation of any of the first four of these is highly improbable. The plant does not, however, match satisfactorily with *S. herbacea* x *S. repens*; the undulate-margined leaves, often with obliquely twisted apices, suggest the presence of *S. aurita.*

c. This hybrid occurs on mountain summits in v.c. 108 and H35, and also in No.

f. BAKER, J. G. (1867). On *Salix grahami*, Borrer. *J. Bot., Lond.,* 5: 157-158. FIG.

13 x 17. *S. aurita* L. x *S. lapponum* L.

a. *S.* x *obtusifolia* Willd.

b. This is a small, spreading shrub with twigs which are at first thinly tomentellous but soon glabrous and dark, shining brown. The leaves are oblong or narrowly obovate, dark green above, rather densely grey-pubescent below, and with entire or subentire margins and rather prominent venation. The rather large, erect catkins develop with the leaves. The ovaries are subsessile and densely tomentose; the style is rather long; and the stigmas are linear and undivided or occasionally cleft to the base. B. Floderus has re-identified as this hybrid specimens previously distributed by E. F. Linton as *S. lapponum* forma and *S. lapponum* x *S. phylicifolia.*

c. *S.* x *obtusifolia* occurs on rock-ledges on mountains in v.c. 88–90 and 96 and in Fe, No, Rs and Su.

e. Hybrid (2*n* = 38).

f. WILKINSON, J. (1944). As above.

13 x 20. *S. aurita* L. x *S. myrsinites* L.

(= *S.* x *angusensis* Druce, *nom. nud., non* Reching. f.) was the parentage attributed by E. F. Linton to specimens from v.c. 90 which he distributed as exsiccata. B. Floderus has redetermined them as *S. nigricans*, which may be correct, though the subglabrous twigs and coriaceous, thinly pubescent leaves suggest *S. phylicifolia* or *S. nigricans* x *S. phylicifolia*. There is no very obvious evidence of either *S. aurita* or *S. myrsinites.*

13 x 21. *S. aurita* L. x *S. herbacea* L.

a. *S.* x *margarita* F. B. White.

b. This is a small, decumbent or ascending shrub with twigs which are at first pubescent but soon glabrous and dark, shining red-brown. The leaves are broadly obovate or suborbicular, dark green and glabrous above, slightly paler and more or less pubescent below, and with prominent and minutely or rather coarsely and bluntly serrate margins. The female catkins develop with the leaves; the catkin-scales are elongate-lingulate and villose. The ovaries are pedicellate, narrowly flask-shaped and tomentose; the style is short; and the stigmas are shortly oblong and more or less bifid. Male catkins have not been seen.

c. *S.* x *margarita* occurs on rock-ledges on mountains in v.c. 89, 90, 101 and 104, and also in No.

14 x 15. *S. nigricans* Sm. x *S. phylicifolia* L.

a. *S.* x *tetrapla* Walker.

b. This is an erect shrub or small tree with twigs which are at first pubescent but soon glabrous or subglabrous and lustrous dark brown. The leaves are ovate, obovate or oblong, usually acute, thinly pubescent but soon glabrescent and dark, shining green above, and more or less glaucous and at first pubescent but later often glabrous or subglabrous below; they become blackish on drying. The ovaries are distinctly pedicellate, narrowly flask-shaped, much longer than the catkin-scales and glabrous or more or less tomentose; the style is distinct; and the stigmas are oblong, spreading and usually deeply bifid. This is probably the commonest and most perplexing willow hybrid, completely linking the two parent species and often making their identification virtually impossible.

c. *S.* × *tetrapla* occurs on damp rocky ground, river-banks and lake-shores in northern Br; it is recorded from v.c. 64–67, 69, 80, 88–90, 92, 96 and 111, and also from Fe, Ge, No, Rs and Su.

19 × 14 × 15. *S. arbuscula* L. × *S. nigricans* Sm. × *S. phylicifolia* L.

(= *S.* × *arbusculoides* Druce, *nom. nud.*) was recorded by Linton (1913) from v.c. 83, but the evidence is unsatisfactory, and the occurrence of this hybrid must be considered questionable. There are apparently no records from elsewhere.

20 × 14 × 15. *S. myrsinites* L. × *S. nigricans* Sm. × *S. phylicifolia* L.

a. (*S.* × *myrsinitoides* Druce, *nom. nud.*).
b. This is an erect or suberect shrub with twigs which are at first very thinly pubescent but soon glabrous and dark, shining brown. The leaves are narrowly obovate, acute, dark, shining green above, glaucescent and glabrous or thinly pubescent near the midrib below, and with a prominent venation and broadly ovate, subacute stipules. The catkins develop with the leaves and are rather large, erect and cylindrical. The ovaries are shortly pedicellate and rather densely tomentose; the style is long; and the stigmas are narrowly oblong and cleft into four linear lobes. Male catkins have not been seen.
c. This hybrid occurs in damp rocky places on mountains in v.c. 88, 90 and 92, and also in Su.

14 × 16. *S. nigricans* Sm. × *S. repens* L.

a. *S.* × *felina* Buser ex G. & A. Camus.
b. This is a suberect or sprawling shrub with twigs which are at first densely pubescent but later glabrous or subglabrous and rather lustrous, dark brown. The leaves are lanceolate, dull green and thinly pubescent or subglabrous above, and greyish and pubescent below; they turn slightly blackish on drying. The erect, cylindrical catkins develop with the leaves; the catkin-scales are elongate-lingulate and mid- or dark brown. The ovaries are distinctly pedicellate and tomentose; the style is short but distinct; and the stigmas are oblong or linear and sometimes cleft to the base into two filiform lobes. Male catkins have not been seen. This description is taken from German and Finnish material.
c. This hybrid possibly occurs in damp rocky places in mountainous areas in v.c. 89. It has also been recorded from Au, Fe, Ge, He and It.

17 × 14. *S. lapponum* L. × *S. nigricans* Sm.

(= *S.* × *dalecarlica* auct., *vix* Rouy) is said to have been collected by E. S. Marshall, in 1892, at Caenlochan, v.c. 90. I have not seen any material of the hybrid, and there are no reports from elsewhere.
MARSHALL, E. S. (1892). Some plants observed in E. Scotland, July and August, 1892. *J. Bot., Lond.*, **31**: 228-236.

19 × 14. *S. arbuscula* L. × *S. nigricans* Sm.

(= *S.* × *breadalbensis* Druce, *nom. nud.*) was recorded and described by Linton (1913) from specimens collected in the neighbourhood of Ben Lawers, v.c. 88. None of the material can be referred with any degree of certainty to the hybrid, and its occurrence in Br must be considered questionable. On the Continent there are reports of this hybrid from He and Su.

20 × 14. *S. myrsinites* L. × *S. nigricans* Sm.

a. *S.* × *punctata* Wahlenb.

b. This is a low, spreading shrub with twigs which are at first thinly pubescent but later glabrous and dark, lustrous or dull brown. The leaves are dark green and glabrescent above, green and at first pubescent but later glabrous or subglabrous below, and with distinctly and rather regularly serrate margins, rather prominent venation and often rather conspicuous stipules; they turn blackish on drying. The erect, bluntly cylindrical catkins develop with the leaves. The ovaries are narrowly flask-shaped, shortly pedicellate and thinly or rather densely hairy (the hairs often distinctly iridescent); the styles are rather long; and the stigmas are long and usually divided to the base into four linear lobes. The anthers are tipped with red or crimson.

c. *S.* × *punctata* has been recorded from damp, rocky slopes at high altitudes in v.c. 88, 90, 92 and 98 and also from No and Su. It also occurred spontaneously in E. F. Linton's garden at Bournemouth, v.c. 11.

f. LINTON, E. F. (1900). *Salix myrsinites* × *nigricans*. *Rep. B.E.C.,* 1: 589.

21 × 14. *S. herbacea* L. × *S. nigricans* Sm.

a. *S.* × *semireticulata* F. B. White.

b. This is a dwarf, creeping shrub with twigs which are at first thinly villose but later glabrous. The leaves are broadly ovate, obtuse, truncate or cordate at the base, thinly villose above and below, and with rather prominent venation and subentire, narrowly revolute margins. The female catkins develop with the leaves; the catkin-scales are narrowly obovate-lingulate and long-villose. The ovaries are narrowly flask-shaped and tomentose; the style is distinct; and the stigmas are narrowly oblong and cleft into linear or filiform lobes. E. F. Linton's somewhat tentative identification of *S.* × *semireticulata* as this hybrid is reinstated here, simply for want of a more satisfactory solution. Linton (1913) cited White's plant under two distinct hybrids: *S. herbacea* × *S. nigricans* and *S. nigricans* × *S. reticulata*. It might in fact be *S. herbacea* × *S. lapponum*.

c. *S.* × *semireticulata* occurs on ledges at high altitudes on Meall Ghaordie, v.c. 88, and is apparently endemic.

14 × 22. *S. nigricans* Sm. × *S. reticulata* L.

is one of the parentages which has been suggested for *S.* × *semireticulata* F. B. White—see under *S. herbacea* × *S. nigricans*.

15 × 16. *S. phylicifolia* L. × *S. repens* L.

a. *S.* × *schraderana* Willd.

b. This is a small, spreading shrub with twigs which are at first thinly pubescent but soon glabrous and dark, shining red-brown. The leaves are small, ovate, acute, at first thinly appressed-sericeous, later glabrous and dark green above, slightly paler or glaucescent below, and with subentire or obscurely and bluntly serrate margins and obscure nervation. I have not seen catkins. This description is based solely upon a sterile specimen collected by Miss U. K. Duncan in 1959 at Delnamer, v.c. 90. W. R. Linton's specimen from Braemar, v.c. 92, is almost certainly *S. nigricans* × *S. phylicifolia*, likewise specimens collected by him and by E. S. Marshall from Glen Shee, v.c. 89. The H. H. Johnston specimens named by J. Fraser appear to be *S. cinerea* × *S. phylicifolia*.

c. *S.* × *schraderana* has been found on a roadside in v.c. 90; records from v.c. 89 and 111 are either doubtful or erroneous. It is apparently endemic, but is said to be cultivated in some Continental botanic gardens.

f. FRASER, J. (1933). As above.

17 × 15. *S. lapponum* L. × *S. phylicifolia* L.

a. *S.* × *gillotii* G. & A. Camus.

b. This is a low, spreading shrub with twigs which are at first thinly villose-pubescent but soon glabrous and lustrous dark reddish-brown. The leaves are lanceolate, acute or narrowly oblong-acute, at first thinly villose on both surfaces but soon subglabrous and shining green above and glaucous and persistently but sparsely villose-pubescent below, and with minutely and rather sharply serrate margins. The erect, cylindrical catkins develop a little in advance of the leaves; the catkin-scales are elongate-lingulate. The ovaries are shortly pedicellate, narrowly flask-shaped and densely tomentose; the style is distinct and rather long; and the stigmas are linear and undivided or cleft into two filiform lobes. Male catkins have not been seen. This description is based on Swedish material.

c. The natural occurrence of this hybrid in Br is questionable; E. F. Linton's material from Ben Lawers, v.c. 88, is probably a form of *S. lapponum*, and his second record from Glen Doll, v.c. 90, looks like *S. phylicifolia* or *S. nigricans* × *S. phylicifolia*. The plant cultivated in this country as *S.* × *gillotii* seems quite distinct from either of its presumed parents and is almost certainly not the true *S.* × *gillotii*, which was described from Ga.

f. GILLOT, F. X. (1890). *Salix . . .? Revue de Botanique, Bull. mens. Soc. franç. Bot., Toulouse*, 8: 517-518.

19 × 15. *S. arbuscula* L. × *S. phylicifolia* L.

was recorded by E. F. Linton (1913) from material collected on Ben Chaisteil, v.c. 98, by E. S. Marshall. I have not been able to trace these specimens, and suspect that, like other E. S. Marshall specimens from the same locality, they may have been no more than small-leaved *S.*

phylicifolia or *S. nigricans* x *S. phylicifolia.* The occurrence of this hybrid in Br must be considered questionable.

20 x 15. *S. myrsinites* L. x *S. phylicifolia* L.

a. *S.* x *notha* Anderss.
b. This is a low, spreading shrub with twigs which are glabrous at first but later glossy, dark brown. The leaves are obovate, acute, lustrous green above and below or faintly glaucescent below, glabrous above and below at maturity, and with finely and regularly serrate margins. The catkins develop with the leaves. The ovaries are pedicellate, narrowly flask-shaped and densely iridescent-tomentose; the style is rather long; and the stigmas are linear and deeply bifid. The anthers are reddish.
c. *S.* x *notha* has been recorded from rocky slopes at high altitudes in v.c. 88 and 90, and also from No and Su.
d. W. R. Linton made artificial hybrids using *S. phylicifolia* as the female parent at Shirley, v.c. 57, and distributed them as exsiccata. E. F. Linton made the cross at Bournemouth, v.c. 11, where spontaneous hybrids also appeared in the garden; both sorts were distributed as exsiccata.
f. LINTON, E. F. (1903b). *S. Myrsinites* x *phylicifolia. Rep. Watson B.E.C.,* 1902-3: 20-21.
LINTON, W. R. (1898). *S. myrsinites* (male) x *phylicifolia* (female). *Rep. B.E.C.,* 1: 566.
PATTON, D. (1924). The vegetation of Beinn Laoigh. *Rep. B.E.C.,* 7: 268-319.

21 x 15. *S. herbacea* L. x *S. phylicifolia* L.

has been suggested as the parentage of *S.* x *grahamii* and of *S.* x *moorei*—see under *S. aurita* x *S. herbacea* x *S. repens.*

17 x 16. *S. lapponum* L. x *S. repens* L.

a. *S.* x *pithoensis* Rouy.
b. This is a low, decumbent shrub with twigs which are at first thinly pubescent but soon glabrous and reddish-brown. The leaves are lanceolate or ovate-elliptic, dark green above, silvery-grey and rather densely appressed-pubescent below, and with rather prominent venation; they turn blackish on drying. The catkins develop in advance of the leaves. The ovaries are rather broadly flask-shaped and thinly tomentose; the style is short but distinct; and the stigmas are linear and apparently undivided. Male catkins have not been seen. The female catkins are described from artificially made hybrids distributed by E. F. Linton.
c. *S.* x *pithoensis* occurs on rocky ground on mountains in v.c. 89, and also in Fe, No and Su.

21 x 17 x 16. *S. herbacea* L. x *S. lapponum* L. x *S. repens* L.

has been recorded from No.

21 × 16. *S. herbacea* L. × *S. repens* L.

a. *S.* × *cernua* E. F. Linton.

b. This is a dwarf, trailing shrub with twigs which are at first rather densely appressed-pubescent but later glabrous and light or dark brown. The leaves are oblong or ovate, glabrous or thinly appressed-pubescent and dark, shining green above, dull green or slightly greyish and often thinly appressed-pubescent but occasionally glabrous below, and with subentire or minutely serrate margins and usually rather prominent venation. The catkins develop with the leaves; the catkin-scales are shortly ovate-lingulate, pale brown, often stained reddish towards the apex and thinly villose or subglabrous. The ovaries are narrowly flask-shaped, usually longer than the catkin-scales and glabrous or more or less densely appressed-pubescent; the style is rather short, but distinct; and the stigmas are short and deeply bifid. The two stamens are free, with yellow or reddish anthers.

c. *S.* × *cernua* occurs on rocky mountain slopes in v.c. 88, 92, 98, 101, 104 and 107, and also in No.

f. LINTON, E. F. (1894). Two new willow-hybrids. *J. Bot., Lond.*, 32: 201-203.

LINTON, E. F. (1897). *Salix cernua* Linton. *J. Bot., Lond.*, 35: 362.

18 × 17. *S. lanata* L. × *S. lapponum* L.

(= *S.* × *stuartii* Hort.) was described from foliage specimens only from v.c. 92 (Linton, 1913); it is most probably a broad-leaved form of *S. lapponum*.

DRUCE, G. C. (1930). *Salix lanata* × *lapponum. Rep. B.E.C.*, 9: 36-37.

19 × 17. *S. arbuscula* L. × *S. lapponum* L.

a. *S.* × *pseudospuria* Rouy.

b. This is a dwarf, erect or spreading shrub with twigs which are at first very thinly appressed-pubescent but soon glabrous and shining red-brown. The leaves are oblong-ovate, acute, dark green and shining or thinly pubescent above, greyish or glaucous and more or less persistently pubescent or thinly tomentellous below, and with obscure venation. The catkins develop with or somewhat in advance of the leaves. The ovaries are rather shortly flask-shaped and densely tomentose; the style is shortish, but distinct; and the stigmas are oblong and undivided or cleft to the base into linear lobes.

c. *S.* × *pseudospuria* occurs on rock-ledges on mountains in v.c. 88, 90 and 110, and also in No, Rs and Su.

19 × 21 × 17. *S. arbuscula* L. × *S. herbacea* L. × *S. lapponum* L.

has been recorded from No and Su.

17 × 20. *S. lapponum* L. × *S. myrsinites* L.

= *S.* × *phaeophylla* Anderss. has been recorded from Br, but all the British material so named has subsequently been re-identified as *S. herbacea* x *S. lapponum*, and the description given by Linton (1913) is based solely upon artificially made hybrids. The hybrid has also been recorded from No and Su, but has been omitted from recent Scandinavian Floras.

21 × 17 × 20. *S. herbacea* L. × *S. lapponum* L. × *S. myrsinites* L.

= *S.* × *eugenes* E. F. Linton was recorded by Linton (1913) from Glen Fiagh, v.c. 90. B. Floderus has subsequently re-identified one of the specimens at K as *S. herbacea* x *S. lapponum*, and the E. S. Marshall material cited by Linton appears to be the same.

21 × 17. *S. herbacea* L. × *S. lapponum* L.

a. *S.* × *sobrina* F. B. White.
b. This is a dwarf, trailing shrub with twigs which are at first thinly spreading-pubescent but soon glabrous and dark, rather lustrous brown. The leaves are ovate or obovate, dark green and often persistently, but very thinly, greyish-pubescent above and below, and with frequently prominent venation. The shortly stipitate, erect, cylindrical catkins develop with the leaves; the catkin-scales are elongate-lingulate or shortly obovate-lingulate, pale brown or tinged reddish and thinly villose. The ovaries are densely pubescent or tomentose; the style is distinct; and the stigmas are narrowly oblong and usually cleft into two linear lobes. The two stamens are free, with yellow or dark red-tipped anthers.
c. *S.* × *sobrina* occurs on ledges and rocky ground at high altitudes in v.c. 88, 90 and 92 and also in Fe, No, Rs and Su.
e. Hybrid ($2n = 38$).
f. WILKINSON, J. (1944). As above.

17 × 22. *S. lapponum* L. × *S. reticulata* L.

a. *S.* × *boydii* E. F. Linton.
b. This is a dwarf, erect shrub with twigs which are rather persistently pubescent. The leaves are suborbicular, rather densely pubescent above, densely grey-tomentose below, and with prominent venation. The catkins appear with or rather before the leaves; the catkin-scales are obovate, rounded and blackened above and with long sericeous hairs. The ovaries are shortly ovoid in the young state, tomentose and sessile; and the style and stigmas are fairly long. The suggestion by Linton (1913) that *S.* × *boydii* has *S. lapponum* and *S. reticulata* as its parents is not wholly satisfactory, as Linton himself admitted, but to date no other guess as to the origin of the hybrid has proved more acceptable. B. Floderus may be correct in regarding it as a form of *S. lapponum*.
c. *S.* × *boydii* occurs on rock-ledges at high altitudes in v.c. 90; it is apparently endemic.

21 × 18. *S. herbacea* L. × *S. lanata* L.

a. *S.* × *sadleri* Syme.

b. This is a dwarf, trailing shrub with twigs which are at first thinly or rather densely lanuginose but later glabrous and dark, shining red-brown. The leaves are suborbicular or very broadly ovate, at first thinly lanuginose but later subglabrous and green on both surfaces. The catkins develop with the leaves; they are shortly cylindrical and have a lanuginose stalk. The ovaries are glabrous and broadly flask-shaped; the style is rather long; and the stigmas are rather shortly oblong and undivided or bifid.

c. *S.* × *sadleri* occurs on rocky ground at high altitudes in v.c. 88, 90 and 92, and also in Fe, Is, No, Rs and Su.

f. SYME, J. T. B. (1875). On *Salix sadleri*, Syme, and *Carex frigida*, Allioni, recently discovered in the highlands of Scotland. *J. Bot., Lond.*, **13**: 33-35, t. 158. FIG.

18 × 22. *S. lanata* L. × *S. reticulata* L.

(= *S.* × *superata* F. B. White, *sec.* Moss) has been recorded from Br, but all the material referred to this hybrid by Linton (1913) has subsequently been re-identified by B. Floderus as *S. herbacea* × *S. lanata*. None of the records for *S. lanata* × *S. reticulata* can be accepted.

19 × 20. *S. arbuscula* L. × *S. myrsinites* L.

= *S.* × *perthensis* Druce (*S.* × *serta* F. B. White, *non* Willd.) was based on specimens in Herb. Hanbury and Herb. Hooker from the Breadalbanes, v.c. 88. A duplicate of the J. D. Hooker specimen (**K**) has larger catkins than is usual in *S. arbuscula*, and could be such a hybrid, but I have not seen wholly convincing material from Br or the Continent, where there is a record from He.

19 × 21. *S. arbuscula* L. × *S. herbacea* L.

a. *S.* × *simulatrix* F. B. White.

b. This is a dwarf, spreading shrub with twigs which are thinly pubescent at first but soon glabrous and dark, rather lustrous red-brown. The leaves are broadly ovate or suborbicular, at first very sparsely pubescent or ciliate, soon glabrous, dark, shining green above, green or slightly glaucous below, and with prominent venation and distinctly serrate margins. The catkins develop with the leaves. The ovaries are very shortly flask-shaped, sessile or subsessile and densely tomentose; the style is distinct; and the stigmas are oblong and usually cleft into four linear lobes. Male catkins have not been seen.

c. *S.* × *simulatrix* occurs on rock-ledges and mountain summits in v.c. 88 and 98 and also in He, No and Su.

19 × 22. *S. arbuscula* L. × *S. reticulata* L.

= *S.* × *ganderi* Huter ex Zahn is according to Enander the identity of a specimen in **BM** collected by R. Brown in 1793 from Ben Lawers, v.c. 88. I have seen no British material, but the hybrid could occur in the Ben Lawers area, where both parents are tolerably common. The hybrid has been recorded from Au, He, No and Su.

21 × 20. *S. herbacea* L. × *S. myrsinites* L.

is one of the parentages which has been suggested for *S.* × *grahamii* Borrer ex Bak.—see under *S. aurita* × *S. herbacea* × *S. repens.*

21 × 22. *S. herbacea* L. × *S. reticulata* L.

= *S.* × *onychiophylla* Anderss. has been recorded by Linton (1913) and others from v.c. 88, but I have seen no authentic British material. The specimens cited by Linton are wrongly identified, but those of D. Patton may be correct. Enander (I think wrongly) considered *S.* × *semireticulata* F. B. White as this hybrid (see under *S. herbacea* × *S. nigricans*). *S.* × *onychiophylla* has been recorded from Au, No and Rs.
PATTON, D. (1924). As above.

351 × 352. *Gaultheria* L. × *Pernettya* Gaudich. = × *Gaulnettya* W. J. Marchant
(by C. A. Stace)

351/1 ×352/1. *Gaultheria shallon* Pursh × *Pernettya mucronata* (L.f.) Gaudich. ex Spreng.

(= × *G. wisleyensis* W. J. Marchant, *nom. nud.*) arose in cultivation at Wisley Gardens, v.c. 17; hybrids between various non-British species of these two genera are found in the wild in Central and South America and in New Zealand.

356 × 357. *Calluna* Salisb. × *Erica* L. = × *Ericalluna* Krüssm.

(by C. A. Stace)

356/1 × 357/4. *Calluna vulgaris* (L.) Hull × *Erica cinerea* L.

= × *E. bealeana* Krüssm. is the parentage almost certainly incorrectly assigned to schizopetalous variants of *Erica cinerea.*

McCLINTOCK, D. (1965). Notes on British heaths, 2. Hybrids in Britain. *Heather Soc. Yr Bk*, **1965**: 9-17.

McCLINTOCK, D. (1971). Recent developments in the knowledge of European Ericas. *Bot. Jb.*, **90**: 509-523.

357. *Erica* L.

(by D. A. Webb)

2 × 1. *E. mackaiana* Bab. × *E. tetralix* L.

a. *E.* × *praegeri* Ostenf.

b. The great variability of *E. tetralix* makes its separation from the hybrid rather difficult (except by the examination of the pollen-grains, which are about 99% full in *E. tetralix* and 99% empty in the hybrid); the hybrid can, however, usually be recognized by its wider leaves, bushier habit, later flowering and more sparsely pubescent sepals. From *E. mackaiana* it can always be distinguished by the presence of some downy hairs on the sepals and ovary. It is completely sterile.

c. In all Irish stations of *E. mackaiana* this hybrid is plentiful; it also occurs with diminishing frequency in a zone a mile or two wide around each colony of *E. mackaiana.* Since the ovule-sterility of *E. mackaiana* in Hb is almost complete, the hybrid is presumably derived from the arrival of pollen of *E. mackaiana* (of which about 30% is good) on the stigma of *E. tetralix.* In Hs the parent species occupy rather different habitats, and the hybrid has not been observed.

d. The production of viable seed by pollinating *E. tetralix* with pollen from *E. mackaiana* has been demonstrated, but the resulting seedlings have not been raised to maturity.

e. Both parents and hybrid $2n = 24$.

f. BANNISTER, P. (1966). *Erica tetralix* L., in Biological Flora of the British Isles. *J. Ecol.*, **54**: 795-813.

CHAPPLE, F. J. (1964). *The heather garden*, p. 77. London. FIG.

McCLINTOCK, D. (1965). Notes on British heaths, 2. Hybrids in Britain. *Heather Soc. Yr Bk*, **1965**: 9-17.

OSTENFELD, C. H. (1912). The International Phytogeographical Excursion in the British Isles, 6. Some remarks on the floristic results of the excursion. *New Phytol.*, **11**: 114-127.

PERRING, F. H. and SELL, P. D. (1968). *Op. cit.*, p. 44.

SMITH, M. H. (1930). Leaf anatomy of British heaths. *Trans. Proc. bot. Soc. Edinb.*, **30**: 199-205.

WEBB, D. A. (1955). *Erica mackaiana* Bab., in Biological Flora of the British Isles. *J. Ecol.*, **43**: 319-330.

3 × 1. *E. ciliaris* L. × *E. tetralix* L.

a. *E.* x *watsonii* Benth.

b. The leaf of the hybrid is rather closer to that of *E. tetralix*. The inflorescence resembles that of *E. ciliaris* in being distinctly racemose, but the raceme is shorter, as is also the corolla. The hybrid is said to be less than 1% pollen-fertile, but apparent backcrosses occur in v.c. 9.

c. The hybrid occurs with moderate frequency almost throughout the range of *E. ciliaris* in v.c. 1 and 9, but has not been recorded from v.c. 3. A single plant was found in 1971 in v.c. H16, immediately adjacent to the small colony of *E. ciliaris*. It has also been recorded from a wide range of localities in western Ga.

d. None.

e. Both parents $2n = 24$.

f. GAY, P. A. (1960). A new method for the comparison of populations that contain hybrids. *New Phytol.*, **59**: 218-226.

McCLINTOCK, D. (1965). As above.

MAXWELL, D. F. (1927). *The low road*, pp. 69 and 71. FIG.

PERRING, F. H. and SELL, P. D. (1968). *Op. cit.*, p. 44.

RILSTONE, F. (1929). *E. ciliaris* x *Tetralix* = *E. Watsoni* Bentham. *Rep. B.E.C.*, **8**: 631.

VIGURS, C. C. and RILSTONE, F. (1935). *Erica ciliaris* L. and *E. watsoni* Benth. in Cornwall. *J. Bot., Lond.*, **73**: 89-90.

4 × 1. *E. cinerea* L. × *E. tetralix* L.

has been reported occasionally from several places, but almost certainly incorrectly identified.

1 × 8. *E. tetralix* L. × *E. vagans* L.

a. *E.* x *williamsii* Druce.

b. In habit, leaf and corolla the plant is closer to *E. vagans*, which it also resembles in the lack of anther-appendages. The anthers are, however, included, as in *E. tetralix*, which it also resembles in some details of the indumentum. The pollen is, at least in part, sterile.

c. The hybrid was described from a single plant found by P. D. Williams in 1910 near the Lizard, v.c. 1; the discoverer stated that it had been seen there nearly 50 years earlier. It has been seen six times since in the same general area, and two different nothomorphs are in cultivation.[*]

d. None.

e. *E. tetralix* $2n = 24$; *E. vagans* ($2n = 24$).

f. DAVEY, F. H. (1910). *Erica vagans* x *cinerea. J. Bot., Lond.,* **48**: 333-334.

DRUCE, G. C. (1912). *Erica tetralix* x *vagans* = *E. Williamsii,* Druce. *Rep. B.E.C.,* **3**: 24.

McCLINTOCK, D. (1965). As above.

TURRILL, W. B. and BOODLE, L. A. (1911). A hybrid heath. *Kew Bull.,* **1911**: 378-379.

7 x 2. *E. erigena* R. Ross (*E. mediterranea* auct., *non* L.) x *E. mackaiana* Bab.

has been suggested as the parentage of the enigmatic plant, *E. stuartii* E. F. Linton, once found near Roundstone, v.c. H16, but the evidence is not convincing.

PRAEGER, R. L. (1934). *The botanist in Ireland,* p. 130. Dublin.

WEBB, D. A. (1954). Notes on four Irish heaths. *Ir. Nat. J.,* **11**: 190-192.

4 x 8. *E. cinerea* L. x *E. vagans* L.

was the parentage originally and incorrectly ascribed to *E. tetralix* x *E. cinerea* = *E.* x *williamsii.*

DRUCE, G. C. (1911). The International Phytogeographical Excursion in the British Isles, 3. The floristic results. *New Phytol.,* **10**: 306-328.

358. *Vaccinium* L.

(by C. A. Stace)

2 x 1. *V. myrtillus* L. x *V. vitis-idaea* L.

a. *V.* x *intermedium* Ruthe.

b. Hybrid plants are intermediate between the parents in a wide range of features, notably stem-pubescence and -shape in section; leaf-persistence, -thickness, -shape and -serration; corolla-shape and -colour; style-length; and berry-colour. They have the upright habit of *V. myrtillus* but are often

[*] See *The Lizard,* **5**: 3-5 (1974).

more vigorous than either parent; they exist either as isolated clumps or extensive patches. In general there is very little morphological variation shown by plants, and most or all hybrids found in the wild are probably F_1s. The hybrid is partially sterile; it produces fewer flowers per plant, a smaller proportion of flowers produce fruits and there are fewer seeds per fruit than in either parent. In tests on Derbyshire material the parents showed roughly 50% seed and 65% pollen germination, and the hybrid approximately 16% and 4% respectively.

c. The hybrid is absent from most of the areas of overlap of the two species, and is common only on the north Derbyshire moors, v.c. 57, and in one area of v.c. 39 (Cannock Chase) where it was first found as a British plant in 1870. In addition it is known in v.c. 62–65 and there are old records for v.c. 69, 97 and 109, the first two probably erroneous. On the Continent it is found in Cz, Da, Fe, Ge, Po, Rs and Su. The two species show some geographical and ecological differences, but are often found intermingled and flowering at the same time. The occurrence of *V.* x *intermedium* is frequently found to be correlated with some sort of human disturbance, such as burning, drainage or path-building.

d. The hybrid has been resynthesized in Br and Fe only when *V. myrtillus* was used as the female parent (7/109 flowers), and the progeny appeared identical with natural *V.* x *intermedium*. Selfed wild *V.* x *intermedium* produced no seed, and of the four possible backcrosses only female *V.* x *intermedium* x male *V. vitis-idaea* was successful. The parents are homogamous and self-fertile. The rarity of wild hybrids and apparent absence of backcrossing is not attributable to any known single factor.

e. Both parents (2n = 24); hybrid 2n = 24.

f. AHOKAS, H. (1971). Notes on polyploidy and hybridity in *Vaccinium* species. *Ann. bot. Fenn.*, **8**: 254-256.

BROWN, N. E. (1887). *Vaccinium intermedium*, Ruthe, a new British plant. *J. Linn. Soc., Bot.*, **24**: 125-128. FIG.

GOURLAY, W. B. (1919). Notes from Cannock Chase on *Vaccinium intermedium*, Ruthe. *Proc. Trans. bot. Soc. Edinb.*, **27**: 327-333.

PERRING, F. H. and SELL, P. D. (1968). *Op. cit.*, p. 44.

RITCHIE, J. C. (1955a). A natural hybrid in *Vaccinium*, 1. The structure, performance and chorology of the cross *Vaccinium intermedium* Ruthe. *New Phytol.*, **54**: 49-67.

RITCHIE, J. C. (1955b). A natural hybrid in *Vaccinium*, 2. Genetic studies in *Vaccinium intermedium* Ruthe. *New Phytol.*, **54**: 320-335.

SHAW, G. A. (1970). *Vaccinium* x *intermedium* Ruthe in v.c. 65. *Naturalist, Hull*, **915**: 132.

5 x 4. *V. microcarpum* (Rupr.) Hook. f. x *V. oxycoccus* L.

= *V.* x *hagerupii* Löve & Löve occurs (as a 'gigas' form with a doubled chromosome number) in Da, Fe and Rs, and intermediates have been reported from He.

359. *Pyrola* L.
(by C. A. Stace)

1 × 3. *P. minor* L. × *P. rotundifolia* L.

= *P.* × *graebnerana* Seeman occurs in Ge, Scandinavia and perhaps Greenland.

360 × 359. *Orthilia* Raf. × *Pyrola* L.
(by C. A. Stace)

360/1 × 359/1. *Orthilia secunda* (L.) House × *Pyrola minor* L.

(= *P.* × *redgrovensis* Druce) was recorded from Elgin, v.c. 95, in 1922, but examination of the specimen at **OXF** by E. F. Warburg showed that it is merely a small form of *P. minor*.

DRUCE, G. C. (1923). *Pyrola secunda × minor. Rep. B.E.C.*, 6: 614-615.

WARBURG, E. F. (1949). *Pyrola minor × secunda = Redgrovensis* Druce. *Watsonia*, 1: 115.

362. *Monotropa* L.
(by D. J. Miller)

1/2 × 1/1. *M. hypophegea* Wallr. × *M. hypopitys* L.

a. None.

b. Plants are found which closely resemble *M. hypophegea* with respect to the shape of the flowers and the fruits, i.e. flowers two to three times as long as broad and the style up to two-thirds the length of the fruit, but which have the inside of the petals, filaments and sometimes ovary and fruit hairy. They have an estimated pollen fertility of over 50%.

c. Such plants occur in most areas where *M. hypophegea* occurs, except on the sand-dunes in v.c. 59 and 60; they have been mainly recorded from v.c. 17, 33 and 34 on chalky soil under beech and pine trees, and more recently from Tentsmuir sand-dunes, v.c. 86. There are a few records of such plants from Au, He, Hu and Su.

d. The Tentsmuir plant had normal meiosis with 24 bivalents and no abnormal pollen (K. Moore *in litt.* 1972).

e. *M. hypopitys* (2*n* = 48); *M. hypophegea* (2*n* = 16); intermediate from v.c. 86 2*n* = 48 (K. Moore *in litt.* 1972).

f. ANGUS, A. (1970). *Monotropa hypopitys* L.—New to Fife. *Watsonia*, **8**: 163.

PERRING, F. H. and SELL, P. D. (1968). *Op cit.*, p. 45 (For *D. J. Wicker* read *D. J. Wicks* = *Mrs. D. J. Miller*).

365. *Limonium* Mill.
(by D. J. Ockendon)

2 × 1. *L. humile* Mill. × *L. vulgare* Mill.

a. *L.* × *neumanii* C. E. Salmon.

b. Plants intermediate between the parents in characters such as scape-length, length and density of spikes, bract-length and seed-length are often considered to be hybrids. Recognition of the hybrids is difficult because of the morphological variability of the parents. British hybrids are reported to be monomorphic with fertile pollen but reduced seed-set; Danish material has some sterile pollen. Introgression is claimed at two Norfolk localities (L. A. Boorman *in litt.*).

c. Hybrids occur only where the two parents grow together, in salt-marshes in v.c. 11, 13, 19, 28 and 48, and also in Da.

d. At meiosis the hybrid possesses 16 bivalents and 2 univalents.

e. *L. humile* 2*n* = 36; *L. vulgare* 2*n* = 32, 36; hybrid 2*n* = 34. Most reports show both parents with 2*n* = 36, but Choudhuri (1942) claimed *L. vulgare* with 2*n* = 32 and the hybrid with 2*n* = 34.

f. BOORMAN, L. A. (1967). *Limonium vulgare* Mill. and *L. humile* Mill., in Biological Flora of the British Isles. *J. Ecol.*, **55**: 221-232.

CHOUDHURI, H. C. (1942). Chromosome studies in some British species of *Limonium. Ann. Bot.*, n.s., **6**: 183-217.

SALMON, C. E. (1904). Notes on *Limonium*, 2. *Limonium Neumani (L. humile × vulgare). J. Bot., Lond.*, **42**: 361-363. FIG.

3 × 5/1. *L. bellidifolium* (Gouan) Dumort. × *L. binervosum* (G. E. Sm.) C. E. Salmon

has been reported from Blakeney Point, v.c. 27/28, but the evidence is not convincing. Fraine and Salisbury, and Choudhuri were unaware that the great morphological variation of *L. binervosum* is associated with apomixis, and it is likely that the supposed hybrid is an apomictic variant of that species, although in 1951 Baker appeared to accept the existence of hybrids.

BAKER, H. G. (1951). Hybridisation and natural gene-flow between higher plants. *Biol. Rev.*, **26**: 302-337.

CHOUDHURI, H. C. (1942). As above.

FRAINE, E. de and SALISBURY, E. J. (1916). The morphology and anatomy of the genus *Statice* as represented at Blakeney Point, 1. *Statice binervosa*, G. E. Smith, and *S. bellidifolia*, DC. (= *S. reticulata*). *Ann. Bot.*, **30**: 239-282. FIG.

5/1 × 5/4. *L. binervosum* (G. E. Sm.) C. E. Salmon × *L. paradoxum* Pugsl.

was recorded from St. David's Head, v.c. 45, by Pugsley, but Baker has shown that both parents are male-sterile apomicts and that the "hybrids" are probably habitat modifications of them.

BAKER, H. G. (1951). As above.

PUGSLEY, H. W. (1932). *Limonium paradoxum* Pugsley. *J. Bot., Lond.*, **70**: 81-82.

366. *Armeria* Willd.

(by D. J. Ockendon)

2 × 1. *A. arenaria* (Pers.) Schult. × *A. maritima* (Mill.) Willd.

a. None.

b. Plants intermediate between the parents in characters such as leaf-width, flowering-head diameter and length of the outer involucral bracts and calyx-teeth are assumed to be hybrids. They are fertile.

c. Hybrids are confined to St Ouen's Bay and St Brelade's Bay, Jersey. The parents are ecologically separated, *A. maritima* occurring on rocks and in salt-marshes and *A. arenaria* on sand-dunes, but hybrids occur where the habitats of the parents meet.

d. Baker produced fertile F_1 hybrids by reciprocal crosses.

e. *A. maritima* $2n = 18$.
f. BAKER, H. G. (1951). Hybridisation and natural gene-flow between higher plants. *Biol. Rev.*, **26**: 302-337.
   SYME, J. T. B., ed. (1867). *English Botany*, 3rd ed., **7**: 159-160, t. 1155. London. FIG.

367. *Primula* L.
(by D. H. Valentine)

4 × 3. *P. elatior* (L.) Hill × *P. veris* L.

a. *P.* × *media* Petermann (*P.* × *christyi* Druce, *nom. nud.*).
b. Hybrids are intermediate between the parents in characters of the leaves, flowers, fruit and indumentum. They are moderately fertile but only occur in very small numbers in Br and introgression has not been recorded.
c. They have been found in a very few localities in wet meadows or woodland in v.c. 26, 27 and 29, sometimes with the parents, sometimes isolated. Hybrids are also recorded from Au, Cz, Ga, Ge, He, Hu and Po.
d. Artificial hybrids can be made but with great difficulty, and only with *P. veris* as the female parent. Both diploid and triploid hybrids have been synthesized, and backcrosses have been made to both parent species. F_1 hybrids have about 30% of the fertility of the parents. At meiosis, artificial diploid hybrids have mostly 11 bivalents or 10 bivalents and 2 univalents, but some trivalents and quadrivalents are occasionally formed.
e. Both parents and natural hybrid $2n = 22$; artificial hybrids, and natural hybrids from Cz ($2n = 22, 33$).
f. VALENTINE, D. H. (1952). Studies in British Primulas, 3. Hybridization between *Primula elatior* (L.) Hill and *P. veris* L. *New Phytol.*, **50**: 383-399. FIG.
   VALENTINE, D. H. (1966). The experimental taxonomy of some *Primula* species. *Trans. bot. Soc. Edinb.*, **40**: 169-180.

4 × 3 × 5. *P. elatior* (L.) Hill × *P. veris* L. × *P. vulgaris* Huds.

= *P.* × *murbeckii* Lindquist has been recorded from Da.

3 × 5. *P. veris* L. × *P. vulgaris* Huds.

a. *P.* × *tommasinii* Gren. & Godron (*P.* × *variabilis* Goupil, *non* Bast.).
b. Hybrids are intermediate between the parents in characters of the leaves, flowers and indumentum; they usually have a pedunculate inflorescence,

as in *P. veris*, but may have some basal flowers as in *P. vulgaris*. They are moderately fertile. Most natural hybrids are probably F_1s, and only in a few localities is there evidence of backcrossing or introgression. The Polyanthus Primulas are developments from crosses between *P. veris* and *P. vulgaris*. They escape from gardens and may occasionally become naturalized; a wild hybrid with *P. veris* has been recorded from v.c. 13.

c. Hybrids are widely distributed in England, round the coast of Wales, in eastern Scotland and in various parts of Hb. They occur usually as scattered individuals, most often in meadows with the *P. veris* parent. Hybrids are also recorded from Au, Da, Ga, Ge, He, It and Su.

d. Artificial hybrids have been made, but only with *P. veris* as the female parent; the reciprocal cross gives empty or imperfect seeds. The viable seeds are small and the germination is low. Artificial F_1 hybrids are vigorous and can be backcrossed to the parents; their fertility is about 30% of that of the parents. In wild hybrids, pollen fertility is about 40% and seed fertility 20%. Artificial hybrids have mainly 11 bivalents or 10 bivalents and 2 univalents at meiosis, but some trivalents and quadrivalents are occasionally formed.

e. Both parents and artificial hybrids $2n = 22$.

f. BUTCHER, R. W. and STRUDWICK, F. E. (1930). *Further illustrations of British plants*, t. 244. Ashford. FIG.

CLIFFORD, H. T. (1958). Studies in British Primulas, 6. On introgression between Primrose (*Primula vulgaris* Huds.) and Cowslip (*P. veris* L.). *New Phytol.*, 57: 1-10.

MOWAT, A. B. (1961). An investigation of mixed populations of *Primula veris* and *P. vulgaris*. *Trans. bot. Soc. Edinb.*, 39: 206-211.

PERRING, F. H. and SELL, P. D. (1968). *Op. cit.*, p. 46.

VALENTINE, D. H. (1955). Studies in British Primulas, 4. Hybridization between *Primula vulgaris* Huds. and *P. veris* L. *New Phytol.*, 54: 70-80.

WOODELL, S. R. J. (1965). Natural hybridization between the Cowslip (*Primula veris* L.) and the Primrose (*P. vulgaris* Huds.) in Britain. *Watsonia*, 6: 190-202.

4 × 5. *P. elatior* (L.) Hill × *P. vulgaris* Huds.

a. *P.* × *digenea* A. Kerner.

b. Hybrids are intermediate between the parents in characters of the leaves, flowers and indumentum; they usually have a pedunculate inflorescence, as in *P. elatior*, but may have some basal flowers as in *P. vulgaris*. They show a fairly high degree of fertility and, although many are of the F_1 type, extensive backcrossing and introgression may occur.

c. Hybrids occur in two well-defined areas in E. Anglia, one to the west and one to the east of Cambridge, in v.c. 19, 20, 25–27 and 29–31. In these areas *P. elatior* replaces *P. vulgaris*, but at the margins of the areas the species meet and interbreed. Hybrid populations are usually in *Quercus robur*–*Fraxinus* woodland on chalky boulder-clay. There is some habitat separation of the parents, *P. elatior* occurring in wetter and *P. vulgaris* in drier parts of woods. Hybrids are also known in Au, Be, Da, Ga, Ge, He and Ju.

d. Artificial hybrids can be made relatively easily with either parent as the female; thus the main natural barrier to crossing in Br is spatial. Backcrosses have been made to both parent species. Artificial F_1 hybrids have about 75% fertility of the parents. At meiosis, artificial hybrids have mostly 11 bivalents or 10 bivalents and 2 univalents, but some trivalents and quadrivalents are occasionally formed.

e. Both parents and natural and artificial hybrids $2n = 22$.

f. CHRISTY, M. (1922). *Primula elatior*: its distribution in Britain. *J. Ecol.*, 10: 200-210.

   VALENTINE, D. H. (1947). Studies in British Primulas, 1. Hybridization between Primrose and Oxlip. *New Phytol.*, 46: 229-253. FIG.

   VALENTINE, D. H. (1948). Studies in British Primulas, 2. Ecology and taxonomy of Primrose and Oxlip. *New Phytol.*, 47: 111-130.

   WOODELL, S. R. J. (1969). Natural hybridization in Britain between *Primula vulgaris* Huds. (the Primrose) and *P. elatior* (L.) Hill (the Oxlip). *Watsonia*, 7: 115-127.

jul × 5. *P. juliae* Kusn. × *P. vulgaris* Huds.

and its derivatives are extensively grown in gardens under a variety of names, of which *Primula* 'Wanda' is one of the best known. They occasionally escape from cultivation and have become naturalized in a few localities.

372. *Anagallis* L.
(by L. F. Ferguson)

2 × 3. *A. arvensis* L. × *A. foemina* Mill.

a. *A.* × *doerfleri* Ronn.

b. Hybrids are robust and sterile; the corollas generally exhibit dulled hues of the dominant parental colour, but rarely may be parti-coloured. *A. arvensis* corollas range from red (dominant) through pink, salmon and lilac to blue (recessive), but those of *A. foemina* are always blue. Both 4-celled and 3-celled glandular hairs have been recorded in hybrids.

c. This hybrid has been greatly over-recorded; it appears to have been definitely and reliably recorded only from between Wilmcote and Billesley, v.c. 38, and from Au, Cz, Ge and Po.

d. Artificial hybrids are made fairly easily; they are sterile except when the salmon-flowered variant (var. *carnea*) of *A. arvensis* is a parent, when they are fertile and on selfing show segregation of colour and other characters. Treatment of a sterile hybrid with colchicine produced a fertile shoot, suggesting sterility in this case was caused by chromosomal factors.

e. *A. arvensis* 2n = 40; *A. foemina* (2n = 40).

f. KOLLMAN, F. and FEINBRUN, N. (1968). A cyto-taxonomic study in Palestinian *Anagallis arvensis* L. *Notes R. bot. Gdn, Edinb.*, **28**: 173-186.

MARSDEN-JONES, E. M. and TURRILL, W. B. (1959). The genetics and pollination of *Anagallis arvensis* subsp. *arvensis* and *Anagallis arvensis* subsp. *foemina. Proc. Linn. Soc. Lond.*, **170**: 27-29.

MARSDEN-JONES, E. M. and WEISS, F. E. (1938). The essential differences between *Anagallis arvensis* Linn. and *Anagallis foemina* Mill. *Proc. Linn. Soc. Lond.*, **150**: 146-155.

SVEREPOVÁ, G. (1964). *Anagallis* x *doerfleri* Ronn. *Preslia*, **36**: 289-293. FIG.

382. *Centaurium* Hill

(by D. M. Moore and R. Ubsdell)

4 x 1. *C. erythraea* Rafn x *C. pulchellum* (Sweet) Druce

a. (*C.* x *wheldonianum* Druce, *nom. nud.*, *C.* x *jolivetinum* P. Fourn., *nom. nud.*).

b. Hybrids are intermediate between the parents in such characters as stem-height, leaf-shape, leaf-size, relative lengths of calyx and corolla-tube, petal-shape, petal-size and pedicel-length. They appear to be highly fertile.

c. This hybrid, growing fairly close to both parents, is recorded from the coast of v.c. 6, 18 and 60. It is also reported from Ga and He.

d. Crosses with Scandinavian parental material were readily made and F_1 plants produced highly fertile seed.

e. *C. erythraea* 2n = 40 (20, 40); *C. pulchellum* 2n = 36.

f. MELDERIS, A. (1932). Genetical and taxonomical studies in the genus *Erythraea* Rich., 1. *Acta Horti bot. Univ. latv.*, **6**: 123-156.

MELDERIS, A. (1972). Taxonomic studies on the European species of the genus *Centaurium* Hill. *Bot. J. Linn. Soc.*, **65**: 224-250.

WHELDON, J. A. and SALMON, C. E. (1925). Notes on the genus *Erythraea. J. Bot., Lond.*, **63**: 345-352.

6 × 1. *C. littorale* (D. Turner) Gilmour × *C. pulchellum* (Sweet) Druce
= *C.* × *aschersonianum* (Seemen) Hegi is reported from Ga, Ge and Ho.

4 × 2. *C. erythraea* Rafn × *C. tenuiflorum* (Hoffmans. & Link) Fritsch
= *C.* × *litardierei* Ronniger is known in Co and Sa.

4 × 6. *C. erythraea* Rafn × *C. littorale* (D. Turner) Gilmour
- a. (*C.* × *klattii* P. Fourn., *nom. nud.*; *C.* × *intermedium* (Wheldon) Druce is in fact a variant of *C. littorale*).
- b. Hybrids are intermediate between the parents in such characters as leaf-shape, inflorescence-density, stigma-shape and the relative lengths of calyx and corolla-tube. They have an irregular meiosis with unequal chromosome segregation at anaphase I, 0–10% pollen fertility and a low seed-set ($2n = 40$). Hybrids whose morphology suggests introgression towards *C. erythraea* have a regular meiosis, 80–90% pollen fertility and high seed-set ($2n = 40$). Hybrids which appear morphologically to be backcrosses to *C. littorale* can have regular or irregular meiosis, but show 80–90% pollen fertility and high seed-set ($2n = 60$); since their progeny is identical to the parent in all features they seem to be stabilized hexaploid hybrid derivatives isolated from the parent species and from the tetraploid hybrids.
- c. Hybrids are known on the coast of v.c. 59 and 60 and have also been reported in v.c. 48. Plants of uncertain status, intermediate between the parents in one or two characters, have been reported from v.c. 27 and 45 where *C. littorale* is not known to occur. The hybrid is also known in Da and Ge.
- d. Artificial hybrids are easily made, in both combinations, and the F_1 has about 60% pollen fertility.
- e. *C. erythraea* $2n = 40$ (20, 40); *C. littorale* $2n = 40$ (38, 40, 42); hybrid $2n = 40, 50–52, 54, 56–60$; artificial hybrid $2n = 40$.
- f. MELDERIS, A. (1932). As above.
  MELDERIS, A. (1972). As above.
  O'CONNOR, W. M. T. (1955). Variation in *Centaurium* in West Lancashire, in LOUSLEY, J. E., ed. *Species studies in the British flora* pp. 119-125. London.
  SALMON, C. E. and THOMPSON, H. S. (1902). West Lancashire notes. *J. Bot., Lond.*, **40**: 293.
  UBSDELL, R. (1972). The status of some intermediates between *Centaurium littorale* and *Centaurium erythraea* from the Lancashire coast. *Watsonia*, **9**: 204.
  UBSDELL, R. (1973). *A study of variation and evolution in Centaurium erythraea Rafn and C. littorale (D Turner) Gilmour.* Ph.D. thesis, University of Reading.
  UBSDELL, R. (1974). A natural hybrid in *Centaurium*. *Watsonia*, **10**: 231-232.

WHELDON, J. A. and SALMON, C. E. (1925). As above.

ZELTNER, L. (1970). Recherches de biosytématique sur les genres *Blackstonia* Huds. et *Centaurium* Hill (Gentianacées). *Bull. Soc. neuchâtel. Sci. nat.*, **93**: 1-164.

385. *Gentianella* Moench
(by N. M. Pritchard)

1 x 2. *G. campestris* (L.) Börner x *G. germanica* (Willd.) Börner

= *G.* x *macrocalyx* Čelak. has been reported from central Europe, but is doubtful.

3/1 x 1. *G. amarella* (L.) Börner *sensu lato* x *G. campestris* (L.) Börner

has been reported from Scandinavia.

3/1 x 2. *G. amarella* (L.) Börner subsp. *amarella* x *G. germanica* (Willd.) E. F. Warb.

a. *G.* x *pamplinii* (Druce) E. F. Warb.

b. Populations are very variable in overall appearance, depending on the extent to which introgression has occurred. Apparent F_1 plants are intermediate in most respects, but have large corollas closer to those of *G. germanica* and the pyramidal habit of *G. amarella*. At the other extreme, thoroughly introgressed colonies may only be recognized by having flowers somewhat larger than those of *G. amarella* combined with rather broader leaves with slightly cordate bases. Introgression seems to occur almost exclusively in the direction of *G. amarella*. Hybrids have a fairly high percentage of good pollen (53.8% and 61.7% in two samples) and the capsules are well-filled with apparently fertile seed.

c. Clearly recognizable hybrids occur on the chalk of the Chilterns and Berkshire Downs, and on the edge of Salisbury Plain, in v.c. 7, 12 and 22–24. The habitat is that of *G. germanica*, so that the hybrids are found in woodland margins and clearings, often in light scrub. In some localities in v.c. 15–17, 22 and 23 only *G. amarella* showing signs of introgression of *G. germanica* can be seen. There is, in fact, direct historical evidence that *G. germanica* and F_1 hybrids between it and *G. amarella* formerly occurred in some of those sites in v.c. 22 and 23, but there is no such evidence for v.c. 15–17. Plants resembling the British hybrids are to be found in Au, Cz, Ga, Ge, Hu and Po; in Hu and Po the other parent is *G. austriaca* (A. & J. Kerner) Holub, which replaces *G. germanica* there.

d. Artificial crosses have resulted in the production of apparently healthy seed. None of this, however, has germinated; the germination of both parent species in cultivation is extremely slow and uncertain.
e. Both parents $2n = 36$.
f. DRUCE, G. C. (1893). *Gentiana germanica*, Willd. *Rep. B.E.C.*, **1**: 379.
   DRUCE, G. C. (1896). The occurrence of a hybrid gentian in Britain. *Ann. Bot.*, **10**: 621-622.
   PRITCHARD, N. M. (1961). *Gentianella* in Britain, 3. *Gentianella germanica* (Willd.) Börner. *Watsonia*, **4**: 290-303.
   WETTSTEIN, R. von (1896). Die europäischen Arten der Gattung *Gentiana* aus der Section *Endotricha* Froel. *Pamphlets Acad. Vienna*, **1896**: 309-382.

3/1 × 5. *G. amarella* (L.) Börner subsp. *amarella* × *G. uliginosa* (Willd.) Börner

a. None.
b. Apparent F_1 plants are intermediate between the parents, most noticeably in calyx-size and -shape and in the number of internodes. Introgressed populations again vary in the direction of *G. amarella*, although the tendency of *G. uliginosa* to show annual duration may be maintained. One introgressed population showed an average pollen fertility (visual) of 70% by comparison with values of 95% and 89% for nearby colonies of *G. amarella* and *G. uliginosa* respectively. The capsules of hybrids are well-filled with apparently fertile seed.
c. All British populations of *G. uliginosa* occur in dune-slacks in South Wales (v.c. 41, 44 and 45). Most of these localities are close to colonies of *G. amarella*, and nearly all the colonies of *G. uliginosa* show signs of hybridization. Some populations of *G. amarella* on Bristol Channel coasts are atypical and may represent introgressed plants of long standing. Plants from Ge and Po often resemble British hybrid plants more closely than British *G. uliginosa*.
d. None.
e. *G. amarella* subsp. *amarella* $2n = 36$.
f. LOUSLEY, J. E. (1950). The habitats and distribution of *Gentiana uliginosa* Willd. *Watsonia*, **1**: 279-282.
   PRITCHARD, N. M. (1959). *Gentianella* in Britain, 1. *Gentianella amarella*, *G. anglica* and *G. uliginosa*. *Watsonia*, **4**: 169-193. FIG.

3/1 × 4c. *G. amarella* (L.) Börner subsp. *amarella* × *G. anglica* (Pugsl.) E. F. Warb. subsp. *cornubiensis* Pritchard

a. None.
b. *G. anglica* subsp. *cornubiensis* is distinguished from *G. amarella* mainly by its few internodes, more obtuse stem-leaves, less branched habit and considerably earlier flowering. It should probably be regarded as a distinct species. In dune-slacks in northern v.c. 1 it frequently occurs with *G.*

amarella and hybridizes with it, forming hybrid swarms in which intermediate plants of all grades appear. In these areas the flowering of the hybrid is more or less continuous from April to August, overlapping the flowering times of the two parents. The plants appear to be fertile. Hybrids probably also occur between *G. amarella* subsp. *amarella* and *G. anglica* subsp. *anglica*, but have yet to be confirmed.

c. The hybrid is known only from dune-slacks in v.c. 1.

d. None.

e. *G. amarella* subsp. *amarella* $2n = 36$.

f. PRITCHARD, N. M. (1959). As above.

389. *Cynoglossum* L.
(by C. A. Stace)

2 × 1. *C. germanicum* Jacq. × *C. officinale* L.
= *C.* × *modorense* Rech. has been recorded from Au, Cz and Rm.

392. *Symphytum* L.
(by F. H. Perring)

2 × 1. *S. asperum* Lepech. × *S. officinale* L.

a. *S.* × *uplandicum* Nyman. (Russian Comfrey).

b. Hybrids form a range of intermediates between the parents in characters of the leaves and flowers. The leaves generally lack the broadly decurrent bases of *S. officinale* but they are never cordate and petiolate as in *S. asperum*; the calyx, in bud, is 4–5 mm with acute segments, not 2–3 mm with obtuse segments, or expanding rapidly at fruiting as in *S. asperum*; the filaments equal or exceed the anthers, never shorter than them as in *S. officinale*, and the stigma is often bent at the tip; the flower-colour varies from reddish-purple to violet, but is never sky-blue as in *S. asperum* or yellowish-white as in some variants of *S. officinale*. Hybrids vary considerably in fertility, but British populations are undoubtedly derived from material which was introduced as a crop and has spread into semi-natural habitats vegetatively, as one of its parents, S. asperum, is an extremely rare introduction.

c. Hybrids occur throughout BI and have been recorded from all vice-counties with the exception of v.c. 74, 99, 108 and 109. It is the most widespread *Symphytum* in BI, characteristically appearing in the absence of either parent. It occurs on roadsides and river-banks, and generally in more obviously man-made habitats than *S. officinale*. The hybrid is also known in Au, Ge, Ho, Scandinavia and the Caucasus.

d. Artificial hybrids have been produced in Ho from Dutch material. Hybrids with purple flowers and $2n = 36$ were made by crossing *S. officinale* subsp. *uliginosum* (Kerner) Nyman ($2n = 40$) and *S. asperum* ($2n = 32$). Other hybrids with pink flowers and $2n = 40$ were made by crossing *S. officinale* ($2n = 48$) and *S. asperum*. Attempts to cross *S. officinale* ($2n = 24$) and *S. asperum* have not been successful.

e. *S. officinale* $2n = 24$ (40), 44, 48; *S. asperum* $2n = 32$; hybrid $2n = 36, 40$.

f. BUCKNALL, C. (1912). Some hybrids of the genus *Symphytum*. *J. Bot., Lond.*, **50**: 332-337.

BUCKNALL, C. (1913). A revision of the genus *Symphytum*, Tourn. *J. Linn. Soc., Bot.*, **41**: 491-556.

BUTCHER, R. W. and STRUDWICK, F. E. (1930). *Further illustrations of British plants*, p. 240. Ashford. FIG.

FAEGRI, K. (1931). Über die in Skandinavien gefunden *Symphytum*-Arten. *Bergens Mus. Årb.*, **1931**, Naturv. No. 4.

GADELLA, W. J. and KLIPHUIS, E. (1969). Cytotaxonomic studies in the genus *Symphytum*, 2. Crossing experiments between *Symphytum officinale* L. and *Symphytum asperum* Lepech. *Acta bot. neerl.*, **18**: 544-549.

PERRING, F. H. and SELL, P. D. (1968). *Op. cit.*, p. 49.

TUTIN, T. G. (1956). The genus *Symphytum* in Britain. *Watsonia*, **3**: 280-281.

WADE, A. E. (1958). The history of *Symphytum asperum* Lepech. and *S.* x *uplandicum* Nyman in Britain. *Watsonia*, **4**: 117-118.

2 x 1 x 6. *S. asperum* Lepech. x *S. officinale* L. x *S. tuberosum* L.

a. None.

b. In size and general appearance this hybrid is intermediate between *S. tuberosum* and *S.* x *uplandicum*, its presumed parents. It has the tuberous rootstock of *S. tuberosum* and the stems are hispid, not scabrid as in *S.* x *uplandicum*; the leaves are 3–4 times as long as broad, not 2–3 times as in *S. tuberosum*; the flowers are more numerous than in *S. tuberosum* but resemble them closely except that they are pale mauve in colour; and the calyx segments are lanceolate, the style is slender and the scales considerably exceed the stamens. The Arbroath plant had pollen with a large number of abortive grains.

c. This hybrid has only been reported reliably from an artificial pool near

Woodbridge, v.c. 33, in 1955, and from the foot of a railway bank near Arbroath, v.c. 90, in 1944. In both cases the two parents were present. A similar plant grows on a roadside verge near Brodie Station, v.c. 95 (*fide* M. McC. Webster and C. A. Stace).

d. None.

e. *S.* x *uplandicum* 2*n* = 36, 40; *S. tuberosum* (2*n* = *c* 72, 144).

f. DUNCAN, U. K. (1960). *Symphytum* x *uplandicum* x *tuberosum. Proc. B.S.B.I.,* 3: 407.

MILNE-REDHEAD, E. (1958). *Symphytum tuberosum* x *uplandicum. Proc. B.S.B.I.,* 3: 46.

1 x 6. *S. officinale* L. x *S. tuberosum* L.

= *S.* x *wettsteinii* Sennholz has been recorded from Au and He.

399. *Pulmonaria* L.
(by D. McClintock)

1 x 2. *P. longifolia* Bast. x *P. obscura* Du Mort. (*P. officinalis* subsp. *obscura* (Du Mort.) Murb.)

a. None.

b. The only examples of this hybrid seen by me (**BM** and **K**) are intermediate between the parents. The leaves are slightly more obtusely cuneate at the base than in *P. longifolia,* but not at all cordate. Its fertility is unknown.

c. These specimens were collected in 1912 by S. H. Bickham from Underdown, Ledbury, v.c. 36, but may have been from a garden. E. S. Marshall's plant (Linton, 1907) certainly was. W. T. Stearn (pers. comm. 1970) tells me that hybrid swarms have occurred freely in his garden, and Hegi (1927) also claimed that hybrids occurred wherever the parents met, and extended beyond the area of the parents, and Encke (1960) and others agree. But H. Merxmüller (*in litt.* 1970) considers what Gams has written in Hegi "pure speculation" and that "very occasionally F_1 hybrids might occur", but "mostly they will be sterile". He has "never seen hybrid swarms". Hybrids have been recorded from Be, Ge and Su.

d. None.

e. Both parents (2*n* = 14).

f. BICKHAM, S. H. and MARSHALL, E. ·S. (1913). *Pulmonaria angustifolia* x *officinalis. Rep. B.E.C.,* 3: 270.

ENCKE, F. (1960). In PAREY, P. *Blumengärtnerei,* 2nd ed., 2: 432. Berlin and Hamburg.

HEGI, G. (1927). *Illustrierte Flora von Mitteleuropa*, 5: 2219. Munich.

LINTON, E. F. (1907). Hybrids among British phanerogams. *J. Bot., Lond.*, 45: 268-276.

MERXMÜLLER, H. and SAUER, W. (1972). *Pulmonaria*, in TUTIN, T. G. *et al.*, eds. *Flora Europaea*, 3: 100-102. Cambridge.

400. *Myosotis* L.
(by P. M. Benoit)

4 × 1. *M. caespitosa* K. F. Schultz × *M. scorpioides* L.

a. *M.* × *suzae* Domin.

b. This hybrid is a vigorous perennial morphologically more or less intermediate between the parents. It is partially fertile and very variable, sometimes forming complex populations. It is most often distinguishable by intermediacy in corolla-size (5–7 mm diam.) and style-length (about equalling the developed nutlets), by the long racemes with small calyces which become brown and shrivelled and contain only 0–2 developed nutlets, by the corolla-tube being shorter than the calyx, and by the pollen grains being very irregular in size and shape and evidently partially sterile. The weak, elongated stems often fall over and produce perennating, vegetative shoots from the leaf axils. *M. secunda* A. Murr., which has similar-sized corollas, is quite different in the dense spreading hairs on its main stems and leaves, and in its longer pedicels, more deeply incised calyces, very short styles, and normal fertility.

c. The hybrid is known with both parents and in quantity in marshland by the River Glaslyn, and more sparingly from Harlech, Dolgellau and Bala, v.c. 48; from Colney Heath and Braughing, v.c. 20; and from Lower Shuckburgh, v.c. 38. It is probably widespread in marshes where the parent species grow together. It is also recorded from Cz and Ge, though Schuster (1967) doubted its existence.

d. None.

e. *M. scorpioides* (2n = 64, 66); *M. caespitosa* 2n = 22.

f. BENOIT, P. M. (1958). *Myosotis caespitosa* × *scorpioides*. *Proc. B.S.B.I.*, 3: 46-47.

MERXMÜLLER, H. and GRAU, J. (1963). Chromosomenzahlen aus der Gattung *Myosotis* L. *Ber. dt. bot. Ges.*, 76: 23-29.

SCHUSTER, R. (1967). Taxonomische Untersuchungen über die Serie *Palustres* M. Pop der Gattung *Myosotis* L. *Reprium nov. Spec. Regni veg.*, 76: 39-98.

1 × 7. *M. scorpioides* L. × *M. sylvatica* Hoffm.

= *M.* × *permixta* Domin has been recorded from Cz, but requires confirmation.

6 × 7. *M. alpestris* Schmidt × *M. sylvatica* Hoffm.

= *M.* × *kablikiana* Domin has been recorded from Cz, but requires confirmation.

6 × 8. *M. alpestris* Schmidt × *M. arvensis* (L.) Hill

= *M.* × *krajinae* Domin has been recorded from Cz, but requires confirmation.

8 × 7. *M. arvensis* (L.) Hill × *M. sylvatica* Hoffm.

= *M.* × *parviflora* (Schur) Domin has been recorded from Cz, but requires confirmation.

10 × 7. *M. ramosissima* Rochel × *M. sylvatica* Hoffm.

= *M.* × *bohemica* Domin has been recorded from Cz, but requires confirmation.

8 × 10. *M. arvensis* (L.) Hill × *M. ramosissima* Rochel

= *M.* × *pseudohispida* (Murr.) Domin has been recorded from Cz, but requires confirmation.

406. *Calystegia* R.Br.
(by C. A. Stace)

2 × 1. *C. pulchra* Brummitt & Heywood (*C. dahurica* auct.) × *C. sepium* (L.) R.Br.

a. None.

b. Plants more or less intermediate between the two parents are very occasionally encountered, and may be hybrids. Intermediacy is marked by flower-colour and -size, pubescence, and bracteole-shape. Some pink-flowered variants of *C. sepium* (e.g. subsp. *spectabilis* Brummitt and subsp. *roseata* Brummitt) may be derived from this parentage. All such intermediates are fertile.

c. Such plants are sparsely scattered in the London area and in the Kent and

Sussex Weald, and have been reported from v.c. 34 and 47 and from Guernsey, CI. They are also known in Scandinavia.

d. Brummitt (1963) found reciprocal crosses at least as successful as intraspecific experimental crosses. The offspring were fertile and more or less intermediate in character, but showed considerable variation which was not related to the direction of the cross.

e. Both parents $2n = 22$.

f. BRUMMITT, R. K. (1963). *A taxonomic revision of the genus Calystegia.* Ph.D. thesis, University of Liverpool.

  McCLINTOCK, D. (1972). Field meetings, 1971: Guernsey. 9th-18th July. *Watsonia*, 9: 184-186.

1 × 3. *C. sepium* × *C. silvatica* (Kit.) Griseb.

a. *C.* × *lucana* (Ten.) G. Don.

b. Hybrids are intermediate in all floral measurements, notably corolla- and stamen-length, and in the degree of inflation of the bracteoles. In many areas the variation of the intermediates is such that there is a continuous range in morphology from one parent to the other. Most intermediates are equally as fertile as the parents, but some well-marked hybrids are consistently largely sterile. Fertility of the parents is, however, also very variable, probably due to their virtually complete self-incompatibility. The hybrids may be found mixed with or close to one or both parents, or (because of their easily effected vegetative dispersal) far from either, usually on waste ground or climbing fences or hedges.

c. In areas where the two parents are common putative hybrids are frequent; this is particularly so in the London area and other parts of south-eastern England. Hybrids have been recorded from v.c. 3, 14, 16–18, 20–23, 31, 33, 39, 41, 43, 44, 46, 48, 51, 58 and 59; they are much less common in the north than in the south. They are also known in CI, Al, Gr, Hs, It, Rs, Algeria and Morocco.

d. Hybrids can quite easily be produced artificially with either parent as female. F_1 plants resemble naturally-occurring intermediates and are apparently equally as fertile as the parents. Backcrosses between natural hybrids and both parents are also easily carried out reciprocally, and in nature are presumably the cause of the wide range of variation of the hybrids.

e. Both parents $2n = 22$.

f. BANGERTER, E. B. (1967). A survey of *Calystegia* in the London area. *Lond. Nat.*, 46: 15-23.

  STACE, C. A. (1961). Some studies in *Calystegia*: Compatibility and hybridisation in *C. sepium* and *C. silvatica. Watsonia*, 5: 88-105. FIG.

1 × 4. *C. sepium* (L.) R.Br. × *C. soldanella* (L.) R.Br.

has been recorded from Chile and New Zealand.

2 × 3. *C. pulchra* Brummitt & Heywood (*C. dahurica* auct.) × *C. silvatica* (Kit.) Griseb.

a. None.
b. Intermediates are occasionally encountered and may be hybrids. They resemble *C. silvatica* very closely, but have pink corollas; they produce at least some seeds. *C. pulchra* itself is of unknown (perhaps recent hybrid) origin, and colonies of it are more frequently largely or wholly sterile than are those of *C. sepium, C. silvatica* or *C.* × *lucana.*
c. Intermediates are found in the London area and in the Kent and Sussex Weald, and in v.c. 56 and 62, but have not been noted elsewhere.
d. Brummitt (1963) found reciprocal crosses at least as successful as intraspecific experimental crosses. The offspring were less fertile than any naturally-occurring species or intermediates (pollen well under 50% fertile) but some seed was produced. The artificial hybrids were mostly more or less intermediate but showed a great deal of variation.
e. Both parents $2n = 22$.
f. BRUMMITT, R. K. (1963). As above.

416. *Verbascum* L.
(by I. K. Ferguson)

The species of this genus hybridize quite freely and in many combinations. Besides the species covered here there are several other alien species sometimes found in BI; many of these are known to hybridize in their native areas.

General References

ARTS-DAMLER, T. (1960). Cytogenetical studies on six *Verbascum* species and their hybrids. *Genetica,* **31:** 241-328.
FRANCHET, M. A. (1868). Essai sur les espèces du genre *Verbascum* et plus particulièrement sur leurs hybrides. *Mém. Soc. acad. Maine-Loire,* **22:** 65-204.
HARTL, D. (1965). *Verbascum* L., in HEGI, G. *Illustrierte Flora von Mitteleuropa,* 2nd ed., **6:** 37-61. Munich.
LAVIER-GEORGE, L. (1937). Le trichome floral et ses rapports avec l'hybridité dans le genre *Verbascum. Bull. Mus. natn. Hist. nat., Paris,* Sér. 2, **9:** 219-226.
MURBECK, S. (1933). Monographie der Gattung *Verbascum. Acta Univ. lund.,* Nov. ser., **29(2):** 1-630.
ROUY, G. (1909). *Flore de France,* **11:** 4-30. Paris.

2 × 1. V. *densiflorum* Bertol. (*V. thapsiforme* Schrad.) × *V. thapsus* L.

= *V.* x *humnickii* Franch. has been found in Ga.

3 × 1. V. *phlomoides* L. × *V. thapsus* L.

a. *V.* x *kerneri* Fritsch.
b. Hybrids are very robust and intermediate between the parents in most characters. The calyx-hairs, which are not stellate, resemble those of *V. phlomoides*. The pollen is mostly infertile and ripe seeds are rarely set.
c. This hybrid has been recorded from v.c. 21, and from Au, Cz, Ge, He, Hu, Po and Rm.
d. Kölreuter found that crosses using either parent as female produced similar, robust, sterile individuals.
e. *V. thapsus* ($2n$ = 34, 36); *V. phlomoides* ($2n$ = 34).
f. FOCKE, W. O. (1881). *Op. cit.*, p. 307.
   LOUSLEY, J. E. and KENT, D. H. (1954). Handlist of plants of the London area, 4: 201. *Lond. Nat.*, 33 (Suppl.).

4 × 1. V. *lychnitis* L. × *V. thapsus* L.

a. *V.* x *thapsi* L. (*V.* x *foliosum* Franch., *V.* x *spurium* Koch).
b. Hybrids are usually very robust. They are intermediate to varying degrees between the parents, some specimens having simple inflorescences as in *V. thapsus* but others having branched ones as in *V. lychnitis*. The hairs of the calyx and filaments, and the shape of the anthers (reniform and mediofixed), resemble those of *V. lychnitis*. The pollen is mostly infertile and the capsules have very few ripe seeds.
c. This hybrid is found occasionally where the two parents grow together, usually on dry soils, in v.c. 14–17, 20, 26, 50 (T. Edmondson *in litt.* 1973) and 52, and from Au, ?Az, Cz, Ga, Ge, Ho, Hs, It, Po and Rs.
d. Experimental reciprocal pollinations yielded 2–3 seeds per capsule, while neighbouring plants of *V. thapsus* self-pollinated produced capsules with some 700 seeds. In the 33 hybrids produced the leaves usually resembled those of *V. lychnitis* but one specimen was recorded with decurrent leaves. Flower-colour and -size was as in *V. thapsus*.
e. *V. thapsus* ($2n$ = 34, 36); *V. lychnitis* ($2n$ = 32, 34).
f. FOCKE, W. O. (1881). *Op. cit.*, p. 305.
   SYME, J. T. B., ed. (1866). *English Botany*, 3rd ed., 6: 117-118, t. 943. London. FIG.

5 × 1. V. *pulverulentum* Vill. × *V. thapsus* L.

a. *V.* x *godronii* Boreau (*V.* x *lamottei* Franch.).
b. This is intermediate between its parents in most characters, but the indumentum is persistent and not conspicuously floccose as in *V. pulverulentum*. The stem-leaves are sessile and scarcely decurrent. The inflorescence has a few branches only, near the base, and the fascicles of

flowers are crowded above but more interrupted below. The hairs of the calyx and filaments resemble those of *V. pulverulentum.* It is usually completely sterile.

c. This hybrid is rare in Br, being recorded only from v.c. 11 (in a garden), 18, 19, 27 and 28 where the two parents grow together. It is also recorded from Au, Ga, Ge, He, Hs and Hu.

d. None.

e. *V. thapsus* (2*n* = 34, 36); *V. pulverulentum* (2*n* = 32).

f. LOUSLEY, J. E. (1934). Two interesting hybrids in the British flora. *J. Bot., Lond.,* **72**: 171-175.

7 × 5 × 1. *V. nigrum* × *V. pulverulentum* Vill. × *V. thapsus* L.

was once, almost certainly incorrectly, suggested as the parentage of some plants of *V.* × *schottianum* (= *V. pulverulentum* × *V. nigrum*) found in v.c. 28.

DRUCE, G. C. and ROBINSON, F. (1919). *Verbascum nigrum* × *pulverulentum. Rep. B.E.C.,* **5**: 511-512.

7 × 1. *V. nigrum* L. × *V. thapsus* L.

a. *V.* × *semialbum* Chaub. (*V.* × *collinum* Schrad., *non* Salisb.).

b. This hybrid is intermediate between its parents in most characters but is variable. The leaves are less hairy than in *V. thapsus,* crenate and usually petiolate as in *V. nigrum.* The inflorescence is simple or weakly branched. The filament-hairs of the upper three stamens are purple, those of the lower two white. The anthers are all reniform and mediofixed as in *V. nigrum.* Hybrids appear to be highly sterile with shrivelled pollen; capsules with ripe seeds are rarely formed.

c. It appears to be fairly frequent in southern and eastern England, being recorded from v.c. 2, 8, 11–13, 17, 20, 22–26, 28, 29, 33, 45, 57 and 73. A plant appeared spontaneously with the parents in a garden in v.c. 17 in 1931. It is widespread in Europe: Au, Cz, Da, Ga, Ge, He, Ho, Hu, It, No, Po, Rm, Rs and Su.

d. Both Gärtner and Kölreuter made reciprocal crosses, and spontaneous hybrids have been recorded from experimental gardens.

e. *V. thapsus* (2*n* = 34, 36); *V. nigrum* (2*n* = 30).

f. BRITTON, C. E. and FRASER, J. (1932). *Verbascum nigrum* × *Thapsus. Rep. B.E.C.,* **9**: 835.

FOCKE, W. O. (1881). *Op. cit.,* p. 303.

SYME, J. T. B., ed. (1866). *English Botany,* 3rd ed., **6**: 118, t. 944. London. FIG.

9 × 1. *V. blattaria* L. × *V. thapsus* L.

= *V.* × *pterocaulon* Franch. has been found in Ga, Ge, He and Hu.

1 × 10. V. *thapsus* L. × V. *virgatum* Stokes

a. V. × *lemaitrei* Boreau.
b. This hybrid is intermediate between its parents. The leaves are crenate, less hairy than in V. *thapsus*, and weakly decurrent. The flowers are large, and solitary or in fascicles. The filament-hairs of the lower two stamens are purple, those of the upper three white and purple. The anthers of the lower stamens are obliquely inserted, and those of the upper ones reniform and mediofixed. The capsules are abortive.
c. There is one record of a casual plant in v.c. 38; elsewhere it occurs in Ga.
d. None.
e. V. *thapsus* ($2n = 34, 36$); V. *virgatum* ($2n = 32, 66$).
f. BAGNALL, J. E. (1893). Notes on the flora of Warwickshire. *Midl. Nat.*, 1892-3: 22.

2 × 3. V. *densiflorum* Bertol. (V. *thapsiforme* Schrad.) × V. *phlomoides* L.

= V. × *trolanderi* Rothm. has been found in central Europe.

2 × 4. V. *densiflorum* Bertol. (V. *thapsiforme* Schrad.) × V. *lychnitis* L.

= V. × *ramigerum* Link has been found in central Europe.

2 × 5. V. *densiflorum* Bertol. (V. *thapsiforme* Schrad.) × V. *pulverulentum* Vill.

= V. × *nothum* Koch has been found in Ga, Ge and He.

2 × 7. V. *densiflorum* Bertol. (V. *thapsiforme* Schrad.) × V. *nigrum* L.

= V. × *ambiguum* Lej. (V. × *adulterinum* Koch) has been found in central Europe and according to Focke as far west as England and Su, but nothing further is known of the British record. The hybrid is intermediate between its parents, particularly in the stamen characters, and, although the pollen appears fertile, ripe seed is usually not set.
FOCKE, W. O. (1881). *Op. cit.*, p. 302.

9 × 2. V. *blattaria* L. × V. *densiflorum* Bertol. (V. *thapsiforme* Schrad.)

= V. × *bastardii* Roem. & Schult. has been found in Au, Ga, Ge, He and Hu.

2 × 10. V. *densiflorum* Bertol. (V. *thapsiforme* Schrad.) × V. *virgatum* Stokes

= V. × *martinii* Franch. has been found in Ga.

4 × 3. *V. lychnitis* L. × *V. phlomoides* L.

= *V.* × *denudatum* Pfund has been found in Au, Cz, Ga, Ge, He, Hu, Po, Rm and Rs.

3 × 5. *V. phlomoides* L. × *V. pulverulentum* Vill.

= *V.* × *murbeckii* Borbás has been found in Cz, Ga, Ge, Gr, He, Hu and Ju.

7 × 3. *V. nigrum* L. × *V. phlomoides* L.

= *V.* × *brockmuelleri* Ruhmer has been found in Au, Cz, Ge, He, Hu and Rs.

9 × 3. *V. blattaria* L. × *V. phlomoides* L.

= *V.* × *fragriforme* Pfund has been found in Bu, Cz, Ga, Ge, Gr, He, Hu and Ju.

3 × 10. *V. phlomoides* L. × *V. virgatum* Stokes

has arisen in cultivation in Ho.

4 × 5. *V. lychnitis* L. × *V. pulverulentum* Vill.

a. *V.* × *regelianum* Wirtg. (*V.* × *euryale* Franch., *V.* × *pulvinatum* auct.).
b. This hybrid is intermediate between its parents; it is very robust and has low pollen-fertility and is apparently sterile. The leaves are usually petiolate and hairy, but not often floccose.
c. In Br it is recorded only from v.c. 3 and 4, but it is widespread on the Continent: Au, Ga, Ge, Gr, He, Hs, Hu, It and Ju.
d. None.
e. *V. lychnitis* (2*n* = 32, 34); *V. pulverulentum* (2*n* = 32).
f. None.

4 × 7. *V. lychnitis* L. × *V. nigrum* L.

a. *V.* × *schiedeanum* Koch.
b. This hybrid is sterile and intermediate between its parents in leaf characters. The petioles vary in length and the stems may be simple or branched. The filament-hairs are usually purple as in *V. nigrum.*
c. There are several records from southern England: v.c. 2, 13–17 and ?29. On the Continent it is widespread: Au, Be, Cz, Ga, Ge, He, Ho, Hs, Hu, It, Po, Rm and Rs.
d. Kölreuter made the cross in both directions and used both white- and yellow-flowered *V. lychnitis*; the F_1 was intermediate and sterile and showed no reciprocal differences.
e. *V. lychnitis* (2*n* = 32, 34); *V. nigrum* (2*n* = 30).
f. FOCKE, W. O. (1881). *Op. cit.*, pp. 301-302.
   SIRKS, M. J. (1916). Sur quelques hybrides artificiels dans le genre *Verbascum* L. *Archs. néerl. Sci.*, Ser. 3B, 3: 32-42.

SYME, J. T. B., ed. (1866). *English Botany*, 3rd ed., 6: 119, t. 946. London. FIG.

9 × 4. *V. blattaria* L. × *V. lychnitis* L.

= *V.* × *muehlenbeckii* Godr. has been found in Au, Cz, Ga, Ge, He and Hs.

4 × 10. *V. lychnitis* L. × *V. virgatum* Stokes

has arisen in cultivation in Ho and is doubtfully recorded from Hs.

7 × 5. *V. nigrum* L. × *V. pulverulentum* Vill.

a. *V.* × *wirtgenii* Franch. (*V.* × *schottianum* Schrad., *V.* × *mixtum* Rom.).
b. This is also intermediate between its parents in most characters; the inflorescence may be simple or branched. The filament-hairs are usually purple as in *V. nigrum*.
c. There are records for v.c. 2, 26–28 and ?38, and it is known in Au, Cz, Ga, Ge, He, Hu and It.
d. None.
e. *V. nigrum* ($2n = 30$); *V. pulverulentum* ($2n = 32$).
f. SYME, J. T. B., ed. (1866). *English Botany*, 3rd ed., 6: 118-119, t. 945. London. FIG.

9 × 5. *V. blattaria* L. × *V. pulverulentum* Vill.

= *V.* × *macilentum* Franch. has been found in Ga, Ge, Gr, He, Hu and Ju.

9 × 7. *V. blattaria* L. × *V. nigrum* L.

a. *V.* × *intermedium* Rupr.
b. This is again intermediate between its parents in many characters. The leaves and stems are more or less glandular-pubescent, the inflorescence is usually simple with flowers solitary or in groups of 2 to 4, and the pedicels are longer than the calyx. The anthers of the lower two stamens are decurrent on their filaments as in *V. blattaria*.
c. An early record for Ingoldmells, v.c. 54, appears to be the only known occurrence in Br. On the Continent it is recorded from Bu, Cz, Ga, Ge, He, Hu, Lu, Po and Rm.
d. Kölreuter made the cross in both directions; the leaves of the F_1 were more obtuse when *V. blattaria* was used as the female parent. The plants were sterile.
e. *V. nigrum* ($2n = 30$); *V. blattaria* ($2n = 30, 32$).
f. FOCKE, W. O. (1881). *Op. cit.*, p. 299.
REYNOLDS, B. (1910). Lincolnshire plants. *J. Bot., Lond.*, 48: 57.

7 × oly. *V. nigrum* L. × *V. olympicum* Boiss.

(= *V.* × *oxoniense* Druce, *nom. nud.*) was recorded by Druce from a garden in Oxford, v.c. 23, close to both parents. It has also been synthesized in

Su. The leaves may be linear-lanceolate and sharply crenate or lanceolate and crenate; both the Oxford and Swedish plants have flowers of intermediate size with pale purple filament-hairs, and are sterile.
DRUCE, G. C. (1918). *Verbascum nigrum* L. x *olympicum* Boiss., *hybr. nov. Rep. B.E.C.*, 5: 39.

9 × 10. *V. blattaria* L. × *V. virgatum* Stokes
has arisen in cultivation in Ho.

424 × 416. *Scrophularia* L. × *Verbascum* L. = × *Scrophulariverbascum* P. Fourn.
(by C. A. Stace)

A plant said to be a hybrid between *Verbascum blattaria* and an unknown species of *Scrophularia* has been reported from Ga, but is unlikely to be correctly identified.

420. *Linaria* Mill.
(by C. A. Stace)

2 × 3. *L. purpurea* (L.) Mill. × *L. repens* (L.) Mill.
a. *L.* × *dominii* Druce.
b. Hybrid plants are intermediate in habit, inflorescence-denseness, flower-colour, and size and shape of the corolla-spur; the rootstock is non-creeping. The corolla is darker in colour than in *L. repens,* but has a distinctly darker venation. The seeds were infertile in Druce's material, but apparently good seeds were produced by a Berkshire specimen (W. M. Keens *in litt.* 1972).
c. Plants of this parentage appeared in Druce's garden at Oxford, v.c. 23, in 1912, alongside the two parents, and were still there at least until 1930

(**BM**). It has been recorded since on waste ground and railways in v.c. 22, 30, 57 and 59 and probably elsewhere. As natives these two species are not sympatric, but Dillemann noted that seed collected from *L. purpurea* in the Paris and Copenhagen Botanic Gardens was in fact seed of this hybrid combination.

  d. Reciprocal crosses have both given rise to similar F_1 hybrid offspring, the description of which matches that given above.
  e. Both parents ($2n = 12$).
  f. DILLEMANN, G. (1951). Notes sur quelques hybridations dans le genre *Linaria* et remarques sur les hybrides obtenus. *Bull. Mus. nat. Hist. nat., Paris*, **23**: 140-145.
     DRUCE, G. C. (1913). *Linaria purpurea* x *repens* = x *L. Dominii, mihi. Rep. B.E.C.*, **3**: 168-169.

3 x 4. *L. repens* (L.) Mill. x *L. vulgaris* Mill.

  a. *L.* x *sepium* Allm. (*L.* x *baxteri* Druce).
  b. Hybrid populations exist either as one form or as several forms variously intermediate between the two parents. Habit and flower characters are the best features by which to determine the degree of intermediacy. Characters of *L. vulgaris* are the large flowers, yellow corolla with orange palate, and long spur; and those of *L. repens* the small flowers, lilac corolla with violet veins and an orange spot on the palate, and short spur. F_1 plants, which are commonest, usually have a pale yellow corolla of intermediate size and with violet veins. They are highly fertile and backcross readily. In some situations the whole series of intermediates between the two species exists, at one extreme differing from *L. repens* only in a pale yellow colour on the corolla, and at the other from *L. vulgaris* only in the pale violet corolla-veins.
  c. The hybrid is scattered throughout England and Wales as far north as v.c. 48 and 53, with isolated localities in v.c. 60 and H6 (Perring and Sell, 1968), v.c. 66 (G. G. Graham *in litt.*, 1971), and v.c. 58 (T. Edmondson *in litt.* 1973), but it is absent from many areas where the two species occur together, especially in northern Br. It is particularly frequent on dry, open banks, especially by roads or railways. It is known in Be, Da, Ga, Ge, Ho, No and Su, but it is apparently only common in parts of Ga.
  d. Both species are self-incompatible, which may account for the frequency of hybrids where one parent is uncommon. The hybrid has been synthesized on the Continent several times from reciprocal crosses, although with some difficulty when *L. repens* is used as the female parent. Reciprocal crosses do not differ in appearance; F_1 plants resemble the commonest (half-way) intermediates found in the wild. F_2 plants and backcrosses have also been obtained.
  e. Both parents ($2n = 12$).
  f. DILLEMANN, G. (1949a). Remarques sur l'hybridation spontanée de *Linaria vulgaris* Mill. et de *L. striata* DC. dans la nature. *Bull. Soc. bot. Fr.*, **96**: 48-49.

DILLEMANN, G. (1949b). Hybrides réciproques des *Linaria vulgaris* Mill. et *L. striata* DC. et identification expérimentale de l'hybride x *L. intermedia* Babey. *Bull. Soc. bot. Fr.*, **96**: 171-172.

DILLEMANN, G. (1951). As above.

DRUCE, G. C. (1893). *L. repens* x *vulgaris*. *Rep. B.E.C.*, **1**: 380-381.

DRUCE, G. C. (1896). The hybrids of *Linaria repens* and *L. vulgaris* in Britain. *Ann. Bot.*, **10**: 622-623.

FOCKE, W. O. (1881). *Op. cit.*, p. 311.

PERRING, F. H. and SELL, P. D. (1968). *Op. cit.*, p. 50.

3 x 6. *L. repens* (L.) Mill. x *L. supina* (L.) Chazelles

a. *L.* x *cornubiensis* Druce.

b. The hybrid appears similar to *L. supina* in stature; the flowers are very pale yellow with pale violet veins, but the herbarium specimens seen are of poor quality and it is impossible to make out further details. It was reported that the seeds were barren. The specimen in herb. Druce (**OXF**) was annotated in 1926 by Thellung: "*Linaria repens* x altera species forsan *L. supina.*"

c. This was collected by L. Medlin from Par, v.c. 2, where it grew with both parents in 1925 (**OXF**) and 1930 (**BM**), but it has apparently not been found there since. It is not known elsewhere.

d. Both parents are self-sterile according to several authors, but Bruun reported that *L. supina* was self-fertile. He divided *Linaria* into four groups which he claimed were inter-sterile. The previous three species fell into one group and *L. supina* into a separate one.

e. Both parents ($2n = 12$).

f. BRUUN, H. G. (1937). Genetical notes on *Linaria*, 1-2. *Hereditas*, **22**: 395-400.

DRUCE, G. C. (1926). *Linaria repens* x *supina*, *nov. hybr. Rep. B.E.C.*, **7**: 998.

424. *Scrophularia* L.
(by C. A. Stace)

2 x 3. *S. auriculata* L. (*S. aquatica* L.) x *S. umbrosa* Dumort. (*S. alata* Gilib.)

a. *S.* x *hurstii* Druce.

b. Hybrid plants have been identified in areas where the two parents grow close together. They are said to be intermediate in a whole range of characters, including the staminode. Putative hybrids from v.c. 8 collected by C. P. Hurst in 1915 were said to be luxuriant and variable, sometimes nearer one parent and sometimes nearer the other. Of the specimens from here in herb. Druce (**OXF**) two appear to be normal *S. umbrosa* and two are affected by mildew and have badly malformed inflorescences. No flowers are present but all the specimens bear well-formed fruit and seeds. Goddijn and Goethart, who annotated the sheets in 1931, considered that the plants were normal or diseased examples of *S. umbrosa*, though they noted the anomalous staminode-shape.

c. This hybrid was described from banks of the River Shalbourne between Hungerford and Shalbourne, v.c. 8 (*fide* Hurst *in litt.*, **OXF**; not v.c. 22 as published by Druce). It was also reported by Praeger from the two main localities of *S. umbrosa* in Hb: River Liffey, v.c. H21; and Lough Erne, v.c. H33.

d. Godijn and Goethart (1931 *in sched.*, **OXF**) mentioned that Hurst's specimens were not at all like the artificial hybrid, which was completely intermediate in morphology and "almost absolutely sterile". Vaarama and Hiirsalmi (1967), however, stated that hybrids of the cross *S. umbrosa* $2n = 26$ x *S. aquatica* $2n = 78$ were fertile, had $2n = 52$, and closely resembled $2n = 52$ *S. umbrosa*. They suggested that plants of the last sort might have arisen in the wild in this way.

e. *S. umbrosa* ($2n = 26, 52$); *S. auriculata* $2n = c$ 40, 78, 80.

f. DRUCE, G. C. (1916). *Scrophularia alata* x *aquatica* = x *S. hurstii nov. hyb. Rep. B.E.C.*, 4: 204-205.

DRUCE, G. C. (1928). *Scrophularia. Rep. B.E.C.*, 9: 568-570.

PRAEGER, W. L. (1951). *Op. cit.*, p. 10.

VAARAMA, A. and HIIRSALMI, H. (1967). Chromosome studies on some Old World species of the genus *Scrophularia. Hereditas*, 58: 333-358.

4 x 3. *S. scorodonia* L. x *S. umbrosa* Dumort. (*S. alata* Gilib.)

= *S.* x *towndrowii* Druce was identified in 1926 in Towndrow's garden at Malvern Wells, v.c. 37, growing near the two putative parents from which it was supposed to have arisen. It was said to be intermediate in many characters, including the staminode (reniform), and to produce fruits very freely; the fruits were smaller than in the parent and mainly sterile. The specimen in herb. Druce (**OXF**) is badly affected with mildew and the inflorescence has an abnormal stunted appearance. The plant appears quite glabrous and scarcely resembles *S. scorodonia* in any vegetative features, but the staminode certainly is reniform in outline. The specimen was annotated in 1931 by Goddijn and Goethart, who saw no evidence of *S. scorodonia* and suggested it was a teratological form *S. ehrhardtii (S. umbrosa)*.

DRUCE, G. C. (1927). *Scrophularia alata* x *Scorodonia* = x *S. towndrowi* Dr. *Rep. B.E.C.*, 8: 312-313.

DRUCE, G. C. (1928). As above.

425. *Mimulus* L.
(by R. H. Roberts)

1 × 2. *M. guttatus* DC. × *M. luteus* L.
a. None.
b. The hybrids are robust and often form extensive clones. They are intermediate between the parents in leaf- and corolla-shapes. The lobes of the corolla are variously spotted with reddish-brown or purple spots or blotches, and the calyx, pedicels and often the upper nodes and leaf-bases are densely to sparsely glandular-puberulent. Their pollen is highly sterile and no seeds are set. Meiosis is very irregular; McArthur studied wild British hybrids and found 6–15 bivalents and an occasional trivalent in each pollen mother cell, the rest of the chromosomes being unpaired.
c. The hybrids occur on the gravelly banks of streams and on river-shingle, often in the absence of either parent. They have been recorded in many places throughout BI, but their distribution is incompletely known because of confusion with *M. luteus*. The parents are not sympatric in their native areas of distribution.
d. Artificial hybrids are readily made with *M. guttatus* as the female parent, much less readily with *M. luteus* as the female parent. No backcrosses have been obtained.
e. *M. guttatus* 2n = 28, (28, 30, 32, 56); *M luteus* (2n = 60, 62, 64); wild hybrid 2n = 45; artificial hybrid (2n = 44, 45, 60).
f. McARTHUR, E. D. (1974). The cytotaxonomy of naturalized British *Mimulus. Watsonia,* 10: 155-158.

MIA, M. M. and VICKERY, R. K. (1968). Chromosome counts in section *Simiolus* of the genus *Mimulus* (Scrophulariaceae), 8. Chromosomal homologies of *M. glabratus* and its allied species and varieties. *Madroño,* 19: 250-256.

MUKHERJEE, B. B. and VICKERY, R. K. (1962). Chromosome counts in the section *Simiolus* of the genus *Mimulus* (Scrophulariaceae), 5. The chromosomal homologies of *M. guttatus* and its allied species and varieties. *Madroño,* 16: 141-155.

ROBERTS, R. H. (1964). *Mimulus* hybrids in Britain. *Watsonia,* 6: 70-75.

cup × 1. *M. cupreus* Dombrain × *M. guttatus* DC.
a. *M.* × *burnetii* S. Arnott.
b. Hybrids are intermediate between the parents in stature, leaf-shape and corolla-shape. The corolla is copper-coloured with a lighter throat which is spotted with small, red dots, and the calyx, pedicels and often the upper nodes and leaf-bases are sparsely to densely glandular-puberulent. Their pollen is highly sterile and no seeds are set. A variant of this hybrid having a corolloid calyx has pollen with a higher proportion of fully-developed grains, and produces a few seeds after selfing, but these fail to germinate.

c. Hybrids occur on stream-banks and river-shingle in a few localities from v.c. 64 northwards to v.c. 112, mostly in the absence of either parent. The parents are not sympatric in their native areas of distribution.

d. Artificial hybrids are readily obtained with *M. guttatus* as the female parent, but not with *M. cupreus* as the female parent. No backcrosses were obtained with the single-flowered hybrids, but the double-flowered plant produced a few viable seeds after pollination with *M. guttatus*. These backcross plants also gave a low seed-set after pollination with *M. guttatus*.

e. *M. guttatus* 2n = 28, (28, 30, 32, 56); *M. cupreus* (2n = 62, 64); wild hybrid 2n = 45.

f. McARTHUR, E. D. (1974). As above.

ROBERTS, R. H. (1968). The hybrids of *Mimulus cupreus*. *Watsonia*, **6**: 371-376.

VICKERY, R. K., COOK, K. W., LINDSAY, D. W., MIA, M. M. and TAI, W. (1968). Chromosome counts in section *Simiolus* of the genus *Mimulus* (Scrophulariaceae), 7. New numbers for *M. guttatus, M. cupreus* and *M. tilingii. Madroño,* **19**: 211-218.

cup × 1 × 2. *M. cupreus* Dombrain × *M. guttatus* DC. × *M. luteus* L. is likely to occur in some places in Br, but would be extremely difficult to distinguish from *M. guttatus* × *M. luteus*. *M. luteus* and *M. cupreus* are completely interfertile, and such hybrids have been successfully crossed with *M. guttatus* to produce a highly variable and sterile triple hybrid.

McARTHUR, E. D. (1974). As above.

ROBERTS, R. H. (1968). As above.

VICKERY, R. K. *et al.* (1968). As above.

426. *Limosella* L.

(by C. A. Stace)

1 × 2. *L. aquatica* L. × *L. australis* R.Br. (*L. subulata* Ives)

a. None.

b. Hybrids are vegetatively more vigorous and produce more flowers than either parent. They are intermediate in leaf-shape and in many flower characters, but in other features closely approach one or the other parent, and most characters show considerable variation between the parental extremes. The hybrid more regularly survives the winter than either parent, but never produces seeds.

c. The hybrid is known only from Morfa Pools, near Port Talbot, v.c. 41, sometimes in great abundance on mud among both parents.
d. None.
e. *L. aquatica* 2*n* = 40; *L. australis* 2*n* = 20; hybrid 2*n* = 30.
f. GLÜCK, H. (1934). *Limosella*-Studien. Beiträge zur Systematik, Morphologie und Biologie der Gattung *Limosella. Bot. Jb.,* 66: 488-566.
VACHELL, E. and BLACKBURN, K. B. (1939). The *Limosella* plants of Glamorgan. *J. Bot., Lond.,* 77: 65-71. FIG.

430. *Veronica* L.
(by S. M. Walters)

2 × 3. *V. anagallis-aquatica* L. × *V. catenata* Pennell
a. *V.* × *lackschewitzii* Keller.
b. Since the differences between the parent species are relatively slight, and the variation, particularly in *V. anagallis-aquatica*, is large, the sterility is the most obvious taxonomic character of the F_1 hybrid. Coupled with the high sterility is a robust and vegetatively vigorous habit. Whilst the easily-recognized hybrids are sterile, in some localities where both species are present much more complex hybridization is found. Marchant (1970) described a topodeme near Barrington, v.c. 29, where fertility of pollen varied from less than 3% to more than 99%.
c. Marchant (1970) found the sterile hybrid common in England, Wales and Hb, but less so in Scotland; he gave records for v.c. 8, 9, 11, 13, 17, 20, 23, 24, 28, 29, 39/57, 41, 59, 64, 66, 68-71, H2 and H6. On the Continent the hybrid is widespread from Scandinavia to Ju and Rm. *V. catenata* is restricted to open, muddy habitats with little or no flowing water, while *V. anagallis-aquatica* is tolerant of a much wider habitat range. Marchant found that sterile F_1 hybrids were at a selective advantage in shallow, flowing streams because of their vegetative vigour.
d. Marchant (1970) produced the F_1 hybrid artificially from seven separate crossings, but a further six attempts failed. No reciprocal differences were found. All the F_1 plants were sterile and resembled the wild hybrids. The sterility was incomplete, however, and in F_2 and later generations fertility was much increased.
e. Both parents 2*n* = 36; hybrid 2*n* = 35, 36.
f. KELLER, J. (1942). Species sect. *Beccabungae* Griseb. genus *Veronica* L. in Hungaria sponte crescentes. *Bot. Közl.,* 39: 137.

MARCHANT, N. G. (1970). *Experimental taxonomy of Veronica section Beccabungae Griseb.* Ph.D. thesis, University of Cambridge.

SCHLENKER, G. (1936). Experimentelle Untersuchungen der Sektion *Beccabungae* Griseb. der Gattung *Veronica. Flora, Jena,* **130**: 305-350. FIG.

WEIMARCK, H. (1963). *Skånes flora,* p. 574. Lund.

WILLIAMS, I. A. (1929). A British *Veronica* hybrid. *J. Bot., Lond.,* **67**: 23-24.

9 × 8. *V. longifolia* L. × *V. spicata* L.

is said to be the parentage of some of the garden plants of the *V. spicata* group which are quite frequently found more or less naturalized as garden escapes or outcasts. This hybrid is known in the wild in Fe, Ru, Su and parts of central Europe.

431. *Hebe* Commers.

(by P. S. Green)

ell × spe. *H. elliptica* (Forst. f.) Pennell × *H. speciosa* (R. Cunn. ex A. Cunn.) Andersen

a. *H.* × *franciscana* (Eastw.) Souster (*H.* × *lewisii* auct.).

b. This is an evergreen shrub, to about 1 m, with narrowly obovate-elliptic, obtuse leaves 4–5 cm long. The flowers are bluish or purplish, 10–12 mm diameter, and borne in terminal axillary racemes. It is apparently fertile.

c. This shrub is naturalized in coastal areas of v.c. 1–3 (including the Isles of Scilly) and CI, where it has escaped from cultivation but is sometimes self-sown.

d. It was first synthesized in a garden in Edinburgh about 100 years ago, and is widely cultivated. It is not known in the wild in New Zealand, where its parents are native. *H. speciosa* is naturalized in Hb.

e. Both parents (2*n* = 40).

f. CLARKE, D. (1973). *Hebe* × *franciscana* (Eastw.) Souster, in BEAN, W. J. *Trees and shrubs hardy in the British Isles,* 8th ed., **2**: 331-332. London.

   GREEN, P. S. (1973). *Hebe* × *franciscana* (Eastwood) Souster, not *H.* × *lewisii,* naturalized in Britain. *Watsonia,* **9**: 371-372.

433. *Rhinanthus* L.
(by A. S. Lean and A. J. E. Smith)

2 × 1. *R. minor* L. × *R. serotinus* (Schönh.) Oborny
(? = *R.* × *fallax* (Wimm. & Grab.) Sterneck has been reported from a few localities in Br (e.g. v.c. 9, 96 and 105) from time to time, but all herbarium specimens examined have proved to be large variants of *R. minor*. There are also records from Au, Ga, Ge and Ho.

2b × 2m. *R. minor* L. subsp. *borealis* (Sterneck) Sell (*R. borealis* (Sterneck) Druce) × *R. minor* L. subsp. *monticola* (Sterneck) O. Schwarz (*R. monticola* (Sterneck) Druce) = *R.* × *gardineri* Druce
2b × 2s. *R. minor* L. subsp. *borealis* (Sterneck) Sell (*R. borealis* (Sterneck) Druce) × *R. minor* L. subsp. *stenophyllus* (Schur) O. Schwarz (*R. stenophyllus* Schur)
These two fertile hybrids occur in Scotland where the parents meet. We prefer to recognize the three species as subspecies of *R. minor*, and Sell suggested that in addition the two hybrids, which form fairly constant populations, should together be considered a further subspecies, subsp. *lintonii* (Wilmott) Sell.
SELL, P. D. (1967). *Rhinanthus* L., in Taxonomic and nomenclatural notes on the British flora. *Watsonia*, 6: 298-301.

435. *Euphrasia* L.
(by P. F. Yeo)

E. salisburgensis Funck is a tetraploid ($2n = 44$) and the only representative in BI of subsection *Angustifoliae* (Wettst.) Joerg. All the other species (*E. officinalis* L. agg.) belong to subsection *Ciliatae* Joerg.; six of them (species 1/19 to 1/24) are diploids ($2n = 22$) and the rest (species 1/1 to 1/18 and *E. stricta* D. Wolff ex J. F. Lehmann) are tetraploids ($2n = 44$). As a general rule, where any two species of *Euphrasia* alike in chromosome number occur together, hybrids are likely to be found. This situation, in fact, contributes largely to the poor definition of species in the genus. It means that it is impracticable to give here full details for every known hybrid, as this would be excessively repetitive and probably incomplete anyway. In addition to hybridization on one chromosome-level, hybridization takes place across the diploid–tetraploid boundary.

Essentially there are three types of field-situation arising from hybridization:

i.   abundant hybrids, showing great variation, occurring in areas of disturbed or intermediate habitats and often broken up into small patches of relatively uniform plants, the parents usually growing nearby (hybrid swarms);

ii. extensive populations of comparatively uniform plants apparently of hybrid origin, occupying an area from which the putative parents are absent and often consistently occupying a particular habitat (incipient speciation);

iii. populations obviously referable to a particular species which appear to have selectively absorbed certain characters of one or more other species (introgression).

Hybrid swarms are most frequent between closely related species in which the F_1 has a fairly high fertility, for example *E. confusa* Pugsl. x *E. tetraquetra* (Bréb.) Arrond. (*E. occidentalis* Wettst.) and *E. nemorosa* (Pers.) Wallr. x *E. pseudokerneri* Pugsl.

Incipient speciation occurs most strikingly through hybridization between diploids and tetraploids. Recognized cases of this are so far all diploid: genes are presumed to pass into a diploid from a tetraploid and modify it in such a way that it comes to occupy a distinct habitat. A species which has probably arisen in this way is *E. vigursii* Davey, from *E. anglica* Pugsl. x *E. micrantha* Reichb. Plants rather similar to *E. vigursii* but probably derived from *E. anglica* x *E. nemorosa* occur in widely scattered areas of southern England, but have not received specific recognition. Incipient species can also arise from the crossing of two species with the same chromosome number. Thus the mountain-flush species *E. scottica* Wettst., which is rather rare in Wales, crosses readily with the short-turf species *E. confusa* and many streamside and flush colonies in Wales are of this hybrid rather than of *E. scottica*. A less definite example is that of *E. confusa* x *E. nemorosa*; it seems that in certain agricultural areas bordering on moorland and heathland where *E. confusa* is common this hybrid is much commoner than, or totally replaces, *E. nemorosa*. These two hybrids represent a slightly different case from those in which the parents are unlike in chromosome number, because they appear to replace one parent rather than invade a new habitat. They are distinguished from introgressants only by being more strictly intermediate between the parents. A species which has probably arisen in this way is *E. campbelliae* Pugsl., its parents being *E. marshallii* Pugsl. and *E. micrantha*.

Introgression seems to occur mainly where there are strong barriers to interbreeding. It is quite frequent in diploids (in Br mostly *E. anglica*); the diploid retains its essential characters but is modified, most often in leaf-shape, in the direction of tetraploids, which generally tend to have finer, deeper leaf-teeth. Frequently the species which is the source of the introgression no longer grows with the diploid, e.g. *E. anglica* in Windsor Great Park, v.c. 22, resembles *E. confusa* in habit and leaf-shape but *E. confusa* is not known in the area; and a strongly introgressed *E. anglica* was found at Farley Heath, v.c. 17, in 1956, unaccompanied by any other *Euphrasia* species. In Hb species from within the area of distribution of *E. salisburgensis* quite often show signs of the leaf-shape and fine, wiry stems of that species, even though there is a strong breeding barrier between it and *E. officinalis* agg. A very small flower-size is also a partial barrier to interbreeding in *Euphrasia*, especially as it is always accompanied by arrangements which promote self-pollination. A probable instance of introgression into a small-flowered species is in *E. scottica*, which in its apparently most typical form has anthocyanin pigmentation more strongly developed on the underside of the leaf than on the upper; many populations lack this character, possibly through hybridization. Similarly, some of the variation in leaf-shape and corolla-size, -shape and -colour which is found in *E. micrantha* may be the result of introgression. In its typical form this species is very distinct, and many of the apparently slightly modified colonies or populations retain enough of the distinctive characters for them to be determined without qualification as *E. micrantha*.

As I have not consistently preserved the information which is now called for I have based the following accounts on my own herbarium, on the herbarium at CGE, and on material collected by Park, Campbell, Warburg and Willmott from northern Br (particularly the Hebrides) and preserved at BM. *Euphrasia curta* (Fr.) Wettst. in the present work is to be construed in approximately the sense of Warburg (1952).

General References

DANDY, J. E. (1958). *Op. cit.*, pp. 98-100.
DRUCE, G. C. (1928). *Op. cit.*, p. 87.
PERRING, F. H. and SELL, P. D. (1968). *Op. cit.*, pp. 55-62.
PRAEGER, R. L. (1951). *Op. cit.*, pp. 10-11.
PUGSLEY, H. W. (1930). A revision of the British Euphrasiae. *J. Linn. Soc., Bot.*, **48**: 467-544.
TOWNSEND, F. (1897). Monograph of the British species of *Euphrasia*. *J. Bot., Lond.*, **35**: 465-477.
WARBURG, E. F. (1952). *Euphrasia*, in CLAPHAM, A. R., TUTIN, T. G., and WARBURG, E. F. *Op. cit.*, pp. 894-911.
WETTSTEIN, R. von (1896). *Monographie der Gattung Euphrasia*. Leipzig.
YEO, P. F. (1954). The cytology of British species of *Euphrasia*. *Watsonia*, **3**: 101-108.
YEO, P. F. (1956). Hybridisation between diploid and tetraploid species of *Euphrasia*. *Watsonia*, **3**: 253-269.
YEO, P. F. (1966). The breeding relationships of some European Euphrasiae. *Watsonia*, **6**: 216-245.

Diploid–diploid hybrids

The diploid species in BI largely replace each other ecogeographically; the principal interspecific contacts are in Wales, between 1/22. *E. anglica* Pugsl. and 1/19. *E. rostkoviana* Hayne, and intergradation between these two sometimes occurs there. Hybrids between these species may also occur in Hb, where, however, *E. anglica* appears to be rather rare and ill-defined. Intergradation, presumably because of hybridization, also occurs between: **1/21.** *E. rivularis* Pugsl. and **1/19.** *E. rostkoviana* Hayne, in v.c. 70; **1/20.** *E. montana* Jord. and **1/21.** *E. rivularis* Pugsl., in v.c. 70; **1/22.** *E. anglica* Pugsl. and **1/23.** *E. vigursii* Davey, in v.c. 2 and 3.

Tetraploid–tetraploid hybrids within subsection Ciliatae

Most British *Euphrasia* hybrids are in this category. Extremely complex situations occur in areas where the number of *Euphrasia* species is large, as on the northern coast of Scotland (v.c. 108 and 109) and in the Outer Hebrides (v.c. 110). As a result of my determinations of Orkney specimens Miss E. R. Bullard states that gatherings of hybrids outnumber gatherings of species from these islands (v.c. 111).

The following list gives all the hybrids of this group which I have personally identified in Br (including those only tentatively determined), together with those that have been reported by other workers, and fuller details of the five most important hybrids. All these hybrids are highly fertile.

1/1 × 1/2. *E. micrantha* Reichb. × *E. scottica* Wettst.
= *E.* x *electa* Towns.

1/4 × 1/1. *E. frigida* Pugsl. × *E. micrantha* Reichb.,
according to Warburg (1952).

1/5 × 1/1. *E. foulaensis* Towns. ex Wettst. × *E. micrantha* Reichb.

1/9 × 1/1. *E. marshallii* Pugsl. × *E. micrantha* Reichb.,
according to Warburg (1952).

1/10 × 1/1. *E. curta* (Fr.) Wettst. × *E. micrantha* Reichb.
= *E.* x *areschougii* Wettst.

1/1 × 1/12. *E. micrantha* Reichb. × *E. tetraquetra* (Bréb.) Arrond. (*E. occidentalis* Wettst.).

1/1 × 1/13. *E. micrantha* Reichb. × *E. nemorosa* (Pers.) Wallr.

1/15 × 1/1. *E. confusa* Pugsl. × *E. micrantha* Reichb.
Artificial hybrids have also been made.

1/17 × 1/1. *E. arctica* Lange ex Rostrup (*E. borealis* auct., *non* Wettst.) × *E. micrantha* Reichb.

1/18 × 1/1. *E borealis* (Towns.) Wettst. (*E. brevipila* auct.) × *E. micrantha* Reichb.
= *E.* x *difformis* Towns.

1/4 × 1/2. *E. frigida* Pugsl. × *E. scottica* Wettst.

1/5 × 1/2. *E. foulaensis* Towns. ex Wettst. × *E. scottica* Wettst.

1/7 × 1/2. *E. campbelliae* Pugsl. × *E. scottica* Wettst.

1/10 × 1/2. *E. curta* (Fr.) Wettst. × *E. scottica* Wettst.

1/13 × 1/2. *E. nemorosa* (Pers.) Wallr. × *E. scottica* Wettst.

1/15 × 1/2. *E. confusa* Pugsl. × *E. scottica* Wettst.

differs from *E. confusa* in its relatively tall, narrow habit, its internodes being usually longer in relation to the leaves and in its short, little-branched branches; and from *E. scottica* in its more flexuous stem and branches, its longer, finer leaf-teeth and sometimes its shorter, broader leaves and larger corollas. It occurs throughout much of the range of *E. scottica* in Wales, northern England and Scotland, except in the Central and Northern Highlands, where *E. confusa* is rare. It is usually found in flushed places or damp stream-edges in hill country, and often shows incipient speciation.

1/18 × 1/2. *E. borealis* (Towns.) Wettst. (*E. brevipila* auct.) × *E. scottica* Wettst.

= *E.* × *venusta* Towns.

1/15 × 1/4. *E. confusa* Pugsl. × *E. frigida* Pugsl.,

according to Warburg (1952).

1/18 × 1/4. *E. borealis* (Towns.) Wettst. (*E. brevipila* auct.) × *E. frigida* Pugsl. (*E. latifolia* auct.)

is likely to occur in Scotland but the only precise record (for v.c. 108) is suspect.

1/5 × 1/8. *E. foulaensis* Towns. ex Wettst. × *E. rotundifolia* Pugsl.

1/5 × 1/9. *E. foulaensis* Towns. ex Wettst. × *E. marshallii* Pugsl.

1/10 × 1/5. *E. curta* (Fr.) Wettst. × *E. foulaensis* Towns. ex Wettst.

1/5 × 1/12. *E. foulaensis* Towns. ex Wettst. × *E. tetraquetra* (Bréb.) Arrond. (*E. occidentalis* Wettst.,

was recorded by Warburg (1952) but is improbable.

1/5 × 1/13. *E. foulaensis* Towns. ex Wettst. × *E. nemorosa* (Pers.) Wallr.

1/17 × 1/5. *E. arctica* Lange ex Rostrup (*E. borealis* auct., *non* Wettst.) × *E. foulaensis* Towns. ex Wettst.

has been recorded but could refer to the next hybrid.

1/18 × 1/5. *E. borealis* (Towns.) Wettst. (*E. brevipila* auct.) × *E. foulaensis* Towns. ex Wettst.,

according to Dandy (1958).

1/7 × 1/9. *E. campbelliae* Pugsl. × *E. marshallii* Pugsl.

1/7 × 1/15. *E. campbelliae* Pugsl. × *E. confusa* Pugsl.

1/9 × 1/8. *E. marshallii* Pugsl. × *E. rotundifolia* Pugsl.,

according to Warburg (1952).

1/18 × 1/8. *E. borealis* (Towns.) Wettst. (*E. brevipila* auct.) × *E. rotundifolia* Pugsl.

was recorded by Warburg (1952) but is improbable.

1/9 × 1/13. *E. marshallii* Pugsl. × *E. nemorosa* (Pers.) Wallr.

1/17 × 1/9. *E. arctica* Lange ex Rostrup (*E. borealis* auct., *non* Wettst.) × *E. marshallii* Pugsl.

has been recorded but could refer to the next hybrid.

1/18 × 1/9. *E. borealis* (Towns.) Wettst. (*E. brevipila* auct.) × *E. marshallii* Pugsl.

1/10 × 1/12. *E. curta* (Fr.) Wettst. × *E. tetraquetra* (Bréb.) Arrond. (*E. occidentalis* Wettst.).

1/10 × 1/13. *E. curta* (Fr.) Wettst. × *E. nemorosa* (Pers.) Wallr.

1/15 × 1/10. *E. confusa* Pugsl. × *E. curta* (Fr.) Wettst.

1/18 × 1/10. *E. borealis* (Towns.) Wettst. (*E. brevipila* auct.) × *E. curta* (Fr.) Wettst.

> = *E.* × *murbeckii* Wettst. was recorded by Warburg (1952) but is improbable.

1/13 × 1/12. *E. nemorosa* (Pers.) Wallr. × *E. tetraquetra* (Bréb.) Arrond. (*E. occidentalis* Wettst.).

1/15 × 1/12. *E. confusa* Pugsl. × *E. tetraquetra* (Bréb.) Arrond. (*E. occidentalis* Wettst.)

> is a variable, low-growing plant differing from *E. tetraquetra* in its greater profusion of primary and higher order branches which bear usually very small leaves at least at the base, and in its more cuneate leaves with longer and finer teeth. It is found on cliff-tops and stabilized dunes on the coasts of western Br from south-western England to south-western Scotland, and inland on limestone pasture in v.c. 6. It forms hybrid swarms.

1/16 × 1/12. *E. pseudokerneri* Pugsl. × *E. tetraquetra* (Bréb.) Arrond. (*E. occidentalis* Wettst.).

> Artificial hybrids have also been made.

1/18 × 1/12. *E. borealis* (Towns.) Wettst. (*E. brevipila* auct.) × *E. tetraquetra* (Bréb.) Arrond. (*E. occidentalis* Wettst.)

> = *E.* × *pratiuscula* Towns.

1/str × 1/12. *E. stricta* D. Wolff ex J. F. Lehmann × *E. tetraquetra* (Bréb.) Arrond. (*E. occidentalis* Wettst.).

1/15 × 1/13. *E. confusa* Pugsl. × *E. nemorosa* (Pers.) Wallr.

> usually has the tall habit of *E. nemorosa* but the stem may be flexuous at the base and it bears slender, flexuous, small-leaved branches, and the leaves are less regularly opposite and sometimes have finer teeth than in *E. nemorosa*. Sometimes the hybrid resembles *E. confusa* but has a stouter stem and less flexuous branches. It shows incipient speciation, occurring in large colonies usually not accompanied by the parents, though usually one or both occur in the district; in v.c. 1 it apparently replaces *E. confusa* and, in v.c. 2, *E. nemorosa*. The hybrid is widespread in Br from Cornwall and East Anglia to northern Scotland; it occurs in pastures, roadsides, patches of disturbed ground and re-colonized ploughed land, and like *E. nemorosa* becomes coastal in northern Scotland.

1/13 × 1/16. *E. nemorosa* (Pers.) Wallr. × *E. pseudokerneri* Pugsl.

> differs from *E. pseudokerneri* in its smaller corollas and sometimes coarser foliage; it is not easily distinguished from *E. nemorosa* but some individuals usually show larger corollas and smaller, more finely toothed

leaves. Other characters depend upon the precise characters of the parental strains involved. This hybrid is widespread in southern England in disturbed areas of chalk and oolite where both species occur, and often forms abundant hybrid swarms. The artificial hybrid has been made on two occasions using different parental stocks. The F_1 was a fertile tetraploid with normal pollen and meiosis; F_2 and backcross generations have been raised.

1/17 × 1/13. *E. arctica* Lange ex Rostrup (*E. borealis* auct., *non* Wettst.) × *E. nemorosa* (Pers.) Wallr.

1/18 × 1/13. *E. borealis* (Towns.) Wettst. (*E. brevipila* auct.) × *E. nemorosa* (Pers.) Wallr.

1/13 × 1/str. *E. nemorosa* (Pers.) Wallr. × *E. stricta* D. Wolff ex J. F. Lehmann.

1/15 × 1/14. *E. confusa* Pugsl. × *E. heslop-harrisonii* Pugsl.

1/17 × 1/14. *E. arctica* Lange ex Rostrup (*E. borealis* auct., *non* Wettst.) × *E. heslop-harrisonii* Pugsl.

1/15 × 1/16. *E. confusa* Pugsl. × *E. pseudokerneri* Pugsl.

1/17 × 1/15. *E. arctica* Lange ex Rostrup (*E. borealis* auct., *non* Wettst.) × *E. confusa* Pugsl.

differs from *E. arctica* in its shorter lower internodes, more profuse branching, generally narrower leaves with fewer, more acute teeth, smaller and less elliptic capsules and sometimes smaller corollas; and from *E. confusa* in its coarser growth, larger leaves, large, usually emarginate capsules and sometimes larger corollas. *E. arctica* occurs in the Orkneys and Shetlands, v.c. 111 and 112, and there it hybridizes with *E. confusa*; at least some of these hybrids have been called *E. borealis* var. *zetlandica* Pugsl. Hybrids usually show incipient speciation but hybrid swarms are probably not infrequent; they occur mainly in pastures and on roadsides and stabilized dunes.

1/18 × 1/15. *E. borealis* (Towns.) Wettst. (*E. brevipila* auct.) × *E. confusa* Pugsl.

1/17 × 1/18. *E. arctica* Lange ex Rostrup (*E. borealis* auct., *non* Wettst.) × *E. borealis* (Towns.) Wettst. (*E. brevipila* auct.)

was recorded when *E. borealis* was a greatly confused name; such records probably refer to variants of *E. borealis* (Towns.) Wettst.

Tetraploid–tetraploid hybrids between subsection Ciliatae and subsection Angustifoliae

1 × 2. *E. officinalis* L. agg. × *E. salisburgensis* Funck.

Plants in this group have the more or less copiously ciliate capsule of *E. officinalis*, but acquire from *E. salisburgensis* a tendency for the stem and branches to be slender and wiry, for the leaves to be narrow and with fewer, longer, finer, more forwardly-directed teeth than usual, and for the leaves near the base of the branches to be conspicuously small. They are all largely but not completely sterile. The characters of *E. salisburgensis* may be so well developed that it is difficult to say which segregate of *E. officinalis* is involved. Artificial crosses have been made between *E. salisburgensis* var. *hibernica* Pugsl. (the taxon found in BI) and both *E. tetraquetra* and *E. nemorosa,* and between *E. salisburgensis* var. *salisburgensis* (not found in BI) and *E. nemorosa.* In the last case backcrosses to *E. nemorosa* were obtained; they showed some increase in fertility. In the first case above the hybrid had $2n = 44$, with many univalents at meiosis.

The following three hybrids have been found within the range of *E. salisburgensis* in Hb:

1/1 × 2. *E. micrantha* Reichb. × *E. salisburgensis* Funck.

1/13 × 2. *E. nemorosa* (Pers.) Wallr. × *E. salisburgensis* Funck.

1/18 × 2. *E. borealis* (Towns.) Wettst. (*E. brevipila* auct.) × *E. salisburgensis* Funck.

The hybrids referred to by Praeger (1951) as "?*borealis* × *salisburgensis*" may well also belong here; they could not refer to *E. arctica* × *E. salisburgensis.*

Diploid–tetraploid hybrids

The distinctive character of British diploid Euphrasiae is the long, glandular indumentum of the leaves, a character usually retained despite gene-flow from tetraploids. On the other hand tetraploids might be expected not to acquire it so readily by gene-flow from diploids, owing to dilution of the genes of the diploid in the larger tetraploid genotype. The same would apply to other characters, so that tetraploids introgressed in this way would be difficult to detect. Thus it is not surprising that none of the reported hybrids in this group is a tetraploid. All attempts to cross diploids with tetraploids have so far failed.

1/1 × 1/19. *E. micrantha* Reichb. × *E. rostkoviana* Hayne

is represented by one gathering from v.c. 73, in CGE.

1/22 × 1/1. *E. anglica* Pugsl. × *E. micrantha* Reichb.

has been found both as diploids and (a single plant) as a triploid. What appears to be this hybrid is established as a fertile diploid with a reasonably uniform morphology and in a distinct habitat in various localities in v.c. 1–3; I regard it as a distinct species, *E. vigursii* Davey. However, in some places plants with different characters but probably of the same parentage occur, and are better not assigned to *E. vigursii.* In 1952, near Withypool, v.c. 5, a population was found which consisted of fertile diploids showing incipient speciation and differing from *E. vigursii* in being smaller in all their parts (except the capsules) and in their narrower leaves and whitish corollas. A single individual of this colony showed a much closer approach to *E. micrantha* and was a highly sterile triploid ($2n = 33$), with 11 bivalents and 11 univalents at meiosis. Plants resembling introgressed *E. anglica* occur at St David's Head, v.c. 45.

1/19 × 1/2. *E. rostkoviana* Hayne × *E. scottica* Wettst.

has been recorded from v.c. 108 but in error, for the second parent is not known north of v.c. 88 and 98.

1/12 × 1/23. *E. tetraquetra* (Bréb.) Arrond. (*E. occidentalis* Wettst.) × *E. vigursii* Davey

probably exists as introgressed *E. vigursii* south of Perranporth, v.c. 1. The plants are dwarfer and have more rounded leaves and denser inflorescences than is usual in *E. vigursii.*

1/13 × 1/19. *E. nemorosa* (Pers.) Wallr. × *E. rostkoviana* Hayne

was recorded (as *E.* × *glanduligera*, q.v.) before *E. anglica* was segregated from *E. rostkoviana.*

1/22 × 1/13. *E. anglica* Pugsl. × *E. nemorosa* (Pers.) Wallr.

= *E.* × *glanduligera* Wettst. has the tall habit of *E. nemorosa* and moreover differs from *E. anglica* in its slightly less broad leaves with longer, finer teeth. It differs from *E. rostkoviana* in its more numerous and more widely divergent branches and in the relatively small upper leaves. It is a fertile incipient species which has been found growing in rather dry, heathy places, often in longish grass, in v.c. 2, 4, 14 and 17. It has $2n = 22$, with 11 bivalents or 10 bivalents plus 2 univalents at meiosis.

1/15 × 1/19. *E. confusa* Pugsl. × *E. rostkoviana* Hayne

resembles *E. anglica* × *E. confusa* but has very large corollas. It has been found in Hobcarton Ghyll, v.c. 70.

1/22 × 1/15. *E. anglica* Pugsl. × *E. confusa* Pugsl.

(= *E.* × *rechingeri* auct., *non* Wettst.) resembles *E. anglica* but has more flexuous and sometimes more numerous branches, and smaller, narrower leaves with all the teeth forwardly directed. It occurs occasionally where the parents meet, particularly in the Mendips, v.c. 6, and on Exmoor, v.c. 5, and sometimes in the absence of *E. confusa,* as in Windsor Great Park, v.c. 22. This hybrid was first misidentified in this country as *E. kerneri* Wettst. × *E. rostkoviana* = *E.* × *rechingeri* Wettst.

1/22 × 1/16. *E. anglica* Pugsl. × *E. pseudokerneri* Pugsl.

was reported by Pugsley (1930) but was an error for *E. anglica* × *E. nemorosa.*

1/18 × 1/19. *E. borealis* (Towns.) Wettst. (*E. brevipila* auct.) × *E. rostkoviana* Hayne

= *E.* × *notata* Towns. could occur but would be difficult to recognize. There are records from v.c. 46, 48 and H27, but Pugsley (1930) treated *E. notata* under "*E. brevipila*", and Townsend's type localities also suggest that it may really apply to *E. borealis.*

1/22 × 1/18. *E. anglica* Pugsl. × *E. borealis* (Towns.) Wettst. (*E. brevipila* auct.)

resembles *E. rostkoviana* but is a larger plant with very large corollas. The leaves are relatively small with fine teeth. It is found on the Somerset Peat Moors near Street, v.c. 6, with *E. borealis*; it exhibits incipient speciation but plants approaching *E. anglica* also occur.

441. *Pinguicula* L.
(by D. A. Webb)

2 × 3. *P. alpina* L. × *P. vulgaris* L.
= *P.* × *hybrida* Wettst. is recorded from Au, Cz and Fe.

4 × 3. *P. grandiflora* Lam. × *P. vulgaris* L.
a. *P.* × *scullyi* Druce.
b. The hybrid is intermediate between the parents, particularly in floral
characters, with a wide range of variation, suggesting segregation in the F_2
and subsequent generations, or backcrossing to the parents, or both these
phenomena. There appears, however, to be no direct evidence as to the
precise fertility of the F_1 hybrid, though Druce recorded that the capsules
were infertile.
c. The hybrid occurs occasionally in the mountains of v.c. H1 and H3; there
is one very recent unpublished record from v.c. H9 and a more doubtful
record for v.c. H35. In southern Hb *P. vulgaris* is unexpectedly rare, and
throughout most of the range of *P. grandiflora* it is virtually confined to
the mountains. The hybrid occurs locally here with *P. vulgaris*, but there is
no very precise information as to its distribution or frequency. It has also
been found in the eastern Pyrenees, Ga.
d. None.
e. *P. vulgaris* ($2n = 64$); *P. grandiflora* ($2n = 32$).
f. CASPER, S. J. (1966). Monographie der Gattung *Pinguicula* L. *Biblitbca
bot.*, **31 (127/128)**.
DRUCE, G. C. (1922). *Pinguicula grandiflora* × *vulgaris mihi* = × *P. Scullyi
mihi. Rep. B.E.C.*, **6**: 301.
PRAEGER, R. L. (1930). Notes on Kerry plants. *J. Bot., Lond.*, **68**:
249-250.
PRAEGER, R. L. (1934). *The botanist in Ireland*, p. 145. Dublin. FIG.
SCULLY, R. W. (1916). *Flora of County Kerry*, pp. 221-224. Dublin.

442. *Utricularia* L.
(by C. A. Stace)

3 × 4. *U. intermedia* Hayne × *U. minor* L.
a. (*U. ochroleuca* auct.).

b. Plants intermediate between these species in several diagnostic characters, e.g. presence of a few bladders on the floating green leaves, corolla sulphur-yellow with a spur half as long as the lower lip, have been recorded as hybrids, but confirmation is difficult as the species of this genus frequently fail to flower. *U. ochroleuca* Hartm., which is widespread in northern and central Europe, is sometimes considered to represent the hybrid, but other authorities consider it a distinct species or a variant of *U. intermedia.*

c. *U. ochroleuca* has been recorded from many parts of northern and western Br, and from Hb, but many of these records refer to *U. intermedia.* Records of hybrids are much scarcer, and come mainly from western Scotland. On the Continent *U. ochroleuca* is widespread, but records of hybrids are extremely rare.

d. None.

e. *U. minor* and *U. ochroleuca* ($2n = c$ 40).

f. HALL, P. M. (1939). The British species of *Utricularia. Rep. B.E.C.,* **12**: 100-117.

445. *Mentha* L.
(by R. M. Harley)

Within Section *Mentha*, which includes all British species of mint except *M. pulegium* L. and the introduced *M. requienii* Benth., the same general characteristics of breeding behaviour and reproductive biology are to be found. In each species a less or greater proportion of individuals are male-sterile and thus biologically female, relying on cross-fertilization for seed production. The remaining individuals are either wholly hermaphrodite, or produce functionally female as well as hermaphrodite flowers – a process which is under both genetic and environmental control. There appear to be no intrinsic barriers to cross-fertilization between different species, and hybrids can be easily produced under experimental conditions. Hybrids are sometimes very frequent in the wild, where the parents are regularly within breeding proximity, although it seems probable that rigorous selection occurs in the seedling stages. Hybrids reaching maturity are vigorous plants, and by means of their well-developed rhizome systems are often able to persist and even spread clonally to new areas in the absence of the parents.

Crossing experiments indicate that whereas hybrids between species of the same ploidy level are generally highly fertile, those between species at different ploidy levels show a high degree of sterility, although viable seed, producing segregant, often abnormal progeny, has been observed in some cases. Furthermore, normally sterile hybrids with irregular meiosis in the anthers have been found on occasions to produce abnormally large pollen grains with stainable cell contents, which may be viable. It is possible that these may sometimes be responsible for the occurrence of backcrosses in which unreduced hybrid gametes are involved.

The situation is further complicated by a history of cultivation of *M. spicata* and its hybrids. In some instances certain clones have achieved taxonomic recognition through being much cultivated because of their desirable characteristics (i.e. scent and glabrousness), and then escaping to become naturalized. However, these are often morphologically very different from individuals derived from the same parent species which have survived the selection pressures imposed by a wild environment.

In the past *M. longifolia* has been erroneously recorded from BI, being misidentified for hairy variants of *M. spicata*. Nevertheless two hybrids involving the true *M. longifolia* have been found in Br, and specimens of this species were found by Druce in 1912 in v.c. 74, where they occurred as relics of cultivation.

General References

FRASER, J. (1927). Menthae Britannicae. *Rep. B.E.C.*, **8**: 213-247.

HARLEY, R. M. (1963). *Taxonomic studies in the genus Mentha*. D.Phil. Thesis, University of Oxford.

HARLEY, R. M. (1967). The spicate mints. *Proc. B.S.B.I.*, **6**: 369-372.

HARLEY, R. M. (1972). *Mentha*, in TUTIN, T. G. *et al.*, eds. *Flora Europaea*, **3**: 183-186. Cambridge.

MORTON, J. K. (1956). The chromosome numbers of the British Menthae. *Watsonia*, **3**: 244-252.

4 × 3. *M. aquatica* L. × *M. arvensis* L.

a. *M.* × *verticillata* L. (*M.* × *sativa* L.). (Whorled Mint).

b. This hybrid is extremely variable, but more or less intermediate between its parents; it is usually distinguishable from these and from *M.* × *gentilis* by calyx characters. The verticillasters are in the axils of leafy bracts and often decrease in size towards the stem apex. It is usually sterile, but occasionally highly fertile plants, sometimes associated with other segregants, occur, probably as a result of backcrossing. Introgressing populations have also been reported.

c. *M.* × *verticillata* is most frequently found in the vicinity of its parents, and occupying a wide range of usually damp habitats including arable fields, track-sides, woodland rides, marshes, river-banks and pond-sides. It is found throughout most of BI as well as the rest of Europe.

d. Studies of flavonoids from the leaves of different cytotypes collected in the wild in Su appear in most instances to confirm the suggested parentage.

e. *M. arvensis* $2n = 72$; *M. aquatica* $2n = 96$; hybrid $2n = 42, 78, 84, (90), (96), 120, 132$.

f. BUTCHER, R. W. (1961). *A new illustrated British flora*, **2**: 303, t. 1100. London. FIG.

OLSSON, U. (1967). Chemotaxonomic analysis of the cytotypes in the *Mentha* x *verticillata* complex (Labiatae). *Bot. Notiser*, **120**: 255-267.

PERRING, F. H. and SELL, P. D. (1968). *Op. cit.*, p. 66.

SYME, J. T. B., ed. (1867). *English Botany*, 3rd ed., **7**: 15-16, t. 1031. London. FIG.

4 × 3 × 5. *M. aquatica* L. × *M. arvensis* L. × *M. spicata* L.

a. *M.* x *smithiana* R. A. Graham (*M.* x *rubra* Sm., *non* Mill.). (Tall Mint).

b. This mint is a morphologically rather uniform, robust perennial similar to *M.* x *gentilis* but differing in the longer calyx and in the upper bracts being usually suborbicular and cuspidate. It is typically glabrous or almost so, and is perhaps represented by only one or a few clones. It is usually sterile, though occasionally produces some viable seed which has been shown to give rise to segregating progeny. Morton (1956) has suggested that the hybrid originated as an amphidiploid involving *M.* x *verticillata* and *M. spicata*.

c. *M.* x *smithiana* is much more frequent in southern Br than in northern Br or in Hb; some of its occurrences are escapes from cultivation. It is generally associated with wet habitats, stream-sides, marshes, etc. It is widespread in central Europe from Ga to Rm.

d. None.

e. *M. aquatica* $2n = 96$; *M. arvensis* $2n = 72$; *M. spicata* $2n = 48$; hybrid $2n = 120$.

f. BUTCHER, R. W. (1961). As above, p. 306, t. 1103. FIG.

GRAHAM, R. A. (1948). Mint Notes, 1. *Mentha rubra*. *Watsonia*, **1**: 88-90.

PERRING, F. H. and SELL, P. D. (1968). *Op. cit.*, p. 66.

3 × 5. *M. arvensis* L. × *M. spicata* L.

a. *M.* x *gentilis* L. (*M.* x *cardiaca* (Gray) Bak.). (Bushy Mint).

b. This is extremely variable, but more or less intermediate between its parents; it is usually distinguishable from them by the shorter, campanulate calyx with more or less subulate teeth. The inflorescence is very like that of *M.* x *verticillata*. Glabrous plants are probably often escapes from cultivation, and usually have the characteristic odour of *M. spicata*. Hairy plants, which have been misidentified as *M.* x *dalmatica* (q.v.) or *M.* x *verticillata*, are probably always spontaneous. Although often sterile, some seed-set occurs with resultant backcrossing and segregation. Morton (1956) reported introgressing populations with some individuals setting well-formed seed.

c. *M.* x *gentilis* is a frequent and widespread hybrid over most of BI, commonly cultivated and escaping and found in a wide range of damp habitats and in disturbed ground. Its spontaneous and introduced distributions are not easily distinguished, but it is possibly native in parts of Br and in Be, Co, Ga and It. It is found as a naturalized plant over most of Europe.

d. Synthetic F_1 hybrids, involving a *M. spicata* parent heterozygous for hairiness, are either glabrous or hairy, and vary in the intensity of the characteristic spearmint odour. They are apparently sterile, and are morphologically comparable with F_1s collected in the field.

e. *M. arvensis* $2n = 72$; *M. spicata* $2n = 48$; wild hybrid $2n = 54, 60, 84, 96, 108, 120$; artificial F_1 $2n = 60, 61$.

f. BUTCHER, R. W. (1961). As above, pp. 304-305, t. 1101-1102. FIG.
GRAHAM, R. A. (1950). Mint notes, 2. *Mentha gracilis* Sole, and its relationship to *Mentha cardiaca* Baker. *Watsonia*, 1: 276-278.
PERRING, F. H. and SELL, P. D. (1968). *Op. cit.*, p. 65.

3 × 6. *M. arvensis* L. × *M. longifolia* (L.) Huds.

= *M.* × *dalmatica* Tausch was erroneously recorded for *M.* × *gentilis*. *M.* × *dalmatica* occurs in the vicinity of its parents on the Continent, but is rare. STILL, A. L. (1938). *Mentha verticillata* L. var. *trichodes* Briq. *Rep. B.E.C.*, 11: 663.

3 × 7. *M. arvensis* L. × *M. suaveolens* Ehrh. (*M. rotundifolia* auct., *non* Huds.)

a. *M.* × *muellerana* F. W. Schultz (*M.* × *wolwerthiana* F. W. Schultz).

b. This hybrid is intermediate between its two parents, but in general appearance closer to *M. arvensis*, with verticillasters more or less remote and in the axils of leafy bracts which decrease in size upwards. The only plants recorded from Br appear to be backcrosses with *M. suaveolens*, and are sterile.

c. It has been found only in v.c. 3 and 9, but is now probably extinct in its original stations, although stock may still persist in cultivation. In Europe the hybrid is of occasional occurrence in the vicinity of its parents.

d. Synthesized F_1 hybrids are almost indistinguishable from *M. arvensis* and are highly sterile, though abnormally large pollen grains, which may represent unreduced gametes, are occasionally produced. These synthesized crosses indicate that, on morphological grounds and in accordance with the cytological evidence, the British hybrid is a backcross with *M. suaveolens* involving an unreduced hybrid gamete, and not a backcross with *M. arvensis*, as originally suggested by Morton (1956).

e. *M. arvensis* $2n = 72$; *M. suaveolens* $2n = 24$; wild hybrid $2n = 60$; artificial F_1 $2n = 48$.

f. PERRING, F. H. and SELL, P. D. (1968). *Op. cit.*, p. 68.
PUGSLEY, H. W. (1935). A new British mint. *J. Bot., Lond.*, 73: 75-78.
ROLES, S. J. (1963). *Flora of the British Isles. Illustrations*, 3: 1174. Cambridge. FIG.

4 × 5. *M. aquatica* L. × *M. spicata* L.

a. *M.* × *piperita* L. (*M.* × *dumetorum* auct. angl., *non* Schultes, *M.* × *citrata* Ehrh.). (Pepper-mint).

b. This is a variable hybrid more or less intermediate between its parents, but usually recognizable by its petiolate, often lanceolate leaves and its terminal, oblong, spike-like inflorescence which is much broader than in *M. spicata*. Both glabrous and hairy plants occur, the latter having been confused with *M.* × *dumetorum* (q.v.). Plants frequently have a characteristic pungent odour, though in some, usually hairy individuals this is replaced by a weak odour reminiscent of that of one or both of the parents. They are apparently highly sterile, though the occurrence of presumed backcrosses or segregants suggests that viable seed can occasionally be set. The most familiar of these is nm. *citrata* (Ehrh.) Briq. (Eau de Cologne Mint), a sterile cultivar which is widely cultivated and occasionally naturalized. It has a strong odour resembling Eau de Cologne, glabrous or subglabrous, ovate, subcordate leaves and a capitate inflorescence.

c. Various cultivated, glabrous clones have become widely naturalized in wet habitats throughout most of BI. Hairy plants have arisen spontaneously either as crosses between the parent species or as mutations of glabrous hybrids; they occur particularly in southern and western Br. Both are widespread throughout Europe. Nm. *citrata* is apparently represented by a few widespread clones, especially in southern England.

d. Synthetic hybrids have been formed on several occasions. Those involving the homozygous recessive menthone-producing genotype of *M. spicata*, with glabrous leaves, give offspring closely comparable with the cultivars of *M.* × *piperita*. Crosses involving other genotypes of *M. spicata* produce offspring comparable with many of the hairy variants of the hybrid, both in general appearance and in odour. All these appear to be sterile.

e. *M. aquatica* $2n = 96$; *M. spicata* $2n = 48$; *M.* × *piperita* $2n = 72, 66$; artificial F_1 $2n = 72$; nm. *citrata* $2n = 84, 108, 120$.

f. BUTCHER, R. W. (1961). As above, pp. 308-309, t. 1105-1106. FIG.

GRAHAM, R. A. (1951). Mint notes, 4. *Mentha piperita* L. and the British peppermints. *Wa onia*, **2**: 30-35.

MURRAY, J. M. (1958). Evolution in the genus *Mentha*. *Proc. 10th International Congress of Genetics*, **2**: 201. Montreal.

PERRING, F. H. and SELL, P. D. (1968). *Op. cit.*, p. 68.

4 × 6. *M. aquatica* L. × *M. longifolia* (L.) Huds.

= *M.* × *dumetorum* Schultes has been misidentified for hairy variants of *M.* × *piperita*. *M.* × *dumetorum* apparently occurs in many places on the Continent in the vicinity of its parents.

4 × 7. *M. aquatica* L. × *M. suaveolens* Ehrh. (*M. rotundifolia* auct., *non* Huds.)

  a. *M.* × *maximilianea* F. W. Schultz.

  b. This is a variable hybrid, but more or less intermediate between its parents. It is perhaps sometimes confused with *M. aquatica,* but it often has somewhat rugose leaves with a cordate base, and the terminal inflorescence forms a blunt spike. It is usually sterile, but backcrosses to both parents and introgression have been recorded, with some individuals producing viable seed.

  c. *M.* × *maximilianea* is a rare spontaneous hybrid found in scattered localities in v.c. 1, 3 and 4 and in Jersey, occurring within the native range of *M. suaveolens* in damp hedge-banks and on stream-sides, etc.

  d. None.

  e. *M. aquatica* $2n = 96$; *M. suaveolens* $2n = 24$; hybrid $2n = 60, 72–78, 120$.

  f. GRAHAM, R. A. (1958). Mint Notes, 7. *Mentha* × *maximilianea* F. Schultz in Britain. *Watsonia,* 4: 72-76. FIG.
    PERRING, F. H. and SELL, P. D. (1968). *Op. cit.,* p. 68.

6 × 5. *M. longifolia* (L.) Huds. × *M. spicata* L.

  a. (*M.* × *villoso-nervata* auct.?, *an* Opiz, *M. longifolia* var. *horridula* auct. angl., *non* Briq.).

  b. This is a variable hybrid, intermediate between its parents, and often confused with *M. spicata* and with *M.* × *villosa* from which it is not readily separable. It differs mainly in its narrower, usually patently-toothed leaves. It is sterile.

  c. It is widely cultivated and often thrown out from gardens and becoming locally naturalized.

  d. Artificial F_1 hybrids are closely comparable with the cultivated plants and are sterile.

  e. *M. longifolia* ($2n = 24$); *M. spicata* $2n = 48$; wild hybrid $2n = 38$; artificial F_1 $2n = 36$.

  f. None.

6 × 5 × 7. *M. longifolia* (L.) Huds. × *M. spicata* L. × *M. suaveolens* Ehrh.
has been recorded, but almost all plants formerly called *M. longifolia* in Br are now known to be variants of *M. spicata.*

5 × 7. *M. spicata* L. × *M. suaveolens* Ehrh. (*M. rotundifolia* auct., *non* Huds.)

  a. *M.* × *villosa* Huds. (*M.* × *nemorosa* Willd., *M.* × *cordifolia* auct.?, *an* Opiz, *M.* × *niliaca* auct., *non* Juss. ex Jacq.). (Large Apple-mint, Bowles' Mint).

  b. This is an extremely variable hybrid, intermediate between the parents though often close to one or the other; it is sometimes almost indistinguishable from *M. spicata,* though it usually has broader, more rugose leaves. In the past glabrous variants have been called *M.* ×

cordifolia, and hairy ones *M.* x *niliaca.* There are many, often widespread cultivated variants; others may be locally frequent and spontaneously produced, and some have been given formal taxonomic status, often of rather dubious value. All are highly sterile, though occasionally some seed has been reported, but its viability is unknown. Probably the most widespread hybrid is nm. *alopecuroides* (Hull) Briq. This is a uniform hybrid, probably represented in Br by a single clone, and often mistaken for *M. suaveolens.* It can be distinguished by its more robust habit, its sweet odour like that of *M. spicata,* the patent teeth on its broadly ovate or orbicular leaves, and its robust spikes of pink flowers. If it is true that *M. suaveolens* is involved in the original parentage of *M. spicata* then *M.* x *villosa* can be looked upon as a kind of backcross, and its variability is not surprising. *M. scotica* R. A. Graham has often been considered a variant of *M.* x *villosa,* but it is a tetraploid, $2n = 48$, and produces some good seed; it seems better to include it under *M. spicata.*

c. Various nothomorphs are widespread throughout Br on roadsides and in waste places and damp ground, but are apparently rare in Hb. The European distribution is uncertain owing to confusion with *M.* x *rotundifolia.*

d. Synthetic F_1 hybrids, made using a *M. spicata* genotype heterozygous for hairiness, gave both hairy and glabrous progeny. Some of these had very narrow leaves and were difficult to distinguish readily from forms of *M. spicata.* All were highly sterile. Plants closely comparable to but not identical with nm. *alopecuroides,* and with the same chromosome number, have been synthesized by crossing *M. suaveolens* with a very broad-leaved, rugose *M. spicata* with $2n = 48$, which is widespread in south-western England.

e. *M. suaveolens* $2n = 24$; *M. spicata* $2n = 48$; *M.* x *villosa* $2n = 36$; nm. *alopecuroides* $2n = 36$; artificial F_1 $2n = 36, 48$–51.

f. BUTCHER, R. W. (1961). As above, p. 313, t. 1110. FIG.
   GRAHAM, R. A. (1958). Mint notes, 8. A new mint from Scotland. *Watsonia,* 4: 119-121.
   PERRING, F. H. and SELL, P. D. (1968). *Op. cit.,* p. 67.

6 x 7. *M. longifolia* (L.) Huds. x *M. suaveolens* Ehrh. (*M. rotundifolia* auct., *non* Huds.)

a. *M.* x *rotundifolia* (L.) Huds. (?*M.* x *niliaca* auct. var. *webberi* Fraser).

b. This hybrid is intermediate between the parents. It is usually fertile, and, within the range of both parents but not in Br, forms fertile introgressing populations. It is probable that *M. spicata* arose from this hybrid by chromosome doubling; *M. spicata* behaves cytologically as a segmental allopolyploid.

c. A single clone cultivated in v.c. 105 and becoming locally naturalized along a small stream appears to be this hybrid. The more widespread Scottish plant known as *M.* x *niliaca* var. *webberi* may perhaps also be referable here, but needs cytological study.

d. Synthetic F_1 hybrids, involving the non-British *M. longifolia*, have been shown to be highly fertile and capable of backcrossing with the parents.

e. *M. suaveolens* $2n = 24$; *M. longifolia* ($2n = 24$); Scottish wild hybrid, non-British wild hybrids, and artificial F_1s $2n = 24$.

f. None.

448. *Thymus* L.
(by C. D. Pigott)

1 × 2. *T. pulegioides* L. × *T. serpyllum* L.

= *T. × oblongifolius* Opiz has been reported from v.c. 11, 17, 28, 32, 49, 62, 88 and H15; but apart from the v.c. 28 record (Swaffham) these are areas where only *T. pulegioides* occurs, or where neither species is known. The Swaffham plant in herb. Druce (**OXF**) appears to be *T. drucei*. Mixed populations of the two species occur on the Breckland in East Anglia but no hybrid plants have been discovered in these localities even after thorough searches. The hybrid is also reported from western and central Europe and Scandinavia but identification appears to be based only on gross morphology. It is possible that hybrids do exist because emasculated hermaphrodite plants of *T. serpyllum* pollinated artificially with pollen from *T. pulegioides* (Pigott, 1954) gave approximately 6.5% fertile nutlets; some of these produced mature plants which flowered sparsely after three years. The reciprocal cross produced some nutlets but none of the plants survived to maturity.

PIGOTT, C. D. (1954). Species delimitation and racial divergence in British *Thymus*. *New Phytol.*, 53: 470-495.

RONNIGER, K. (1924). Contribution to the knowledge of the genus *Thymus*. The British species and forms. *Rep. B.E.C.*, 7: 226-239.

RONNIGER, K. (1928). The distribution of *Thymus* in Britain. *Rep. B.E.C.*, 8: 509-517.

3 × 1. *T. drucei* Ronn. × *T. pulegioides* L.

= *T. × henryi* Ronn. (*T. × lansdowneiae* Druce & Ronn., *T. × jacksonii* Ronn.) has been recorded from v.c. 3, 6, 7, 10, 14, 17, 22, 23, 36, 41, 48, 74 and 80, but all these determinations are likely to be errors. No authentic herbarium specimens have been seen and thorough searches of localities where the two species occur together have revealed no hybrids. Moreover attempts to make the hybrid artificially (Pigott, 1954) have been unsuccessful, although the experiments were too limited in scale to be conclusive. In southern England *T. drucei* commences flowering a month before *T. pulegioides.* There appear to be no foreign records for this hybrid.

PIGOTT, C. D. (1954). As above.
RONNIGER, K. (1924). As above.
RONNIGER, K. (1928). As above.

1 × **vul.** *T. pulegioides* L. × *T. vulgaris* L.

= *T.* × *citriodorus* Pers. (Lemon-thyme) was found as a garden outcast or escape in an old quarry in v.c. 32 in 1873. The parentage suggested for this garden plant is based purely on its intermediate morphology. Plants commonly have the stamens abortive and are sterile, but no cytological or genetical studies have been made, so that their true nature remains uncertain.

RONNIGER, K. (1924). As above.

455. *Salvia* L.

(by C. A. Stace)

4 × **2.** *S. horminoides* Pourr. × *S. pratensis* L.

has been recorded from Ga.

nem × **2.** *S. nemorosa* L. × *S. pratensis* L.

= *S.* × *sylvestris* L. occurred on waste ground at Burton-on-Trent, v.c. 39, in 1937, but like the first parent is not an established member of the British flora. It has often been confused with *S. pratensis,* but certainly occurs frequently in central and south-eastern Europe from Ge to Rm as a native, and it is recorded as an alien in No and Su.

BURGES, R. C. L. (1938). *Salvia sylvestris* L. *Rep. B.E.C.,* **11**: 498.

457. *Prunella* L.
(by J. E. Lousley)

2 × 1. *P. laciniata* (L.) L. × *P. vulgaris* L.

a. *P.* × *hybrida* Knaf (*P.* × *intermedia* Link (*partim?*)).

b. Plants intermediate between the parents occur freely in places where the parents grow, or have grown, together. They usually attract attention first by strange and varied colours of the corolla, or by a combination of large, creamy-white flowers with undivided leaves, or purple flowers with pinnatifid or deeply toothed leaves. According to Hegi the pollen is usually sterile.

c. The hybrid occurs in usually calcareous grassland where *P. laciniata* has been introduced. The cross occurs freely and often *P. laciniata* is replaced by a series of hybrids which persist after the introduced species is lost. It has been recorded from v.c. 6, 9, 12, 14, 16–18, 20, 23, 24, 29, 32 and 37. On the Continent the hybrid is widespread from Ga to Gr.

d. None.

e. Both parents $2n = 28$ (32).

f. FOUILLADE, A. (1927). *Brunella (hybrida) Knafi,* in GUÉTROT, M. *Plantes hybrides de France,* **1** and **2**: 54. Lille.

HEGI, G. (1927). *Prunella,* in *Illustrierte Flora von Mittel-Europa,* **5**: 2377-2384. Munich.

LAPORTE-CRU, J. (1970). Étude des variations morphologiques de l'espèce *Prunella vulgaris* et de l'hybride *Prunella hybrida* en Gironde. *Botaniste, Bordeaux,* **53**: 63-115. FIG.

MORTON, J. K. (1973). A cytological study of the British Labiatae (excluding *Mentha*). *Watsonia,* **9**: 239-246.

MOSS, C. E. (1915). Notes on British plants, 4. *Brunella laciniata* × *vulgaris. J. Bot., Lond.,* **53**: 8-13.

PANINI, F. (1926). Ibridi naturali nel genere *Brunella,* I., *Archo bot. Sist. Fito-geogr. Genet.,* **2**: 63-78.

PERRING, F. H. and SELL, P. D. (1968). *Op. cit.,* p. 69.

459. *Stachys* L.
(by P. S. Green)

5 × 4. *S. alpina* L. × *S. germanica* L.

a. *S.* × *digenea* Legué.

b. This is a variable hybrid said to be more or less intermediate between the parental species.

c. It is recorded as occurring spontaneously in cultivation from parents of British origin, and as a wild plant in Au, Cz, Ga, Ge, He and Hu.

d. Lang (1940) crossed *S. germanica* with *S. alpina* and obtained vigorous, partially fertile hybrids. From the results of comparative studies he suggested that *S. germanica* may have originally arisen from hybridization between *S. alpina* and *S. byzantina* (*S. lanata*) at a time when, hypothetically, their now disjunct ranges were contiguous.

e. *S. alpina* $2n = 30$; *S. germanica* ($2n = 30$).

f. LANG, A. (1940). Untersuchungen über einige Verwandtschafts- und Abstammungsfragen in der Gattung *Stachys* L. auf cytogenetischer Grundlage. *Biblithca bot.*, **29**(118): 7-10, 75-92.

LOUSLEY, J. E. (1934). *Stachys alpina* L. x *S. germanica* L. *Rep. B.E.C.*, **10**: 539.

SALMON, C. E. (1919). A hybrid *Stachys*. *J. Linn. Soc., Bot.*, **44**: 357-362. FIG.

4 × 6. *S. germanica* L. × *S. palustris* L.

= *S. x mirabilis* Rouy has been reported from Ga, Ge and Hu.

4 × 7. *S. germanica* L. × *S. sylvatica* L.

(? = *S. x pannonica* Láng.) has been reported from Hu.

5 × 6. *S. alpina* L. × *S. palustris* L.

has been reported from He.

5 × 7. *S. alpina* L. × *S. sylvatica* L.

= *S. x medebachensis* Feld & Koenen has been reported from Ga, Ge, He and Hu.

6 × 7. *S. palustris* L. × *S. sylvatica* L.

a. *S. x ambigua* Sm.

b. Hybrids are more or less intermediate between the parents in characters of the leaves, flowers and rhizome; the leaves are distinctly petiolate yet narrower in proportion to their length than in *S. sylvatica*. The hybrid frequently resembles *S. palustris* more than *S. sylvatica* which, it has been suggested, indicates that occasional backcrossing to *S. palustris* perhaps takes place. So-called petiolate varieties of *S. palustris* (var. *petiolata* Čelak. and var. *pseudo-ambigua* Mejer) have been described and need investigation. Hybrids show a high degree of sterility both in abortive pollen and lack of seed-set, but care should be taken to avoid confusion with male-sterile plants of the parental species.

c. The hybrid occurs throughout much of BI, often in the absence of the parental species, especially *S. sylvatica,* and is presumably distributed by man by fragmentation of the brittle rhizome; in the north and west *S.* x *ambigua* is often a weed of cultivated ground. On the Continent the hybrid is recorded from Au, Be, Cz, Da, Fe, Ga, Ge, He, Ho, Hs, Hu, It, No, Po, Rm and Su.

d. Attempts to produce artificial hybrids have proved very difficult. Using *S. sylvatica* as seed parent, Wilcock (1973) obtained 0–50% seed-set with different genotypes. However, only 7% of the seeds germinated and all died in the seedling stage. Using *S. palustris* as the seed parent (135 pollinations) no seed was obtained.

e. *S. palustris* $2n = (64, 96)$, 97–103, mostly 102; *S. sylvatica* $2n = (48)$, 62–68, mostly 64 and 66; hybrid $2n = 78$–86, mostly 83 and 84.

f. BORNMÜLLER, J. (1920). Bemerkungen über den Formenkreis von *Stachys palustris* x *sylvatica* in Thüringen. *Beih. bot. Zbl.,* **37**: 310-315.

CLOS, M. D. (1889). Le *Stachys ambigua* Sm. est-il espèce, variété ou hybride? *Bull. Soc. bot. Fr.,* **36**: 66-71.

EDEES, E. S. (1945). *Stachys* x *ambigua* Sm. *NWest. Nat.,* **19**: 275.

MORTON, J. K. (1973). A cytological study of the British Labiatae (excluding *Mentha*). *Watsonia,* 9: 239-246.

PERRING, F. H. and SELL, P. D. (1968). *Op. cit.,* p. 69.

SYME, J. T. B., ed. (1867). *English Botany,* 7: 58-59, t. 1070. London. FIG.

WILCOCK, C. C. (1972). Is *Stachys* x *ambigua* Sm. always distinguishable? *Watsonia,* 9: 62.

WILCOCK, C. C. (1973). *The experimental taxonomy of Stachys ambigua Sm., S. palustris L. ana S. sylvatica L. in Britain.* Ph.D. thesis, University of London.

WILCOCK, C. C. and JONES, B. M. G. (1974). The identification and origin of *Stachys* x *ambigua* Sm. *Watsonia,* **10**: 139-147.

462. *Lamium* L.
(by P. S. Green)

1 x 4. *L. amplexicaule* L. x *L. purpureum* L.

has been treated in many Floras as the parentage of *L. hybridum* Vill., which is somewhat intermediate in appearance. This last species is an allotetraploid, and Bernström (1955) suggested that it is derived from the diploids *L. purpureum* and *L. bifidum* Cyr. Bernström has in fact synthesized another allotetraploid, *L. moluccellifolium* Fr., from L. *amplexicaule* x *L. purpureum*.

BERNSTRÖM, P. (1955). Cytogenetic studies on relationships between annual species of *Lamium*. *Hereditas*, **41**: 1-122.

FOCKE, W. O. (1881). *Op. cit.*, p. 340.

JONES, S. B. and JONES, C. A. (1965). Status of *Lamium hybridum* Vill. (Labiatae). *Am. Midl. Nat.*, **74**: 503-506.

LITTLE, J. E. and WARBURG, E. F. (1953). *Lamium hybridum* Vill. *Watsonia*, **2**: 361-368.

2 x 4. *L. moluccellifolium* Fr. x *L. purpureum* L.

a. *L.* x *boreale* Druce.
b. This plant was said to show a mixture of the characters of the parental species and to be sterile. *L. purpureum* is thought to be involved in the parentage of *L. moluccellifolium* (see above).
c. The hybrid was reported once from Tongue, v.c. 108, in 1917.
d. Crosses using Swedish and Danish parental material have failed to produce hybrids.
e. *L. moluccellifolium* $2n = 36$; *L. purpureum* $2n = 18$.
f. DRUCE, G. C. (1924). *Lamium mollucellifolium* x *purpureum, nova hybr. Rep. B.E.C.*, **7**: 53.

DRUCE, G. C. (1929). x *Lamium boreale* Dr. *Rep. B.E.C.*, **8**: 873.

JÖRGENSEN, C. A. (1927). Cytological and experimental studies in the genus *Lamium*. *Hereditas*, **9**: 126-136.

MÜNTZING, A. (1926). Ein Art-Bastard in der Gattung *Lamium*. *Hereditas*, **7**: 215-228.

3 x 4. *L. hybridum* Vill. x *L. purpureum* L.

has been reported on a few occasions, e.g. from v.c. 29 and 69, but always with doubt attached to the identification.

BENNETT, A., HOSKING, A. and MARSHALL, E. S. (1904). *Lamium hybridum*, Vill. x *purpureum* var. *decipiens*, Sonder.? *Rep. Watson B.E.C.*, **1903-4**: 15.

LUMB, D. and WHELDON, J. A. (1914). *Lamium purpureum* L., var.? *Rep. B.E.C.*, **3**: 491.

5 x 4. *L. album* L. x *L. purpureum* L.

= *L.* x *schroeteri* Gams is reported from Ga, He and Hu.

6 x 4. *L. maculatum* L. x *L. purpureum* L.

has been reported from Ge.

5 x 6. *L. album* L. x *L. maculatum* L.

= *L.* x *holsiaticum* E. H. L. Krause is reported from Au, Cz, Ga, Ge, Hu and Rm.

465. *Galeopsis* L.
(by P. M. Benoit and C. A. Stace)

1 × 2. *G. angustifolia* Ehrh. ex Hoffm. × *G. ladanum* L.

has been reported from the Continent and, although the identity of these putative hybrids has been doubted, artificial hybrids have also been obtained.

1 × 3. *G. angustifolia* Ehrh. ex Hoffm. × *G. segetum* Neck.

= *G.* × *wirtgenii* Ludw. has been found on the Continent (e.g. Be, Ga, Ge and He), and it arose spontaneously in a Botanic Garden in Su.

2 × 3. *G. ladanum* L. × *G. segetum* Neck.

= *G.* × *ochrerythra* E. H. L. Krause has been recorded from Ge.

4/2 × 4/1. *G. bifida* Boenn. × *G. tetrahit* L. *sensu stricto*

a. *G.* × *ludwigii* Hausskn.
b. The hybrid can be distinguished from its parents by its intermediate corolla and by having a high proportion of sterile pollen grains and few developed nutlets. It occurs as occasional individuals with the parent species and does not form complex populations, probably owing to the preponderance of autogamy and the low pollen-fertility of the hybrid.
c. The hybrid has been found as single plants at Arthog, v.c. 48, in 1960; at Peniarth, Towyn, v.c. 48, in 1963; and at Flitwick, v.c. 30, in 1971. It has also been found in Au, Ge, Ho and Su.
d. The two species are self-compatible. Hybrid plants have been obtained with ease on several occasions by reciprocal crosses. The F_1 plants resemble the wild putative hybrids in morphology and sterility (pollen fertility varies between c 25 and 70%), and show some hybrid vigour. The F_2 shows a great deal of variation in morphology and fertility, but F_3 and subsequent generations show less morphological range and increased fertility.
e. Both parents, putative wild hybrids and all artificial hybrids ($2n = 32$); *G. tetrahit* agg. $2n = 32$.
f. BENOIT, P. M. (1965). *Galeopsis bifida*: a species. *Proc. B.S.B.I.*, **6**: 169.
HAGBERG, A. (1952). Heterosis in F_1 combinations in *Galeopsis*, 1 and 2. *Hereditas*, **38**: 33-82, 221-245.
MÜNTZING, A. (1930). Outlines to a genetic monograph of the genus *Galeopsis* with special reference to the nature and inheritance of partial sterility. *Hereditas*, **13**: 185-341. FIG.

5 × 4/1. *G. speciosa* Mill. × *G. tetrahit* L. *sensu stricto*

has been occasionally recorded from Br, and also from Au and Ge, but such records probably referred to *G. tetrabit sensu lato,* i.e. including *G. bifida* (q.v.).

4/2 × 5. *G. bifida* Boenn. × *G. speciosa* Mill.

= *G.* × *sulfurea* Druce has been recorded from v.c. 45 and 92, but Dandy (1958) included *G. sulfurea* under *G. bifida.* No specimen of *G. sulfurea* can be traced in herb. Druce (**OXF**). The experiments of Müntzing (1930) have shown that *G. speciosa* cannot be artificially hybridized with *G. tetrabit* or *G. bifida,* although with either of the latter two as the female parent some embryo development does occur. The plants recorded as hybrids are probably variants of *G. bifida* or *G. tetrabit sensu stricto* with rather large, cream- or yellow-coloured corollas.

DANDY, J. E. (1958). *Op. cit.,* p. 105.

DRUCE, G. C. (1904). Notes on the flora of Westerness. *Ann. Scot. nat. Hist.,* **49**: 36-42.

DRUCE, G. C. (1929). Notes on the second edition of the "British Plant List". *Rep. B.E.C.,* **8**: 867-877.

MÜNTZING, A. (1930). As above.

469. *Scutellaria* L.
(by F. H. Perring)

1 × 2. *S. galericulata* L. × *S. minor* Huds.

a. *S.* × *hybrida* Strail (*S.* × *nicholsonii* Taub.).

b. This hybrid is more or less intermediate between the parents in characters of the leaves and flowers, but is rather closer to *S. minor.* The upper leaves have two or three distant crenations on each margin, not *c* 6 as in *S. galericulata* or one as in *S. minor*; and the corolla is 8–10 mm as in *S. minor,* but usually blue-violet as in *S. galericulata,* though with a tinge of pink. Plants are usually infertile but some ripe seeds are occasionally produced; they reproduce vigorously vegetatively, persisting in the absence, locally, of either parent.

c. Hybrids occur in only a limited area of southern England and Hb, in v.c. 1–4, 9, 11, 13, 14, 16, 17, 22, H1, H2 and H6. The area of overlap of the distribution of the two parents is much greater. The hybrid is also recorded from Ga and Su.

d. None.

e. *S. galericulata* $2n = 32$; *S. minor* $2n = c$ 32.

f. NICHOLSON, G. (1885). *Scutellaria galericulata* x *minor. Rep. B.E.C.*, **1**: 93.

PERRING, F. H. and SELL, P. D. (1968). *Op. cit.*, p. 70.

SALMON, C. E. (1931). *Flora of Surrey*, p. 525. London.

TAUBERT, P. (1886). *Scutellaria minor* x *galericulata* (*S. Nicholsoni* Taubert), ein neuer Bastard. *Verh. bot. Ver. Prov. Brandenb.*, **28**: 25-28.

471. *Ajuga* L.
(by C. A. Stace)

3 x 2. *A. genevensis* L. x *A. reptans* L.

= *A.* x *hybrida* Kerner occurs in much of Europe from Ga to the Balkan peninsula.

4 x 2. *A. pyramidalis* L. x *A. reptans* L.

a. *A.* x *hampeana* Braun & Vatke (*A.* x *hybrida* Druce, *nom. nud.*, *non* Kerner).

b. Hybrids are vegetatively vigorous; they apparently produce normal *A. reptans*-like runners but only later in the season. The earlier stolons are suberect (*A. pyramidalis* lacks stolons). The flowers, leaves and inflorescence characters are more or less intermediate. It is reported to be sterile.

c. The hybrid is known from Betty Hill, v.c. 108, and from Ballyvaughan, v.c. H9, where it was discovered at the start of this century. There are also apparently authentic specimens in **BM** from v.c. 106 (1888) and v.c. 111 (1920). It is also found in much of Europe from the Balkan peninsula to Scandinavia and Ga.

d. None.

e. Both parents $2n = 32$.

f. DRUCE, G. C. (1916). *Ajuga pyramidalis* x *reptans. Rep. B.E.C.*, **4**: 207.
DRUCE, G. C. (1924). *Ajuga pyramidalis* x *reptans. Rep. B.E.C.*, **7**: 401-402.

3 x 4. *A. genevensis* L. x *A. pyramidalis* L.

= *A.* x *adulterina* Wallr. occurs in Au, Cz and Ge, and probably elsewhere, and is reported to be cultivated in Br in rock-gardens, etc.

472. *Plantago* L.
(by D. M. Moore)

3 × 2. *P. lanceolata* L. × *P. media* L.

has been reported by Woodruffe-Peacock. Herbarium material of this presumed parentage at **OXF** and **K** has been found by Sagar and Harper to have the typical inflorescence of *P. media* and the attenuated, more or less lanceolate-spathulate leaves which this species shows when growing in tall herbage. Repeated attempts to produce artificial hybrids have been unsuccessful and the species seem to be very strongly isolated genetically.

DRUCE, G. C. (1912). The check list of Lincolnshire plants, by E. A. Woodruffe-Peacock. *Rep. B.E.C.,* **3**: 43-44.

RAHN, K. (1957). Chromosome numbers in *Plantago. Bot. Tidsskr.,* **53**: 369-378.

SAGAR, G. R. and HARPER, J. L. (1964). *Plantago major* L., *P. media* L. and *P. lanceolata* L., in Biological Flora of the British Isles. *J. Ecol.,* **52**: 189-221.

3 × 4. *P. lanceolata* L. × *P. maritima* L.

was suggested as the parentage of *P. edmonstonii* Druce, from v.c. 111 and 112. This taxon is not of hybrid origin, however; it appears to be nothing more than a variant of *P. maritima* occurring on serpentine soils.

475. *Campanula* L.
(by D. W. Shimwell)

1 × 2. *C. latifolia* L. × *C. trachelium* L.

a. None.
b. This hybrid is apparently recorded in error for products of continuous selfing in both species which leads to wide ranges of variation in such characters as height, stem-indumentum and calyx- and corolla-length in *C. trachelium,* and leaf-shape and calyx-tube indumentum in *C. latifolia.*
c. Records exist for v.c. 36 and 57, but are presumably erroneous.
d. The cross made with *C. trachelium* as female parent produces some good seed, but the F_1 dies in the seedling stage; the cross made with *C. latifolia* as female parent fails to produce viable seed. This work was

originally carried out by Gadella (1964) but has been repeated by me after a report of the hybrid from a locality in v.c. 57.

e. Both parents and artificial hybrid ($2n = 34$).

f. GADELLA, T. W. J. (1964). Cytotaxonomic studies in the genus *Campanula. Wentia*, **11**: 1-104.

TOWNDROW, R. F. (1909). *Campanula trachelium* x *latifolia. J. Bot., Lond.*, **47**: 386.

3 × 2. *C. rapunculoides* L. × *C. trachelium* L.
    = *C.* x *chevalieri* Sennen is recorded from Hs.

6 × 2. *C. glomerata* L. × *C. trachelium* L.
    is recorded from various parts of central Europe.

475 × 478. *Campanula* L. × *Phyteuma* L. (= × *Fockeanthus* Wehrh.)
(by C. A. Stace)

A few hybrids between various species (none British) of these genera have been reported from central Europe, mainly the Alps.

484 × 485. *Cruciata* Mill. × *Galium* L.
(by K. M. Goodway)

484/1 × 485/6. *Cruciata laevipes* Opiz (*C. chersonensis* (Willd.) Ehrend.) × *Galium pumilum* Murr.
    was recorded from Walton Downs, v.c. 17, in 1923, but the specimens examined are *G. mollugo* x *G. verum*.
    BRITTON, C. E. and FRASER, J. (1924). *Galium. Rep. B.E.C.*, **7**: 391.

485. *Galium* L.

(by K. M. Goodway)

3b × 3a. *G. erectum* auct. angl. (*G. mollugo* subsp. *erectum* Syme) × *G. mollugo* L.

has been recorded from v.c. 6, 15, 22 and 24 and probably elsewhere, but the variation found in these taxa is now considered better summarized by placing them all in *G. album* Mill. subsp. *album*.

BEEBY, W. H., DRUCE, G. C. and MARSHALL, E. S. (1895). *Galium Mollugo. Rep. B.E.C.*, 1: 484-485.

PUGSLEY, H. W. and THOMPSON, H. S. (1934). *Galium Mollugo × erectum? Rep. Watson B.E.C.*, 4: 220.

3 × 4. *G. album* Mill. subsp. *album* (*G. mollugo* L. incl. *G. erectum* auct. angl.) × *G. verum* L.

a. *G. × pomeranicum* Retz. (*G. × ochroleucum* Wolf ex Schweigg. & Koerte, *G. × rothschildii* Druce, *nom. nud.*, *G. × hillardiae* Druce).

b. Hybrids range between the parental extremes in flower-colour and inflorescence- and leaf-shape. Although the commonest form appears, morphologically, to be an F_1 hybrid, in some populations the range of plants present suggests that segregation and backcrossing to the parents is occurring. This is more usual on the Continent than in BI. The hybrid is somewhat less fertile than the parents.

c. This hybrid is frequent along the southern coast and on inland chalk and limestone in England, but is rare in Scotland, Wales and Hb. It occurs in CI and is widespread on the Continent. Hybrids identified as *G. erectum* × *G. verum* occur mainly where the former parent replaces *G. mollugo sensu stricto*.

d. Fagerlind hand-pollinated 747 flowers (giving a possible 1494 seeds). 146 seeds developed but only 22 ripened; of 12 F_1 plants produced only five developed.

e. Both parents and hybrid $2n = 44$.

f. ARMITAGE, E. (1909). Hybrids between *Galium verum* and *G. mollugo. New Phytol.*, 8: 351-353. FIG.

DRUCE, G. C. (1929). Notes on the second edition of the "British Plant List". *Rep. B.E.C.*, 8: 867-877.

FAGERLIND, F. (1937). Embryologische, zytologische und bestäubungsexperimentelle Studien in der Familie Rubiaceae nebst Bemerkungen über einige Polyploidistätsprobleme. *Acta Horti Bergiani*, 11: 195-470.

HUNNYBUN, E. W. (1910). *Galium erectum × verum. Rep. Watson B.E.C.*, 2: 236-237.

PERRING, F. H. and SELL, P. D. (1968). *Op. cit.*, p. 70.

8 × 4. *G. palustre* L. × *G. verum* L.

was recorded by Praeger from v.c. H14, but this is probably a typographical error for *G. mollugo* × *G. verum*, which was not mentioned. PRAEGER, R. L. (1951). *Op. cit.*, p. 10.

6 × 5. *G. pumilum* Murr. × *G. saxatile* L.

was recorded in 1920 from Cheddar Gorge, v.c. 6, growing with both the parents. The specimens seen are *G. saxatile* with no trace of *G. pumilum*, as suggested at the time by Salmon, and it has proved impossible to hybridize these species.

SALMON, C. E. and THOMPSON, H. S. (1921). *G. saxatile* × *asperum? Rep. Watson B.E.C.*, 3: 142.

5 × 7. *G. saxatile* L. × *G. sterneri* Ehrend.

a. None.
b. Plants are intermediate between the parents, particularly in the leaf-shape and in the marginal prickles on the leaves. Hybrids are highly sterile in the field and in cultivation, but there are sometimes fertile plants of *G. saxatile* in the population showing a few morphological characters of *G. sterneri*, which suggests occasional backcrossing.
c. The hybrids have only been detected in a few upland areas, where they are associated with rich ledge-vegetation, although the parents do grow in close proximity in many other localities. Hybrids have been identified by morphology and chromosome count in v.c. 49, 88, 96, 108, and by morphology alone in v.c. 90.
d. It has proved impossible to hybridize *G. saxatile* and tetraploid *G. sterneri*. One plant has been produced by crossing *G. saxatile* and diploid *G. sterneri*. This was a dwarf plant whose buds aborted at an early stage.
e. *G. saxatile* $2n = 44$; *G. sterneri* $2n = 22, 44$. Chromosome numbers for the hybrid (together with those for *G. sterneri* in the same area) are $2n = 33$ (v.c. 49, *G. sterneri* $2n = 22$); $2n = 55$ (v.c. 88 and 108, *G. sterneri* $2n = 44$).
f. None.

487. *Sambucus* L.

(by C. A. Stace)

2 × 3. *S. nigra* L. × *S. racemosa* L.

has been reported from Da.

494. *Valerianella* Mill.
(by C. A. Stace)

5 x 3. *V. dentata* (L.) Poll. x *V. rimosa* Bast.
= *V.* x *zoltanii* Borbás has been recorded from Rm, but needs checking.

498 × 500. *Knautia* L. × *Succisa* Haller = × *Succisoknautia* Baksay
(by C. A. Stace)

Hybrids between *S. pratensis* Moench and two different non-British species of *Knautia* have been found in He and Hu respectively.

499 × 500. *Scabiosa* L. × *Succisa* Haller
(by C. A. Stace)

499/1 x 500/1. *Scabiosa columbaria* L. x *Succisa pratensis* Moench
was recorded from Dovedale, v.c. 57, in 1903 ("both species growing plentifully at the place"), and according to F. A. Lees in Yorkshire in 1909. It is not a likely hybrid and no hybrids between these genera have been found elsewhere.
BICKHAM, S. H. (1910). *Scabiosa Columbaria* x *Succisa. Rep. B.E.C.*, **2**: 415.
LINTON, W. R. (1903). *Flora of Derbyshire*, p. 177. London.

502. *Bidens* L.
(by T. G. Tutin)

1 × 2. B. *cernua* L. × *B. tripartita* L.

(= *B.* × *peacockii* Druce, *nom. nud.*) was recorded from v.c. 17 and 32 by Woodruffe-Peacock, but no specimens have been traced. It was perhaps recorded without sufficient regard to the variability of these two species, which have different chromosome numbers.

DRUCE, G. C. and WOODRUFFE-PEACOCK, E. A. (1918). ?*Bidens cernua* × *tripartita*. *Rep. B.E.C.*, 5: 33-34.

503. *Galinsoga* Ruiz. & Pav.
(by C. A. Stace)

2 × 1. G. *ciliata* (Raf.) Blake × *G. parviflora* Cav.

= *G.* × *mixta* J. Murr has been reported from Au and Ga, but needs checking.

506. *Senecio* L.
(by P. M. Benoit, P. C. Crisp and B. M. G. Jones)

2 × 1. S. *aquaticus* Hill × *S. jacobaea* L.

a. *S.* × *ostenfeldii* Druce.

b. This hybrid is partially fertile and morphologically very variable, sometimes forming complex populations through which the two species intergrade. Hybrid individuals can be recognized by their obviously more or less intermediate characters and lower fertility. They are showy plants through having wide-spreading inflorescences of numerous large capitula, and late in the summer they often produce fresh axillary flowering branches which overtop the primary inflorescence. The disc-achenes are distinctly pubescent, though more minutely so than in *S. jacobaea*, and foliage characters are intermediate. The hybrids further differ from *S. aquaticus* in their small fruiting involucres (5.5–7.0 mm diam.; 9–10 mm in *S. aquaticus*) and from *S. jacobaea* in their more numerous florets (90–120 per capitulum; 50–75 in *S. jacobaea*). Fertility is variable, but the most obvious hybrids (F_1s?) have a high proportion of sterile pollen and have less than 15% developed achenes. The hybrid has often been collected or recorded as *S. aquaticus* var. *pinnatifidus* Gren. & Godr., but authentic material of that taxon has not been seen by us.

c. *S. aquaticus* x *S. jacobaea* is widespread in BI. Meikle remarked on its rarity in southern England (Praeger, 1951), but at least in Wales and Ireland, where agriculture is poorer and wet habitats are more frequent, it is common and sometimes forms large populations in neglected marshy pastures; it is also known in Scotland. Occasionally it occurs with *S. jacobaea* alone when, apparently, *S. aquaticus* has been swamped by hybridization. The hybrid undoubtedly occurs on the Continent, but its distribution is not known in detail.

d. None.

e. Both parents ($2n = 40$).

f. DRUCE, G. C. (1915). *Senecio aquaticus* x *Jacobaea*. *Rep. B.E.C.*, **4**: 17.
DRUCE, G. C. (1924). *Senecio aquaticus* x *Jacobaea* Druce. *Rep. B.E.C.*, **7**: 39.
HARPER, J. L. and WOOD, W. A. (1957). *Senecio jacobaea* L., in Biological Flora of the British Isles. *J. Ecol.*, **45**: 617-637.
PRAEGER, R. L. (1951). *Loc. cit.*

3 x 1. *S. erucifolius* L. x *S. jacobaea* L.

= *S.* x *liechtensteinensis* Murr. (*S.* x *whitwellianus* Lees ex Cheetham, *nom. nud.*) has been recorded from Farming (?) Woods, v.c. 32 (**BIRM**), and from several places in Yorkshire. The three sheets at **BIRM**, collected in 1910, bear a mixture of plants with acute leaf-lobes and pubescent ray-achenes and with obtuse leaf-lobes and glabrous ray-achenes. They may represent both species and possibly a hybrid, but all the plants have normal pollen grains which would not be expected in such a hybrid. No specimens supporting the Yorkshire records have been seen, and the few records for the Continent also remain unconfirmed. The two species often grow close together and hybrids might well occur with them.

LEES, F. A. (1887). *Senecio Jacobaea* x *erucifolius*. *Rep. bot. Record Club*, **1884-1886**: 129.
LEES, F. A. (1938). In CHEETHAM, C. A., ed. The vegetation of Yorkshire and supplement to the Floras of the county. *Naturalist, Hull*, **981**: 297.

1 x 4. *S. jacobaea* L. x *S. squalidus* L.

was recorded from railway banks in v.c. H4 in 1902 (**BM**) and at Maryborough, v.c. H14, in 1927 (**DBN**). The v.c. H4 specimen appears to

be a depauperate *S. squalidus.* The plant from v.c. H14 has glabrous ray-achenes and is sterile, and is either *S. aquaticus* x *S. jacobaea* or an abnormal plant of *S. jacobaea.* The hybrid has not been found abroad.
PRAEGER, R. L. (1951). *Loc. cit.*

1 x 8. *S. jacobaea* L. x *S. vulgaris* L.

was recorded in 1931 from Aberdovey, v.c. 48 (**OXF**), but the specimen is almost certainly a plant of *S. squalidus* which had been infested with caterpillars.
CRISP, P. C. (1972). *Cytotaxonomic studies in the section Annui of Senecio.* Ph.D. thesis, University of London.

18 x 1. *S. cineraria* DC. x *S. jacobaea* L.

a. *S.* x *albescens* Burbidge & Colgan.
b. The hybrid is partially fertile and variable in its characters. Most often it resembles *S. jacobaea* in being biennial without a profusion of barren shoots at the base in summer, and in its erect stems with leafy inflorescences. It is intermediate between the parents in its indumentum, the dissection of its leaves, and its achenes. Thus the underside of the leaves is grey with tomentum but this is distinctly thinner than the white felt-like tomentum of *S. cineraria,* and the disc-achenes are minutely pubescent. Some specimens have small, apparently sterile pollen grains and wholly or almost wholly sterile achenes, but others, probably backcrosses, have more numerous developed achenes. The frequency, variability and persistence for 70 years of this short-lived hybrid at Killiney Bay suggests that freely-interbreeding populations occur.
c. *S. cineraria* x *S. jacobaea* is reliably recorded from the coast at Newquay, v.c. 1, Budleigh Salterton, v.c. 3, Saunton, v.c. 4, Ventnor, v.c. 10, and Killiney Bay, v.c. H21, and inland in 1909 as a spontaneous garden weed at Brampton Abbotts, v.c. 36. It is also reported from Torquay, v.c. 3, Swanage, v.c. 9, Newport, v.c. 10, Horton Cliff, Gower, v.c. 41, and Bishop's Castle, v.c. 40, where it is said to have been first noted in BI by W. Borrer in 1836, but no voucher material from these localities has been traced.
d. None.
e. Both parents ($2n = 40$).
f. BURBIDGE, F. W. and COLGAN, N. (1902). A new *Senecio* hybrid. *Ir. Nat.,* **11**: 311-317; also in *J. Bot., Lond.,* **40**: 401-406. FIG.
HARPER, J. L. and WOOD, W. A. (1957). As above.
PRAEGER, R. L. (1951). *Loc. cit.*

3 x 7. *S. erucifolius* L. x *S. viscosus* L.

has been recorded from rubbish dumps in v.c. 18 and 31 (**BM**). The former specimen is *S.* x *londinensis* and the latter *S. squalidus.* There are no foreign records.

DONY, J. G. (1950). Excursions, 1948. Huntingdonshire. *Year Book B.S.B.I.*, **1950**: 51-53.

4 × 7. *S. squalidus* L. × *S. viscosus* L.

a. *S.* × *londinensis* Lousley.

b. Hybrids are intermediate between the parents in shapes of leaf and capitulum, in ligule dimensions, and in branching habit. They resemble *S. squalidus* in the black tips to their outer involucral bracts and in being polycarpic. They are near to *S. viscosus* in their degree of viscidity. Ripe achenes have not been reported from the hybrid under natural conditions, and wild hybrids are highly pollen sterile. No evidence exists of introgression between the species.

c. Hybrids occur fairly regularly (although not commonly) in BI where the parental species grow together, usually on waste ground or similar disturbed sites. Most records have been made during the last three decades in southern England and Wales, where both parental species are common. The rarity or absence of the hybrid in Scotland and Hb correlates with the relative rarity there of the parents.

d. Hybrids are easily produced, although with a low frequency, by artificial cross-pollination. Hybrids are highly seed- and pollen-sterile; achenes when set are of the dimensions of those of *S. viscosus*, but are brown and hirsute as are those of *S. squalidus*. Successful backcrosses have been made only to *S. viscosus*. Increase in fertility together with considerable segregation has been noted in F_2 plants, and experimentally-produced allopolyploids are large, vigorous and fertile.

e. *S. squalidus* $2n = 20$; *S. viscosus* $2n = 40$; hybrid $2n = 30$.

f. CRISP, P. C. (1972). As above.

LOUSLEY, J. E. (1946). A new hybrid *Senecio* from the London area. *Rep. B.E.C.*, **12**: 869-874. FIG.

LOUSLEY, J. E. (1947). A new hybrid *Senecio* (*S. squalidus* L. × *S. viscosus* L.) from the London area. *Proc. Linn. Soc. Lond.*, **158**: 21-22.

4 × 8. *S. squalidus* L. × *S. vulgaris* L.

a. *S.* × *baxteri* Druce. *S. cambrensis* Rosser is the fertile allohexaploid.

b. Doubts exist about the status of many of the reported intermediates between these two species. *S.* × *baxteri* refers to the F_1 hybrid, which should be triploid ($2n = 30$), but no natural hybrid with this number has been recorded. Several specimens have been found with intermediate characters and sterile pollen and achenes, but these characters also occur as segregants in more or less tetraploid natural hybrid swarms. The most striking characteristic of these hybrid swarms is the intermediate (although segregating) nature of the ligule character. Probably authentic records of the F_1 hybrid refer, as do experimental observations, to large plants with very numerous, ligulate, pendulous capitula. Introgression from *S. squalidus* into *S. vulgaris* has given rise to the f. *radiatus* Hegi. Hybrids between f. *radiatus* and f. *vulgaris* occur commonly in mixed populations, and can be readily detected by their very small, inconspicuous ligules. The allohexaploid *S. cambrensis* arose within the last century by hybridization between the two species, and it still persists in segregant forms in northern Wales. Usually both tetraploid hybrid swarms and *S. cambrensis* have lower seed- and pollen-fertility than the parent species, but fertile achenes are often larger, and stainable pollen-grains are larger and often have four rather than three pores.

c. *S. vulgaris* f. *radiatus* is rapidly becoming ubiquitous, if ephemeral. *S. cambrensis* is restricted to v.c. 50, 51 and 58. Possible F_1 hybrids have been recorded from at least 18 English and Welsh vice-counties. Hybrid swarms, probably tetraploid and distinct from f. *radiatus,* have been found in at least 20 English and Welsh and in three Irish vice-counties. The hybrids typically occur on waste ground, disturbed soils, railway embankments and similar habitats.

d. There are strong reproductive barriers between the two species. The two synthesized hybrids reported have been triploid. In one case colchicine treatment produced a synthetic allohexaploid analogous to *S. cambrensis.* In the other a single F_2 achene probably backcrossed into native *S. vulgaris* to form populations resembling f. *radiatus* by the F_4. Experimental crosses have successfully been made between *S. cambrensis* and *S. vulgaris,* producing a fairly highly sterile hybrid.

e. *S. vulgaris* (all forms) $2n = 40$; *S. squalidus* $2n = 20$; artificial F_1 hybrid $2n = 30$; wild fertile hybrid swarms $2n = 36-42$; *S. cambrensis* $2n = 58-60$.

f. BRENAN, J. P. M. (1948). *Senecio squalidus* L. x *vulgaris* L. *Rep. B.E.C., 13*: 364.

CRISP, P. C. (1972). As above.

CRISP, P. C. and JONES, B. M. G. (1970). *Senecio squalidus* L., *S. vulgaris* L. and *S. cambrensis* Rosser. *Watsonia, 8*: 47-48.

HARLAND, S. C. (1954). The genus *Senecio* as a subject for cytogenetical investigation. *Proc. B.S.B.I., 1*: 256-257.

PERRING, F. H. and SELL, P. D. (1968). *Op. cit.,* p. 71.

ROSSER, E. M. (1955). A new British species of *Senecio. Watsonia, 3*: 228-232. FIG.

STEPHENSON, T. (1946). A new *Senecio* hybrid. *Naturalist, Hull,* **819**: 137-138. FIG.

TROW, A. H. (1916). On the numbers of nodes and their distribution along the main axis in *Senecio vulgaris* and its segregates. *J. Genet.,* **6**: 1-63.

6 x 7. *S. sylvaticus* L. x *S. viscosus* L.

a. *S.* x *viscidulus* Scheele.

b. Hybrids are intermediate between the parents in leaf-colour and -shape, in capitulum-shape and -size, in the lengths of the outer involucral bracts and ligules, and in the degree of viscidity. The hybrids have lower seed- and pollen-fertility than the parents; the occasional achenes set are of intermediate size and are black or dark brown as in *S. viscosus* but hirsute as in *S. sylvaticus.* Introgression may take place, but the species are sharply delimited.

c. The parental species often grow together, especially on sandy heaths, and here the hybrids sometimes occur; records exist for v.c. 6, 16, 17, 37, 82 and H4. The hybrid has also been found in Au, Cz, Ga, Ge, Ho and Su.

d. None.

e. Both parents and hybrid $2n = 40$.

f. CRISP, P. C. (1972). As above.
   LOUSLEY, J. E. (1954). *Senecio* x *viscidulus* Scheele. *Proc. B.S.B.I.*, **1**: 37-39.

6 x 8. *S. sylvaticus* L. x *S. vulgaris* L.

has been doubtfully reported from Ge and He.

7 x 8. *S. viscosus* L. x *S. vulgaris* L.

has been recorded from v.c. 17 (in 1942), 33 (in 1923) and 57 (in 1949), but none of the records has been confirmed. Putative hybrids have smaller capitula and are less viscid than *S. viscosus,* and they have involucral bracts which are intermediate in morphology. Seed-set is poor. Attempts at experimental hybridization have resulted in failure due to hybrid embryo inviability. There is also a doubtful record from Su.

GIBBS, P. E. (1971). Studies on synthetic hybrids of British species of *Senecio*, 1. *Senecio viscosus* L. x *S. vulgaris* L. *Trans. Proc. bot. Soc. Edinb.*, **41**: 213-218.

LOUSLEY, J. E. (1946). As above.

LOUSLEY, J. E. (1950). Exhibition Meeting 1949. *Year Book B.S.B.I.*, **1950**: 61.

RIDDELSDELL, H. J. (1923). ?*Senecio viscosus* x *vulgaris.* *J. Bot., Lond.*, **61**: 176-177.

ver x 8. *S. vernalis* Waldst. & Kit. x *S. vulgaris* L.

= *S.* x *pseudovernalis* Zabel has been reported from Ge, He, Po and Rm, and robust plants occurring with the parents near Market Harborough, v.c. 55, in 1968 and 1969 (K. G. Messenger *in litt.* 1972) also might well have been this hybrid. The British plants had intermediate leaf-shapes, rayed capitula as large as or larger than those of *S. vernalis,* and deformed pollen; they occurred on road-verges newly sown with seed mixtures obtained from Da and Ho (K. G. Messenger *in litt.* 1972). At least some of the hybrids found on the Continent have been sterile and, since *S. vernalis* has $2n = 20$, were presumably triploid. Earlier reports of the hybrid from BI (and several of *S. vernalis* itself) were based on specimens of *S. squalidus* x *S. vulgaris.*

CRISP, P. C. (1972). As above.

JACOBASCH, E. (1894). *Senecio vulgaris* L. und *Senecio vernalis* W. K.

sind nur Endglieder zweier Entwickelungsreihen einer Urform. *Verh. bot. Ver. Prov. Brandenb.*, **36**: 78-87.

MESSENGER, K. G. (1970). *Senecio vernalis* Waldst. & Kit. in Leicestershire (v.c. 55). *Watsonia*, **8**: 90.

VATKE, W. (1874). *Senecio vulgaris* x *vernalis* Ritschl., forma *Weylii* nebst. Allgemeinen Bemerkungen über Pflanzenbastarde. *Verh. bot. Ver. Prov. Brandenb.*, **14**: 45-51.

509. *Petasites* Mill.
(by C. A. Stace)

2 x 1. *P. albus* (L.) Gaertn. x *P. hybridus* (L.) Gaertn., Mey. & Scherb.

= *P.* x *rechingeri* Hayek has been reported from central Europe.

514. *Filago* L.
(by C. A. Stace)

1 x 5. *F. germanica* (L.) L. (*F. vulgaris* Lamarck) x *F. minima* (Sm.) Pers.

was recorded from v.c. 22 in 1894, but W. R. Linton considered it to be a form or variety of *F. germanica*. It is not recorded elsewhere.

DRUCE, G. C. and LINTON, W. R. (1895). *Filago germanica* x *minima?* *Rep. B.E.C.*, **1**: 450-451.

2 x 3. *F. apiculata* G. E. Sm. (*F. lutescens* Jord.) x *F. spathulata* C. Presl (*F. pyramidata* L.)

= *F.* x *costei* Revol. is reported from Ga.

4 x 3. *F. gallica* L. x *F. spathulata* C. Presl (*F. pyramidata* L.)

= *F.* x *schultzeana* P. Fourn. is reported from Ga and Ge.

514 × 515. *Filago* L. × *Gnaphalium* L.
(by C. A. Stace)

514/4 × 515/4. *Filago gallica* L. x *Gnaphalium uliginosum* L.
has been reported from Ga, but is doubtful.

515. *Gnaphalium* L.
(by C. A. Stace)

2 x 1. *G. norvegicum* Gunn. x *G. sylvaticum* L.
= *G.* x *traunsteineri* J. Murr has been recorded from Au, but needs checking.

518. *Solidago* L.
(by C. A. Stace)

2 x 1. *S. canadensis* L. x *S. virgaurea* L.
= *S.* x *niederederi* E. Khek has been recorded from Au.

2 x 3. *S. canadensis* L. x *S. gigantea* Ait.
has been recorded from North America.

519 × 518. *Aster* L. × *Solidago* L.
= × *Solidaster* Wehrhahn
(by C. A. Stace)

Three hybrids between these genera are known in North America, and one (involving *S. canadensis* L.) has arisen in a European botanic garden.

519. *Aster* L.
(by P. F. Yeo)

7 × 6. *A. laevis* L. × *A. novi-belgii* L.

a. *A.* x *versicolor* Willd.

b. *A.* x *versicolor* differs from *A. laevis* in its non-pruinose foliage and its less regularly imbricated involucral bracts with more extensive herbaceous tips, and from *A. novi-belgii* in its more regular and more appressed involucral bracts without long, more or less leafy tips. The ray-florets are coloured, or are white at first and become coloured later.

c. The hybrid is recorded from waste-ground as an escape from cultivation, the parents being North American. Britton reported it from v.c. 11, 17 and 55, and there are also records from v.c. 13, 29 and 71. It is not known in North America.

d. None.

e. *A. laevis* ($2n = 48, 54$); *A. novi-belgii* ($2n = 18, 48, 49, 54$).

f. BRITTON, C. E. (1932). The naturalised and alien Asters of the British Plant List, ed. II. *Rep. B.E.C.*, 9: 710-718.

8 × 6. *A. lanceolatus* Willd. × *A. novi-belgii* L.

a. *A.* x *salignus* Willd. (?*A. lanceolatus* sensu Clapham (1962), *non* Willd.; *A. longifolius* auct., *non* Lam.).

b. *A.* x *salignus* is here conceived as corresponding to the plant figured and described under this name by Wagenitz (1964, pp. 54-55). Specimens of this kind at W have been determined as *A. novi-belgii* x *A. simplex* Willd. by Cronquist, and Wagenitz (1964, pp. 56-57) states that *A. lanceolatus* Willd. is the correct name for *A. simplex*. However, some British material assigned to *A.* x *salignus* is really *A. novi-belgii*; this includes the well-known colony at Wicken Fen, v.c. 29. *A.* x *salignus* differs from *A.*

novi-belgii in having the cauline and inflorescence leaves less distinctly auriculate (sometimes not at all) and the involucral bracts not, or less distinctly, loose, enlarged and herbaceous, and from *A. lanceolatus* in its typical state by having the leaves sometimes slightly auriculate, the involucral bracts sometimes slightly dilated and the ligules blue-violet, not white.

c. The hybrid occurs in water-side situations, and on railway-banks and waste-ground, where it is widespread and probably rather common as an escape from cultivation. There are records from v.c. 23, 29, 69, 75 and 80.

d. None.

e. *A. lanceolatus* ($2n = 32, 64$); *A. novi-belgii* ($2n = 18, 48, 49, 54$); hybrid ($2n = 18$).

f. CLAPHAM, A. R. (1962). *Aster*, in CLAPHAM, A. R., TUTIN, T. G. and WARBURG, E. F. *Op. cit.*, pp. 841-844.

CRONQUIST, A. J. (1952). *Aster*, in GLEASON, H. A. *The new Britton and Brown illustrated Flora of the Northeastern United States and adjacent Canada*, 3: 440-467. New York.

FERNALD, M. L. (1950). *Gray's Manual of botany*, 8th ed., pp. 1416-1442. New York, etc.

WAGENITZ, G. (1964). *Aster*, in HEGI, G. *Illustrierte Flora von Mitteleuropa*, 2nd ed., 6(3): 35-71. Munich. FIG.

WHELDON, J. A. (1920). *Aster novi-belgii?* x *A. salignus. Rep. B.E.C.*, 5: 561-562.

522 × 521. *Conyza* Less × *Erigeron* L.
(by G. Halliday)

522/1 × 521/1. *Conyza canadensis* (L.) Cronquist x *Erigeron acer* L.

a. (*E.* x *huelsenii* Vatke).

b. Hybrids are intermediate between the parents, particularly in the involucral bracts, which are slightly pubescent, 4–5 mm long and tinged with lilac at the tips. The hyaline margins to the outer involucral bracts are also intermediate in width. The plants are usually rather small and weak, with relatively few capitula, and appear to be sterile. The ligules exceed the outer involucral bracts by 1–2 mm and are lilac, appreciably paler than those of *E. acer*.

c. Hybrids are very rare and irregular in occurrence and always associated with the parents, mostly in neglected sand-pits and similar habitats. They have been reported from v.c. 4–6, 13, 15–17 and 26, and from Ge, Ho, Po, Rm and Su.

d. None.

e. Both parents ($2n = 18$).

f. MARSHALL, E. S. (1907). A hybrid *Erigeron. J. Bot., Lond.,* **45**: 164.

ROPER, I. M. (1911). *Erigeron acre* x *canadense* = *E. Hulsenii* Kerner. *J. Bot., Lond.,* **49**: 348.

VATKE, W. (1871). *Erigeron Huelsenii* Vatke. Ein neuer Bastart aus der Posener Flora. *Öst. bot. Z.,* **21**: 346-347.

526. *Anthemis* L.
(by Q. O. N. Kay)

2 x 1. *A. cotula* L. x *A. tinctoria* L.
= *A.* x *bollei* Schultz Bip. has been reported from Ge.

3 x 1. *A. arvensis* L. x *A. tinctoria* L.
= *A.* x *adulterina* Wallr. has been reported from Cz and Ge.

3 x 2. *A. arvensis* L. x *A. cotula* L.
has been reported from He.

526 × 531. *Anthemis* L.
× *Tripleurospermum* Schultz Bip.
(by Q. O. N. Kay)

526/1 x 531/1b. *Anthemis tinctoria* L. x *Tripleurospermum inodorum* (L.) Schultz Bip. (*T. maritimum* (L.) Koch subsp. *inodorum* (L.) Hyland. ex Vaarama)
(= x *Anthematricaria hampeana* Geisenh.) has been reported from Ge.

526/2 × 531/1b. *Anthemis cotula* L. × *Tripleurospermum inodorum* (L.) Schultz Bip. (*T. maritimum* (L.) Koch subsp. *inodorum* (L.) Hyland. ex Vaarama)

a. (× *Anthematricaria celakovskyi* Geisenh.).

b. F_1 hybrids are intermediate between the parents in the characters of the leaves, indumentum, and capitulum, but are less vigorous and completely sterile. Their mature achenes (which do not contain embryos) have five to ten smooth whitish ribs of varying thickness, with traces of "oil-glands" between the ribs at the upper end, below a well-developed spreading membranous pappus. Receptacular bracts are present only on some capitula on some plants, and are intermediate in structure between involucral bracts and the receptacular bracts of *A. cotula*.

c. I have found hybrids on two occasions in BI, growing in fairly dense mixed populations of the parents: a single plant in a ley meadow near Moulsford, v.c. 22, in August 1966; and two plants in barley-stubble near Bridgnorth, v.c. 40, in September 1969. I have also found hybrid plants in Ga, and apparent hybrids have been reported from Cz and Ge.

d. Both parents are strongly self-incompatible and artificial hybrids can be made surprisingly easily with either parent as the female; the achene-set is low (below 10%) but most of the achenes contain hybrid embryos. The characters of F_1 hybrid plants are as described above.

e. *A. cotula* $2n = 18$; *T. inodorum* $2n = 18$, (18, 36); wild and artificial hybrids $2n = 18$. Artificial *A. cotula* × tetraploid *T. inodorum* $2n = 27$.

f. KAY, Q. O. N. (1971). *Anthemis cotula* L., in Biological Flora of the British Isles. *J. Ecol.*, **59**: 623-636.

526/3 × 531/1b. *Anthemis arvensis* L. × *Tripleurospermum inodorum* Schultz Bip. (*T. maritimum* (L.) Koch subsp. *inodorum* (L.) Hyland. ex Vaarama)

a. (× *Anthematricaria gruetterana* Aschers.).

b. Presumed F_1 hybrids are generally morphologically intermediate between the parents, rather weak, and completely sterile. Their receptacular bracts resemble the involucral bracts and are arranged irregularly among the disk-florets.

c. The hybrid has not been observed in the field in BI, but presumed F_1 hybrids were grown from achenes collected from an isolated plant of *A. arvensis* found growing in a population of *T. inodorum* near Oxford, v.c. 23, in 1964. Apparent hybrids have also been reported from Ge.

d. Both parents are strongly self-incompatible.

e. Both parents and presumed hybrids $2n = 18$.

f. GRÜTTER, M. (1891). *Anthemis arvensis* × *Matricaria inodora* nov. hyb. *Dt. bot. Mschr.*, **9**: 5-7.

   KAY Q. O. N. (1971). *Anthemis arvensis* L., in Biological Flora of the British Isles. *J. Ecol.*, **59**: 637-648.

526 × 532. *Anthemis* L. × *Matricaria* L. = × *Anthematricaria* (Geisenh.) Dom.
(by Q. O. N. Kay)

526/2 × 532/1. *Anthemis cotula* L. × *Matricaria recutita* L.
= × *A. dominii* Rohlena has been reported from Cz.

528 × 533. *Achillea* L. × *Chrysanthemum* L. = × *Chrysanthemoachillea* Prodan
(by C. A. Stace)

Two hybrids between species of these genera have been reported from Rm; none of the species involved is British.

531. *Tripleurospermum* Schultz Bip.
(by Q. O. N. Kay)

1a × 1b. *T. maritimum* (L.) Koch (*T. maritimum* (L.) Koch subsp. *maritimum*) × *T. inodorum* (L.) Schultz Bip. (*T. maritimum* (L.) Koch subsp. *inodorum* (L.) Hyland. ex Vaarama)

a. None.

b. F_1 hybrids are intermediate between the parents in the characters of the mature achenes, leaves, and capitula, and are vigorous and of moderate to high (20–80%) fertility. They are usually intermediate in habit, and annual to biennial. *T. maritimum* shows marked geographic variation in BI and F_1 hybrids vary correspondingly. F_2 and later generations, and backcross plants, are more variable and generally of lower fertility than the F_1; most populations of hybrid origin include 5% to 20% of sterile plants, of which some are vegetatively vigorous but have abnormal and distorted capitula.

c. Morphologically and ecologically intermediate plants can often be found in BI in localities where the parental species grow in close proximity to one another, especially where arable land, with *T. inodorum* as a weed, occurs within a few metres of the drift-line habitat of *T. maritimum*. Hybrid populations of this type are usually small, apparently consisting mainly of F_1 plants restricted to a narrow strip of intermediate habitat, but some evidence of introgression between the parental populations can usually be observed. Larger intermediate populations, typically varying around the F_1 morphological type but sometimes closer to *T. maritimum*, and often separated by distances of some kilometres from the parents, are not infrequent in inland ruderal habitats in the maritime areas of western and southern BI, where they occur on railway-ballast, slag-heaps (e.g. in the lower Swansea valley) and waste-ground, and around quarries (e.g. china-clay workings in Cornwall). Hybrids or presumed hybrids have also been found in Da, Fe, Ga, Ge, ?No, Rs and Su.

d. Both parents are usually strongly self-incompatible, and artificial diploid hybrids can be made easily with either parent as the female; their characters are as described above. Isolation appears to be mainly ecological in BI and western Europe, where both parents are diploid. In the Baltic region, where *T. inodorum* is tetraploid and *T. maritimum* diploid, hybrids are probably triploid and isolation mainly genetic. Artificial triploid F_1 hybrids, which can be made easily with either species as the female parent, are of very low fertility, though not completely sterile..

e. *T. maritimum* $2n$ = 18; *T. inodorum* $2n$ = 18, (18, 36); hybrid $2n$ = 18.

f. HÄMET-AHTI, L. (1967). *Tripleurospermum* (Compositae) in the northern parts of Scandinavia, Finland, and Russia. *Acta bot. fenn.*, **75**: 3-16.

KAY, Q. O. N. (1969). The origin and distribution of diploid and tetraploid *Tripleurospermum inodorum* (L.) Schultz Bip. *Watsonia*, **7**: 130-141.

KAY, Q. O. N. (1972). Variation in sea mayweed (*Tripleurospermum maritimum* (L.) Koch) in the British Isles. *Watsonia*, **9**: 81-107.

NEHOU, J. (1954). Étude comparative de *Matricaria inodora* L. et de *M. maritima* L. (Composées Radiées). *Bull. Soc. scient. Bretagne*, **28**: 133-153.

532 × 531. *Matricaria* L.
× *Tripleurospermum* Schultz Bip.
(= × *Pseudomatricaria* Domin)
(by Q. O. N. Kay)

532/1 × 531/1b. *Matricaria recutita* L. × *Tripleurospermum inodorum* (L.) Schultz Bip. (*T. maritimum* (L.) Koch subsp. *inodorum* Hyland. ex Vaarama)

(= x *P. roblenae* Domin) has been reported from Cz and Ge.

535. *Artemisia* L.
(by C. A. Stace)

6 × 8. *A. absinthium* L. × *A. campestris* L.

has been recorded from He.

538. *Arctium* L.
(by F. H. Perring and P. D. Sell)

1 × 2. *A. lappa* L. × *A. nemorosum* Lej.
1 × 3. *A. lappa* L. × *A. pubens* Bab. = *A.* × *debrayi* Senay
1 × 4. *A. lappa* L. × *A. minus* Bernh. = *A.* × *nothum* (Ruhmer) J. Weiss
2 × 3. *A. nemorosum* Lej. × *A. pubens* Bab.
4 × 2. *A. minus* Bernh. × *A. nemorosum* Lej.
4 × 3. *A. minus* Bernh. × *A. pubens* Bab.

All four species of *Arctium*, *A. lappa* L., *A. nemorosum* Lej., *A. pubens* Bab., and *A. minus* Bernh. which are at present recognized in BI appear to be interfertile, as numerous fertile intermediates occur between all of them. Experiments at Cambridge showed that self-pollination gives about 100% good achenes even in intermediates, and that after emasculation no good achenes are formed; thus apomixis is unlikely. So far no hybridization experiments have been undertaken, and there appear to be no detailed chromosome studies.

Intermediates between *A. minus* and *A. nemorosum* are infrequent as the former is mainly southern in distribution whilst the latter is northern. *A. minus* occurs in pure populations in woodlands where other species do not encroach, but, in open ground in the south and east, intermediates with *A. pubens* and *A. lappa* are frequent. *A. lappa,* in fact, appears to be at one end and *A. minus* at the other of a continuum showing a complete range of intermediates. *A. pubens* itself may well be the result of hybridization between *A. lappa* and *A. minus,* as it has the characters which would be expected from such a cross. In the past *A. lappa* x *A. pubens* and *A. lappa* x *A. minus* have been the most frequently reported hybrids.

General References

ARÈNES, J. (1950). Monographie du genre *Arctium* L. *Bull. Jard. bot. État Brux.,* **20**: 67-156.

BEEBY, W. H. (1908). The British species of *Arctium. J. Bot., Lond.,* **46**: 380-382.

EVANS, A. H. (1913). The British species of *Arctium. J. Bot., Lond.,* **51**: 113-119.

PERRING, F. H. (1960). Report of the survey of *Arctium* L. agg. in Britain, 1959. *Proc. B.S.B.I.,* **4**: 33-37.

539. *Carduus* L.
(by W. A. Sledge)

4 x 3. *C. acanthoides* L. x *C. nutans* L.

a. *C.* x *orthocephalus* Wallr.

b. The capitula are intermediate in size, often 2–4 on the branches, and normally somewhat drooping; the involucral bracts are narrower and less strongly spinous than in *C. nutans* but more spreading than in *C. acanthoides;* and the florets are intermediate in size. The achenes are mostly abortive but variable numbers of good achenes are formed. In Canada hybrid swarms occur and plants show a complete range of variation from one parent to the other. F_2 plants mostly resemble *C. acanthoides* closely and have $2n = 22$.

c. This is probably the most frequently occurring thistle hybrid in Br, the parents often growing near one another. It is recorded from 25 vice-counties in England and Wales northwards to northern Yorkshire, and is widespread on the Continent. The two species are naturalized in Canada and hybrid swarms occur in Ontario.

d. None.

e. *C. nutans* ($2n = 16$); *C. acanthoides* ($2n = 22$); hybrids ($2n = 16-22$).

f. MOORE, R. J. and MULLIGAN, G. A. (1956). Natural hybridisation between *Carduus acanthoides* and *Carduus nutans* in Ontario *Can. J. Bot.*, **34**: 71-85. FIG.

MULLIGAN, G. A. and MOORE, R. J. (1961). Natural selection among hybrids between *Carduus acanthoides* and *C. nutans* in Ontario. *Can. J. Bot.*, **39**: 269-279.

SYME, J. T. B., ed. (1866). *English Botany*, 3rd ed., 5: 9, t. 685. London. FIG.

cri × 3. *C. crispus* L. × *C. nutans* L.

= *C.* × *polyacanthus* Schleich. has been recorded both in Br and on the Continent, but the Br records involve *C. crispus sensu lato*, i.e. including *C. acanthoides*.

4 × cri. *C. acanthoides* L. × *C. crispus* L.

= *C.* × *leptocephalus* Peterm. has been recorded from the Continent, but these two taxa have been much confused in the past and in any case are regarded by many workers as belonging to a single species.

539 × 540. *Carduus* L. × *Cirsium* Mill = × *Carduocirsium* Sennen
(by W. A. Sledge)

Carduus nutans L. × *Cirsium arvense* (L.) Scop. and *Carduus nutans* × *Cirsium vulgare* (Savi) Ten. have been tentatively recorded from v.c. 36 and 53/54 respectively, but both are almost certainly errors. A few hybrids between mostly non-British species of these genera have occasionally been doubtfully reported on the Continent.

540. *Cirsium* Mill
(by W. A. Sledge)

Virtually no experimental work has been done on hybridization in this genus, and many of the earlier records of *Cirsium* hybrids are unreliable. Most hybrids show low fertility, with few or no ripe achenes produced amongst a prevalence of shrivelled ones. ·

General Reference

NÄGELI, C. (1857). Dispositio specierum generis Cirsii, in KOCH, G. D. J. *Synopsis florae Germanicae et Helveticae*, 3rd ed., **2**: 741-760. Leipzig.

1 x 2. *C. eriophorum* (L.) Scop. x *C. vulgare* (Savi) Ten.

a. *C.* x *gerhardtii* Schultz Bip.

b. The leaves have a slightly clasping base forming shortly decurrent, spinous wings on the stem; the capitula are intermediate in shape with arachnoid webbing less dense than in *C. eriophorum*; and the involucral bracts are not dilated below the spinous tips. This hybrid is partially fertile and apparently backcrosses to *C. eriophorum* in v.c. 7 to give plants with non-decurrent leaves very similar to *C. eriophorum* but with capitula intermediate in size and with very sparse webbing on the involucral bracts.

c. The parents often grow together and flower at the same time but hybrids occur rarely and then solitarily or in small numbers. Examples have been seen by me from v.c. 7, 9, 19, 53 and 61, and there are also records from v.c. 6, 23 and 29. The hybrid is known from Au, Cz, Ga, Ge, He, Hu, It and Ju.

d. None.

e. *C. eriophorum* ($2n = 34$); *C. vulgare* ($2n = 68$).

f. BRENAN, J. P. M. (1946). Notes on the flora of Oxfordshire and Berkshire. *Rep. B.E.C.*, **12**: 781-802.

LOUSLEY, J. E. (1934). Two interesting hybrids in the British flora. *J. Bot., Lond.*, **72**: 171-173.

1 x 3. *C. eriophorum* (L.) Scop. x *C. palustre* (L.) Scop.

= *C.* x *dominii* M. Schulze has been reported from Au and Cz.

4 x 1. *C. arvense* (L.) Scop. x *C. eriophorum* (L.) Scop.

= *C.* x *sennenii* Rouy has been reported from Ga.

3 x 2. *C. palustre* (L.) Scop. x *C. vulgare* (Savi) Ten.

= *C.* x *subspinuligerum* Peterm. has been reported from v.c. 11 (wrongly) and v.c. 55 (no specimens seen); it is known from Au, Ga, Ge and He.

4 × 2. *C. arvense* (L.) Scop. × *C. vulgare* (Savi) Ten.

= *C.* × *csepeliense* Borbás has been reported from v.c. 13, 17, 18/19 and 24, but no genuine British examples of this hybrid have been seen. A plant so-named from v.c. 23 in herb. Druce (**OXF**) is *C. vulgare.* The hybrid is known from Be, Cz, Ga, Ge, He and Hu.

5 × 2. *C. oleraceum* (L.) Scop. × *C. vulgare* (L.) Scop.

= *C.* × *bipontinum* F. Schultz has been reported from Au, Ga, Ge and He.

6 × 2. *C. acaulon* (L.) Scop. × *C. vulgare* (Savi) Ten.

a. *C.* × *sabaudum* Löhr.
b. This hybrid grows to *c* 30 cm and has strongly spinous leaves, normally with a few spinules on the upper surface of the lobes and with a shortly decurrent base forming a spinous wing on the stem. The indumentum of the stem and leaves consists of jointed and some arachnoid hairs. The stem bears *c* 5 solitary, glabrous or feebly arachnoid, ovoid capitula of the colour of those of *C. acaulon,* with adpressed involucral bracts which have squarrose, spinous tips. The achenes are mostly shrivelled but a few are apparently fertile.
c. Hybrids have occurred as solitary plants in v.c. 6, 11 and 34, and unconfirmed records exist for v.c. 18/19, 53 and 54. The hybrid is also known in Ga, Ge, He, It and Su.
d. None.
e. *C. acaulon* ($2n = 34$); *C. vulgare* ($2n = 68$).
f. BRENAN, J. P. M. (1946). × *Cirsium sabaudum* Löhr (*C. acaule* × *vulgare* = *lanceolatum*). *Rep. B.E.C.,* **12**: 682-683.

7 × 2. *C. heterophyllum* (L.) Hill × *C. vulgare* (Savi) Ten.

= *C.* × *breunium* Goller & Huter has been reported from Au. It was also recorded for Br by Druce (with doubt) and by Dandy, but I have not been able to trace the source of these records.
DANDY, J. E. (1958). *Op. cit.,* p. 117.
DRUCE, G. C. (1928). *Op. cit.,* p. 63.

8 × 2. *C. dissectum* (L.) Hill × *C. vulgare* (Savi) Ten.

has been recorded from v.c. 46, 53 and 54, but almost certainly in error. A specimen so-named from v.c. 53 is *C. vulgare.* This combination is not recorded on the Continent.

4 × 3. *C. arvense* (L.) Scop. × *C. palustre* (L.) Scop.

a. *C.* × *celakovskianum* Knaf (*C.* × *mixtum* Druce, *nom. nud.*).
b. The stem is winged below with decurrent, sinuate-lobate leaves, but more or less naked above. The capitula are small and clustered or shortly stalked; the corolla-tube is longer than the corolla-limb which is 5-partite to three-quarters of its length.

c. *C. palustre* often grows on dry ground with or near *C. arvense,* but the hybrid is uncommon. It is recorded from scattered localities northwards to Yorkshire in v.c. 4–6, 8, 15, 17, 22–24, 33, 36–38, 41, 55 and 64, and also in Au, Cz, Ge, He and It.

d. None.

e. Both parents (2*n* = 34).

f. None.

5 × 3. *C. oleraceum* (L.) Scop. × *C. palustre* (L.) Scop.

= *C.* × *hybridum* Koch has been found in Au, Cz, Ga, Ge, He, No, Rm and Su.

6 × 3. *C. acaulon* (L.) Scop. × *C. palustre* (L.) Scop.

a. *C.* × *kirschlegeri* Schultz Bip.

b. The stems are erect, to 40 cm, and the upper part bears arachnoid and jointed hairs; the leaves are pinnatifid, some forming short, decurrent, spinous wings. The few capitula are intermediate in size and have shortly spinous, feebly arachnoid involucral bracts and florets which are shorter than in *C. acaulon* and have a limb equalling the tube.

c. Hybrids have occurred as solitary plants in v.c. 6, 9 and 23, and an unconfirmed record exists for v.c. 15. They are also known in Au, Ga, Ge, He, It and Su.

d. None.

e. Both parents (2*n* = 34).

f. YOUNG, D. P. (1961). *Cirsium acaulon* × *palustre* = *C.* × *kirschlegeri* Sch.-Bip. *Proc. B.S.B.I.,* 4: 275-276, where the figures in the table of corolla-, tube- and limb-lengths have been transposed. Those for *C.* × *kirschlegeri* should be 20–23, 10–11 and 10–12 mm respectively.

7 × 3. *C. heterophyllum* (L.) Hill × *C. palustre* (L.) Scop.

a. *C.* × *wankelii* Reichardt.

b. The stem is branched above and bears leaves which are white-tomentose beneath, irregularly serrate, lobed or pinnatifid, shortly decurrent and weakly spinous; the radical leaves are entire. The capitula are solitary or clustered, 0.5–2.0 cm diam., feebly arachnoid, and bear purple involucral bracts with short spinous tips.

c. Hybrids have been noted in v.c. 88, 89, 96, 97, 104 and 105 and probably occur elsewhere in mixed populations of the parents. It is one of the common thistle hybrids on the Continent, being recorded from Au, Cz, Ge, He, It, No, Rs and Su.

d. None.

e. Both parents (2*n* = 34).

f. WILMOTT, A. J. (1933). *Cirsium heterophyllum* × *palustre* in Inverness. *J. Bot., Lond.,* 71: 17-18.

8 × 3. *C. palustre* (L.) Scop. × *C. dissectum* (L.) Hill

a. *C.* × *forsteri* (Sm.) Loud. (*C.* × *spurium* Del.).

b. The leaves are decurrent, sinuate-pinnatifid or pinnatipartite with lobed, weakly spinous segments and arachnoid-pubescent below, and the upper part of the branched stem, which is not continuously winged, is cottony. The capitula are intermediate in size, solitary or 2–3 together on the branches and bear shortly spinous, webbed outer involucral bracts. Druce mistakenly applied the name *C.* × *forsteri* to *C. arvense* × *C. dissectum.*

c. The species commonly occur together and hybrids are sometimes numerous in mixed populations of the parents. It is certainly the most frequently occurring hybrid in BI, being recorded from v.c. 3, 4, 5/6, 9–11, 13, 14, 15/16, 17, 22, 29, 34, 37, 54, 55, 61, 102; H9, 10, 13, 14, 16, 19, 21, 23, 26, 27, 33, 35 and 40. It is also reported from Ga and Ho.

d. None.

e. *C. palustre* (2*n* = 34).

f. DRUCE, G. C. (1930). *Cirsium palustre* × *pratense* = × *C. spurium* Delastre. *Rep. B.E.C.,* 9: 23.

   PRAEGER, R. L. (1951). *Op. cit.,* p. 10.

   SYME, J. T. B., ed. (1866). *English Botany,* 3rd ed., 5: 19, t. 695. London. FIG.

3 × 9. *C. palustre* (L.) Scop. × *C. tuberosum* (L.) All.

a. *C.* × *semidecurrens* Richt.

b. This hybrid has more or less thickened roots, the stems are *c* 60 cm and unbranched, and the cauline leaves are pinnatipartite and decurrent and form spinous wings on the stem, which is naked above. The 6–15 capitula are subsolitary, ovoid, intermediate in size and borne on felted peduncles. The involucral bracts are appressed, spinous-tipped, glandular on the backs and the inner ones are purplish; the limb of the corolla exceeds the tube; and the fruits are all abortive.

c. Hybrids have occurred as isolated plants or in very small numbers at Great Ridge, v.c. 8 ("six flowering stems", site now destroyed) and at Nash Point, v.c. 41. They are also known in Ga, Ge, He and It.

d. None.

e. *C. palustre* (2*n* = 34).

f. GROSE, J. D. (1949). A hybrid thistle from Wiltshire. *Watsonia,* 1: 91.

4 × 5. *C. arvense* (L.) Scop. × *C. oleraceum* (L.) Scop.

= *C.* × *reichenbachianum* Löhr has been reported from Au and Ga.

6 × 4. *C. acaulon* (L.) Scop. × *C. arvense* (L.) Scop.

a. *C.* × *boulayi* Camus.

b. This hybrid has a creeping rootstock, branched stems 20–60 cm, and leaves resembling those of *C. arvense.* The indumentum of the stem and leaves is of jointed and some arachnoid hairs, the latter especially below the capitula. The capitula are numerous, on long peduncles, intermediate in size and bear strongly spinous involucral bracts; the limb of the corolla is split nearly to the base.

c. *C.* x *boulayi* has occurred in v.c. 19 and unconfirmed records exist for v.c. 6, 12, 33/34, 35, 37, 53 and 54; it is also known in Au, Cz, Ga, Ge and Su.

d. None.

e. Both parents (2*n* = 34).

f. GIBSON, G. S. (1844). Notice of a *Carduus* found near Saffron Walden, Essex. *Phytol.*, 1: 902-903.

SYME, J. T. B., ed. (1866). *English Botany,* 3rd ed., 5: 20-21, t. 697. London. FIG.

4 x 7. *C. arvense* (L.) Scop. x *C. heterophyllum* (L.) Hill

= *C.* x *discolor* Goller has been recorded from Au.

4 x 8. *C. arvense* (L.) Scop. x *C. dissectum* (L.) Hill

has been recorded for v.c. 6, doubtless in error; it is unknown on the Continent. It was also listed by Druce, who mistakenly took it for the parentage of *C.* x *forsteri.*

DRUCE, G. C. (1928). *Op. cit.,* p. 63.

MURRAY, R. P. (1896). *The flora of Somerset,* p. 202. Taunton.

4 x 9. *C. arvense* (L.) Scop. x *C. tuberosum* (L.) All.

= *C.* x *prantlii* Gremli has been recorded from Ge. A few British records exist, but these refer to *C. acaulon* x *C. tuberosum.*

6 x 5. *C. acaulon* (L.) Scop. x *C. oleraceum* (L.) Scop.

= *C.* x *rigens* Wallr. has been found in Au, Cz, Ga and Ge.

7 x 5. *C. heterophyllum* (L.) Hill x *C. oleraceum* (L.) Scop.

= *C.* x *affine* Tausch has been found in Au, Cz, Ge, No and Su.

5 x 9. *C. oleraceum* (L.) Scop. x *C. tuberosum* (L.) All.

= *C.* x *braunii* F. Schultz has been found in Ga, Ge and He.

6 x 7. *C. acaulon* (L.) Scop. x *C. heterophyllum* (L.) Hill

= *C.* x *alpestre* Näg. has been found in Au, Ga and He.

6 x 8. *C. acaulon* (L.) Scop. x *C. dissectum* (L.) Hill

a. *C.* x *woodwardii* (H. C. Wats.) Nyman.

b. The creeping rootstock bears unthickened roots and branched stems. The leaves are deeply pinnatifid, less spinous than in *C. acaulon* and arachnoid beneath, and the upper part of the stem has both arachnoid and jointed hairs. The capitula are solitary and terminal with lanceolate-acuminate, mucronate, feebly arachnoid involucral bracts. Hybrid plants are functionally female but set some seed (?parthenogenetically); the resulting plants show no segregation.

c. This is an extremely rare hybrid since the parents differ widely both in their ecology and time of flowering, but it formerly occurred near Swindon, v.c. 7, where it persisted for a century in its original station. The locality has now been destroyed but plants are maintained in cultivation. Records from v.c. 41 are almost certainly errors; that from Nash Point is probably *C. acaulon* x *C. tuberosum*. Elsewhere it is only recorded from Ga (no specimens seen).

d. None.

e. *C. acaulon* ($2n = 34$).

f. GROSE, J. D. (1957). *Flora of Wiltshire*, p. 356. Devizes.

SYME, J. T. B., ed. (1866). *English Botany*, 3rd ed., **5**: 19-20, t. 696. London. FIG.

6 x 9. *C. acaulon* (L.) Scop. x *C. tuberosum* (L.) All.

a. *C.* x *zizianum* Koch (*C.* x *medium* All., *C.* x *fraseranum* Druce, *nom. nud.*).

b. The non-creeping rootstock bears more or less thickened roots and 20–60 cm, branched stems, the upper parts of which bear arachnoid and jointed hairs. The leaves are intermediate in shape, narrower in outline and with shorter, broader and closer lobes than in *C. tuberosum*. The capitula are more elongated than in *C. tuberosum*. F_1 hybrids are intermediate between the parents but backcrossing gives plants intergrading with and scarcely distinguishable from both. The most reliable character for recognizing plants of hybrid origin is the nature of the pubescence—jointed hairs only in *C. acaulon*, arachnoid hairs only in *C. tuberosum* and a mixture in the hybrids. Hybrid plants are functionally female.

c. Hybrids occur in several places on the Wiltshire Downs, v.c. 7 and 8, where introgression is common; they also occur in v.c. 41, and are known from Au, Ga, Ge, He and It.

d. None.

e. *C. acaulon* ($2n = 34$).

f. GROSE, J. D. (1942). The tuberous thistle in Wiltshire. *Wilts. archaeol. nat. Hist. Mag.*, **49**: 557-561. FIG.

GROSE, J. D. (1957). As above, pp. 357-358.

PIGOTT, C. D. (1968). *Cirsium acaulon* (L.) Scop., in Biological Flora of the British Isles. *J. Ecol.*, **56**: 597-612.

8 x 9. *C. dissectum* (L.) Hill x *C. tuberosum* (L.) All.

is listed for Br by Dandy but I can trace no British record, nor can Dandy trace the evidence for its inclusion in his list. It is unknown on the Continent.

DANDY, J. E. (1958). *Op. cit.*, p. 117.

544. *Centaurea* L.
(by D. J. Ockendon)

7 × 1. *C. nemoralis* Jord. × *C. scabiosa* L.

= *C.* × *cantiana* C. E. Britton (*C.* × *brittonii* Druce) was reported by Britton from near Luddesdown, v.c. 16. The record appears to be based on a single specimen which is probably an aberrant plant of *C. nigra* L. (incl. *C. nemoralis*). Confirmation is needed for the existence of such a hybrid.
BRITTON, C. E. (1923). *Centaurea scabiosa* L. varieties and a hybrid. *Rep. B.E.C.*, 6: 767-773.

5 × 6. *C. jacea* L. × *C. nigra* L. (incl. *C. nemoralis* Jord.)

a. *C.* × *drucei* C. E. Britton (*C.* × *moncktonii* C. E. Britton).
b. Hybrids are intermediate between the parents and recognized by their deeply and irregularly laciniate bract-appendages. The similarity and great variability of the parents makes certain recognition of the hybrids very difficult. Characters such as bract-colour and the presence or absence of ray-florets are probably not reliable indicators of hybridity. The hybrids are fully fertile and introgression is claimed to be common in southern England (Marsden-Jones and Turrill, 1954), but the evidence is not very convincing. A comparable, but reverse situation, has been described in Fe, where *C. jacea* is native and *C. nigra* occurs as an introduction. Pure *C. nigra* does not persist for long, but hybridizes readily with adjacent *C. jacea* and may introgress into neighbouring populations of it.
c. Hybrids have been recorded from about 20 vice-counties in Br, chiefly from central and southern England, with scattered occurrences as far north as the Orkneys. Hybrids probably have not persisted for very long at many of these localities and little reliable current information is available. As *C. nigra* is common and widespread in Br, hybrids may well be found wherever *C. jacea* becomes established. The latter is probably less often introduced now than formerly and seldom persists for very long. *C. pratensis* Thuill., which is morphologically intermediate between *C. jacea* and *C. nigra,* may be a hybrid. It is reported from v.c. 14, 16, 17, 22, 24, 34, 58 and 88/89, and from Be, Ga and Ge.
d. Artificial hybrids can be made easily with either parent as the female. Backcrosses to both parents have also been made.
e. Both parents and hybrid 2*n* = 44.
f. BRITTON, C. E. (1921). British forms of *Centaurea jacea. Rep. B.E.C.*, 6: 163-173.
BRITTON, C. E. (1922). British Centaureas of the *nigra* group. *Rep. B.E.C.*, 6: 406-417.
BRITTON, C. E. (1927). *Centaurea pratensis. Rep. B.E.C.*, 8: 149-152.
MARSDEN-JONES, E. M. and TURRILL, W. B. (1954). *British knapweeds.* London.

SAARISALO-TAUBERT, A. (1966). A study of hybridisation in *Centaurea*, section *Jacea*, in eastern Fennoscandia. *Annls bot. fenn.*, **3**: 86-95.

9 × 5. *C. calcitrapa* L. × *C. jacea* L.

= *C.* × *jaceiformis* Rouy has been recorded from Ga, as has *C. calcitrapa* × *C. pratensis* Thuill. = *C.* × *nouelii* Franchet (see under *C. jacea* × *C. nigra*).

7 × 6. *C. nemoralis* Jord. × *C. nigra* L.

is widespread in Br, but the current consensus of opinion in this country is that these two taxa are not worthy of recognition as separate species. They are completely interfertile and a complete range of intermediates between the two is frequently found; moreover the characters used to separate the two taxa are found in varying combinations often in one population.

ELKINGTON, T. T. and MIDDLEFELL, L. C. (1972). Population variation within *Centaurea nigra* L. in the Sheffield region. *Watsonia*, **9**: 109-116.

MARSDEN-JONES, E. M. and TURRILL, W. B. (1954). As above.

OCKENDON, D. J., WALTERS, S. M. and WHIFFEN, T. P. (1969). Variation within *Centaurea nigra* L. *Proc. B.S.B.I.*, **7**: 549-552.

8 × 6. *C. aspera* L. × *C. nigra* L.

= *C.* × *pagesii* Coste & Soul. has been recorded from Ga.

8 × 9. *C. aspera* L. × *C. calcitrapa* L.

= *C.* × *pouzinii* DC. has been recorded from Ga, Hs and It.

544 × 545. *Centaurea* L. × *Serratula* L. = × *Centauserratula* Arènes
(by C. A. Stace)

A hybrid between two non-British species of these genera has been reported from the Lebanon.

549. *Hypochoeris* L.
(by J. S. Parker)

2 × 1. *H. glabra* L. × *H. radicata* L.
a. None.
b. Hybrids are short-lived perennials with many, much-branched, floriferous scapes arising from a basal leaf-rosette. The leaves are usually hairier than in *H. glabra,* but the degree of hairiness depends largely on the *H. radicata* parent. The capitula are intermediate in size between those of the parents; the ligules are longer than the involucral bracts and the capitula open in dull weather, as in *H. radicata.* Hybrid-fertility is less than 5% of that of the parents, most mature capitula having no viable achenes. The achenes produced are usually backcrosses to *H. radicata.*
c. Hybrids probably occur occasionally wherever the parents grow together in sandy fields and on brecks and sand-dunes, etc. They have so far been recorded only from v.c. 17, 26, 28 and 48. They are also known in Ge, He and Su.
d. Artificial hybrids are easily made using *H. glabra* as the female parent. At meiosis in F_1 hybrids maximum pairing is a ring-bivalent and a chain of seven chromosomes. The two species differ by at least three chromosome interchanges. Viable gametes may contain four, five or rarely nine chromosomes. Backcrosses with *H. radicata* as the male parent may have $2n = 8$, 9 or 13; those with *H. glabra* as the female parent all have $2n = 10$, but are deformed, non-flowering plants with numerous, linear leaves bearing a few, irregular teeth.
e. *H. glabra* $2n = 10$; *H. radicata* $2n = 8$; wild and artificial F_1 hybrids $2n = 9$.
f. BENOIT, P. M. (1959). Two interesting botanical discoveries in Merioneth-shire. *Nature Wales,* **5**: 726-728.
HOLMBERG, O. R. (1930). *Hypochoeris glabra* L. × *radicata* L., *nova hybr. Bot. Notiser,* **1930**: 413-416. FIG.
PARKER, J. S. (1971). *The control of recombination.* D.Phil. thesis, University of Oxford.
SANDWITH, N. Y. (1954). A *Hypochoeris* hybrid. *Proc. B.S.B.I.,* **1**: 99.

550. *Leontodon* L.
(by R. A. Finch)

1 × 2. *L. autumnalis* L. × *L. hispidus* L.
= *L.* × *ambiguus* Fleischer (*L.* × *hispidaster* Beauv.) has been recorded from Cz, Ga, Ge and He, but requires confirmation.

2 × 3. L. hispidus L. × **L. taraxacoides** (Vill.) Mérat

a. None.

b. F₁ hybrids are as variable as the two species but may usually be recognized in the field by their intermediacy between accompanying parental forms, especially in size, indumentum, leaf-shape and pappus development, and low (<0.1%) pollen- and achene-fertility. Most F₁ hybrids have nearly glabrous involucres and mainly the *L. taraxacoides* pappus-type on the marginal achenes. All F₁ hybrids show unusually great variability in pollen grain size.

c. F₁ hybrids have been found with the parental species in two places near Wytham and near Oxford, v.c. 22, between Bicester and Aylesbury, v.c. 24, and at Seaton Sluice, v.c. 67, and a presumed backcross with *L. hispidus* (2*n* = 14) was found near Arundel, v.c. 13. P. M. Benoit (*in litt.* 1971) has found putative hybrids in v.c. 40.

d. Vigorous F₁ hybrids were obtained easily from reciprocal crosses. Backcrosses were obtained reciprocally with both species, but seed-set was very low. First generation backcrosses varied greatly in vigour, morphology and fertility. Some were identical, except chromosomally, with the pure species. Hybridization and a little backcrossing probably occur in many mixed populations, for the species are genetically separated mainly by the very low fertility in the F₁ hybrid due to its meiotic irregularity.

e. *L. hispidus* 2*n* = 14, 14 + 1B (rare), 21 (rare, autotriploids); *L. taraxacoides* 2*n* = 8, 12 (uncommon, autotriploids); F₁ hybrid 2*n* = 11.

f. BLACKBURN, K. B. (1945). A new hybrid plant. *The Vasculum*, **30**: 55.

　ELLIOT, E. (1950). A new phase of amphiplasty in *Leontodon*. *New Phytol.*, **49**: 344-349.

　FINCH, R. A. (1967). Natural chromosome variation in *Leontodon*. *Heredity, Lond.*, **22**: 359-386.

552. *Tragopogon* L.
(by S. M. Walters)

2 × 1. T. porrifolius L. × **T. pratensis** L.

a. *T.* × *mirabilis* Rouy.

b. The F₁ hybrid is recognized by its flower-colour, which is usually rather dull purple in general appearance; in detail the individual ligules have a yellow base but are distally suffused with purple to a variable extent. In other characters, such as the leaf-base and the degree of inflation of the peduncle, hybrids are also intermediate. Ellis (1931, 1932) gave details of a complex hybrid swarm established between 1924 and 1931 on waste ground at Great Yarmouth, v.c. 27. Different individuals showed varied combinations of characters, and clearly represented later generations derived from partially fertile F_1 plants, most of which, however, generally show greatly reduced fertility. Typical *T.* x *mirabilis* is the hybrid involving *T. pratensis* subsp. *pratensis* but British hybrids probably all involve subsp. *minus* (Mill.) Wahlenb.

c. Hybrids are very occasionally recorded on road-sides or waste ground where the alien *T. porrifolius* has become established, but they are not known to persist for many years. There are definite records for v.c. 15, 17, 27 and 29, and for Da, Ga, Ho and Su.

d. Britton (1917) raised the hybrid in his garden at Merton, v.c. 17, and distributed specimens via the Botanical Exchange Club. He reported that the plants were largely sterile but that "most capitula produced one or two fertile seeds". Winge (1938) undertook extensive studies using Danish material, and analysed progenies as far as the F_7. The F_1 was only 2% to 3% seed-fertile, but chromosome pairing at meiosis was regular and complete; F_1 sterility is caused by death of the embryo at an early stage of development. Artificial crosses have been made using *T. pratensis* subsp. *pratensis*, subsp. *minus* and subsp. *orientalis* (L.) Vollmann.

e. Both parents and artificial hybrids ($2n = 12$).

f. BRITTON, C. E. (1917). *T. porrifolium* x *pratensis. Rep. B.E.C.,* **4**: 576-577.

BRITTON, C. E. and TODD, W. A. (1910). *Tragopogon porrifolium* x *pratense. J. Bot., Lond.,* **48**: 203-204.

CLAUSEN, J. (1966). Stability of genetic characters in *Tragopogon* species through 200 years. *Trans. Proc. bot. Soc. Edinb.,* **40**: 148-158.

ELLIS, A. E. (1931). *Tragopogon minor* x *porrifolius. Rep. B.E.C.,* **9**: 272-274.

ELLIS, A. E. (1932). *Tragopogon porrifolius* x *pratensis* (*minus* Mill.). *Rep. B.E.C.,* **9**: 566-567.

MERTENS, T. R. (1972). Student investigations of speciation in *Tragopogon. J. Hered.,* **63**: 39-41. FIG.

OWNBEY, M. (1950). Natural hybridisation and amphiploidy in the genus *Tragopogon. Am. J. Bot.,* **37**: 487-499.

WINGE, O. (1938). Inheritance of species characters in *Tragopogon. C. r. Trav. Lab. Carlsberg,* Sér. physiol., **22**: 155-193.

554. *Lactuca* L.

(by C. A. Stace)

3 x **1.** *L. saligna* L. x *L. serriola* L.

= *L.* x *dichotoma* Simonk. has been recorded from Rm.

556. *Sonchus* L.
(by R. A. Lewin)

2 × 1. *S. arvensis* L. × *S. palustris* L.

was recorded tentatively from a gravel-pit in v.c. 31, but no specimen has been seen and such a hybrid is unknown elsewhere..

DONY, J. G. (1950). Excursions, 1948. Huntingdonshire.. *Year Book B.S.B.I.*, **1950**: 51-53.

2 × uli. *S. arvensis* L. × *S. uliginosus* Bieb. (*S. arvensis* var. *uliginosus* (Bieb.) Béguinot)

occurs in north-eastern North America. The *S. arvensis* there has $2n = 54$ (cf. British material $2n = 64$), *S. uliginosus* has $2n = 36$ and the hybrid swarms have $2n = 37\text{-}43$.

4 × 3. *S. asper* (L.) Hill × *S. oleraceus* L. em. Gouan

a. (*S. × picquetii* Druce has been determined by me as *S. oleraceus*).

b. Although the parental species are both very common and widespread as weeds in Europe and elsewhere, recognizable authentic hybrids are extremely rare. In true hybrids, which are sterile, the auricles of the upper leaves are rounded and dentate, more or less as in *S. asper*, with a single elongate tooth directed downwards or backwards. The achenes are abortive (small and white), but possibly inclined to the *S. asper* type (with wings and lacking transverse rugosity).

c. Barber (1941) found that 9 out of 12 seedlings raised from achenes obtained from a single, typical *S. asper* plant, open-pollinated in a field at Merton, v.c. 17, were undoubted hybrids, as indicated by their intermediate leaf-form and by their chromosome number and complete sterility. Lewin (1948), despite extensive field observations throughout Br, found only two such hybrids, one in Cambridge, v.c. 29, in 1941 and the other in Kinnerley, v.c. 40, in 1943. Doubtful records include those of Bennett from v.c. 99, Druce from v.c. 8 and Jersey, and Harrison from v.c. 104. A record of Simpson from v.c. H2 also requires checking. Elsewhere there are records from several places on the Continent, but all require confirmation.

d. The parental species are self-fertile and, presumably, readily self-pollinated. Barber (1941) tried to cross the two species artificially, but without success. Hybrids have recently been synthesized by Tsun-Shi *et al.*

e. *S. oleraceus* $2n = 32$; *S. asper* $2n = 18$; hybrid $2n = 25$.

f. BARBER, H. N. (1941). Spontaneous hybrids between *Sonchus asper* and *S. oleraceus*. *Ann. Bot.*, n.s., **5**: 375-377.

DRUCE, G. C. (1924). *Sonchus asper × oleraceus = S. Piquetii mihi. Rep. B.E.C.*, **7**: 43-44.

LEWIN, R. A. (1948). *Sonchus* L., in Biological Flora of the British Isles. *J. Ecol.*, **36**: 203-223.

TSUN-SHI, H., SCHOOLER, A. B., BELL, A. and NALEWAJA, J. D. (1972). Cytotaxonomy of three *Sonchus* species. *Am. J. Bot.*, **59**: 789-796.

ZENARI, S. (1921). Forme ereditarie e variabilità nei cicli di *Sonchus oleraceus* L. et di *Sonchus asper* Hill. *Riv. Biol.*, **3**: 6.

558. *Hieracium* L.
(by P. D. Sell and C. West)

All British species of *Hieracium*, except *H. umbellatum* L., are apparently apomictic. Gustafsson (1946) reported both apomictic and amphimictic plants in *H. umbellatum*, and the behaviour of this species in BI suggests this is so. All the hybrids reported from BI, with the exception of the last of those listed below, are errors of identification.

General References

GUSTAFSSON, A. (1946). Apomixis in higher plants. *Acta Univ. lund.*, Ser. 2, **42**.

PUGSLEY, H. W. (1948). A prodromus of the British Hieracia. *J. Linn. Soc., Bot.*, **54**.

1/33 × 1/65. *H. anglicum* Fries × *H. hypochoeroides* Gibson

was reported from two areas near Settle, v.c. 64, by Hanbury (1893). The specimens on which these reports were based were included in *H. hypochoeroides* var. *lancifolium* W. R. Linton by Linton (1905) and Pugsley (1948). In our opinion they are referable to *H. rubiginosum* F. J. Hanb. (1/139).

HANBURY, F. J. (1893). Further notes on Hieracia new to Britain. *J. Bot., Lond.*, **31**: 16-19.

LINTON, W. R. (1905). *An account of the British Hieracia*. London.

1/164 × 1/219. *H. strumosum* (W. R. Linton) A. Ley (*H. lachenalii* auct., *non* C. C. Gmel.) × *H. perpropinquum* (Zahn) Druce

(as *H. sciaphilum* (Uechtr.) F. J. Hanb. × *H. boreale* Fries) was reported by Hanbury (1894) from a railway-bank near Rhayader, v.c. 43, where it was collected in 1888 by A. Ley, and mentioned by Linton (1905) and by

Druce (1928) (as *H. boreale* x *H. vulgatum* Fries = *H.* x *luescheri sensu* Druce, *non* Zahn). The specimens on which the record was based were described by Pugsley (1948) as a new species, *H. rhayaderense* Pugsl. Similar plants still grow in quantity on the railway-bank near Rhayader. In our opinion they are all ' referable to *H. scabrisetum* (Zahn) Roffey (1/199).

DRUCE, G. C. (1928). *Op. cit.*, p. 71.

HANBURY, F. J. (1894). Notes on British Hieracia. *J. Bot., Lond.*, **32**: 225-233.

LINTON, W. R. (1905). As above.

1/219 x **1/209.** *H. perpropinquum* (Zahn) Druce x *H. subcrocatum* (E. F. Linton) Roffey

(as *H. commutatum* Beck x *H. eupatorium* Griseb.) was reported by Hanbury (1893) from Torpantau Station, v.c. 42, where it was collected by A. Ley in 1892, and mentioned by Hanbury and Ley (1894) and Linton (1905). Pugsley (1948) described the specimens on which the record was based as a new species, *H. argutifolium* Pugsl. (1/219). In our opinion they are referable to *H. perpropinquum.*

HANBURY, F. J. (1893). As above.

HANBURY, F. J. and LEY, A. (1894). *H. Eupatorium* x *boreale. Rep B.E.C.*, **1**: 419.

LINTON, W. R. (1905). As above.

1/218 x **1/217.** *H. umbellatum* L. subsp. *bichlorophyllum* (Druce & Zahn) Sell & West (*H. bichlorophyllum* (Druce & Zahn) Pugsl.) x *H. umbellatum* L. subsp. *umbellatum*

is the probable parentage of intermediates between these two taxa which occur within the range of the latter in Wales, Devon and Cornwall. For this reason we prefer to consider the two taxa as subspecies of *H. umbellatum.*

558P. *Pilosella* Hill
(by P. D. Sell and C. West)

The genus *Pilosella* is largely sexual in BI, and intermediates have been recorded between all taxa that are known to grow close together. On the Continent the situation is much more complicated because about 18 species occur as natives, and in addition to the hybrids given here about 12 others are known, each of which involves three or four of the species mentioned in this account. In BI only *P. officinarum*, *P. peleterana* and the isolated *P. flagellaris* subsp. *bicapitata* are certainly native.

General References

NAEGELI, C. and PETER, A. (1885). *Die Hieracien Mittel-Europas — Piloselloiden.* Munich.

PETER, A. (1884). Über spontane und künstliche Gartenbastarde der Gattung *Hieracium* sect. *Piloselloidea. Bot. Jb.*, 5: 239-286, 448-496; 6: 111-136.

PUGSLEY, H. W. (1948). A prodromus of the British Hieracia. *J. Linn. Soc., Bot.*, 54.

2/1 × 2/2. *P. officinarum* C. H. & F. W. Schultz (*Hieracium pilosella* L.) × *P. peleterana* (Mérat) C. H. & F. W. Schultz (*H. peleteranum* Mérat)

a. (*H.* × *longisquamum* Peter; *P.* × *pachylodes* (Naegeli & Peter) Sojak, *nom. illegit.*).

b. Plants intermediate in hair-clothing and in the shape of the involucral bracts appear to be hybrids between these two species. Those from v.c. 39 have short stolons, those from elsewhere longer ones.

c. Putative hybrids were recognized by us in 1954 at Wetton Mill, v.c. 39. In 1958 C. West found intermediates at Plemont and Quennevais, Jersey, and on a roadside bank between St. Peter Port and Fort George, Guernsey. There is also a similar specimen at **CGE** collected by F. R. Tennant at Folkestone, v.c. 15, in 1890, but *P. peleterana* has never been recorded from there. A specimen collected by A. J. Wilmott in 1933 near Jubilee Hill Road, Jersey (**BM**), was determined by Pugsley (1948, p. 313) as this hybrid, and we agree with this. In v.c. 39 and Jersey the *P. peleterana* parent is referable to subsp. *tenuiscapa* (Pugsl.) Sell & West, but in Guernsey to subsp. *peleterana.* The *P. officinarum* parent is probably subsp. *concinnata* (F. J. Hanb.) Sell & West in the case of the Wilmott specimen, but subsp. *officinarum* in all other cases. Hybrids have also been recorded from Fe, Ga, Ge, He, It, No, Rs and Su.

d. Hybrids synthesized by R. A. Finch were similar in appearance to the wild putative hybrids.

e. *P. officinarum* $2n = 18, 36, 45, 54, 63$; *P. peleterana* $2n = 18, 27, 36, 45$; artificial hybrid $2n = 27$ (parents $2n = 36$ and $2n = 18$ respectively).

f. None.

2/3 × 2/1. *P. lactucella* (Wallr.) Sell & West (*Hieracium lactucella* Wallr.) × *P. officinarum* C. H. & F. W. Schultz (*H. pilosella* L.)

= *P.* × *schultesii* (F. W. Schultz) C. H. & F. W. Schultz has been recorded from Au, Be, Co, Cz, Fe, Ga, Ge, He, Ho, Hu, It, Ju, Po, Rs, Sa and Su.

2/6 × 2/1. *P. caespitosa* (Dumort.) Sell & West (*Hieracium collini-forme* (Naegeli & Peter) Roffey) × *P. officinarum* C. H. & F. W. Schultz (*H. pilosella* L.)

> (= *H.* × *duplex* Peter, *P.* × *prussica* (Naegeli & Peter) Sojak) has been recorded from Au, Cz, Fe, Ge, Hu, Ju, Po, Rm, Rs and Su.

2/7 × **2/1.** *P. aurantiaca* (L.) C. H. & F. W. Schultz (*H. aurantiacum* L., incl. *H. brunneocroceum* Pugsl.) × *P. officinarum* C. H. & F. W. Schultz (*Hieracium pilosella* L.)

> a. *P.* × *stoloniflora* (Waldst. & Kit.) C. H. & F. W. Schultz.
> b. The putative hybrids in v.c. 95 show considerable morphological variation which suggests that backcrossing occurs. Plants of several sorts are apparent: some differing from *P. officinarum* only in having long black hairs; some similar to the last but with reddish flowers; some with two reddish-flowered capitula on a stem; and some with several reddish-flowered capitula and which may differ from *P. aurantiaca* only in the larger size of the capitula. The vegetative parts are nearly always intermediate in character.
> c. Such plants were first found in BI by Miss M. McC. Webster in 1957 on a grassy bank by Cromdale Railway Station, v.c. 95, where they have been seen several times since. The parents growing with the hybrid are the type subspecies in each case. In 1959 Miss U. K. Duncan collected a rather poor specimen from Lintrose near Newtyle, v.c. 90, which seems to be referable to this hybrid. An apparent hybrid swarm is also locally abundant on old lime waste near Northwich, v.c. 58, where it was first found by Mrs G. Mackie in 1967; in this case the *P. aurantiaca* parent is subsp. *brunneocrocea* (Pugsl.) Sell & West (J. N. Mills pers. comm. 1972). Plants apparently of this parentage are grown by nurserymen, and they sometimes escape from gardens and spread vegetatively. The hybrid has also been recorded from Au, Cz, Ga, Ge, He, It, Ju, Po, Rm and Rs.
> d. Hybrids have been produced artificially; they were very variable, some being nearer one parent and some the other.
> e. *P. officinarum* $2n$ = 18, 36, 45, 54, 63; *P. aurantiaca* ($2n$ = 18, 27, 36, 45, 54, 63, 72).
> f. OSTENFELD, C. H. and ROSENBERG, O. (1906). Experimental and cytological studies in the Hieracia. *Bot. Tidsskr.*, **27**: 225-248. FIG.

2/1 × **2/9.** *P. officinarum* C. H. & F. W. Schultz (*Hieracium pilosella* L.) × *P. praealta* (Vill. ex Gochnat) C. H. & F. W. Schultz (*H. praealtum* Vill. ex Gochnat)

> = *P.* × *brachiata* (Bertol. ex Lam.) C. H. & F. W. Schultz has been recorded from Al, Au, Bu, Cz, Ga, Ge, He, Hu, It, Ju, Po, Rm and Rs.

2/3 × 2/2. *P. lactucella* (Wallr.) Sell & West (*Hieracium lactucella* Wallr.) × *P. peleterana* (Mérat) C. H. & F. W. Schultz (*H. peleteranum* Mérat)

(= *H.* × *auriculiforme* Fries) has been recorded from Fe, Ga, He, It, No and Su.

2/6 × 2/2. *P. caespitosa* (Dumort.) Sell & West (*Hieracium colliniforme* (Naegeli & Peter) Roffey) × *P. peleterana* (Mérat) C. H. & F. W. Schultz (*H. peleteranum* Mérat)

(= *H.* × *chaetocephalum* Hofmann) has been recorded from Ge.

2/7 × 2/2. *P. aurantiaca* (L.) C. H. & F. W. Schultz (*Hieracium aurantiacum* L.) × *P. peleterana* (Mérat) C. H. & F. W. Schultz (*H. peleteranum* Mérat)

(= *H.* × *bryhnii* Blytt ex Omang) has been recorded from No.

2/2 × 2/9. *P. peleterana* (Mérat) C. H. & F. W. Schultz (*Hieracium peleteranum* Mérat) × *P. praealta* (Vill. ex Gochnat) C. H. & F. W. Schultz (*H. praealtum* Vill. ex Gochnat)

(= *H.* × *longistolonosum* Vollman; *H.* × *promeces* Peter) has been recorded from Ge.

2/6 × 2/3. *P. caespitosa* (Dumort.) Sell & West (*Hieracium colliniforme* (Naegeli & Peter) Roffey) × *P. lactucella* (Wallr.) Sell & West (*H. lactucella* Wallr.)

= *P.* × *floribunda* (Wimmer & Graebner) Arvet Touvet has been recorded from Au, Cz, Fe, Ge, Ho, Hu, No, Po, Rs and Su.

2/7 × 2/3. *P. aurantiaca* (L.) C. H. & F. W. Schultz (*Hieracium aurantiacum* L.) × *P. lactucella* (Wallr.) Sell & West (*H. lactucella* Wallr.)

(= *H.* × *fuscum* Vill.). has been recorded from Au, Cz, Fe, Ga, Ge, He, It, Ju, No, Po, Rm and Rs.

2/3 × 2/9. *P. lactucella* (Wallr.) Sell & West (*Hieracium lactucella* Wallr.) × *P. praealta* (Vill. ex Gochnat) C. H. & F. W. Schultz (*H. praealtum* Vill. ex Gochnat)

= *P.* × *koernickeana* (Naegeli & Peter) Sojak has been recorded from Au, Cz, Ge, Hu, Po, Rm and Rs.

2/7 × 2/5. _P. aurantiaca_ (L.) C. H. & F. W. Schultz (_Hieracium aurantiacum_ L.) × **_P. flagellaris_** (Willd.) Sell & West (_H. flagellare_ Willd.)

= _P._ × _rubra_ (Peter) Sojak has been recorded from Cz.

2/7 × 2/6. _P. aurantiaca_ (L.) C. H. & F. W. Schultz (_Hieracium aurantiacum_ L.) × **_P. caespitosa_** (Dumort.) Sell & West (_H. colliniforme_ (Naegeli & Peter) Roffey)

= _P._ × _fuscatra_ (Naegeli & Peter) Sojak has been recorded from Au, Bu, Cz, Fe, Gr, He, It, Ju, No, Po, Rm, Rs and Su.

2/6 × 2/9. _P. caespitosa_ (Dumort.) Sell & West (_Hieracium colliniforme_ (Naegeli & Peter) Roffey) × **_P. praealta_** (Vill. ex Gochnat) C. H. & F. W. Schultz (_H. praealtum_ Vill. ex Gochnat)

(= _H._ × _polymastix_ Peter; _P._ × _obornyana_ (Naegeli & Peter) Sojak, _nom. illegit._) has been recorded from Au, Cz, Fe, Ge, Hu, Po, Rm and Rs.

2/7 × 2/9. _P. aurantiaca_ (L.) C. H. & F. W. Schultz (_Hieracium aurantiacum_ L.) × **_P. praealta_** (Vill. ex Gochnat) C. H. & F. W. Schultz (_H. praealtum_ Vill. ex Gochnat)

= _P._ × _calomastix_ (Peter) Sojak has been recorded from Au, Cz, ?Rm and Rs.

559. _Crepis_ L.
(by C. A. Stace)

1 × 3. _C. foetida_ L. × **_C. setosa_** Haller f.

is reported from Cz.

5 × 2. _C. biennis_ L. × **_C. vesicaria_** L. subsp. **_taraxacifolia_** (Thuill.) Thell.

was reported doubtfully from Ge.

5 × 6. _C. biennis_ L. × **_C. capillaris_** (L.) Wallr.

= _C._ × _druceana_ Murr. ex Druce was reported from v.c. 22 in 1923, but Druce later placed _C. druceana_ as a synonym of _C. capillaris_ var. _anglica_ Druce & Thell., and J. B. Marshall (after examining the type of _C._

druceana) seemed to conclude that it should be placed under *C. capillaris*. There is a doubtful record of this hybrid from Au.

DRUCE, G. C. (1926). *Crepis Druceana* Murr. *Rep. B.E.C.*, **7**: 774.
DRUCE, G. C. (1928). *Op. cit.*, p. 65.
MARSHALL, J. B. (1964). Notes on British *Crepis*, 2. Variants of *Crepis capillaris* (L.) Wallr. *Proc. B.S.B.I.*, **5**: 325-333.

559 × 560. *Crepis* L. × *Taraxacum* Weber
(by C. A. Stace)

559/2 × 560/1. *Crepis vesicaria* L. subsp. *taraxacifolia* (Thuill.) Thell. × *Taraxacum officinale* Weber agg.

was reported from v.c. 37 in 1908, but no specimen has been traced. The existence of such a hybrid is extremely unlikely.

TOWNDROW, R. F. (1908). Worcestershire hybrids. *J. Bot., Lond.*, **46**: 364-365.

560. *Taraxacum* Weber
(by A. J. Richards)

Richards (1972) recognized 132 agamospecies of dandelion in BI, and so far 15 more have been detected since that publication appeared. Traditionally all these agamospecies have been grouped into four or five aggregates, but a clear understanding of the hybridization potential of British dandelions can only be obtained by considering the agamospecies separately.

In BI sexuality is known only in four of the agamospecies, and it is probable that throughout the world at least 90% of the agamospecies are obligate apomicts (Richards, 1972). Of these four species, two are probably introduced: *T. austriacum* v. Soest in v.c. 66 and *T. obtusilobum* Dahlst. in v.c. 15 and 66. Both hybridize extensively in their native areas of Europe, and hybrids of *T. obtusilobum* have been found in both its British sites. The other two species (*T. brachyglossum* Dahlst. and *T. subcyanolepis* M.P. Chr.) are more widespread and hybrids involving both of them have been detected here. The following composite account covers all these hybrids.

   b. It is thought that sexual species habitually hybridize with polliniferous agamospecies whenever they meet. So far the following 13 hybrid combinations have been found in Br:

   T. ancistrolobum Dahlst. x *T. subcyanolepis.*
   T. anglicum Dahlst. x *T. subcyanolepis.*
   T. brachyglossum x *T. hamatum* Raunk.
   T. brachyglossum x *T. insigne* Ekm.
   T. brachyglossum x *T. marklundii* Palmgr.
   T. brachyglossum x *T. polyodon* Dahlst.
   T. brachyglossum x *T. rubicundum* Dahlst.
   T. brachyglossum x *T. subcyanolepis.*
   T. cophocentrum Dahlst. x *T. subcyanolepis.*
   T. dahlstedtii H. Lindb. f. x *T. subcyanolepis.*
   T. hamatiforme Dahlst. x *T. obtusilobum.*
   T. marklundii x *T. subcyanolepis.*
   T. oxoniense Dahlst. x *T. subcyanolepis.*

   T. anglicum belongs to section *Palustria; T. oxoniense, T. rubicundum* and *T. brachyglossum* to section *Erythrosperma;* and the other ten to section *Vulgaria.* Although critical populations showing extensive morphological intergrading can indicate the presence of sexuality, it can be difficult to identify parents in a population with several species. Widespread back-crossing seems to occur in some instances, and hybrids are fertile and can be at least partially sexual. Identification of hybrids is usually possible in cases of populations with only two or three polliniferous species present, or when very distinctive species are involved. Hybridization can be suspected when morphological intergrading and partial or complete seed-sterility (indicating self-sterile sexuality) are noted. Sexual plants can also be detected by the presence of small pollen of regular size. Hybrids are usually intermediate in character, but dominance is exhibited in some important characters such as the *Erythrosperma*-type fruit-shape and -colour.

   c. Records of hybrids are limited to v.c. 12, 15, 17, 22, 23, 41, 56, 60, 62 and 66–68, which have been the most thoroughly worked. The vice-comital distribution of the two most frequently hybridizing sexual species is given below; hybrids are likely to be found in these, and indeed in all other areas in which sexual species are discovered. *T. brachyglossum*: v.c. 1–3, 5, 6, 9, 10, 12–17, 20–26, 28, 29, 32–42, 44, 45, 47–50, 52, 55–60, 62, 64, 66–71, 73, 82, 83, 85, 88–91, 93–98, 101, 103, 105, 106, 109,

H9, H16 and CI. *T. subcyanolepis*: v.c. 9, 15, 22, 23, 25, 26, 29, 41, 48, 58–60, 66, 67, 89, 90, 99, 101, 102, 107, H9, H12, H15, H16, H17, H38 and CI. None of the 13 hybrids listed above is known from outside BI at present, although all may occur. Hybrids are found in association with sexual species, and there is no evidence that they inhabit a more open or otherwise different habitat. *T. brachyglossum* is a plant of dry places, especially short, calcareous grassland, sandy heathland and dunes, always at low altitude. *T. subcyanolepis* occurs locally in rough, closed grassland, especially in sand-dunes, and on sandy road-verges. The two records of *T. obtusilobum* are from grassland.

d. *T. brachyglossum*, *T. subcyanolepis* and *T. austriacum* have been success-fully crossed with a number of other species. Of the naturally-occurring hybrids that have been reported in BI only *T. brachyglossum* x *T. subcyanolepsis* and *T. brachyglossum* x *T. polyodon* have been attempted, and these set good seed. *T. subcyanolepis* has set seed in artificial crosses with six species, acting as the female parent, and *T. brachyglossum* with five, but only as the male parent. There is no evidence of sterility between any British species, except in so far as agamosperms can only act as male parents, and then only if they produce pollen (many lack it) sufficiently small to effect stylar penetration.

e. *T. brachyglossum* $2n = 16, 18, 20, 24, 25, 26, 27, 28$; *T. subcyanolepis* $2n = 16, 17, 18, 20, 24, 25, 26, 27$; *T. obtusilobum* $2n = 16, 17$; *T. rubicundum*, *T. marklundii*, *T. ancistrolobum*, *T. hamatiforme*, *T. polyodon* and *T. hamatum* $2n = 24$; *T. dahlstedtii* $2n = 24, 27$; *T. insigne* $2n = 25$; *T. oxoniense* and *T. anglicum* $2n = 32$. In the first two species the numbers $2n = 16$ to 20 refer to sexual plants; $2n = 24$ and above to apomictic plants. None of the naturally-occurring hybrids has been examined cytologically, nor have either of the two synthesized hybrids. Extensive analyses of hybrids in natural populations, through the examination of pollen-type and seed-set (Richards 1970a, b, c), suggest that hybrids with diploid male and female parents are diploid sexuals. Those with one or both parents as triploid facultative agamosperms are diploid, hyperdiploid or triploid facultative agamosperms, dependent on the chance course of the irregular meiosis of the facultatively agamo-spermous parent(s). If the male parent is an obligate agamosperm, the hybrid offspring is (always?) a near-triploid obligate agamosperm.

f. RICHARDS, A. J. (1970a). Eutriploid facultative agamospermy in *Taraxacum*. *New Phytol.*, **69**: 761-774.

RICHARDS, A. J. (1970b). Hybridisation in *Taraxacum*. *New Phytol.*, **69**: 1103-1121.

RICHARDS, A. J. (1970c). Observations on *Taraxacum* sect. *Erythrosperma* Dt. em. Lindb. fil. in Slovakia. *Acta Fac. Rerum nat. Univ. comen., Bratisl., Bot.*, **18**: 81-120.

RICHARDS, A. J. (1972). The *Taraxacum* flora of the British Isles. *Watsonia*, **9** (Suppl.).

563 × 561. *Alisma* L. × *Baldellia* Parl.
(by C. A. Stace)

563/1 × 561/1. *Alisma plantago-aquatica* L. × *Baldellia ranunculoides* (L.) Parl.

a. (*Alisma* x *glueckii* Druce, *nom. nud.*).
b. Hybrids were described in detail by Glück from British and Irish plants. He recognized two variants, one more similar to *B. ranunculoides* and one more like *A. plantago-aquatica*, and he considered them to be the reciprocal crosses. They were somewhat intermediate in habit, leaf-shape, pedicel-length and other characters, but they were apparently quite fertile and resembled one or other parent closely in many features. Björkqvist examined the specimens and concluded that those resembling *B. ranunculoides* were in fact abnormal plants of that species and those resembling *A. plantago-aquatica* were plants of *A. lanceolatum*, a species not mentioned by Glück.
c. Glück identified hybrids from around Tuam, v.c. H17 (coll. Praeger), near the Holland Arms, v.c. 52, between Fearn and Balintore, v.c. 106, and near Taunton, v.c. 5. Praeger added Pontoon, v.c. H27, but there appear to be no other records.
d. Björkqvist was unable to obtain hybrids between *Alisma* and *Baldellia*.
e. *B. ranunculoides* $2n = (14, 16, 18, 22), 30$; *A. plantago-aquatica* $2n = 14$ (10, 12, 14, 16).
f. BJÖRKQVIST, I. (1968). Studies in *Alisma* L., 2. Chromosome studies, crossing experiments and taxonomy. *Op. bot. Soc. bot. Lund.*, **19**.
GLÜCK, H. (1913). Gattungs-Bastarde innerhalb der Familie Alismaceen. *Beih. bot. Zbl.*, **30**(2): 124-137. FIG.
PRAEGER, R. L. (1951). *Op. cit.*, p. 12.

563. *Alisma* L.
(by C. A. Stace)

2 × 1. *A. lanceolatum* With. × *A. plantago-aquatica* L.
a. *A.* x *rhicnocarpum* Schotsman.
b. The hybrid is intermediate in most characters, particularly leaf-shape, bract-size, style-length and stomatal length, but in others, especially certain floral characters, it more closely resembles one or other parent. It is usually stated to be highly or completely sterile, but Pogan (1971)

reported a plant from Po which had 20% viable pollen and Allen (1970) found his putative hybrids from the Isle of Man fertile.

c. Two putative hybrids, described as "feeble and infertile", have been found in Hb in recent years (P. J. Boyle *in litt.* 1971), and Allen has found intermediates widespread in the Isle of Man, v.c. 71. The reported differing flowering times of the two parents (*A. lanceolatum* in the morning, *A. plantago-aquatica* in the afternoon) might partly explain the rarity of the hybrid in the wild. Elsewhere cytologically-confirmed hybrids have been found in Ho, Po and Su.

d. Artificial hybrids have been obtained by Schotsman, and (reciprocally) by Björkqvist, who used both $2n = 26$ and 28 cytodemes of *A. lanceolatum*. The artificial hybrids were intermediate and highly sterile, but further generations and backcrosses were obtained. F_2 plants grown by Pogan (1971) from wild F_1 hybrids had $2n = 21-25$. Pogan earlier suggested that *A. lanceolatum* might be an allotetraploid derived partly or wholly from *A. plantago-aquatica.*

e. *A. plantago-aquatica* $2n = 14$ (also unconfirmed 10, 12, 16); *A. lanceolatum* $2n = 26$, 28 (also unconfirmed 24); wild hybrid ($2n = 20$, 21, according to *A. lanceolatum* parent).

f. ALLEN, D. E. (1964). An *Alisma* hybrid. *Proc. B.S.B.I.*, **5**: 373.

ALLEN, D. E. (1970). *The flowering plants and ferns of the Isle of Man.* Douglas.

BJÖRKQVIST, I. (1968). Studies in *Alisma* L., 2. Chromosome studies, crossing experiments and taxonomy. *Op. bot. Soc. bot. Lund.*, **19**. FIG.

POGAN, E. (1961). Odrebność gatunkowa i próba wyjaśnienia genezy *Alisma lanceolatum* With. *Acta Soc. bot. Pol.*, **30**: 667-727. FIG.

POGAN, E. (1965). Badania embriologiczne nad triploidalnyn mieszańcem *Alisma. Acta biol. cracov., Bot.*, **8**: 11-19.

POGAN, E. (1971). Karyological studies in a natural hybrid of *Alisma lanceolatum* With. x *Alisma plantago-aquatica* L. and its progeny. *Genet. pol.*, **12**: 219-225.

SCHOTSMAN, H. D. (1949). Korte mededeling betreffende het geslacht *Alisma* in Nederland. *Ned. kruidk. Archf*, **56**: 199-203.

576. *Zostera* L.

(by T. G. Tutin)

1 x 3. *Z. marina* L. x *Z. noltii* Hornem.

(= *Z.* x *hagstromii* Druce, *nom. nud.*) has been recorded from Hayling Island, v.c. 11; v.c. 67/68; Montrose Basin, v.c. 90; near Kirkwall, v.c. 111;

and Lough Ine, v.c. H3. Putative hybrids examined cytologically by Tutin showed no evidence of hybridity, and it is likely that specimens of *Z. angustifolia* (Hornem.) Reichb., which is intermediate in size, have been misidentified as the hybrid.

BUTCHER, R. W. (1934). Notes on the variation of the British species of *Zostera. Rep. B.E.C.*, **10**: 592-597.

DRUCE, G. C. (1920). *Zostera marina* x *nana. Rep. B.E.C.*, **5**: 684-685.

HALL, P. M. and PEARSALL, W. H. (1934). *Zostera nana* Roth. *Rep. B.E.C.*, **10**: 773.

TUTIN, T. G. (1936). New species of *Zostera* from Britain. *J. Bot., Lond.*, **74**: 227-230.

TUTIN, T. G. (1942). *Zostera* L., in Biological Flora of the British Isles. *J. Ecol.*, **30**: 217-226.

577. *Potamogeton* L.
(by J. E. Dandy)

Hybridization is frequent in the genus *Potamogeton,* and is important ecologically as well as taxonomically because the hybrids often form vigorous, long-persistent, clonal populations and, especially in moving waters like rivers and canals, may become dispersed over long distances by means of vegetative winter-buds and detached regenerative fragments.

The two hydrophilous species, *P. pectinatus* and *P. filiformis* (subgen. *Coleogeton*) hybridize with each other but not with the anemophilous species (subgen. *Potamogeton*) which form the remainder of the genus. Within subgen. *Potamogeton* numerous hybrid combinations occur, some of them surprising in view of the great disparity in the characters of the parents. Thus some of the broad-leaved species (sect. *Potamogeton*) hybridize with narrow-leaved "pusil-loid" species (sect. *Graminifolii*), and the very distinct *P. crispus* (which forms the monotypic sect. *Batrachoseris*) hybridizes freely with both broad-leaved and narrow-leaved species. With the possible exception of *P. gramineus* x *P. lucens* (in which well-formed fruits are often produced), all the hybrids are sterile, though very occasionally there is partial development of a malformed drupelet. Thus if a *Potamogeton* plant bears well-formed fruits (and is not *P. gramineus* x *P. lucens*) it is most unlikely to be a hybrid.

In the following account the statements of distribution within BI are based entirely on specimens seen during the preparation of the work. Study of relevant material has shown that many of the previously published records are erroneous; these are not included in the stated distributions.

1 × 2. *P. natans* L. × *P. polygonifolius* Pourr.

a. *P. × gessnacensis* G. Fisch.

b. The plants are intermediate between the parent species but resemble *P. natans* rather than *P. polygonifolius*. They are most easily recognized by the form of the submerged leaves, but these, as in both the parent species, often disappear early, and in their absence identification is difficult. The submerged leaves resemble the narrow, bladeless phyllodes of *P. natans* except that they are mostly more or less expanded towards the apex into a narrowly elliptic or oblanceolate blade which shows the influence of *P. polygonifolius*. The floating leaves resemble those of either parent, but usually show evidence of a discoloured joint at the top of the petiole, derived from *P. natans*; and the stipules in their nervation resemble *P. natans* rather than *P. polygonifolius*.

c. British plants which appear to be this hybrid have been found in Llyn Anafon, v.c. 49 (C. Bailey, 1884, and later collectors), and on the Hill of Nigg, v.c. 106 (U. K. Duncan, 1970). The hybrid is surprisingly rare, and outside BI is known only from the Gessnach stream in Ge. Records previously published for BI are errors; they refer to plants of *P. natans* (v.c. 17 and H2) or *P. polygonifolius* (v.c. 102).

d. None.

e. *P. natans* ($2n = 52$); *P. polygonifolius* ($2n = 26$).

f. HAGSTRÖM, J. O. (1916). *Critical researches on the Potamogetons*, pp. 192-193 (excluding the Irish f. *hibernicus*). Stockholm.

3 × 1. *P. coloratus* Hornem. × *P. natans* L.

has been wrongly claimed as the parentage of plants of *P. gramineus* × *P. natans* from v.c. 110.

1 × 4. *P. natans* L. × *P. nodosus* Poir.

= *P. × schreberi* G. Fisch. has been recorded from Ge and He.

5 × 1. *P. lucens* L. × *P. natans* L.

a. *P. × fluitans* Roth (*P. crassifolius* Fryer, *P. sterilis* Hagstr.).

b. The plants are clearly intermediate between the parent species. They differ from *P. natans* in having laminate submerged leaves with a more or less oblanceolate, translucent blade, and floating leaves in which the petiole is not longer than the blade and lacks a discoloured, flexible joint at the top. They differ from *P. lucens* in having floating leaves, and submerged leaves with longer petioles.

c. This hybrid is surprisingly rare in BI, being known only from the fens of Cambridgeshire, v.c. 29 and 31, from the Moors River and tributary in v.c. 9 and 11, and, formerly, from the old Wey and Arun Canal in v.c. 13 and 17. It is also recorded from Au, Da, Ge, Rs and Su.

d. None.

e. Both parents ($2n = 52$).

f. DANDY, J. E. and TAYLOR, G. (1939a). Some new county records. *J. Bot., Lond.,* **77**: 253-259.

DANDY, J. E. and TAYLOR, G. (1939e). Another record of x *Potamogeton fluitans* from South Hants. *J. Bot., Lond.,* **77**: 342.

FRYER, A. (1890). On a new hybrid *Potamogeton* of the *fluitans* group. *J. Bot., Lond.,* **28**: 321-326, t. 299. FIG.

FRYER, A. (1898). *The Potamogetons (Pond Weeds) of the British Isles,* t. 4-7. London. FIG.

HAGSTRÖM, J. O. (1916). As above, pp. 216, 238-241, fig. 112. FIG.

PERRING, F. H. and SELL, P. D. (1968). *Op. cit.,* p. 136.

6 x 5 x 1. *P. gramineus* L. x *P. lucens* L. x *P. natans* L.

has been wrongly claimed as the parentage of *P. crassifolius* Fryer (= *P. lucens* x *P. natans*) from v.c. 29.

6 x 1. *P. gramineus* L. x *P. natans* L.

a. *P.* x *sparganifolius* Laest. ex Fries (*P. kirkii* (Hook f.) Syme ex Hook f., *P. tiselii* K. Richt.). (Ribbon-leaved Pondweed).

b. The plants resemble *P. natans* rather than *P. gramineus,* but differ from *P. natans* in the submerged leaves, which are laminate with a translucent blade which varies in shape from narrowly oblanceolate to linear and ribbon-like. These submerged leaves also distinguish the hybrid plants from *P. gramineus,* being narrower in shape than in that species and being tapered at the base into a petiole. The floating leaves of the hybrid plants sometimes show evidence of the discoloured, flexible joint at the top of the petiole which characterizes *P. natans.*

c. This hybrid was formerly thought to be a very rare plant confined in BI to a locality in western Hb. It is certainly rare, but is now known from a number of localities in Hb and in Br, where it ranges very sporadically from south-eastern England northwards to Scotland. On the Continent the hybrid is recorded from Ho and Rs and is especially frequent in Scandinavia; it is recorded also from Siberia.

d. None.

e. Both parents (2*n* = 52).

f. FRYER, A. (1898). As above, t. 8, 9. FIG.

PERRING, F. H. and SELL, P. D. (1968). *Op. cit.,* p. 137.

SYME, J. T. B., ed. (1869). *English Botany,* 3rd ed., **9**: 31-32, t. 1403. London. FIG.

7 x 1. *P. alpinus* Balb. x *P. natans* L.

was formerly accepted as the parentage of *P. drucei* Fryer (= *P. nodosus* Poir.) in the mistaken belief that it was a hybrid.

15 x 1. *P. berchtoldii* Fieb. x *P. natans* L.

a. *P.* x *variifolius* Thore.

b. This remarkable hybrid suggests *P. natans* in miniature, having small floating leaves which often show an indication of the discoloured, flexible joint at the top of the petiole characteristic of that species. The influence of *P. berchtoldii* is seen, however, in the numerous, narrowly linear submerged leaves which, though resembling on a small scale the submerged, bladeless phyllodes of *P. natans,* are more delicate in texture and have a venation suggesting *P. berchtoldii.* The influence of *P. berchtoldii* is also evident in the small flower-spikes.

c. This hybrid is known in BI only from the Glenamoy River in v.c. H27, and elsewhere only in south-western Ga.

d. None.

e. *P. berchtoldii* ($2n = 26$); *P. natans* ($2n = 52$).

f. DANDY, J. E. and TAYLOR, G. (1967). Taxonomic and nomenclatural notes on the British flora. *Watsonia,* 6: 315-316.
HAGSTRÖM, J. O. (1916). As above, pp. 193-195, fig. 97. FIG.

3 × 2. *P. coloratus* Hornem. × *P. polygonifolius* Pourr.

was proposed as the parentage of *P. anglicus* Hagstr. (= *P. polygonifolius*). The plant came from Woking Heath, v.c. 17, and is not a hybrid.

5 × 2. *P. lucens* L. × *P. polygonifolius* Pourr.

was at one time wrongly suggested as the parentage of *P. drucei* Fryer (= *P. nodosus* Poir.).

6 × 2. *P. gramineus* L. × *P. polygonifolius* Pourr.

a. (*P. gramineus* f. *lanceolatifolius* Tisel.).

b. In general aspect the hybrid resembles *P. gramineus* rather than *P. polygonifolius*; indeed Swedish plants were originally named *P. gramineus* forma *lanceolatifolius* by Tiselius. It differs, however, from *P. gramineus* most obviously in the form of the submerged leaves, which are narrowly oblanceolate, being gradually tapered below into a petiolate or sub-petiolate (narrow-winged) base. The shape of the submerged leaves shows the influence of *P. polygonifolius,* but they are more acute at the apex, and some of them (especially the lower ones) are more shortly petiolate than in that species, while the stipules resemble those of *P. gramineus.* Long-petiolate floating leaves occur, as in both the parent species.

c. This is a very rare hybrid, in BI seen so far only from two stations in Scotland: the River Bladnoch at Spittal, v.c. 74, and a small loch east of Nairn, v.c. 96; it was collected in both these localities by G. Taylor in 1953. It is also known from Su, and there is a doubtful record from Ge under the name *P. seemenii* Aschers. & Graebn. Other plants from Scotland and Hb recorded as this hybrid are all referable to *P. gramineus.*

d. None.

e. *P. gramineus* ($2n = 52$); *P. polygonifolius* ($2n = 26$).

f. HAGSTRÖM, J. O. (1916). As above, pp. 231-232, fig. 109. FIG.

7 × 2. *P. alpinus* Balb. × *P. polygonifolius* Pourr.

=*P.* × *spathulatus* Schrad. ex Koch & Ziz has been erroneously recorded from Jersey, CI, the plant concerned being *P. polygonifolius*, not a hybrid. What is apparently the true hybrid was described from Ge and is reported also from Su.

2 × 8. *P. polygonifolius* Pourr. × *P. praelongus* Wulf.

has been wrongly claimed as the parentage of *P. macvicarii* (= *P. alpinus* × *P. praelongus*) from v.c. 97.

2 × 13. *P. polygonifolius* Pourr. × *P. pusillus* L.

= *P.* × *miguelensis* Dandy has been recorded from Az.

15 × 2. *P. berchtoldii* Fieb. × *P. polygonifolius* Pourr.

= *P.* × *rivularis* Gillot has been recorded from Ga.

3 × 6 × 5. *P. coloratus* Hornem. × *P. gramineus* L. × *P. lucens* L.

has been wrongly claimed as the parentage of *P.* × *billupsii* (= *P. coloratus* × *P. gramineus*) from v.c. 29.

3 × 6. *P. coloratus* Hornem. × *P. gramineus* L.

a. *P.* × *billupsii* Fryer.
b. The plants resemble *P. gramineus* rather than *P. coloratus*, and in particular the floating upper leaves have longer petioles than those of the latter species. The influence of *P. coloratus* is obvious, however, especially in the submerged leaves which are narrowed at the base into a petiole, whereas those of *P. gramineus* are mostly sessile.
c. This is known in BI only from Benwick in the fens of the Isle of Ely, v.c. 29, and the island of Benbecula, v.c. 110. The two parent species occur in both these areas. The hybrid is recorded also from Su.
d. None.
e. *P. coloratus* (2*n* = 26); *P. gramineus* (2*n* = 52).
f. DANDY, J. E. and TAYLOR, G. (1941). Further records of *Potamogeton* from the Hebrides. *J. Bot., Lond.,* 79: 97-101.
FRYER, A. (1893). Notes on pondweeds. A new hybrid *Potamogeton. J. Bot., Lond.,* 31: 353-355, t. 337, 338. FIG.
FRYER, A. (1898). As above, t. 16, 17. FIG.
HAGSTRÖM, J. O. (1916). As above, p. 181.

15 × 3. *P. berchtoldii* Fieb. × *P. coloratus* Hornem.

a. *P.* × *lanceolatus* Sm. (*P. perpygmaeus* Hagstr. ex Druce). (Lanceolate Pondweed).
b. Although the plants are clearly intermediate between *P. berchtoldii* and *P. coloratus* they cannot be confused with either. The numerous, small,

narrow submerged leaves show the influence of *P. berchtoldii* but differ from those of that species in being broader and oblanceolate, tapering below to a sessile or subsessile base, and more obtuse at the apex; they are distinguished from those of *P. coloratus* by the smaller size, narrower shape, and lack of a distinct petiole. The influence of *P. coloratus* is most evident in the occurrence of small, shortly petiolate floating leaves. The small flower-spikes are a further indication of the influence of *P. berchtoldii*. The parentage *P. berchtoldii* x *P. coloratus* (in the form *P. coloratus* x *P. "pusillus"*) was first suggested by Hagström for his *P. perpygmaeus*, based on plants from v.c. 29 and Hb.

c. This hybrid is known only from Burwell Fen in v.c. 29, the Afon Lligwy in v.c. 52, the Caher River in v.c. H9, and the Clonbruck and Grange Rivers in v.c. H17.

d. None.

e. Both parents ($2n = 26$).

f. BENNETT, A. (1881). On *Potamogeton lanceolatus* of Smith. *J. Bot., Lond.,* **19**: 65-67, t. 217. FIG.

  DRUCE, G. C. (1923). *P. coloratus* x *pusillus* = *P. perpygmaeus* Hagström. *Rep. B.E.C.,* **6**: 630.

  FRYER, A. (1913). *The Potamogetons (Pond Weeds) of the British Isles,* t. 39. London. FIG.

  HAGSTRÖM, J. O. (1916). As above, pp. 149-151, fig. 68 (excluding the French *P.* x *rivularis*). FIG.

  SYME, J. T. B., ed. (1869). *English Botany,* 3rd ed., **9**: 34-35, t. 1405. London. FIG.

6 x 5. *P. gramineus* L. x *P. lucens* L.

a. *P.* x *zizii* Koch ex Roth (*P. coriaceus* (Mert. & Koch) A. Benn., *P. babingtonii* A. Benn., *P. angustifolius* auct., *P. longifolius sensu* Bab.). (Long-leaved Pondweed).

b. The plants are obviously intermediate between the parent species but vary considerably, occurring with or without floating leaves (or "semi-floating" leaves with lacunar tissue) which, when present, show the influence of *P. gramineus* and at once separate the hybrid plants from *P. lucens*. In their submerged leaves (apart from occasional bladeless phyllodes) the hybrid plants resemble *P. lucens* rather than *P. gramineus,* but are distinguished from *P. lucens* by the leaf-base, this being sessile or narrowed into a short petiole which is flattened and is often bordered on one side by a downward production of the blade. These submerged leaves differ from those of *P. gramineus* in being broader in shape and tapered to the sessile or petiolate base. The plants often produce fully formed fruits, and this is the only European *Potamogeton* hybrid which does so; the drupelets are intermediate in size between those of *P. gramineus* and *P. lucens,* but to what extent they are viable is not known.

c. This hybrid has a wide but sporadic distribution in BI northwards to

Zetland, and is widespread on the Continent; it is also recorded from
Siberia.
d. None.
e. Both parents ($2n = 52$).
f. BUTCHER, R. W. (1961). *A new illustrated British flora*, 2: fig. 1398.
London. FIG.
DANDY, J. E. and TAYLOR, G. (1939b). The identity of *Potamogeton
babingtonii. J. Bot., Lond.*, 77: 161-164, t. 617. FIG.
FRYER, A. (1913). As above, t. 42, 43. FIG.
PERRING, F. H. and SELL, P. D. (1968). *Op. cit.*, p. 137.
SYME, J. T. B., ed. (1869). *English Botany*, 3rd ed., 9: 40-41, t. 1410.
London. FIG.

6 x 5 x 9. *P. gramineus* L. x *P. lucens* L. x *P. perfoliatus* L.

has been wrongly claimed as the parentage of *P. involutus* (= *P. gramineus*
x *P. perfoliatus*) from v.c. 29 and 31.

7 x 5. *P. alpinus* Balb. x *P. lucens* L.

a. *P.* x *nerviger* Wolfg.
b. The hybrid is clearly intermediate between *P. alpinus* and *P. lucens*. In
shape and general appearance the submerged leaves resemble those of *P.
lucens*, but (apart from an occasional bladeless phyllode) they differ in
having a sessile and slightly amplexicaul base and a blunter apex. The
influence of *P. alpinus* also appears in the rufescent tinge of the upper
parts of the plant, where there is sometimes a suggestion of floating leaves.
The apex of the submerged leaves is not so broadly rounded or obtuse as
in *P. alpinus*.
c. The only known station in BI is the River Fergus in v.c. H9; elsewhere
there is one other locality in Lithuania, Rs.
d. None.
e. Both parents ($2n = 52$).
f. DANDY, J. E. and TAYLOR, G. (1967). As above, p. 316.

5 x 8. *P. lucens* L. x *P. praelongus* Wulf.

has been wrongly claimed as the parentage of *P. babingtonii* (= *P.
gramineus* x *P. lucens*) from v.c. H16.

5 x 9. *P. lucens* L. x *P. perfoliatus* L.

a. *P.* x *salicifolius* Wolfg. (*P. decipiens* Nolte ex Koch, *P. upsaliensis* Tisel.,
P. salignus Fryer, *P. kupfferi* A. Benn.).
b. This hybrid, like both its parent species, varies greatly in leaf-shape
according to the type of water in which it is growing, being narrower-
leaved in stronger currents. All the hybrid plants are readily distinguishable
from *P. perfoliatus*; they more or less resemble *P. lucens*, the broad-leaved
ones often very closely. The leaf-base, however, always shows the

influence of *P. perfoliatus* and serves to distinguish the hybrid plants from *P. lucens*. Usually the leaves are sessile and more or less amplexicaul, but sometimes the base is tapered into a short petiole which is flattened and canaliculate, slightly clasping the stem, whereas in *P. lucens* the petiole is thick, non-canaliculate and not at all amplexicaul.

c. This hybrid is widely though sporadically distributed over Br northwards to Angus, and there are scattered stations in Hb. The plants are usually found in rivers, canals, lakes and fen-drains; sometimes, as in the River Wye (v.c. 34 and 36) and the lower River Tweed (v.c. 68, 80 and 81), they form large clonal populations. The hybrid is also widely distributed on the Continent.

d. None.

e. Both parents ($2n = 52$).

f. BAKER, J. G. and TRIMEN, H. (1867). Report of the London Botanical Exchange Club for the year 1866. *J. Bot., Lond.*, 5: 65-73, t. 61. FIG.

FRYER, A. (1913, 1915). *The Potamogetons (Pond Weeds) of the British Isles*, t. 48, 49. London. FIG.

PERRING, F. H. and SELL, P. D. (1968). *Op. cit.*, p. 136.

SYME, J. T. B., ed. (1869). *English Botany*, 3rd ed., 9: 39-40, t. 1409. London. FIG.

19 × 5. *P. crispus* L. × *P. lucens* L.

a. *P.* × *cadburyae* Dandy & Taylor.

b. The hybrid is manifestly intermediate between the parent species. It is easily distinguished from *P. crispus* by the shape of the leaves, which are narrowly elliptic-oblong and tapered to the base, and by the more elongated stipules. It differs from *P. lucens* in the sessile leaves which are rounded-obtuse at the apex and have only 5–7 longitudinal veins.

c. This is known only from a single plant which was found growing with both the parent species in a pool near Nuneaton, v.c. 38. Attempts to find more plants of the hybrid in this pool have so far been unsuccessful.

d. None.

e. Both parents ($2n = 52$).

f. DANDY, J. E. and TAYLOR, G. (1957). Two new British hybrid pondweeds. *Kew Bull.*, 12: 332.

7 × 6. *P. alpinus* Balb. × *P. gramineus* L.

a. *P.* × *nericius* Hagstr.

b. The plants are clearly intermediate between the parent species. Though they resemble *P. gramineus* rather than *P. alpinus* the influence of the latter species is evident especially in the submerged leaves.

c. This hybrid is known in BI only from the River Don in v.c. 92. Elsewhere it is reported from Is, No and Su.

d. None.

e. Both parents ($2n = 52$).

f. DRUCE, G. C. (1921). *Potamogeton alpinus* x *gramineus* = x *P. nericius*
Hagst. *Rep. B.E.C.*, 6: 49.
HAGSTRÖM, J. O. (1916). As above, pp. 145-146, fig. 65. FIG.

6 x 8. *P. gramineus* L. x *P. praelongus* Wulf.

was said to be the parentage of *P. lundii* K. Richt. from Su and of similar
plants from the Lunan Burn, v.c. 89, but *P. lundii* is in fact *P. gramineus* x
P. perfoliatus.

6 x 9. *P. gramineus* L. x *P. perfoliatus* L.

a. *P.* x *nitens* Weber (*P. involutus* (Fryer) H. & J. Groves, *P. lundii* K. Richt.).
b. Though always more or less intermediate between the parent species the
plants are very variable in the shape of the submerged leaves, and also in
the presence or absence of petiolate floating leaves which, when
developed, show the influence of *P. gramineus*. Most plants resemble *P.
gramineus* (sometimes very closely) but differ from that species in that the
submerged leaves are more lanceolate in shape, with a more or less
amplexicaul base, while the apex is blunter and the secondary venation not
so regularly ascending. The plants with broad submerged leaves show more
resemblance to *P. perfoliatus*, but differ from that species in having more
conspicuous stipules, a less rounded leaf-apex, and a different leaf-
venation; when floating leaves are present they at once distinguish hybrid
plants from *P. perfoliatus*.
c. This hybrid has a wide distribution in BI, occurring frequently in waters
where *P. gramineus* and *P. perfoliatus* grow in association but sometimes
where one or both appear to be absent. It is widely distributed elsewhere
in Europe and has been found in temperate Asia; variants of it are also
reported from North America.
d. None.
e. Both parents ($2n = 52$).
f. BUTCHER, R. W. (1961). As above, fig. 1400. FIG.
FRYER, A. (1913). As above, t. 37, 38, 44. FIG.
MOORE, D. (1864). On *Potamogeton nitens*, Weber, as an Irish plant. *J.
Bot., Lond.*, 2: 325-326, t. 23. FIG.
PERRING, F. H. and SELL, P. D. (1968). *Op. cit.*, p. 137.
SYME, J. T. B., ed. (1869). *English Botany*, 3rd ed., 9: 36-37, t. 1407.
London. FIG.

15 x 6 x 9. *P. berchtoldii* Fieb. x *P. gramineus* L. x *P. perfoliatus* L.

was a parentage attributed (in the form *P. gramineus* L. x *P. perfoliatus* L.
x *P. millardii* H.-Harrison) to *P. heslop-harrisonii* Clark, based on material
from North Uist, v.c. 110. The proposed hybrid formula is scarcely
feasible, as it means that one of the parents of *P. heslop-harrisonii* must
have been a sterile hybrid. No voucher specimen has been seen, and the
identity of the plant remains in doubt.

11 × **6.** *P. friesii* Rupr. × *P. gramineus* L.

has been wrongly claimed as the parentage of *P.* × *lanceolatus* (= *P. berchtoldii* × *P. coloratus*) from v.c. 29.

6 × **13.** *P. gramineus* L. × *P. pusillus* L.

has been wrongly claimed as the parentage of *P.* × *lanceolatus* (= *P. berchtoldii* × *P. coloratus*).

15 × **6.** *P. berchtoldii* Fieb. × *P. gramineus* L.

was the parentage originally suggested for *P. heslop-harrisonii* Clark, but later changed to *P. berchtoldii* × *P. gramineus* × *P. perfoliatus* (q.v.).

7 × **8.** *P. alpinus* Balb. × *P. praelongus* Wulf.

a. *P.* × *griffithii* A. Benn. (*P. macvicarii* A. Benn.).
b. The plants combine the characters of the parent species. They show strong resemblance to *P. praelongus* in the stipules and submerged leaves, but the influence of *P. alpinus* is evident in the less amplexicaul base of the submerged leaves and in the occurrence of petiolate floating leaves, which are never found in *P. praelongus*. Hagström, while accepting the parentage *P. alpinus* × *P. praelongus*, equated *P.* × *griffithii* with *P.* × *nerviger* (= *P. alpinus* × *P. lucens*).
c. This is known only from Llyn Anafon, v.c. 49, and from two hill lochs in Moidart and Ardnamurchan, v.c. 97.
d. None.
e. Both parents (2*n* = 52).
f. BENNETT, A. (1883). Two new Potamogetons. *J. Bot., Lond.,* **21**: 65-67, t. 235. FIG.
   DANDY, J. E. and TAYLOR, G. (1939c). *Potamogeton griffithii* and *P. macvicarii. J. Bot., Lond.,* **77**: 277-282.
   FRYER, A. (1898). As above, t. 22, 23. FIG.

7 × **9.** *P. alpinus* Balb. × *P. perfoliatus* L.

a. *P.* × *prussicus* Hagstr. (*P. johannis* H.-Harrison).
b. The British specimens of this hybrid are strikingly intermediate between the parent species. The vegetative shoots appear rufescent towards the top as in *P. alpinus*, and the rather elongated leaf-shape also recalls that species, although in some states of *P. perfoliatus* the leaves are similarly elongated. At the base, however, the leaves of the hybrid differ from those of *P. alpinus* in being definitely amplexicaul, approaching those of *P. perfoliatus* but not so strongly clasping. In leaf-venation and stipules the hybrid plants are intermediate, the stipules being larger and more conspicuous than in *P. perfoliatus* but more delicate than in *P. alpinus*. Floating leaves are not present in the specimens seen.

c. This is known only from two islands off the west coast of Scotland: Colonsay, v.c. 102, and Benbecula, v.c. 110. It is also recorded from Ge, No and Su.

d. None.

e. Both parents ($2n = 52$).

f. DANDY, J. E. and TAYLOR, G. (1941). As above.

7 × 15. *P. alpinus* Balb. × *P. berchtoldii* Fieb.

(as *P. alpinus* × *P. "pusillus"*) has been wrongly claimed as the parentage of *P.* × *lanceolatus* (= *P. berchtoldii* × *P. coloratus*).

7 × 19. *P. alpinus* Balb. × *P. crispus* L.

a. *P.* × *olivaceus* Baagöe ex G. Fisch. (*P. venustus* Baagöe ex A. Benn.).

b. In general appearance this hybrid suggests a narrow-leaved form of *P. alpinus*, and it is easily distinguishable from *P. crispus* by the subentire margins of the leaves and by the occasional presence of petiolate floating leaves. The influence of *P. crispus* is, however, apparent, chiefly in the compressed stem, the fewer (usually 7) longitudinal veins of the leaves, and the shorter flower-spikes. Floating leaves, when developed, taper to a short petiole.

c. This hybrid is known in BI only from the River Tweed and some of its tributaries in v.c. 68, 78, 79, 80 and 81, the River Earn in v.c. 88, and the River Ythan in v.c. 93. It also occurs in Da.

d. None.

e. Both parents ($2n = 52$).

f. DANDY, J. E. and TAYLOR, G. (1943). × *Potamogeton olivaceus* (*P. alpinus* × *crispus*). *J. Bot., Lond.*, **80**: 117-120.
HAGSTRÖM, J. O. (1916). As above, p. 144, fig. 64. FIG.
PERRING, F. H. and SELL, P. D. (1968). *Op. cit.*, p. 138.

9 × 8. *P. perfoliatus* L. × *P. praelongus* Wulf.

a. *P.* × *cognatus* Aschers. & Graebn.

b. The plants bear a strong resemblance to *P. perfoliatus*, and, as in that species but not in *P. praelongus*, the leaves are denticulate at least when young. The influence of *P. praelongus*, however, appears in a number of characters, most noticeably in the stipules, which are larger than those of *P. perfoliatus* and are persistent, at least as a row of fibres, and in the leaf-apex, which is distinctly cucullate and splits under pressure.

c. This hybrid is known in BI from the North Idle Drain and Double Rivers, v.c. 54, and from Loch Borralie, v.c. 108. On the Continent the hybrid has been recorded from Da, Ge, No, Rs and Su.

d. None.

e. Both parents ($2n = 52$).

f. TAYLOR, J. M. and SLEDGE, W. A. (1944). × *Potamogeton cognatus* Asch. and Graeb. in Britain. *Naturalist, Hull*, **1944**: 121-123, t. 1. FIG.

19 × 8. *P. crispus* L. × *P. praelongus* Wulf.

a. *P.* × *undulatus* Wolfg.

b. In foliage the plants resemble *P. praelongus* rather than *P. crispus,* but the influence of the latter species is shown in the fewer longitudinal leaf-veins as well as in the compressed stem. The leaf-margins are entire, as in *P. praelongus,* or show only slight indication of the serrulation characteristic of *P. crispus.*

c. In Br this hybrid is known only from Llyn Hilyn, v.c. 43, growing in company with both the parent species. In Hb it has been found in the River Lagan and the Six Mile Water, v.c. H39. On the Continent it is recorded from Da, Po and Rs.

d. None.

e. Both parents (2*n* = 52).

f. DANDY, J. E. and TAYLOR, G. (1967). As above, p. 316.

19 × 9. *P. crispus* L. × *P. perfoliatus* L.

a. *P.* × *cooperi* (Fryer) Fryer (*P. undulatus* var. *cooperi* Fryer).

b. The plants at first sight resemble *P. perfoliatus,* often very closely, but are distinguished from that species by the compressed stem and by the fewer veins of the leaves, there being only one or sometimes two fine, secondary longitudinal veins between the midrib and the nearest primary veins; these characters show the influence of *P. crispus,* as also do the short flower-spikes. In dentition the leaf-margins are intermediate between the parent species, but do not have the regular serrulation of *P. crispus.*

c. This is one of the more frequent *Potamogeton* hybrids, with a wide though sporadic distribution in BI. The known localities in Br range from Hertfordshire and Wiltshire to Stirlingshire, and there are a few stations in Hb. On the Continent the hybrid is recorded from Cz, Da, Ge and Rm.

d. None.

e. Both parents (2*n* = 52).

f. FRYER, A. (1891). On a new British *Potamogeton* of the *nitens* group. *J. Bot., Lond.,* **29**: 289-292, t. 313. FIG.

FRYER, A. (1900). *The Potamogetons (Pond Weeds) of the British Isles,* t. 31, 32. London. FIG.

PERRING, F. H. and SELL, P. D. (1968). *Op. cit.,* p. 139.

11 × 14. *P. friesii* Rupr. × *P. obtusifolius* Mert. & Koch

has been wrongly claimed as the parentage of some plants of *P. obtusifolius* from v.c. 17 and 89.

15 × 11. *P. berchtoldii* Fieb. × *P. friesii* Rupr.

(as *P. friesii* × *P. "pusillus"*) has been wrongly claimed as the parentage of some plants of *P. friesii* (v.c. 34), *P. berchtoldii* (v.c. 40 and 86) and *P. pusillus* (v.c. 86).

18 × 11. *P. acutifolius* Link × *P. friesii* Rupr.

  a. *P.* × *pseudofriesii* Dandy & Taylor.
  b. The hybrid clearly combines the characters of the parent species. At first sight it recalls *P. friesii,* but it is distinguished from that species by the presence in the leaves of scattered vein-like longitudinal sclerenchymatous strands between the primary veins, these showing the influence of *P. acutifolius.* It differs from *P. acutifolius* in the stipules, which are tubular towards the base as in *P. friesii,* and in the sclerenchymatous strands, which are much less numerous than in *P. acutifolius.*
  c. This hybrid is known only from a single station near Strumpshaw, v.c. 27, where it was found growing with both the parent species.
  d. None.
  e. Both parents (2*n* = 26).
  f. DANDY, J. E. and TAYLOR, G. (1957). As above.

19 × 11. *P. crispus* L. × *P. friesii* Rupr.

  a. *P.* × *lintonii* Fryer.
  b. The plants in general appearance resemble *P. crispus,* and are also rather similar to *P. crispus* × *P. trichoides,* with which some of them (from v.c. 17 and H37) have been confused. The influence of *P. friesii* is apparent in the leaves, which are narrower than in mature plants of *P. crispus* and less conspicuously serrulate along the margins, the serrulation being usually confined to the apical part of the leaf. Another character derived from *P. friesii* is seen in the stipules, which are tubular towards the base, though sometimes very shortly so.
  c. This hybrid is sporadically distributed in Br from Kent to Lancashire, and there is a single known Irish locality in v.c. H37. Most of the British localities are in the River Trent and the associated canal-systems of the midlands and north, and it seems likely that the hybrid has, to some extent at least, been distributed through these waters by means of its winter-buds and the effects of boat-traffic. Outside BI the hybrid is known from Ho.
  d. None.
  e. *P. crispus* (2*n* = 52); *P. friesii* (2*n* = 26).
  f. DANDY, J. E. and TAYLOR, G. (1939d). × *Potamogeton bennettii* and × *P. lintonii. J. Bot., Lond.,* 77: 304-311.
     PERRING, F. H. and SELL, P. D. (1968). *Op. cit.,* p. 138.

14 × 13. *P. obtusifolius* Mert. & Koch × *P. pusillus* L.

  (as *P. obtusifolius* × *P. panormitanus*) has been wrongly claimed as the parentage of some plants of *P. obtusifolius* (*P. sturrockii,* v.c. 89), *P. berchtoldii* (v.c. 89) and *P. pusillus* (v.c. 33).

15 × 13. *P. berchtoldii* Fieb. × *P. pusillus* L.

(as *P. panormitanus* x *P. "pusillus"*) has been wrongly claimed as the parentage of some plants of *P. berchtoldii* (v.c. 40 and 64) and *P. pusillus* (v.c. 2).

13 × 16. *P. pusillus* L. × *P. trichoides* Cham. & Schlecht.

a. *P.* x *grovesii* Dandy & Taylor (*P. trinervius* auct.).
b. The plants are clearly intermediate between the parent species. In general appearance they more closely resemble *P. trichoides* and have been mistaken for that species, but they differ from it in having the stipules tubular towards the base as in *P. pusillus*; moreover the leaves are more evidently 3-veined than in *P. trichoides*, the lateral veins being more conspicuous than in that species. The plants differ from *P. pusillus*, which has 4-carpellate flowers, in having the number of carpels reduced to 1–3 as in *P. trichoides*.
c. This hybrid is known only from a single locality near Palling, v.c. 27, where it was collected in 1897 and 1900 growing in company with the parent species. It has been sought unsuccessfully in recent years.
d. None.
e. Both parents ($2n = 26$).
f. DANDY, J. E. and TAYLOR, G. (1967). As above, pp. 316-317.

19 × 14. *P. crispus* L. × *P. obtusifolius* Mert. & Koch

has been wrongly claimed as the parentage of *P.* x *bennettii* (= *P. crispus* x *P. trichoides*).

15 × 16. *P. berchtoldii* Fieb. × *P. trichoides* Cham. & Schlecht.

(as *P. "pusillus"* x *P. trichoides*) has been wrongly claimed as the parentage of some plants of *P. berchtoldii* (v.c. 6, 12, 16, 24, 25, 27 and 32), *P. trichoides* (v.c. 17) and *P. pusillus* (v.c. 22).

18 × 15. *P. acutifolius* Link × *P. berchtoldii* Fieb.

a. *P.* x *sudermanicus* Hagstr.
b. The plants are obviously intermediate between the parent species, but in general appearance suggest a robust state of *P. berchtoldii* rather than *P. acutifolius*. The leaves distinguish the hybrid from both parents: they are 3-veined but here and there between the primary veins are short, broken longitudinal sclerenchymatous strands which show the influence of *P. acutifolius*. The parentage was given by Hagström in the form *P. acutifolius* x *P. "pusillus"*.
c. This hybrid is known in BI only from a single locality near Arne, v.c. 9. Both the parent species are known from the same locality. Outside BI the hybrid is recorded from Su.
d. None.
e. Both parents ($2n = 26$).

458 HYBRIDIZATION

f. GOOD, R. (1948). *A geographical handbook of the Dorset flora*, p. 227. Dorchester.
HAGSTRÖM, J. O. (1916). As above, p. 73, fig. 28 A-E. FIG.

15 × 19. *P. berchtoldii* Fieb. × *P. crispus* L.
(as *P. crispus* × *P. "pusillus"*) has been wrongly claimed as the parentage of *P.* × *bennettii* (= *P. crispus* × *P. trichoides*).

17 × 16. *P. compressus* L. × *P. trichoides* Cham. & Schlecht.
= *P.* × *ripensis* Baagöe ex G. Fisch. has been recorded from Da.

19 × 16. *P. crispus* L. × *P. trichoides* Cham. & Schlecht.
a. *P.* × *bennettii* Fryer.
b. In appearance the hybrid suggests a narrow-leaved form of *P. crispus,* and the leaves, especially towards their apex, have traces of a minute serrulation obviously derived from that species. The influence of *P. trichoides,* however, is clearly shown in the flowers, which mostly have the number of carpels reduced to 2 or 3. The leaves of the hybrid are much broader than those of *P. trichoides.*
c. This hybrid is known only from the Forth and Clyde Canal in v.c. 77 and 86. It was originally described from Grangemouth, v.c. 86, where it grew in association with both *P. crispus* and *P. trichoides.*
d. None.
e. *P. crispus* (2*n* = 52); *P. trichoides* (2*n* = 26).
f. DANDY, J. E. and TAYLOR, G. (1939d). As above.
FRYER, A. (1895). *Potamogeton Bennettii. J. Bot., Lond.,* **33**: 1-3, t. 348. FIG.
FRYER, A. (1900). As above, t. 33. FIG.
HAGSTRÖM, J. O. (1916). As above, pp. 63-64, fig. 24 B, E. FIG.

20 × 21. *P. filiformis* Pers. × *P. pectinatus* L.
a. *P.* × *suecicus* K. Richt.
b. The hybrid is variously intermediate between the parent species. Some plants resemble *P. pectinatus* in having acute leaves and stigmas with a stylar "neck", but differ from that species in the leaf-sheaths which are tubular towards the base as in *P. filiformis.* In other plants, notably those from the Rivers Wharfe, Ure and Tweed, the leaf-sheaths are less obviously tubular and on cursory examination often appear to be quite open as in *P. pectinatus;* but these plants have leaves which are more or less obtuse to rounded at the apex as in *P. filiformis,* and have broad, sessile stigmas which also show the influence of the latter species.
c. This hybrid is known from scattered localities in Scotland in areas where both the parent species occur, and there are also populations in the lower River Tweed and its tributary the Till, v.c. 68 and 81, and the Rivers Ure and Wharfe, v.c. 64 and 65, which are outside the present-day range of *P.*

filiformis and may possibly represent a clone of ancient origin. Outside BI it is recorded from Da, Fe, Ge, Is, No, Rs and Su, and has also been reported from Siberia and North America.

d. None.

e. Both parents ($2n = 78$).

f. DANDY, J. E. and TAYLOR, G. (1946). An account of x *Potamogeton suecicus* Richt. in Yorkshire and the Tweed. *Trans. Proc. bot. Soc. Edinb.,* **34**: 348-360, t. 4-8, fig. 1. FIG.

PERRING, F. H. and SELL, P. D. (1968). *Op. cit.,* p. 139.

589. *Polygonatum* Mill.
(by D. McClintock)

3 x 2. *P. multiflorum* (L.) All. x *P. odoratum* (Mill.) Druce

a. *P.* x *hybridum* Brügger.

b. *P.* x *hybridum* is intermediate between its two putative parents in the number of flowers per leaf axil, the length of and degree of constriction of the perianth, the size of the stamens, and in several other characters. The most complete details are given by Brügger (1886) and Vestergren (1925). The hybrid possesses to a variable degree the ridged stem characteristic of *P. odoratum*. No detailed observations on fertility have been made; most hybrid plants are sterile but fruit is occasionally produced.

c. The hybrid is not known to have arisen in the wild in Br, but Warburg (1962) concluded that it was probably the commonest member of the genus in gardens. Thus at least a good proportion of the records of garden escapes is likely to be of the hybrid (see Perring and Walters, 1962). On the Continent it is also frequent in gardens, but wild hybrids have been recorded in scattered localities from Scandinavia to Ga and Ge. There appear to be more than one clone of the hybrid in cultivation.

d. Meiosis in the hybrid is fairly regular; usually eight bivalents and a trivalent are formed.

e. *P. multiflorum* ($2n = 18$); *P. odoratum* ($2n = 20$); hybrid ($2n = 19$); but in all three taxa other numbers up to ($2n = 30$) are known.

f. BRÜGGER, C. G. (1886). *Polygonatum hybridum. Jber. naturf. Ges. Graubündens,* **29**: 115-116.

PERRING, F. H. and WALTERS, S. M. (1962). *Op. cit.,* p. 313.

THERMAN, E. (1953). Chromosomal evolution in the genus *Polygonatum. Hereditas,* **39**: 277-288.

VESTERGREN, T. (1925). *Polygonatum multiflorum* (L.) All. x *officinale* All. i Sverige. *Svensk bot. Tidskr.*, **19**: 495-519. FIG.

WARBURG, E. F. (1962). *Polygonatum,* in CLAPHAM, A. R., TUTIN, T. G. and WARBURG, E. F. *Op. cit.*, pp. 966-967.

600. *Endymion* Dumort.
(by P. M. Smith)

2 × **1**. *E. hispanicus* (Mill.) Chouard × *E. non-scriptus* (L.) Garcke

a. (?*Scilla patula* Lam. ex DC.).

b. A complete range of hybrid intermediates exists between the parental conditions, notably in characters of the inflorescence, which is erect to nodding, and the perianth, which is companulate to parallel-sided, pale to dark blue in colour, and with straight to recurved tips to the segments. The stamens are inserted fairly close together near the middle of the perianth in many hybrid plants, and the anthers vary in colour from cream to blue. Hybrids seem as fertile as the parents, and backcrossing quickly leads to the formation of hybrid swarms.

c. Hybrid bluebells are common in gardens where both parents are grown, and occasionally escape, or they may arise in the wild where *E. hispanicus* has escaped. They occur throughout western Europe wherever both parents occur, either native or introduced. The parents are probably not sympatric in their native areas of distribution.

d. Hybrids have been made artificially, and have been commercially exploited. In BI some obstacle to crossing is offered by the slightly different flowering times, which overlap in May, but the main isolation barrier is geographical.

e. Both parents and hybrid $2n = 16$. Certain large plants of both parents are triploid, $2n = 24$; these are quite common in gardens.

f. BAKER, J. G. (1872). A study of Wood Hyacinths. *Gdnrs' Chron.*, **31**: 1038-1039.

OULD LOOSDRECHT, Mol van (1950). A giant *Scilla. Gdnrs' Chron.*, **128**: 214.

TURRILL, W. B. (1952). *Endymion hispanicus. Curtis bot. Mag.*, **169**: t. 176. FIG.

601. *Muscari* Mill.
(by C. A. Stace)

1 × 2. *M. atlanticum* Boiss. & Reut. × *M. comosum* (L.) Mill.

= *M.* × *rocheri* P. Fourn. has been recorded from Ga, but needs checking.

605. *Juncus* L.
(by C. A. Stace)

4 × 5. *J. compressus* Jacq. × *J. gerardii* Lois.

a. *J.* × *royeri* P. Fourn. (*J.* × *transiens* Druce, *nom. nud.*).

b. The occurrence of these hybrids is very uncertain and their supposed characters vary according to the plant under consideration. They are variously intermediate in capsule-shape and length of anthers, styles and perianth. They are said to produce occasional capsules, but most or all British specimens seen by me appear to be immature or partially sterile plants of one or other species.

c. Hybrids have been reported from v.c. 6, 15, 16 and 28, and from Ge, Ho, Rm and Su.

d. None.

e. *J. gerardii* ($2n = 80, 84$); *J. compressus* ($2n = 40, 44$).

f. LOUSLEY, J. E. (1935). *Juncus compressus* × *Gerardi. Rep. B.E.C.,* **10**: 986-987.

WHITE, J. W. (1889). *Juncus Gerardi* Lois. *J. Bot., Lond.,* **27**: 49-50.

9 × 8. *J. effusus* L. × *J. inflexus* L.

a. *J.* × *diffusus* Hoppe.

b. This hybrid has the densely tufted habit of its parents; the inflorescence shape resembles that of *J. inflexus* but the floral characters and stem-texture and -anatomy are intermediate. The pith is continuous or nearly so. The pollen appears perfect. The capsule is distinctly smaller than in either parent and seed production is much reduced but very variable. Well-formed seeds germinate readily and F_2 generations showing segregation in diagnostic characters have been raised. The variation of the hybrid found in the wild may be due to the presence of F_2s or backcrosses.

c. Hybrid plants are widespread but nowhere common. They are recorded from most parts of BI where both parents occur (Perring and Sell, 1968; Stace, 1972). Some of these records require confirmation, but it is clear that the hybrid is more frequent in England than elsewhere in BI. It is very widespread in Europe, at least from Su to Rm, usually in close proximity to both parents.

d. Attempts to re-synthesize the hybrid have so far failed. *J. inflexus* flowers a little later than *J. effusus* and prefers more base-rich conditions, but the two species overlap considerably in both these characters.

e. *J. effusus* $2n = 40$ (40, 42); *J. inflexus* $2n = (40)$ 42.

f. CLIFFORD, H. T. (1958). On putative hybrids between *Juncus inflexus* and *J. effusus*. *Kew Bull.*, **13**: 392-395.

NILSSON, Ö. and SNOGERUP, S. (1971). Drawings of Scandinavian plants, 53. *Bot. Notiser*, **124**: 183-184. FIG.

PERRING, F. H. and SELL, P. D. (1968). *Op. cit.*, p. 140.

REICHENBACH, L. and REICHENBACH, H. G. (1847). *Icones Florae Germanicae et Helveticae*, t. 921. Leipzig. FIG.

STACE, C. A. (1972). The history and occurrence in Britain of hybrids in *Juncus* subgenus *Genuini*. *Watsonia*, **9**: 1-11.

SYME, J. T. B. (1870). *English Botany*, ed. 3, **10**: t. 1562. London. FIG.

10 × 8. *J. conglomeratus* L. × *J. inflexus* L.

a. *J.* × *ruhmeri* Aschers. & Graeb.

b. Sterile plants, supposedly intermediate between these two species, have been considered to represent hybrids, but considerable doubt must be attached to all these identifications.

c. Such plants have been reported from v.c. 15/16 and 22, and from Cz, Da, Ga, Ge, Ho and Algeria.

d. *J. inflexus* usually flowers several weeks later than *J. conglomeratus*.

e. *J. inflexus* $2n = (40)$ 42; *J. conglomeratus* $2n = 42$ (40, 42).

f. ASCHERSON, P. and GRAEBNER, P. (1904). *Synopsis der Mitteleuropäischen Flora*, **2**: 450-451. Leipzig.

FOUILLADE, A. (1927). *Juncus Ruhmeri*, in GUÉTROT, M., ed. *Plantes hybrides de France*, **1** and **2**: 34. Lille.

STACE, C. A. (1972). As above.

13 × 8. *J. balticus* Willd. × *J. inflexus* L.

a. None.

b. The hybrids are more robust (to 2 m) than either parent and possess extensively creeping rhizomes. In general habit the plants do not closely resemble any other *Juncus*. The inflorescence-morphology and stem-anatomy are intermediate between those of the parents, but the anthers are larger than in either and possess apparently perfect pollen. Seed-set has, however, not been detected. There is much variation in the size and degree of branching of the inflorescence. The pith is interrupted.

c. Two large patches were independently discovered in 1952 on the coast of v.c. 59, and a smaller one on the coast of v.c. 60 in 1966. All these patches lie in base-rich dune-slacks close to the sea and within about 20 km of each other, one of them lying within 1 km of the only English site of *J. balticus. J. inflexus* is frequent in the area. The hybrid has been found nowhere else.

d. Hybridization attempts have so far been unsuccessful, as have attempts at backcrossing.

e. *J. inflexus* $2n = (40)$ 42; *J. balticus* $2n = c$ 80 (40, 80); hybrid $2n = c$ 82.

f. STACE, C. A. (1970). Unique *Juncus* hybrids in Lancashire. *Nature, Lond.,* **226**: 180.

STACE, C. A. (1972). As above.

8 × pal. *J. inflexus* L. × *J. pallidus* R.Br.

a. None.

b. In areas where *J. pallidus* has become fairly well established, plants have appeared which seem intermediate between this species and native *J. inflexus,* particularly in stem-ridging and inflorescence-shape.

c. Such intermediates have been found in v.c. 30, but now, like most established *J. pallidus,* are extinct.

d. None.

e. *J. inflexus* $2n = 42$ (40).

f. DONY, J. G. (1953). *Flora of Bedfordshire,* p. 398. Luton.

EDGAR, E. (1964). The leafless species of *Juncus* in New Zealand. *N.Z.J. Bot.,* **2**: 177-204.

10 × 9. *J. conglomeratus* L. × *J. effusus* L.

a. *J.* × *kern-reichgeltii* Janch. & Wacht. ex van Ooststr.

b. Both species are variable and because of their overall similarity plants difficult to identify with one species or the other are not rare. Such plants have sometimes been recorded as hybrids, but often wrongly so. The characters which can best be used to identify the parents (and hybrids) are the ridging of the stems and the degree of inrolling shown by the base of the main bract. Genuine hybrids found in Br are apparently completely fertile; Agnew reported introgression into *J. effusus* in one Welsh and one Scottish locality. Snogerup claimed that Swedish hybrids were sterile.

c. Agnew found the hybrid only at higher altitudes in Wales and Scotland, but I have found it in lowland localities in northern England and eastern Scotland. Records exist for scattered localities throughout Br, but all need careful checking. The hybrid has also been recorded from Cz, Ga, Ge, Ho, Rm, Su and Newfoundland.

d. *J. conglomeratus* flowers earlier than *J. effusus* in the same locality, but flowering times usually overlap, especially at higher altitudes. Attempts to produce hybrids artificially have so far failed. Agnew (1968) reported the formation of germinable seed following artificial cross-pollination, but there is no conclusive evidence that this was hybrid seed. A segregating F_2 generation has been grown from wild F_1 plants.

e. *J. effusus* $2n = 40$ (40, 42); *J. conglomeratus* $2n = 42$ (40, 42).

f. AGNEW, A. D. Q. (1968). The interspecific relationships of *Juncus effusus* and *J. conglomeratus* in Britain. *Watsonia*, **6**: 377-388.

KRISA, B. (1962). Relations of the ecologico-phenological observations on the taxonomy of the species *Juncus effusus* L. *s.l. Preslia*, **34**: 114-126.

NILSSON, Ö. and SNOGERUP, S. (1971). Drawings of Scandinavian plants, 54. *Bot. Notiser*, **124**: 184-186. FIG.

SNOGERUP, S. (1970). Studies in the genus *Juncus*, 4. The typification of *J. conglomeratus* L. *Bot. Notiser*, **123**: 425-429. FIG.

STACE, C. A. (1972). As above.

13 × 9. *J. balticus* Willd. × *J. effusus* L.

a. *J.* × *obotritorum* Rothm. (non *J.* × *scalovicus* Aschers. & Graeb.).

b. The hybrid has rather weak, slender stems but spreads rapidly by extensive rhizomes. The plants are intermediate in stem-anatomy and inflorescence characters, but the anthers are larger than in either parent and produce apparently perfect pollen. The inflorescence varies a great deal in size and branching, but seed-set has not been detected.

c. One large patch discovered in a wet dune-slack in v.c. 59 in 1933 was eradicated by building in 1968. In 1966 a second very small colony was found 9 km further south, but this was similarly destroyed in 1974. The former was 3 km from the only English site of *J. balticus*; *J. effusus* is common in the neighbourhood. The hybrid has also been reported from Da, Ge, No and Rs, but only the second record has been confirmed and the first and last are erroneous.

d. Attempts to re-synthesize the hybrid and to backcross it to both parents have so far failed.

e. *J. effusus* $2n = 40$ (40, 42); *J. balticus* $2n = c$ 80 (40, 80); hybrid $2n = c$ 80.

f. STACE, C. A. (1970). As above.
   STACE, C. A. (1972). As above.

9 × pal. *J. effusus* L. × *J. pallidus* R.Br.

has been reported from v.c. 30 in similar circumstances to the putative *J. inflexus* × *J. pallidus* hybrids, but they are more difficult to characterize due to the lack of qualitative differences between the parents.

DONY, J. G. (1953). As above.

EDGAR, E. (1964). As above.

13 × 12. *J. balticus* Willd. × *J. filiformis* L.

= *J.* × *inundatus* Drej. is found in Da, Fe, Ge, Is, No, Rs and Su.

15 × 14. *J. acutus* L. × *J. maritimus* Lam.

has been occasionally reported from Europe and once from New Zealand, but the records have not been confirmed.

18 × 17. *J. acutiflorus* Ehrh. ex Hoffm. × *J. subnodulosus* Schrank

was recorded from Freshwater, v.c. 10, by E. Drabble in 1924. The specimen in **BM** seems too young to determine satisfactorily, but this hybrid has apparently never been discovered anywhere else. The record must therefore be treated as very doubtful.

DRABBLE, E. and LONG, J. W. (1932). A list of plants from the Isle of Wight. *Rep. B.E.C.*, **9**: 734-757.

19 × 17. *J. articulatus* L. × *J. subnodulosus* Schrank

= *J.* × *degenianus* Boros was described from Hu in 1922.

18 × 19. *J. acutiflorus* Ehrh. ex Hoffm. × *J. articulatus* L.

a. *J.* × *surrejanus* Druce.

b. The hybrid is variable in habit but on the whole intermediate between its parents in this feature as well as in flowering time, inflorescence-branching, and perianth and other floral characters. The capsules of the hybrid do not swell and they produce no seed. Because of the variability of *J. articulatus* hybrids are most easily recognized in the field where they can be compared with the variant of *J. articulatus* occupying that area. An erect, fertile jointed-rush with the chromosome number of *J. articulatus* ($2n = 80$) but characters intermediate between the two above species was described by Timm and Clapham (1940); it might be a hybrid originating from the fusion of $n = 40$ gametes from both parents.

c. The two species show a small separation in ecological preference (*J. acutiflorus* is commoner in more acid, less trampled places) but they are commonly found in flower together. Hybrids have probably been greatly overlooked because of the overall similarity of the parents; they are reported as very common in western Wales and around Oxford, and are almost certainly the commonest British rush hybrids. Records exist for v.c. 1–6, 9, 11, 17, 20, 22, 23, 28, 45–49, 70, 87, 94, 96, 101, 111, H28, H33 and H35, and for Au, Ga, Ge, He and Ho.

d. None.

e. *J. acutiflorus* $2n = 40$; *J. articulatus* $2n = 80$; hybrid $2n = 60$.

f. CHAPPLE, J. F. G. (1948). *Juncus acutiflorus* (Ehrh.) Hoffm. × *J. articulatus* L. *Rep. B.E.C.*, **13**: 370-371.

CLAPHAM, A. R. (1949). Taxonomic Problems in *Galium* and *Juncus*, in WILMOTT, A. J., ed. *British flowering plants and modern systematic methods*, pp. 72-74. Arbroath. FIG.

NILSSON, Ö. and SNOGERUP, S. (1972). Drawings of Scandinavian plants, 76. *Bot. Notiser*, **125**: 203-204. FIG.

TIMM, E. W. and CLAPHAM, A. R. (1940). Jointed rushes of the Oxford district. *New Phytol.,* **39**: 1-11. FIG.

18 × **20. *J. acutiflorus*** Ehrh. ex Hoffm. × *J. alpinoarticulatus* Chaix
= *J.* × *langei* Erdner has been recorded from Au, Ga, Ge and He.

20 × **19. *J. alpinoarticulatus*** Chaix × *J. articulatus* L.
  a. *J.* × *buchenaui* Dörfl.
  b. The hybrid varies somewhat in appearance between the two parents, i.e. the culms may be erect to decumbent and the perianth segments acute to obtuse, and the inflorescence varies in dimensions, branching and the number of heads and flowers per head. Elsewhere in Europe the variation is greater due to the large number of variants of the parents involved in hybridization. The capsule is usually seedless, but may possess a few seeds; their germinability is untested.
  c. The hybrid was discovered by G. C. Druce in 1930 with *J. alpino-articulatus* in Upper Teesdale, v.c. 66, and there are old records from Inchnadamph, v.c. 108, but it appears that there are no recent records from anywhere in BI. It occurs in Au, Cz, Da, Fe, Ga, Ge, He, Ho, Is, No, Rs, Su and Newfoundland where the parents are sympatric.
  d. Artificial hybrids have been made by Snogerup.
  e. *J. articulatus* 2*n* = 30; *J. alpinoarticulatus* 2*n* = 40 (Fern, 1971); (hybrid 2*n* = 60).
  f. DRUCE, G. C. (1931). *Juncus nodulosus* Wahl.. *Rep. B.E.C.,* **9**: 372-373.
     FEARN, G. M. (1971). *Biosystematic studies of selected species in the Teesdale flora.* Ph.D. thesis, University of Sheffield.
     HÄMET-AHTI, L. (1966). Some races of *Juncus articulatus* in Finland. *Acta bot. fenn.,* **72**: 3-22. FIG.
     NILSSON, Ö. and SNOGERUP, S. (1972). Drawings of Scandinavian plants, 78. *Bot. Notiser,* **125**: 206. FIG.

19 × **21. *J. articulatus*** L. × *J. nodulosus* Wahlenb.
  a. None.
  b. Sterile plants intermediate in inflorescence-form and in size, shape and colour of the perianth segments have been found with both species and are thought to be hybrids. Their identity with cytologically-confirmed Scandinavian hybrids has yet to be proved.
  c. Intermediates have only been found in one of the few British localities of *J. nodulosus,* on the shores of Loch Ussie, v.c. 106. The first specimens (det. Lindquist) were collected by E. S. Marshall in 1892, when he first discovered *J. nodulosus,* and they have been recently refound. The areas of *J. nodulosus* are close to the common *J. articulatus,* and the hybrids occur in similar wet, stony, lake-shore conditions. They are frequent in Su.
  d. None.

e. *J. articulatus* 2*n* = 80; *J. nodulosus* 2*n* = 40, (40, 80) (Fearn, 1971); hybrid (2*n* = 60).

f. FEARN, G. M. (1971). As above.
HÄMET-AHTI, L. (1966). As above. FIG.
NILSSON, Ö. and SNOGERUP, S. (1972). Drawings of Scandinavian plants, 79. *Bot. Notiser*, **125**: 206-207. FIG.
PUGSLEY, H. W. (1931). A new *Juncus* in Scotland. *J. Bot., Lond.*, **69**: 278-284.
WEBSTER, M. McC. (1970). *Juncus nodulosus* Wahlenb. var. *marshallii* (Pugsley) P. W. Richards refound at Loch Ussie, Ross and Cromarty. *Trans. Proc. Bot. Soc. Edinb.*, **40**: 621-622.

19 × 22. *J. articulatus* L. × *J. bulbosus* L.
was discovered in Su in 1957.

22 × koc. *J. bulbosus* L. × *J. kochii* F. W. Schultz.
There is a disagreement as to whether or not these two species are distinct, and on the characters which should be used to separate them. Hybrids or intermediates between them have been noted, but there are no firm records and no voucher specimens have been seen. The two taxa produce a fertile F_1 when crossed with *J. bulbosus* as the female.
BENOIT, P. M. and ALLEN, D. E. (1968). Synthesized *Juncus bulbosus* × *kochii*. *Proc. B.S.B.I.*, **7**: 504.

606. *Luzula* DC.

(by C. A. Stace)

2 × 1. *L. forsteri* (Sm.) DC. × *L. pilosa* (L.) Willd.

a. *L. × borreri* Bromf. ex Bab.

b. Hybrid plants are morphologically intermediate in basal leaf-width, cauline leaf-length, degree to which the inflorescence branches spread out, presence of a mucro at the leaf-apex, and the shape and size of the seed-caruncle. The anthers are longer than in either parent and possess normal-looking pollen. The seeds are, however, almost always abortive. In England some hybrid specimens have been detected which fall very close to *L. pilosa*, and, coupled with the fact that apparently good seeds are occasionally found, this tends to suggest that backcrossing at least to that parent occasionally occurs.

c. The hybrid is frequent and sometimes common where the two species are found together in southern England, from Cornwall and Kent to Herefordshire. There is also a record for v.c. H20 (specimen in **CGE**) despite the absence of *L. forsteri* in Hb (Perring and Sell, 1968). Elsewhere it is very scarce; records exist for Ga and Ge.

d. Hybrids are easily synthesized with either parent as female. They are completely sterile but all or nearly all of the 33 small chromosomes from *L. pilosa* pair at meiosis with the 12 large ones from *L. forsteri*, indicating a fairly close homology.

e. *L. forsteri* ($2n = 24$); *L. pilosa* ($2n = 66$); hybrid ($2n = 45$).

f. EBINGER, J. E. (1962). *Luzula* x *borreri* in England. *Watsonia*, **5**: 251-254.

   EBINGER, J. E. (1964). Taxonomy of the subgenus *Pterodes*, genus *Luzula. Mem. N.Y. bot. Gdn*, **10**: 279-304.

   NORDENSKIOLD, H. (1957). Hybridisation experiments in the genus *Luzula*, 3. The subgenus *Pterodes. Bot. Notiser*, **110**: 1-16.

   PERRING, F. H. and SELL, P. D. (1968). *Op. cit.*, p. 140.

   SYNNOTT, D. (1973). *Luzula* x *borreri* in Ireland. *Ir. Nat J.*, **17**: 327-330.

1 x 3. *L. pilosa* (L.) Willd. x *L. sylvatica* (Huds.) Gaudin

= *L.* x *buchenaui* Fourn. has been recorded from He.

4 x 1. *L. luzuloides* (Lam.) Dandy & Wilmott x *L. pilosa* (L.) Willd.

= *L.* x *cechica* Domin is recorded from Cz.

4 x 3. *L. luzuloides* (Lam.) Dandy & Wilmott x *L. sylvatica* (Huds.) Gaudin

= *L.* x *hermanni-muelleri* Aschers. & Graebn. has been recorded from Ga and Ge.

9 x 3. *L. multiflora* (Retz) Lejeune x *L. sylvatica* (Huds.) Gaudin

= *L.* x *johannis-principis* Murr has been recorded from Au and Ge.

4 x 5. *L. luzuloides* (Lam.) Dandy & Wilmott x *L. nivea* (L.) DC.

= *L.* x *favratii* Richter has been recorded from Au, Ge and He.

8 x 6. *L. campestris* (L.) DC. x *L. spicata* (L.) DC.

has been recorded doubtfully from Abyssinia, central Europe, the Balkans and Lapland.

9 × 6. *L. multiflora* (Retz) Lejeune × *L. spicata* (L.) DC.

= *L.* × *winderiae* Murr has been recorded from Au and Ge.

8 × 9. *L. campestris* (L.) DC. × *L. multiflora* (Retz) Lejeune

= *L.* × *intermedia* Figert has been recorded from Au, Ga and Ge.

8 × 10. *L. campestris* (L.) DC. × *L. pallescens* Sw.

occurs in Su.

9 × 10. *L. multiflora* (Retz) Lejeune × *L. pallescens* Sw.

has been mentioned as occurring on the Continent. Lousley (1950) noted what "seemed to be intermediates" in Denton Fen, v.c. 31, in 1948, and a specimen in **BM** similarly labelled was collected at Woodwalton Fen, v.c. 31, by Mr Tams in 1960. In both these places the two species occur close together, but present evidence is far from sufficient to prove the existence of the hybrid.

LOUSLEY, J. E. (1950). *Luzula pallescens* (Wahlenb.) Wahlenb. *Year Book B.S.B.I.*, **1950**: 96.

con × 9. *L. congesta* (Thuill.) Lejeune × *L. multiflora* (Retz) Lejeune

a. None.
b. These taxa are usually considered varieties or subspecies of one species, but they differ at least in Br in habit, inflorescence-form, seed-size, length of perianth segments, chromosome number and ecological preferences. Intermediates are sometimes encountered in mixed populations; such plants often also have an intermediate chromosome number and are wholly or partially sterile. Less often plants with an intermediate chromosome number fall into the normal morphological range of *L. multiflora* or *L. congesta.*
c. Intermediates, presumed hybrids, were found by Buchanan (1959) in unspecified areas of Scotland where the two parents grew together. Hybrid swarms are reported from Da.
d. Using Scandinavian material the two taxa were easily crossed and the hybrids were fairly fertile and gave rise to a progeny with varying chromosome numbers. The existence of hybrid swarms suggests back-crossing.
e. *L. multiflora* $2n = 36$; *L. congesta* $2n = 48$; intermediates $2n = 42$ (and 39, 40, 44, 47 in Da).
f. BUCHANAN, J. (1959). *Luzula campestris* (L.) DC. *Proc. Linn. Soc. Lond.*, **170**: 126-128.
NORDENSKIOLD, H. (1956). Cyto-taxonomical studies in the genus *Luzula*, 2. Hybridization experiments in the *campestris-multiflora* complex. *Hereditas*, **42**: 7-73.

614. *Narcissus* L.
(by D. McClintock)

In addition to the following three groups of hybrids all manner of garden clones, mostly of obscure origin, nameable and unnameable, may be found freely naturalized in many parts of BI (particularly south-western Br and CI); over 50 are listed for Guernsey alone by McClintock (1975). In the following account one of the parents of the first two hybrids is given as 1–3. *Narcissus* sect. *Ajax* (Salisb.) Spach. It seems likely that both of these groups of hybrids have originated from crosses involving more than one species of sect. *Ajax*, e.g. *N. N. pseudonarcissus, N. hispanicus* and *N. abscissus.*

General References

FERNANDEZ, A. (1967). Keys to the identification of native and naturalised taxa of the genus *Narcissus. Royal Horticultural Society's Daffodil and Tulip Year Book*, **1968**: 37-66.

McCLINTOCK, D. (1975). *The wild flowers of Guernsey*, pp. 230-233. London.

WARBURG, E. F. (1962). *Narcissus* L., in CLAPHAM, A. R., TUTIN, T. G. and WARBURG, E. F. *Op. cit.*, pp. 999-1002.

1–3 × 6. *Narcissus* section *Ajax* (Salisb.) Spach × *N. poeticus* L. agg.

a. *N.* × *incomparabilis* Mill., *sensu lato* (incl. *N.* × *barrii* Baker, *N.* × *burbidgei* Baker and *N.* × *leedsii* Baker).

b. The nothomorphs of this parentage are conveniently divided into two groups. The first is what is usually meant by *N.* × *incomparabilis*, viz. the large-cupped cultivars of Division 2 of the Royal Horticultural Society's 1950 Classification, with the corona more than one-third but less than equal to the length of the perianth. Very numerous examples have been grown for at least three or four hundred years. 'Carlton', 'Fortune' and 'Sir Watkin' are among the many which have become naturalized. 'Butter and Eggs' is a long-known double-flowered cultivar. Fernandez (1967) gave the parentage of this hybrid in the restricted sense as *N. hispanicus* Gouan × *N. poeticus*; Jefferson-Brown (1969) as *N. pseudonarcissus* × *N. poeticus*. The second group is what is usually meant by *N.* × *barrii*, viz. small-cupped cultivars of Division 3 of the R.H.S. Classification, which are backcrosses of *N.* × *incomparabilis* to *N. poeticus*. These originated about 1880. They have the corona not more than one-third the length of the perianth segments. Cultivars known to be naturalized include 'Barri Conspicuus' and 'Edward Buxton'. In the wild in BI seed-set is rare, depending on climatic factors, and seedlings apparently even rarer (McClintock, 1970).

c. These plants are liable to persist wherever they have been thrown out or left after commercial or decorative planting, and may flourish vegetatively indefinitely. Such naturalized Narcissi have hardly ever been accurately recorded; only very few local Floras even mention this hybrid, but it is doubtless widespread. On the Continent it is recorded from Au, Ga, He, Hs and It.

d. Hybrids of this origin have been made countless times by breeders.

e. *N. hispanicus* (2*n* = 14, 21, 42); *N. pseudonarcissus* (2*n* = 14); *N. poeticus* (2*n* = 14, 21); hybrid (2*n* = 14).

f. BAKER, J. G. (1875). In BURBIDGE, F. W. *The Narcissus: its history and culture*, pp. 38-41, pl. XVIII-XXII. London. FIG.

BAKER, J. G. (1886). New garden plants. *Gdnrs' Chron.*, 25: 648.

BOWLES, E. A. (1934). *A handbook of Narcissus*, pp. 91-116, pl. I, IV, VI-X. London. FIG.

JEFFERSON-BROWN, M. J. (1969). *Daffodils and Narcissi*, pp. 92-111, 194, 201; pl. 9-22, III-V. London. FIG.

McCLINTOCK, D. (1970). The purity of naturalised Narcissi. *Royal Horticultural Society's Daffodil and Tulip Year Book*, 1971: 181-183.

1–3 × jon. *Narcissus* sect. *Ajax* (Salisb.) Spach × *N. jonquilla* L.

a. *N. × infundibulum* Poir., *sensu lato* (*N. × odorus* L., *N. × heminalis* (Salisb.) Schultes f.).

b. This is a group of elegant, fragrant hybrids of Division 7 of the R.H.S. 1950 Classification, some double-flowered. They have 1–3 bright yellow flowers on shortish scapes, and narrow leaves. The parentage of *N. × infundibulum sensu stricto* was given by Fernandez (1967) as *N. abscissus* Schultes f. × *N. jonquilla*, and that of *N. × odorus* as *N. hispanicus* × *N. jonquilla*, both naturalized in several places. Jefferson-Brown (1969) recorded that the number and forms of the chromosomes show that *N. × odorus* is a hybrid of *N. pseudonarcissus* and *N. jonquilla*.

c. The small-flowered taxon called *N. × heminalis* was known in plenty 2 or 3 miles south of St Austell, v.c. 2, by 1870 (Davey, 1909), but the colony was later nearly extirpated (Thurston and Vigurs, 1922). It still survives, however (Mrs B. E. M. Garratt, pers. comm. 1973). Hybrids are also recorded from Ga, He, ?Hs, It, Ju and Lu.

d. Hybrids of this parentage have been made by horticulturists.

e. *N. hispanicus* (2*n* = 14, 21, 42); *N. pseudonarcissus*, *N. abscissus* and *N. jonquilla* (2*n* = 14); *N. × odorus* (2*n* = 14, 28).

f. BAKER, J. G. (1875). As above, pp. 41-42, pl. XXIV. FIG.

BOWLES, E. A. (1934). As above, pp. 139-143.

DAVEY, F. H. (1909). *Flora of Cornwall*, pp. 431-432. Penryn.

D. K. (1887). The great and common Jonquils. *The Garden*, 32: 394-395, pl. 620. FIG.

JEFFERSON-BROWN, M. J. (1969). As above, pp. 143-145, 195, 205; pl. 34. FIG.

THURSTON, E. and VIGURS, C. C. (1922). *Supplement to the Flora of Cornwall*, p. 131. Truro.

6 × **taz.** *N. poeticus* L. agg. × *N. tazetta* L.

   a. *N.* × *medioluteus* Mill. (*N.* × *biflorus* Curt.).
   b. This comprises the Poetaz Narcissi, Division 8 of the R.H.S. 1950 Classification. The best-known clone is 'Primrose Peerless', named thus and common even in Gerard's time (*c* 1600). It has 1–3 fragrant flowers on each scape, each flower up to 2 inches across with a cream perianth and a small, pale-yellow corona. It is said never to perfect its ovules and pollen (Baker, 1875).
   c. 'Primrose Peerless' is recorded from several southern English and Irish counties, and from Au, He, It, Ju and Lu.
   d. This was chiefly worked on in the first half of the present century (Gray, 1970), when many cultivars were produced.
   e. *N. poeticus* ($2n = 14, 21$); *N. tazetta* ($2n = 20$); hybrid ($2n = 17$).
   f. BAKER, J. G. (1875). As above, pp. 83-84, pl. XLI. FIG.
      BOWLES, E. A. (1934). As above, pp. 174-178.
      GRAY, A. (1970). *Narcissus* Poetaz. Royal Horticultural Society's *Daffodil and Tulip Year Book*, **1971**: 43-45.
      JEFFERSON-BROWN, M. J. (1969). As above, pp. 128-129, 195; pl. 33. FIG.

616. *Iris* L.
(by J. R. Ellis)

2 × **vir.** *I. versicolor* L. × *I. virginica* L.

   a. *I.* × *robusta* E. Anderson.
   b. Hybrids are similar to *I. versicolor* with extensive vegetative increase by rhizomes. They are appreciably but not completely sterile, setting 0–2 capsules per flowering shoot. The capsules are irregularly cylindrical with an attenuated conical apex and 0–12 (mean 5.7) rounded seeds which often have degenerated embryos.
   c. Hybrids occur in reed-swamps and rough pasture on the northern fringes of Lake Windermere, v.c. 69. Neither parent occurs in the immediate vicinity, but *I. versicolor* is naturalized at Ullswater and Esthwaite; no naturalized colonies of *I. virginica* are known in this country. In North America natural hybrid colonies occur in Ontario and Michigan, where the two species are sympatric.
   d. Anderson obtained vigorous hybrids from reciprocal pollinations between *I. versicolor* and *I. virginica*. The F_1 hybrids were similar to the natural American hybrids and were highly but not completely sterile. British wild

hybrids have an irregular meiotic division with 35 bivalents and 19 univalents, suggesting that *I. virginica* is one of the progenital species of the allopolyploid *I. versicolor*. In spite of the irregular meiosis, some viable seeds result from self-pollinations and from backcross pollinations with *I. versicolor*.

e. *I. virginica* ($2n = 70, 71$); *I. versicolor* $2n = 108$; wild hybrid $2n = 89$; artificial F_2 $2n = 86, 89, 93, 96$; artificial backcross to *I. versicolor* $2n = 98, 99, c 102$.

f. ANDERSON, E. (1928). The problem of species in the northern blue flags, *I. versicolor* L. and *I. virginica* L. *Ann. Mo. bot. Gdn*, **15**: 241-332.

CHIMPHAMBA, B. B. (1971). *Cytogenetic studies in the genus Iris*. Ph.D. thesis, University of London.

624–645. Orchidaceae
(by P. F. Hunt, R. H. Roberts and D. P. Young)

Extensive natural hybridization in both temperate and tropical species, together with the vast number (over 45,000) of artificially produced hybrids involving, on occasions, up to five genera in one plant, has made the Orchidaceae a very complex family with which to deal. Nomenclatural details are provided in Willis (1973) and in Garay and Sweet (1974), where all orchid hybrid-genera are listed.

Records of hybrids between British native orchids are very scattered; whenever the distribution ranges of two species overlap and these species are known to hybridize a careful examination of populations should be carried out. Hybrids are not always immediately recognizable as such as they can show all stages of intermediacy between the two parents. In some cases, although the two species may cross elsewhere in their geographical range, they may not have overlapping flowering times in Br, or one or both parents may be very rare or they may be separated by different ecological preferences.

The identification of the parents of British orchid hybrids can rarely be more than tentative; it depends on a sound knowledge of the possible parents and their ranges of morphological variation, and on the ability to recognize certain diagnostic characters. Study of the details of the reproductive parts of the flowers of suspected hybrids to deduce their possible parents is not always worth while. Summerhayes (1968) recorded that, in plants of x *Dactylogymnadenia* that he examined, the bursicles (which are present in *Dactylorhiza* but absent from *Gymnadenia*) were present, absent or imperfectly formed in different flowers of the same spike. Apart from the extensive studies made in *Dactylorhiza* very little accurate experimental or biometrical work has been carried out on orchid hybrids. Surprisingly little observational work has been carried out on the behaviour and specificity of the insect pollinators of British orchids; such a study would ensure a greater predictability of hybridization patterns.

General References

CAMUS, A. and CAMUS, E. G. (1929). *Iconographie des Orchidées d'Europe et du Bassin Méditerranéen.* Paris. FIG.

GARAY, L. A. and SWEET, H. R. (1974). Hybrid generic names, in WITHNER, C. L., ed. *The orchids. Scientific studies.* London.

GODFERY, M. J. (1933). *Monograph and iconograph of native British Orchidaceae.* Cambridge.

HUNT, P. F. (1971). Taxonomic and nomenclatural notes on European and British orchid hybrids. *Orchid Rev.,* **79**: 138-142.

SUMMERHAYES, V. S. (1968). *Wild orchids of Britain,* 2nd ed. London.

WILLIS, A. J. (1973). *Dictionary of flowering plants and ferns,* 8th ed. by AIRY-SHAW, H. K. Cambridge.

624. *Cephalanthera* Rich.
(by P. F. Hunt)

1 × 2. C. damasonium (Mill.) Druce × *C. longifolia* (L.) Fritsch

= *C.* × *schulzei* G. Camus, Berg. & A. Camus has been reported from Ga and Ge.

1 × 3. C. damasonium (Mill.) Druce × *C. rubra* (L.) Rich.

= *C.* × *mayeri* Camus has been reported from Ge.

2 × 3. C. longifolia (L.) Fritsch × *C. rubra* (L.) Rich.

= *C.* × *otto-hechtii* Keller has been reported from Su.

624 × 625. *Cephalanthera* Rich.
× *Epipactis* Sw.
= × *Cephalopactis* Aschers. & Graebn.
(by D. P. Young)

624/1 × 625/2. *Cephalanthera damasonium* (Mill.) Druce × *Epipactis helleborine* (L.) Cr.

= × *C. hybrida* Domin has been once reported from central Europe, but is doubtful.

624/1 × 625/7. *Cephalanthera damasonium* (Mill.) Druce × *Epipactis atrorubens* (Hoffm.) Schult.

= × *C. speciosa* (Wettst.) Aschers. & Graebn. has been once reported from central Europe, but is doubtful.

624 × 640. *Cephalanthera* Rich.
× *Ophrys* L.
= × *Cephalophrys* Cox
(by P. F. Hunt)

624/3 × 640/1. *Cephalanthera rubra* (L.) Rich. × *Ophrys apifera* Huds.

= × *C. integra* Cox was reported from It in 1912, but is most unlikely to be correctly identified.

625. *Epipactis* Sw.
(by D. P. Young)

7 × 1. *E. atrorubens* (Hoffm.) Schult. × *E. palustris* (L.) Cr.

= *E.* × *pupplingensis* K. Bell was described from Ge in 1969.

2 × 3. *E. helleborine* (L.) Cr. × *E. purpurata* Sm.

   a. *E.* × *schulzei* P. Fourn.

   b. Plants with intermediate characters are occasionally seen, but, because of the wide variability of *E. helleborine* particularly, it is difficult to be sure whether these are hybrids or merely variants of one or other putative parent. The most convincing hybrids have leaves longer than in *E. purpurata* but retaining their lanceolate shape and violet suffusion, and flowers with the dull purple colour of *E. helleborine* rather than the clear pale green and mauve of *E. purpurata*. Other morphological characters are more or less intermediate between the parents. Such plants are apparently fertile.

   c. Probable hybrids of this parentage have been found, sometimes in numbers, in various parts of southern England and the southern Midlands where the two species coexist. They have also been found on the Continent within the area of *E. purpurata*.

   d. None.

   e. *E. helleborine* ($2n = 40$, but also aneuploid variants).

   f. HALL, P. M. (1932). Notes on British Orchidaceae. *Rep. B.E.C.*, **9**: 724-725.

     KNIGHT, J. T. H. (1959). Notes on a colony of *Epipactis* in the Sussex Weald. *Proc. B.S.B.I.*, **3**: 279-280.

2 × 4. *E. helleborine* (L.) Cr. × *E. leptochila* (Godf.) Godf.

   = *E.* × *stephensonii* Godf. was reported from v.c. 17 and 33/34 by Godfery, but the evidence was not adequate for such an identification. Plants with characters of both putative parents have been seen in Be and Ge, but require further study.

   GODFERY, M. J. (1926). *Epipactis dunensis* Godf. *J. Bot., Lond.*, **64**: 65-68.

7 × 2. *E. atrorubens* (Hoffm.) Schult. × *E. helleborine* (L.) Cr.

   a. *E.* × *schmalhousenii* K. Richt. (*Helleborine* × *crowtheri* (Druce) Druce, nom. nud.).

   b. Plants with characters intermediate between the parents have been regarded as hybrids. There is nothing against this view, but the wide and overlapping ranges of variation of each parent make it difficult to be certain whether a plant is a genuine hybrid. One credible example from Da was sterile.

   c. Reconsideration of the reports of this hybrid from northern Br leads to the conclusion that most were shade forms of *E. atrorubens*. The evidence that it has ever occurred here is unconvincing at present. There are records from v.c. 50, 64 and 108. The hybrid has been rarely reported on the Continent.

d. The parents are to some degree genetically isolated by non-overlapping times of flowering, besides different ecological preferences.

e. Both parents ($2n = 40$, but *E. helleborine* also has aneuploid variants).

f. DRUCE, G. C. (1910). *Helleborine atrorubens* Druce var. *Crowtheri nov. var. Naturalist, Hull,* **638**: 128.

GODFERY, M. J. (1933). As above, p. 83, pl. 12. FIG.

LEES, F. A. (1910). Note on *Helleborine atrorubens* var. *Crowtheri. Naturalist, Hull,* **638**: 129-131.

627. *Spiranthes* Rich.
(by P. F. Hunt)

2 × 1. *S. aestivalis* (Poir.) Rich. × *S. spiralis* (L.) Chevall.

= *S.* × *zahlbruckneri* H. Fleischmann has been reported from Au.

634 × 637. *Herminium* R.Br.
× *Pseudorchis* Séguier (*Leucorchis* E. Mey.) = × *Pseudinium* P. F. Hunt
(by P. F. Hunt)

634/1 × 637/1. *Herminium monorchis* (L.) R.Br. × *Pseudorchis albida* (L.) A. & D. Löve (*Leucorchis albida* (L.) E. Mey. ex Schur)

= × *P. aschersonianum* (Brügg. & Killias) P. F. Hunt has been reported from the Continent.

644 × 634. *Aceras* R.Br. × *Herminium* R.Br.
= × *Aceraherminium* Camus
(by P. F. Hunt)

644/1 × 634/1. *Aceras anthropophorum* (L.) Ait. f. × *Herminium monorchis* (L.) R.Br.

has been reported from the Continent.

635 × 636. *Coeloglossum* Hartm.
× *Gymnadenia* R.Br.
= × *Gymnaglossum* Rolfe
(by P. F. Hunt)

635/1 × 636/1. *Coeloglossum viride* (L.) Hartm. × *Gymnadenia conopsea* (L.) R.Br.

a. × *G. jacksonii* (Quirk) Rolfe.
b. Plants with varying degrees of parental influence have been occasionally seen but usually the hybrids resemble *Gymnadenia* in most respects but with the otherwise pink flowers tinged green. The labellum is intermediate in shape but the spur is much shorter than in *Gymnadenia*. The leaves are intermediate between the relatively long and narrow ones of *Gymnadenia* and the shorter, ovate ones of *Coeloglossum*. Plants are usually apparently sterile, which is unusual in orchid hybrids, but they often exhibit hybrid vigour, with one plant from v.c. 40 reaching 74 cm in height.
c. Scattered records exist for this hybrid throughout most of BI, and it is to be expected wherever the two species coexist, mostly on chalk and limestone. It also occurs over much of the Continent, at least from Ga to Rm.
d. None.
e. *G. conopsea* $2n = 20$ (also 40, 80); *C. viride* ($2n = 40$).
f. GODFERY, M. J. (1933). As above, pp. 132-133, pl. 23 and 29. FIG.
McKECHNIE, H. (1918). Notes on some new hybrid orchids. *Rep. B.E.C.*, **5**: 180-183, pl. 12 and 14A. FIG.
QUIRK, R. (1911). *Gymplatanthera jacksonii*. *Rep. Winchester Coll. nat. Hist. Soc.*, **1909-1911**: 5-6. FIG.

STEPHENSON, T. and STEPHENSON, T. A. (1922). Hybrids of *Gymnadenia conopsea* and *Coeloglossum viride. Orchid Rev.*, **30**: 101-103. FIG.

SUMMERHAYES, V. S. (1968). As above, p. 213, pl. V. FIG.

635 × 638. *Coeloglossum* Hartm.
× *Platanthera* Rich.
= × *Coeloplatanthera* Ciferri & Giacomini
(by P. F. Hunt)

635/1 × 638/1. *Coeloglossum viride* (L.) Hartm. × *Platanthera chlorantha* (Custer) Reichb.

(= × *C. brueggeri* Ciferri & Giacomini, *nom. nud.*) has been reported from the Continent.

635/1 × 638/2. *Coeloglossum viride* (L.) Hartm. × *Platanthera bifolia* (L.) Rich.

was reported by Heslop-Harrison from South Uist, v.c. 110, in 1949. No specimen has been seen and nothing further is known of the discovery. HESLOP-HARRISON, J. (1949). A new orchid hybrid. *Vasculum*, **34**: 22.

635 × 643. *Coeloglossum* Hartm.
× *Dactylorhiza* Nevski
= × *Dactyloglossum* Hunt & Summerh.
(by P. F. Hunt)

635/1 × 643/1. *Coeloglossum viride* (L.) Hartm. × *Dactylorhiza fuchsii* (Druce) Soó

a. × *D. mixtum* (Aschers. & Graebn.) P. F. Hunt (× *Orchicoeloglossum mixtum* Aschers. & Graebn.)

b. Plants with intermediate characters are regularly seen in several areas. Most are smaller than *D. fuchsii* and the size and arrangement of the leaves are much more like those of *C. viride*, but the leaves can be faintly spotted. The flower is usually of the *D. fuchsii* colour tinged and overlaid with the pale green of *C. viride* to give a distinct and curious mottled effect. The labellum is intermediate and variable; it is basically the shape of that of *C. viride* with a well-marked, trilobed apical half, an almost tooth-like mid-lobe and broad side-lobes. The spur is about as long as that of *D. fuchsii* but of the width of that of *C. viride*.

c. It is reported from many places in Br where the distribution of the two species overlaps, but is most frequently found on the chalk downs. It is also found on the Continent.

d. None.

e. *C. viride* ($2n = 40$); *D. fuchsii* $2n = 40$.

f. GODFERY, M. J. (1933). As above, pp. 132-133, pl. 22. FIG.

GROSE, J. D. (1950). *Coeloglossum viride* x *Orchis fuchsii* on the Wiltshire Downs. *Watsonia*, **1**: 207-208.

HALL, P. M. (1929). Hybrids between *Habenaria viridis* (L.) Br. and the palmate orchids. *Rep. B.E.C.*, **8**: 789-790.

HALL, P. M. (1930). Hybrid orchids. *Rep. B.E.C.*, **9**: 37.

McKECHNIE, H. (1918). Notes on some new hybrid orchids. *Rep. B.E.C.*, **5**: 180-183, pl. 15. FIG.

SUMMERHAYES, V. S. (1968). As above, p. 213.

635/1 x 643/1 x 643/3. *Coeloglossum viride* (L.) Hartm. x *Dactylorhiza fuchsii* (Druce) Soó x *Dactylorhiza praetermissa* (Druce) Soó

(= x *Orchicoeloglossum ullmanii* P. M. Hall) was reported from near Winchester, v.c. 11/12, in 1912-1914, but apparently no specimens or detailed descriptions exist. At first other species such as *D. incarnata, D. maculata* or '*D. latifolia*' were suggested as possible parents, but these are rarer in the area concerned.

HALL, P. M. (1913). New hybrid orchid. *Rep. Winchester Coll. nat. Hist. Soc.*, **1912-1913**: 6-8. FIG. (This is summarized in *Rep. B.E.C.*, **5**: 158-159, 172-173 (1918)).

HALL, P. M. (1929). As above.

635/1 x 643/2. *Coeloglossum viride* (L.) Hartm. x *Dactylorhiza maculata* (L.) Soó

a. x *D. drucei* (Camus) Soó (*Habenaria* x *websteri* Druce).

b. First assumed to be a hybrid between *C. viride* and a marsh orchid (*D. incarnata, D. praetermissa* or *D. majalis*), the consensus of opinion today is that x *Dactyloglossum drucei* is applicable to hybrids between *C. viride* and *D. maculata*. The overall resemblance of the plants is to *D. maculata* but the flower is suffused green and the labellum is narrower than in *D. maculata*.

c. This hybrid is apparently very rare but there are scattered records in Br; its rarity may be due to the ecological separation of the two species and to the possibility that some records of x *D. mixtum* may be referable to this hybrid instead. It is also reported from the Continent and from Is.

d. None.

e. *C. viride* (2n = 40); *D. maculata* 2n = 80.

f. GODFERY, M. J. (1933). As above, pp. 130-131, pl. 23. FIG.

ROLFE, R. A. (1892). On *Habenari-orchis viridi-maculata*, Rolfe, *hyb. nat. Ann. Bot.*, 6: 325-327, pl 18. FIG.

635/1 x 643/3. *Coeloglossum viride* (L.) Hartm. x *Dactylorhiza incarnata* (L.) Soó

= x *D. guilhotii* (Camus) Soó (*Habenaria* x *vaughanii* Druce, *nom. nud.*) has been reported from Br and from Ga. No authentic British specimen has been found and the French plant is very similar to *C. viride*, it being difficult to discern any influence of *D. incarnata*.

GODFERY, M. J. (1933). As above, p. 130.

635/1 x 643/4. *Coeglossum viride* (L.) Hartm. x *Dactylorhiza praetermissa* (Druce) Soó

(= x *D. dominianum* auct., x *Orchicoeloglossum dominianum* auct.). No British specimen or verified record of this hybrid can be traced but it has several times been referred to in botanical literature. Much confusion has occurred regarding the correct name for it and it is quite probable that x *D. dominianum* is in fact applicable to *C. viride* x *D. fuchsii*.

635/1 x 643/5. *Coeloglossum viride* (L.) Hartm. x *Dactylorhiza purpurella* (T. & T. A. Steph.) Soó

a. x *D. viridella* (H.-Harrison f.) Soó (*Orchis* x *viridella* H.-Harrison f.).

b. The plants are usually small (up to 11 cm) with erect, narrow, light yellowish-green, unspotted or very faintly spotted leaves. The labellum is intermediate in shape between the rhomboidal one of *D. purpurella* and the strap-shaped one of *C. viride*, and the mid-lobe is shorter than the side-lobes. The flowers are a rich, rosy purple tinged green and have irregular, darker purple spots and loops on the labellum. *Coeloglossum viride* x *Dactylorhiza latifolia* x *D. purpurella* has been reported from Harris, v.c. 110, but the identity of the second parent is unknown.

c. This hybrid is reported from the Hebridean islands of Iona, Eigg and Harris, v.c. 103, 104 and 110, and there is also a record for v.c. 66.

d. None.

e. *C. viride* (2n = 40); *D. purpurella* 2n = 80.

f. HESLOP-HARRISON, J. (1949). As above.

635/1 × 643/6. *Coeloglossum viride* (L.) Hartm. × *Dactylorhiza majalis* (Reichb.) Hunt & Summerh.

> has been reported from Harris, v.c. 110, by Heslop-Harrison. No specimen has been seen and nothing further is known of the discovery.
> HESLOP-HARRISON, J. (1949). As above.

636. *Gymnadenia* R.Br.
(by P. F. Hunt)

1 × 2. *G. conopsea* (L.) R.Br. × *G. odoratissima* (L.) Rich.

> = *G.* × *intermedia* Peterm. has been reported from the Continent.

636 × 637. *Gymnadenia* R.Br.
× *Pseudorchis* Séguier (*Leucorchis* E. Mey.)
= × *Pseudadenia* P. F. Hunt
(by P. F. Hunt)

636/1 × 637/1. *Gymnadenia conopsea* (L.) R.Br. × *Pseudorchis albida* (L.) A. & D. Löve (*Leucorchis albida* (L.) E. Mey. ex Schur)

a. × *P. schweinfurthii* (Hegelm. ex A. Kerner) P. F. Hunt (× *Gymleucorchis schweinfurthii* (Hegelm. ex A. Kerner) T. & T. A. Steph.).

b. Plants with intermediate characters are occasionally to be found but are more often nearer to one or other of the parents. Most plants, on critical examination, resemble *P. albida* in foliage and labellum and *G. conopsea* in sepals and petals, but the spur is truly intermediate, being shorter and stouter than in *G. conopsea*. This means that the flowers are usually pink but the labellum more or less pure white.

c. This hybrid has been recorded in various parts of Br where the distribution of the two species overlaps, particularly in Scotland. It is recorded also from Au, Ge, He, No and Su.

d. None.

e. *G. conopsea* 2*n* = 20 (also 40, 80); *P. albida* (2*n* = 42).
f. GODFERY, M. J. (1933). As above, p. 150.
  ROLFE, R. A. (1898). *Gymnadenia* x *Conopseo-albida. Orchid Rev.,* 6:
  238-239.
  SUMMERHAYES, V. S. (1968). As above, p. 227.

636/2 x 637/1. *Gymnadenia odoratissima* (L.) Rich. x *Pseudorchis albida* (L.) A. & D. Löve (*Leucorchis albida* (L.) E. Mey. ex Schur)
  = x *P. strampfii* (Aschers.) P. F. Hunt has been reported from Ga and Ge.

636 × 638. *Gymnadenia* R.Br. × *Platanthera* Rich. = × *Gymnaplatanthera* Lambert
(by P. F. Hunt)

636/1 x 638/2. *Gymnadenia conopsea* (L.) R.Br. x *Platanthera bifolia* (L.) Rich.
  = x *G. chodatii* (Leudner) Lambert has been reported from Su.

636/2 x 638/1. *Gymnadenia odoratissima* (L.) Rich. x *Platanthera chlorantha* (Custer) Reichb.
  = x *G. borelii* Lambert has been reported from the Continent.

636 × 642. *Gymnadenia* R.Br. × *Orchis* L. = × *Orchigymnadenia* Camus
(by P. F. Hunt)

636/1 x 642/7. *Gymnadenia conopsea* (L.) R.Br. x *Orchis mascula* (L.) L.
  = x *O. robsonii* H.-Harrison pat. was recorded from Blackhall Rocks, v.c.

66, in 1928. No authenticating specimen can be traced, and it is not very likely that species of these genera could hybridize.

GODFERY, M. J. (1933). As above, p. 146.

GRAHAM, G. G., SAYERS, C. D. and GAMAN, J. H. (1972). *A check list of the vascular plants of County Durham*, p. 58. Durham.

643 × 636. *Dactylorhiza* Nevski × *Gymnadenia* R.Br. = × *Dactylogymnadenia* Soó
(by P. F. Hunt)

643/1 × 636/1. *Dactylorhiza fuchsii* (Druce) Soó × *Gymnadenia conopsea* (L.) R.Br.

a. × *D. cookei* (H.-Harrison f.)‡ Soó (*Orchis* × *hibernica* Druce, *nom. nud.*) (× *Orchigymnadenia heinzeliana* (Reichardt) Camus may be correctly placed here, but is considered by Vermeulen to be a form of *Dactylorhiza maculata*).

b. Plants with various features of each of the parents have been reported but usually some of the characters are truly intermediate. Both subsp. *fuchsii* and subsp. *hebridensis* (Wilmott) Soó of *D. fuchsii* hybridize with *G. conopsea*. In most plants the short, rounded basal leaf so characteristic of *D. fuchsii* is absent, but the leaf-spots are generally evident and occasionally there is very little green between them. The labellum is usually more reminiscent of *D. fuchsii*, although in some flowers there are 3 more or less equal, truncated lobes. The influence of *G. conopsea* is evidenced by the shapes of the lateral sepals and petals and especially by the long, slender spur and the characteristic sweet scent.

c. Scattered records exist for this hybrid mainly in southern England, but it could well be much more frequent than hitherto reported. It also occurs northwards to the Outer Hebrides, and in Hb.

d. None.

e. *G. conopsea* 2n = 20 (also 40, 80); *D. fuchsii* 2n = 40.

f. GODFERY, M. J. (1933). As above, pp. 144-145, pl. 26 and 28. FIG.

HALL, P. M. (1933). Three hybrid orchids: 1931. *Rep. B.E.C.*, **9**: 364-366, pl. 49. FIG.

643/2 × 636/1. *Dactylorhiza maculata* (L.) Soó × *Gymnadenia conopsea* (L.) R.Br.

a. × *D. legrandiana* (Camus) Soó (× *Orchigymnadenia evansii* (Druce) T. & T. A. Steph.).

b. Hybrids purporting to be of this parentage have the leaves variable in shape and size but usually faintly spotted. The flowers are intermediate in shape and size; the labellum bears a superficial resemblance to that of *D. maculata*, being pale lilac with a few very fine spots and occasional loops and the spur is as slender and long as that of *G. conopsea*. The flowers are usually fragrant.

c. This hybrid is recorded from scattered localities throughout most of BI. It could well occur at any suitable site where both species exist, but ecologically they occupy distinct habitats. It is also reported from the Continent.

d. None.

e. *G. conopsea* $2n = 20$ (also 40, 80); *D. maculata* $2n = 80$.

f. GODFERY, M. J. (1933). As above, p. 145, pl. 30. FIG.

STEPHENSON, T. and STEPHENSON, T. A. (1921a). The forms of *Orchis maculata. J. Bot., Lond.,* 59: 121-128, pl. 559, fig. 24. FIG.

STEPHENSON, T. and STEPHENSON, T. A. (1921b). Natural hybrid orchids from Arran. *Orchid Rev.,* 29: 131-133. FIG.

STEPHENSON, T. and STEPHENSON, T. A. (1922). Hybrids of *Orchis purpurella. J. Bot., Lond.,* 60: 33-35. FIG.

643/3 × 636/1. *Dactylorhiza incarnata* (L.) Soó × *Gymnadenia conopsea* (L.) R.Br.

= × *D. vollmannii* (M. Schulze) Soó is possibly represented by a specimen at K collected at the Lizard, v.c. 1, in 1950. The leaves of the preserved specimen are small and unspotted, the flowers bright rose-pink and of the general structure of those of *G. conopsea*, and fragrant. The spur is twisted and malformed and there is some doubt as to whether this specimen is of the parentage claimed. The hybrid has also been reported from Su.

643/3d × 636/1. *Dactylorhiza cruenta* (O. F. Müll.) Soó × *Gymnadenia conopsea* (L.) Br.

= × *D. raetica* Paroz & Reinhard has been reported from Ge.

643/4 × 636/1. *Dactylorhiza praetermissa* (Druce) Soó × *Gymnadenia conopsea* (L.) R.Br.

a. × *D. wintoni* (Quirk) Soó (*Habenaria* × *wintoni* Quirk, *H.* × *quirkiana* Druce).

b. Hybrid plants of this parentage are robust and the leaves are variable in shape and size and usually faintly spotted. The flowers, subtended by large bracts, are intermediate in size but morphologically are closer to those of

D. praetermissa. The labellum is more trilobed than *D. praetermissa* but with the faint radiating lines of violet dots as in that species. The spur is slender but shorter than in *G. conopsea,* and the scent faint but indicative of the latter species.

c. Although the geographical ranges of the parents overlap the two species are usually separated ecologically. There are records of the hybrid from v.c. 3, 11 and 12.

d. None.

e. *G. conopsea* $2n = 20$ (also 40, 80); *D. praetermissa* $2n = 80$.

f. DRUCE, G. C. (1910). *Habenaria Gymnadenia* x *Orchis praetermissa. Rep. B.E.C.,* **5**: 157-158, pl. 10. FIG.
   GODFERY, M. J. (1933). As above, pp. 146-147, pl. G. FIG.
   HALL, P. M. (1933). As above, pl. 47 and 48. FIG.
   SUMMERHAYES, V. S. (1968). As above, p. 292.

643/5 x 636/1. *Dactylorhiza purpurella* (T. & T. A. Steph.) Soó x *Gymnadenia conopsea* (L.) R.Br.

a. x *D. varia* (T. & T. A. Steph.) Soó (x *Orchigymnadenia varia* T. & T. A. Steph.).

b. The few plants of this hybrid that have been reported are so similar to *D. purpurella* that doubts have been cast as to their authenticity, but the long, slender spike, small, relatively pale flowers and long, slender spur all suggest the influence of *G. conopsea.* It is surprising, however, that the latter parent has not expressed itself in the shape or markings of the labellum, which closely resembles that of the paler variants of *D. purpurella.* Vegetatively the plants often resemble a fairly robust *D. purpurella* with finely spotted leaves.

c. The hybrid is recorded from Upper Teesdale in v.c. 70, from various parts of Scotland (especially the western islands), and from v.c. H38.

d. None.

e. *G. conopsea* $2n = 20$ (also 40, 80); *D. purpurella* $2n = 80$.

f. GODFERY, M. J. (1933). As above, pp. 147-148, pl. 26 and K. FIG.
   HESLOP-HARRISON, J. (1955). Orchid hybrids in North Down. *Ir. Nat. J.,* **11**: 342-346. FIG.
   STEPHENSON, T. and STEPHENSON, T. A. (1921b). As above. FIG.
   STEPHENSON, T. and STEPHENSON, T. A. (1922). As above. FIG.

643/6 x 636/1. *Dactylorhiza majalis* (Reichb.) Hunt & Summerh. x *Gymnadenia conopsea* (L.) R.Br.

= x *D. lebrunii* (Camus) Soó has been reported from the Continent.

643/7 × 636/1. *Dactylorhiza traunsteineri* (Sauter) Soó × *Gymnadenia conopsea* (L.) R.Br.

> = × *D. fuchsii* (G. Keller & Soó) Soó has been reported from the Continent.

643/2 × 636/2. *Dactylorhiza maculata* (L.) Soó × *Gymnadenia odoratissima* (L.) Rich.

> has been reported from He.

645 × 636. *Anacamptis* Rich. × *Gymnadenia* R.Br. = × *Gymnanacamptis* Aschers. & Graebn.

(by P. F. Hunt)

645/1 × 636/1. *Anacamptis pyramidalis* (L.) Rich. × *Gymnadenia conopsea* (L.) R.Br.

a. × *G. anacamptis* (Wilms) Aschers. & Graebn.

b. Although the floral structure of these two species is very similar and the few observations made suggest that both are pollinated by the same species of moth, chances of hybridization are probably very limited because in most localities the flowering seasons only just overlap. In the reported English examples the spike is oblong, not pyramidal, and the flowers are much duller than in *A. pyramidalis* but with the two converging plates at the base of the labellum which are a feature of that species. The scent is that of *G. conopsea*.

c. This hybrid has been recorded from v.c. 11/12, 33/34 and 66_x and from Ge.

d. None.

e. *G. conopsea* $2n = 20$ (also 40, 80); *A. pyramidalis* ($2n = 40$).

f. GODFERY, M. J. (1933). As above, p. 156.

645/1 × 636/2. *Anacamptis pyramidalis* (L.) Rich. × *Gymnadenia odoratissima* (L.) Rich.

> has been recorded from He.

643 × 637. *Dactylorhiza* Nevski × *Pseudorchis* Séguier (*Leucorchis* E. Mey.) = × *Pseudorhiza* P. F. Hunt

(by P. F. Hunt)

643/1 × **637/1.** *Dactylorhiza fuchsii* (Druce) Soó × *Pseudorchis albida* (L.) A. & D. Löve (*Leucorchis albida* (L.) E. Mey. ex Schur) = × *P. nieschalkii* (Senghas) P. F. Hunt has been reported from Ge.

643/2 × **637/1.** *Dactylorhiza maculata* (L.) Soó × *Pseudorchis albida* (L.) A. & D. Löve (*Leucorchis albida* (L.) E. Mey. ex Schur) = × *P. bruniana* (Brügg.) P. F. Hunt has been reported from the Continent.

638. *Platanthera* Rich.

(by P. F. Hunt)

2 × **1.** *P. bifolia* (L.) Rich. × *P. chlorantha* (Custer) Reichb.
= *P.* × *hybrida* Brügg. has been reported on several occasions from Br (e.g. v.c. 70, 101 and 104), but it is probable that the specimens are aberrant forms of one or other of the putative parents, in which peloric and other abnormalities are quite frequent. Some specimens from the Continent (e.g. Fe and Su) are perhaps a little more convincing, but nevertheless most or all are probably odd forms of *P. chlorantha* or *P. bifolia*.
GODFERY, M. J. (1933). As above, pp. 138-139.

643 × 638. *Dactylorhiza* Nevski
× *Platanthera* Rich.
= × *Rhizanthera* Hunt & Summerh.
(by P. F. Hunt)

643/1 x 638/2. *Dactylorhiza fuchsii* (Druce) Soó x *Platanthera bifolia* (L.) Rich.

= x *R. somersetensis* (Camus) Soó has been reported from v.c. 6, originally as x *Orchiplatanthera chevallierana.* Close examination of the preserved specimen, together with the comments of R. A. Rolfe and M. J. Godfery, suggest very strongly that the plant is in fact not a hybrid but a variant of *D. fuchsii.* In any case the true parentage, if it were a hybrid, would most probably be *P. chlorantha* x *D. fuchsii,* since *P. bifolia* is not known in the locality.

ROLFE, R. A. (1913). *Orchiplatanthera chevallieriana. Orchid Rev.,* **21**: 235.

643/2 x 638/2. *Dactylorhiza maculata* (L.) Soó x *Platanthera bifolia* (L.) Rich.

= x *R. chevallierana* (Camus) Soó (x *Orchiplatanthera chevallierana* (Camus) Camus) has been reported from near Perth, v.c. 88/89, and Isle of Harris, v.c. 104. The plants apparently had the overall appearance of *P. bifolia* but the flower, although green and white, had a shape similar to that of *D. maculata,* having a marked trilobed labellum and short spur. The leaves were faintly spotted. Nevertheless more careful study of further specimens is necessary to show that the plants were correctly identified. This hybrid is also recorded from Ga.

GODFERY, M. J. (1933). As above, p. 217.
ROLFE, R. A. (1897). A new British orchid. *Orchid Rev.,* **5**: 234-235.

645 × 638. *Anacamptis* Rich.
× *Platanthera* Rich.
= × *Anacamptiplatanthera* P. Fourn.
(by P. F. Hunt)

645/1 x 638/2. *Anacamptis pyramidalis* (L.) Rich. x *Platanthera bifolia* (L.) Rich.

= x *A. payotii* P. Fourn. has been reported from the Continent.

640. *Ophrys* L.
(by P. F. Hunt)

1 × 2. *O. apifera* Huds. × *O. fuciflora* (Crantz) Moench

   a. *O.* × *albertiana* Camus.

   b. Vegetatively the plants of this hybrid do not suggest its hybrid origin but the flowers exhibit an admixture of the characters of both parents. The influence of *O. apifera* is shown by the general appearance of the flowers which have a strongly convex labellum, long, narrow, deflexed sepals and a long, curved anther-beak. The *O. fuciflora* characters transmitted to the hybrid are the short triangular petals and the undivided labellum with a short apical tooth. The caudicles are ribbon-like at the base as in *O. fuciflora*, and filamentous at the apex as in *O. apifera*.

   c. This hybrid is recorded only from v.c. 15 where there is the only substantial colony of *O. fuciflora* in BI. It is also reported from Ga, He and It.

   d. None.

   e. Both parents and hybrid ($2n = 36$).

   f. DANESCH, E. (1972). *Orchideen Europas. Ophrys-Hybriden*, p. 134. Bern and Stuttgart. FIG.

      GODFERY, M. J. (1933). As above, p. 241, pl. 56. FIG.

      SUMMERHAYES, V. S. (1968). As above, pl. 6. FIG.

1 × 3. *O. apifera* Huds. × *O. sphegodes* Mill.

   = *O.* × *pseudoapifera* Caldesio has been recorded once in Br from southern v.c. 15 at the end of the last century. However, no accurate description or preserved specimen is extant and it is not possible therefore to check the authenticity of this record. It is reported from the Continent.

1 × 4. *O. apifera* Huds. × *O. insectifera* L.

   = *O.* × *pietzschii* Kümpel was described from Ge in 1971.

2 × 3. *O. fuciflora* (Crantz) Moench × *O. sphegodes* Mill.

   = *O.* × *obscura* G. Beck has been reported from v.c. 15 but the specimens seen by me from the same locality exhibit no evidence of *O. sphegodes* and fall well within the range of variation shown by *O. fuciflora*. The flowering times of these two species scarcely overlap in Br and hybridization seems unlikely; nevertheless the hybrid is reported from Ga, Ge and It.

   DANESCH, E. (1972). As above, p. 135. FIG.

   GODFERY, M. J. (1933). As above, pp. 234-235, pl. 57. FIG.

2 × 4. *O. fuciflora* (Crantz) Moench × *O. insectifera* L.

   = *O.* × *devenensis* Reichb. f. has been reported from Ge and He.

4 × 3. *O. insectifera* L. × *O. sphegodes* Mill.

   a. *O.* × *hybrida* Pokorny.

   b. Plants undoubtedly of this parentage bear an overall resemblance to *O. sphegodes* but the influence of *O. insectifera* is clearly visible in the long and narrow, dark, velvety petals with incurled edges, the marked side-lobes, and the shiny, blue marking of the labellum.

   c. This hybrid has been reported from v.c. 15, and from Au, Ga and Ge.

   d. None.

   e. Both parents ($2n = 36$).

   f. DANESCH, E. (1972). As above, pp. 144-145. FIG.

     GODFERY, M. J. (1933). As above, pp. 230-231, pl. 55. FIG.

     GUÉTROT, M. (1927). *Ophrys (hybrida) Pokornyi*, in *Plantes hybrides de France*, **1** and **2**: 60. Lille. FIG.

     REINHARDT, H. R. (1969) *Ophrys* × *apicula* J. Schmidt and *Ophrys* × *hybrida* Pokorny. *Orchidee (Hamb.)*, **20**: 131-135.

     RENDLE, A. B. (1906). *Ophrys* × *hybrida*. *J. Bot., Lond.*, **44**: 347-349. FIG.

641 × 642. *Himantoglossum* Spreng. × *Orchis* L. = × *Orchimantoglossum* Aschers. & Graebn.
(by P. F. Hunt)

641/1 × 642/3. *Himantoglossum hircinum* (L.) Spreng. × *Orchis simia* Lam.

   = × *O. lacazei* Aschers. & Graebn. has been reported from the Continent.

642. *Orchis* L.
(by P. F. Hunt)

2 × 1. *O. militaris* L. × *O. purpurea* Huds.

   = *O.* × *jacquinii* Godr. has been reported from Ga and Rm.

1 × 3. *O. purpurea* Huds. × *O. simia* Lam.

= *O.* × *weddellii* G. Camus has been reported from Ga and He.

5 × 1. *O. morio* L. × *O. purpurea* Huds.

= *O.* × *perrettii* K. Richt. has been reported from the Continent.

7 × 1. *O. mascula* (L.) L. × *O. purpurea* Huds.

= *O.* × *wilmsii* K. Richt. has been reported from Ge.

2 × 3. *O. militaris* L. × *O. simia* Lam.

a. *O.* × *beyrichii* A. Kerner.

b. The early British and modern Continental specimens of this hybrid are very variable in appearance but the flower-colour and especially the details of the labellum-shape are truly intermediate and instantly recognizable.

c. Up to the middle of the last century both parents were relatively common in parts of the middle Thames valley and hybrids between them were reported on several occasions. However, now that the two species are among our rarest native plants (two localities in Br for *O. simia* and three for *O. militaris*) the chances of hybridization are virtually non-existent. The hybrid is also reported from Au, Ga and Ge.

d. None.

e. Both parents and hybrid (2*n* = 42).

f. GODFERY, M. J. (1933). As above, pp. 170-171, pl. 34. FIG.

GUÉTROT, M. (1927). *Orchis Beyrichii*, in *Plantes hybrides de France*, 1 and 2: 61. Lille. FIG.

SUMMERHAYES, V. S. (1951). As above, p. 248.

2 × 5. *O. militaris* L. × *O. morio* L.

= *O.* × *ladurneri* Murr. has been reported from the Continent.

7 × 2. *O. mascula* (L.) L. × *O. militaris* L.

has been reported from He.

3 × 4. *O. simia* Lam. × *O. ustulata* L.

= *O.* × *dallii* Zimmerman has been reported from the Continent.

5 × 4. *O. morio* L. × *O. ustulata* L.

has been reported from It.

6 × 5. *O. laxiflora* Lam. × *O. morio* L.

a. *O.* × *alata* Fleury.

b. The plants of this hybrid recorded from its sole British locality are more or less intermediate in most characters but the sepals are always prominently veined in green or deep greenish-purple. There are fewer flowers than in *O. laxiflora* but their overall appearance is that of one of the paler-flowered forms of that species.

c. This hybrid is known in BI only from Guernsey; it is recorded also from Ga, He and It.

d. None.

e. *O. morio* ($2n = 36$); *O. laxiflora* ($2n = 42$).

f. DRUCE, G. C. (1915). x *Orchis alata* Fleury. *Rep. B.E.C.*, **4**: 25.
GODFERY, M. J. (1933). As above, p. 185, pl. 42. FIG.

7 x 5. *O. mascula* (L.) L. x *O. morio* L.

a. *O.* x *morioides* Brand.

b. The green-veined lateral sepals and broadly truncated labellum mid-lobe of *O. morio* in an otherwise typical *O. mascula* flower is perhaps the best description of this hybrid. The leaves can be spotted or unspotted depending, presumably, on the leaf-type of the *O. mascula* parent; the spur is shorter than in *O. mascula*.

c. This hybrid is surprisingly rare; in many cases the parents grow in very close juxtaposition, they have the same pollinators, and their flowering seasons overlap considerably. There are scattered records from England and Wales and on the Continent from Su southwards.

d. None.

e. *O. morio* ($2n = 36$); *O. mascula* ($2n = 42$).

f. GODFERY, M. J. (1933). As above, p. 181.
SUMMERHAYES, V. S. (1951). As above, p. 259.

6 x 7. *O. laxiflora* Lam. x *O. mascula* (L.) L.

= *O.* x *largei* K. Richt. has been reported from Ge and Hs.

643 × 642. *Dactylorhiza* Nevski × *Orchis* L.
= × *Orchidactyla* Hunt & Summerh.

(by P. F. Hunt)

Hybrids involving several species of both genera, including *Dactylorhiza incarnata*, *D. maculata* and *D. majalis*, and *Orchis laxiflora*, *O. mascula*, *O. militaris*, *O. morio* and *O. purpurea*, have been reported from the Continent, but all are most likely to be erroneously identified.

644 × 642. *Aceras* R.Br. × *Orchis* L.
= × *Orchiaceras* Camus
(by P. F. Hunt)

644/1 × 642/1. *Aceras anthropophorum* (L.) Ait. f. x *Orchis purpurea* Huds.

> = x *O. macra* (Lindl.) Camus (x *O. meilsheimeri* Rouy) has been reported from Ga and Ge.

644/1 × 642/2 × 642/1. *Aceras anthropophorum* (L.) Ait. f. x *Orchis militaris* L. x *Orchis purpurea* Huds.

> = x *O. bispuria* Keller (x *O. verdunensis* Peitz) has been reported from Ga.

644/1 × 642/2. *Aceras anthropophorum* (L.) Ait. f. x *Orchis militaris* L.

> = x *O. spuria* (Reichb. f.) Camus (x *O. weddellii* Camus) has been reported from Be, Ga, Ge, He and Ho.

644/1 × 642/3. *Aceras anthropophorum* (L.) Ait. f. x *Orchis simia* Lam.

> = x *O. bergonii* Camus has been reported from Ga, He, Hs, It and North Africa.

644/1 × 642/7. *Aceras anthropophorum* (L.) Ait. f. x *Orchis mascula* (L.) L.

> = x *O. orphanidesii* Camus has been reported from Ga and Gr.

645 × 642. *Anacamptis* Rich. × *Orchis* L.
= × *Anacamptorchis* Camus
(by P. F. Hunt)

645/1 × 642/4. *Anacamptis pyramidalis* (L.) Rich. x *Orchis ustulata* L.

> = x *A. fallax* Camus has been reported from the Continent.

645/1 x 642/5. *Anacamptis pyramidalis* (L.) Rich. x *Orchis morio* L.
= x *A. laniccae* Br.-Blanquet has been reported from Ga and He.

645/1 x 642/6. *Anacamptis pyramidalis* (L.) Rich. x *Orchis laxiflora* Lam.
has been reported from Ga.

643. *Dactylorhiza* Nevski (*Dactylorchis* (Klinge) Vermeul.)
(by R. H. Roberts)*

Hybrids occur very widely in this genus, although not all of them are as common or widespread as some observers have claimed. In this account eight species are recognized, the seven given by Dandy (1958) plus *D. cruenta*, which was included by Dandy as a subspecies of *D. incarnata*. In general the synonymy under the genus *Dactylorchis* (as given by Dandy) has not been included, and the different subspecies involved in hybridization have been mentioned only where of particular relevance.

General References (in addition to those listed at the start of the Orchidaceae)

DANDY, J. E. (1958). *Op. cit.*, pp. 146-148.

HAGERUP, O. (1938). Studies in the significance of polyploidy, 2. *Orchis. Hereditas*, **24**: 258-264.

HESLOP-HARRISON, J. (1954). A synopsis of the dactylorchids of the British Isles. *Ber. geobot. ForschInst. Rübel*, **1953**: 53-82.

HESLOP-HARRISON, J. (1968). Genetic system and ecological habit as factors in dactylorchid variation, in SENGHAS, K. and SUNDERMANN, H., eds. As below, pp. 20-27.

HEUSSER, C. (1938). Chromosomenverhältnisse bei schweizerischen basitonen Orchideen. *Ber. schweiz. bot. Ges.*, **48**: 562-605.

PUGSLEY, H. W. (1939). Recent work on dactylorchids. *J. Bot., Lond.*, **77**: 50-56.

SENGHAS, K. and SUNDERMANN, H., eds (1968). Probleme der Orchideengattung *Dactylorhiza*. *Jber. naturw. Ver. Wuppertal*, **21** and **22**.

---

* With assistance from J. Heslop-Harrison.

SENGHAS, K. and SUNDERMANN, H. eds (1972). Probleme der Orchideengattung *Orchis* mit Nächtragen in *Ophrys, Dactylorhiza, Epipactis* und Hybriden. *Jber. naturw. Ver. Wuppertal*, **25**.

SOÓ, R. v. (1960). Synopsis generis *Dactylorhiza (Dactylorchis). Annls Univ. Scient. bpest. Rolando Eötvös, Biol.*, **3**: 335-357.

SOÓ, R. v. (1962). *Nomina nova generis Dactylorhiza.* Budapest.

VERMEULEN, P. (1938). Chromosomes in *Orchis. Chronica bot.*, **4**: 107-108.

1 × 2. *D. fuchsii* (Druce) Soó × *D. maculata* (L.) Soó

a. *D. × transiens* (Druce) Soó (*Orchis × transiens* Druce).

b. Hybrids have the robust habit of *D. fuchsii*, but their leaves are narrower, folded upwards and all oblong-lanceolate, with the lowest leaves not so bluntly rounded at the tip. The flowers have a heavily marked, trilobed labellum with broader, more crenate lateral lobes than in *D. fuchsii*, and a long, straight, cylindrical spur, which is not so slender as in *D. maculata*. They are highly sterile. In BI *D. maculata* occurs as subsp. *ericetorum* (Linton) Hunt & Summerh. and subsp. *rhoumensis* (H.-Harrison f.) Soó. No hybrids of the latter are recorded.

c. Hybrids occur with the parents where soil conditions vary from acid to base-rich, but even then only in small numbers. They have been recorded from 21 vice-counties in Br northwards to the Outer Hebrides, and from v.c. H9, H16 and H38; they are much rarer than some authors have assumed and their distribution is imperfectly known because they have often been recorded in error. They have also been recorded from Ga.

d. Artificial cross-pollination of the parent species gives a very low seed-set. F_1 hybrids have been raised to flowering from crosses between German parental material, but their fertility has not been tested. Isolation is both genetical and ecological. No experimental crosses have been tried with subsp. *rhoumensis*.

e. *D. fuchsii* $2n = 40$; *D. maculata* subsp. *ericetorum* $2n = 80$; wild hybrid $2n = 60$; subsp. *rhoumensis* $2n = 40$.

f. DRUCE, G. C. (1915). *Orchis maculata* L. and *O. Fuchsii*, 2. *O. maculata* L. *Rep. B.E.C.*, **4**: 99-108.

GODFERY, M. J. (1933). As above. FIG.

HARBECK, M. (1972). Künstliche Hybriden bei *Dactylorhiza*, in SENGHAS, K. and SUNDERMANN, H., eds. As above, pp. 140-141.

STEPHENSON, T. and STEPHENSON, T. A. (1921). The forms of *Orchis maculata. J. Bot., Lond.*, **59**: 121-128.

1 × 3. *D. fuchsii* (Druce) Soó × *D. incarnata* (L.) Soó

a. *D. × kernerorum* (Soó) Soó (*Orchis × curtisiana* Druce).

b. Hybrids are usually taller than *D. incarnata*, with a narrower, less hollow stem and yellowish green, often strict leaves which are broadest near the middle, narrowly hooded at the tip and marked with rather pale spots or are unspotted. The bracts exceed the flowers, but are narrower than in *D. incarnata* and often tinged with purple. The flowers vary from a pale salmon-pink to a bluish-purple, with a flat, trilobed labellum marked with a pattern of broken loops and dots. The spur is intermediate between those of the parents. The hybrids are highly sterile.

c. The hybrids usually occur as single plants in mixed populations of the parents. There are records from 12 vice-counties in Br northwards to the Inner Hebrides, in base-rich marshes, dune-slacks and fens, where *D. fuchsii* occurs in the drier areas. They are also recorded from Au, Ga, Ge, He and Rm.

d. Artificial cross-pollination of the parent species gives a low and erratic seed-set, with embryos varying a good deal in size. The few seeds set may possibly result from occasional haploid parthenogenesis; their viability has not been tested.

e. Both parents $2n = 40$.

f. GODFERY, M. J. (1933). As above. FIG.

HAGERUP, O. (1944). On fertilisation, polyploidy and haploidy in *Orchis maculatus* L. *sens. lat. Dansk bot. Ark.*, **11**: 1-25.

HALL, P. M. (1933). Three hybrid orchids. *Rep. B.E.C.*, **10**: 364-366. FIG.

PETTERSSON, B. (1947). On some hybrid populations of *Orchis incarnata* x *maculata* in Gotland. *Svensk bot. Tidskr.*, **41**: 115-140. FIG.

1 x 3 x 4. *D. fuchsii* (Druce) Soó x *D. incarnata* (L.) Soó x *D. praetermissa* (Druce) Soó

has been tentatively recorded from Rudley, v.c. 11.

3d x 1. *D. cruenta* (O. F. Müll.) Soó (*D. incarnata* subsp. *cruenta* (O. F. Müll.) Sell) x *D. fuchsii* (Druce) Soó

a. None.

b. The hybrids are highly variable, but usually have the tall, slender habit and the larger number of leaves of *D. fuchsii*, with the lower leaves elliptical and blunt-tipped. The stem is hollow, but the upper leaves are bract-like and both these and the bracts have heavy purple spots and blotches on both surfaces. The flowers are dark purplish in colour, with the lateral lobes of the labellum reflexed; the spur is intermediate between those of the parents. Although many plants seem to be of the F_1 type, there is some indication that backcrosses and second, or later, generation segregates occur.

c. In BI hybrids have been recorded only from near the southern end of Lough Carra, v.c. H26. In this area *D. cruenta* grows in Schoenetum occupying the highly calcareous peat near the lake-shore, and *D. fuchsii* grows on the drier ground nearby. Most of the hybrids are found on the drier ground. They have also been recorded from Su.

d. None.

e. Both parents $2n = 40$.

f. HESLOP-HARRISON, J. (1950). Notes on some Irish dactylorchids. *Ir. Nat. J.*, **10**: 82.

1 × 4. *D. fuchsii* (Druce) Soó × *D. praetermissa* (Druce) Soó

a. *D.* × *grandis* (Druce) P. F. Hunt (*Orchis* × *mortonii* Druce).

b. Hybrids are usually tall and the stems have a narrow cavity. The leaves are whitish beneath, the upper narrower and more acute, but the lower shorter and blunter than those of *D. praetermissa*; all of them are marked with faint, transversely-elongated spots, or may be unspotted. The spike is conical, many-flowered and dense, with bracts which are much longer than the flowers, but narrower than those of *D. praetermissa*; the flowers are intermediate between those of the parents. Their pollen is highly sterile and seed production very low.

c. Hybrids are often abundant with the presumed parents in marshes, wet meadows, fens and occasionally on chalk downs. In these places *D. fuchsii* occupies the drier and *D. praetermissa* the wetter parts. Hybrids are found throughout the distributional area of *D. praetermissa* in Br, and have also been recorded from Ga.

d. Artificial cross-pollination of the parent species results in a good seed-set and a viable F_1, but such plants have not been grown to maturity. The natural triploid hybrids produce no seed after selfing, but pollination with the presumed parents results in a very low set of good seed with triploid embryos. There is considerable evidence that these seeds are the result of parthenogenesis.

e. *D. fuchsii* $2n = 40$; *D. praetermissa* $2n = 80$; wild hybrid $2n = 60$.

f. BROOKE, J. (1950). *The wild orchids of Britain.* London. FIG.

DRUCE, G. C. (1918). Notes on the British orchids chiefly the palmate section. *Rep. B.E.C.*, **5**: 149-180. FIG.

GODFERY, M. J. (1933). As above. FIG.

HARBECK, M. (1972). As above.

HESLOP-HARRISON, J. (1953a). Microsporogenesis in some triploid dactylorchid hybrids. *Ann. Bot.*, n.s., **17**: 539-549.

HESLOP-HARRISON, J. (1957). On the hybridisation of the common spotted orchid, *Dactylorchis fuchsii* (Druce) Vermln., with the marsh orchids, *D. praetermissa* (Druce) Vermln. and *D. purpurella* (T. & T. A. Steph.) Vermln. *Proc. Linn. Soc. Lond.*, **167**: 176-185.

HESLOP-HARRISON, J. (1959). Apomictic potentialities in *Dactylorchis*. *Proc. Linn. Soc. Lond.*, **170**: 174-178.

1 × 5. *D. fuchsii* (Druce) Soó × *D. purpurella* (T. & T. A. Steph.) Soó

a. *D.* × *venusta* (T. & T. A. Steph.) Soó (*Orchis* × *hebridella* Wilmott).

b. Hybrids are often very robust, but extremely variable, with stems which are either solid or have a narrow hollow. The leaves are whitish beneath and marked with pale to deep purple, transversely-elongated, large spots, or may be unspotted. The flowers are in a dense, pointed spike and intermediate in characters between those of the parents. Their pollen is highly sterile and seed production is very low.

c. Hybrids are often abundant where the parents grow together in wet meadows, marshes and dune-slacks, throughout the distributional area of *D. purpurella* in BI.

d. Artificial cross-pollination of the parent species gives a good seed-set and the F_1 is viable, but such plants have not been raised to maturity. The natural triploid hybrids produce no seed after selfing, but pollination with the presumed parents results in a very low set of good seed with triploid embryos. There is considerable evidence that these seeds are the result of parthenogenesis.

e. *D. fuchsii* $2n = 40$; *D. purpurella* $2n = 80$; wild hybrid $2n = 60$.

f. GODFERY, M. J. (1933). As above. FIG.

   HESLOP-HARRISON, J. (1953a). As above.

   HESLOP-HARRISON, J. (1955). Orchid hybrids in North Down. *Ir. Nat. J.*, **11**: 342-346. FIG.

   HESLOP-HARRISON, J. (1957). As above.

   HESLOP-HARRISON, J. (1959). As above.

   STEPHENSON, T. and STEPHENSON, T. A. (1922). Hybrids of *Orchis purpurella*. *J. Bot., Lond.*, **60**: 33-35.

   WILMOTT, A. J. (1939). Notes on a short visit to Barra. *J. Bot., Lond.*, **77**: 189-195.

1 x 6. *D. fuchsii* (Druce) Soó x *D. majalis* (Reichb.) Hunt & Summerh.

a. *D.* x *braunii* (Halácsy) Borsos & Soó.

b. Hybrids are robust. Their leaves are whitish beneath and usually heavily marked with transversely-elongated, dark purple spots, which are occasionally ring-like. The flowers are in a large, pointed spike and are intermediate in characters between those of the parents. Their pollen is highly sterile and their seed-set very low.

c. Hybrids have been recorded from western Scotland (the mainland and the Hebrides) and Anglesey, v.c. 52. They occur where the parents grow together in base-rich marshes. They have also been recorded from Au, Cz, Ga, Ge, He and Po.

d. Artificial cross-pollination of the parent species gives a good set of seed, but the viability of such seed has not been tested.

e. *D. fuchsii* $2n = 40$; *D. majalis* $2n = 80$; wild hybrid $2n = 60$.

f. GODFERY, M. J. (1933). As above. FIG.

1 x 7. *D. fuchsii* (Druce) Soó x *D. traunsteineri* (Sauter) Soó

a. *D.* x *kellerana* P. F. Hunt.

b. Hybrids are nearer *D. traunsteineri* in their dwarf habit and few, narrow leaves, the lowest two of which are recurved and blunt at the tip. The leaves are whitish beneath and often marked with dark purple, narrow, transverse bars or transversely-elongated spots or rings. The flowers have a trilobed labellum with a pattern nearer that of *D. fuchsii* and slightly reflexed lateral lobes. The spur is intermediate between those of the parents. The pollen appears to be highly sterile, and no seeds are set.

c. This hybrid has been recorded from five localities in BI: Scraw Bog, v.c. H23; Keel Bridge, near Ballinrobe, v.c. H26; Cors Erddreiniog and Rhos-y-Gad, v.c. 52; and Helmsley, v.c. 62. At three of these places it occurs with the parents in base-rich fen. It has also been recorded from No.

d. Artificial cross-pollination of the parent species gives a good seed-set, but the viability of this seed has not been tested. Plants of the presumed F_1 hybrid set no seed after pollination with the supposed parents or after selfing.

e. *D. fuchsii* $2n = 40$; *D. traunsteineri* $2n = 80$; wild hybrid $2n = 60$.

f. GODFERY, M. J. (1933). As above. FIG.

HESLOP-HARRISON, J. (1953b). Studies in *Orchis* L., 2. *Orchis Traunsteineri* Saut. in the British Isles. *Watsonia*, **2**: 371-391.

3 × 2. *D. incarnata* (L.) Soó × *D. maculata* (L.) Soó

a. *D.* x *claudiopolitana* (Soó) Soó.

b. Hybrids have the strict habit of *D. incarnata*, with a slightly hollow stem and oblong-lanceolate, yellowish-green leaves usually marked with large round, pale spots. The upper leaves are bract-like, and the bracts of the oblong spike are rather broad and longer than the flowers. The flowers have a nearly flat labellum, with broadly rounded, often crenate lateral lobes and a small mid-lobe; their colour varies from pale salmon-pink to bright reddish-purple and the labellum is marked with a pattern of dots arranged in parallel loops. The spur has a wide throat, but is slender and distinctly curved. The pollen is highly sterile and no seed seems to be set after selfing.

c. Hybrids are rare and usually occur as single plants where the parents grow in close proximity. There are records from 13 vice-counties in Br northwards to the Outer Hebrides, and from v.c. H1 and H35, in marshes and bogs where soil conditions vary from acid to base-rich. Hybrids are also recorded from Da, Ga, Ge, No, Rm and Su.

d. Artificial cross-pollination of the parent species produces few seeds, but some of these are viable. F_1 plants have been raised to the flowering stage from crosses between Continental parental material, but their fertility has not yet been tested.

e. *D. maculata* $2n = 80$; *D. incarnata* $2n = 40$; wild hybrid ($2n = 60$).

f. GODFERY, M. J. (1933). As above. FIG.

HARBECK, M. (1968). Versuche zur Samenvermehrung einiger *Dactylo-rhiza*-Arten, in SENGHAS, K. and SUNDERMANN, H., eds. As above pp. 112-118.

PUGSLEY, H. W. (1939). As above.

3 × **2** × **6.** *D. incarnata* (L.) Soó × *D. maculata* (L.) Soó × *D. majalis* (Reichb.) Hunt & Summerh.

has been reported from Ge and He.

3d × **2.** *D. cruenta* (O. F. Müll.) Soó (*D. incarnata* subsp. *cruenta* (O. F. Müll.) Sell) × *D. maculata* (L.) Soó

= *D.* × *ampolai* Hautzinger (*D.* × *samnaunensis* (Gsell) Soó, *nom. nud.*) has been recorded from He and It.

2 × **4.** *D. maculata* (L.) Soó × *D. praetermissa* (Druce) Soó

a. *D.* × *hallii* (Druce) Soó (*Orchis* × *hallii* Druce, *O.* × *scotica* Druce, *nom. nud.*).

b. Hybrids are somewhat similar to *D.* × *grandis,* but their leaves are longer, narrower, and more erect. The flower spike is cylindrical, and the lilac-purple flowers have a broad, shallowly trilobed labellum with a pattern of lines and dots and with its lateral lobes rounded and crenulate. The spur is usually straight, long and intermediate in thickness between those of the parents. The presumed hybrids show normal pollen and seed fertility.

c. The distribution of these hybrids is incompletely known because of confusion with other taxa, but they may be expected wherever the parent species occur together. They are also recorded from Ho.

d. Artificial cross-pollination of the parent species gives a good seed-set. F_1 hybrid plants have been raised from crosses with Continental parental material, but their fertility has not been tested. Presumed F_1 hybrids have shown normal seed-fertility after pollination with the presumed parents or with other presumed F_1 plants.

e. Both parents and wild putative hybrids $2n = 80$.

f. BROOKE, J. (1950). As above. FIG.
DRUCE, G. C. (1918). As above.
HARBECK, M. (1972). As above.

2 × **5.** *D. maculata* (L.) Soó × *D. purpurella* (T. & T. A. Steph.) Soó

a. *D.* × *formosa* (T. & T. A. Steph.) Soó (*Orchis* × *formosa* T. & T. A. Steph.).

b. Hybrids are extremely variable, but their leaves are usually long, narrow, and marked with large, round spots, or are unspotted. The flowers are in a small, oblong spike and vary from pale lilac to dull purple, with narrow, spreading lateral sepals; the labellum has rounded, crenate lateral lobes, a small mid-lobe, and is marked with a pattern of purple spots and broken lines. The spur is intermediate between those of the parents. These hybrids appear to have normal fertility.

c. Such plants are often abundant where the parents grow together, in wet meadows on soils showing a transition from base-rich to more acid conditions. They are recorded from most of the localities where *D. purpurella* occurs in BI, and also from No.

d. Artificial cross-pollination of the parent species gives a normal seed-set and a viable F_1, but the fertility of such plants has not been tested. Presumed F_1 hybrids give a normal seed-set after pollination with the supposed parents or with other presumed F_1 plants.

e. Both parents and wild putative hybrids $2n = 80$.

f. GODFERY, M. J. (1933). As above. FIG.

HESLOP-HARRISON, J. (1955). As above.

STEPHENSON, T. and STEPHENSON, T. A. (1920). The British palmate orchids. *J. Bot., Lond.*, **58**: 257-262, pl. 556, fig. 12. FIG.

STEPHENSON, T. and STEPHENSON, T. A. (1922). As above.

2 x 6. *D. maculata* (L.) Soó x *D. majalis* (Reichb.) Hunt & Summerh.

a. *D.* x *townsendiana* (Rouy) Soó (*Orchis* x *dinglensis* Wilmott).

b. Hybrids are usually taller than *D. majalis* and have narrower, more acute, spotted leaves. The flowers have a broad, flat labellum marked with broken lines and dots, slightly crenate lateral lobes, and a small mid-lobe; they vary in colour from pale to dark reddish-purple. The spur is intermediate between those of the parents, but has a narrow throat closer to that of *D. maculata*. They show a high degree of fertility.

c. They have been recorded from v.c. H2 and H27, and from Au, Da, Ho and Su.

d. Artificial cross-pollination of the parent species gives a normal seed-set. F_1 hybrid plants, from crosses with Continental *D. maculata* and *D. majalis* from Ireland, have been raised to flowering, but their fertility has not been tested. Presumed F_1 hybrids give a normal seed-set after pollination with the supposed parent or with other presumed F_1 plants.

e. Both parents and wild putative hybrids $2n = 80$.

f, HARBECK, M. (1968). As above.

HARBECK, M. (1972). As above.

WILMOTT, A. J. (1936). New British marsh orchids. *Proc. Linn. Soc. Lond.*, **148**: 126-130.

2 x 7. *D. maculata* (L.) Soó x *D. traunsteineri* (Sauter) Soó

a. *D.* x *jenensis* (Brand.) Soó.

b. Hybrids have a slender, flexuous stem with 4–6 narrowly linear-lanceolate, subacute leaves more or less faintly marked with round spots like those of *D. maculata*. The spike is short, and the bracts are as long as the flowers and broader than in *D. maculata*. The lilac flowers have a large, trilobed labellum with the mid-lobe slightly longer than the lateral lobes, which are broad, crenulate and reflexed; the labellum is lightly marked with a pattern of reddish flecks and dots. The spur is cylindrical and intermediate in thickness between those of the parents. The pollen appears to be sterile and no seed-set has been observed.

c. They have been recorded from localities in v.c. 49 and 52 where *D. traunsteineri* occurs in base-rich fen and *D. maculata* is found in the more acid, raised areas around its edge. These hybrids are also recorded from Da, Fe, Ge, He, No, Po, Rs and Su, but some of these records may refer to *D.* x *kellerana* as Continental authorities have not always distinguished *D. fuchsii* from *D. maculata.*

d. Artificial cross-pollination of the presumed parents gives a good seed-set. F_1 hybrids from crosses with Continental parental material, made by M. Harbeck, have been raised to flowering, but their fertility has not yet been tested.

e. Both parents $2n = 80$.

f. ROBERTS, R. H. (1962). *Dactylorchis maculata* subsp. *ericetorum* x *D. traunsteineri. Proc. B.S.B.I.,* **4**: 418.

3d x 3. *D. cruenta* (O. F. Müll.) Soó (*D. incarnata* subsp. *cruenta* (O. F. Müll.) Sell) x *D. incarnata* (L.) Soó

= *D.* x *krylowii* (Soó) Soó has been recorded from Rs.

3 x 4. *D. incarnata* (L.) Soó x *D. praetermissa* (Druce) Soó

a. *D.* x *wintoni* (A. Camus) P. F. Hunt.

b. Hybrids are closer in habit to *D. praetermissa*; they have a moderately hollow stem, with erect, yellowish-green, more or less hooded leaves which are broadest near the middle. The spike is oblong and narrow with long, often incurved bracts. The flowers vary from pale lilac to bright violet (depending on the colour of the *D. incarnata* parent) and are smaller than those of *D. praetermissa,* with erect outer perianth segments and a more distinctly trilobed, narrower labellum whose lateral lobes are often reflexed. The spur is wide-throated and curved. The hybrids are highly sterile.

c. Such hybrids are generally rare and usually occur as single plants in base-rich marshes and fens. They have been recorded from 18 vice-counties in southern Br, northwards to v.c. 45 and 59.

d. Artificial cross-pollination of the presumed parents gives a poor seed-set, and artificial F_1 hybrids have not been raised.

e. *D. incarnata* $2n = 40$; *D. praetermissa* $2n = 80$.

f. DRUCE, G. C. (1918). As above.
   GODFERY, M. J. (1933). As above. FIG.

3 x 5. *D. incarnata* (L.) Soó x *D. purpurella* (T. & T. A. Steph.) Soó

a. *D.* x *latirella* (P. M. Hall) Soó (*Orchis* x *latirella* P. M. Hall).

b. Hybrids are intermediate in habit between the parents. Their leaves are narrower and a yellower green than in *D. purpurella,* and are sometimes marked with small indistinct spots. The flowers are in a cylindrical spike and nearer the colour of those of *D. purpurella,* but the lateral lobes of the labellum are somewhat reflexed, and the spur is intermediate in size and shape between those of the parents.

c. Hybrids are uncommon and are usually found as single plants in base-rich marshes and dune-slacks where the parents grow together. There are records from v.c. 46, 48, 52, 64, 65, 68, 96, 100, 101, 110 and 111.

d. Artificial cross-pollination of the parents gives few seeds whose viability has not been tested.

e. *D. incarnata* 2n = 40; *D. purpurella* 2n = 80.

f. HALL, P. M. (1936). New hybrid marsh orchid. *J. Bot., Lond.,* **74**: 329.

HALL, P. M. and STEPHENSON, T. (1942). A new variety of x *Orchis latirella. J. Bot., Lond.,* **80**: 131.

WILMOTT, A. J. (1938). In CAMPBELL, M. S. Further botanising in the Outer Hebrides. *Rep. B.E.C.,* **11**: 534-560. FIG.

3 × 6. *D. incarnata* (L.) Soó × *D. majalis* (Reichb.) Hunt & Summerh.

a. *D.* x *aschersoniana* (Hausskn.) Soó.

b. Hybrids have a somewhat slenderer and less hollow stem than *D. incarnata*. The leaves are rather broad, yellowish-green, usually tapering from the base and marked with faint spots, or they may be unspotted. The flowers vary from pale pink to reddish-purple, but are larger and more vividly coloured than in *D. incarnata*; the labellum shows varying degrees of intermediacy between those of the parents in shape and marking, but its lateral lobes are usually reflexed; the lateral perianth segments are often faintly spotted, and broader and blunter than those of *D. majalis*.

c. Such plants have been recorded from v.c. 110 and H8 and from Au, Da, Ge, He and Su.

d. Artificial cross-pollination of the parent species gives rather few seeds whose viability has not been tested.

e. *D. incarnata* 2n = 40; *D. majalis* 2n = 80.

f. SENGHAS, K. and SUNDERMANN, H., eds (1968). As above. FIG (only).

3 × 7. *D. incarnata* (L.) Soó × *D. traunsteineri* (Sauter) Soó

a. *D.* x *lehmanii* (Klinge) Soó.

b. Hybrids of this presumed parentage have the habit of *D. traunsteineri*, but their leaves are narrowly lanceolate, widest near the base, a somewhat yellowish green and sometimes marked with a few, small spots. The upper leaves exceed the base of the spike, and the bracts are long and suffused with purple. The flowers are intermediate between those of the parents. They seem to have a very low seed-set, but the viability of these seeds has not been tested.

c. Hybrids are very rare in areas of rich-fen, and in BI have been recorded only from v.c. 49, 52 and H20. They have also been reported from Au, Fe, Ga, Ge, He, Po, Rs and Su.

d. None.

e. *D. incarnata* 2n = 40; *D. traunsteineri* 2n = 80; hybrid 2n = 60.

f. HESLOP-HARRISON, J. (1953b). As above.

LACEY, W. S. and ROBERTS, R. H. (1958). Further notes on *Dactylorchis traunsteineri* (Saut.) Vermeul. in Wales. *Proc. B.S.B.I.,* **3**: 22-27.

3d × 6. *D. cruenta* (O. F. Müll.) Soó (*D. incarnata* subsp. *cruenta* (O. F. Müll.) Sell) × *D. majalis* (Reichb.) Hunt & Summerh.

= *D.* × *predaënsis* (Gsell) Soó has been recorded from He and Rs.

3d × 7. *D. cruenta* (O. F. Müll.) Soó (*D. incarnata* subsp. *cruenta* (O. F. Müll.) Sell) × *D. traunsteineri* (Sauter) Soó

(= *D.* × *engadinensis* (Ciferri & Giacomini) Soó, *nom. nud.*) has been recorded from He, Rs and Su.

4 × 5. *D. praetermissa* (Druce) Soó × *D. purpurella* (T. & T. A. Steph.) Soó

a. *D.* × *insignis* (T. & T. A. Steph.) Soó (*Orchis* × *salteri* T. Steph.).

b. The putative hybrids are intermediate between the parents in habit and vegetative characters, but nearer *D. purpurella* in flower characters. They have normal pollen- and seed-fertility, and introgression may occur, but there is no conclusive evidence of this.

c. Such plants occur at Morfa Harlech, v.c. 48, where they appear to be frequent with the presumed parents in calcareous dune-slacks. They have also been recorded from v.c. 46. They are generally uncommon because the parents are largely vicarious in BI.

d. Artificial cross-pollination of the parents gives a normal seed-set, from which a viable F_1 has been raised. The fertility of F_1 hybrids has not been tested.

e. Both parents and wild putative hybrid $2n = 80$.

f. ROBERTS, R. H. (1966). Studies on Welsh orchids, 3. The coexistence of some of the tetraploid species of marsh orchids. *Watsonia*, **6**: 260-267.
STEPHENSON, T. (1942). A new hybrid *Dactylorchis*. *J. Bot., Lond.*, **80**: 104.
STEPHENSON, T. and STEPHENSON, T. A. (1922). As above.

4 × 7. *D. praetermissa* (Druce) Soó × *D. traunsteineri* (Sauter) Soó

a. None.

b. Plants morphologically intermediate between these species have been considered to be hybrids. They are fully fertile and appear to backcross to form a complete range of intermediates linking the parent species.

c. Such plants have been recorded from areas of rich-fen in v.c. 28 and 29.

d. None.

e. Both parents and putative hybrid $2n = 80$.

f. None.

6 × 5. *D. majalis* (Reichb.) Hunt & Summerh. × *D. purpurella* (T. & T. A. Steph.) Soó

a. None.

b. The presumed hybrids are intermediate between the parents in leaf and flower characters. They are fully fertile.

c. Some populations in western Hb and in Scotland appear to show extensive hybridization between these two species. On the other hand a biometric study of a mixed population of *D. majalis* and *D. purpurella* in v.c. 52 indicated that they coexist there without intergrading; only a few plants of the presumed F$_1$ hybrid have been found there. Such hybrids have also been recorded from v.c. 100 and 110. They occur in wet meadows, base-rich marshes and dune-slacks.

d. Artificial cross-pollination of the parents gives a good seed-set, but the viability of these seeds has not been tested.

e. Both parents and putative hybrid $2n = 80$.

f. ROBERTS, R. H. (1966). As above.

5 x 7. *D. purpurella* (T. & T. A. Steph.) Soó x ***D. traunsteineri*** (Sauter) Soó

has been recorded from v.c. 49, but subsequent studies have shown that this identification was almost certainly erroneous. Artificial cross-pollination results in a good seed-set, but the viability of these seeds has not been tested. Biometric studies of mixed populations of *D. purpurella* and *D. traunsteineri* have shown that they coexist without intergrading; even F$_1$ hybrids seem not to be produced.

ROBERTS, R. H. (1966). As above.

6 x 7. *D. majalis* (Reichb.) Hunt & Summerh. x ***D. traunsteineri*** (Sauter) Soó

= ***D. x dufftiana*** (Schulze) Soó has been recorded from Au, Ga, Ge, He, Po and Rs.

644 × 643. *Aceras* R.Br. × *Dactylorhiza* Nevski = × *Dactyloceras* Garay & Sweet

(by P. F. Hunt)

A hybrid between *"Dactylorhiza latifolia"* (a name referable to one of several species of marsh-orchid) and *Aceras anthropophorum* (L.) Ait. f. (= x *D. helvetica* (Ciferri & Giacomini) Garay & Sweet, *nom. nud.*) has been reported form the Continent.

645 × 643. *Anacamptis* Rich.
× *Dactylorhiza* Nevski
= × *Dactylocamptis* Hunt & Summerh.
(by P. F. Hunt)

645/1 × 643/1. *Anacamptis pyramidalis* (L.) Rich. x *Dactylorhiza fuchsii* (Druce) Soó

> has been reported from v.c. 7/8, but its identity has not been authenticated and the different flowering times and spur-lengths of the two species make it unlikely that such a hybrid could arise.
> SUMMERHAYES, V. S. (1968). As above, p. 236.

645/1 × 643/2. *Anacamptis pyramidalis* (L.) Rich. x *Dactylorhiza maculata* (L.) Soó

> = x *D. weberi* (M. Schulze) Soó has been reported from the Continent.

649. *Arum* L.
(by J. D. Lovis and C. T. Prime)

2 × 1. *A. italicum* Miller x *A. maculatum* L.

a. None.
b. This hybrid is approximately intermediate between the parents in morphology, but is closer to *A. italicum* in phenology, and therefore most likely to be mistaken for that species; it is difficult to distinguish even in the field. Known hybrid examples have a pale purple spadix. The spotted-leaf character (common in *A. maculatum* but rare in *A. italicum*) is dominant to the non-spotted one. Pollen from hybrid plants has been found to be capable of germination, but backcross hybrids have not been detected. The difference in flowering season of the two parent species undoubtedly restricts the opportunities for cross-pollination, but their flowering periods do overlap in some seasons. Hybrids involving both subspecies of *A. italicum* have been found.
c. Hybrids have so far been positively identified only from Helston v.c. 1, and Jersey (involving *A. italicum* subsp. *italicum*); and from Purbeck, v.c. 9, Arundel, v.c. 13, Jersey and Guernsey (involving *A. italicum* subsp. *neglectum*). Foreign and other British records require confirmation.

d. Attempts to cross the species have so far failed.

e. *A. italicum* 2*n* = 83, 84; *A. maculatum* 2*n* = (28), 56, ?84; hybrid 2*n* = 69 or 70.

f. LOVIS, J. D. (1954). A wild *Arum* hybrid. *Proc. B.S.B.I.,* **1**: 97.

PRIME, C. T. (1954). *Arum neglectum* (Towns.) Ridley, in Biological Flora of the British Isles. *J. Ecol.,* **42**: 241-248.

PRIME, C. T. (1961). Taxonomy and nomenclature in some species of the genus *Arum* L. *Watsonia,* **5**: 106-109.

PRIME, C. T. (1973). Arums — Lords-and-Ladies, in GREEN, P. S., ed. *Plants: wild and cultivated,* pp. 167-171, pl. VII. London. FIG.

SOWTER, F. A. (1949). *Arum maculatum* L., in Biological Flora of the British Isles. *J. Ecol.,* **37**: 207-219.

652. *Sparganium* L.
(by C. D. K. Cook)

1 × neg. *S. erectum* L. × *S. neglectum* Beeby

has been reported occasionally, but these two taxa are now considered to be subspecies of *S. erectum* differing only in fruit-shape. It is possible that subsp. *oocarpum* (Čelak.) Domin could be the hybrid between subsp. *erectum* and subsp. *neglectum* (Beeby) Schinz & Thell.

2 × 1. *S. emersum* Rehman × *S. erectum* L.

(= *S.* × *aschersonianum* Hausskn. (*S. erectum* subsp. *erectum* × *S. emersum*) or *S.* × *engleranum* Aschers. & Graebn. (*S. erectum* subsp. *neglectum* (Beeby) Schinz & Thell. × *S. emersum*)). It is doubtful that this hybrid, although widely reported, really exists. Attempts have been made to synthesize it but the F_1 is apparently inviable. Authentic herbarium material of *S.* × *aschersonianum* is referable to *S. erectum* subsp. *microcarpum* (Neuman) Domin and of *S.* × *engleranum* to depauperate *S. erectum* subsp. *neglectum*.

COOK, C. D. K. (1962). *Sparganium erectum* L., in Biological Flora of the British Isles. *J. Ecol.,* **50**: 247-255.

3 × 2. *S. angustifolium* Michx. × *S. emersum* Rehman

a. (*S. angustifolium* subsp. *borderi* Aschers. & Graebn.; *S.* × *zetlandicum* Druce, *nom. nud.*).

b. Hybrids are intermediate between the parents in vegetative and floral characteristics. The F_1 is highly fertile. Segregants and backcrosses are frequent and an almost continuous variation pattern between the parents is found. The F_1 has the inflated leaf-bases of *S. angustifolium* (even when growing terrestrially) but the remote male capitula of *S. emersum.*

c. Hybrids occur in western Scotland in v.c. 74, 104 and 105. Hybrid populations are found in disturbed or previously oligotrophic but now eutrophic lakes, tarns and streams. Near Stoer, v.c. 105, a hybrid population was found on a shingle beach, a habitat not exploited by either parent. Hybrids are frequent in northern Europe and are known in Au, Da, Fe, Ga, Ge, No, Rs and Su.

d. None.

e. Both parents ($2n = 30$).

f. COOK, C. D. K. (1961). *Sparganium* in Britain. *Watsonia*, 5: 1-10.

3 × 4. *S. angustifolium* Michx. × *S. minimum* Wallr.

= *S.* × *oligocarpon* Ångstr. is found in Fe, No and Su.

653. *Typha* L.
(by C. A. Stace)

2 × 1. *T. angustifolia* L. × *T. latifolia* L.

a. *T.* × *glauca* Godr.

b. Plants variously intermediate between the two species and found in the vicinity of one or both of them are probably hybrids. The leaf-width and the degree of separation of the male and female parts of the inflorescence are usually more or less intermediate. Other characters, however, seem to vary between those of the two parents. The stigmas may be filiform or flattened, the apex of the gynophore hairs enlarged and brown or linear and colourless, the staminate bracteoles linear and brown or hair-like and colourless, the pistillate bracteoles present or absent, and the stalks of the pistillate flowers short and stubby or long and hairlike, as in *T. angustifolia* and *T. latifolia* respectively. The pollen grains (single in *T. angustifolia* and in tetrads in *T. latifolia*) are usually found partly single and partly in tetrads, with a good proportion misshapen or of unusual size. In North America most putative hybrids set very little good seed, if any, but in Europe fertility and introgression have not been properly investigated. In North America the hybrid exists both as obviously intermediate populations and as introgressed populations apparently of one species but possessing one or two characteristics of the other.

c. Specimens which appear to be authentic hybrids have been found in v.c. 8, 9, 28, 32 and 53, and also in Cz, Da, Fe, Ga, Ge, He, Ho, No, Po and Rs, and in U.S.A. and Canada. Like the parents they may form large clones and become dispersed by vegetative means.

d. Both species are protogynous and self-compatible. In North America the hybrid has been synthesized only when *T. angustifolia* was used as female; such hybrids resembled the wild *T.* x *glauca*. Pistillate bracts were present, the stigmas were intermediate in character, the pollen was largely abortive and consisted of from 5 to 36% tetrads, and meiosis was variously irregular with up to 6 univalents at each metaphase I. Both parents of this cross were $2n = 30$, but the hybrid was highly sterile.

e. *T. latifolia* ($2n = 30$); *T. angustifolia* ($2n = 30, 60$).

f. ALM, C. G. and WEIMARCK, H. (1933). *Typha angustifolia* L. x *latifolia* L. funnen i Skåne. *Bot. Notiser*, **1933**: 279-284. FIG.

BAYLY, I. L. and O'NEILL, T. A. (1971). A study of introgression in *Typha* at Point Pelee marsh, Ontario. *Can. Field Nat.*, **85**: 309-314.

FASSETT, N. C. and CALHOUN, B. (1952). Introgression between *Typha latifolia* and *T. angustifolia*. *Evolution, Lancaster, Pa.*, **6**: 367-379.

LOUSLEY, J. E. and TUTIN, T. G. (1947). *Typha angustifolia* L. x *latifolia* L. *Rep. B.E.C.*, **13**: 173-174.

SMITH, S. G. (1967). Experimental and natural hybrids in North American *Typha* (Typhaceae). *Am. Midl. Nat.*, **78**: 257-287. FIG.

VAN OOSTSTROOM, S. J. and REICHGELT, T. J. (1962). *Typha angustifolia* L. x *T. latifolia* L. (*T.* x *glauca* Godr.) in Nederland. *Gorteria*, **1**: 90-92. FIG.

655. *Scirpus* L. (*Schoenoplectus* (Reichb.) Palla)
(by J. E. Lousley)

8 x 6. *S. lacustris* L. x *S. triquetrus* L.

a. *S.* x *carinatus* Sm. (*Schoenoplectus* x *carinatus* (Sm.) Palla).

b. Hybrids are usually intermediate between the parents in height but readily recognized by the stem being rounded below but bluntly trigonous above. They resemble *S. triquetrus* in having usually three stigmas, and a rigid, triangular upper bract. Limited variation occurs, but there is usually local uniformity owing to vegetative reproduction. Seed is rarely ripened and there is no evidence of backcrossing.

c. This hybrid has occurred in BI only by the River Tamar, v.c. 2 and 3, and by the River Thames, v.c. 16, 17 and 21, where it was formerly plentiful and associated with *S. triquetrus*. It is probably now extinct. It is scattered in Europe, from Ho and Ga to Ge and Au.

d. None.

e. Both parents and hybrid (2*n* = 42).

f. JACKSON, A. B. and DOMIN, K. (1908). *S. lacustris* x *triqueter* (*S. carinatus*, Sm.). *Rep. B.E.C.*, 2: 314-316.

   LOUSLEY, J. E. (1931). *Schoenoplectus* group of the genus *Scirpus* in Britain. *J. Bot., Lond.*, 69: 151-163.

   OTZEN, D. (1962). Chromosome studies in the genus *Scirpus* L., section *Schoenoplectus* Benth. et Hook., in the Netherlands. *Acta bot. neerl.*, 11: 37-46.

   PERRING, F. H. and SELL, P. D. (1968). *Op. cit.*, p. 145.

9 x 6. *S. tabernaemontani* C. C. Gmel. x *S. triquetrus* L.

a. *S.* x *scheuchzeri* Brügg. (*S.* x *kuekenthalianus* Junge, *S.* x *arunensis* Druce).

b. Hybrids are usually intermediate between the parents in height but have the stem rounded below (as in *S. tabernaemontani*) and bluntly trigonous above. The influence of *S. triquetrus* is often also evident in a rigid, triangular upper bract. This hybrid is distinguished from *S.* x *carinatus* by the slenderer, more glaucous stems and asperous glumes, but these characters are sometimes difficult to apply to old herbarium specimens. It is thought that fertility is low but plants reproduce freely vegetatively. Thus field observations suggest that the limited variation is more likely to be explained by independent crosses and vegetative increase than by fertile hybrid swarms.

c. This hybrid occurs by rivers near the upper limits of the tide, by the River Tamar, v.c. 2, the River Arun, v.c. 13, and the River Medway, v.c. 15 and 16. The record from the River Thames, v.c. 17/21, given in Perring and Sell (1968), requires confirmation. It has become reduced in recent years, but still occurs in some quantity by the Arun. It is very rare elsewhere in Europe: Ga, Ge, He and Ho.

d. None.

e. Both parents and hybrid (2*n* = 42).

f. BAKKER, D. (1968). *Scirpus lacustris* L. ssp. *glaucus* (Sm.) Hartm. x *Scirpus triqueter* L. (*S.* x *scheuchzeri* Bruegg.) in Nederland. *Gorteria*, 4: 76-79.

   BRÜGGER, C. G. (1882). *S. scheuchzeri. Ber. natur. Ges. Graubündens*, 25: 108-111.

   JUNGE, P. (1905). *S. kuekenthalianus. Jb. hamb. wiss. Anst.*, 22(3): 73.

   LOUSLEY, J. E. (1931). As above.

   OTZEN, D. (1962). As above.

   PERRING, F. H. and SELL. P. D. (1968). *Op. cit.*, p. 145.

7 x 8. *S. americanus* Pers. x *S. lacustris* L.

= *S.* x *schmidtianus* Junge has been recorded from Ge.

8 × **9.** *S. lacustris* L. × *S. tabernaemontani* C. C. Gmel.
has been recorded from Da, Ge and Ho.

656. *Eleocharis* R.Br.
(by S. M. Walters)

5b × **5a.** *E. palustris* (L.) Roem. & Schult. subsp. *microcarpa* S. M.
Walters × *E. palustris* subsp. *vulgaris* S. M. Walters
  a. None. It is not certain which of the above two is the type subspecies.
  b. Lewis and John (1961) described a topodeme of *E. palustris* which
     consisted, on cytological evidence, of a mixture of the two subspecies and
     sterile hybrids with intermediate chromosome numbers and abnormal
     meiosis. One plant, which had $2n = 27$, showed a stomatal size at the lower
     end of the range of that of subsp. *vulgaris*; such plants are putative F_1
     hybrids. On the other hand, since *E. uniglumis* is also present in the
     locality, Lewis and John were uncertain whether to attribute plants with
     chromosome numbers varying around that of subsp. *vulgaris* (i.e.
     $n = 16$–21) to backcrossing to that subspecies or to hybridization with *E.
     uniglumis*.
  c. The only records to date are from Port Meadow, Oxford, v.c. 23, and from
     Su.
  d. Strandhede made artificial hybridizations between the two subspecies;
     subsp. *microcarpa* is self-incompatible, whilst subsp. *vulgaris* is self-
     compatible. Using the former as female parent almost normal fruit-set was
     obtained with pollen from subsp. *vulgaris*. He suggested that the reported
     sterility of Lewis and John's putative hybrid might have been due to
     self-incompatibility, since his own artificial hybrids were reasonably
     fertile.
  e. *E. palustris* subsp. *vulgaris* $2n = 38$; subsp. *microcarpa* $2n = 16$; putative
     hybrids $n = 11$–18, $2n = 27$ (plants with $n = 16$–21 could be hybrid
     derivatives).
  f. LEWIS, K. R. and JOHN, B. (1961). Hybridisation in a wild population of
     Eleocharis palustris. *Chromosoma*, **12**: 433-468.
     STRANDHEDE, S.-O. (1965). Chromosome studies in *Eleocharis*, subser.
     Palustres, 3. Observations on western European taxa. *Opera Botanica*,
     9(2): 1-86.
     STRANDHEDE, S.-O. (1966). Morphological variation and taxonomy in
     European *Eleocharis*, subser. *Palustres*. *Opera Botanica*, **10(2)**: 1-187.
     WALTERS, S. M. (1949). *Eleocharis*, in Biological Flora of the British
     Isles. *J. Ecol.*, **37**: 192-206.

WALTERS, S. M. (1950). *Variation in Eleocharis palustris agg.* Ph.D. thesis, University of Cambridge.

5a × 6. *E. palustris* (L.) Roem. & Schult. subsp. *vulgaris* S. M. Walters × *E. uniglumis* (Link) Schult.

a. None.

b. Morphologically intermediate plants can be found which are of putative hybrid origin between these two taxa. Such plants do not show any significant sterility.

c. There are no definite records of hybrids, but intermediates are found where the two parents meet, mainly on the western coast of Br, and in Su.

d. Artificial hybrids have been made by Strandhede and showed almost normal fruit-set.

e. *E. palustris* subsp. *vulgaris* $2n = 38$; *E. uniglumis* $2n = 46$; putative hybrids $2n = 38\text{-}46$.

f. STRANDHEDE, S.-O. (1965). As above.
STRANDHEDE, S.-O. (1966). As above.
WALTERS, S. M. (1949). As above.
WALTERS, S. M. (1950). As above.

659. *Schoenus* L.
(by C. A. Stace)

2 × 1. *S. ferrugineus* L. × *S. nigricans* L.
= *S.* × *intermedius* Čelak. has been reported from Au, Cz, Ga, Ge, He and Su.

663. *Carex* L.
(by E. C. Wallace)*

References to *Carex* hybrids are widely scattered in the literature, but many of the records are based on mistaken identifications, frequently of immature

* With assistance from P. M. Benoit, A. O. Chater, A. C. Jermy and C. A. Stace.

specimens. Accurate identification can only be guaranteed by the examination of freshly gathered, mature material, along with specimens of the suspected parents from the same locality.

Although very many *Carex* hybrids have been claimed for BI, only about 34 can be said to have been identified with any certainty. This is fewer than in the genera *Salix* or *Epilobium,* each of which has far fewer species than *Carex* in this country. Unlike the case in the two former genera, there are strong genetical barriers between many groups of *Carex,* and in general hybrids are only formed between species which are obviously quite closely related and fairly similar in appearance. There are some exceptions to this rule. In most cases the hybrids are sterile, although this is not true of hybrids within *C. flava* agg. and *C. nigra* agg., where hybrid swarms may arise.

Greatest uncertainty in the identity of putative hybrids arises in certain critical groups, particularly *C. flava* agg., *C. nigra* agg., and *C. spicata* agg. The names *C. flava* and *C. vulpina* in particular have been used in different senses in the past, and references to these names in the literature often implicate a species different from the one associated with that name nowadays. I have been unable to discover any factual basis for the inclusion of several of the hybrids listed by Pearsall (1934) for BI.

General References

JERMY, A. C. and TUTIN, T. G. (1968). *British sedges.* London.

KÜKENTHAL, G. (1909). Cyperaceae-Caricoideae-*Carex,* in ENGLER, A., ed. *Das Pflanzenreich,* **38(IV,20):** 67-767. Leipzig.

PEARSALL, W. H. (1934). Some hybrid Carices. *Rep. B.E.C.,* **10:** 682-685.

5 × 1. *C. binervis* Sm. × *C. laevigata* Sm.

a. *C.* × *deserta* Merino.

b. The Welsh plants are intermediate between the parents in most of the diagnostic characters and are sterile, with empty, green utricles 4.0–5.5 mm long and lacking conspicuous stripes or dots. The female glumes are 4.0–4.5 mm, obovate-acuminate and with an excurrent, pale-brown midrib and a very narrow, scarious margin (A. O. Chater *in litt.* 1973).

c. Two plants were found in 1961 by A. O. Chater, growing with both parents on a damp roadside slope east of Llyn Du, near Tremadoc, v.c. 49. The hybrid was described from Hs in 1909.

d. None.

e. *C. binervis* 2n = 74; *C. laevigata* 2n = 72.

f. None.

8 × 1. *C. demissa* Hornem. × *C. laevigata* Sm.

a. None.

b. This hybrid has intermediate characters and is highly sterile. It differs conspicuously from *C. demissa* in its tall stems, lowest female spike exserted on a long peduncle, and larger female glumes and utricles; and from *C. laevigata* in its curved stem-base, narrower leaves, less acuminate female glumes, and utricles with a less deeply notched beak. It resembles the common *C. demissa* x *C. hostiana* but is taller and has larger female glumes and utricles and more distant female spikes, the lowest of which has a long-exserted peduncle (P. M. Benoit *in litt.* 1971).

c. *C. demissa* x *C. laevigata* is known only as a single plant found in 1970 by P. M. Benoit with the parents near Egryn Abbey, v.c. 48.

d. None.

e. *C. demissa* 2n = 70; *C. laevigata* 2n = 72.

f. None.

1 x 24. *C. laevigata* Sm. x *C. pallescens* L.

is the parentage assigned by A. O. Chater, R. W. David and A. C. Jermy to a plant collected in 1973 by J. E. Raven by the road between Ardgour and Trislaig, v.c. 97, on the northern side of Loch Linnhe. It was growing near both putative parents and other Carices, and was sterile. The broad leaves and long-beaked utricles suggested *C. laevigata* as one parent, while the few, long, white hairs on the leaf-sheath indicated *C. pallescens* as the other.

DAVID, R. W. (1974). A *Carex* hybrid previously unknown. *Watsonia*, **10**: 165-166.

2 x 4. *C. distans* L. x *C. hostiana* DC.

a. *C.* x *muellerana* F. W. Schultz.

b. This hybrid is recognized by its sterility, the 3–4 female spikes which are cylindrical and fuscous when mature, and the male and female glumes having a silvery margin. It would be difficult to diagnose correctly if *C. demissa* or *C. lepidocarpa* (which could hybridize with *C. distans* and *C. hostiana*) were also present in the area.

c. This is a rare hybrid known in BI only in v.c. 12 and H21, in base-rich, damp meadows and a valley bog. It is also recorded from Au, Be, Cz, Ga, Ge and Rm.

d. None.

e. *C. distans* 2n = 74; *C. hostiana* 2n = 56.

f. None.

2 x 6. *C. distans* L. x *C. flava* L.

= *C.* x *luteola* Sendtn. has been recorded from Au, Ge and Ho.

2 x 7. *C. distans* L. x *C. lepidocarpa* Tausch

a. *C.* x *binderi* Podp.

b. Sterile plants, more or less intermediate in morphology between these two species, have been identified as this hybrid; more detailed study and diagnosis is required.

c. The plant from Totternhoe, v.c. 30, identified as this hybrid by Davies (1955), was considered by Nelmes (1955) to be *C. hostiana* x *C. lepidocarpa*, and A. C. Jermy (*in litt.* 1973) agrees. A plant collected by A. G. Kenneth and A. McG. Stirling on the Isle of Danna, v.c. 101 (A. G. Kenneth *in litt.* 1971) was identified as *C. distans* x *C. lepidocarpa* (presumably subsp. *scotica* E. W. Davies) by A. G. Kenneth and P. D. Sell. There are also records from Au and Cz.

d. None.

e. *C. distans* $2n = 74$; *C. lepidocarpa* $2n = 68$.

f. DAVIES, E. W. (1955). The cytogenetics of *Carex flava* and its allies. *Watsonia*, 3: 129-137.

   NELMES, E. (1955). *Carex hostiana* in Bedfordshire. *Proc. B.S.B.I.*, 1: 314-315.

2 × 10. *C. distans* L. × *C. serotina* Mérat

= *C.* x *gogelana* Podp. has been recorded from Cz and Ge.

2 × 11. *C. distans* L. × *C. extensa* Gooden.

a. *C.* x *tornabenii* Chiov.

b. Hybrids are intermediate between the parents in characters of the leaves, bracts and spikes, and are sterile.

c. This is known for certainty only as a few plants at Mochras and at Morfa Harlech, v.c. 48, where it has persisted for over 20 years on salt-marshes close to both parents. Records from v.c. 6, 9 and 15 are errors for *C. extensa* and/or *C. distans*. The hybrid is also recorded from Su.

d. None.

e. *C. distans* $2n = 74$; *C. extensa* $2n = 60$.

f. BRENAN, J. P. M. and SIMPSON, N. D. (1946). A hybrid sedge new to Britain. *NWest. Nat.*, 20: 202-206.

   NELMES, E. (1952). A hybrid *Carex* from Wales. *Watsonia*, 2: 148-150.

   SANDWITH, C. I. (1947). *C. distans* L. x *extensa* Good. *Proc. Bristol Nat. Soc.*, 27: 155.

5 × 3. *C. binervis* Sm. × *C. punctata* Gaud.

a. None.

b. This sterile hybrid resembles *C. binervis* in its robust habit, very distant female spikes, purplish-brown female glumes, and mature utricles spreading not more than 70°. It differs from that species in its slenderer, paler male spike (2.5–3.0 mm diam.), and slender utricle-beak with a shortly bifid tip.

c. *C. binervis* x *C. punctata* was known in BI only from a wet rock by the shore near Barmouth, v.c. 48, where it grew with both parents. One plant was found in 1954, two more in 1959, but none has been seen since 1960 (P. M. Benoit *in litt.* 1971). Elsewhere the hybrid was reported from Gironde, Ga, in 1967 (E. Contré *in litt.* 1970).

d. None.
e. *C. binervis* 2n = 74; *C. punctata* 2n = 68.
f. BENOIT, P. M. (1958). A new hybrid sedge. *Watsonia*, 4: 122-124.

8 × 3. *C. demissa* Hornem. × *C. punctata* Gaud.
has been recorded from Ge.

3 × 10. *C. punctata* Gaud. × *C. serotina* Mérat
was recorded by Pearsall for BI, but nothing further is known of this. It
has also been recorded from He.
PEARSALL, W. H. (1934). As above.

24 × 3. *C. pallescens* L. × *C. punctata* Gaud.
has been recorded from He.

5 × 4. *C. binervis* Sm. × *C. hostiana* DC.
was recorded by Praeger for v.c. H39, but nothing further is known of this.
PRAEGER, R. L. (1951). *Op. cit.*, p. 13.

6 × 4. *C. flava* L. × *C. hostiana* DC.
= *C.* x *xanthocarpa* Degland has been recorded from Au, Cz, Da, Fe, Ga,
Ju, No, Rm and Su, but British records refer to segregates of *C. flava* other
than *C. flava sensu stricto*.

4 × 7. *C. hostiana* DC. × *C. lepidocarpa* Tausch
a. *C.* x *fulva* Gooden.
b. Hybrids are easily recognized by their intermediacy and complete sterility.
The bracts are long, the spikes longer than in *C. lepidocarpa* and fuscous
when mature, and the female spikes are frequently male at the top.
c. Hybrids occur throughout most of BI in wet valley pasture, heathland
bogs, wet moorland and loch-side fens on base-rich soil, and involve both
subsp. *lepidocarpa* and subsp. *scotica* E. W. Davies of *C. lepidocarpa*. The
latter hybrid has been recorded from v.c. 88/89, 109 and 111. There are
records from Au, Cz, Da, Fe, Ge, He, No and Su.
d. *C. hostiana* (from Teesdale) x *C. lepidocarpa* (from the Pyrenees) was
synthesized by Davies (1955) and grown at least to the seedling stage.
e. *C. hostiana* 2n = 56; *C. lepidocarpa* (both subspp.) 2n = 68.
f. DAVIES, E. W. (1955). As above.
NELMES, E. (1955). As above.
NELMES, E. and ROB, C. M. (1948). *Carex Hostiana* DC. x *lepidocarpa*
Tausch. *Rep. B.E.C.*, 13: 373.
REITER, M. (1950). Der Formenkreis von *Carex flava* L. *s. lat.* und seine
Bastarde im Lande Salzburg. *Mitt. naturw. ArbGemein. Haus Nat.
Salzb., Bot.*, 1950: 42-46.

8 × 4. *C. demissa* Hornem. × *C. hostiana* DC.

a. None.

b. This hybrid is intermediate between the parents and highly sterile. It is taller than *C. demissa* and has longer bracts than in *C. hostiana*; the female spikes taper from the base and are pale fuscous, the lowest often being long-stalked. It may be distinguished from *C. hostiana* x *C. lepidocarpa*, which it closely resembles, by its shorter and slightly smaller spikes and narrower and longer male glumes.

c. *C. demissa* x *C. hostiana* is frequent with the parents, and sometimes in the absence of *C. hostiana*, in damp pastures and heathy ground, especially bordering base-rich swamps, throughout most of BI; it was said by E. S. Marshall to be by far the commonest hybrid sedge in Br, and by R. D. Meikle (Praeger, 1951) to be the only common *Carex* hybrid in Hb. It is also recorded from Da, Fa, Fe, Ho, No and Su.

d. Davies (1955) reported a plant from Loch Tummel, v.c. 88/89, with a highly irregular meiosis with several chains of chromosomes and numerous univalents.

e. *C. demissa* $2n = 70$; *C. hostiana* $2n = 56$.

f. DAVIES, E. W. (1955). As above.

  MARSHALL, E. S. (1914). *A supplement to the Flora of Somerset,* p. 204. Taunton.

  PRAEGER, R. L. (1951). *Op. cit.,* p. 13.

4 × 10. *C. hostiana* DC. × *C. serotina* Mérat

a. *C.* x *appeliana* Zahn.

b. This is a sterile hybrid with the habit of *C. hostiana*, long bracts, 1–2 short and slender female spikes, and utricles shorter than the glumes.

c. This hybrid has been recorded from v.c. 33, 88, 90, 101 (in a dune-slack), 110 and 111, and from Au, Cz, Da, Fa, Fe, Ho, No, Rm and Su. Some of the older records of *C. flava* x *C. hostiana* from Hb may well refer to this hybrid.

d. Meiosis is irregular, with 5–16 univalents.

e. *C. hostiana* $2n = 56$; *C. serotina* $2n = 70$; hybrid ($2n = 61–71$).

f. HEILBORN, O. (1928). Chromosome studies in Cyperaceae. *Hereditas,* 11: 182-192.

  LINTON, E. F. and LINTON, W. R. (1890). *C. fulva* x *Oederi*, Ehrh. *Rep. B.E.C.,* 1: 273-274, 321.

  REITER, M. (1950). As above.

5 × 6. *C. binervis* Sm. × *C. flava* L.

has been recorded from Ge.

5 × 7. *C. binervis* Sm. × *C. lepidocarpa* Tausch

has been recorded from Ge.

5 × 8. *C. binervis* Sm. × *C. demissa* Hornem.

a. *C.* x *corstorphinei* Druce.

b. This hybrid has the habit of *C. binervis,* but the mature female spikes are fuscous in colour and the utricles are shorter, more shortly beaked and lack the green lines. Some of the utricles are short and inflated, showing evidence of *C. demissa.* A note by E. S. Marshall (*in sched.,* **OXF**) remarks that a cross between *C. binervis* and the alpine representative of the *C. flava* agg. from the Clova Mountains would, he thought, produce something very like the plant in question.

c. *C.* x *corstorphinei* was recorded from Glen Phee, v.c. 90, in 1915. A "mittelform" between the two parental species was recorded from Stornoway, v.c. 110, by Krause (1898), but the evidence for this is inconclusive.

d. None.

e. *C. binervis* $2n = 74$; *C. demissa* $2n = 70$.

f. DRUCE, G. C. (1916). *Carex binervis* Sm. x *C. flava,* var. *oedocarpa* Anders. *Rep. B.E.C.,* 4: 216.

KRAUSE, E. H. L. (1898). Floristische Notizen. *Bot. Zbl.,* 75: 38.

5 × 16. *C. binervis* Sm. × *C. rostrata* Stokes

= *C.* x *catteyensis* A. Benn. was recorded from Winless, v.c. 109, in 1911, but the evidence is not convincing and no specimens have been seen.

BENNETT, A. (1911). *C. catteyensis, mihi. Ann. Scot. nat. Hist.,* 77: 49.

52 × 5. *C. bigelowii* Torr. ex Schwein. × *C. binervis* Sm.

was recorded from Ben Wyvis, v.c. 106, in 1900 by E. S. Marshall, but surely in error.

MARSHALL, E. S. (1901). Plants of north Scotland, 1900. *J. Bot., Lond.,* 39: 266-275.

6 × 7a. *C. flava* L. × *C. lepidocarpa* Tausch subsp. *lepidocarpa*

a. *C.* x *pieperana* P. Junge.

b. Plants with utricles intermediate in size and beak-length between those of the parents have been considered hybrids. Italian material showed 29% pollen fertility and backcrosses occurred in the field.

c. Such plants have been reported from a calcareous marsh in v.c. H17, where *C. flava* was discovered in 1968, new to Hb, growing with *C. lepidocarpa.* Records from v.c. 110–112 must be errors, since *C. flava* does not occur in these areas. It is possible that plants of *C. flava* from Malham Tarn, v.c. 64, have suffered introgression from *C. lepidocarpa* in the past, and the situation there may parallel that in v.c. H17. There are also records of the hybrid from Au, Da, Fe, Ge, Ho, It, No, Rm and Su.

d. Davies obtained seedlings from reciprocal crosses using British and Swiss material and *C. lepidocarpa* subsp. *lepidocarpa* and subsp. *scotica.*

e. *C. flava* $2n = 60$: *C. lepidocarpa* $2n = 68$.

f. DAVIES, E. W. (1955). As above.

PERRING, F. H. (1970). Vascular plants new to Ireland. *Watsonia*, **8**: 91.

REITER, M. (1950). As above.

SENAY, P. (1951). Le groupe des *Carex flava* et *C. Oederi. Bull. Mus. nat. Hist. nat.*, Sér. 2, **22**: 618-624, 790-796; **23**: 146–152.

8 × 6. *C. demissa* Hornem. × *C. flava* L.

a. (?*C.* × *alsatica* Zahn).

b. Hybrid plants are intermediate between the parents to a varying degree. At Roudsea Wood they showed 22% pollen fertility and appeared to backcross to both parents.

c. A hybrid swarm exists in Roudsea Wood, v.c. 69; elsewhere this hybrid is recorded from Au, Fe, Ho, No and Su.

d. Davies obtained reciprocal crosses, some of which had 20% pollen fertility. The wild hybrid from Roudsea Wood had a very irregular meiosis.

e. *C. demissa* $2n = 70$; *C. flava* $2n = 60$.

f. DAVIES, E. W. (1955). As above.

SENAY, P. (1951). As above.

6 × 10. *C. flava* L. × *C. serotina* Mérat

= *C.* × *ruedtii* Kneuck. (?*C.* × *alsatica* Zahn) has been recorded from Au, Fe, Ga, Rm and Su. The few records from Br must either be errors or refer to segregates of *C. flava* other than *C. flava sensu stricto*, as the latter and *C. serotina* do not grow together anywhere in BI.

6 × 15. *C. flava* L. × *C. pseudocyperus* L.

was mentioned by Druce (1928), with a query, but nothing more is known of it.

DRUCE, G. C. (1928). *Op. cit.*, p. 121.

6 × 19. *C. flava* L. × *C. saxatilis* L.

= *C.* × *marshallii* A. Benn. has been recorded from v.c. 87/88/89, 90, 92/93, 98 and 105, but *C. flava sensu stricto* does not occur in these areas. Druce (1926) said his plant from v.c. 98 was *C. demissa* × *C. saxatilis*, and probably Bennett also intended this combination. A. C. Jermy (*in litt.* 1973) believes that *C.* × *marshallii* was in any case a variant of *C. saxatilis*.

BENNETT, A. (1925). Notes on British Carices. *Trans. Proc. bot. Soc. Edinb.*, **29**: 127-129.

DRUCE, G. C. (1926). *Carex flava* × *saxatilis. Rep. B.E.C.*, **7**: 789.

31 × 6. *C. flacca* Schreb. × *C. flava* L.

was recorded by Beeby from v.c. 112, but almost certainly in error and no voucher specimen has been found.

BEEBY, W. H. (1907). *Carex glauca* × *flava. Ann. Scot. nat. Hist.*, **64**: 235-236.

8 × 7. *C. demissa* Hornem. × *C. lepidocarpa* Tausch

a. None.
b. Hybrids have female spikes intermediate in size and utricle characters between those of the parents. *C. lepidocarpa* subsp. *lepidocarpa* and subsp. *scotica* both frequently hybridize with *C. demissa*; the hybrids show 22–35% pollen fertility and often backcross to form a hybrid swarm.
c. Hybrids are frequent in scattered localities throughout BI, in base-rich, wet pastures, moorland and heathland fens and base-rich, mountain flushes and rock-ledges. They are also recorded from Da, Fe and Su.
d. Davies obtained crosses between *C. demissa* (female) and *C. lepidocarpa* subsp. *lepidocarpa*; some of the F_1 plants showed 25% pollen fertility.
e. *C. demissa* $2n = 70$; *C. lepidocarpa* (both subspp.) $2n = 68$.
f. DAVIES, E. W. (1955). As above.

7a × 10. *C. lepidocarpa* Tausch. subsp. *lepidocarpa* × *C. serotina* Mérat

a. *C.* × *schatzii* Kneuck.
b. This hybrid is intermediate between the parents in stature, bract-length and female spike-length. The pollen shows 20–25% fertility.
c. It is known from a few base-rich, wet places in England, e.g. v.c. 29 (Wicken Fen), 33, 65 and 70, and from Au, Cz, Fe, Ge, He, Rm and Su.
d. Davies obtained hybrids using foreign material (*C. serotina* as female), and the F_1 showed 24% pollen fertility. In the wild the late flowering-time of *C. serotina* undoubtedly acts as a strong isolating factor.
e. *C. lepidocarpa* $2n = 68$; *C. serotina* $2n = 70$.
f. DAVIES, E. W. (1955). As above.
   REITER, M. (1950). As above.

8 × 10. *C. demissa* Hornem. × *C. serotina* Mérat

a. None.
b. This hybrid differs from *C. demissa* in its shorter, curved stems, and its utricles with a more or less tapered beak which is, however, not so abruptly contracted as in *C. serotina*. The female spikes are pale fuscous when mature and most utricles are empty; pollen fertility has been found to be 26–29%.
c. It has been found in open places in fens, in dune-slacks and on sandy flats by Scottish sea-lochs, in v.c. 6 (Shapwick peat-moor), 28, 97 and 101. It is also recorded from Fe and Ho.
d. Davies obtained artificial crosses with *C. demissa* as the female parent; some of the F_1 plants showed 29–30% pollen fertility. The two species are isolated in the wild by their different flowering times.
e. Both parents $2n = 70$.
f. DAVIES, E. W. (1955). As above.

8 × 11. *C. demissa* Hornem. × *C. extensa* Gooden.
has been recorded from Ge.

8 × 19. *C. demissa* Hornem. × *C. saxatilis* L.
see under *C. flava* x *C. saxatilis*.

12 × 23. *C. sylvatica* Huds. × *C. strigosa* Huds.
= *C.* x *strigulosa* Chat. has been recorded from Ga.

32 × 12. *C. hirta* L. × *C. sylvatica* Huds.
= *C.* x *cetica* Rech. pat. has been recorded from Au.

15 × 16. *C. pseudocyperus* L. × *C. rostrata* Stokes
  a. *C.* x *justi-schmidtii* Junge.
  b. This hybrid differs from *C. pseudocyperus* in the partly glaucous leaves, less pendulous and aggregated spikes which are, however, longer and laxer, and the more inflated and divergent utricles. It differs from *C. rostrata* in its flatter, wider and greener leaves, longer and more pendulous spikes, and longer utricles tapering gradually into the beak.
  c. There is so far only one known locality in BI, at Cranberry Rough, Hockham, v.c. 28, where it was found in 1955 on peat in carr. It is also known in Fe, Ge, No, Po and Su.
  d. None.
  e. *C. pseudocyperus* ($2n = 66$); *C. rostrata* ($2n = 60, 76, 82$).
  f. PETCH, C. P. and SWANN, E. L. (1956). A hybrid sedge from West Norfolk. *Proc. B.S.B.I.*, **2**: 1-3. FIG.

15 × 17. *C. pseudocyperus* L. × *C. vesicaria* L.
= *C.* x *wolteri* R. Gross has been recorded from Cz, Ge and Su. Pearsall noted it from BI, but apparently in error.
PEARSALL, W. H. (1934). As above.

16 × 17. *C. rostrata* Stokes × *C. vesicaria* L.
  a. *C.* x *involuta* (Bab.) Syme. (Involute-leaved Sedge).
  b. This is a variable hybrid, some variants being nearer one parent and some the other. It differs from *C. rostrata* in the flatter, greener leaves, the longer and less inflated utricles and the female glumes more conspicuous in the mature spikes; and from *C. vesicaria* in the narrower leaves and smaller utricles which are somewhat more crowded and patent. It is highly sterile.
  c. This hybrid grows in wet places and by water in scattered localities over most of BI. It is also recorded from Au, Da, Fe, Ga, Ho, No, Rm, Rs and Su.
  d. None.
  e. *C. rostrata* ($2n = 60, 76, 82$); *C. vesicaria* ($2n = 82$).

f. BENNETT, A., LINTON, E. F. and MARSHALL, E. S. (1910). *Carex inflata* x *vesicaria. Rep. B.E.C.,* **2**: 479-480.

PEARSALL, W. H. (1934). As above.

SYME, J. T. B., ed. (1870). *English Botany,* 3rd ed., **10**: 169-170, t. 1681. London. FIG.

16 x 19. *C. rostrata* Stokes x *C. saxatilis* L.

has been recorded from No and Su. Some robust plants determined as *C. saxatilis, C.* x *grahamii* or *C.* x *ewingii* from v.c. 88 and 97 might be better attributed to the above parentage (A. C. Jermy *in litt.* 1972).

JERMY, A. C. and TUTIN, T. G. (1968). As above, p. 94.

20 x 16. *C. riparia* Curt. x *C. rostrata* Stokes

(? = *C.* x *beckmanniana* Figert) has been recorded from Da, Fe, Ge, He, Ho, No, Po, Rs and Su. It was mentioned for BI by Pearsall (1934) and Jermy and Tutin (1968), but probably in error (A. C. Jermy *in litt.* 1973).

PEARSALL, W. H. (1934). As above.

21 x 16. *C. acutiformis* Ehrh. x *C. rostrata* Stokes

= *C.* x *beckmanniana* Figert has been recorded from BI by Jermy and Tutin (1968), who said it is "intermediate and sterile". The only record traced is of a specimen collected by C. Waterfall in v.c. 33 in 1902, considered to be this hybrid by Bennett and Marshall, but the specimen in BM appears to me to be *C. acuta. C.* x *beckmanniana* is considered by most Continental authors to be *C. riparia* x *C. rostrata.*

BENNETT, A., MARSHALL, E. S. and WATERFALL, C. (1903). *C. acuta,* L. *Rep. Watson B.E.C.,* **1902-1903**: 23.

32 x 16. *C. hirta* L. x *C. rostrata* Stokes

has been recorded from Ga. It was mentioned for BI by Pearsall, but probably erroneously so.

PEARSALL, W. H. (1934). As above.

33 x 16. *C. lasiocarpa* Ehrh. x *C. rostrata* Stokes

= *C.* x *prahliana* P. Junge has been recorded from Fe, Ge, Po and Su. It was mentioned for BI by Pearsall, but probably erroneously so.

PEARSALL, W. H. (1934). As above.

19 x 17. *C. saxatilis* L. x *C. vesicaria* L.

is the parentage ascribed to *C.* x *grahamii* Boott (*C.* x *stenolepis* auct., *non* Less.) by Jermy and Tutin (1968). They described it as being sterile and intermediate in stature, leaf-shape and -anatomy, and glume- and utricle-shape, and stated that it was not the same as the true *C. stenolepis* Less. from Scandinavia. It is recorded from mountainous areas (above 600 m) of

Scotland, in v.c. 88, 90 and 98, and shows considerable variation. Whatever the identity of the disputed Scottish plant I do not consider that *C. vesicaria*, a lowland species, is involved. Hybrids are recorded in Scandinavia between *C. stenolepis* Less. and both *C. saxatilis* and *C. vesicaria*.

BUTCHER, R. W. and STRUDWICK, F. E. (1930). *Further illustrations of British plants*, p. 411. Ashford. FIG.

EWING, P. (1910). On some Scottish alpine forms of *Carex. Ann. Scot. nat. Hist.*, **75**: 174-181.

JERMY, A. C. and TUTIN, T. G. (1968). As above, pp. 94-95. FIG.

MARSHALL, E. S. (1911). Dalmally plants, 1910. *J. Bot., Lond.*, **49**: 191-198.

SYME, J. T. B., ed. (1870). *English Botany*, 3rd ed., **10**: 173-174, t. 1684. London. FIG.

TUTIN, T. G. (1962). In CLAPHAM, A. R., TUTIN, T. G. and WARBURG, E. F. *Op. cit.*, p. 1091. FIG.

20 × 17. *C. riparia* Curt. × *C. vesicaria* L.

a. *C.* × *csomadensis* Simonk.

b. This hybrid differs from *C. riparia* in its much narrower leaves, more slender male spikes, longer and more slender peduncles to the female spikes, glumes with a paler margin, and a more slender beak to the utricle. From *C. vesicaria* it differs in the glaucous leaves, and darker and smaller utricles. It is sterile.

c. The only certain record is from a marsh south of Mizen Head, v.c. H20, where it was found by A. W. Stelfox in 1946. There are also records from v.c. 13, 20 and 22–24, but Nelmes (1949) examined all of them except that from v.c. 20 and concluded they were *C. riparia*. The hybrid is recorded from Cz, Da, Fe, Ge, Ho, No, Rm and Su.

d. None.

e. *C. riparia* ($2n = 72$); *C. vesicaria* ($2n = 82$).

f. DRUCE, G. C. and MARSHALL, E. S. (1912). *C. csomadensis*, Simonkai. *Rep. B.E.C.*, **3**: 133.

NELMES, E. (1949). Another hybrid *Carex* from Ireland. *Watsonia*, **1**: 86-87.

21 × 17. *C. acutiformis* Ehrh. × *C. vesicaria* L.

= *C.* × *ducellieri* Beauv. has been recorded from He.

32 × 17. *C. hirta* L. × *C. vesicaria* L.

a. *C.* × *grossii* Fiek.

b. This is intermediate between the parents. The utricles are much less hairy and with shorter hairs than in *C. hirta*, and the fruits do not develop. There is a well-developed rhizome system.

c. In BI it is only known in a marsh among sand-dunes, growing with *C.*

vesicaria, 4 miles south of Wicklow, v.c. H20, where it was found by J. P. Brunker in 1944. According to Brunker (1946) the marsh was brackish, but A. C. Jermy (*in litt.* 1972) informs me it is a freshwater marsh. The colony had spread to form a large patch by the mid 1960s (M. J. P. Scannell *in litt.* 1973). Elsewhere it is known from Ge, Po and Su.

  d. None.
  e. *C. hirta* 2*n* = 112; *C. vesicaria* (2*n* = 82).
  f. BRUNKER, J. P. (1946). A hybrid sedge new to the British Isles. *Ir. Nat. J.,* 8: 339.
  NELMES, E. (1947). A hybrid sedge new to the British Isles. *Rep. B.E.C.,* 13: 93-94.

33 × 17. *C. lasiocarpa* Ehrh. × *C. vesicaria* L.

= *C. × kohtsii* K. Richt. has been recorded from Ge, No, Po and Su. It was mentioned for BI by Pearsall, but probably in error.
PEARSALL, W. H. (1934). As above.

47 × 17. *C. acuta* L. × *C. vesicaria* L.

= *C. × prostii* Chassagne has been recorded from Ga.

18 × 19. *C. × grahamii* Boott (*C. × stenolepis* auct., *non* Less.) × *C. saxatilis* L.

= *C. × ewingii* E. S. Marshall has been recorded from bogs on the eastern side of Ben More, v.c. 88, but Jermy and Tutin (1968) considered *C. × grahamii* to be the hybrid *C. saxatilis* × *C. vesicaria,* and *C. × ewingii* therefore a synonym of it. There is much *C. saxatilis* in the area, where it is very variable, but despite frequent searches I have not found *C. × grahamii* on Ben More, although it has been recorded there.
EWING, P. (1910). As above.
MARSHALL, E. S. (1911). As above.

21 × 20. *C. acutiformis* Ehrh. × *C. riparia* Curt.

has been recorded from BI on several occasions, e.g. v.c. 17, 22–24, 27, 28 and 41, but remains unconfirmed and I have seen no specimens of it. Bowen (1968) commented that Druce's record for v.c. 22 was not represented by a specimen at **OXF**.
BENNETT, A., DRUCE, G. C., HORWOOD, A. R. and MARSHALL, E. S. (1916). *Carex acutiformis × riparia. Rep. B.E.C.,* 4: 379-380.
BOWEN, H. J. M. (1968). *The flora of Berkshire,* p. 292. Oxford.
DRUCE, G. C. (1916a). *Carex acutiformis × riparia. Rep. B.E.C.,* 4: 215.

31 × 20. *C. flacca* Schreb. × *C. riparia* Curt.

= *C. × lausii* Podp. has been recorded from Cz and Ge.

33 × 20. *C. lasiocarpa* Ehrh. × *C. riparia* Curt.

  a. *C.* × *evoluta* Hartm.
  b. This hybrid is intermediate between its parents. It differs from *C. lasiocarpa* in its broader, flatter leaves, more robust spikes, and less hairy utricles, which are sterile. From *C. riparia* it differs in its narrower leaves, smaller spikes and hairy utricles.
  c. *C.* × *evoluta* is known in BI only from Sharpham peat-moor, v.c. 6, where it was found in 1915 by H. S. Thompson and apparently last seen in 1955 by A. J. Willis. It is also recorded from Fe, Ga, Ge, Ho, No, Rm, Rs and Su.
  d. None.
  e. *C. riparia* (2n = 72).
  f. THOMPSON, H. S. (1915). *Carex evoluta* Hartm. in Britain. *J. Bot., Lond.,* 53: 309.
     THOMPSON, H. S. (1917). *Carex evoluta* Hartm. *Rep. Watson B.E.C.,* 3: 37-38.
     VOO, E. E. van der (1962). *Carex* × *evoluta* Hartm. *Gorteria,* 1: 19-20.

46 × 20. *C. elata* All. × *C. riparia* Curt.

  = *C.* × *nicoloffii* Pampanini has been recorded from It.

21 × 31. *C. acutiformis* Ehrh. × *C. flacca* Schreb.

  = *C.* × *jaegeri* F. Schultz has been recorded from Cz and Ge, and it was mentioned for BI by Pearsall (1934) and Druce (1928). There is no evidence of its occurrence in BI.
  DRUCE, G. C. (1928). *Op. cit.,* p. 121.
  PEARSALL, W. H. (1934). As above.

21 × 33. *C. acutiformis* Ehrh. × *C. lasiocarpa* Ehrh.

  = *C.* × *uechtritziana* K. Richt. has been recorded from Ge. Pearsall mentioned it for BI, but probably in error.
  PEARSALL, W. H. (1934). As above.

21 × 46. *C. acutiformis* Ehrh. × *C. elata* All.

  = *C.* × *felixii* Lambert has been recorded from Ga.

47 × 21. *C. acuta* L. × *C. acutiformis* Ehrh.

  = *C.* × *subgracilis* Druce has been recorded from v.c. 22, 23 and 27 and from Ga. It is said to be intermediate between the parents, with female flowers with two or three stigmas in the same spike, and Jermy (1967) suggested that it is at least partly fertile and more common than generally thought. I have seen no specimens that might be attributed to this hybrid, but Bowen (1968) cited Druce's specimen from v.c. 22 at **OXF**, determined by E. S. Marshall.

BOWEN, H. J. M. (1968). As above, p. 293.
DRUCE, G. C. (1916b). *Carex acutiformis* x *gracilis*. *Rep. B.E.C.*, 4: 215-216.
JERMY, A. C. (1967). *Carex* section *Carex* (= *Acutae* Fr.). *Proc. B.S.B.I.*, 6: 375-379.

25 x 31. *C. filiformis* L. x *C. flacca* Schreb.

= *C.* x *bouchardii* Genty was described from Ga in 1956. Lousley (1937) reported plants of *C. flacca* growing at Chertsey, v.c. 17 and 21, accompanied by *C. filiformis*, which closely resembled *C. filiformis*, but he was unable to demonstrate a hybrid origin for them. There are also doubtful records from Ge.

LOUSLEY, J. E. (1937). *C. tomentosa* L. and *diversicolor* Crantz. *Rep. B.E.C.*, 11: 230.

26 x 27. *C. panicea* L. x *C. vaginata* Tausch

has been recorded from Su.

26 x 29. *C. panicea* L. x *C. paupercula* Michx.

has been recorded from Fe.

31 x 26. *C. flacca* Schreb. x *C. panicea* L.

= *C.* x *fontis-sancti* Podp. has been recorded from Cz, Ge and He.

28 x 29. *C. limosa* L. x *C. paupercula* Michx.

= *C.* x *connectens* Holmb. has been recorded from Cz, Ge, Is, No, Rm and Su.

28 x 30. *C. limosa* L. x *C. rariflora* (Wahlenb.) Sm.

= *C.* x *firmior* (Norm.) Holmb. has been recorded from Fe, Is, No, Rs and Su.

29 x 30. *C. paupercula* Michx. x *C. rariflora* (Wahlenb.) Sm.

= *C.* x *stygia* Fr. has been recorded from No and Su.

31 x 37. *C. flacca* Schreb. x *C. montana* L.

has been recorded from Ge.

31 x 50. *C. flacca* Schreb. x *C. nigra* (L.) Reichard

= *C.* x *winkelmannii* Aschers. & Graebn. has been recorded from Ge and ?Po. Plants from BI suggested as this hybrid have turned out to be variants of *C. nigra*.

JERMY, A. C. and TUTIN, T. G. (1968). As above, p. 114.

35 × 34. *C. ericetorum* Poll. × *C. pilulifera* L.

= *C.* × *lackowitziana* A. R. Paul has been recorded from Ge.

36 × 34. *C. caryophyllea* Latourr. × *C. pilulifera* L.

= *C.* × *paulii* Aschers. & Graebn. has been recorded from Ge and Po.

37 × 34. *C. montana* L. × *C. pilulifera* L.

= *C.* × *ginsiensis* Waisbecker has been recorded from Hu.

36 × 35. *C. caryophyllea* Latourr. × *C. ericetorum* Poll.

= *C.* × *sanionis* K. Richt. has been recorded from Ge, Rs and Su.

35 × 37. *C. ericetorum* Poll. × *C. montana* L.

has been recorded from Au.

39 × 40. *C. digitata* L. × *C. ornithopoda* Willd.

= *C.* × *dufftii* Hausskn. has been recorded from Au, Fe, No, Rm and Su.

43 × 44. *C. atrata* L. × *C. norvegica* Retz.

= *C.* × *candriani* Kneuck. has been recorded from Au, Fe, Is, No and Su.

47 × 46. *C. acuta* L. × *C. elata* All.

   a. *C.* × *prolixa* Fr. (*C.* × *curtisii* Druce, *nom. nud.*).

   b. Putative hybrids are intermediate between the parents in their diagnostic characters, and show disturbed pollen mother cell meiosis which leads to a variable percentage of inviable pollen.

   c. There are old records for v.c. 1, 27 and 41; it has recently been confirmed from Walton Common, v.c. 28, and a putative hybrid from Reach, v.c. 29, has been studied by Faulkner. Jermy (1967) suggested that hybrids are at least partly fertile and more common than generally thought. The hybrid has also been recorded from Au, Cz, Da, Ge, Ho and Su.

   d. Faulkner synthesized hybrids, which resembled the wild putative hybrids in morphology and chromosome behaviour, and also made backcrosses.

   e. *C. acuta* $2n = 82–85$; *C. elata* $2n = 74–77$.

   f. FAULKNER, J. S. (1972). Chromosome studies on *Carex* section *Acutae* in north-west Europe. *Bot. J. Linn. Soc.*, **65**: 271-301.

     FAULKNER, J. S. (1973). Experimental hybridization of north-west European species in *Carex* section *Acutae* (Cyperaceae). *Bot. J. Linn. Soc.*, **67**: 233-253.

     JERMY, A. C. (1967). As above.

48 × 46. *C. aquatilis* Wahlenb. × *C. elata* All.

(= *C.* × *hibernica* auct., *non* A. Benn.) has been recorded on a number of occasions, but most or all of the records refer to *C.* × *hibernica* A. Benn.

(*C. aquatilis* x *C. nigra*, q.v.). *C. aquatilis* x *C. elata* has been recorded from Su.

BENNETT, A. (1897). Notes on British plants, 2. *Carex. J. Bot., Lond.*, 35: 244-252, 259-264.

JERMY, A. C. and TUTIN, T. G. (1968). As above, p. 138.

46 × 50. *C. elata* All. × *C. nigra* (L.) Reichard

a. *C.* x *turfosa* Fr.

b. This hybrid differs from *C. elata* in its non-tufted, more slender habit, narrower leaves, more slender spikes and shorter glumes and utricles. From *C. nigra* it differs in its taller stems, broader leaves and less slender spikes. The utricles are usually empty but some viable fruits are produced, despite irregular meiosis.

c. *C. elata* x *C. nigra* occurs mainly in fens and fen-carr in East Anglia and parts of Hb. There are records from near Winchester, v.c. 11, Wheatfen Broad, v.c. 27, Foulden Common and Breckles Heath, v.c. 28, Flemingstone Moors, v.c. 41, Newtown Butler, v.c. H33, and v.c. H38 and H39. Other records need confirming (e.g. v.c. 97) or are errors (e.g. v.c. 22), but Jermy (1967) suggested that the hybrid is more common than generally thought. Elsewhere the hybrid has been recorded from Au, Cz, Fe, Ga, Ge, He, Ho, No, Rm, Rs and Su.

d. Faulkner synthesized hybrids, which resembled the wild hybrids in morphology and chromosome behaviour, and also made reciprocal backcrosses to both parents. Seed-set in parental crosses varied from 0 to 93%, but there were no reciprocal differences. In the backcrosses seed-set was higher when either parent was used as female and the F_1 as male.

e. *C. elata* $2n = 74–77$; *C. nigra* $2n = 83–85$ (also 82); hybrid $n = 37–42$.

f. BOWEN, H. J. M. (1968). As above, p. 293.

FAULKNER, J. S. (1972). As above.

FAULKNER, J. S. (1973). As above.

JALAS, J. and HIRVELÄ, U. (1964). Notes on the taxonomy and leaf anatomy of *Carex elata* All., *C. omskiana* Meinsch. and *C.* x *turfosa* Fr. *Annls bot. fenn.*, 1: 47-54. FIG.

JERMY, A. C. (1967). As above.

47 × 48. *C. acuta* L. × *C. aquatilis* Wahlenb.

has been recorded from No, Rs and Su, and Jermy (1967) suggested that it is at least partly fertile and more common than generally thought.

JERMY, A. C. (1967). As above.

47 × 50. *C. acuta* L. × *C. nigra* (L.) Reichard

a. (? *C.* x *elytroides* Fr.).

b. This is said to be very distinct in appearance when fresh, but I have never seen it so. It is described as having rather glaucous, semi-cylindrical leaves, pale green utricles, and brown, orbicular, punctulate, flat fruits, some of which are apparently fertile.

c. It has been recorded on good authority from v.c. 8, 9, 11, 23, 37, 41, 49, 58, 64 and 108. In the localities in v.c. 58 and 108 no *C. acuta* was observed, but it usually occurs close to its parents. Jermy (1967) suggested it is more common than generally thought. It has also been recorded from Au, Cz, Fe, Ga, Ge, He, Ho, No, Rs and Su.

d. Faulkner synthesized hybrids, which agreed with the wild hybrids in their morphology and their irregular pollen mother cell meiosis, and also made reciprocal backcrosses to both parents. In the backcrosses seed-set was higher when either parent was used as female and the F_1 as male.

e. *C. acuta* $2n = 82–85$; *C. nigra* $2n = 83–85$ (also 82).

f. FAULKNER, J. S. (1972). As above.
FAULKNER, J. S. (1973). As above.
JERMY, A. C. (1967). As above.
LINTON, E. F. and MARSHALL, E. S. (1898). *Carex* sp. *Rep. B.E.C.*, **1**: 572-573.

47 × 51. *C. acuta* L. × *C. trinervis* Degl.

has been recorded from Ho.

47 × 52. *C. acuta* L. × *C. bigelowii* Torr. ex Schwein.

(?*C.* × *elytroides* Fr.) has been recorded from Rm.

48 × 49. *C. aquatilis* Wahlenb. × *C. recta* Boott

a. *C.* × *grantii* A. Benn.

b. This hybrid is intermediate between its parents to varying degrees, and many variants of it occur in its single known locality, some approaching one parent and some the other. It has shorter bracts and spikes than *C. aquatilis,* and the midrib of the female glumes is shorter. The hybrid is largely sterile, with a disturbed pollen mother cell meiosis. Faulkner (1972) found that *C. recta* itself shows a range of chromosome numbers and a degree of meiotic irregularity which suggest that it is not a true species, but of hybrid origin, at least in Br.

c. *C. aquatilis* × *C. recta* is known only on banks of the Wick River, v.c. 109.

d. Faulkner obtained hybrids from reciprocal crosses; seed-set was higher when *C. recta* was the male parent.

e. *C. aquatilis* $2n = 76–77$; *C. recta* $2n = 74$ (*c* 70, *c* 80, 84).

f. BENNETT, A. (1897). As above.
BURGES, R. C. L., NELMES, E. and WALLACE, E. C. (1948). *Carex aquatilis* Wahl. × *C. recta* Boott. *Rep. B.E.C.*, **13**: 373.
FAULKNER, J. S. (1972). As above.
FAULKNER, J. S. (1973). As above.

48 × 50. *C. aquatilis* Wahlenb. × *C. nigra* (L.) Reichard

a. *C.* × *hibernica* A. Benn.

b. This hybrid is intermediate between its parents and sterile. The stem-leaves and bracts are longer than in *C. nigra*, the leaves narrower and more glaucous than in *C. aquatilis*, and the spikes longer than in *C. nigra* but more slender and less tapering at the base than in *C. aquatilis*. The name *C.* x *hibernica* was wrongly attributed by Bennett to *C. aquatilis* x *C. elata*.

c. *C. aquatilis* x *C. nigra* occurs on river-banks, loch-shores and mountain mires (as on the Angus-Aberdeen tableland). There are records from v.c. 76, 88–90, 92, 96, 108, 109, H1, H2 and H16. It is also recorded from Fe, No, Rs and Su.

d. Faulkner obtained hybrids from reciprocal crosses, and also made backcrosses.

e. *C. aquatilis* $2n = 76–77$; *C. nigra* $2n = 83–85$ (also 82); hybrid $2n =$ approximately the mean of the two parents, but variable, and one triploid, $n = 56–70$, $2n = 120$, was found by Faulkner (1972) at Loch Dochart, v.c. 88.

f. BENNETT, A. (1897). As above.
FAULKNER, J. S. (1972). As above.
FAULKNER, J. S. (1973). As above.
JERMY, A. C. (1967). As above.
SCULLY, R. W. (1916). *Flora of County Kerry*, p. 329. Dublin.

48 × 52. *C. aquatilis* Wahlenb. × *C. bigelowii* Torr. ex Schwein.

a. *C.* x *limula* Fr.

b. Scottish plants involve the small variant of *C. aquatilis* that grows in alpine bogs at 2000–3000 ft. The stems are taller than in *C. bigelowii*, but curved as in that species. The floral and fruiting characters are intermediate between those of the parents; the spikes are more slender than those of *C. bigelowii* and the fruits sterile. In the wild the hybrid has a fairly distinctive habit.

c. The hybrids occur at around 2600–3000 ft in open stony ground with *C. bigelowii* in the Scottish mountains from near Dalwhinnie across to Lochnagar, v.c. 89, 90, 92 and 96. They are also recorded from Fe, No, Rs and Su.

d. Faulkner obtained hybrids from reciprocal crosses.

e. *C. aquatilis* $2n = 76–77$; *C. bigelowii* $2n = 68–71$.

f. FAULKNER, J. S. (1973). As above.

50 × 49. *C. nigra* (L.) Reichard × *C. recta* Boott

= *C.* x *spiculosa* Fr. has been recorded from Fe, No, Rs and Su. Plants resembling the Scandinavian material, but differing in that the glumes are not serrulate, were found on Harris, v.c. 110, by a stream at Loch Langavat, by W. S. Duncan in 1895, but have apparently not been seen since. The Scottish plants were described as *C.* x *spiculosa* var. *hebridensis* A. Benn., but Wilmott (1938) later decided they were not hybrids but a variety of *C. nigra* (*C. goodenoughii* var. *hebridensis* (A. Benn.) Wilmott). A. C. Jermy (*in litt.* 1973) agrees with Wilmott.

BENNETT, A. (1897). As above.
BENNETT, A. (1922). *Carex spiculosa* Fries. *Rep. B.E.C.* 6: 319.
DRUCE, G. C. (1932). x *C. spiculosa* var. *hebridensis* (A. Benn.). *Rep. B.E.C.*, 9: 675.
WILMOTT, A. J. (1938). *Carex spiculosa* var. *hebridensis* A. Benn. *J. Bot., Lond.,* 76: 137-141.

50 x 51. *C. nigra* (L.) Reichard x *C. trinervis* Degl.

= *C.* x *timmiana* P. Junge has been recorded from Da and Ho.

52 x 50. *C. bigelowii* Torr. ex Schwein. x *C. nigra* (L.) Reichard

a. *C.* x *decolorans* Wimm.
b. This hybrid is intermediate between the parents and sterile. Many of the Carices found in the area of Scotland occupied by this hybrid are attacked by gall-forming flies and the utricles are abnormal. This at times makes the identification of hybrids difficult.
c. *C. bigelowii* x *C. nigra* occurs in the Scottish mountains from near Dalwhinnie across to Lochnagar, v.c. 89, 90, 92 and 96, along with the two parents and often *C. aquatilis* x *C. bigelowii* also. It is also recorded from Au, Cz, Ge, No, Po and Su.
d. Faulkner synthesized hybrids, which agreed with the wild hybrids in morphology and their irregular pollen mother cell meiosis, and also made backcrosses. Most chromosome associations in the wild hybrids are bivalents, but univalents and chain-trivalents or even chains of more than three chromosomes have been found.
e. *C. bigelowii* $2n = 68-71$; *C. nigra* $2n = 83-85$ (also 82).
f. FAULKNER, J. S. (1972). As above.
FAULKNER, J. S. (1973). As above.

52 x 72. *C. bigelowii* Torr. ex Schwein. x *C. curta* Gooden.

was recorded from Ben Wyvis, v.c. 106, by E. S. Marshall in 1900, but on cultivation proved to be a variety of *C. curta*.
MARSHALL, E. S. (1901). As above.
MARSHALL, E. S. (1909). Notes on *Carex canescens* Lightf. *J. Bot., Lond.,* 47: 107-108.

55 x 54. *C. appropinquata* Schumach. x *C. paniculata* L.

a. *C.* x *solstitialis* Figert.
b. This hybrid differs from *C. appropinquata* in its tufted habit and in having broader leaves, stouter stems and denser (more contiguous but not larger) spikes. From *C. paniculata* it is distinguished by its more slender stems, less rough leaves and smaller inflorescence of the dark fuscous colour of that of *C. appropinquata*. The utricles are empty.
c. *C.* x *solstitialis* occurs with the parents in Upton and Wheatfen Broads, v.c.

27, and between Barton Mills and Icklingham, v.c. 26, and is also recorded from Cz, Da, Ge, Rm and Su.

d. None.

e. *C. appropinquata* (2n = 64); *C. paniculata* (2n = 60, 62, 64).

f. LOUSLEY, J. E. (1937). *C. paniculata* L. x *C. paradoxa* Willd. *Rep. B.E.C.*, **11**: 231-232.

56 x 54. *C. diandra* Schrank x *C. paniculata* L.

= *C.* x *beckmannii* Keck ex F. Schultz has been recorded from Au, Cz, Fe, Ge, Ho, Rm and Su.

57 x 54. *C. otrubae* Podp. x *C. paniculata* L.

has been recorded from near Tiverton, v.c. 4, near Chiddingstone, v.c. 16, and near Witley, v.c. 17, but none of these records can be substantiated. The latter two were confirmed by Kükenthal as *C. paniculata* x *C. vulpina*, before *C. otrubae* and *C. vulpina sensu stricto* were recognized as distinct species in Br, but the Chiddingstone plant (**CGE**) was later determined by Nelmes (1939) as *C. vulpina sensu stricto*. Further study is required of specimens supposedly of this hybrid. There are reliable records from Ho and Su.

BAKKER, D. and PLOEG, D. T. E. van der (1972). *Carex otrubae* Podp. x *Carex paniculata* L. nieuw voor Nederland. *Gorteria*, **6**: 21-24. FIG.

MARSHALL, E. S. (1897). A new British hybrid sedge from Surrey. *J. Bot., Lond.*, **35**: 491-492.

NELMES, E. (1939). Notes on British Carices, 4. *Carex vulpina* L. *J. Bot., Lond.*, **77**: 259-266.

54 x 58. *C. paniculata* L. x *C. vulpina* L. *sensu stricto*

= *C.* x *pseudovulpina* K. Richt. has been recorded from Ge and Rs. British records probably referred to *C. otrubae* x *C. paniculata* (q.v.).

60 x 54. *C. disticha* Huds. x *C. paniculata* L.

has been recorded from v.c. 37 but no specimen has been traced; the record is probably erroneous.

AMPHLETT, J. and REA, C. (1909). *The botany of Worcestershire*, p. 384. Birmingham.

69 x 54. *C. elongata* L. x *C. paniculata* L.

= *C.* x *fussii* Simonk. has been recorded from Ga, Ge and Rm.

70 x 54. *C. echinata* Murr. x *C. paniculata* L.

has been recorded from Ga.

54 × 71. *C. paniculata* L. × *C. remota* L.

a. *C.* × *boenninghausiana* Weihe. (Bönninghausen's Sedge).

b. This sterile hybrid varies in appearance between its two parents, but can always be recognized by its long, lax, densely tufted stems and long, narrow, dull green leaves. The spikes are numerous, with a few remote ones below and more congested ones above; they resemble those of *C. curta* rather than those of *C. remota*.

c. *C.* × *boenninghausiana* occurs in wet alder and willow swamps, often among the tussocks of *C. paniculata* or at the edge of runnels in the swamp, and tolerates shade. It is one of the most frequent British sedge hybrids, being scattered over most of BI (but not in CI). It is also recorded from Au, Cz, Da, Ga, Ge, Ho and Su.

d. None.

e. *C. paniculata* ($2n = 60, 62, 64$); *C. remota* $2n = 60, 62$.

f. PEARSALL, W. H. (1933). The British species of *Carex. Rep. B.E.C.*, **10**: 166-196.

   PEARSALL, W. H. (1934). As above.

   PERRING, F. H. and SELL, P. D. (1968). *Op. cit.*, p. 147.

   SYME, J. T. B., ed. (1870). *English Botany*, 3rd ed., **10**: 98-99, t. 1629. London. FIG.

72 × 54. *C. curta* Gooden. × *C. paniculata* L.

= *C.* × *ludibunda* Gay (*C.* × *silesiaca* Figert) has been recorded from Da, Fe, Ga and Ge. A single tuft growing with the supposed parents at Newbridge Bog, Ashdown Forest, v.c. 14, was found by F. Rose and J. R. Wallis in 1944 and identified by them as this hybrid. This determination was tentatively confirmed by E. Nelmes, but the plant could not be refound and the original specimen has not been traced (F. Rose *in litt.* 1970).

55 × 56. *C. appropinquata* Schumach. × *C. diandra* Schrank

= *C.* × *limnogena* Appel ex Kneuck. has been recorded from Au, Cz, Da, Fe and Su.

55 × 57. *C. appropinquata* Schumach. × *C. otrubae* Podp.

has been recorded from Su.

55 × 69. *C. appropinquata* Schumach. × *C. elongata* L.

has been recorded from Ge.

55 × 71. *C. appropinquata* Schumach. × *C. remota* L.

= *C.* × *rieseana* Figert has been recorded from Ge and Su.

55 × 72. *C. appropinquata* Schumach. × *C. curta* Gooden.
= *C.* × *schuetzeana* Figert has been recorded from Cz, Ge and No.

72 × 56. *C. curta* Gooden. × *C. diandra* Schrank
= *C.* × *limicola* H. Gross has been recorded from Ge.

57 × 58. *C. otrubae* Podp. × *C. vulpina* L. *sensu stricto*
was tentatively identified by Nelmes from the River Medway, near
Tonbridge, v.c. 16, and from Amberley Wild Brooks, v.c. 13, but the
records have not been confirmed. Such hybrids are likely to be found by
careful searching.
NELMES, E. (1939). As above.

65 × 57. *C. divulsa* Stokes × *C. otrubae* Podp.
(often as *C. divulsa* × *C. vulpina*) has been recorded from Br on several
occasions, but more study of the specimens is needed before the records
can be confirmed. There are several early records from v.c. 37 and one
from v.c. 35, but Druce (1910) and others doubted their correctness, and
G. Kükenthal identified at least one of them as *C. otrubae*. A recent
collection by R. W. David from v.c. 23 (**CGE**) is perhaps this hybrid (A. O.
Chater *in litt.* 1971).
BENNETT, A., BICKHAM, S. H., DRUCE, G. C., MARSHALL, E. S. and
    THELLUNG, A. (1915). *Carex divulsa* × *vulpina*. *Rep. B.E.C.*, **4**: 170.
DRUCE, G. C. (1910). Notes on British Carices. *J. Bot., Lond.*, **48**:
    98-101.
LINTON, E. F. (1907). Hybrids among British phanerogams. *J. Bot.,
    Lond.*, **45**: 296-304.
MARSHALL, E. S. (1903). *Carex divulsa* × *vulpina*. *Rep. B.E.C.*, **2**: 60-61.

57 × 67. *C. otrubae* Podp. × *C. spicata* Huds. (*C. contigua* Hoppe)
= *C.* × *haussknechtii* Senay (as *C. contigua* × *C. vulpina*) has been
occasionally recorded from Br, e.g. v.c. 6, 17 and 33, but the identity of
the voucher specimens has in each case been doubted. Specimens at **K**
from v.c. 17 and 33 are in my opinion examples of *C. spicata* and *C.
otrubae* respectively. *C.* × *haussknechtii* was described from Ga.
LITTLE, J. E., SALMON, C. E. and THOMPSON, H. S. (1926). *C.
    contigua* × *vulpina*. *Rep. Watson B.E.C.*, **3**: 355.
PUGSLEY, H. W. and THOMPSON, H. S. (1933). *Carex contigua* ×
    vulpina? Rep. Watson B.E.C., **4**: 191.

68 × 57. *C. muricata* L. (*C. pairaei* F. W. Schultz) × *C. otrubae* Podp.
(as *C. muricata* × *C. vulpina*) was thought by E. F. Linton to be the
parentage of a plant collected by E. S. Marshall in 1892 at Clymping, v.c.
13, but G. Kükenthal named it *C. otrubae*.
LINTON, E. F. (1907). As above.

57 × 71. *C. otrubae* Podp. × *C. remota* L.

    a. *C.* × *pseudoaxillaris* K. Richt. (*C.* × *axillaris* auct., *non* L.). (Axillary Sedge).

    b. This hybrid is intermediate between its parents and generally sterile. It differs from *C. otrubae* in the longer, laxer, narrower leaves, and distant spikes, of which the lower are often compound; and from *C. remota* in the stouter stems, longer, broader leaves and more robust, pale brown spikes.

    c. *C.* × *pseudoaxillaris* is locally frequent on ditch-sides and road-side banks near both parents throughout Br north to v.c. 83. It occurs in CI and is scattered in Hb. It is also reliably recorded from Ge and Ho.

    d. Faulkner studied meiosis in a wild hybrid from v.c. 23, and found up to 50 "chromosome bodies", many of which were obviously univalents.

    e. *C. otrubae* $2n = 58$ (60); *C. remota* $2n = 60, 62$.

    f. FAULKNER, J. S. (1972). As above.
       PEARSALL, W. H. (1933). As above.
       PEARSALL, W. H. (1934). As above.
       PERRING, F. H. and SELL, P. D. (1968). *Op. cit.*, p. 147.
       SYME, J. T. B., ed. (1870). *English Botany*, 3rd ed., **10**: 97-98, t. 1628. London. FIG.

71 × 58. *C. remota* L. × *C. vulpina* L. *sensu stricto*

    = *C.* × *axillaris* L. has been recorded from Au, Be, Cz, Ga and Ge. British records refer to *C. otrubae* × *C. remota* (q.v.).

61 × 71. *C. arenaria* L. × *C. remota* L.

    has been recorded from Ge.

67 × 62. *C. spicata* Huds. (*C. contigua* Hoppe) × *C. divisa* Huds.

    was recorded from an alluvial meadow by the River Avon, south of Sea Mills, v.c. 34, in 1921, but the specimens were not mature and there was disagreement as to the correct identity.
    BENNETT, A. and THOMPSON, H. S. (1922). ?*C. divisa* Huds. × *contigua* Hoppe. *Rep. Watson B.E.C.*, **3**: 186.

63 × 72. *C. chordorrhiza* L.f. × *C. curta* Gooden.

    = *C.* × *lidii* Flatb. was described from No in 1972.

81 × 64. *C. dioica* L. × *C. maritima* Gunn.

    = *C.* × *deinbolliana* Gay has been recorded from Fe, Is, No and Rs.

65 × 66. *C. divulsa* Stokes × *C. polyphylla* Kar. & Kir. (*C. muricata* subsp. *leersii* Aschers. & Graebn.)

    was recorded from Unhill Wood, v.c. 22, by Bowen (1968). The record was based on a specimen (**OXF**) collected by G. C. Druce in 1890 and identified as this hybrid by E. Nelmes. In view of the uncertainty of the nomenclature and taxonomy of *C. polyphylla* this record should be regarded as doubtful.

BOWEN, H. J. M. (1968). As above, p. 294.

65 × 67. *C. divulsa* Stokes × *C. spicata* Huds. (*C. contigua* Hoppe)

has been recorded from Ga and Ge. A specimen collected from near West Monkton, v.c. 5, was identified *in situ* as this hybrid by E. S. Marshall, and another from Wimbotsham, v.c. 28, by J. E. Little, but neither has been confirmed. Little (1931) later considered that the latter plant was in fact *C. divulsa* (written *C. divisa* in error).

BENNETT, A. and MARSHALL, E. S. (1915). *Carex contigua* Hoppe × *divulsa* Good.? *Rep. B.E.C.*, **4**: 170.

LITTLE, J. E. (1928). *Carex contigua* × *divulsa? Rep. B.E.C.*, **8**: 930.

LITTLE, J. E. (1931). *Carex contigua* × *divulsa? Rep. Watson, B.E.C.*, **4**: 56 (and see p. 104 for erratum).

LITTLE, J. E., THOMPSON, H. S. and WOLLEY-DOD, A. H. (1929). *Carex contigua* × *divulsa? Rep. Watson B.E.C.*, **3**: 494.

65 × 68. *C. divulsa* Stokes × *C. muricata* L. (*C. pairaei* F. W. Schultz)

was in the opinion of some workers the identity of a plant collected in 1919 in v.c. 28 by J. E. Little, but this was never confirmed.

LITTLE, J. E. *et al.* (1920). *Carex. Rep. Watson B.E.C.*, **3**: 120-121.

65 × 71. *C. divulsa* Stokes × *C. remota* L.

a. *C.* × *emmae* L. Gross.

b. The putative hybrid from Waldron differs from *C. remota* in the more rigid stems which are more scabrid above, and the stiffer leaves. Only the lowest spike has a long bract, and the glumes are longer in proportion to the utricles than in *C. remota,* but more serrulate than in *C. divulsa.* Fruits were not formed, and its identity was confirmed by G. Kükenthal.

c. The single British plant was on a roadside verge near Waldron, Heathfield, v.c. 14, where it was found in 1924 by C. E. Salmon. It persisted until at least 1935, when it was in imminent danger of being built upon (Wolley-Dod, 1937). There are also records from Cz and Ge.

d. None.

e. *C. divulsa* $2n = 58$; *C. remota* $2n = 60, 62$.

f. BENNETT, A. and SALMON, C. E. (1925). *C. remota* × *divulsa. Rep. Watson B.E.C.*, **3**: 315.

BENNETT, A., SALMON, C. E. and WOLLEY-DOD, A. H. (1925). *Carex divulsa* × *remota. Rep. B.E.C.*, **7**: 743.

SALMON, C. E. (1925). *Carex remota* × *divulsa. J. Bot., Lond.*, **63**: 140-141.

WOLLEY-DOD, A. H. (1937). *Flora of Sussex*, p. 482. Hastings.

65 × 74. *C. divulsa* Stokes × *C. ovalis* Gooden.
   has been recorded from Ge.

68 × 67. *C. muricata* L. (*C. pairaei* F. W. Schultz) × *C. spicata* Huds. (*C. contigua* Hoppe)
   has been recorded from Au.

71 × 67. *C. remota* L. × *C. spicata* Huds. (*C. contigua* Hoppe)
   (= *C.* × *pseudoaxillaris* auct., *non* K. Richt.) has been recorded from Br on a number of occasions, but all records are likely to be errors for *C. otrubae* × *C. remota*. There are records for *C. remota* × *C. spicata* as distinct from *C. otrubae* × *C. remota* from v.c. 3, 4 and 24, and from Cz and Ge, but all require confirmation.
   LINTON, E. F., MARSHALL, E. S. and RIDDELSDELL, H. J. (1910). *C. contigua* × *remota*? *Rep. B.E.C.*, 2: 477.
   MARTIN, W. K. and FRASER, G. T. (1939). *Flora of Devon*, p. 665. Arbroath.

68 × 71. *C. muricata* L. (*C. pairaei* F. W. Schultz) × *C. remota* L.
   has been recorded from central Europe, but requires confirmation; the occasional older British records referred to *C. otrubae* × *C. remota*.

69 × 71. *C. elongata* L. × *C. remota* L.
   = *C.* × *ploettnerana* R. Beyer has been recorded from Ge.

72 × 69. *C. curta* Gooden. × *C. elongata* L.
   (= *C.* × *helvola* Wimm., *non* Blytt) has been recorded from Ge.

70 × 71. *C. echinata* Murr. × *C. remota* L.
   = *C.* × *gerhardtii* Figert has been recorded from Au and Ge.

72 × 70. *C. curta* Gooden. × *C. echinata* Murr.
   a. *C.* × *biharica* Simonk.
   b. This hybrid is intermediate between the parents and sterile. The spikes are stouter and fewer-flowered than in *C. curta*, and the utricles slightly larger and more spreading, with a serrated beak. The hybrid differs from *C. echinata* in the narrower but longer spikes and the slightly smaller utricles. In the past it has been confused with *C. curta* × *C. lachenalii*, from which it is best distinguished by the serrated beak of the utricles.
   c. *C.* × *biharica* occurs on the higher ground on the Scottish mountains in v.c. 88, 90 and 92. It has also been recorded from Au, Cz, Fe, Ge, No, Po, Rm and Su.
   d. None.
   e. *C. curta* ($2n = 56$); *C. echinata* ($2n = 56, 58$).

f. DRUCE, G. C. (1898). On the occurrence of *Carex helvola*, Blytt, in Britain. *J. Linn. Soc.*, **33**: 458-464.

DRUCE, G. C. (1900). *C. helvola*, Blytt. *Ann. Scot. nat. Hist.*, **36**: 232-233.

DRUCE, G. C. (1909). *Carex helvola*, Blytt, on Ben Lawers. *Ann. Scot. nat. Hist.*, **72**: 238-241.

DRUCE, G. C., LINTON, E. F. and MARSHALL, E. S. (1898). *C. helvola*, Blytt. *Rep. B.E.C.*, **1**: 571-572.

MARSHALL, E. S. (1909). As above.

81 × 70. *C. dioica* L. × *C. echinata* Murr.

a. *C.* × *gaudiniana* Guthn.

b. This hybrid is intermediate between the parents and sterile. It is a tall plant with very narrow leaves and usually one male spike with two female spikes below it.

c. It is known in a marsh 2.5 miles from Louisburgh, v.c. H27, where it was found in 1942 by A. W. Stelfox (**BM, K**), and in a bog at Mynydd Hiraethog, v.c. 50 (I. R. Bonner, 1970, conf. P. M. Benoit). It has also been recorded from Au, Cz, Ge, He, No, Po and Su.

d. None.

e. *C. dioica* $2n = 52$; *C. echinata* ($2n = 56, 58$).

f. None.

82 × 70. *C. davalliana* Sm. × *C. echinata* Murr.

= *C.* × *paponii* Muret ex Durand & Pitt. has been recorded from Au and He.

72 × 71. *C. curta* Gooden. × *C. remota* L.

= *C.* × *arthuriana* Beckm. & Figert has been recorded from Cz, Da, Ge, He, No, Rs and Su. A plant from near Chertsey, v.c. 17, was queried as this hybrid by Marshall (1909).

MARSHALL, E. S. (1909). As above.

74 × 71. *C. ovalis* Gooden. × *C. remota* L.

= *C.* × *ilseana* Ruhm. has been recorded from Ge and He.

72 × 73. *C. curta* Gooden. × *C. lachenalii* Schkuhr

a. *C.* × *helvola* Blytt ex Fr.

b. This hybrid is sterile and intermediate between its parents. It is somewhat variable in its Scottish stations, and has been confused with *C. curta* x *C. echinata* (q.v.). It differs from *C. lachenalii* in its more numerous, somewhat longer spikes which are more slender and less congested, and from *C. curta* in its more robust, darker coloured spikes. The utricles have a smooth beak.

c. *C.* x *helvola* occurs for certain only in mires on rocky slopes on the mountains around Lochnagar, v.c. 92, and there is a very old unconfirmed record from Ben Macdhui. Specimens named *C.* x *helvola* and collected on Ben Lawers, v.c. 88, are probably *C. curta* x *C. echinata*, since *C. lachenalii* is not known in that area. *C.* x *helvola* has also been recorded from Fe, Is, No, Rs and Su.

d. None.

e. *C. curta* ($2n = 56$); *C. lachenalii* ($2n = 58, 64$).

f. BENNETT, A. (1886). *Carex helvola* Blytt, in Scotland. *Trans. bot. Soc. Edinb.*, **16**: 361-362.

BENNETT, A., LINTON, E. F. and MARSHALL, E. S. (1910). *C. helvola* Blytt. *Rep. Watson B.E.C.*, **2**: 263.

DRUCE, G. C. (1898). As above.

DRUCE, G. C. (1900). As above.

DRUCE, G. C. (1909). As above.

LOUSLEY, J. E., MACKECHNIE, R., WALLACE, E. C. and WILMOTT, A. J. (1935). x *Carex helvola* Blytt. *Rep. B.E.C.*, **9**: 994.

MARSHALL, E. S. (1909). As above.

72 x 81. *C. curta* Gooden. x *C. dioica* L.

= *C.* x *microstachya* Ehrh. has been recorded from Au, Fe, Ge, Is, No, Po, Rs and Su.

81 x 73. *C. dioica* L. x *C. lachenalii* Schkuhr

(= *C.* x *gaudiniana* Blytt, *non* Guthn.) has been recorded from Fe, No and Su.

82 x 81. *C. davalliana* Sm. x *C. dioica* L.

= *C.* x *figertii* Aschers. & Graebn. has been recorded from Au, Cz, Ge and Po.

664–720. Gramineae

There is far more literature concerning hybrids in the Gramineae than those in any other group of plants, partly because of the widespread occurrence of hybridization but largely due to the great economic significance of grasses (including hybrids). Hybrid grasses are extensively used for grain (especially wheat) and fodder (e.g. *Festuca* and *Lolium*), and many have been bred and studied in agricultural establishments. Those grown commercially often escape into the wild, or remain as relics of cultivation, and in some cases the natural distribution of the hybrids has become obscured.

Apart from numerous interspecific hybrids within most of the larger (and many smaller) genera, hybrids occur between British species in six intergeneric combinations, although none of the hybrids found in the wild embraces more than one tribe. The great majority of grass hybrids are wholly or largely sterile; certain *Bromus* and *Lolium* hybrids are notable exceptions.

Of the many extensive reviews of grass hybrids, both natural and artificial, the following are of great value. The most complete work (Knobloch, 1968) should be used with caution as no distinction is made between spontaneous and artificial crosses; even so the book is indispensable to students of grass hybrids.

General References

CAMUS, A. (1958). Graminées hybrides de la flore française (genre *Bromus* excepté). *Bull. Jard. bot. Etat Brux.*, **28**: 337-374.

HUBBARD, C. E. (1968). *Grasses*, 2nd ed. Harmondsworth.

KNOBLOCH, I. W. (1968). *A check list of crosses in the Gramineae.* E. Lansing, Michigan, U.S.A.

ULLMANN, W. (1936). Natural and artificial hybridisation of grass species and genera. *Herbage Rev.*, **4**: 105-127.

669. *Glyceria* R.Br.

(by M. Borrill)

1 × 2. G. *fluitans* (L.) R.Br. × G. *plicata* Fr.

a. *G. × pedicillata* Townsend. (Hybrid Sweet-grass).

b. This hybrid shows a wide range of plants intermediate between the parents; indeed some of the more extreme variants may resemble one or other parent very closely and are usually recognized by the fact that they are completely sterile, with persistent spikelets on the mature culms. The anthers are shrunken and the pollen grains completely empty. No caryopsis development occurs. To identify *G. × pedicillata* with certainty the inflorescences should be young enough to contain anthers, or old enough for the persistent nature of the spikelets to be obvious.

c. The hybrid is stoloniferous, like *G. plicata*, and occurs sporadically in aquatic or paludal habitats over most of BI. It ranges from small colonies by ponds or at bends of streams to large stands in shallow water; and can flourish in swiftly flowing streams. It is often found in the absence of both

parents, and is a classic example of a successful, vegetatively-propagated, sterile hybrid. On the Continent it occurs from No and Su to Ga and Cz.
d. Attempts to synthesize this hybrid have not yet been successful. Morphological variation between collections indicates that it is polytopic in origin, rather than a single, sterile clone.
e. Both parents and hybrid $2n = 40$.
f. BORRILL, M. (1956). A biosystematic study of some *Glyceria* species in Britain, 1. Taxonomy. *Watsonia*, 3: 299-306.
   HUBBARD, C. E. (1968). As above, pp. 120-121. FIG.
   PERRING, F. H. and SELL, P. D. (1968). *Op. cit.*, p. 148.
   TOWNSEND, F. (1853). On a supposed new species of *Glyceria*. *Trans. bot. Soc. Edinb.*, 4: 27-30.

3 × 1. *G. declinata* Bréb. × *G. fluitans* (L.) R.Br.

a. None.
b. This little-known hybrid is intermediate in features between the parents, and may be recognized by its persistent spikelets, blunt lemmas 5.0-5.5 mm long, and sterile anthers 0.5-1.8 mm long. The hybrid would be expected to be a triploid, $2n = 30$, but this number has not yet been found in any *Glyceria*; until it has the existence of this hybrid must remain doubtful.
c. According to Hubbard (1968) this hybrid occurs in Br but is "rather rare". There are specimens in K determined by him from v.c. 1, 8, 17, 22 and 64 (C. E. Hubbard *in litt.* 1973). It has also been reported from Su.
d. Attempts to synthesize this hybrid have not yet been successful.
e. *G. declinata* $2n = 20$; *G. fluitans* $2n = 40$.
f. BORRILL, M. (1958). A biosystematic study of some *Glyceria* species in Britain, 4. Breeding systems, fertility relationships and general discussion. *Watsonia*, 4: 89-100.
   HUBBARD, C. E. (1968). As above, p. 117.

1 × 4. *G. fluitans* (L.) R.Br. × *G. maxima* (Hartm.) Holmberg

= *G.* x *digenea* Domin has been reported from Cz, but needs checking. The plant collected by Druce in v.c. 22 (Druce, 1932) and said by W. O. Howarth to have been "possibly a hybrid" between *G. maxima* (which it resembled) and "one of the forms of *G. fluitans*, such as *G. plicata*," has been examined by C. E. Hubbard (Lambert, 1947) and found to show no evidence of hybrid origin.
DRUCE, G. C. (1932). *Glyceria. Rep. B.E.C.*, 9: 678.
LAMBERT, J. M. (1947). *Glyceria maxima* (Hartm.) Holmb., in Biological Flora of the British Isles. *J. Ecol.*, 34: 310-344.

669 × 671. *Glyceria* R.Br. × *Lolium* L.
(by M. Borrill)

669/1 × 671/1. *Glyceria fluitans* (L.) Br. × *Lolium perenne* L.
has been reported from Su, but probably erroneously so.

669/2 × 671/1. *Glyceria plicata* Fr. × *Lolium perenne* L.
was recorded from v.c. 13 by Druce, and E. S. Marshall thought the identification "most likely correct". Specimens examined by M. H. Sutton, however, were identified as x *Festulolium loliaceum*, and it is most unlikely that hybrids between species of *Glyceria* and *Lolium* could occur. DRUCE, G. C., MARSHALL, E. S. and SUTTON, M. H. (1918). *L. perenne* x *Glyceria plicata. Rep. B.E.C.*, 5: 260-261.

670. *Festuca* L.
(by M. Borrill)

2 × 1. *F. arundinacea* Schreb. × *F. pratensis* Huds.
a. *F.* x *aschersoniana* Dörfl.
b. This rarely-reported hybrid may vary a little in the direction of *F. pratensis*, but mostly resembles *F. arundinacea*. Plants apparently of *F. arundinacea* but with sterile anthers might, after careful study, be assigned to this hybrid. The hybrids are sterile and exhibit a complex pairing of chromosomes at meiosis.
c. The characteristic habitats of *F. pratensis* (old pastures, water-meadows, and old, low-lying grassland) are shared by certain robust races of *F. arundinacea*. The hybrid has been found in Br by C. E. Hubbard (*in litt.* 1970), and there is a specimen in K from v.c. 98 determined by him (C. E. Hubbard, *in litt.* 1973). There are also records from Cz, Da, Fe, Ge, It, No, Rs and Su.
d. Artificial hybrids were made by Jenkin (1955a) and Malik (1967). Very variable numbers of hybrids were obtained in both directions of crossing. All the 44 hybrids obtained were completely male-sterile, and set virtually no seed on backcrossing or out-crossing. Hybrid morphology is unaffected by the direction of cross, and more definitely resembles the *F. arundinacea* parent in the coarseness of the leaf and in the inflorescence branching. Auricle-pubescence is somewhat less than in typical *F. arundinacea*. Jenkin

concluded that "If found in nature these plants would inevitably be classified as *F. arundinacea* or a form of that species".

e. *F. pratensis* $2n = 14$; *F. arundinacea* $2n = 42$; artificial hybrids ($2n = 28$, 35).

f. JENKIN, T. J. (1955a). Interspecific and intergeneric hybrids in herbage grasses, 9. *F. arundinacea* with some other *Festuca* species. *J. Genet.*, **53**: 81-93.

   MALIK, C. P. (1967). Hybridisation of *Festuca* species. *Canad. J. Bot.*, **45**: 1025-1029.

   MALIK, C. P. and THOMAS, P. T. (1967). Cytological relationship and genome structure in some *Festuca* species. *Caryologia*, **20**: 1-39.

2 × 3 × 1. *F. arundinacea* Schreb. × *F. gigantea* (L.) Vill. × *F. pratensis* Huds.

arose in Su from open-pollination of a plant of *F. arundinacea* x *F. gigantea* which had been collected in the wild and grown in cultivation.

3 × 1. *F. gigantea* (L.) Hill × *F. pratensis* Huds.

a. *F. x schlickumii* Grantz.

b. This hybrid is said to be intermediate between the parents, although the results of artificial crosses indicate that there is often a stronger resemblance to *F. gigantea*. Experience shows that such hybrids look like a morphologically "diluted" *F. gigantea*. The plants are male-sterile with non-dehiscent anthers, and set very little or no viable caryopses.

c. The rarity of the hybrids is not surprising in view of the distinct ecological preferences of the parents. *F. gigantea* is found mostly in damp, open woodlands and shady places, while *F. pratensis* is often abundant in water-meadows, low-lying ground, old pastures, and by road-sides. Two or three plants collected in v.c. 28 in 1918 by F. Robinson were identified as this hybrid by G. C. Druce, and C. E. Hubbard in 1952 also found the hybrid at two places in v.c. 28. Elsewhere it has been reported from Au, Cz, Da, Ga, Ge, Ho, Po, No, Rm and Su.

d. Jenkin (1955b) has obtained artificial hybrids and backcross derivatives. With *F. gigantea* as female the cross was easily made (seed-set 61%, germination 43%), but with *F. pratensis* as female few hybrids were obtained. Jenkin found that "Owing to their stout stems, broad leaves, drooping inflorescences, and awned florets the F_1 hybrids were all more suggestive of *F. gigantea* than *F. pratensis*, yet all these characters were less pronounced than in *F. gigantea*". Fertility was low in the F_1 hybrids, which had non-dehiscent anthers. Backcrossing of the parents to F_1 plants used as females gave a few hybrid-derivatives which also had a low fertility. Even when the species meet the chances of successful cross-fertilization are low due to self-fertility in the parents, particularly *F. gigantea* which was the more successful female in the artificial crosses.

e. *F. pratensis* $2n = 14$; *F. gigantea* $2n = 42$.

f. DRUCE, G. C. (1919). *F. elatior* x *gigantea* = x *F. schlickumi* Grantzow. *Rep. B.E.C.*, 5: 315-316.

JENKIN, T. J. (1955b). Interspecific and intergeneric hybrids in herbage grasses, 1. Some of the breeding interactions of *F. gigantea*. *J. Genet.*, 53: 94-99.

PETCH, C. P. and SWANN, E. L. (1968). *Flora of Norfolk*, p. 250. Norwich.

5 x 1. *F. heterophylla* Lam. x *F. pratensis* Huds.

= *F.* x *wippraensis* K. Wein has been reported from Ge, but needs careful checking.

1 x 6. *F. pratensis* Huds. x *F. rubra* L.

= *F.* x *hercynica* K. Wein has been reported from Ga, Ge and Rm, but needs careful checking.

8 x 1. *F. ovina* L. x *F. pratensis* Huds.

= *F.* x *pseudofallax* K. Wein has been reported from Ge, but needs careful checking.

2 x 3. *F. arundinacea* Schreb. x *F. gigantea* (L.) Vill.

a. *F.* x *gigas* Holmberg.

b. The parents differ appreciably in appearance, and the hybrids are variably intermediate. Hybridization tends to produce a "diluted" *F. gigantea* in appearance, particularly with respect to the shorter awns, but the *F. arundinacea* parent contributes a positive feature, i.e. minute hairs on the auricles. Consequently a sterile plant suspected to be hybrid on account of general intermediacy of panicle structure and habit might be confirmed as such by presence of these hairs.

c. Both parents are common grasses, but since they usually occupy quite distinct habitats the scarcity of reputed hybrids is probably to be expected. The hybrid is probably reliably reported from v.c. 31 and 101 (det. C. E. Hubbard) and v.c. 95 and 96 (det. A. Melderis), and there are also records from Au, Cz, Da, Ge, Ho, Po and Su.

d. Jenkin (1933) crossed a single plant of each species. With *F. gigantea* as the female parent 74% of the florets developed caryopses; of the 77 caryopses obtained 90.9% were fully developed and 89.5% germinated. There was only a small seed-set with *F. arundinacea* as the female parent but seven plants were obtained. All the hybrids were vigorous but sterile. It appears that these species can be readily intercrossed by artificial pollination following emasculation, but that in the wild crossing is curtailed by the appreciable self-fertility of *F. gigantea*. Among the progeny of a wild-collected hybrid from Su Nilsson found an amphidiploid with $2n = 84$ and over 50% pollen fertility. Five artificial F_1 hybrids were examined by Peto, who concluded that there was "a fairly consistent failure of pairing between seven chromosomes from each parent."

e. Both parents and hybrid $2n = 42$.

f. JENKIN, T. J. (1933). Interspecific and intergeneric hybrids in herbage grasses. Initial crosses. *J. Genet.*, **28**: 205-264.

MALIK, C. P. and THOMAS, P. T. (1966). Chromosomal polymorphism in *Festuca arundinacea. Chromosoma*, **18**: 1-18.

NILSSON, F. (1935). Amphipolyploidy in the hybrid *Festuca arundinacea x gigantea. Hereditas,* **20**: 181-198. FIG.

PETO, F. H. (1933). The cytology of certain intergeneric hybrids between *Festuca* and *Lolium. J. Genet.*, **28**: 113-156.

3 × 6. *F. gigantea* (L.) Vill. × *F. rubra* L.

= *F.* x *haussknechtii* Torges has been reported from Cz, but needs careful checking.

5 × 6. *F. heterophylla* Lam. × *F. rubra* L.

= *F.* x *napecae* Prodan has been reported from Rm.

5 × 8. *F. heterophylla* Lam. × *F. ovina* L.

= *F.* x *osswaldii* K. Wein has been reported from Ge.

7 × 6. *F. juncifolia* St. Amans (*F. arenaria* Osbeck, *sec.* Kjellqvist) × *F. rubra* L.

has been occasionally reported from the Atlantic coast of Europe, including Br, but the determinations are all very doubtful.

8 × 6. *F. ovina* L. × *F. rubra* L.

= *F.* x *zobelii* K. Wein has been reported from Au, Ga and Ge.

11 × 6. *F. longifolia* Thuill. × *F. rubra* L.

has been reported from Rs.

12 × 6. *F. glauca* Lam. × *F. rubra* L.

= *F.* x *wettsteinii* Vetter has been reported from Au.

8 × 9. *F. ovina* L. (incl. *F. vivipara* (L.) Sm.) × *F. tenuifolia* Sibth.

a. None.

b. Hybrids are intermediate between the parents, which are themselves often difficult to separate. They are either sexual, in which case the inflorescences are usually male-sterile, or viviparous. Watson (1958) studied the concordance of taxa with chromosome races, and concluded that *F. tenuifolia* is typically diploid, and that the tetraploids should be assigned to *F. ovina.* Wild plants can intergrade morphologically, although in cultivation a quantitative separation on the basis of general appearance does provide a fairly accurate guide to their chromosome number.

c. Natural sexual triploid hybrids were found on cliffs at the southernmost tip of the Mull of Kintyre, v.c. 101. Viviparous triploid hybrids were also observed there (P. J. Watson *in litt.*1971). The only sexual plant reported in that part of Scotland is diploid, though tetraploid viviparous plants occur in the vicinity.

d. Hybridization experiments were carried out by bagging together unemasculated diploid and tetraploid sexual plants. Self-fertility was assessed by bagging heads separately at the same time. Diploids were self-sterile, tetraploids moderately self-fertile. With a diploid as female all the progeny were triploid hybrids. These grew well and tended to resemble the tetraploid parent; they produced apparently normal sexual panicles, but few viable pollen grains. With a tetraploid as female all progeny were tetraploid, and could have been selfs or hybrids. There was some evidence that the triploids could produce caryopses by backcrossing to either parent. The presence of triploid hybrids in areas containing sexual diploids but only viviparous tetraploids might be explained by the occasional production of incomplete flowers on some viviparous plants and the formation of some tetraploid pollen in an unusual season (Watson, 1958).

e. *F. ovina* $2n = 28$; *F. tenuifolia* $2n = 14$; wild hybrid $2n = 21$; artificial hybrid $2n = 21$, ?28.

f. HORANSZKY, A. (1972). Problems in biosystematic studies of Hungarian *Festuca ovina sens. lat.* representatives. *Symp. Biol. Hung.,* **12**: 177-182.

WATSON, P. J. (1958). Distribution in Britain of diploid and tetraploid races within the *Festuca ovina* group. *New Phytol.,* **57**: 11-18.

12 × 8. *F. glauca* Lam. × *F. ovina* L.

= *F.* x *duernsteinensis* Vetter has been reported from Au.

670 × 671. *Festuca* L. × *Lolium* L.
= × *Festulolium* Aschers. & Graebn.

(by E. J. Lewis)

670/1 × 671/1. *Festuca pratensis* Huds. × *Lolium perenne* L.

a. x *F. loliaceum* (Huds.) P. Fourn. (Hybrid Fescue).

b. Hybrids, while variable in their degree of resemblance to either parent, exhibit some of the morphological characters of both species and are fairly readily recognizable in the reproductive stage. In general they resemble *L. perenne* in having the spikelets arranged alternately in two rows on opposite sides of an axis which is more frequently unbranched than branched. The spikelets, which are borne on short pedicels or may be nearly sessile, resemble those of *F. pratensis* but are more compressed and have markedly unequal glumes, the lower one being much reduced and occasionally absent. The anthers are non-dehiscent and contain mostly empty pollen grains.

c. This hybrid is the commonest x *Festulolium,* and may be found in old pastures and meadows, in water-meadows and on road-sides through old grassland, usually on rich, heavy soils. It has been recorded from most lowland districts of England but is less frequent in Wales and Hb and is rather rare in Scotland, being unknown north of v.c. 85. It is also widely found in Europe from No and Su to It and central Europe.

d. Results from the extensively investigated artificial cross indicate that there are large genotypic, and reciprocal, differences in crossability and that compatible combinations can give rise to large numbers of hybrids. These have non-dehiscent anthers containing very few viable pollen grains. They are also highly female-sterile although some workers have reported a low degree of success on backcrossing to both parents. Hybrids with artificially doubled chromosome numbers are relatively fertile.

e. Both parents $2n = 14$; wild hybrid $2n = 14, 21$.

f. CAMUS, A. (1958). As above.

GYMER, P. T. (1971). *The nature of hybrids between Lolium perenne L. and Festuca pratensis Huds.* Ph.D. Thesis, University of Nottingham.

GYMER, P. T. and WHITTINGTON, W. J. (1973a). Hybrids between *Lolium perenne* L. and *Festuca pratensis* Huds., 1. Crossing and incompatibility. *New Phytol.,* **72**: 411-424.

GYMER, P. T. and WHITTINGTON, W. J. (1973b). Hybrids between *Lolium perenne* and *Festuca pratensis,* 2. Comparative morphology. *New Phytol.,* **72**: 861-865.

HUBBARD, C. E. (1968). As above, pp. 148-149. FIG.

PERRING, F. H. and SELL, P. D. (1968). *Op. cit.,* p. 148.

TERRELL, E. E. (1966). Taxonomic implications of genetics in ryegrasses (*Lolium*). *Bot. Rev.,* **32**: 138-164.

670/1 x 671/2. *Festuca pratensis* Huds. x *Lolium multiflorum* Lam.

a. x *F. braunii* (K. Richt.) A. Camus.

b. This rarely reported hybrid can be distinguished from x *F. loliaceum* by its awned lemmas. Experience with the artificial hybrid shows that it can be variably intermediate between the parents in gross morphology. In general, the inflorescence resembles that of *L. multiflorum* in structure but the spikelets tend to be more widely spaced on the rhachis and some branching may occur in the basal region. The spikelets are subtended by two glumes, the lower one being much reduced and occasionally absent. The hybrids are functionally male-sterile.

c. The hybrid has been recorded only in v.c. 23 and 28 (both det. C. E. Hubbard), and must be assumed to be of rare occurrence. Specimens are most likely to be found where the two species are cultivated. It has also been recorded in Ga, Ge, He, Ho, Lu and Su.

d. Experimental hybridization of the two species shows their breeding affinity to be on a par with that found between *F. pratensis* and *L. perenne*. I have found appreciable genotypic and reciprocal differences in crossability, with better seed-setting and viability being obtained when *L. multiflorum* was the female parent. Similar results were also obtained when artificial tetraploid plants of the species were used. Hybrids from the former cross are male-sterile whereas the tetraploid hybrids are relatively fertile.

e. Both parents $2n = 14$; artificial hybrid ($2n = 14$).

f. CAMUS, A. (1958). As above.

   HUBBARD, C. E. (1968). As above, p. 149.

   TERRELL, E. E. (1966). As above.

670/2 × 671/1. *Festuca arundinacea* Schreb. × *Lolium perenne* L.

a. × *F. holmbergii* (Dörfl.) P. Fourn.

b. Hybrids resemble *F. arundinacea* in gross vegetative and reproductive morphology. The panicle may be less branched and the spikelets almost sessile giving the former a slightly more compact appearance than in *F. arundinacea*. The auricles are ciliate as in the latter species. The hybrids are male-sterile.

c. Hybrids have only been recorded in v.c. 6, 11, 17, 23 (all det. A. Melderis or C. E. Hubbard) and 33 (det. D. M. Barling). They are possibly fairly rare but their resemblance to *F. arundinacea* would make them difficult to recognize in nature. The two species are frequently sympatric. Hybrids have also been recorded in Su.

d. Experimental work shows that while relatively good seed-set can be obtained when the cross is performed in either direction, viable caryopses can only be obtained with *L. perenne* as female parent. Initially the seedlings tend to lack vigour but when well established they are quite robust. They are functionally male-sterile. A low degree of backcrossing to both parental species is possible.

e. *F. arundinacea* $2n = 42$; *L. perenne* $2n = 14$; artificial hybrid ($2n = 28$).

f. HUBBARD, C. E. (1968). As above, pp. 145, 149.

   JENKIN, T. J. (1955a). Interspecific and intergeneric hybrids in herbage grasses, 17. Further crosses involving *Lolium perenne*. *J. Genet.*, **53**: 442-466.

   ULLMANN, W. (1936). As above.

670/2 × 671/2. *Festuca arundinacea* Schreb. × *Lolium multiflorum* Lam.

a. None.

b. This hybrid differs from other × *Festulolium* hybrids in having both ciliate auricles and markedly awned lemmas. Natural hybrids, obtained from caryopses collected in Ga, resembled *F. arundinacea* in general habit but their panicles had a more compact appearance due to reduced branching and shorter internodes. The mean number of florets per spikelet and the spikelet-length were almost identical to those of *F. arundinacea*. All the hybrids had non-dehiscent anthers.

c. The hybrid is possibly not as rare as the paucity of records would suggest; it has been found in v.c. 10 and v.c. 23 (C. E. Hubbard, *in litt.* 1970). Beddows (1964) reported that, amongst inflorescences of *L. multiflorum* collected for breeding purposes in south-western Ga, two gave families of seedlings which consisted mainly of this hybrid (80% and 95%). It has also been recorded in Au and Da.

d. This hybrid has been extensively investigated by plant breeders on account of its potential agricultural value. Pronounced genotypic and reciprocal differences in crossability are found, with compatible combinations giving rise to large numbers of progeny when *L. multiflorum* is used as female parent. The hybrids are sterile but relatively fertile material can be obtained by artificially doubling either the chromosome number of the hybrids or that of the parents prior to crossing.

e. *F. arundinacea* $2n = 42$; *L. multiflorum* $2n = 14$; artificial hybrid ($2n = 28$).

f. BEDDOWS, A. R. (1964). *Lolium multiflorum* Lam. × *Festuca arundinacea* Schreb. Natural and artificial hybrids. *J. Linn. Soc., Bot.,* **59**: 89-98.

HUBBARD, C. E. (1968). As above, p. 149.

LEWIS, E. J. (1966). The production and manipulation of new breeding material in *Lolium—Festuca. Proc 10th Int. Grassland Congr.,* pp. 688-693. Helsinki.

670/3 × 671/1. *Festuca gigantea* (L.) Vill. × *Lolium perenne* L.

a. × *F. brinkmannii* (A. Braun) Aschers. & Graebn.

b. This rare hybrid is described as having *Lolium*-like racemes, awned lemmas and non-dehiscent anthers. Jenkin's artificial hybrids resembled their *F. gigantea* pollen-parent in habit, the stems being stout and the leaves broad and thin. The inflorescences were large, lax and branched, but were denser and less elaborately branched than in the male parent. The spikelets resembled those of *F. gigantea* but the awns were shorter. The anthers were non-dehiscent.

c. This hybrid is rare in Br and has been reported only on a few occasions elsewhere in Da, Ga, Ge and Su. Generally, the habitat preferences of *F. gigantea* and *L. perenne* are quite distinct, which thus reduces the opportunities for interbreeding. The hybrid has been recorded from Haverfordwest, v.c. 45.

d. Data on artificial crosses are insufficient for generalizations on the breeding affinity of the two species. Jenkin (1955) obtained caryopses from the cross made in both directions, but only when *L. perenne* was the female parent were these viable. The seedlings were quite vigorous and developed into mature plants. No caryopses were obtained from attempts to backcross one of these plants to both parental species. The self-fertile and inbreeding nature of *F. gigantea* helps to ensure the rarity of hybrids in the wild.

e. *F. gigantea* $2n = 42$; *L. perenne* $2n = 14$.

f. ANONYMOUS (1972). Plant Records. *B.S.B.I. News,* 1: 17.
   CAMUS, A. (1958). As above.
   HUBBARD, C. E. (1968). As above, pp. 147, 149.
   JENKIN, T. J. (1955a). As above.
   ULLMANN, W. (1936). As above.

670/3 × 671/2. *Festuca gigantea* (L.) Vill. × *Lolium multiflorum* Lam.

= × *F. nilssonii* Cugnac & A. Camus has been recorded once from Su. It was included for BI by Dandy, but I know of no other reports of its occurrence in this country.
DANDY, J. E. (1958). *Op. cit.,* p. 158.

670/6 × 671/1. *Festuca rubra* L. × *Lolium perenne* L.

a. × *F. frederici* Cugnac & A. Camus.

b. Of the two subspecies of *F. rubra* which are indigenous to this country the rhizomatous subsp. *rubra* is the more common and more likely to be involved. According to Jenkin (1955) artificial hybrids, which are male-sterile, agree in every morphological feature with *F. rubra* except that their rhizomes are much less well developed.

c. The single reported British wild specimen was found in a meadow at Borrowdale, v.c. 70, by Miss G. A. Hayes in 1956 (det. A. Melderis). The two parental species frequently occur together, particularly in old grazed pastures, and on the basis of results from artificial crosses it would be expected that they would occasionally hybridize in nature. Elsewhere there is a single record from Su.

d. In artificial crosses where *L. perenne* was the female parent the seed-set and -viability was variable but relatively good. Seed-set was lower in the reciprocal cross and there was no germination. All the hybrids were both male- and female-sterile, the anthers in some plants being very weakly developed and sometimes abortive.

e. *F. rubra* L. subsp. *rubra* $2n = 42$, 56; *L. perenne* $2n = 14$; wild hybrid ($2n = 28$).

f. BANGERTER, E. B. (1957). *Festuca rubra* x *Lolium perenne. Proc. B.S.B.I.,* 2: 381.

JENKIN, T. J. (1955b). Interspecific and intergeneric hybrids in herbage grasses, 15. The breeding affinities of *Festuca rubra. J. Genet.,* 53: 125-130.

NILSSON, F. (1933). Ein spontaner Bastard zwischen *Festuca rubra* und *Lolium perenne. Hereditas,* 18: 1-15.

670 × 672. *Festuca* L. × *Vulpia* C. C. Gmel. = × Festulpia Melderis ex Stace & Cotton

(by A. J. Willis)*

670/6a × 672/1. *Festuca rubra* L. subsp. *rubra* × *Vulpia membranacea* (L.) Dumort.

a. x *F. hubbardii* Stace & Cotton

b. Hybrids, which flower freely, are variously intermediate between the parents, but are distinguished by their perennial and variably tufted or shortly creeping rhizomatous habit, one-sided inflorescences, and long awns. The lower glume of hybrids is much longer than that of *V. membranacea*, being about half the length of the upper. The lemmas are usually 6–9.5 mm long and the awns 2–6.5 mm long. The anthers are often very small and slender and do not open; pollen is mostly imperfect. The ovary is glabrous at the apex, as in *F. rubra*. Seed-set is less than 1%, and viability very low, but F_2 plants can be raised.

c. Hybrids occur among or near both parents on coastal sand-dunes, mostly in rather bare areas but occasionally in tall grass and *Hippophaë rhamnoides* scrub, in v.c. 1, 3, 4, 6, 9, 13, 15, 41, 44, 48, 49, 52 and 59 and in Guernsey. Most records date from 1954 or later, when myxomatosis had reduced the number of rabbits (which graze *V. membranacea* selectively), but the hybrid was found in Guernsey in 1928 and 1953, and there are a few very old British specimens in **BM** and **K.** There is an unconfirmed record for Hs.

d. Crosses between the parents have produced healthy F_1 plants, but only with *V. membranacea* as the female (Stace and Cotton, 1974). Isolation seems largely cytogenetical, and the parents appear homozygous for a series of chromosome inversions, but the semi-cleistogamous flowering

---

* With assistance from R. Cotton and C. A. Stace

habit of *V. membranacea* probably contributes to the scarcity of hybrids. At meiosis the hybrid often shows about 14 bivalents and 7 univalents.

e. *F. rubra* 2*n* = 42, 56; *V. membranacea* 2*n* = 28, 42; natural and artificial hybrid 2*n* = 35.

f. BENOIT, P. M. (1960). *Festuca rubra* x *Vulpia membranacea* at Harlech and Newborough. *Nature Wales*, **6**: 59-60. FIG.

MELDERIS, A. (1955). A hybrid between *Festuca rubra* and *Vulpia membranacea.* Proc. B.S.B.I., **1**: 390-391.

PATZKE, E. (1970). Untersuchungen über Wurzelfluoreszenz von Schwingelarten zur Gliederung der Verwandtschaftsgruppe *Festuca* Linné. *Senckenberg. biol.*, **51**: 255-276.

STACE, C. A. and COTTON, R. (1974). Hybrids between *Festuca rubra* L. *sensu lato* and *Vulpia mumbranacea* (L.) Dum. *Watsonia*, **10**: 119-138.

WILLIS, A. J. (1967). The genus *Vulpia* in Britain. *Proc. B.S.B.I.*, **6**: 386-388.

670/6a x 672/2. *Festuca rubra* L. subsp. *rubra* x *Vulpia bromoides* (L.) Gray

a. None.

b. The hybrid is a rhizomatous perennial, intermediate between the parents, generally resembling *F. rubra*, but approaching *V. bromoides* in the linear-lanceolate lemma, the length of the awn of the lemma, and the unequal glumes. It resembles some plants of *F. rubra* x *V. membranacea* closely, but may be distinguished by the slightly longer lower glumes and thinner spikelet pedicels. It is highly male-sterile.

c. The hybrid has occurred with both parents, in open grassy sward on sandy sites, at Littlehampton, v.c. 13 (1961, coll. and det. A. Melderis), and Hollesley Bay, v.c. 25 (1969 and 1973, coll. P. J. O. Trist, det. C. E. Hubbard).

d. None.

e. *F. rubra* 2*n* = 42, 56; *V. bromoides* 2*n* = 14.

f. MELDERIS, A. (1965). *Festuca rubra* x *Vulpia bromoides*, a new hybrid in Britain. *Proc. B.S.B.I.*, **6**: 172-173.

TRIST, P. J. O. (1971). *Festuca rubra* L. x *Vulpia bromoides* (L.) Gray. *Watsonia*, **8**: 311.

670/6a x 672/3. *Festuca rubra* L. subsp. *rubra* x *Vulpia myuros* (L.) C. C. Gmel.

a. None.

b. The hybrid is a densely caespitose perennial, without creeping rhizomes. It is intermediate between the parents in the exsertion of the panicle, the number of nodes of the panicle (mostly 11–16), the width of the lemmas (flattened, 1.5–1.8 mm) and the length of the awns (about equalling the lemmas). The anthers are small and sterile, pollen being wholly imperfect, and no caryopses are formed (P. M. Benoit, *in litt.* 1970).

c. This hybrid has been found by P. M. Benoit in two localities at Arthog, v.c. 48, by a road-side and on a railway line, where it occurred in 1957 and 1972. In 1974 it was found on cinders by a railway track at Stockport, v.c. 59, by R. Cotton and C. A. Stace, and on a sandy shingle bank at Snettisham, v.c. 28, by R. P. Libbey.

d. None.

e. *F. rubra* $2n$ = 42, 56; *V. myuros* $2n$ = 42.

f. BENOIT, P. M. (1958). A new hybrid grass. *Proc. B.S.B.I.*, **3**: 85-86.

670/7 × 672/1. *Festuca juncifolia* St Amans (*F. arenaria* Osbeck, *sec.* Kjellqvist) × ***Vulpia membranacea*** (L.) Dumort.

a. x *F. melderisii* Stace & Cotton.

b. This hybrid resembles *F. rubra* x *V. membranacea* extremely closely, but may be distinguished by the lemmas which are 9.5–10.5 mm and which may be conspicuously hairy. It is sterile.

c. Cytologically confirmed hybrids are frequent with both parents (and *F. rubra* x *V. membranacea*) at Littlehampton, v.c. 13 (*fide* A. Melderis and C. A. Stace, 1972). Hybrids from Sandwich Bay, v.c. 15, reported to have $2n$ = 42 (M. D. Hooper, *in litt.* 1967), probably represent the same taxon, as may some plants from Kenfig Burrows, v.c. 41 (P. L. Thomas, *in litt.* 1973). Plants from Dawlish Warren, v.c. 3, found in 1956 by J. F. and P. C. Hall (**BM**), and from Littlehampton, v.c. 13, found in 1958 by E. B. Bangerter, P. C. Hall and J. E. Lousley (**BM**), and in both cases determined by A. Melderis as *F. rubra* var. *arenaria* x *V. membranacea,* differ from *F. rubra* x *V. membranacea* only in their slightly hairy lemmas. They may represent the latter hybrid or *F. juncifolia* x *V. membranacea.*

d. None.

e. *F. juncifolia* $2n$ = 56; *V. membranacea* $2n$ = 28, 42; hybrid $2n$ = 42.

f. MELDERIS, A. (1957). *Festuca rubra* var. *arenaria* x *Vulpia membranacea. Proc. B.S.B.I.*, **2**: 243.

STACE, C. A. and COTTON, R. (1974). As above.

683 × 670. *Bromus* L. × *Festuca* L. = × *Bromofestuca* Prodan

(by P. M. Smith)

683/1 × 670/2. *Bromus erectus* Huds. × ***Festuca arundinacea*** Schreb.

= x *B. cojocnensis* Prodan has been described from Rm, but is very doubtful.

671. *Lolium* L.
(by E. J. Lewis)

2 × 1. *L. multiflorum* Lam. × *L. perenne* L.

a. *L.* × *hybridum* Hausskn.

b. Hybrids are generally more or less intermediate in morphology between the parents and may be annuals or short-lived perennials. They have mostly awned, but occasionally awnless, lemmas and the leaf-blades are rolled in the young shoot. The range of variability in morphological detail, however, can be appreciable and is probably due to introgression, bearing in mind the complete interfertility and largely sympatric distribution of the two species. Pollen is usually over 70% viable.

c. Hybrids are frequently found, especially where both species are cultivated and have been reported from v.c. 7, 8, 17, 22, 23, 28, 30, 32–34 and 41 and from Da, Ga, Ge, He, Ho, Rm and Su.

d. Extensive experimental work shows the two species to be completely interfertile and cultivars of the hybrid have been developed by plant breeders. The validity of their separation into distinct species has been questioned on account of their ability to interbreed freely.

e. Both parents and artificial hybrids $2n = 14$.

f. HUBBARD, C. E. (1968). As above, p. 153.

JENKIN, T. J. (1931). The inter-fertility of *Lolium perenne* and *L. perenne* var. *multiflorum. Bull. Welsh Pl. Breed. Stn*, Ser. H., **12**: 121-125.

JENKIN, T. J. and THOMAS, P. T. (1938). The breeding affinities and cytology of *Lolium* species. *J. Bot., Lond.*, **76**: 10-12.

TERRELL, E. E. (1966). Taxonomic implications of genetics in ryegrasses (*Lolium*). *Bot. Rev.*, **32**: 138-164.

1 × 3. *L. perenne* L. × *L. temulentum* L.
has been recorded from Ge.

2 × 3. *L. multiflorum* Lam. × *L. temulentum* L.

a. None.

b. The Blackmoor hybrid was a robust specimen which was intermediate in several of the morphological details in which the parental species differ, including thickness of culm and length of upper glume (A. C. Leslie, *in litt.* 1971). The florets were smaller and more numerous than those of *L. temulentum*, bearing anthers which contained mostly imperfect pollen. Experimental results indicate that hybrids can be morphologically very variable, particularly in details such as awn-length, bearing in mind the existence of the awnless *L. temulentum* var. *arvense* Lilj.

c. Hybrids are probably fairly rare. The specimen described above, det. C. E. Hubbard and A. Melderis, was found by A. C. Leslie in 1970 at

Blackmoor, v.c. 12, as a shoddy-alien. Another plant, det. C. E. Hubbard, was found by Miss M. McC. Webster in 1970 on distillery refuse (from imported barley grain) at the Longman tip, Inverness, v.c. 96. The limited occurrence of *L. temulentum* in Br and the generally weak nature of the hybrid seedlings, at least from experimental evidence, make the establishment of the hybrid in any quantity unlikely; the two British records almost certainly originated from imported hybrid caryopses. The hybrid is recorded from Ge.

d. Jenkin (1954) made the cross in both directions and obtained slightly better seed-set with *L. temulentum* as female parent (43% compared with 25%). However, these caryopses were inviable, whereas from the reciprocal cross he obtained 55 mature but semi-sterile plants. Backcrosses to both parental species were moderately successful giving progeny which was variable in morphology and fertility. D. H. Hides (pers. comm. 1971), using artificial tetraploid parents and employing the technique of embryo culture, successfully raised F_1 plants from a cross with *L. temulentum* as female parent. He found an appreciable degree of self- and cross-fertility in these and their derivatives.

e. Both parents $2n = 14$.

f. JENKIN, T. J. (1954). Interspecific and intergeneric hybrids in herbage grasses, 6. *Lolium italicum* A. Br. intercrossed with other *Lolium* types. *J. Genet.,* **52**: 282-299.

JENKIN, T. J. and THOMAS, P. T. (1938). As above.

WEBSTER, M. McC. (1973). *Lolium multiflorum* Lam. x *L. temulentum* L. *Watsonia,* **9**: 390.

2 x rig. *L. multiflorum* Lam. x *L. rigidum* Gaud.

a. None.

b. The only specimen of this hybrid to be recorded in Br was annual or biennial, and intermediate in several respects between its putative parental species. It differed from *L. multiflorum* in having a longer upper glume and a much shorter awn on the lemma, and from *L. rigidum* in having a shorter upper glume and a short-awned lemma.

c. The above specimen was collected as a shoddy-alien at Blackmoor, v.c. 12, in 1960 by J. E. Lousley. It was originally identified as *L. rigidum* but later discovered by E. E. Terrell to be this hybrid. Like *L. rigidum* the hybrid is probably of rare occurrence in Br. Despite the strong interfertility of the two species, British records are likely to originate from imported hybrid caryopses.

d. Artificial hybrids, which are completely fertile, can be obtained without difficulty. Results of crosses among the three cross-pollinated and largely self-sterile species, *L. multiflorum*, *L. perenne* and *L. rigidum*, indicate that they are more or less completely interfertile.

e. Both parents $2n = 14$.

f. JENKIN, T. J. (1954). As above.

MELDERIS, A. (1971). *Lolium multiflorum* Lam. x *L. rigidum* Gaud. *Watsonia*, **8**: 299-300.

TERRELL, E. E. (1966). As above.

rig x 3. *L. rigidum* Gaud. x *L. temulentum* L.

a. None.

b. Jenkin's artificially-produced hybrid resembled its pollen-parent, *L. temulentum*, in general habit; the culms were tall and erect and the spikelets large. As in *L. rigidum*, the spikelets were longer than their glumes and had a high number of florets. The latter were also smaller than those of *L. temulentum* and the lemmas less broad. The anthers were non-dehiscent.

c. This alien hybrid has been found twice by Miss M. McC. Webster (*in litt.* 1973). It was first collected in Galashiels, v.c. 79, in 1965 (det. C. E. Hubbard, E). Plants (tentatively det. C. E. Hubbard, **CGE, E, K**) were also collected on distillery refuse at the Longman tip, Inverness, v.c. 96, in 1971.

d. Jenkin (1954, 1955) made the cross in both directions and obtained variable (11–91%), but mostly high, seed-sets. While caryopsis development was generally better with *L. temulentum* as female parent the only viable plant was obtained from the reciprocal cross. Its florets opened very irregularly and the anthers and stigmas were poorly exserted. Five established plants were obtained from a backcross to *L. temulentum*; one was examined and this was male-sterile.

e. Both parents $2n = 14$.

f. JENKIN, T. J. (1954). As above.

JENKIN, T. J. (1955). Interspecific and intergeneric hybrids in herbage grasses, 18. Various crosses including *Lolium rigidum sens. ampl.* with *L. temulentum* and *L. loliaceum* with *Festuca pratensis* and with *F. arundinacea. J. Genet.*, **53**: 467-486.

sub x 3. *L. subulatum* Vis. (*L. loliaceum* (Bory & Chaub.) Hand. Mzt.) x *L. temulentum* L.

was found by Miss M. McC. Webster (*in litt.* 1973) at Blackmoor, v.c. 12, in 1960 (det. C. E. Hubbard, E). Jenkin (1954) made reciprocal crosses and obtained good seed-set (57–79%) and germination (66–100%). The seedlings were slow in establishment and development and the emergence of the third and subsequent leaves from their sheaths was abnormal. Inflorescences were badly deformed but some backcrossing to *L. loliaceum* was possible. Such abnormalities would reduce the success of natural hybridization.

JENKIN, T. J. (1954). As above.

683 × 671. *Bromus* L. × *Lolium* L.
(by P. M. Smith)

683/15 × 671/1. *Bromus commutatus* Schrad. × *Lolium perenne* L.
was reported in error by Linton from Christchurch, v.c. 11. The plants
were variants of *L. perenne* with unusually swollen, protruding spikelets.
Artificial hybridization between these genera has not been successful.

DRUCE, G. C. and LINTON, E. F. (1901). *Bromus commutatus*, Schrad. ×
    Lolium perenne, L. *Rep. B.E.C.*, **1**: 651-652.

LINTON, E. F. (1901). *Bromus commutatus*, Schrad. × *Lolium perenne*,
    L. *Rep. Watson B.E.C.*, **1900-1901**: 35.

673. *Puccinellia* Parl.
(by B. M. G. Jones and C. A. Stace)

2 × 1. *P. distans* (L.) Parl. × *P. maritima* (Huds.) Parl.

a. *P.* × *hybrida* Holmberg (*Glyceria* × *jansenii* P. Fourn., *G.* × *salina* Druce,
    nom. nud., P. × *kattegatensis* (Neum.) Holmberg).

b. Plants more or less intermediate between *P. maritima* and *P. distans* in
    morphology have been reported as hybrids. Their status is unconfirmed;
    some at least are fertile. Holmberg described the lemmas as 2.3–2.8 mm
    and the pollen as very unequally developed. Both species are very variable
    and the putative hybrids may be ecads or extreme genetic variants.

c. There are records from v.c. 13–15 and 28, and Ga, Ho, No and Su.

d. None.

e. *P. maritima* $2n = 49, 56, 63, 77$ (also 60, 70); *P. distans* $2n = 42$ (28, 42).

f. BENNETT, A., DRUCE, G. C., LITTLE, J. E. and SALMON, C. E. (1920).
    G. borreri Bab. *Rep. Watson B.E.C.*, **3**: 125.

   DRUCE, G. C. (1918). *G. maritima* × *distans. Rep. B.E.C.*, **5**: 61-62.

   HOLMBERG, O. R. (1920). Einige *Puccinellia*-Arten und -Hybriden. *Bot.
       Notiser*, **1920**: 103-111.

   JANSEN, P. (1951). *Puccinellia* Parl., in WEEVERS, T. *et al. Flora
       Neerlandica*, **1(2)**: 64-76. Amsterdam.

   LOUSLEY, J. E. (1936). *Puccinellia pseudodistans* (Crépin) Jans. &
       Wacht. in Britain. *J. Bot., Lond.*, **74**: 260-266.

4 × 1. *P. fasciculata* (Torr.) Bicknell × *P. maritima* (Huds.) Parl.
(= *Glyceria* × *burdonii* Druce) was reported from Pagham, v.c. 13, in 1917

and later, but opinions as to its identity were divided. W. B. Turrill considered it to be a variant of *P. maritima*, and Dandy also included it under that species.

BENNETT, A., DRUCE, G. C., LITTLE, J. E. and SALMON, C. E. (1920). As above.

DANDY, J. E. (1958). *Op. cit.*, p. 159.

DRUCE, G. C. (1918). *G. maritima* x *Borreri* = x *G. Burdoni. Rep. B.E.C.*, 5: 61.

1 x 5. *P. maritima* (Huds.) Parl. x *P. rupestris* (With.) Fernald & Weatherby

a. *P.* x *krusemaniana* Jans. & Wacht.

b. Plants with the habit of *P. rupestris*, but with an open panicle whose floral parts approach those of *P. maritima* in size, have been found once. The anthers were non-dehiscent and the pollen grains were sterile.

c. Plants were collected from Chichester Harbour, v.c. 13, in 1920; they are also known in Ho.

d. None.

e. *P. maritima* $2n$ = 49, 56, 63, 77 (also 60, 70); *P. rupestris* $2n$ = 42.

f. JANSEN, P. (1951). As above.

JANSEN, P. and WACHTER, W. H. (1932). Grassen langs de Zuider-zeekust, 2. *Puccinellia*, 2. *Ned. kruidk. Archf*, **1932**: 301-305. FIG.

HUBBARD, C. E. (1968). As above, p. 205.

DRUCE, G. C. (1921). *Glyceria maritima* Wahl. x *procumbens* Dum. *Rep. B.E.C.*, 6: 256.

cap x 1. *P. capillaris* (Liljebl.) Jans. x *P. maritima* (Huds.) Parl.

= *P.* x *mixta* Holmberg has been recorded from Da, Ho, Is, No and Su.

2 x 3. *P. distans* (L.) Parl. x *P. pseudodistans* (Crép.) Jans. & Wacht.

= *P.* x *harmseniana* Jans. & Wacht. has been recorded from Ho (see *P. distans* x *P. fasciculata*).

2 x 4. *P. distans* (L.) Parl. x *P. fasciculata* (Torr.) Bicknell

a. None.

b. Putative hybrids are said to be more or less intermediate between these two species, and sterile. All wild plants which on first examination appeared to be intermediate have on closer study proved to be fertile and have the chromosome number of *P. fasciculata* ($2n$ = 28). They are referable to *P. pseudodistans* (Crép.) Jans. & Wacht., which is most likely to be an ecad of *P. fasciculata*. The very small anthers of *P. pseudodistans* and *P. fasciculata* may account for the reports of sterility.

c. *P. pseudodistans*, or intermediates between *P. distans* and *P. fasciculata*, have been recorded from v.c. 13–16, 18, 19, 25 and 27, but none has been confirmed as a hybrid.

d. *P. fasciculata* is notably plastic in its response to soil-water availability; plants rooted in wet mud become more luxuriant, with laxer panicles, and identical with plants known as *P. pseudodistans.*

e. *P. distans* 2*n* = 42 (28, 42); *P. fasciculata* 2*n* = 28; *P. pseudodistans* 2*n* = 28.

f. JONES, B. M. G. and NEWTON, L. E. (1970). The status of *Puccinellia pseudodistans* (Crép.) Jansen & Wachter in Great Britain. *Watsonia,* **8:** 17-26.

LOUSLEY, J. E. (1936). As above.

NEWTON, L. E. (1965). *Taxonomic studies in the British species of Puccinellia.* M.Sc. thesis, University of London.

2 × 5. *P. distans* (L.) Parl. × *P. rupestris* (With.) Fernald & Weatherby

a. *P.* × *pannonica* (Hackel) Holmberg.

b. Putative hybrids are said to be sterile perennials and to be intermediate in morphology between the parents, with panicles similar to but looser than those of *P. rupestris,* very short pedicels, and lemmas 2.0–3.3 mm. Such plants examined by Newton (1965) were fertile and on the basis of panicle characters he determined them as luxuriant examples of *P. rupestris.*

c. Putative hybrids have been reported from the coasts of south-eastern England from v.c. 3 to v.c. 27, and from Ho and No.

d. Attempts at experimental hybridization have failed.

e. *P. distans* 2*n* = 28 (28, 42); *P. rupestris* 2*n* = 42.

f. HOLMBERG, O. R. (1920). As above.

HUBBARD, C. E. (1968). As above, p. 205.

JANSEN, P. (1951). As above.

LOUSLEY, J. E. (1937). *P. distans* (L.) Parl. × *rupestris* (With.) Fernald & Weatherby. *Rep. B.E.C.,* **11:** 234.

NEWTON, L. E. (1965). As above.

cap × 2. *P. capillaris* (Liljebl.) Jans. × *P. distans* (L.) Parl.

= *P.* × *elata* (Holmberg) Holmberg has been recorded from Fe, Ho and Su.

cap × 3. *P. capillaris* (Liljebl.) Jans. × *P. pseudodistans* (Crép.) Jans. & Wacht.

= *P.* × *feekesiana* Jans. & Wacht. has been recorded from Ho.

674. *Catapodium* Link.

(by P. M. Benoit)

2 × 1. *C. marinum* (L.) C. E. Hubbard × *C. rigidum* (L.) C. E. Hubbard

  a. None.
  b. This hybrid is intermediate between the parent species in habit, length of spikelet-stalks, shape of spikelets, size of glumes, and size, shape and imbrication of florets. It has slender, yellow (not whitish), indehiscent anthers containing imperfect pollen.
  c. A single plant (**K**) was discovered with the parents on railway ballast at Aberdovey Station, v.c. 48, in 1960. The hybrid has not been found since, and is unknown elsewhere.
  d. None.
  e. Both parents $2n = 14$.
  f. BENOIT, P. M. (1961). *Catapodium marinum* x *rigidum. Proc. B.S.B.I.*, 4: 276.

676. *Poa* L.

(by T. G. Tutin)

Many of the hybrids which have been reported outside BI from morphological characters alone are very doubtful and require careful re-examination.

1 × 2. *P. annua* L. × *P. infirma* Kunth

  a. None.
  b. Hybrids are intermediate between the parents in their diagnostic floral characters, and largely sterile.
  c. They have been found only in Guernsey, CI, and there only detected amongst caryopses collected from a wild population.
  d. The hybrid has been made artificially, using *P. annua* as the female parent, and as expected proved to be triploid ($2n = 21$). At meiosis seven bivalents and seven univalents were formed and no good pollen was observed.
  e. *P. annua* $2n = 28$; *P. infirma* $2n = 14$; artificial hybrid $2n = 21$; wild putative hybrids $2n = 15–21$.
  f. HUBBARD, C. E. (1968). As above, p. 167.
     TUTIN, T. G. (1957). A contribution to the experimental taxonomy of *Poa annua* L. *Watsonia*, 4: 1-10.

1 × 7. *P. annua* L. × *P. glauca* Vahl

   was identified by H. Fisher (Linton, 1907) from Corrie Ardran, v.c. 88,
   but no voucher specimen has been seen.
   LINTON, E. F. (1907). Hybrids among British phanerogams. *J. Bot.,*
      Lond., **45**: 296-304.

1 × 10. *P. annua* L. × *P. pratensis* L.

   has been recorded from Rs, and a grass collected from Corrie Ardran, v.c.
   88, was identified as this by H. Fisher (Linton, 1907). Nothing further is
   known of the latter.
   LINTON, E. F. (1907). As above.

4 × 5. *P. alpina* L. × *P. flexuosa* Sm.

   a. *P.* × *jemtlandica* (Almq.) K. Richt.
   b. Readily distinguished from *P. flexuosa* by its viviparous spikelets, but at
      times resembling *P. alpina* rather closely, it can usually be distinguished
      from the latter by its narrow leaves, which taper gradually from the base
      to tip. The sheaths of the basal leaves are less persistent than in *P. alpina,*
      and the lemmas have only a few long hairs at the base instead of
      throughout the lower half. The hybrid is remarkably uniform throughout
      its range and is always sterile, reproducing by bulbils (vivipary).
   c. It is known from rock-ledges and damp, stony slopes on Ben Nevis, v.c. 97,
      and Lochnagar and the Cairntoul, v.c. 92. It also occurs on mountains in
      southern and central No and central Su.
   d. None.
   e. *P. alpina* $2n = 28$, 35 (also 36–38, 42–44, 46, 48, 51, 52, 54–57); *P.
      flexuosa* $2n = 42$ (also 43, *c* 78, 81); hybrid ($2n = 37$, 39).
   f. NANNFELDT, J. A. (1937). On *Poa jemtlandica* (Almqv.) Richt., its
      distribution and possible origin. *Bot. Notiser,* **1937**: 1-27.
      NYGREN, A. (1950). Cytological and embryological studies in Arctic
         Poae. *Symb. bot. upsal.,* **10(4)**: 1-64.
      PERRING, F. H. and SELL, P. D. (1968). *Op. cit.,* p. 149.

4 × 10. *P. alpina* L. × *P. pratensis* L.

   = *P.* × *berjedalica* H. Sm. has been recorded from Fe, Ge, He, No and Su.

7 × 6. *P. glauca* Vahl × *P. nemoralis* L.

   = *P.* × *jurassica* Chrtek & Jirasek has been recorded from He.

9 × 6. *P. compressa* L. × *P. nemoralis* L.

   = *P.* × *figertii* Gerhardt has been recorded from Ga and Ge.

6 × 10. *P. nemoralis* L. × *P. pratensis* L.

= *P.* × *coarctata* Hall. f. ex Gaud. has been recorded from Ga, Ge and He, and a specimen from Ben Creachan, v.c. 98, was identified as this by H. Fisher (Linton, 1907), but nothing more is known of the latter.
LINTON, E. F. (1907). As above.

11 × 6. *P. angustifolia* L. × *P. nemoralis* L.

has been recorded from Fe.

6 × 13. *P. nemoralis* L. × *P. trivialis* L.

has been recorded from Ga and Ge.

6 × 14. *P. nemoralis* L. × *P. palustris* L.

= *P.* × *intricata* K. Wein has been recorded from Ge.

7 × 10. *P. glauca* Vahl × *P. pratensis* L.

has been recorded from He.

9 × 10. *P. compressa* L. × *P. pratensis* L.

= *P.* × *complanata* Schur has been recorded from Cz, Ga, Ge and He.

9 × 13. *P. compressa* L. × *P. trivialis* L.

has been recorded from Ge.

9 × 14. *P. compressa* L. × *P. palustris* L.

= *P.* × *fossaerusticorum* K. Wein has been recorded from Ge and Su, and arose spontaneously in cultivation in Scandinavia.

10 × 13. *P. pratensis* L. × *P. trivialis* L.

= *P.* × *sanionis* Aschers. & Graebn. was identified by W. O. Howarth from Pyrford, v.c. 17, in 1929. The specimens differed from *P. pratensis* in having scabrid leaf-sheaths and a ligule nearly 5 mm long. It is also recorded from Cz, Ga, Ho, Po and Rs.
ASCHERSON, P. and GRAEBNER, P. (1900). *Synopsis der Mittel-europäischen Flora*, 2: 434. Leipzig.
BIDDISCOMBE, W. and HOWARTH, W. O. (1929). *Poa pratensis* L. var. *Rep. Watson B.E.C.,* 3: 496.

14 × 10. *P. palustris* L. × *P. pratensis* L.

has been recorded from Da.

15 × 10. *P. chaixii* Vill. × *P. pratensis* L.

= *P.* × *wippraensis* K. Wein has been recorded from Ge.

15 × 13. *P. chaixii* Vill. × *P. trivialis* L.
   = *P.* x *austrohercynica* K. Wein has been recorded from Ge.

678. *Dactylis* L.

(by C. A. Stace)

1 × 2. *D. glomerata* L. × *D. polygama* Horvat. (*D. aschersoniana* Graebn.)
   has been reported from Da, Fe, No and Su.

681. *Melica* L.

(by C. A. Stace)

2 × 1. *M. nutans* L. × *M. uniflora* Retz.
   = *M.* x *weinii* Hempel was described from Ge in 1970.

683. *Bromus* L.

(by P. M. Smith)

1 × 2. *B. erectus* Huds. × *B. ramosus* Huds.
   has been reported from Ge.

6 ×5. *B. madritensis* L. × *B. sterilis* L.
   = *B.* x *fischeri* Cugnac & A. Camus has been reported from Ga.

5 × 8. *B. sterilis* L. × *B. rigidus* Roth

has been suggested as the origin of *B. diandrus* Roth.

5 × 9. *B. sterilis* L. × *B. tectorum* L.

= *B.* × *guetrotii* A. Camus has been reported from Ga and Ge.

6 × 8. *B. madritensis* L. × *B. rigidus* Roth

= *B.* × *husnotii* A. Camus has been reported from Ga and Su.

6 × 9. *B. madritensis* L. × *B. tectorum* L.

= *B.* × *rosettae* A. Camus has been reported from Ga.

10 × 12. *B. hordeaceus* L. subsp. *hordeaceus* (*B. mollis* L.) × *B. hordeaceus* subsp. *thominii* (Hardouin) Hylander (*B. thominii* Hardouin, *non sensu* Tutin)

= *B.* × *jansenii* A. Camus has been reported from Br, Be, Ga, Ho and Su. Some reduction of fertility in such hybrids has occasionally been noted, but considering all available evidence it seems better to include both these taxa within a single species.

CAMUS, A. (1957a). *Bromus* hybrides de la Flore française. *Bull. Jard. bot. État Brux.*, **27**: 479-485.

SMITH, P. M. (1968). The *Bromus mollis* aggregate in Britain. *Watsonia*, **6**: 327-344.

10 × 13. *B. hordeaceus* L. (*B. mollis* L.) × *B. lepidus* Holmberg

a. *B.* × *pseudothominii* P. M. Smith (*B. thominii sensu* Tutin, *non* Hardouin).

b. Hybrids, the true identity of which have only recently been recognized, are intermediate in general morphology between the parents. Lemmas of hybrid plants are 6.5–8.0 mm long, bluntly angled and with a narrow hyaline margin. They are highly fertile and somewhat variable; introgression occurs. This hybrid taxon is not to be confused with *B. hordeaceus* subsp. *thominii* (Hardouin) Hylander, which is a small, sometimes prostrate plant of sand-dunes.

c. *B.* × *pseudothominii* is common and widely distributed in BI and is found more often than *B. lepidus.* Typically it occurs on or near sown grassland, e.g. road-side verges and sown pastures. It occurs in most ruderal habitats. Hybrids are common in Europe, and are found in western Asia and Mediterranean Africa. They have been introduced, or have arisen independently, in North America.

d. Artificial crosses closely resemble naturally occurring material referred to the hybrid. F_1 and F_2 hybrids are highly fertile. In Br the incidence of hybridization may be lower than elsewhere, where the autogamous habit is less frequent. The native area of *B. lepidus,* where hybridization may be common, is not known.

e. Both parents and hybrid $2n = 28$.

f. CAMUS, A. (1957b). Un *Bromus* hybride des dunes du Cotentin. *Bull.*
   Mus. Hist. nat., Paris, sér. 2, **29**: 184-185.
   SMITH, P. M. (1968). As above. FIG.

10 × 14. *B. bordeaceus* L. (*B. mollis* L.) × *B. racemosus* L.

= *B.* × *hannoveranus* K. Richt. has been reported from Ge and Su.

15 × 10. *B. commutatus* Schrad. × *B. bordeaceus* L. (*B. mollis* L.)

= *B.* × *brevieri* Chassagne has been reported from Ga and Hu.

10 × 18. *B. bordeaceus* L. (*B. mollis* L.) × *B. secalinus* L.

has been reported from Ge, but distinct species such as *B. commutatus*
were considered as part of the hybrid progeny.

15 × 14. *B. commutatus* Schrad. × *B. racemosus* L.

a. None.

b. Hybridization between these two species is manifested more by the
   apparent introgression of genes of each into the other than by specimens
   of exactly intermediate morphology, though these are occasionally found.
   Intermediate plants have the lowest internode of the rachilla 1–1.3 mm
   long, lemmas about 8 mm long, and anthers about 1.5–2 mm long.
   Backcrossing is presumably responsible for the occurrence of some
   characters of one parent in what would otherwise be regarded as a typical
   specimen of the other. Part of the anomalous variation in anther-length in
   B. racemosus may be related to cleistogamy, but this does not account for
   the occasional presence of long anthers in *B. commutatus.* Hybridization is
   probably uncommon in these mainly self-pollinating plants, but the high
   fertility of the hybrids and their facultative autogamous habit would
   enable recombinations to persist and accumulate.

c. Hybrids are to be expected where the parents grow together. They are not
   uncommon, but have rarely been reported, partly due to the long-standing
   confusion between the parents. Occasionally the parents may be found
   growing together on road-sides, but more commonly they co-habit lowland
   water-meadows, especially in southern and south-western England.
   Hybrids have been recorded from much of western Europe, but a record
   from Da of a sterile hybrid may be erroneous.

d. The hybrids appear to be highly fertile, as indicated by Wilson (1956)
   from the behaviour of artificial hybrids.

e. Both parents and hybrid $2n = 28$.

f. HYLANDER, N. (1953). *Nordisk Kärlväxtflora,* **1**: 356-357. Stockholm.
   SMITH, P. M. (1973). Observations on some critical bromegrasses.
   Watsonia, **9**: 319-332.

WILSON, D. (1956). *Cytogenetic studies in the genus Bromus.* Ph.D. thesis, University of Wales, Aberystwyth. FIG.

17 × 14. B. arvensis L. × **B. racemosus** L.

has been reported from Su.

17 × 15. B. arvensis L. × **B. commutatus** Schrad.

has been reported from Su.

684. *Brachypodium* Beauv.

(by C. A. Stace)

2 × 1. B. pinnatum (L.) Beauv. × **B. sylvaticum** (Huds.) Beauv.

a. *B. × cugnacii* A. Camus.
b. Putative hybrids are intermediate in awn-length (awns from half as long to as long as lemmas), pubescence of glumes, culms and leaves, and general habit. They have been little investigated, however, and much doubt remains concerning the existence of hybrids of this parentage. Praeger said they were "intermediate in all respects" and considered them "without question hybrids", whereas St.-Yves, who treated the two parents as varieties of one species, equated *B. cugnacii* with *B. pinnatum* var. *glaucovirens.*
c. Intermediates have been recorded from v.c. H11 and H39, and from Cz, Da and Ga.
d. None.
e. *B. sylvaticum* (2n = 18), *B. pinnatum* (2n = 28).
f. CAMUS, A. (1931). Quelques hybrides des genres *Cistus, Bromus* et *Brachypodium. Bull. Soc. bot. Fr.,* **78**: 97-102.
CAMUS, A. (1958). As above.
PRAEGER, R. L. (1937). *Brachypodium pinnatum* in Ireland. *Ir. Nat. J.,* **6**: 159-161.
SAINT-YVES, A. (1934). Contribution à l'étude des *Brachypodium* (Europe et Région méditerranéenne). *Candollea,* **5**: 427-493.
SIMPSON, N. D. (1942). *Brachypodium pinnatum* (L.) Beauv. × *sylvaticum* (Huds.) Beauv. *Rep. B.E.C.,* **12**: 427-428.

685. *Agropyron* Gaertn.
(by A. Melderis)

The hybrids described below are usually intermediate in their leaf and spikelet characters between their parental species and are characterized by high male-sterility, having narrow, indehiscent anthers with angular or shrivelled, translucent pollen grains. Seed-set has not been observed.

1 × 2. *A. caninum* (L.) Beauv. × *A. donianum* F. B. White

> was said by Tutin (1962) to occur with the parents in Scotland. Specimens so-labelled from Inchnadamph, v.c. 108, collected by J. E. Raven and S. M. Walters in 1953 (CGE), appear to me to be fertile and typical *A. caninum*, with long awns. Hybrids between *A. caninum* and Scottish *A. donianum*, with awns of intermediate length, arose spontaneously in my garden and were also fertile, thus throwing doubt on the specific distinctness of *A. donianum*.
>
> TUTIN, T. G. (1962). *Agropyron*, in CLAPHAM, A. R., TUTIN, T. G. and WARBURG, E. F. *Op. cit.*, p. 1156.

1 × 3. *A. caninum* (L.) Beauv. × *A. repens* (L.) Beauv.

> was reported from Da in 1931.

4 × 3. *A. pycnanthum* (Godr.) Gren. & Godr. (*A. littorale* auct., *Triticum littorale* Host, *non* Pallas, *A. pungens* auct., *non Triticum pungens* Pers.) × *A. repens* (L.) Beauv.

> a. *A.* × *oliveri* Druce.
> b. Hybrids have more prominent and closer ribs on the upper surface of the leaf-blades than in *A. repens* and possess sometimes shorter cilia on the margin of the leaf-sheaths than in *A. pycnanthum*. They differ from the two hybrids involving *A. junceiforme* in having a tough inflorescence-axis with rigid scabridules on the angles and usually acute glumes.
> c. There are many records from Br but most are unconfirmed. Authentic records exist from the coasts of v.c. 2, 28 and 60, and the hybrid is reported also from Da, Fe, Ga, Ge, Ho, No and Su.
> d. Artificial crosses between *A. repens* and *A. pycnanthum* carried out by Cauderon were unsuccessful.
> e. Both parents 2*n* = 42.
> f. CAUDERON, Y. (1958). *Étude cytogénétique des Agropyrum français et de leurs hybrides avec les blés*, p. 30. Paris.
> HANSEN, A. (1960). *Elytrigia (Agropyron)*-hybrider i Danmark. *Bot. Tidsskr.*, 55: 296-312. FIG.
> HYLANDER, N. (1953). *Nordisk Kärlväxtflora*, 1: 368. Stockholm.

JANSEN, P. (1951). *Agropyron* Gaertn., in WEEVERS, T. *et al. Flora Neerlandica,* **1(2):** 112-119. Amsterdam.

MELDERIS, A. (1953). Hybrids of the British *Elytrigia. Year book B.S.B.I.,* 1953: 61.

VESTERGREN, T. (1925). *Agropyron litorale* (Host) Dum., en mediterran-atlantisk art vid Nordeuropas kuster. *Svensk bot. Tidskr.,* **19:** 263-288. FIG.

5 × 3. *A. junceiforme* (A. & D. Löve) A. & D. Löve × *A. repens* (L.) Beauv.

a. *A.* × *laxum* (Fr.) Tutin.

b. Hybrids may usually be recognized by the glabrous margin of the leaf-sheaths, somewhat prominent, scabrous or shortly hairy ribs on the upper surface of the leaf-blades, the slightly fragile inflorescence-axis with smooth angles, and the usually obtuse glumes.

c. In Br this hybrid is fairly rare and local, occurring on coasts, particularly on sand-dunes, where the two parental species grow in close proximity, particularly in northern Br. There are confirmed records from Jersey, Guernsey, v.c. 15 and v.c. 41; from numerous localities from v.c. 59 and 61 northwards to the Outer Hebrides, v.c. 110, and Orkneys, v.c. 111; and from v.c. H38. It is known also from Da, Fa, Fe, Ga, Ge, Ho, Is, No, Po and Su.

d. Cauderon obtained caryopses in a cross between *A. repens* and *A. junceiforme.*

e. *A. repens* $2n = 42$; *A. junceiforme* $2n = 28$; hybrid $2n = 35$ (35, 49).

f. CAUDERON, Y. (1958). As above.

GODLEY, E. J. (1951). Two natural *Agropyron* hybrids occurring in the British Isles. *Ann. Bot.,* n.s., **15:** 535-544. FIG.

HANSEN, A. (1960). As above. FIG.

HENEEN, W. K. (1962). Karyotype studies in *Agropyron junceum, A. repens* and their spontaneous hybrids. *Hereditas,* **48:** 471-502.

HYLANDER, N. (1953). As above, p. 369.

JANSEN, P. (1951). As above.

MELDERIS, A. (1953). As above.

ÖSTERGREN, G. (1940). On the morphology of *Agropyron junceum* (L.) PB., *A. repens* (L.) PB. and their spontaneous hybrids. *Bot. Notiser,* **1940:** 133-143. FIG.

5 × 4. *A. junceiforme* (A. & D. Löve) A. & D. Löve × *A. pycnanthum* (Godr.) Gren. & Godr. (*Triticum littorale* Host, *non* Pallas, *A. littorale* auct., *A. pungens* auct., *non Triticum pungens* Pers.)

a. *A.* × *obtusiusculum* Lange (*A. hackelii* Druce, *Triticum acutum* auct., *A. acutum* auct.). (Hybrid Sea-Couch).

b. In the characters of the inflorescence-axis and glumes this hybrid resembles *A. junceiforme* × *A. repens,* but it may be distinguished from the latter by the ciliate margin of the leaf-sheaths, and by the more prominent ribbing on the upper surface of the leaf-blades. In some hybrids, probably originated from a cross between awned *A. pycnanthum* (*A. pungens* var. *setigerum* Dum.) and *A. junceiforme*, the lemmas are shortly awned.

c. Hybrids in Br are not uncommon but local, growing in a zone between loose sand and fixed sand-dunes or at the margins of salt-marshes, in CI and throughout Br northwards to v.c. 66 and 70, and from southern Hb. They are reported also from Be, Co, Da, Fe, Ga, Ge, Ho, Hs, Po and Su.

d. None.

e. *A. pycnanthum* 2*n* = 42; *A. junceiforme* 2*n* = 28; hybrid 2*n* = 35.

f. GODLEY, E. J. (1951). As above. FIG.

HANSEN, A. (1960). As above. FIG.

HUBBARD, C. E. (1968). As above, pp. 102-103. FIG.

HYLANDER, N. (1953). As above, p. 369.

JANSEN, P. (1951). As above.

MELDERIS, A. (1953). As above.

SIMONET, M. (1935). Sur la valeur taxonomique de l'*Agropyrum acutum* Roehm. et S. – Contrôle cytologique. *Bull. Soc. bot. Fr.*, **81**: 801-814. FIG.

VESTERGREN, T. (1925). As above. FIG.

685 × 686. *Agropyron* Gaertn. × *Elymus* L. = × *Agroelymus* Camus ex A. Camus
(by C. E. Hubbard)

685/3 × 686/1. *Agropyron repens* (L.) Beauv. × *Elymus arenarius* L.
= × *A. bergrothii* (H. Lindb.) Rousseau has been recorded from Rs.

685/4 × 686/1. *Agropyron pycanthum* (Godr.) Gren. & Godr. (*A. littorale* auct., *Triticum littorale* Host, *non* Pallas, *A. pungens* auct., *non Triticum pungens* Pers.) × *Elymus arenarius* L.

was reported from sand-dunes between Heacham and Hunstanton, v.c. 28, by Hubbard (1936), but closer examination has shown that the specimen is a depauperate plant of *E. arenarius* with narrower leaf-blades, more slender culms and inflorescences, and fewer-flowered spikelets.

HUBBARD, C. E. (1936). *Elymus arenarius* L. x *Agropyron pungens* R. & S. *Proc. Linn. Soc. Lond.*, **148**: 109.

685/5 × 686/1. *Agropyron junceiforme* (A. & D. Löve) A. & D. Löve
× *Elymus arenarius* L.

> = × *A. strictus* (Dethard.) Çamus ex A. Camus has been recorded from Da,
> Ge, Po and Su.

685 × 687. *Agropyron* Gaertn.
× *Hordeum* L.
= × *Agrohordeum* Camus ex A. Camus
(by C. E. Hubbard)

685/3 × 687/1. *Agropyron repens* (L.) Beauv. × *Hordeum secalinum*
Schreb.

a. × *A. langei* (K. Richt.) Camus ex A. Camus.
b. This hybrid is more or less intermediate between its parents, differing from
   A. repens in the inflorescence-axis being articulated just above each
   spikelet so that the latter falls with an internode of the axis attached, and
   from *H. secalinum* in possessing rhizomes and mostly single spikelets, at
   least at the upper nodes of the spike. There appear to be two nothomorphs
   (A & B) of the hybrid. Nothomorph A has narrower, fewer (3–4)-nerved,
   awned glumes and fewer (2–4) florets per spikelet, and also occasionally
   2(–3) spikelets at the lower nodes of the inflorescence. Nothomorph B is
   easily confused with *A. repens* var. *aristatum* Baumg. Among specimens
   referred to this variety are some with an articulated inflorescence-axis and
   with the internode remaining attached to the spikelet-base. In these the
   awned glumes are wider and 5–7-nerved, and the spikelets 3–5-flowered.
   Nothomorph A is male-sterile, but nothomorph B, while usually male-
   sterile, has been found with perfect pollen and occasionally a well-
   developed caryopsis.
c. Nothomorph A was known only from a single Danish locality for a
   number of years (1865–1877 or later) but in 1961 it was rediscovered in
   salt-meadows elsewhere in Da. It has also been found at Shirehampton, v.c.
   34, in a brackish pasture by the R. Avon (1945–54), with both parent
   species. Nothomorph B occurs in various parts of lowland England from
   the Thames Valley to Northumberland, in meadows, field- and road-
   margins, often with or near awnless *Agropyron repens*, but its distribution
   is imperfectly known.
d. None.
e. *A. repens* $2n = 42$; *H. secalinum* $2n = 28$ (14, 28); hybrid ($2n = 49$).

f. ASCHERSON, P. and GRAEBNER, P. (1902). *Synopsis der Mittel-europäischen Flora*, **2(1)**: 747-748. Leipzig.

HANSEN, A. (1960). *Elytrigia (Agropyron)*-hybrider in Danmark. *Bot. Tidsskr.*, **55**: 296-312. FIG.

HUBBARD, C. E. and SANDWITH, N. Y. (1955). An intergeneric grass hybrid new to Britain. *Proc. B.S.B.I.*, **1**: 323, 387.

HYLANDER, N. (1953). *Nordisk Kärlväxtflora*, **1**: 369. Stockholm.

VESTERGREN, T. (1925). En hybrid mellan *Agropyron repens* (L.) PB. och *Hordeum nodosum* L. *Svensk bot. Tidskr.*, **19**: 412-418. FIG.

685/4 × 687/1. Agropyron pycnanthum (Godr.) Gren. & Godr. (*A. littorale* auct., *Triticum littorale* Host, *non* Pallas, *A. pungens* auct., *non Triticum pungens* Pers.) × *Hordeum secalinum* Schreb.

= × *A. rouxii* (Gren. & Duval-Jouve) Camus ex A. Camus has been recorded from Ga.

686 × 687. *Elymus* L. × *Hordeum* L. = × *Elymordeum* Lepage

(by C. A. Stace)

About seven combinations between species of these two genera are known in North America, but none involves British species.

689. *Koeleria* Pers.

(by R. S. Callow)

1a × 1g. K. albescens DC. × *K. gracilis* Pers.

= *K.* × *supraarenaria* Domin was recorded from various parts of Br before these three taxa were considered by British botanists to be variants of one species (*K. cristata* (L.) Pers.).

DRUCE, G. C. (1905). Notes on the British Koelerias. *J. Bot., Lond.*, **43**: 354-357.

HOWARTH, W. O. (1933). Notes on *Koeleria* Pers. *Rep. B.E.C.*, **10**: 37-41.

1 × 2. *K. cristata* (L.) Pers. × *K. vallesiana* (Honck.) Bertol.

a. None.

b. The hybrid is intermediate between the parents in most diagnostic characters, of which the most useful is the nature of the leaf-sheaths. In *K. vallesiana* leaf-sheaths are persistent due to interconnecting vascular strands between the parallel sheath-veins; these interconnections are not found in *K. cristata*. In the hybrid cross-strands are found infrequently but the leaf-sheaths are persistent. Hybrid fertility is probably low but so is that of the parents in natural populations. A chromosome count is essential to confirm hybridity.

c. In BI *K. vallesiana* is found only on limestone hills on the seaward end of the Mendips, in v.c. 6. Hybrids probably occur infrequently in mixed populations in this area; so far they have been detected on Crook Peak and Uphill. There are no confirmed records from outside BI. Plants determined as *K. mixta* Domin, and said to be probably *K. britannica* (Domin) Ujhelyi (a segregate of *K. cristata*) × *K. vallesiana*, were collected from Uphill, v.c. 6, by G. C. Druce in 1905, but the specimens (and the taxon *K. mixta*) are almost certainly variants of *K. cristata*.

d. The pentaploid hybrid shows a considerable amount of chromosome pairing at meiosis. Mostly bivalents are formed, but with up to 3 chains of five chromosomes and 3–6 univalents per pollen mother cell. The pairing pattern indicates a considerable degree of chromosome homology between the parental species.

e. *K. cristata* $2n = 28$; *K. vallesiana* $2n = 42$: natural F_1 hybrid $2n = 35$.

f. HOWARTH, W. O. (1933). As above.

692. *Avena* L.

(by C. A. Stace)

1 × sat. *A. fatua* L. × *A. sativa* L.

a. (*A. × marquandii* Druce, ?*A. hybrida* Peterm., ?*A. intermedia* Lindgr.).

b. Naturally-occurring putative hybrids are recognized by their intermediacy in the main diagnostic characters. At least the lowest lemma and rhachilla is hairy, the florets disarticulate tardily, and the awn-form and -length varies between that of the parents. Hybrids are usually highly fertile and appear to give rise by backcrossing to aberrant variants (fatuoids) of cultivated oats.

c. Hybrids usually appear sporadically in or around oat-fields or where oats have been previously grown, but there are not many British records. They have been found in v.c. 4, 21, 24, 25/26, 34 and 111, sporadically across Europe from Su to Bu and Tu, and throughout much of Canada and the U.S.A. They are rarely persistent in any one place.

d. Both species are mainly inbreeding. Reciprocal crosses have been made on several occasions and grown on to at least the F_3 generation. Artificial F_1 plants resemble *A. sativa* in most morphological characters, and subsequent generations show segregation of the diagnostic characters as would be expected if most of the *A. sativa* characters were dominant. Thus the natural intermediates are more likely to be backcrosses or subsequent segregants than F_1 hybrids, and some are possibly mere variants of *A. fatua*. F_1 hybrids can be distinguished from *A. sativa* by the lodicule (with no or a short, rounded side-lobe) and epiblast (apex irregularly truncate) being of the *A. fatua* type. F_1 plants show from 0% to about 35% meiotic abnormalities, but many intervarietal hybrids of *A. sativa* are equally irregular.

e. *A. sativa* $2n = 42$; *A. fatua* and hybrid ($2n = 42$).

f. AAMODT, O. S., JOHNSON, L. P. V. and MANSON, J. M. (1934). Natural and artificial hybridisation of *Avena sativa* with *A. fatua* and its relation to the origin of fatuoids. *Can. J. Res.*, **11**: 701-727. FIG.

BAUM, B. R. (1968). On some relationships of *Avena sativa* and *A. fatua* (Gramineae) as studied from Canadian material. *Can. J. Bot.*, **46**: 1013-1024.

BAUM, B. R. (1969). The role of the lodicule and epiblast in determining natural hybrids of *Avena sativa* x *fatua* in cultivated oats. *Can. J. Bot.*, **47**: 85-91.

COFFMAN, F. A. and MACKEY, J. (1959). Hafer (*Avena sativa* L.), in KAPPERT, H. and RUDORF, W. eds. *Handbuch der Pflanzenzuchtung*, **2**: 427-531. Berlin and Hamburg.

JENSEN, N. F. (1961). Genetics and inheritance in oats: inheritance of morphological and other characters, in COFFMAN, F. A., ed. *Oats and oat improvement*, pp. 125-206. American Society of Agronomy.

PHILP, J. (1933). The genetics and cytology of some interspecific hybrids of *Avena. J. Genet.*, **27**: 133-179.

sat x **3.** *A. sativa* L. x *A. strigosa* Schreb.

has been found in Au.

693. *Helictotrichon* Bess.
(by C. A. Stace)

1 × 2. *H. pratense* (L.) Pilg. × *H. pubescens* (Huds.) Pilg.
    was recorded from v.c. 53/54 by E. A. Woodruffe-Peacock in 1910, but
    the record has never been confirmed and this hybrid has not been found
    elsewhere.
    DRUCE, G. C. (1912). The check list of Lincolnshire Plants. *Rep. B.E.C.*,
        3: 43-44.

695. *Holcus* L.
(by K. Jones)

1 × 2. *H. lanatus* L. × *H. mollis* L.
  a. *H.* × *hybridus* K. Wein.
  b. In general appearance the hybrids resemble *H. mollis* but can be
    distinguished by a combination of characters. The spikelets differ in that
    the glumes are somewhat obtuse, with a more or less abruptly exserted
    nerve, and have some hairs. The awns are somewhat shorter than is usual in
    H. mollis, *c* 3.0–3.5 mm as against 4.0–4.5 mm. The pollen fertility is very
    low and the anthers are indehiscent and shrunken. Whereas in *H. mollis*
    hairs occur only at the nodes of the flag-leaf-sheath, in the hybrid
    pubescence can extend upwards to cover variable amounts of the sheath.
    Leaf-margins may also appear more hairy than in typical *H. mollis*. Some
    hybrids may resemble *H. lanatus* more closely; any plant which seems to
    approach *H. lanatus* but has rhizomes is likely to be a hybrid. On the
    Continent two nothomorphs have been described by Wein, one (which
    resembles *H. lanatus*) was called f. *superlanatus* and the other (which is
    more like *H. mollis*) was called f. *supermollis*.
  c. In Br hybrids are known from only three localities but may well be more
    frequent. They have been found in the Aberystwyth area, v.c. 46, at
    Boghall near Pentlandfield, v.c. 83, and on a sandy heath near Brandon,
    v.c. 26. In each locality both parents were present and the habitats were
    open, though in Aberystwyth the site was at the edge of a wood. All of
    these hybrids tend towards *H. mollis* in appearance. Elsewhere hybrids
    have been recorded only from Ge.
  d. Artificial hybrids between the two parents have been made by Beddows

using *H. mollis* with $2n = 28$, the cross succeeding in both directions. F_1 hybrids, being triploid, are highly sterile. In appearance they resemble *H. mollis* more than *H. lanatus*, but progenies can vary in appearance and may be influenced by the direction of the cross. It is possible that in nature ploidy levels of *H. mollis* higher than tetraploid could hybridize with *H. lanatus*, but these would be much more difficult to detect both morphologically and chromosomally. It is, in fact, likely that at least the pentaploid variant of *H. mollis* has arisen by past hybridization with *H. lanatus*.

e. *H. lanatus* $2n = 14$; *H. mollis* $2n = 28, 35, 42$ and 49; wild hybrid $2n = 21$.

f. BEDDOWS, A. R. (1971). The inter- and intra-specific relationships of *Holcus lanatus* L. and *H. mollis* L. *sensu lato* (Gramineae). *Bot. J. Linn. Soc.*, **64**: 183-198.

CARROLL, C. P. and JONES, K. (1962). Cytotaxonomic studies in *Holcus*, 3. A morphological study of the triploid F_1 hybrid between *Holcus lanatus* L. and *H. mollis* L. *New Phytol.*, **61**: 72-84. FIG.

JONES, K. (1958). Cytotaxonomic studies in *Holcus*, 1. The chromosome complex in *Holcus mollis* L. *New Phytol.*, **57**: 191-210.

JONES, K. and CARROLL, C. P. (1962). Cytotaxonomic studies in *Holcus*, 2. Morphological relationships in *Holcus mollis* L. *New Phytol.*, **61**: 63-71.

WEIN, K. (1913). *Holcus lanatus* x *mollis* (x *Holcus hybridus*) K. Wein, nov. hybr. *Reprium Spec. nov. Regni veg.*, **13**: 36-37.

699 × 700. *Ammophila* Host.
× *Calamagrostis* Adans.
= × *Ammocalamagrostis* P. Fourn.

(by C. E. Hubbard)

699/1 × 700/1. *Ammophila arenaria* (L.) Link x *Calamagrostis epigejos* (L.) Roth

a. x *A. baltica* (Schrad.) P. Fourn. (Purple Marram, Hybrid Marram).

b. This is a vigorous, sterile plant with long, stout rhizomes and stout culms up to 1.5 m high. The leaf-blades are flat and have the lower surface uppermost. The very dense, tapering, elongated panicles have spikelets intermediate in structure between those of the parents. Westergaard found three distinct hybrids on the Danish coast: nm. *subarenaria* Marss. (nm. *baltica*), nm. *intermedia* Westerg., and nm. *epigeoidea* Westerg. As their names suggest, the first and third are nearer *A. arenaria* and *C. epigejos* respectively, while the second is more or less intermediate in structure. The inflorescence are normally purplish or pale green, but a plant with greenish-yellow panicles has been recorded from Dunkerque, Ga. The British representatives are all nm. *subarenaria* but show some variation, the Norfolk plants having a 5–7-nerved lemma and 2-nerved palea, and Northumberland and Sutherland plants a 3–5-nerved lemma and 4-nerved palea.

c. In BI this hybrid is known to occur naturally only from Horsey to Yarmouth, v.c. 27, on Ross Links, v.c. 68 and on Handa Island, v.c. 108, but it has been planted on sand-dunes in v.c. 26–28 on account of its considerable importance as a sand-binder. It occurs on the coasts of north-western Europe from Boulogne, Ga, to No and Su and the Baltic Sea coasts of Po and Rs.

d. Nm. *subarenaria* ($2n = 42$, AAAACC) appears on cytological and morphological evidence to have arisen from an unreduced gamete ($n = 28$) of tetraploid *A. arenaria* and a reduced gamete ($n = 14$) of tetraploid *C. epigejos*, nm. *intermedia* ($2n = 28$, AACC) from reduced gametes of the same parents, and nm. *epigeoidea* ($2n = 42$, AACCCC) from reduced gametes of tetraploid *A. arenaria* and octoploid *C. epigejos*. Meiosis in the hybrids is very irregular, with many univalents and multivalents.

e. *A. arenaria* ($2n = 28, 56$); *C. epigejos* $2n = 28$ ($28, 42, 56$); hybrid $2n = 42$ (D. Ranwell, pers. comm.); nm. *subarenaria* ($2n = 42$); nm. *intermedia* ($2n = 28$); nm. *epigeoidea* ($2n = 42$).

f. ELLIS, E. A. (1960). The purple (hybrid) marram, *Ammocalamagrostis baltica* (Fluegge) P. Fourn. in East Anglia. *Trans. Norfolk Norwich Nat. Soc.*, **19**: 49-51.

HUBBARD, C. E. (1968). As above, pp. 286-287. FIG.

HYLANDER, N. (1953). *Nordisk Kärlväxtflora*, **1**: 311-312. Stockholm. FIG.

KUBIÉN, E. (1964). Badania cytologiczne, morfologiczne i anatomiczne nad *Ammophila baltica* (Flügge) Link i jej domniemanymi gatunkami rodzicielskimi. *Acta Soc. bot. Pol.*, **33**: 527-546. FIG.

KUBIÉN, E. (1970). Cytological and embryological studies in *Ammophila baltica* (Flügge) Link. *Acta biol. cracov., Ser. bot.*, **13**: 99-110.

PERRING, F. H. and SELL, P. D. (1968). *Op. cit.*, p. 152.

WESTERGAARD, M. (1941). *Calamagrostis epigejos* (L.) Roth, *Ammophila arenaria* Link og deres Hybrider (*Ammophila baltica* (Flügge) ·Link). *Bot. Tidsskr.*, **45**: 338-351. FIG.

WESTERGAARD, M. (1943). Cytotaxonomical studies on *Calamagrostis epigejos* (L.) Roth, *Ammophila arenaria* (L.) Link and their hybrids (*Ammophila baltica* (Flügge) Link). *K. danske Vidensk. Selsk., Biol. Skr.*, **2(4)**. FIG.

700. *Calamagrostis* Adans.

(by F. E. Crackles)

2 × 1. *C. canescens* (Weber) Roth × *C. epigejos* (L.) Roth
= *C.* × *rigens* Lindgr. has been reported from Da, Ga, Ge and Su.

1 × 3. *C. epigejos* (L.) Roth × *C. stricta* (Timm) Koel.
= *C.* × *strigosa* (Wg.) Hartm. has been reported from Fe, No, Rs and Su.

2 × 3. *C. canescens* (Weber) Roth × *C. stricta* (Timm) Koel.
a. *C.* × *gracilescens* Blytt (*C.* × *conwentzii* Ulbrich).
b. Octoploid hybrids (2n = 56) are intermediate between the parents in bract-position, glume-shape, length of callus hairs in relation to floret-length, awn-length and the point of origin of the awn on the lemma. The ligule is long and the upper leaf-surface may have long white hairs as in *C. canescens,* while the culm is rough just below the panicle and the top of lower leaf-sheaths usually hairy as in *C. stricta.* The presence of some good pollen and viable caryopses in these plants certainly indicates a degree of fertility, but the plants also spread vegetatively and can persist for many years. Tetraploid plants (2n = 28) of hybrid origin are like *C. stricta* in general appearance but have a long basal inflorescence-branch and some intermediate features, e.g. awn-length and -position, and callus hair-length; they are possibly backcrosses to *C stricta.* They appear to be largely sterile. Past hybridization and introgression is believed to be responsible for much of the variability of *C. stricta* in its various British stations.
c. Both variants of the hybrid are known in Br only from a canal bank and margins at Leven, v.c. 61, where they were first noted in 1951. Both parents are present, the octoploid hybrid tending to be intermediate in its ecological requirements. The plants which might be backcrosses to *C. stricta* occur on the canal-edge, as does *C. stricta.* Hybrids are also known in Da, Fe, Ge, No and Su.
d. Artificial crosses have been produced with ease using Swedish material. The F_1 (2n = 28) had a low fertility but all F_2 plants examined produced viable caryopses and all hybrids produced pollen, though sometimes of poor quality. Fertility of 2n = 56 hybrids is due at least partly to apomixis, but might be also the result of amphiploidy. The two species are normally isolated by geographical and ecological factors and different flowering times.
e. Both parents 2n = 28; wild hybrid 2n = c 28, 56 (28, 56).
f. CRACKLES, F, E. (1953). *Calamagrostis stricta* in S. E. Yorkshire. *Year Book B.S.B.I.,* **1953**: 54.
   CRACKLES, F. E. (1972). *Calamagrostis stricta* and its hybridization with *C. canescens. Watsonia,* **9**: 181-182.

CRACKLES, F. E. (1974). Seeking to understand the flora of the East Riding of Yorkshire. *Naturalist, Hull,* **1974**: 1-17.

HOLMBERG, O. R. (1922). *Hartmans Handbok i Skandinaviens Flora,* **1**: 151. Stockholm.

NYGREN, A. (1946). The genesis of some Scandinavian species of *Calamagrostis. Hereditas,* **32**: 131-262.

NYGREN, A. (1962). Artificial and natural hybridization in European *Calamagrostis. Symb. bot. upsal.,* **17(3)**: 1-105.

ULBRICH, E. (1910). Ein für Mitteleuropa neuer *Calamagrostis*-Bastard: x *Calamagrostis conwentzii* Ulbrich (= *C. neglecta* x *lanceolata* S. Almqvist). *Reprium nov. Spec. Regni veg.,* **8**: 52-54.

701 × 700. *Agrostis* L. × *Calamagrostis* Adans.

(by C. A. Stace)

A few hybrids between species of *Agrostis* (including *A. stolonifera*) and *Calamagrostis* (all non-British species) are known in Europe.

701. *Agrostis* L.

(by A. D. Bradshaw)

2a × 5. *A. canina* L. subsp. *canina* × *A. stolonifera* L.

has been reported from Da, Fe, Ga and Su, but requires confirmation.

2b × 3. *A. canina* L. subsp. *montana* (Hartm.) Hartm. (*A. stricta* J. F. Gmel.) × *A. tenuis* Sibth. (*A. capillaris* L.?)

a. (? *A.* x *sanionis* Aschers. & Graebn.; ? *A.* x *mercieri* Aschers. & Graebn.).

b. Hybrids appear to be intermediate between the parents in all characters, the most useful being the ligule, palea and closing of the panicle. But all these characters are plastic, making the determination of hybrids only easy with plants in cultivation. The situation is confused also by the occurrence of individuals which could be presumed to be backcrosses, etc. The pollen fertility of the few putative hybrids that have been examined is about 80%, and caryopsis fertility about 50%.

c. Because of the difficulty of recognition the distribution of hybrids is not at all clear. They have been reported in many different parts of BI, and it is possible that they occur wherever the two parents coexist, which is usually on poor, acid soils. In Europe properly defined hybrids are known only from Da, Fe, Ga, He, Ho, Is, No, Rs and Su, but may well be much more common.

d. Evidence on the ease of crossing between the parents is confused. Although crossing has been reported (Davies, 1953) later evidence suggests that crossing is difficult or not successful at all (Widén, 1971), and that plants presumed to be hybrids are in fact selfs of *A. canina*. This could equally apply to the putative natural hybrids. The putative artificial and natural hybrids which have been examined tend to form about eight bivalents and three quadrivalents at meiosis, suggesting they might well be *A. canina*.

e. Both parents and hybrid $2n = 28$.

f. CAMUS, A. (1958). As above.

DAVIES, W. E. (1953). The breeding affinities of some British species of *Agrostis*. *Br. agric. Bull.*, 6: 313-315.

JONES, K. (1956a). Species differentiation in *Agrostis*, 2. The significance of chromosome pairing in the tetraploid hybrids of *Agrostis canina* subsp. *montana* Hartm., *A. tenuis* Sibth., and *A. stolonifera* L. *J. Genet.*, 54: 377-393.

PHILIPSON, W. R. (1937). A revision of the British species of the genus *Agrostis* Linn. *J. Linn. Soc., Bot.*, 51: 73-151.

WIDÉN, K.-G. (1971). The genus *Agrostis* L. in eastern Fennoscandia. Taxonomy and distribution. *Flora fenn.*, 5: 1-209. FIG.

2b × 5. *A. canina* L. subsp. *montana* (Hartm.) Hartm. (*A. stricta* J. F. Gmel.) × *A. stolonifera* L.

a. None.

b. The putative hybrid is apparently intermediate between its parents. It will be difficult to recognize, but artificially-produced hybrids are completely sterile. Little is known about the wild plant, and its occurrence requires confirmation.

c. It is reported only from Truro, v.c. 1 (1932), and from Fe, Ga, Ho and Su. Its parents are unlikely to coexist, though they may occur adjacently where heathland and damp, fertile ground meet.

d. The two parents can be crossed to produce a small number of hybrid caryopses. At meiosis the artificial hybrid forms 7–24 univalents and 3–9 bivalents.

e. Both parents and hybrid $2n = 28$.

f. PHILIPSON, W. R. (1937). As above.
WIDÉN, K.-G. (1971). As above. FIG.

4 × 3. *A. gigantea* Roth × *A. tenuis* Sibth. (*A. capillaris* L.?)

a. *A.* × *bjoerkmanii* Widén.

b. The difference between the two parents lies mainly in metrical characters, particularly the size of the leaves and panicles, the length of the ligule, and the size and abundance of rhizomes, *A. gigantea* being bigger and coarser. The hybrid lies in between. It is vegetatively vigorous but infertile (pollen-fertility *c* 45% and caryopsis-fertility *c* 50%). Backcross and F_2 individuals have been found abroad in the wild but are aneuploid and of low vigour.

c. The hybrid does not appear to be very common in BI. This may be due to the restriction of *A. gigantea,* and therefore of hybrids, to disturbed sandy soils, but is also due to lack of critical observation. It may be widespread in weedy situations. The hybrid is widespread in Fennoscandia on the sandy shores of rivers and lakes and also occurs on roadsides. Elsewhere in Europe its distribution is not clear because of taxonomic confusion.

d. The hybrid can be produced readily by crosses between the parents, which give a very high seed-set. Backcross and F_2 individuals can also be produced without great difficulty, but are aneuploid and of low vigour. The hybrid forms 14 bivalents and 7 univalents at meiosis.

e. *A. tenuis* $2n = 28$; *A. gigantea* $2n = 42$; hybrid $2n = 35$.

f. JONES, K. (1956b). Species differentiation in *Agrostis,* 3. *Agrostis gigantea* Roth and its hybrids with *A. tenuis* Sibth. and *A. stolonifera* L. *J. Genet.,* 54: 394-399.
STUCKEY, I. H. and BANFIELD, W. G. (1946). The morphological variations and occurrence of aneuploids in some species of *Agrostis* in Rhode Island. *Am. J. Bot.,* 33: 185-190.
WIDÉN, K.-G. (1971). As above. FIG.

5 × 3. *A. stolonifera* L. × *A. tenuis* Sibth. (*A. capillaris* L.?)

a. *A.* × *murbeckii* Fouillade ex Fournier (*A.* × *intermedia* C. A. Weber, *non* Balbis).

b. The parents are very similar which makes the recognition of the hybrid difficult. In the vegetative state the hybrid can only be recognized by its intermediate ligule. The inflorescence is intermediate in shape and usually half-closed when mature. In cultivation the hybrid is as vigorous as the parents and its growth-habit combines the high tiller density of *A. tenuis* with the ability to spread by stolons of *A. stolonifera*; both stolons and rhizomes may be present. Pollen- and caryopsis-fertilities are low, 0–15%. Most individuals appear to be F_1s, but occasionally individuals are found which may be backcrosses or F_2s.

c. The hybrid occurs wherever the two parents are to be found growing together, particularly in damp neutral pastures and near the coast.

throughout BI. In very old pastures, where vegetative growth is sufficient for survival, the hybrid may be the only *Agrostis* in the sward; in these situations, however, a number of different individual hybrid plants make up the population. It is widespread throughout BI, probably much more common than previously suspected, and is widespread in northern Europe especially on coasts in the spray zone of Fe, Ga, Ge, Ho, No, Rs and Su.

d. Crosses between the parents can be made easily giving a very high seed-set and vigorous F_1 plants. Subsequent generations can be produced only with difficulty because of the sterility of the F_1, which forms 7 bivalents and 14 univalents at meiosis. This forms the main barrier between the species. However, a second barrier is the different times of anthesis in the two species (1000–1130 hrs in *A. stolonifera* and 1300–1700 hrs in *A. tenuis*).

e. Both parents and hybrid $2n = 28$.

f. BRADSHAW, A. D. (1958). Natural hybridisation of *Agrostis tenuis* Sibth. and *A. stolonifera* L. *New Phytol.*, **57**: 66-84. FIG.

  CAMUS, A. (1958). As above.

  FOUILLADE, M. (1932). Sur les *Agrostis alba, vulgaris, castellana* et leurs hybrides. *Bull. Soc. bot. Fr.*, **79**: 789-804.

  JONES, K. (1956a). As above.

  WIDÉN, K.-G. (1971). As above. FIG.

4 × 5. *A. gigantea* Roth × *A. stolonifera* L.

a. None.

b. This hybrid is intermediate between the parents, most easily recognizable by the presence of both stolons and rhizomes, large obtuse ligules, and panicles remaining partially open after flowering. It is vigorous vegetatively but only about 25% pollen- and caryopsis-fertile.

c. The distribution in BI is not clear; plants are known for certain from v.c. 23 but probably are quite widespread where the parents coexist. Elsewhere there are records from Fe and Su.

d. Crosses between the parents can be made easily.

e. *A. stolonifera* $2n = 28$; *A. gigantea* $2n = 42$; hybrid $2n = 35$.

f. WIDEN, K.-G. (1971). As above. FIG.

8 × 5. *A. semiverticillata* (Forsk.) C. Chr. × *A. stolonifera* L.

a. *A.* x *robinsonii* Druce (?*A.* x *densissima* (Hack.) Druce).

b. The hybrid is intermediate and difficult to recognize because of the similarity of the parents. It differs from *A. stolonifera* by its larger palea, more compact inflorescence with shorter pedicels, and tendency to shed its spikelets as a whole at maturity. It is most easily recognized by its complete pollen-sterility and lack of seed-set. *A. densissima* was included by Dandy in *A. stolonifera*.

c. *A. semiverticillata* is only a casual in Br, but is naturalized in Guernsey which is the only recorded locality for the hybrid. The distribution of the hybrid elsewhere is not clear but it is probably to be found in southern Europe wherever *A. semiverticillata* occurs.

d. The ease of crossing is not certain but five caryopses were obtained from a single cross. The parents are highly self-incompatible. The hybrid forms 7 bivalents and 14 univalents at meiosis.

e. *A. stolonifera* 2*n* = 28; *A. semiverticillata* and hybrid (2*n* = 28).

f. BJÖRKMAN, S. O. (1960). Studies in *Agrostis* and related genera. *Symb. bot. upsal.* **17(1):** 1-112. FIG.

DANDY, J. E. (1958). *Op. cit.,* p. 165.

DRUCE, G. C. (1925). *Agrostis verticillata* Vill. x *palustris* Huds. (vel *alba* L.) *nov. hybr. Rep. B.E.C.,* **7:** 457.

DRUCE, G. C., HOWARTH, W. O. and ROBINSON, F. (1925). *Agrostis alba* x *verticillata. Rep. B.E.C.,* **7:** 744.

701 × 703. *Agrostis* L. × *Polypogon* Desf. = × *Agropogon* P. Fourn.

(by A. D. Bradshaw)

701/5 × 703/1. *Agrostis stolonifera* L. × *Polypogon monspeliensis* (L.) Desf.

a. x *A. littoralis* (Sm.) C. E. Hubbard. (Perennial Beard-grass).

b. The hybrid is intermediate between the parents in floral characters, with conspicuous awns yet persistent spikelets. In vegetative characters it more nearly resembles *A. stolonifera.* It is perennial and individual plants can persist for many years but they are not so vigorous as *A. stolonifera.* It varies in vigour perhaps because of hybridization with different ecotypes of *A. stolonifera* or because of habitat conditions. It is almost completely pollen- and caryopsis-sterile, with non-dehiscent anthers. Subsequent generations are not known.

c. The hybrid occurs temporarily and sporadically in coastal areas where the two species coexist, usually on damp sand and silty mud, in v.c. 9, 11, 13, 14, 16, 18, 27, 28, 34 and 41. Like *P. monspeliensis* the hybrid occurs occasionally as a casual on rubbish-tips, e.g. in v.c. 59, probably originating from discarded bird-seed. It has been reported from the coast of Europe from Da to Hs and North Africa and eastwards to Asia, and also from North America, but Hubbard (1954) stated that it is confined to Br and Ga.

d. The hybrid forms 7 bivalents and 14 univalents at meiosis.

e. Both parents and hybrid 2*n* = 28.

f. BJÖRKMAN, S. O. (1960). Studies in *Agrostis* and related genera. *Symb. bot. upsal.*, **17**(1): 1-112. FIG.

CAMUS, A. (1958). As above.

HUBBARD, C. E. (1968). As above, pp. 308-309. FIG.

PERRING, F. H. and SELL, P. D. (1968). *Op. cit.*, p. 152.

PHILIPSON, W. R. (1937). A revision of the British species of the genus *Agrostis* Linn. *J. Linn. Soc., Bot.*, **51**: 73-151.

707. *Phleum* L.

(by C. A. Stace)

1 × 2. *P. bertolonii* DC. × *P. pratense* L.
has been detected in Su.

708. *Alopecurus* L.

(by C. E. Hubbard)

1 × 2. *A. myosuroides* Huds. × *A. pratensis* L.
= *A.* × *turicensis* Brügger has been recorded from He, but is doubtful.

3 × 2. *A. geniculatus* L. × *A. pratensis* L.
a. *A.* × *hybridus* Wimm.
b. Hybrids are generally more or less intermediate between the parents in spikelet- and anther-size, although some plants may resemble one or the other parent more in habit and vigour. Such nothomorphs have been named f. *subgeniculatus* Holmb. and f. *subpratensis* Holmb. Anthers are generally sterile with mainly imperfect pollen but occasionally a few apparently good caryopses may be present. In a plant from v.c. 23 (coll. R. C. Palmer, 1969), with the general characteristics of the hybrid, there is

abundant good pollen. The flowers of the parents are usually markedly protogynous, in *A. pratensis* the stigmas protruding 2–5 days before the anthers begin to be exserted, so that hybridization is possible where species grow in close proximity.

c. The hybrid has been found in the lowlands of England northwards to v.c. 39 and 56 in marshy meadows and by streams and ponds, but is no doubt generally overlooked. It is also recorded from Au, Cz, Da, Fe, Ga, Ge, Ho, Hu, No, Po, Rs and Su.

d. Experimental hybridization in plants of *A. pratensis* has proved difficult on account of the densely crowded inflorescence.

e. Both parents ($2n = 28$).

f. ASCHERSON, P. and GRAEBNER, P. (1899). *Synopsis der Mitteleuropäischen Flora*, **2(1)**: 138-139. Leipzig.

HEIDENREICH, F. A. (1866). *Alopecurus pratensis* x *geniculatus*, beobachtet bei Tilsit in Östpreussen. *Öst. bot. Z.*, **16**: 277-281.

JACKSON, A. B. (1901). *Alopecurus hybridus* in Britain. *J. Bot., Lond.*, **39**: 232-234.

4 × 2. *A. aequalis* Sobol. × *A. pratensis* L.

= *A.* x *winkleranus* Aschers. & Graebn. has been recorded from Ga, Ge, Po and Su.

4 × 3. *A. aequalis* Sobol. × *A. geniculatus* L.

a. *A.* x *haussknechtianus* Aschers. & Graebn.

b. This hybrid is similar in general structure to its parents, but intermediate in spikelet-, awn- and anther-lengths. The anthers do not open and the pollen is mostly imperfect.

c. This rare hybrid was found by C. E. Hubbard in 1936 (**K**) at Appleton, v.c. 28, on the margin of a shallow pond, with both parents. It is also known from Au, Da, Fe, Ga, Ge, No and Su.

d. None.

e. *A. geniculatus* ($2n = 28$); *A. aequalis* ($2n = 14$).

f. ASCHERSON, P. and GRAEBNER, P. (1899). As above, p. 138.

PETCH, C. P. and SWANN, E. L. (1968). *Flora of Norfolk*, p. 258. Norwich.

5 × 3. *A. bulbosus* Gouan × *A. geniculatus* L.

= *A.* x *plettkei* Mattfeld has been recorded from Be, Ge and Ho.

716. *Spartina* Schreb.

(by C. J. Marchant)

2 × 1. *S. alterniflora* Lois. × *S. maritima* (Curt.) Fernald

a. *S.* × *townsendii* H. & J. Groves. (Townsend's Cord-Grass).

b. The F_1 hybrid (*S.* × *townsendii* H. & J. Groves) has given rise in Br to the amphidiploid *S. anglica* Hubbard, *nom. nud.*, which is now by far the commonest *Spartina* in BI. The sterile F_1 in general tends to be as large or larger than *S. alterniflora*, the more robust parent, demonstrating hybrid vigour. Many of its morphological characters tend towards those of *S. alterniflora*, such as the ligule-size, leaf-length and -width, number of spikes per culm, and strongly rhizomatous habit. Yet a weak F_1 individual can easily be confused with *S. maritima*. The high F_1 tiller density is like that of *S. maritima*. The leaf-blades make a less acute angle with the stem than in either parent. The anthers are often incompletely exserted and do not dehisce, with completely aborted grains, and there is no seed-set. The amphidiploid generally resembles the F_1 hybrid but is bigger in most respects (though not necessarily taller) with a lower tiller density. In contrast it is at least partially seed- and pollen-fertile.

c. The natural distribution is confused by introductions. The F_1 hybrid is abundant in and near Southampton Water, v.c. 9–11 and 13, its original locality and the only known British site where *S. alterniflora* was introduced and originally contacted *S. maritima*. Elsewhere there are sporadic patches among the large swards of the amphidiploid on the southern, western and eastern coasts of England and Wales, and in v.c. H21. Here it may be introduced or of secondary polyhaploid derivation. The amphidiploid occurs in most suitable coastal areas of England and Wales, and sporadically in Scotland and Hb. The F_1 and/or amphidiploid are also widespread in Da, Ga, Ge and Ho, and have been successfully introduced to New Zealand, Australia and north-western America. The Bay of Biscay, south-western Ga, may be a second location of the natural hybrid origin of *S.* × *townsendii* (as *S.* × *neyrautii* Foucaud).

d. Artificial hybrids have never been made despite repeated attempts. Chromosome doubling of the F_1 hybrid has also been unsuccessful. A polyhaploid was raised from amphidiploid caryopses of wild origin.

e. *S. alterniflora* $2n = 62$; *S. maritima* $2n = 60$; F_1 hybrid $2n = 62$; amphidiploid clones $2n = 120, 122, 124$. Other plants in Southampton Water suspected as backcrosses from the amphidiploid to *S. alterniflora* have $2n = c\ 90$ and $2n = 76$.

f. GOODMAN, P. J., BRAYBROOKS, E. M., LAMBERT, J. M. and MARCHANT, C. J. (1969). *Spartina* Schreb., in Biological Flora of the British Isles. *J. Ecol.*, 57: 285-313.

HUBBARD, C. E. (1968). As above, pp. 356-359. FIG.

HUSKINS, C. L. (1930). The origin of *Spartina townsendii. Genetica*, **12**: 531-538.

MARCHANT, C. J. (1967). Evolution in *Spartina* (Gramineae), 1. The history and morphology of the genus in Britain. *J. Linn. Soc., Bot.*, **60**: 1-24.

MARCHANT, C. J. (1968). Evolution in *Spartina* (Gramineae), 2. Chromosomes, basic relationships and the problem of *S.* x *townsendii* agg. *J. Linn. Soc., Bot.*, **60**: 381-409.

PERRING, F. H. and SELL, P. D. (1968). *Op. cit.*, p. 152.

PERRING, F. H. and WALTERS, S. M. (1962). *Op. cit.*, p. 405.

720. *Setaria* Beauv.

(by C. A. Stace)

2 x **1.** *S. verticillata* (L.) Beauv. x *S. viridis* (L.) Beauv.

= *S.* x *ambigua* Guss. has been reported from Co, Ga, Hs, It, Ju and parts of central Europe.

3. General bibliography

ANON. (1966). Interspecific incompatibility in *Lathyrus. Rep. John Innes hort. Instn*, **56**: 33-34.

ADAMS, H. and ANDERSON, E. (1958). A conspectus of hybridization in the Orchidaceae. *Evolution, Lancaster, Pa.*, **12**: 512-518.

ADAMS, R. P. and TURNER, B. L. (1970). Chemosystematic and numerical studies of natural populations of *Juniperus ashei* Buch. *Taxon*, **19**: 728-751.

AGNEW, A. D. Q. (1968). The interspecific relationships of *Juncus effusus* and *J. conglomeratus* in Britain. *Watsonia*, **6**: 377-388.

ALLARD, R. W. (1960). *Principles of plant breeding.* Wiley, New York.

ALLAN, H. H. (1937). Wild species-hybrids in the phanerogams, 1. *Bot. Rev.*, **3**: 593-615.

ALLAN, H. H. (1940). Natural hybridization in relation to taxonomy, in HUXLEY, J., ed. *The new systematics*, pp. 515-528. The Systematics Association, Oxford.

ALLAN, H. H. (1949). Wild species-hybrids in the phanerogams, 2. *Bot. Rev.*, **15**: 77-105.

ALLEN, D. E. (1957). G. C. Druce's discovery of introgression. *Proc. B.S.B.I.*, **2**: 292.

ALMQVIST, E. (1926). Newbred plant species. *Rep. B.E.C.*, **7**: 917-919.

ALSTON, R. E., RÖSLER, H., NAIFEH, K. and MABRY, T. J. (1965). Hybrid compounds in natural interspecific hybrids. *Proc. natn. Acad. Sci. U.S.A.*, **54**: 1458-1465.

ALSTON, R. E. and TURNER, B. L. (1962). Comparative chemistry of *Baptisia*: problems of intraspecific hybridization, in LEONE, C. A., ed. *Taxonomic biochemistry and serology*, pp. 225-238. Ronald Press, New York.

ALSTON, R. E. and TURNER, B. L. (1963a). *Biochemical systematics.* Prentice-Hall, Englewood Cliffs, New Jersey.

ALSTON, R. E. and TURNER, B. L. (1963b). Natural hybridization among four species of *Baptisia* (Leguminosae). *Am. J. Bot.*, **50**: 159-173.

ALSTON, R. E., TURNER, B. L., LESTER, R. N. and HORNE, D. (1962). Chromatographic validation of two morphologically similar hybrids of different origins. *Science, N.Y.*, **137**: 1048-1050.

ANDERSON, E. (1936). Hybridization in American Tradescantias. *Ann. Mo. bot. Gdn*, **23**: 511-525.

ANDERSON, E. (1939). Recombination in species crosses. *Genetics, Princeton*, **24**: 668-698.

ANDERSON, E. (1941). The technique and use of mass collections in plant taxonomy. *Ann. Mo. bot. Gdn*, **28**: 287-292.

ANDERSON, E. (1948). Hybridization of the habitat. *Evolution, Lancaster, Pa.*, **2**: 1-9.

ANDERSON, E. (1949). *Introgressive hybridization.* Wiley, New York.

ANDERSON, E. (1951). Concordant versus discordant variation in relation to introgression. *Evolution, Lancaster, Pa.*, **5**: 133-141.

ANDERSON, E. (1953a). Introgressive hybridization. *Biol. Rev.*, **28**: 280-307.

ANDERSON, E. (1953b). The analysis of suspected hybrids, as illustrated by *Berberis* x *gladwynensis*. *Ann. Mo. bot. Gdn*, **40**: 73-78.

ANDERSON, E. and HUBRICHT, L. (1938). Hybridization in *Tradescantia*, 3. The evidence for introgressive hybridization. *Am. J. Bot.*, **25**: 396-402.

ANDERSON, E. and STEBBINS, G. L. (1954). Hybridization as an evolutionary stimulus. *Evolution, Lancaster, Pa.*, **8**: 378-388.

ANDERSON, E. and TURRILL, W. B. (1935). Biometrical studies on herbarium material. *Nature, Lond.*, **136**: 986.

ANDERSON, S. (1931). Graes-hybrider i Danmark. *Bot. Tidsskr.*, **41**: 424-430.

ANDERSSON-KOTTO, I. and GAIRDNER, A. E. (1931). Interspecific crosses in the genus *Dianthus*. *Genetica*, **13**: 77-112.

BAKER, H. G. (1947). Criteria of hybridity. *Nature, Lond.*, **159**: 221-223.

BAKER, H. G. (1951). Hybridization and natural gene-flow between higher plants. *Biol. Rev.*, **26**: 302-337.

BAKER, H. G. (1958). Hybridization between dioecious and hermaphrodite species in the Caryophyllaceae. *Evolution, Lancaster, Pa.*, **12**: 423-427.

BARBER, H. N. (1970). Hybridization and the evolution of plants. *Taxon*, **19**: 154-160.

BARSKI, G., SORIEUL, S. and CORNEFERT, F. (1960). Production dans des cultures *in vitro* de deux souches cellulaires en association, de cellules de caractère "hybride". *C. r. hebd. Séanc. Acad. Sci., Paris*, **251**: 1825-1827.

BAUM, B. R. (1969). On the application of nomenclature to the taxonomy of hybrids. *Taxon*, **18**: 670-671.

BEAN, W. J. (1911). Graft hybrids. *Kew Bull.*, **1911**: 267-269.

BEASLEY, J. O. (1940). Hybridization of American 26-chromosome and Asiatic 13-chromosome species of *Gossypium*. *J. agric. Res.*, **60**: 175-181.

BEEBY, W. H. (1892). On natural hybrids. *J. Bot., Lond.*, **30**: 209-212.

BLACKBURN, K. B. and HARRISON, J. W. H. (1924). Genetical and cytological studies in hybrid roses, 1. The origin of a fertile hexaploid form in the *Pimpinellifoliae–Villosae* crosses. *Br. J. exp. Biol.*, **1**: 557-570.

BLAKESLEE, A. F. (1945). Removing some of the barriers to crossability in plants. *Proc. Am. phil. Soc.*, **89**: 561-574.

BLAKESLEE, A. F. and SATINA, S. (1949). Differences in crossability between species of *Datura* due to individual races used in the cross. *Am. J. Bot.*, **36**: 795.

BORRILL, M. (1961). The pattern of morphological variation in diploid and tetraploid *Dactylis*. *J. Linn. Soc., Bot.*, **56**: 441-452.

BOSIO, M. G. (1940). Ricerche sulla fecondazione intraovarica in *Helleborus* & *Paeonia*. *Nuovo G. bot. ital.*, **47**: 591-598.

BRIGGS, D. and WALTERS, S. M. (1969). *Plant variation and evolution*. Weidenfeld and Nicolson, London.

BRINK, R. A. and COOPER, D. C. (1947). The endosperm in seed development. *Bot. Rev.*, **13**: 423-541.

BRINK, R. A., COOPER, D. C. and AUSCHERMAN, L. E. (1944). A hybrid between *Hordeum jubatum* and *Secale cereale* reared from an artificially cultivated embryo. *J. Hered.*, **35**: 67-75.

BROBOV, E. G. (1973). Introgressive hybridization, Sippenbildung und Vegetationanderung. *Reprium Spec. nov. Regni veg.*, **84**: 273-294.

BROWN, P. and STRATTON, G. B. (1963–65). *World list of scientific periodicals published in the years 1900–1960*, 4th ed. Butterworths, London.

BRÜGGER, G. (1880). Wildwachsende Pflanzenbastarde in der Schweiz und deren Nachbarschaft. *Jber. naturf. Ges. Graubündens*, **23-24**: 47-123.

BUCHHOLZ, J. T., DOAK, C. C. and BLAKESLEE, A. F. (1932). Control of gametophytic selection in *Datura* through shortening and splicing of styles. *Bull. Torrey bot. Club*, **59**: 109-118.

BUNYARD, E. A. (1907). On xenia, in WILKS, W., ed. *Report of the third international conference 1906 on genetics*, pp. 297-300. Royal Horticultural Society, London.

CAMUS, A. (1957). *Bromus* hybrides de la flore française. *Bull. Jard. bot. État Brux.*, **27**: 479-485.

CAMUS, A. (1958). Graminées hybrides de la flore française (genus *Bromus* excepté). *Bull. Jard. bot. État Brux.*, **28**: 337-374.

CAMUS, E. G. (1907). A contribution to the study of spontaneous hybrids in the European flora, in WILKS, W., ed. *Report of the third international conference 1906 on genetics*, pp. 150-154. Royal Horticultural Society, London.

CANNON, W. A. (1909). Studies in heredity as illustrated by the trichomes of species and hybrids of *Juglans, Oenothera, Papaver* and *Solanum. Publs Carnegie Instn*, **117**.

CARLILE, M. J. (1973). Cell fusion and somatic incompatibility in myxomycetes. *Ber. dt. bot. Ges.*, **86**: 123-140.

CARLISLE, A. and BROWN, A. H. F. (1965). The assessment of the taxonomic status of mixed oak (*Quercus* spp.) populations. *Watsonia*, **6**: 120-127.

CARLSON, P. S., SMITH, H. H. and DEARING, R. D. (1972). Parasexual interspecific plant hybridisation. *Proc. natn. Acad. Sci. U.S.A.*, **69**: 2292-2294.

CARNAHAN, H. L. and HILL, H. D. (1961). Cytology and genetics of forage grasses. *Bot. Rev.*, **27**: 1-162.

CASPARI, E. (1948). Cytoplasmic inheritance. *Adv. Genet.*, **2**: 1-66.

CHHEDA, H. R. and HARLAN, J. R. (1962). Mode of chromosome association in *Bothriochloa* hybrids. *Caryologia*, **15**: 461-476.

CLAPHAM, A. R., TUTIN, T. G. and WARBURG, E. F. (1952). *Flora of the British Isles.* Cambridge University Press.

CLAPHAM, A. R., TUTIN, T. G. and WARBURG, E. F. (1962). *Flora of the British Isles*, 2nd ed. Cambridge University Press.

CLARKE, C. B. (1891). *Epilobium Duriaei* J. Gay, a new (?) English plant. *J. Bot., Lond.*, **29**: 225-228.

CLARKE, C. B. (1892a). On *Epilobium Duriaei* J. Gay. *J. Bot., Lond.*, **30**: 78-81.

CLARKE, C. B. (1892b). On *Holoschoenus. J. Bot., Lond.*, **30**: 321-323.

CLAUSEN, J. (1951). *Stages in the evolution of plant species.* Cornell University Press, Ithaca, New York.

CLAUSEN, J., KECK, D. D. and HIESEY, W. M. (1945). Experimental studies on the nature of species, 2. Plant evolution through amphiploidy and autoploidy, with examples from the Madiinae. *Publs Carnegie Instn*, **564**.

CLELAND, R. E. (1972). *Oenothera: cytogenetics and evolution.* Academic Press, London and New York.

CLIFFORD, H. T. (1954). Analysis of suspected hybrid swarms in the genus *Eucalyptus. Heredity*, **8**: 259-269.

CLIFFORD, H. T. (1955). An index for use in quantitative taxonomic problems. *New Phytol.*, **54**: 132-137.

CLIFFORD, H. T. (1961). Factors affecting the frequencies of wild plant hybrids. *Bot. Rev.*, **27**: 561-579.

CLIFFORD, H. T. (1963). Angiosperm hybrids in the Australian flora. *Qd Nat.*, **17**: 32-34.

CLIFFORD, H. T. and BINET, F. E. (1954). A quantitative study of a presumed hybrid swarm between *Eucalyptus elaeophora* and *E. goniocalyx. Aust. J. Bot.*, **2**: 325-336.

COCKAYNE, L. (1923). Hybridism in the New Zealand flora. *New Phytol.*, **22**: 105-127.

COCKAYNE, L. and ALLAN, H. H. (1934). An annotated list of groups of wild hybrids in the New Zealand flora. *Ann. Bot.*, **48**: 1-55.

COCKAYNE, L. and ATKINSON, E. (1926). On the New Zealand wild hybrids of *Nothofagus. Genetica*, **8**: 1-43.

COOK, C. D. K. (1970). Hybridization in the evolution of *Batrachium. Taxon*, **19**: 161-166.

COUSENS, J. E. (1965). The status of the pedunculate and sessile oaks in Britain. *Watsonia*, **6**: 161-176.

CRANE, M. B. and LAWRENCE, W. J. C. (1956). *The genetics of garden plants*, 4th ed. Macmillan, London.

CRANE, M. B. and MARKS, E. (1952). Pear–apple hybrids. *Nature, Lond.*, **170**: 1017.

CUGNAC, A. de (1937). Tendences et possibilités actuelles de la génétique végétale. *Annls Sci. nat., Bot.*, Sér. X, **19**: 113-123.

CULBERSON, C. F. and HALE, M. E. (1973). Chemical and morphological evolution in *Parmelia* sect. *Hypotrachyna*: product of ancient hybridization? *Brittonia*, **25**: 163-173.

CUTLER, D. F. (1972). Leaf anatomy of certain *Aloe* and *Gasteria* species and their hybrids, in GHOUSE, A. K. M. and YUNIS, M., eds. *Research trends in plant anatomy*, pp. 103-122. Tata McGraw-Hill, Delhi.

DANDY, J. E. (1958). *List of British vascular plants.* B.S.B.I., London.

DANDY, J. E. (1969a). *Nomenclatural changes in the list of British vascular plants. Watsonia*, **7**: 157-178.

DANDY, J. E. (1969b). *Watsonian vice-counties of Great Britain.* Ray Society, London.

DANIEL, L. (1914–1915). L'hybridation asexuelle ou variation spécifique chez les plantes greffées. *Revue gén. Bot.*, **26**: 305-341; **27**: 22-29, 33-49.

DANSER, B. H. (1929). Über die Begriffe Komparium, Kommiskuum und Konvivium und über die Entstehungsweise der Konvivien. *Genetica*, **11**: 399-450.

DARLINGTON, C. D. (1937). What is a hybrid? *J. Hered.*, **28**: 308.

DAVIES, A. J. S. (1957). Successful crossing in the genus *Lathyrus* through stylar amputation. *Nature, Lond.*, **180**: 612.

DAVIES, D. R. (1974). Chromosome elimination in interspecific hybrids. *Heredity*, **32**: 267-270.

DAVIS, P. H. and HEYWOOD, V. H. (1963). *Principles of angiosperm taxonomy.* Oliver and Boyd, Edinburgh.

DE WET, J. M. J. and HARLAN, J. R. (1972). Chromosome pairing and phylogenetic affinities. *Taxon*, **21**: 67-70.

DIELS, L. (1921). Die Methoden der Phytographie und der Systematik der Pflanzen. *Handb. der Biol. Arbeitsmethoden*, Abt. 11, I(2): 67-190. Berlin.

DILLEMANN, G. (1948). Remarques sur l'hybridation spontanée des Linaires dans les jardins botaniques. *Bull. Mus. Hist. nat., Paris*, **20**: 546-547.

DILLEMANN, G. (1954). L'hybridation interspécifique naturelle. *Bull. Soc. bot. Fr.*, **101**: 36-87.

DOBZHANSKY, T. (1937). *Genetics and the origin of species.* Columbia University Press, New York.

DOBZHANSKY, T. (1951). *Genetics and the origin of species,* 3rd ed. Columbia University Press, New York.

DRUCE, G. C. (1896). The hybrids of *Linaria repens* and *L. vulgaris* in Britain. *Ann. Bot.*, **10**: 622-624.

DRUCE, G. C. (1897). *The flora of Berkshire,* pp. 368-370. Clarendon, Oxford.

DRUCE, G. C. (1928). *British plant list.* Buncle, Arbroath.

DRUCE, G. C. (1929a). Notes on the second edition of the "British plant list." *Rep. B.E.C.*, **8**: 867-877.

DRUCE, G. C. (1929b). British plant list (Edition II). Additions and corrections. *Rep. B.E.C.*, **8**: 878-883.

DRUERY, C. T. (1900). Fern crossing and hybridising. *Jl R. hort. Soc.*, **24**: 288-297.

DRUERY, C. T. (1907). Fern breeding, in WILKS, W., ed. *Report of the third international conference 1906 on genetics,* pp. 273-277. Royal Horticultural Society, London.

EAST, E. M. (1935). Genetic reactions in *Nicotiana*, 3. Dominance. *Genetics, Princeton*, **20**: 443-451.

EAST, E. M. and PARK, J. B. (1917). Studies of self-sterility, 1. The behaviour of self-sterile plants. *Genetics, Princeton*, **2**: 505-609.

EATON, R. D. (1973). The evolution of seed incompatibility in *Primula*. *New Phytol.*, **72**: 855-860.

EBINGER, J. E. (1962). *Luzula* x *borreri* in England. *Watsonia*, **5**: 251-254.

EDWARDSON, J. R. (1970). Cytoplasmic male sterility. *Bot. Rev.*, **36**: 341-420.

ELLENGORN, J. E. and PETROVA, K. A. (1948). The differential staining of hybrid chromosomes. *Bot. Zhurn.*, **33**: 40-44.

ELLIOT, E. (1950). A new phase of amphiplasty in *Leontodon*. *New Phytol.*, **49**: 344-349.

ELLIS, J. R. (1962). *Fragaria-Potentilla* intergeneric hybridization and evolution in *Fragaria*. *Proc. Linn. Soc. Lond.*, **173**: 99-106.

EMERSON, R. and WILSON, C. M. (1954). Interspecific hybrids and the cytogenetics and cytotaxonomy of *Euallomyces*. *Mycologia*, **46**: 393-434.

EPHRUSSI, B. (1972). *Hybridization of somatic cells*. Princeton University Press, New Jersey.

EPLING, C. (1947). Natural hybridization of *Salvia apiana* and *S. mellifera*. *Evolution, Lancaster, Pa.*, **1**: 69-78.

ERNST, A. (1953). Muttergleiche Nachkommen nach interspezifischen Kreuzungen bei Blütenpflanzen. *Experientia*, **9**: 7-16.

EXELL, A. W. and STACE, C. A. (1972). Patterns of distribution in the Combretaceae, in VALENTINE, D. H., ed. *Taxonomy, phytogeography and evolution*, pp. 307-323. Academic Press, London and New York.

FAGERLIND, F. (1937). Embryologische, zytologische und bestäubungsexperimentelle Studien in der Familie Rubiaceae nebst Bemerkungen über einige Polyploidistätsprobleme. *Acta Horti Bergiani*, **11**: 195-470.

FOCKE, W. O. (1881). *Die Pflanzenmischlinge. Ein Beitrag zur Biologie der Gewächse*. Borntraeger, Berlin.

FOTHERGILL, P. G. (1938). Studies in *Viola*, 1. The cytology of a naturally-occurring population of hybrids between *Viola tricolor* L. and *Viola lutea* Huds. *Genetica*, **20**: 159-186.

GAJEWSKI, W. (1957). A cytogenetic study of the genus *Geum* L. *Monographiae bot.*, **4**.

GARAY, L. A. and SWEET, H. R. (1974). Hybrid generic names, in WITHNER, C. L., ed. *The orchids. Scientific studies*. London.

GARDOU, C. (1967). Sur quelques tentatives d'hybridation expérimentale intraspécifique chez *Centaurea jacea* L. s. lat. *C.r. hebd. Séanc. Acad. Sci., Paris*, **265**: 1195-1198.

GAY, P. A. (1960). A new method for the comparison of populations that contain hybrids. *New Phytol.*, **59**: 219-226.

GILMOUR, J. S. L. *et al.*, eds (1969). International code of nomenclature for cultivated plants. *Regnum veg.*, **64**.

GILMOUR, J. S. L. and GREGOR, J. W. (1939). Demes: a suggested new terminology. *Nature, Lond.*, **144**: 333.

GILMOUR, J. S. L. and HESLOP-HARRISON, J. (1954). The deme terminology and the units of microevolutionary change. *Genetica*, **27**: 147-161.

GOODMAN, M. M. (1967). The identification of hybrid plants in segregating populations. *Evolution, Lancaster, Pa.*, **21**: 334-340.

GOODSPEED, T. H. and BRADLEY, M. V. (1942). Amphidiploidy. *Bot. Rev.*, **8**: 271-316.

GORSHKOV, I. S. (1962). Wide hybridization in tree- and soft-fruit crops, in TSITSIN, N. V., ed. *Wide hybridization in plants*, pp. 31-38. Israel Programme of Scientific Translations, Jerusalem.

GORYUNOV, D. V. (1962). Notes on the history of wide hybridization, in TSITSIN, N. V., ed. *Wide hybridization in plants*, pp. 62-76. Israel Programme of Scientific Translations, Jerusalem.

GOTTLIEB, L. D. (1974). Levels of confidence in the analysis of hybridization in plants. *Ann. Mo. bot. Gdn*, **59**: 435-446.

GOWEN, J. W., ed. (1952). *Heterosis.* State College Press, Ames, Iowa.

GRAHAM, R. A. (1958). Mint notes, 8. A new mint from Scotland. *Watsonia,* 4: 119-121.

GRANT, V. (1952). Isolation and hybridization between *Aquilegia formosa* and *A. pubescens. Aliso,* 2: 341-360.

GRANT, V. (1953). The role of hybridization in the evolution of the leafy stemmed Gilias. *Evolution, Lancaster, Pa.,* 7: 51-64.

GRANT, V. (1956). The influence of breeding habit on the outcome of natural hybridization in plants. *Am. Nat.,* 90: 319-322.

GRANT, W. F. (1965). A chromosome atlas and interspecific hybridization index for the genus *Lotus* (Leguminosae). *Can. J. Genet. Cytol.,* 7: 457-471.

GRELL, R. F. (1965). Chromosome pairing, crossing over and segregation in *Drosophila melanogaster. Natn. Cancer Inst. Monogr.,* 18: 215-242.

GUÉTROT, M., ed. (1927-1931). *Plantes hybrides de France.* (Private publication.) 1 and 2 (1927) Lille; 3 and 4 (1929) Paris; 5-7 (1931) Gap.

GYMER, P. T. and WHITTINGTON, W. J. (1973a). Hybrids between *Lolium perenne* L. and *Festuca pratensis* Huds., 1. Crossing and incompatibility. *New Phytol.,* 72: 411-424.

GYMER, P. T. and WHITTINGTON, W. J. (1973b). Hybrids between *Lolium perenne* and *Festuca pratensis,* 2. Comparative morphology. *New Phytol.,* 72: 861-865.

HAARTMAN, J. (1751). Plantae hybridae. *Amoenitates academicae,* 3: 28-62.

HALL, M. T. (1952). Variation and hybridization in *Juniperus. Ann. Mo. bot. Gdn,* 39: 1-64.

HANBURY, F. J., ed. (1895). *The London catalogue of British plants,* 9th ed. G. Bell, London.

HANBURY, F. J., ed. (1925). *The London catalogue of British plants,* 11th ed. G. Bell, London.

HANSEN, A. (1959). Die Gras-Hybriden in der Flora Frankreichs. Kritik und Ergänzungen. *Bull. Jard. bot. État Brux.,* 29: 61-68.

HARBERD, D. J. and MACARTHUR, E. D. (1972). The chromosome constitution of *Diplotaxis muralis* (L.) DC. *Watsonia,* 9: 131-135.

HARBORNE, J. B. (1968). The use of 27 chemical characters in the systematics of higher plants, in HAWKES, J. G., ed. *Chemotaxonomy and serotaxonomy,* pp. 173-191. Academic Press, London and New York.

HARLEY, R. M. (1972). *Mentha* L., in TUTIN, T. G. *et al.,* eds. *Flora Europaea,* 3: 183-186. Cambridge University Press.

HARRIS, H. (1970). *Cell Fusion.* Clarendon, Oxford.

HARRIS, H. and WATKINS, J. F. (1965). Hybrid cells derived from mouse and man: artificial heterokaryons of mammalian cells from different species. *Nature, Lond.,* 205: 640-646.

HARRISON, B. J. and DARBY, L. (1955). Unilateral hybridization. *Nature, Lond.,* 176: 982.

HATHEWAY, W. H. (1962). Weighted hybrid index. *Evolution, Lancaster, Pa.,* 16: 1-10.

HAUSSKNECHT, C. (1884). *Monographie der Gattung Epilobium.* Fischer, Jena.

HAYES, H. K., IMMER, F. R. and SMITH, D. C. (1955). *Methods of plant breeding*, 2nd ed. Prentice-Hall, New York.

HEISER, C. B. (1949). Natural hybridization with particular reference to introgression. *Bot. Rev.*, **15**: 645-687.

HEISER, C. B. (1961). Natural hybridization and introgression with particular reference to *Helianthus*. *Recent Adv. Bot.*, **1**: 874-877.

HEISER, C. B. (1973). Introgression re-examined. *Bot. Rev.*, **39**: 347-366.

HEMINGWAY, J. S., SCHOFIELD, H. J. and VAUGHAN, J. G. (1961). Volatile mustard oils of *Brassica juncea* seeds. *Nature, Lond.*, **192**: 993.

HEYWOOD, V. H. (1967). *Plant taxonomy*. Edward Arnold, London.

HILL, J. B. (1929). Matrocliny in flower size in reciprocal F_1 hybrids between *Digitalis lutea* and *D. purpurea*. *Bot. Gaz.*, **87**: 548-555.

HOLLINGS, E. and STACE, C. A. (1974). Karyotype variation and evolution in the *Vicia sativa* aggregate. *New Phytol.*, **73**: 195-208.

HOOKER, J. D. (1853). *The botany of the Antarctic voyage*, 2. *Flora Novae-Zelandiae*, 1. *Flowering Plants*. Lovell Reeve, London.

HUBBARD, C. E. (1968). *Grasses*, 2nd ed. Penguin, Harmondsworth.

HUSKINS, C. L. (1929). Criteria of hybridity. *Science, N.Y.*, **69**: 399-400.

IVANOV, M. A. (1928). Experimental production of hybrids in *Nicotiana rustica* L. *Genetica*, **20**: 295-397.

JEFFREY, E. C. (1914). Spore conditions in hybrids and the mutation hypothesis of De Vries. *Bot. Gaz.*, **58**: 322-336.

JENKIN, T. J. (1924). The artificial hybridization of grasses. *Bull. Welsh Pl. Breed. Stn*, Ser. H, **2**: 1-18.

JENKIN, T. J. (1931). The method and technique of selection, breeding and strain-building in grasses. *Bull. imp. Bur. Pl. Genet. Herb.*, **3**: 5-34.

JOHNSON, L. and HALL, O. (1965). Analysis of phylogenetic affinities in the Triticinae by protein electrophoresis. *Am. J. Bot.*, **52**: 506-513.

JOHNSON, L. P. V. (1939). A descriptive list of natural and artificial interspecific hybrids in North American forest-tree genera. *Can. J. Res.*, Ser. C, **17**: 411-444.

JOHNSON, L. P. V. and BRADLEY, E. C. (1946). Hybridisation technique for forest trees. *Can. J. Res.*, Ser. C, **24**: 305-307.

JORDAN, H. D. (1957). Hybridisation of rice. *Trop. Agric., Trin.*, **34**: 133-136.

JØRGENSEN, C. A. (1928). The experimental formation of heteroploid plants in the genus *Solanum*. *J. Genet.*, **19**: 133-210.

KAO, K. N. and MICHAYLUK, M. R. (1974). A method for high-frequency intergeneric fusion of plant protoplasts. *Planta*, **115**: 355-367.

KARPECHENKO, G. D. (1927). Polyploid hybrids of *Raphanus sativus* L. x *Brassica oleracea* L. *Trudy prikl. Bot. Genet. Selek.*, **17**: 305-410.

KENT, D. H. (1958). *British herbaria*. B.S.B.I., London.

KIHARA, H. (1940). How to make difficult crosses successful—a suggestion. *Pl. Breed. Abstr.*, **10**: 1009.

KIMBER, G. and RILEY, R. (1963). Haploid angiosperms. *Bot. Rev.*, **29**: 480-531.

KIRKPATRICK, J. B., SIMMONS, D. and PARSONS, R. F. (1973). The relationship of some populations involving *Eucalyptus cypellocarpa* and *E. globulus* to the problem of phantom hybrids. *New Phytol.*, **72**: 867-876.

KLÁŠTERSKÝ, I. (1968). *Rosa* L., in TUTIN, T. G. *et al.*, eds. *Flora Europaea*, **2**: 25-32. Cambridge University Press.

KNOBLOCH, I. W. (1968). *A check list of crosses in the Gramineae.* (Private publication.) E. Lansing, Michigan.

KNOBLOCH, I. W. (1972). Intergeneric hybridization in flowering plants. *Taxon*, **21**: 97-103.

KNOBLOCH, I. W. (1973). The present status of hybridity among the pteridophytes. *Taiwania*, **18**: 29-34.

KOHNE, D. E. (1968). Taxonomic applications of DNA hybridization techniques, in HAWKES, J. G., ed. *Chemotaxonomy and serotaxonomy*, pp. 117-130. Academic Press, London and New York.

KOSHY, T. K. (1968). Evolutionary origin of *Poa annua* L. in the light of karyotypic studies. *Can. J. Genet. Cytol.*, **10**: 112-118.

KRUCKEBERG, A. R. (1962). Intergeneric hybrids in the Lychnideae (Caryophyllaceae). *Brittonia*, **14**: 311-321.

LAIBACH, F. (1925). Das Taubwerden von Bastardsamen und die künstliche Aufzucht früh absterbender Bastardembryonen. *Z. Bot.*, **17**: 417-459.

LANJOUW, J. and STAFLEU, F. A. (1964). Index herbariorum, Part I. The herbaria of the world, 5th ed. *Regnum veg.*, **31**.

LASCH, W. (1831). Beitrag zur Kenntnis der Varietäten und Bastardformen einheimischer Gewächse. *Linnaea*, **6**: 484-500.

LASCH, W. (1857). Aufzählung der in der Provinz Brandenburg, besonders in der Gegend um Driesen, wildwachsenden Bastard-Pflanzen, nebst kurzen Notizen zur Erkennung solcher Gewächse. *Bot. Ztg*, **15**: 505-517.

LAWRENCE, W. J. C. (1936). The origin of new forms in *Delphinium*. *Genetica*, **18**: 109-115.

LEHMANN, E. (1941). Zur Genetik der Entwicklung in der Gattung *Epilobium*, 4. Das Plasmon in der Gattung *Epilobium*. *Jb. wiss. Bot.*, **89**: 687-753; **90**: 49-98.

LEVEILLÉ, H. (1917). Les hybrides de France. *Bull. Géogr. bot.*, **27**: 34-68, 100.

LEVIN, D. A. (1970). Hybridization and evolution—a discussion. *Taxon*, **19**: 167-171.

LEVIN, D. A. (1971). The origin of reproductive isolating mechanisms in flowering plants. *Taxon*, **20**: 91-114.

LEWIS, D. and CROWE, L. K. (1958). Unilateral interspecific incompatibility in flowering plants. *Heredity*, **12**: 233-256.

LEWIS, H. (1967). The taxonomic significance of autopolyploidy. *Taxon*, **16**: 267-271.

LEWIS, H. and EPLING, C. (1959). *Delphinium gypsophilum*, a diploid species of hybrid origin. *Evolution, Lancaster, Pa.*, **13**: 511-525.

LING, H. and LING, M. (1974). Genetic control of somatic cell fusion in a myxomycete. *Heredity*, **32**: 95-104.

LINNAEUS, C. (1786). *A dissertation on the sexes of plants*, trans. SMITH, J. E. (Private publication). London.

LINTON, E. F. (1907). Hybrids among British phanerogams. *J. Bot., Lond.*, **45**: 268-276, 296-304.

LITTLE, E. L. (1960). Designating hybrid forest trees. *Taxon*, **9**: 225-231.

LJUNGDAHL, H. (1924). Über die Herkunft der in der meiosis konjugierenden Chromosomen bei *Papaver*-hybriden. *Svensk bot. Tidskr.*, **18**: 279-291.

LOTSY, J. P. (1916). *Evolution by means of hybridization.* Martinus Nijhoff, The Hague.

/LOTSY, J. P. and GODDIJN, W. A. (1928). Voyages and exploration to judge of the bearing of hybridisation upon evolution. *Genetica*, **10**: 1-315.

LYNCH, R. I. (1907). Natural hybrids, in WILKS, W., ed. *Report of the third international conference 1906 on genetics*, pp. 159-177. Royal Horticultural Society, London.

MACDOUGAL, D. T. (1907). Hybridization of wild plants. *Bot. Gaz.*, **43**: 45-58.

McNAUGHTON, I. H. and HARPER, J. L. (1960a). The comparative ecology of closely related species living in the same area, 1. External breeding-barriers between *Papaver* species. *New Phytol.*, **59**: 15-26.

McNAUGHTON, I. H. and HARPER, J. L. (1960b). The comparative ecology of closely related species living in the same area, 2. Aberrant morphology and a virus-like syndrome in hybrids between *Papaver rhoeas* and *P. dubium. New Phytol.*, **59**: 27-41.

MANGELSDORF, A. J. and EAST, E. M. (1927). Studies on the genetics of *Fragaria. Genetics, Princeton*, **12**: 307-339.

MANGELSDORF, P. C. and REEVES, R. G. (1931). Hybridization of maize, *Tripsacum* and *Euchlaena. J. Hered.*, **22**: 329-343.

MANTON, I. (1935). The cytological history of watercress (*Nasturtium officinale* R.Br.). *Z. indukt. Abstamm.- u. VererbLehre*, **69**: 132-157.

MANTON, I. and WALKER, S. (1954). Induced apogamy in *Dryopteris dilatata* (Hoffm.) A. Gay and *D. filix-mas* (L.) Schott. emend. and its significance for the interpretation of the two species. *Ann. Bot.*, n.s., **18**: 377-383.

MARSDEN-JONES, E. M. (1930). The genetics of *Geum intermedium* Willd. haud Ehrh., and its backcrosses. *J. Genet.*, **23**: 377-395.

MARSHALL, E. S. (1891). On the supposed occurrence of *Epilobium Duriaei* J. Gay in England. *J. Bot., Lond.*, **29**: 296-298.

MARSHALL, E. S. (1892). *Epilobium Duriaei*: a rejoinder. *J. Bot., Lond.*, **30**: 106-108.

MARSHALL, E. S. (1893). Do natural hybrids exist? *J. Bot., Lond.*, **31**: 20.

MATHER, K. (1947). Species crosses in *Antirrhinum*, 1. Genetic isolation of the species *majus, glutinosum* and *orontium. Heredity*, **1**: 175-186.

MATTFELD, J. (1930). Über hybridogene Sippen der Tannen. *Biblthca bot.*, **25**: 1-84.

MAYR, E. (1942). *Systematics and the origin of species.* Columbia University Press, New York.

MELVILLE, R. (1939). The application of biometrical methods to the study of elms. *Proc. Linn. Soc. Lond.*, **151**: 152-159.

MELVILLE, R. (1955). Morphological characters in the discrimination of species and hybrids, in LOUSLEY, J. E., ed. *Species studies in the British flora*, pp. 55-64. B.S.B.I., London.

MILLARDET, M. A. (1894). Note sur l'hybridation sans croisement ou fausse hybridation. *Mém. Soc. Sci. phys. Inst. Bordeaux*, **4**: 347-372.

MIROV, N. T. (1956). Composition of turpentine of lodgepole x jack pine hybrids. *Can. J. Bot.*, **34**: 443-457.

MOORE, D. M. (1959). Population studies on *Viola lactea* Sm. and its wild hybrids. *Evolution, Lancaster, Pa.*, **13**: 318-332.

MÜNTZING, A. (1929). Cases of partial sterility in crosses within a Linnean species. *Hereditas*, **12**: 297-319.

MÜNTZING, A. (1930). Outlines to a genetic monograph of the genus *Galeopsis* with special reference to the nature and inheritance of partial sterility *Hereditas*, **13**: 185-341.

MYERS, W. M. (1947). Cytology and genetics of forage grasses. *Bot. Rev.*, **13**: 319-421.

NAVASCHIN, M. (1934). Chromosome alterations caused by hybridization and their bearing upon certain general genetic problems. *Cytologia*, **5**: 169-203.

NEILREICH, A. (1852). Über hybride Pflanzen der Wiener Flora. *Verh. zool.-bot. Ges. Wien*, **1**: 114-131.

NEILSON-JONES, W. (1969). *Plant chimeras*, 2nd ed. Methuen, London.

NEWTON, W. C. F. and PELLEW, C. (1929). *Primula kewensis* and its derivatives. *J. Genet.*, **20**: 405-467.

NICHOLSON, W. E. (1905). Notes on two forms of hybrid *Weisia*. *Revue bryol.*, **32**: 19-25.

NICHOLSON, W. E. (1906). *Weisia crispa*, Mitt. ♀ x *W. microstoma* C.M.♂. *Revue bryol.*, **33**: 1-2.

NICHOLSON, W. E. (1910). A new hybrid moss. *Revue bryol.*, **37**: 23-24.

NILSSON, N. H. (1928). *Salix laurina. Acta Univ. lund.*, Avd. 2, **24**(6): 1-88.

NILSSON, N. H. (1954). Über Hochkomplexe Bastardverbindungen in der Gattung *Salix. Hereditas*, **40**: 517-522.

NORDENSKIÖLD, H. (1956). Cyto-taxonomical studies in the genus *Luzula*, 2. Hybridization experiments in the *campestris-multiflora* complex. *Hereditas*, **42**: 7-73.

NYGREN, A. (1948). Some interspecific crosses in *Calamagrostis* and their evolutionary consequences. *Hereditas*, **34**: 387-413.

NYGREN, A. (1949). Apomictic and sexual reproduction in *Calamagrostis purpurea. Hereditas*, **35**: 285-300.

OSTENFELD, C. H. (1928). The present state of knowledge on hybrids between species of flowering plants. *Jl R. hort. Soc.*, **53**: 31-44.

OSTENFELD, C. H. (1929). Genetic studies in *Polemonium*, 2. Experiments with crosses of *P. mexicanum* Cerv. and *P. pauciflorum* Wats. *Hereditas*, **12**: 33-40.

OWNBEY, M. and McCOLLUM, G. D. (1954). The chromosomes of *Tragopogon. Rhodora*, **56**: 7-21.

PAGE, C. N. (1973). Two hybrids in *Equisetum* new to the British flora. *Watsonia*, **9**: 229-237.

PARKER-RHODES, A. F. (1950). The Basidiomycetes of Skokholm Island, 4. A case of hybridisation in *Psilocybe (Deconica)*. *New Phytol.*, **49**: 335-343.

PARSONS, R. F. and KIRKPATRICK, J. B. (1972). Possible phantom hybrids in *Eucalyptus*. *New Phytol.*, **71**: 1213-1219.

PASSMORE, S. F. (1934). Hybrid vigour and reciprocal crosses in *Cucurbita Pepo*. *Ann. Bot.*, **48**: 1029-1030.

PERRING, F. H. (1964). Mapping the hybrids. *Proc. B.S.B.I.*, **5**: 379.

PERRING, F. H. and SELL, P. D., eds (1968). *Critical supplement to the atlas of the British flora*. Nelson, London.

PERRING, F. H. and WALTERS, S. M., eds (1962). *Atlas of the British flora*. Nelson, London.

PHILIPSON, W. R. (1937). A revision of the British species of the genus *Agrostis* Linn. *J. Linn. Soc., Bot.*, **51**: 73-151.

PODDUBNAYA-ARNOL'DI, V. A. (1962). Embryology of the hybrids of some angiosperm plants, in TSITSIN, N. V., ed. *Wide hybridization in plants*, pp. 124-133. Israel Programme of Scientific Translations, Jerusalem.

POEHLMAN, J. M. (1959). *Breeding field crops*. Holt, Rinehart and Winston, New York.

POWER, J. B., CUMMINS, S. E. and COCKING, E. C. (1970). Fusion of isolated plant protoplasts. *Nature, Lond.*, **225**: 1016-1018.

PRAEGER, R. L. (1951). Hybrids in the Irish flora: a tentative list. *Proc. R. Ir. Acad.*, Sect. B, **54**: 1-14.

PRITCHARD, N. M. (1961). *Gentianella* in Britain, 3. *Gentianella germanica* (Willd.) Börner. *Watsonia*, **4**: 290-303.

PRYOR, L. D. and BRYANT, L. H. (1958). Inheritance of oil characters in *Eucalyptus*. *Proc. Linn. Soc. N.S.W.*, **83**: 55-64.

RATTENBURY, J. A. (1962). Cyclic hybridization as a survival mechanism in the New Zealand forest flora. *Evolution, Lancaster, Pa.*, **16**: 348-363.

REES, H. (1961). Genotypic control of chromosome form and behaviour. *Bot. Rev.*, **27**: 288-318.

RENNER, O. (1929). Artbastarde bei Pflanzen. *Handb. Vererbungswissenschaft*, Lief. **7(11A)**, 2. Berlin.

RICHENS, R. H. (1945). Forest tree breeding and genetics. *Imperial Forestry Bureaux Joint Publ.*, **8**.

RICKETT, H. W. and CAMP, W. H. (1948). The nomenclature of hybrids. *Bull. Torrey bot. Club*, **75**: 496-501.

RILEY, H. P. (1952). Ecological barriers. *Am. Nat.*, **86**: 23-32.

RILEY, R. (1966). Genetics and the regulation of meiotic chromosome behaviour. *Sci. Prog., Lond.*, **54**: 193-207.

RILEY, R. and CHAPMAN, V. (1958). Genetic control of the cytologically diploid behaviour of hexaploid wheat. *Nature, Lond.*, **183**: 713-715.

RITCHIE, J. C. (1955b). A natural hybrid in *Vaccinium*, 2. Genetic studies in *Vaccinium intermedium* Ruthe. *New Phytol.*, **54**: 320-335.

ROBERTS, H. F. (1929). *Plant hybridization before Mendel*. Princeton University Press.

ROLFE, R. A. (1900). Hybridization viewed from the standpoint of systematic botany. *Jl R. hort. Soc.*, **24**: 181-202.

ROSE, F. and GÉHU, J. M. (1960). Comparaison floristique entre les comtés anglais du Kent et du Sussex et le département français du Pas de Calais. *Bull. Soc. bot. Fr.*, **13**: 125-139.

ROSENBERG, O. (1907). Cytological investigations in plant hybrids, in WILKS, W., ed. *Report of the third international conference 1906 on genetics*, pp. 289-291. Royal Horticultural Society, London.

RUENESS, J. (1973). Speciation in *Polysiphonia* (Rhodophyceae, Ceramiales) in view of hybridization experiments: *P. hemisphaerica* and *P. boldii*. *Phycologia*, **12**: 107-110.

SANTAMOUR, F. S. (1972). Interspecific hybridisation with fall- and spring-flowering elms. *Forest Sci.*, **18**: 283-289.

SATINA, S. (1944). Periclinal chimaeras in *Datura* in relation to development and structure (A) of the style and stigma, (B) of calyx and corolla. *Am. J. Bot.*, **31**: 493-502.

SAX, K. (1935). The cytological analysis of species hybrids. *Bot. Rev.*, **1**: 100-117.

SCHWARTZ, D. (1960). Genetic studies on mutant enzymes in maize: synthesis of hybrid enzymes by heterozygotes. *Proc. natn. Acad. Sci. U.S.A.*, **46**: 1210-1215.

SCHWARZE, P. (1959). Untersuchungen über die gesteigerte Flavonoid-produktion in *Phaseolus*-Artbastarden (*Phaseolus vulgaris* × *Phaseolus coccineus*). *Planta*, **54**: 152-161.

SEARS, E. R. (1944). Inviability of intergeneric hybrids involving *Triticum monococcum* and *T. aegilopoides*. *Genetics, Princeton*, **29**: 113-127.

SELL, P. D. (1967). Taxonomic and nomenclatural notes on the British flora. *Rhinanthus* L. *Watsonia*, **6**: 298-301.

SKALÍNSKA, M., JANKUN, A. and WCISŁO, H. (1971). Studies in chromosome numbers of Polish angiosperms, eighth contribution. *Acta biol. cracov., Bot.*, **14**: 55-102.

SMITH, D. M. and LEVIN, D. A. (1963). A chromatographic study of reticulate evolution in the Appalachian *Asplenium* complex. *Am. J. Bot.*, **50**: 952-958.

SMITH, W. K. (1948). Transfer from *Melilotus dentata* to *M. alba* of the genes for reduction in coumarin content. *Genetics, Princeton*, **33**: 124-125.

SNEATH, P. H. A. (1968). Numerical taxonomic study of the graft chimaera + *Laburnocytisus adamii* (*Cytisus purpureus* + *Laburnum anagyroides*). *Proc. Linn. Soc. Lond.*, **179**: 83-96.

SNYDER, L. A. (1951). Cytology of inter-strain hybrids and the probable origin of variability in *Elymus glaucus*. *Am. J. Bot.*, **38**: 195-202.

SOLBRIG, O. T. (1968). Fertility, sterility and the species problem, in HEYWOOD, V. H., ed. *Modern methods in plant taxonomy*, pp. 77-96. Academic Press, London and New York.

SOLBRIG, O. T. (1970). *Principles and methods of plant biosystematics*. Macmillan, London.

STACE, C. A. (1961). Some studies in *Calystegia*. Compatibility and hybridization in *C. sepium* and *C. silvatica*. *Watsonia*, **5**: 88-105.

STACE, C. A. (1970). Anatomy and taxonomy in *Juncus* subgenus *Genuini*, in ROBSON, N. K. B., CUTLER, D. F. and GREGORY, M., eds. *New research in plant anatomy*, pp. 75-81. Academic Press, London and New York.

STACE, C. A. (1972). The history and occurrence in Britain of hybrids in *Juncus* subgenus *Genuini. Watsonia*, 9: 1-41.

STACE, C. A. and COTTON, R. (1974). Hybrids between *Festuca rubra* L. *sensu lato* and *Vulpia membranacea* (L.) Dum. *Watsonia*, 10: 119-138.

STAFLEU, F. A., ed. (1972). International code of botanical nomenclature. *Regnum veg.*, 82.

STEBBINS, G. L. (1945). Cytological analysis of species hybrids, 2. *Bot. Rev.*, 11: 463-486.

STEBBINS, G. L. (1947). Types of polyploids: their classification and significance. *Adv. Genet.*, 1: 403-429.

STEBBINS, G. L. (1950). *Variation and evolution in plants.* Oxford University Press, London and Columbia University Press, New York.

STEBBINS, G. L. (1956). Taxonomy and the evolution of genera with special reference to the family Gramineae. *Evolution, Lancaster, Pa.*, 10: 235-245.

STEBBINS, G. L. (1958). The inviability, weakness and sterility of interspecific hybrids. *Adv. Genet.*, 9: 147-215.

STEBBINS, G. L. (1959). The role of hybridization in evolution. *Proc. Am. phil. Soc.*, 103: 231-251.

STEBBINS, G. L. (1969). The significance of hybridization for plant taxonomy and evolution. *Taxon*, 18: 26-35.

STEBBINS, G. L. and ZOHARY, D. (1959). Cytogenetic and evolutionary studies in the genus *Dactylis*, 1. Morphology, distribution, and interrelationships of the diploid subspecies. *Univ. Calif. Publs Bot.*, 31(1): 1-40.

STRAW, R. M. (1955). Hybridization, homogamy and sympatric speciation. *Evolution, Lancaster, Pa.*, 9: 441-444.

SVESCHNIKOVA, I. N. (1940). Cytogenetical analysis of heterosis in hybrids of *Vicia. J. Hered.*, 31: 349-360.

SWANSON, C. P. (1958). *Cytology and cytogenetics.* Macmillan, London.

TERRELL, E. E. (1963). Symbols and terms for morphological intergradation and hybridization. *Taxon*, 12: 105-108.

TERRELL, E. E. (1966). Taxonomic implications of genetics in ryegrass (*Lolium*). *Bot. Rev.*, 32: 138-164.

THOMPSON, W. P. (1940). The causes of hybrid sterility and incompatibility. *Trans. R. Soc. Can.*, Ser. III, Sect. V, 34: 1-13.

TOBGY, H. A. (1943). A cytological study of *Crepis fuliginosa, C. neglecta*, and their F_1 hybrid, and its bearing on the mechanism of phylogenetic reduction in chromosome number. *J. Genet.*, 45: 67-111.

TSITSIN, N. V., ed. (1962). *Wide hybridization in plants.* Israel Programme of Scientific Translations, Jerusalem.

TURESSON, G. (1922). The genotypical response of the plant species to the habitat. *Hereditas*, 3: 211-350.

TURRILL, W. B. (1934). The correlation of morphological variation with distribution in some species of *Ajuga. New Phytol.*, 33: 218-230.

TUTIN, T. G. (1957). A contribution to the experimental taxonomy of *Poa annua* L. *Watsonia,* 4: 1-10.

TUTIN, T. G. *et al.,* eds (1964–). *Flora Europaea.* Cambridge University Press.

ULLMANN, W. (1936). Natural hybridization of grass species and genera. *Herb. Rev.,* 4: 105-142.

VALENTINE, D. H. (1941). Variation in *Viola riviniana* Rchb. *New Phytol.,* 40: 189-209.

VALENTINE, D. H. (1947). Studies in British Primulas, 1. Hybridization between primrose and oxlip (*Primula vulgaris* Huds. and *P. elatior* Schreb.). *New Phytol.,* 46: 229-253.

VALENTINE, D. H. (1952). Studies on British Primulas, 3. Hybridization between *Primula elatior* (L.) Hill and *P. veris* L. *New Phytol.,* 50: 383-399.

VALENTINE, D. H. (1955). Studies in the British Primulas, 4. Hybridization between *Primula vulgaris* Huds. and *P. veris* L. *New Phytol.,* 54: 70-80.

VALENTINE, D. H. (1956). Studies in British Primulas, 5. The inheritance of seed compatibility. *New Phytol.,* 55: 305-318.

VALENTINE, D. H. (1961). Evolution in the genus *Primula,* in WANSTALL, P. J., ed. *A Darwin centenary,* pp. 71-87. B.S.B.I., London.

VALENTINE, D. H. (1963). The treatment of hybrids in Flora Europaea. *Webbia,* 18: 47-55.

VALENTINE, D. H. (1970). Evolution at zones of vegetational transition. *Reprium nov. Spec. Regni veg.,* 81: 33-39.

VALENTINE, D. H. and WOODELL, S. R. J. (1961). Studies in British Primulas, 9. Seed incompatibility in diploid-autotetraploid crosses. *New Phytol.,* 60: 282-294.

WADDINGTON, C. H. (1939). *An introduction to modern genetics.* Allen and Unwin, London.

WAGNER, W. H. (1954). Reticulate evolution in the Appalachian Aspleniums. *Evolution, Lancaster, Pa.,* 8: 103-118.

WAGNER, W. H. (1968). Hybridization, taxonomy and evolution, in HEYWOOD, V. H., ed. *Modern methods in plant taxonomy,* pp. 113-138. Academic Press, London and New York.

WALKER, S. (1955). Cytogenetic studies in the *Dryopteris spinulosa* complex, 1. *Watsonia,* 3: 193-208.

WALTERS, S. M. (1953). *Montia fontana* L. *Watsonia,* 3: 1-6.

WARBURG, E. F. and KÁRPÁTI, Z. E. (1968). *Sorbus* L., in TUTIN, T. G. *et al.,* eds. *Flora Europaea,* 2: 67-71. Cambridge University Press.

WATKINS, A. E. (1932). Hybrid sterility and incompatibility. *J. Genet.,* 25: 125-162.

WATSON, H. C., ed. (1874). *The London catalogue of British plants,* 7th ed. R. Hardwicke, London.

WEBB, D. A. (1951). Hybrid plants in Ireland, *Ir. Nat. J.,* 10: 201-204.

WEBB, D. A. (1972). *Antirrhinum* L., in TUTIN, T. G. *et al.,* eds. *Flora Europaea,* 3: 221-224. Cambridge University Press.

WESTERGAARD, M. (1943). Cytotaxonomical studies on *Calamagrostis epigeios* (L.) Roth, *Ammophila arenaria* (L.) Link and their hybrids

(*Ammophila baltica* (Flügge) Link). *K. danske Vidensk. Selsk.. Biol. Skr.,* 2(4).

WHITEHEAD, F. H. (1954). An example of taxonomic discrimination by biometric methods. *New Phytol.,* 53: 496-510.

WIEGLAND, K. M. (1935). A taxonomist's experience with hybrids in the wild. *Science, N.Y.,* 81: 161-166.

WILKS, W., ed. (1900). International conference on hybridization (the cross-breeding of species) and on the cross-breeding of varieties. *Jl R. hort. Soc.,* 24.

WILKS, W., ed. (1907). *Report of the third international conference 1906 on genetics; hybridization (the cross-breeding of genera or species), the cross-breeding of varieties, and general plant breeding.* Royal Horticultural Society, London.

WILLIAMS, W. (1959). Pear-apple hybrids. *Rep. John Innes hort. Instn,* 49: 8.

WILSON, J. H. (1900). The structure of certain new hybrids. *Jl R. hort. Soc.,* 24: 146-180.

WINGE, O. (1917). The chromosomes, their number and general importance. *C.r. Trav. Lab. Carlsberg,* 13: 131-275.

WINGE, O. (1932). On the origin of constant species hybrids. *Svensk bot. Tidskr.,* 26: 107-122.

WINGE, O. (1944). The *Sambucus* hybrid *S. nigra* x *S. racemosa. C.r. Trav. Lab. Carlsberg,* 24: 73-78.

WIT, F. (1974). Cytoplasmic male sterility in ryegrasses (*Lolium* spp.) detected after intergeneric hybridization. *Euphytica,* 23: 31-38.

WOLLEY-DOD, A. H. (1910). The British roses. *J. Bot., Lond..* 48 (Suppl.).

WOLLEY-DOD, A. H. (1930-31). A revision of the British roses. *J. Bot., Lond.,* 68-69 (Suppl.).

WOODELL, S. R. J. (1965). Natural hybridization between the Cowslip (*Primula veris* L.) and the Primrose (*P. vulgaris* Huds.) in Britain. *Watsonia,* 6: 190-202.

YABLOKOV, A. S. (1962). Wide hybridization in silviculture and green-belt work, in TSITSIN, N. V., ed. *Wide hybridization in plants,* pp. 48-61. Israel Programme of Scientific Translations, Jerusalem.

YAMAMOTO, K. (1966). Studies on the hybrids among *Vicia sativa* L. and its related species. *Mem. Fac. Agric. Kagawa Univ.,* 21: 1-104.

YARNELL, S. H. (1956). Cytogenetics of the vegetable crops, 2. Crucifers. *Bot. Rev.,* 22: 81-166.

YENIKEYEV, K. H. (ENIKEEV, H. H.) (1966). The method of pollination with a pollen mixture to obtain interspecific hybrids of plum and cherry. *Genetica,* 36: 301-306.

YEO, P. F. (1956). Hybridisation between diploid and tetraploid species of *Euphrasia. Watsonia,* 3: 253-269.

ZIRKLE, C. (1935). *The beginnings of plant hybridization.* University of Pennsylvania Press, Philadelphia.

Author Index

This index refers only to the names mentioned in Section A and in the first chapter (Explanation of text) of Section B.

Subject Index

This covers Section A and the first chapter (Explanation of text) of Section B in some detail, but includes only the genus headings of the Systematic accounts (Section B, Chapter 2). The latter references are asterisked.

Y

yield, 90

Z

Zea, 57 (*see also* maize)

Zostera, 443–444*

zygote, 3, 34–36, 60, 61

zygotic

    failure of F_2, 40

    lethality, 16, 61

    sterility, 41